AF754374

ACOUSTIC AND ELASTIC WAVE FIELDS IN GEOPHYSICS, III

Methods in Geochemistry and Geophysics
(Volumes 1–28 are out of print)

1. A.S. RITCHIE — CHROMATOGRAPHY IN GEOLOGY
2. R. BOWEN — PALEOTEMPERATURE ANALYSIS
3. D.S. PARASNIS — MINING GEOPHYSICS
4. I. ADLER — X-RAY EMISSION SPECTROGRAPHY IN GEOLOGY
5. THE LORD ENERGLYN AND L. BREADLY — ANALYTICAL GEOCHEMISTRY
6. A.J. EASTON — CHEMICAL ANALYSIS OF SILICATE ROCKS
7. E.E ANGINO AND G.K. BILLINGS — ATOMIC ABSORPTION SPECTROMETRY IN GEOLOGY
8. A. VOLBORTH — ELEMENTAL ANALYSIS IN GEOCHEMISTRY, A: MAJOR ELEMENTS
9. P.K. BHATTACHARYA AND H.P. PATRA — DIRECT CURRENT GEOELECTRIC SOUNDING
10. J.A.S. ADAMS AND P. GASPARINI — GAMMA-RAY SPECTROMETRY OF ROCKS
11. W. ERNST — GEOCHEMICAL FACIES ANALYSIS
12. P.V. SHARMA — GEOPHYSICAL METHODS IN GEOLOGY
13. C.H. CHEN (Editor) — COMPUTER-AIDED SEISMIC ANALYSIS AND DISCRIMINATION
14A. O. KOEFOED — GEOSOUNDING PRINCIPLES, 1. RESISTIVITY SOUNDING MEASUREMENTS
14B. H.P. PATRA AND K. MALLICK — GEOSOUNDING PRINCIPLES, 2. TIME-VARYING GEOELECTRIC SOUNDINGS
15. A.A. KAUFMAN AND G.V. KELLER — THE MAGNETOTELLURIC SOUNDING METHOD
16. A.A. KAUFMAN AND G.V. KELLER — FREQUENCY AND TRANSIENT SOUNDINGS
17. C.H. CHEN (Editor) — SEISMIC SIGNAL ANALYSIS AND DISCRIMINATION
18. J.E. WHITE — UNDERGROUND SOUND APPLICATION OF SEISMIC WAVES
19. M.N. BERDICHEVSKY AND M.S. ZHDANOV — ADVANCED THEORY OF DEEP GEOMAGNETIC SOUNDINGS
20A. A.A. KAUFMAN AND G.V. KELLER — INDUCTIVE MINING PROSPECTING, PART I: THEORY
21. A.W. WYLIE — NUCLEAR ASSAYING OF MINING BOREHOLES - AN INTRODUCTION
22. C.H. CHEN (Editor) — SEISMIC SIGNAL ANALYSIS AND DISCRIMINATION III
23. R.P. PHILP — FOSSIL FUEL BIOMARKERS
24. R.B. JOHNS (Editor) — BIOLOGICAL MARKERS IN THE SEDIMENTARY RECORD
25. J.C. D' ARNAUD GERKENS — FOUNDATION OF EXPLORATION GEOPHYSICS
26. P. TYGEL AND P. HUBRAL — TRANSIENT WAVES IN LAYERED MEDIA
27. A.A. KAUFMAN AND G.V. KELLER — INDUCTION LOGGING
28. J.G. NEGI AND P.D. SARAF — ANISOTROPY IN GEOELECTROMAGNETISM
29. V.P. DIMRI — DECONVOLUTION AND INVERSE THEORY - APPLICATION TO GEOPHYSICAL PROBLEMS
30. K.-M STRACK — EXPLORATION WITH DEEP TRANSIENT ELECTROMAGNETICS
31. M.S. ZHDANOV AND G.V. KELLER — THE GEOELECTRICAL METHODS IN GEOPHYSICAL EXPLORATION
32. A.A. KAUFMAN AND A.L. LEVSHIN — ACOUSTIC AND ELASTIC WAVE FIELDS IN GEOPHYSICS, I
33. A.A. KAUFMAN AND P.A. EATON — THE THEORY OF INDUCTIVE PROSPECTING
34. A.A. KAUFMAN AND P. HOEKSTRA — ELECTROMAGNETIC SOUNDINGS
35. M.S. ZHDANOV AND P.E. WANNAMAKER — THREE-DIMENSIONAL ELECTROMAGNETICS
36. M.S. ZHDANOV — GEOPHYSICAL INVERSE THEORY AND REGULARIZATION PROBLEMS
37. A.A. KAUFMAN, A.L. LEVSHIN AND K.L. LARNER — ACOUSTIC AND ELASTIC WAVE FIELDS IN GEOPHYSICS, II
38. A.A. KAUFMAN AND YU. A. DASHEVSKY — PRINCIPLES OF INDUCTION LOGGING
39. A.A. KAUFMAN AND A.L. LEVSHIN — ACOUSTIC AND ELASTIC WAVE FIELDS IN GEOPHYSICS, III

Methods in Geochemistry and Geophysics, 39

ACOUSTIC AND ELASTIC WAVE FIELDS IN GEOPHYSICS, III

A.A. KAUFMAN
Colorado School of Mines
Golden, CO, U.S.A.

and

A.L. LEVSHIN
University of Colorado
Boulder, CO, U.S.A.

2005

ELSEVIER

Amsterdam - Boston - Heidelberg - London - New york - Oxford
Paris - San Diego - San Francisco - Singapore - Sydney - Tokyo

ELSEVIER B.V.
Radarweg 29
P.O. Box 211, 1000 AE
Amsterdam, The Netherlands

ELSEVIER Inc.
525 B Street, Suite 1900
San Diego, CA 92101-4495
USA

ELSEVIER Ltd.
The Boulevard, Langford Lane
Kidlington, Oxford OX5 1GB
UK

ELSEVIER Ltd.
84 Theobalds Road
London WC1X 8RR
UK

© 2005 Elsevier B.V. All rights reserved.

This work is protected under copyright by Elsevier B.V., and the following terms and conditions apply to its use:

Photocopying
Single photocopies of single chapters may be made for personal use as allowed by national copyright laws. Permission of the Publisher and payment of a fee is required for all other photocopying, including multiple or systematic copying, copying for advertising or promotional purposes, resale, and all forms of document delivery. Special rates are available for educational institutions that wish to make photocopies for non-profit educational classroom use.

Permissions may be sought directly from Elsevier's Rights Department in Oxford, UK: phone (+44) 1865 843830, fax (+44) 1865 853333, e-mail: permissions@elsevier.com. Requests may also be completed on-line via the Elsevier homepage (http://www.elsevier.com/locate/permissions).

In the USA, users may clear permissions and make payments through the Copyright Clearance Center, Inc., 222 Rosewood Drive, Danvers, MA 01923, USA; phone: (+1) (978) 7508400, fax: (+1) (978) 7504744, and in the UK through the Copyright Licensing Agency Rapid Clearance Service (CLARCS), 90 Tottenham Court Road, London W1P 0LP, UK; phone (+44) 20 7631 5555; fax: (+44) 20 7631 5500. Other countries may have a local reprographic rights agency for payments.

Derivative Works
Tables of contents may be reproduced for internal circulation, but permission of the Publisher is required for external resale or distribution of such material. Permission of the Publisher is required for all other derivative works, including compilations and translations.

Electronic Storage or Usage
Permission of the Publisher is required to store or use electronically any material contained in this work, including any chapter or part of a chapter.

Except as outlined above, no part of this work may be reproduced, stored in a retrieval system or transmitted in any form or by any means, electronic, mechanical, photocopying, recording or otherwise, without prior written permission of the Publisher.
Address permissions requests to: Elsevier's Rights Department, at the fax and e-mail addresses noted above.

Notice
No responsibility is assumed by the Publisher for any injury and/or damage to persons or property as a matter of products liability, negligence or otherwise, or from any use or operation of any methods, products, instructions or ideas contained in the material herein. Because of rapid advances in the medical sciences, in particular, independent verification of diagnoses and drug dosages should be made.

First edition 2005

Library of Congress Cataloging in Publication Data
A catalog record is available from the Library of Congress.

British Library Cataloguing in Publication Data
A catalogue record is available from the British Library.

ISBN: 0 444 51955 6

∞ The paper used in this publication meets the requirements of ANSI/NISO Z39.48-1992 (Permanence of Paper).
Printed in The Netherlands.

Contents

Introduction

This monograph is the last volume in the series *Acoustic and elastic wave fields in geophysics*. The previous two volumes published by Elsevier (A. Kaufman and A. Levshin, 2000; A. Kaufman, A. Levshin and K. Larner, 2002) dealt mostly with wave propagation in liquid media. Here we consider waves in elastic media, and their description is based mainly on the classical papers of Stokes, Love, Lamb, Rayleigh, Stoneley, and others.

The first chapter is devoted entirely to propagation of longitudinal, torsional, and bending waves along a thin bar. Several examples illustrate a reflection of these waves and the deformations they cause. Some attention is paid to the transition from the dynamic stage to equilibrium. In the second chapter, proceeding from Newton's laws, we derive an equation of motion of an elementary volume of the elastic medium. Then, by means of Hooke's law, we obtain the equation for a displacement field. To solve it, we introduce scalar and vector potentials, formulate boundary conditions for wavefields, and derive wave equations for potentials. By analogy with acoustic waves (Parts I and II), the concepts of potential and kinetic energies, as well as the Poynting vector, are described. Hooke's law in the Cartesian system of coordinates is described in Chapter 1. Here we derive relationships between stress and strain in any curvilinear orthogonal system of coordinates.

Behavior of waves in a homogeneous medium for several types of sources is studied in Chapter 3. First we investigate longitudinal and shear waves caused by elementary spherical sources in the near, intermediate, and far zones. Then the field generated by the point force is described in detail. In the last section we consider longitudinal and shear plane waves, which serves as a preparation for the next chapter.

In the fourth chapter we describe the reflection and transmission of plane waves, starting from an analysis of strains and stresses that accompany them. We discuss in detail behavior of reflected and transmitted waves caused by different incident plane waves, including discussion of the dependence of wavefields on parameters of a medium and the angle of incidence. In conclusion, recursive expressions for reflection and transmission coefficients describing plane waves in the $n-$layered medium are derived.

Chapter 5 is devoted to surface waves. First, we consider Rayleigh waves in a homogeneous half-space. We discuss such topics as the characteristic equation for the velocity of propagation, the dependence of wave amplitudes on the depth below the free boundary, and elliptic polarization of particle motion. Then we study Stoneley waves, which may appear at the interface between fluid and elastic media and between two elastic media. Further, we describe Love waves, which arise as a result of the constructive interference of plane SH waves traveling up and down inside a layer of finite thickness overlaying the half-space with the higher shear velocity. Finally, in the last section, behavior of Rayleigh waves in this medium is considered.

Chapter 6 is devoted to the study of waves generated by linear and point sources in a homogeneous half-space, when a source is located either at or beneath the free boundary. Asymptotic behavior of waves in the far zone is studied using integration in a complex wavenumber plane. In addition, we investigate reflection and transmission of waves caused by a linear source in the presence of the boundary between fluid and elastic media.

Chapter 7 describes waves in the borehole that are generated by an elementary spherical source. Main features of the normal modes, Stoneley waves, and as head waves are described.

Finally, in the last chapter we focus on plane wave propagation in a transversely isotropic medium. Influence of the angle of incidence on the velocities of different plane waves, orientation of rays with respect to the phase surface, and other questions are the subject of this Chapter.

Acknowledgments

We express our thanks to Drs. M. Barmin, C. Cheng, I. Fuks, O. Godin, K. Larner, P. Martin, K. Naugolnykh, and L. Ostrovsky for very useful comments and suggestions. We express our sincere thanks to Dr. T. Young for the permanent support of this work.

We also want to thank Marianna Borukaeva for her technical assistance. We deeply appreciate the tremendous editorial help of Linda Bevard.

List of Symbols

a	radius of a borehole
a_n, b_n	coefficients of Fourier's series
A_n, B_n	amplitudes of normal modes, coefficients of Debye's expansion
C_i	path of integration in a complex variable plane
c	phase velocity
c_1	velocity of a compressional wave in a fluid
c_l	velocity of a longitudinal wave in a solid
c_s	velocity of a shear wave in a solid
c_{St}	velocity of a Stoneley wave
c_R	velocity of a Rayleigh wave in a solid half-space
c_{pn}, c_{gn}	phase and group velocities of the nth mode.
c_{ij}	components of an elastic tensor
d_i, d_j	thicknesses of layers in a periodic system
E	Young modulus
E_k	kinetic energy
e_{ij}	component of a strain tensor
$\mathbf{F}$	force
$\underset{\sim}{G}$	propagator matrix
g	gravitational acceleration
g_{ij}	elements of propagator matrix
$H_n^{(1)}(mr)$, $H_n^{(2)}(mr)$	Hankel functions of the nth order
$h(t)$	step-function
$i = \sqrt{-1}$	imaginary unit
$\mathbf{I}$	polar moment of inertia
$I_n(mr)$, $K_n(mr)$	modified Bessel functions of the nth order
$\mathbf{i}$, $\mathbf{j}$, $\mathbf{k}$	unit vectors in a Cartesian system of coordinates
$\mathbf{i}_r$, $\mathbf{i}_\theta$, $\mathbf{i}_\varphi$	unit vectors in a spherical system of coordinates
$\mathbf{i}_z$, $\mathbf{i}_r$, $\mathbf{i}_\varphi$	unit vectors in a cylindrical system of coordinates

$J_n(mr)$, $Y_n(mr)$	Bessel functions of the nth order
J	Jacobian
k	wavenumber
k_n	wavenumbers of normal modes
l	length of a bar
$\mathbf{M}$	torque
M	total mass, bulk modulus
m	variable of integration
N	impulse of force
$\mathbf{n}$	unit normal
$\mathbf{p}$	slowness vector
p	slowness, observation point
$\mathbf{P}$	momentum of motion
r	distance from the z-axis
$\mathbf{r}$	radius-vector
r_n	roots of dispersion equation
R	distance from the source, radius of curvature
$\mathcal{R}_{ij}$	reflection coefficient
S	bar cross-section
$\mathbf{s}$	displacement vector
t	time
T	period
$\mathcal{T}_{ij}$	transmission coefficient
$\mathcal{U}$	complex amplitude of potential
V	volume
u, v, w	components of displacement
$\mathbf{v}$	velocity vector
V_0	elementary volume
W	total energy
x, y, z	Cartesian coordinates
$\mathbf{Y}$	Poynting vector
Z	impedance
δS	elementary area
$\delta(t)$	Delta function
ϵ	small number

Λ	wavelength
λ	Lame's constant
μ	modulus of rigidity, Lame's constant
Ψ	phase shift
$\boldsymbol{\psi}$	vector potential
ρ	density of mass
ω	circular frequency
$\tilde{\boldsymbol{\psi}}$	complex amplitude of vector potential
φ	scalar potential, azimuthal angle
$\tilde{\varphi}$	complex amplitude of scalar potential
Θ	dilatation
θ_i	angle of incidence
θ_r	angle of reflection
θ_2	angle of refraction
θ_c	critical angle
σ	Poisson's ratio
τ_{ij}	component of a stress tensor

Chapter 1

Hooke's law, Poisson's relation and waves along thin bars

Under an action of forces, all real bodies experience deformation, and either their size or shape or both change. This means that the relative position of body particles vary, and this effect becomes more noticeable with an increase of force. Some features of the dependence of deformation on a force magnitude are qualitatively shown in Fig. 1.1a. The initial portion of the curve is practically a straight line. Within this range Hooke's law is valid, and each element of the stress tensor is a linear function of strains (Appendix E). In other words, expanding stresses as functions of strains in Taylor's series, we discard terms that are relatively small. Thus, in this range we can apply the linear theory of elasticity to study a deformation. The latter displays two important features, namely

a. Deformation disappears as soon as forces are removed, that is, we ignore any free vibrations that may arise.

b. The relative change of the position of particles is usually very small.

With further increase of force, the rate of change of deformation becomes greater, and finally a body is broken. The value of the force magnitude corresponding to the breaking point varies for different materials. It also depends on the type of force. For example, in the case of chalk, the breaking force, causing a stretch, is smaller than the twisting one.

There is another interesting feature of deformation. Suppose that a force corresponds to the bending portion of the curve. Then when we begin to decrease the force a change of a deformation may occur along the curve, which differs from that shown in Fig. 1.1a, and the hysteresis effect is observed.

In principle, the process of deformation may take place at different rates. For instance, if the force varies slowly (quasi-statically), there is sufficient time for heat exchange

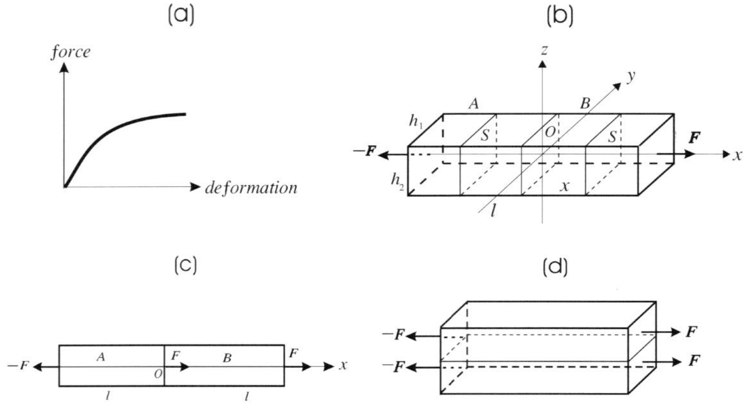

Figure 1.1: (a) Dependence of deformation on force (b) Rectangular bar (c) Influence of the bar length (d) Influence of the bar cross-section

between a body and a surrounding medium, that is, a deformation is isothermal. In such a case the influence of vibrations arising in the body is insignificant. In contrast, wave propagation is accompanied by relatively rapid motions of elementary volumes (particles). Correspondingly, heat exchange between these volumes of a medium is almost absent, and the process of deformation is adiabatic (Appendix E). In both cases we use the same linear theory of elasticity, but there is a very small difference in values of elastic constants (Appendix E).

1.1 Hooke's law and Poisson's relation

Two approaches allow one to establish physical laws of the linear theory of elasticity. One of them requires a knowledge of the atomic-molecular lattice of a medium, in particular, such parameters as mass and the charge of positive and negative ions. Also it is essential to have an information about internal electromagnetic forces acting between charges, because it is due to them that elastic waves exist. We will follow the second approach, which is entirely based on experiments performed by R. Hooke, T. Young, S. Poisson, and others.

Hooke's law

In order to derive Hooke's law, we will proceed from the experimental studies of bar deformation in an equilibrium, when either stretching or compressional forces are applied.

Hooke's experiments Consider a bar of length l having the rectangular cross-section S, and suppose that two extensional forces with equal magnitudes but opposite directions act on the bar faces, Fig. 1.1b. It is implied that these forces are uniformly distributed over each face. From the physical point of view, it is obvious that at the instant when these forces are applied, waves arise inside the bar, and they travel between faces. However, with an increase of time, the effect of propagation becomes less noticeable due to attenuation, and finally waves disappear. Then, equilibrium is observed, since external forces are constants (Part I). For now we pay attention only to this last stage, but later we will study both propagation and equilibrium, as well as the transition from one stage to another. Experiments performed by R. Hooke with bars in equilibrium, that have a different length, l, and the same cross-section demonstrated the following. An expansion of the bar, δl, is directly proportional to the force magnitude:

$$\delta l \sim F \tag{1.1}$$

This is the essence of Hooke's law, and it states that with an increase of force, $|F|$, an expansion of the elastic body, δl, also linearly increases. Of course, such behavior takes place only when the force magnitude corresponds to the initial portion of the curve in Fig. 1.1a. It may be appropriate to notice that δl is a sum of displacements of both faces of the bar, and Hooke's law is also valid in the case of compression.

The first generalization of eq. 1.1 is related to a distribution of internal forces arising due to deformation. Let us mentally draw the cross-section S at any place on the bar and consider portion A, Fig. 1.1b. This portion is surrounded by the lateral surface of the bar, where external forces are absent. The force $-\mathbf{F}$ acts on its left face and on section S. Since the bar is in equilibrium, the resultant force, acting on portion A, has to be equal to zero. This means that the force applied to the surface S coincides with $\mathbf{F}$. In contrast, portion A creates the force $-\mathbf{F}$, which acts on portion B. Because our conclusion is independent of a position of S, we can say that internal forces are uniformly distributed inside the bar, and their magnitude is equal to $|\mathbf{F}|$ at each of the bar's points. In other words, external forces $-\mathbf{F}$ and $\mathbf{F}$ are transmitted inside the bar.

Until now we have discussed expansion of the bar. As was already mentioned, the Hooke's law, eq. 1.1, is also valid when both forces are directed toward the middle point O, Fig. 1.1b, and compression takes place. Now we are prepared to demonstrate that there is a relationship between the displacement δl and the original length of the bar, l. This task can be solved in two ways. One of them is an analysis of measurements of displacements with bars having different lengths. The second approach follows from

Hooke's law and the condition of an equilibrium. First, in accordance with experiments, the bar expansion, δl, is directly proportional to its original length:

$$\delta l \sim l \tag{1.2}$$

This shows that bars with the same cross-section and different lengths experience different expansions, provided that in all cases the external forces are the same. Moreover, between them there is a linear relationship, given by eq. 1.2. For instance, if the bar length is $2l$, an extension is equal to $2\delta l$, and in general, it becomes $n\delta l$ when the bar length is nl. In particular, in the limit, an extension of an infinitely long bar also tends to an infinity. Such dependence between the length l and an expansion δl is a remarkable feature of a deformation, that has an interesting explanation (Part I). As follows from experiments, the expansion δl depends on forces $\mathbf{F}$ and $-\mathbf{F}$ as well as length l. However, the relative change of the length, $\delta l/l$, is defined by the force. Correspondingly, in place of eq. 1.1 we can write

$$\frac{\delta l}{l} \sim F \tag{1.3}$$

It is useful to derive eq. 1.3, using a different approach, which does not require any measurements. In fact, consider two identical bars of the length l and cross-section S, rigidly connected together, Fig. 1.1c. The length of the new bar is $2l$, and, as follows from the condition of an equilibrium, the force, applied to the portion A at points of the middle section, is equal to $\mathbf{F}$. In accordance with eq. 1.1 an expansion of this portion of the bar is equal to δl. In the same manner, the stretching of the portion B is δl, too. Thus, the total expansion of the bar with the length $2l$ is $2\delta l$. By analogy, we obtain eq. 1.3 for bars of an arbitrary length. It is proper to emphasize that we consider only cases when expansion or contraction, δl, is much smaller than the length l:

$$\delta l \ll l \tag{1.4}$$

Until now we paid attention to a displacement of the bar faces. In order to apply eq. 1.3 to any cross-section of the bar, consider its portion of the length x, confined by the middle cross-section and $S(x)$, as well as the lateral surface, Fig. 1.1b. Because of a symmetry, a pair of forces $-\mathbf{F}$ and $\mathbf{F}$ does not move the middle section, and the displacement $\delta l(x)$ is directly proportional to the length x of this portion. This gives

$$\frac{\delta l(x)}{x} \sim F, \tag{1.5}$$

where x is the distance from the origin O.

Next, we will make one more step in a generalization of eq. 1.1, which allows us to transform the relationship 1.3 into the equation. Let us imagine that two identical bars are connected, as is shown in Fig. 1.1d. As before, the same forces are applied to faces of each bar. It is clear that the new bar has an extension δl in spite of the fact that forces acting on its faces are twice as great. However, areas of the cross-sections are also increased by two. Therefore, the ratio F/S remains the same for the single and combined bars. This consideration suggests that the relative extension of the bar is defined by the force per unit area, that is, a traction (Appendix C), and we can write

$$\frac{\delta l}{l} \sim \frac{F}{S} \tag{1.6}$$

The last step is an introduction of the coefficient of proportionality, which gives

$$\frac{F}{S} = E\frac{\delta l}{l} \tag{1.7}$$

Here E is called the Young modulus, (Appendix E). Note that F/S and $\delta l/l$ describe normal stress and strain in the bar, respectively.

Poisson's relation

As we already know, experimental studies performed by R. Hooke allowed others to obtain eq. 1.7. A series of measurements, carried out much later by S. Poisson discovered another important relation describing a deformation. Consider the bar with length l and cross-section $S = h_1 h_2$ that is subjected to an action of external forces, Fig. 1.1b. The experiments showed that the bar extension, δl, is accompanied by a contraction of the cross-section. In the opposite case, the bar compression leads to an increase of S. This phenomenon was studied by S. Poisson, who discovered that

$$\frac{\delta h_1}{h_1} = \frac{\delta h_2}{h_2} = -\sigma\frac{\delta l}{l} \tag{1.8}$$

The coefficient σ is called Poisson's ratio. For all imaginable materials, σ has a positive value. The sign "—" shows that if $\delta l > 0$, (expansion), then $\delta h_1 < 0$ and $\delta h_2 < 0$. On the contrary, in the case of contraction, $\delta l < 0$, we have $\delta h_1 > 0$ and $\delta h_2 > 0$. As follows from eq. 1.8, relative changes of the bar dimensions, normal to the external force, are the same:

$$\frac{\delta h_1}{h_1} = \frac{\delta h_2}{h_2} \tag{1.9}$$

At the same time, displacements themselves, δh_1 and δh_2, differ from each other, if $h_1 \neq h_2$. In essence, eq. 1.8, which we will further call Poisson's relation, describes the second law of elasticity. Both the Hooke's and Poisson's relation are experimental. They provide the foundation of the linear theory of elasticity and play the same role in the theory of elasticity as Newton's laws in classical mechanics and Maxwell's equations in electrodynamics.

Differential form of Hooke's law and Poisson's relation

It is useful to represent eqs. 1.7 and 1.8 in a different form that allows us to study deformation in the vicinity of any point of an elastic medium. Let us consider two cross-sections of the bar, $S(x)$ and $S(x+dx)$, located close to each other, Fig. 1.2a. As a result of deformation, they are displaced at distances $u(x)$ and $u(x+dx)$, respectively. Taking into account that forces applied to each surface are the same, we conclude that relative displacements are equal, that is

$$\frac{u(x)}{x} = \frac{u(x+dx)}{x+dx} = \frac{F_x}{ES} \tag{1.10}$$

Here x is the distance from the origin O, which does not move during deformation, and F_x is the scalar component of the force. From the last equality we have

$$(x+dx)\; u(x) = x\; u(x+dx)$$

Since the distance between two cross-sections is very small, it is natural to assume that displacements between them change linearly. Then, expanding the function $u(x+dx)$ in the Taylor series and discarding all terms except the first two, we obtain

$$x\; u(x) + dx\; u(x) = x\; u(x) + x\frac{du}{dx}dx$$

Comparison with eq. 1.10 shows that the relative expansion of the bar is characterized by the first derivative du/dx. The position of the cross-section $S(x)$ was chosen arbitrarily, and, correspondingly, du/dx describes a deformation of the bar at each point. In our case, a value of du/dx is independent of the point coordinates, and we are dealing with homogeneous deformation. However, in general, this function, du/dx, may change from point to point, and inhomogeneous deformation is observed. By definition, du/dx is called the strain at a point, or more precisely, the diagonal element of the strain tensor (Appendix D). Note that if a displacement u is a function of several coordinates,

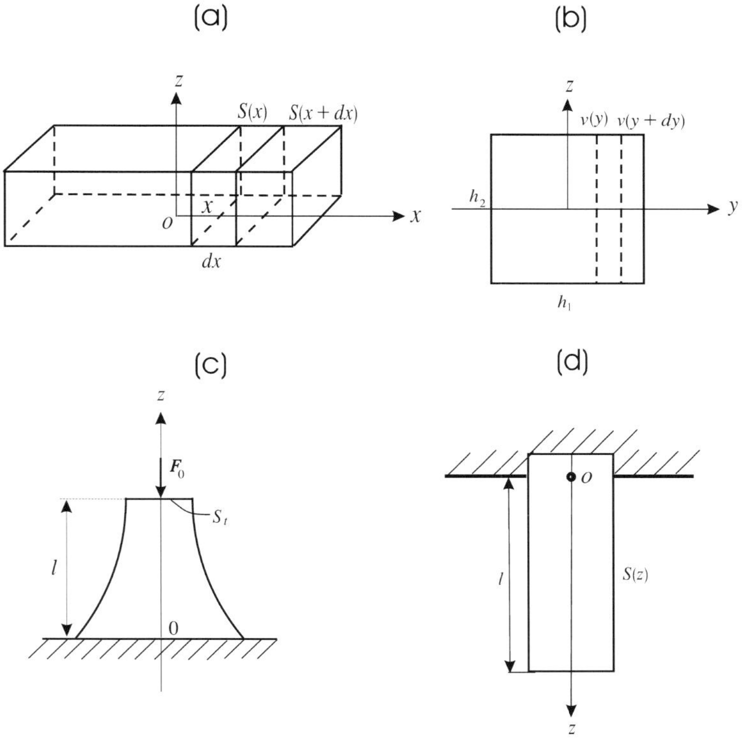

Figure 1.2: (a) and (b) Illustration of eqs. 1.10, 1.13 (c) Deformation of an elastic column (d) Bar extension due to gravitational field

we have to use the partial derivatives, and the strain is written as

$$e_{xx} = \frac{\partial u}{\partial x} \tag{1.11}$$

Then, in place of eq. 1.10 we have

$$e_{xx} = \frac{1}{E}\frac{F_x}{S} \qquad \text{or} \qquad \frac{F_x}{S} = E\, e_{xx} \tag{1.12}$$

As follows from eq. 1.11, the strain e_{xx} characterizes the rate of change of the displacement, as well as a type of deformation. For instance, if $\partial u/\partial x > 0$ expansion occurs, whereas compression is observed when $\partial u/\partial x < 0$. It is clear that the strain is dimensionless.

Next, we perform similar transformations with eq. 1.8 and consider a cross-section of an elementary parallelepiped of the bar in the plane YOZ, Fig. 1.2b. Taking into account that a deformation is homogeneous, we have

$$\frac{\delta h_1}{h_1} = \frac{v(y)}{y} = \frac{v(y+dy)}{y+dy}, \tag{1.13}$$

where $v(y)$ is the displacement of the cross-section $S(y)$ along the y–axis and y is the distance from the origin. Again, using the Taylor series, eq. 1.13 gives

$$\frac{\delta h_1}{h_1} = \frac{\partial v}{\partial y},$$

and, by analogy,

$$\frac{\partial h_2}{h_2} = \frac{\partial w}{\partial z}$$

Here w is the displacement along the z–axis. Respectively, Poisson's ratio, eq. 1.8, is written in the form

$$\frac{\partial v}{\partial y} = \frac{\partial w}{\partial z} = -\sigma \frac{\partial u}{\partial x} \tag{1.14}$$

Thus, the rate of change of the corresponding components of the displacement vector

$$\mathbf{s} = u\mathbf{i} + v\mathbf{j} + w\mathbf{k} \tag{1.15}$$

is the same along the y– and z–axes. By definition,

$$e_{yy} = \frac{\partial v}{\partial y} \quad \text{and} \quad e_{zz} = \frac{\partial w}{\partial z} \tag{1.16}$$

are also the diagonal elements of the strain tensor and, in accordance with eq. 1.8, they are related to each other,

$$e_{yy} = e_{zz} = -\sigma e_{xx} \tag{1.17}$$

provided that forces are applied only along the x–axis. By analogy, if they are directed either along the y– or z–axes, we have

$$e_{xx} = e_{zz} = -\sigma e_{yy} \quad \text{and} \quad e_{xx} = e_{yy} = -\sigma e_{zz}$$

respectively.

Ratio F_x/S and stress

Now let us discuss ratio F_x/S in eq. 1.12, which describes the x–component of the traction $\mathbf{t}$. The latter can be introduced in the following way (Appendix C).

$$t_x = (\mathbf{X} \cdot \mathbf{n}) = \frac{F_x}{S} \tag{1.18}$$

Here $\mathbf{n}$ is the unit vector normal to the surface $S(x)$ and directed toward the portion of the bar that creates the surface force. In our case, the vector $\mathbf{X}$ has only one component X_x:

$$\mathbf{X} = X_x \mathbf{i} \tag{1.19}$$

As follows from eq. 1.18, if $\mathbf{n} = \mathbf{i}$ then $X_x = F_x/S$, but $X_x = -F_x/S$ when $\mathbf{n} = -\mathbf{i}$. Since forces applied to bar faces have opposite directions, the function X_x has the same sign inside the bar. Correspondingly, eqs. 1.12 can be rewritten as

$$e_{xx} = \frac{1}{E} X_x \qquad \text{or} \qquad X_x = E e_{xx} \tag{1.20}$$

This shows that in the case of an expansion, $(e_{xx} > 0)$, X_x is positive, whereas it is negative if a compression takes place. Note that X_x is the diagonal element of the stress tensor. In the same manner we have

$$e_{yy} = \frac{1}{E} Y_y \qquad \text{or} \qquad Y_y = E e_{yy} \tag{1.21}$$

$$\text{and} \qquad e_{zz} = \frac{1}{E} Z_z \qquad \text{or} \qquad Z_z = E e_{zz}$$

when force is oriented either along the y– or z–axis. As is shown in Appendix C, functions Y_y and Z_z, as well as X_x, are the diagonal elements of the stress tensor. At the beginning we will use these notations, but later, throughout almost all monograph the notion τ_{ij} is applied. Thus, the differential forms of Hooke's law and Poisson's relation are

$$X_x = E e_{xx} \qquad \text{and} \qquad e_{yy} = e_{zz} = -\sigma e_{xx} \tag{1.22}$$

As was pointed out, this system is used to describe numerous phenomena in an elastic medium. First, we will study simple cases of homogeneous and inhomogeneous deformation.

Example one Consider an elastic column placed on an ideally rigid foundation, and suppose that force F_0 acts on its top surface S_t, Fig. 1.2c. Because of this and the weight, a cross-section of the column $S(z)$ varies, and our goal is to find its value, provided that the stress is constant inside the body:

$$Z_z = \text{const} \tag{1.23}$$

Since the external force F_0 is transmitted through the elastic column and its mass between the top S_t and the cross-section $S(z)$ is equal to

$$\rho \int_z^l S(z)dz,$$

the condition 1.23 can be written as

$$\frac{F_0}{S(z)} + \frac{\rho\, g}{S(z)} \int_z^l S(\eta)d\eta = \frac{F_0}{S_t}, \tag{1.24}$$

where η is a variable of integration, l is column length, and g is gravitational acceleration. Two terms at the left side of eq. 1.24 describe stresses caused by the external force and the weight of the column above the cross-section $S(z)$. Multiplication of this equation by $S(z)$ and then differentiation by z yield

$$-\rho\, gS(z) = \frac{F_0}{S_t}\frac{dS(z)}{dz} \qquad \text{or} \qquad \ln S(z) = -\frac{\rho\, gS_t}{F_0}z + C$$

Since

$$\ln S_t = -\frac{\rho\, gS_t}{F_0}l + C,$$

we finally obtain

$$S(z) = S_t \exp \frac{\rho\, gS_t}{F_0}(l - z) \tag{1.25}$$

Thus, in approaching the origin, $z = 0$, the cross-section of the column exponentially increases, and it provides a constancy of the stress, eq. 1.23. As follows from eq. 1.25 with a decrease of the density ρ, the change of the function $S(z)$ also becomes smaller. In other words, if an influence of the weight would be absent then a homogeneous deformation takes place with the constant cross-section of the column. From eq. 1.25 we see that

$$S(z) \to \infty \qquad \text{if} \qquad F_0 \to 0 \tag{1.26}$$

This means that the stress caused by a weight cannot be constant along the z–axis. In accordance with Hooke's law, the displacement of the surface S_t is

$$\delta l = \frac{F_0}{S_t E} l$$

Taking into account that a deformation is homogeneous, the strain at points of the column is constant, and it is equal to

$$e_{zz} = \frac{F_0}{S_t E}, \tag{1.27}$$

where F_0 is the scalar component of the force and is negative.

Example two Suppose that the bar is suspended, as is shown in Fig. 1.2d. Its cross-section S, length l, density ρ, and the Young modulus E are given. Our goal is to determine a bar extension due to the gravitational field. First, consider a portion of the bar, bounded by surfaces $S(O)$ and S, located at distance z from origin O. The force applied to $S(z)$ is equal to the weight of the lower portion of the bar:

$$F_z = g\rho(l - z)S \tag{1.28}$$

Since this force is uniformly distributed over the cross-section, the stress is

$$Z_z = \frac{F_z}{S} = g\rho(l - z), \tag{1.29}$$

and it varies linearly within the range

$$0 \leq \frac{F_z}{S} \leq g\ \rho l$$

It is clear that force F_z is external with respect to the upper part of the bar and, correspondingly, eq. 1.28 gives

$$\frac{dw}{dz} = \frac{Z_z}{E} = \frac{g\rho}{E}(l - z) \tag{1.30}$$

This shows that we are dealing with an inhomogeneous deformation. Integration of eq. 1.30 yields

$$w(z) = \frac{\rho g}{E}(lz - \frac{z^2}{2} + C)$$

Inasmuch as $w(0) = 0$, we obtain

$$w(z) = \frac{\rho g z}{E}(l - \frac{z}{2}) \tag{1.31}$$

In particular, an extension of the lower end, $(z = l)$, is

$$w(l) = \frac{g\ \rho l^2}{2E}, \tag{1.32}$$

that is, it is directly proportional to a square of the original length.

Superposition of displacements

To this point we have studied deformation caused by a single force. Taking into account that eqs. 1.12 are linear, we can apply the principle of superposition (Appendix D), which is formulated in the following way. If a deformation e_n corresponds to the force $\mathbf{F}_n$, then the resultant deformation, caused by a sum

$$\mathbf{F} = \mathbf{F}_1 + \mathbf{F}_2 + ... + \mathbf{F}_n,$$

is equal to

$$e = e_1 + e_2 + ... + e_n \tag{1.33}$$

Note that we have already discovered two kinds of deformation, namely, homogeneous and inhomogeneous deformations. The first corresponds to the case in which the strain remains the same inside a body, whereas, the second kind of strain is a function of a point. Of course, within an elementary volume, deformation is homogeneous (Appendix D). To illustrate the principle of superposition consider three more examples in which forces act on different faces of the rectangular body.

Example three Suppose that a body is surrounded by a fluid and the pressure is equal to P. In such case there are only forces that are normal to the body faces and, by definition

$$X_x = Y_y = Z_z = -P, \tag{1.34}$$

where $P > 0$. The presence of the minus sign is related to the fact that the force caused by a fluid is directed inside a body. Our task is to find displacements of faces along the coordinate axes caused by all three forces. Applying Hooke's law and Poisson's relation (eq. 1.8), we determine the relative displacement due to each force. Then, using the principle of superposition, we add them together. First, consider a compression along the x–axis due to the stress X_x. As follows from Hooke's law,

$$\frac{\partial u_1}{\partial x} = -\frac{1}{E}P \tag{1.35}$$

Because there are forces acting on two other faces of the volume, expansion along x is observed. In accordance with Poisson's relation and Hooke's law, we have

$$\frac{\partial u_2}{\partial x} = -\sigma\frac{\partial v}{\partial y} = \frac{\sigma}{E}P \qquad \text{and} \qquad \frac{\partial u_3}{\partial x} = -\sigma\frac{\partial w}{\partial z} = \frac{\sigma}{E}P \tag{1.36}$$

Therefore, the total strain, $\partial u/\partial x$, is

$$\frac{\partial u}{\partial x} = \frac{\partial u_1}{\partial x} + \frac{\partial u_2}{\partial x} + \frac{\partial u_3}{\partial x}$$

or

$$\frac{\partial u}{\partial x} = -\frac{1}{E}(1-2\sigma)P \tag{1.37}$$

In the same manner we obtain

$$\frac{\partial v}{\partial y} = -\frac{1}{E}(1-2\sigma)P \quad \text{and} \quad \frac{\partial w}{\partial z} = -\frac{1}{E}(1-2\sigma)P \tag{1.38}$$

Since a deformation is homogeneous, strains describe a change of the volume size along coordinate axes, that is

$$\frac{\delta l_1}{l_1} = \frac{\delta l_2}{l_2} = \frac{\delta l_3}{l_3} = -\frac{1}{E}(1-2\sigma)P \tag{1.39}$$

Here l_1, l_2, and l_3 are the initial lengths. In particular, when the volume is a cube, displacements of faces are the same.

It is easy to find a relationship between pressure and a relative change of volume caused by compression. The initial and new volumes are

$$V = l_1 l_2 l_3$$

and

$$V_1 = (l_1 + \delta l_1)(l_2 + \delta l_2)(l_3 + \delta l_3) = l_1 l_2 l_3 (1 + \frac{\delta l_1}{l_1})(1 + \frac{\delta l_2}{l_2})(1 + \frac{\delta l_3}{l_3})$$

Discarding higher-order terms we obtain

$$V_1 = V(1 + \frac{\delta l_1}{l_1} + \frac{\delta l_2}{l_2} + \frac{\delta l_3}{l_3}) \tag{1.40}$$

or

$$\frac{\delta V}{V} = -\frac{3(1-2\sigma)}{E}P \tag{1.41}$$

Here $\delta V/V = \Theta =$ div**s** is a dilatation (Appendix D), and, correspondingly,

$$P = -M\frac{dV}{V} = -M \text{div } \mathbf{s}, \tag{1.42}$$

where

$$M = \frac{E}{3(1-2\sigma)} \tag{1.43}$$

is the bulk modulus.

As follows from eq. 1.41 Poisson's ratio, σ, cannot exceed $1/2$; otherwise, with an increase of pressure, volume would increase too. Note that the velocity of propagation in a fluid is defined by the density ρ and the bulk modulus M (Part I).

Example four Assume that there are forces oriented along the x–axis that produce a bar extension. There are also forces applied to faces perpendicular to the y–axis, so that these faces of the bar cannot move in this direction. At the same time, faces normal to the z–axis are not subjected to an action of forces. Applying again the principle of superposition, we have

$$\frac{\delta l_1}{l_1} = \frac{1}{E}\frac{F_x}{S_x} - \frac{\sigma}{E}\frac{F_y}{S_y}, \tag{1.44}$$

where F_y is an unknown force. In order to find it, we take into account that $\delta l_2 = 0$, and this gives

$$\frac{\delta l_2}{l_2} = \frac{1}{E}\frac{F_y}{S_y} - \frac{\sigma}{E}\frac{F_x}{S_x} = 0$$

or

$$\frac{F_y}{S_y} = \sigma\frac{F_x}{S_x} \qquad \text{or} \qquad Y_y = \sigma X_x \tag{1.45}$$

Substituting the latter into eq. 1.44 we have

$$e_{xx} = \frac{\partial u}{\partial x} = \frac{1}{E}(1-\sigma^2)X_x \qquad \text{or} \qquad X_x = \frac{E}{1-\sigma^2}e_{xx} = E_* e_{xx} \tag{1.46}$$

Here $E_* = E/1-\sigma^2$ is usually called the effective Young modulus.

Example five Unlike in the previous case, the faces of a bar normal to the y and z–axes cannot move, but force $\mathbf{F}_x$ produces a deformation along the x–axis. This means that there are normal stresses at faces S_y and S_z. Then, for strains e_{xx}, e_{yy}, and e_{zz}, we can write

$$e_{xx} = \frac{1}{E}X_x - \frac{\sigma}{E}Y_y - \frac{\sigma}{E}Z_z$$

$$e_{yy} = -\frac{\sigma}{E}X_x + \frac{1}{E}Y_y - \frac{\sigma}{E}Z_z = 0 \tag{1.47}$$

$$e_{zz} = -\frac{\sigma}{E}X_x - \frac{\sigma}{E}Y_y + \frac{1}{E}Z_z = 0$$

This system contains three unknowns: e_{xx}, Y_y, and Z_z. First of all we find Y_y. Eliminating Z_z we obtain two equations

$$e_{xx} = \frac{1-\sigma^2}{E}X_x - \frac{\sigma(1+\sigma)}{E}Y_y$$

and

$$0 = -\frac{\sigma(1+\sigma)}{E}X_x + \frac{(1-\sigma^2)}{E}Y_y$$

The last equation gives

$$Y_y = \frac{\sigma}{1-\sigma}X_x \tag{1.48}$$

and, correspondingly,

$$e_{xx} = [1-\sigma^2 - \frac{\sigma^2(1+\sigma)}{1-\sigma}]\frac{X_x}{E}$$

or

$$X_x = E_{**}e_{xx} \tag{1.49}$$

Here

$$E_{**} = \frac{1-\sigma}{(1+\sigma)(1-2\sigma)}E \tag{1.50}$$

is another effective Young modulus. As will be shown later this modulus defines the velocity of the longitudinal waves. By analogy with eq. 1.48 we also have

$$Z_z = \frac{\sigma}{1-\sigma}X_x \tag{1.51}$$

1.2 Longitudinal waves in a thin bar

In the previous section we studied deformation, when a body is in equilibrium, and, correspondingly, its particles do not move. At the same time, as we know, such a state does not occur instantly; it is preceded by wave propagation and attenuation. Now we turn our focus to the wave phenomena and start from the simplest case, when the longitudinal wave propagates along a slender bar, Fig. 1.3a.

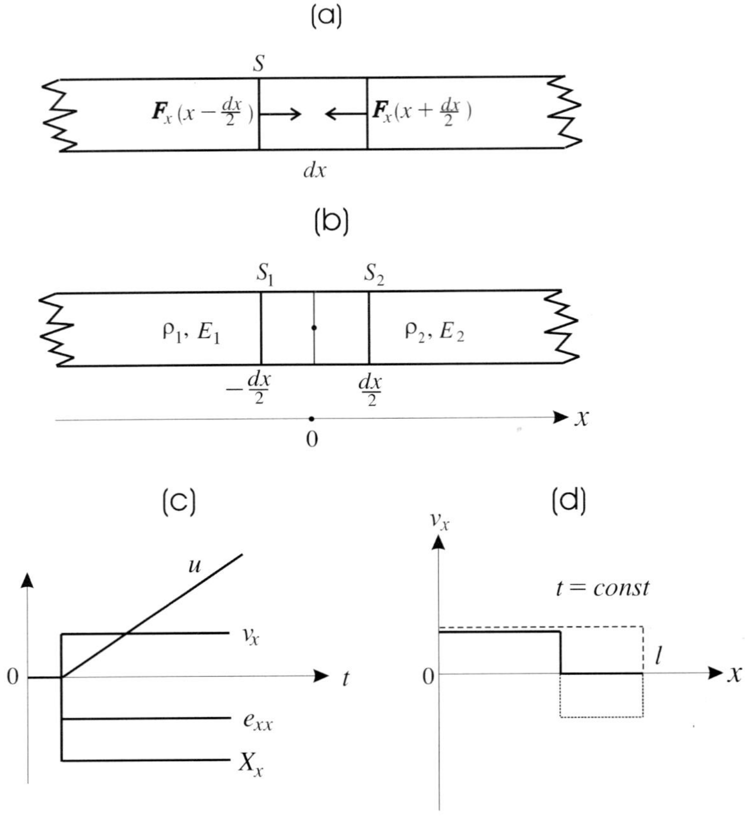

Figure 1.3: (a) Derivation of wave equation (b) Illustration of eq. 1.75 (c) Wave fields of the step function (d) Superposition of two compressional waves

One-dimensional wave equation

By analogy with acoustic waves (Part I), we first derive the equation describing wave propagation along a bar. With this purpose in mind consider the bar element, bounded by two cross-sections, $S(x-dx/2)$ and $S(x+dx/2)$, and the lateral surface, Fig. 1.3a. Suppose that the wave propagates along the x–axis and reaches the cross-section $S(x-dx/2)$. From this moment we begin to observe deformation of this bar element, $dV = Sdx$. With some time delay, the wave arrives at the front face of the volume, $S(x+dx/2)$. Force $F(x,t)$, which accompanies the wave, has the same direction at both cross-sections, but may differ in magnitude. For instance, in the case of compression, this force is directed along the x–axis, whereas it moves in the opposite direction when expansion takes place. The force at points of the section $S(x + dx/2)$ acts on a medium located in front of the bar element. In accordance with Newton's third law, force $\mathbf{F}_x(x+dx/2)$, having the same magnitude and opposite direction, acts on the surface $S(x + dx/2)$ of the elementary volume. Thus, the resultant force, applied to this bar element at any instant t, is

$$\mathbf{F}_x(x - \frac{dx}{2}, t) + \mathbf{F}_x(x + \frac{dx}{2}, t)$$

It is essential that both forces are considered at the same moment and that they always have opposite directions. In general, their magnitudes are different. In accordance with Newton's second law, the equation of motion of the bar element with the mass $m = \rho S dx$ is

$$m\frac{\partial^2 u(x,t)}{\partial t^2} = F_x(x - \frac{dx}{2}, t) + F_x(x + \frac{dx}{2}, t) \tag{1.52}$$

Here $u(x, t)$ is the x–component of the displacement of the center of mass, m. Eq. 1.52 contains three unknowns, $u(x,t)$, $F_x(x - dx/2, t)$, and $F_x(x + dx/2, t)$. In order to derive an equation with respect to only one of these functions, we use the concept of traction, $\mathbf{t}$, and Hooke's law. As was shown in the previous section and in Appendix C, the x–component of the traction $\mathbf{t}$ is defined as

$$t_x = \mathbf{X} \cdot \mathbf{n}$$

Respectively, we have

$$F_x(x - \frac{dx}{2}, t) = t_x(x - \frac{dx}{2}, t)S = -X_x(x - \frac{dx}{2}, t)\ S \tag{1.53}$$

and $$F_x(x + \frac{dx}{2}, t) = t_x(x + \frac{dx}{2}, t)S = \ X_x(x + \frac{dx}{2}, t)\ S$$

Substitution of eqs. 1.53 into eq. 1.52 yields

$$\rho\ dx\frac{\partial^2 u(x,t)}{\partial t^2} = X_x(x + \frac{dx}{2}, t) - X_x(x - \frac{dx}{2}, t) \tag{1.54}$$

Here X_x is the normal stress at the front and back faces of the volume element. As the distance dx is very small, we can assume that the stress X_x linearly changes between the volume faces. It allows us to replace the right side of eq. 1.54 as

$$X_x(x + \frac{dx}{2}, t) - X_x(x - \frac{dx}{2}, t) = \frac{\partial X_x(x,t)}{\partial x}dx, \tag{1.55}$$

and it becomes

$$\rho\,\frac{\partial^2 u(x,t)}{\partial t^2} = \frac{\partial X_x(x,t)}{\partial x}, \tag{1.56}$$

where derivatives are taken at the same point. The next step involves the use of Hooke's law:

$$X_x = E\ e_{xx} = E\frac{\partial u}{\partial x} \tag{1.57}$$

From the last two equations, we obtain

$$\rho\frac{\partial^2 u}{\partial t^2} = E\frac{\partial^2 u}{\partial x^2} \qquad \text{or} \qquad \frac{\partial^2 u}{\partial x^2} = \frac{1}{c_l^2}\frac{\partial^2 u}{\partial t^2} \tag{1.58}$$

As we know (Part I), the latter is the wave equation in an one-dimensional case and

$$c_l = \sqrt{\frac{E}{\rho}} \tag{1.59}$$

describes its velocity of propagation. We derive the same equation for strain and stress. In fact, taking the derivative with respect to x from both sides of eq. 1.58, we have

$$\frac{\partial^2}{\partial x^2}(\frac{\partial u}{\partial x}) = \frac{1}{c_l^2}\frac{\partial^2}{\partial t^2}(\frac{\partial u}{\partial x})$$

or

$$\frac{\partial^2 e_{xx}}{\partial x^2} = \frac{1}{c_l^2}\frac{\partial^2 e_{xx}}{\partial t^2} \tag{1.60}$$

Then, the use of Hooke's law gives

$$\frac{\partial^2 X_x}{\partial x^2} = \frac{1}{c_l^2}\frac{\partial^2 X_x}{\partial t^2} \tag{1.61}$$

It is natural that all three quantities characterizing wave propagation satisfy the same wave equation. It is a partial differential equation of the second order, and its general solution for displacement (Part I) is

$$u(x,t) = A\ f\ [a(t-\frac{x}{c_l})] + Bg[a(t+\frac{x}{c_l})] \tag{1.62}$$

Here f and g are continuous functions that have first and second derivatives with respect to both distance x and time t. Their behavior is defined by the primary source, which generates the wave. A and B are some constants, as well as the parameter a.

Since an argument of any function is dimensionless, this constant is measured in $\sec^{-1}$. Now it may be appropriate to remind ourselves the following.

a. A solution of the wave equation was first obtained by D'Alembert who applied in essence the trial-and-error method. It does not determine functions f and g, but defines a structure of arguments, that is, a relationship between distance x and time t.

b. In deriving eqs. 1.58–1.61, it was assumed that $u(x,t)$, $e_{xx}\ (x,t)$, and $X_x(x,t)$ are continuous functions. However, it turns out that even a discontinuous function can be a solution of the wave equation.

c. It is easy to show that $u(x,t)$, given by eq. 1.62, satisfies the wave equation. In fact, performing a differentiation, we have

$$\frac{\partial^2 f}{\partial x^2} = A\frac{a^2}{c_l^2} f''[a(t-\frac{x}{c_l})] \quad \text{and} \quad \frac{\partial^2 f}{\partial t^2} = A\ a^2 f''[a(t-\frac{x}{c_l})],$$

where derivatives are taken with respect to the argument of the function, which is $a(t-x/c_l)$. It is clear that

$$\frac{\partial^2 f}{\partial x^2} = \frac{1}{c_l^2}\frac{\partial^2 f}{\partial t^2},$$

and the same is valid for the function $g[a(t+x/c_l)]$.

d. Since with an increase of time the same value of the argument $t-x/c_l$ is observed at greater distances, the function $f\ [a(t-x/c_l)]$ describes the outgoing wave that is traveling away from the origin, if $x>0$. In contrast, the function $g[a(t+x/c_l)]$ characterizes the incoming wave, if $x>0$, because with a decrease of distance the same value of the argument $t+x/c_l$ takes place later. It may happen that the wave field is described by either by the outgoing or incoming waves or by the superposition of them.

Now consider the main properties of these waves in some detail.

Outgoing wave By definition, expressions for the displacement, the particle velocity, strain, and stress are

$$u(x,t) = A\ f[a(t-\frac{x}{c_l})], \qquad \dot{u}(x,t) = A\ af'[a(t-\frac{x}{c_l})], \tag{1.63}$$

$$e_{xx}(x,t) = -\frac{Aa}{c_l} f'[a(t-\frac{x}{c_l})], \qquad X_x(x,t) = -\frac{AaE}{c_l} f'[a(t-\frac{x}{c_l})]$$

As we can see, the last three quantities have the same dependence on distance and time. From eqs. 1.63 it also follows that

$$e_{xx} = -\frac{\dot{u}}{c_l}, \tag{1.64}$$

that is, the ratio of velocities of the particle and wave defines the strain with an accuracy of a sign. As was mentioned earlier and in Appendix D the strain e_{xx} is usually very small: $(e_{xx} \ll 1)$. Therefore, the particle velocity is many orders smaller than the wave velocity. It is clear that in the case of the compressional wave, $e_{xx} < 0$ and both velocities have the same direction. Due to this fact, compression takes place. On the contrary, when they have opposite directions, $e_{xx} > 0$, the extensional wave propagates. Let us imagine that at the instant $t = 0$, the wave approaches the left face of some portion of the bar, which has a length equal to $c_l \delta t$. Then during this brief time interval, δt, the bar end moves at the distance $\dot{u}\delta t$, and, correspondingly, the ratio $\dot{u}/c_l$ describes the strain. As follows from Hooke's law, eq. 1.64 can written in the form:

$$\frac{\dot{u}}{c_l} = -\frac{X_x}{E} \qquad \text{or} \qquad \frac{X_x}{\dot{u}} = -\frac{E}{c_l} = -\sqrt{E\rho} \tag{1.65}$$

By analogy with electrodynamics, the right side of eq. 1.65 is called the impedance of a medium for longitudinal waves:

$$Z = \frac{E}{c_l} = \sqrt{E\rho} = c_l \rho \tag{1.66}$$

As is also the case in acoustics, impedance plays an important role in describing reflection and transmission of waves.

Incoming wave In accordance with eq. 1.62 we have in this case

$$u(x,t) = Bg[a(t + \frac{x}{c_l})], \qquad \dot{u}(x,t) = Bag'[a(t + \frac{x}{c_l})] \tag{1.67}$$

$$e_{xx}(x,t) = \frac{Ba}{c_l} g'[a(t + \frac{x}{c_l})], \qquad X_x(x,t) = \frac{BaE}{c_l} g'[a(t + \frac{x}{c_l})]$$

$$e_{xx} = \frac{\dot{u}}{c_l} \qquad \text{and} \qquad \frac{\dot{u}}{c_l} = \frac{X_x}{E} \qquad \text{or} \qquad \frac{X_x}{\dot{u}} = Z$$

Since the wave propagates toward the origin, $(x > 0)$, both the wave and particle velocities have the same direction in places where compression occurs. However, the directions are opposite to each other in places where tension occurs. The similarity for outgoing waves is obvious.

Displacement field

In accordance with Poisson's relation

$$\frac{\partial v}{\partial y} = \frac{\partial w}{\partial z} = -\sigma \frac{\partial u}{\partial x}, \tag{1.68}$$

wave propagation along the x–axis is accompanied by motion of particles along all axes. In the approximation of a thin bar, it is assumed that all functions describing the wave

$$u(x,t), \qquad e_{xx}(x,t), \qquad \dot{u}(x,t), \qquad X_x(x,t)$$

are the same at all points of any cross-section of the bar. Besides, the strains e_{yy} and e_{zz}, caused by the stress X_x, are uniformly distributed over each cross-section S. Because of symmetry, components of the displacement v and w are equal to zero along the middle line of the bar (x–axis), and then they linearly increase toward the lateral surface. Because of all these assumptions, we can only approximately describe the displacement field **s**. As follows from eq. 1.63, all strains vary synchronously, but the vector field **s**:

$$\mathbf{s} = u\mathbf{i} + v\mathbf{j} + w\mathbf{k} \tag{1.69}$$

can be rather complicated. In order to study its behavior, we use eqs. 1.63 and 1.14. They give

$$\frac{\partial u(x,t)}{\partial x} = -\frac{A\,a}{c_l} f'[a(t-\frac{x}{c_l})], \tag{1.70}$$

$$\frac{\partial v(x,t)}{\partial y} = \frac{A\,a\,\sigma}{c_l} f'[a(t-\frac{x}{c_l})], \qquad \frac{\partial w}{\partial z} = \frac{A\,a\,\sigma}{c_l} f'[a(t-\frac{x}{c_l})]$$

After integration of the last two equations, we have for the field **s**

$$u(x,t) = Af\,[a(t-\frac{x}{c_l})], \qquad v(x,t) = \frac{A\,a\,\sigma}{c_l} f'[a(t-\frac{x}{c_l})]\,y, \tag{1.71}$$

$$w(x,t) = \frac{A\,a\,\sigma}{c_l} f'[a(t-\frac{x}{c_l})]\,z,$$

since $v = w = 0$ at points of the x–axis. As is seen from set 1.71, the displacements along the y and z axes are defined by the rate of change of the component $u(x,t)$. Taking into account that the bar is thin, we usually have

$$v \ll u \qquad \text{and} \qquad w \ll u \tag{1.72}$$

However, the strains e_{xx}, e_{yy}, and e_{zz} are comparable, eqs. 1.70, and therefore, the dilatation, Θ, is equal to

$$\Theta = \text{div } \mathbf{s} = -\frac{Aa}{c_l}(1-2\sigma)\ f'[a(t-\frac{x}{c_l})] \tag{1.73}$$

As was shown earlier, the dilatation defines the relative change of an elementary volume, $\Theta = \Delta V/V$, and wave propagation is accompanied by either compression or expansion of the volume. At the beginning, we assumed that the displacement $u(x,t)$ is uniformly distributed over the cross-section. If, in addition, we neglect by components v and w , it is easy to see that curl $\mathbf{s} = 0$. In other words, wave propagation is not accompanied by rotation of elementary volumes of the bar. At the same time, wave causes a change of volumes. This is why these waves are called dilatational or irrotational waves.

Reflection and transmission

Now, suppose that the bar consists of two homogeneous portions and S is the boundary between them. In general, they differ in terms of both the Young modulus and density ρ:

$$E_1 \neq E_2 \qquad \text{and} \qquad \rho_1 \neq \rho_2$$

Since at points of the cross-section S, $(x=0)$, the velocity c_l is usually a discontinuous function, we cannot apply the wave equation. Therefore, it has to be replaced on S by boundary conditions. One of them is a continuity of the displacement u

$$u_1(0,t) = u_2(0,t) \tag{1.74}$$

Here $u_1(x,t)$ and $u_2(x,t)$ are scalar components of displacement along the x-axis at each portion of the bar. It is clear that if condition 1.74 is invalid, we would observe either overlapping of two portions of the bar or a gap between them. In order to derive the second condition, consider an elementary volume, confined by surfaces S_1 and S_2, Fig. 1.3b, which are located at distance dx from each other. In accordance with the Newton's second law, we have

$$m\frac{\partial^2 u}{\partial t^2} = F_x(S_1) + F_x(S_2) \tag{1.75}$$

Here m is the mass of the volume and $F_x(S_1)$ and $F_x(S_2)$ are external forces acting on its faces. Since $F_x = (\mathbf{X}\cdot\mathbf{n})S$, we have

$$m\frac{\partial^2 u}{\partial t^2} = [X_{2x}(\frac{dx}{2}) - X_{1x}(-\frac{dx}{2})]S$$

or

$$\rho\, dx \frac{\partial^2 u}{\partial t^2} = X_{2x}(\frac{dx}{2}) - X_{1x}(-\frac{dx}{2}), \tag{1.76}$$

where X_{2x} and X_{1x} are normal stresses at opposite sides of the boundary.

As S_1 and S_2 approach the boundary S, $(dx \to 0)$, the left side of eq. 1.76 tends to zero, since an acceleration cannot be infinitely large. Therefore, the right side also tends to zero, and we conclude that the normal stress is a continuous function at the boundary

$$X_{1x}(0,t) = X_{2x}(0,t) \tag{1.77}$$

Thus, boundary conditions are

$$u_1(0,t) = u_2(0,t) \qquad \text{and} \qquad X_{1x}(0,t) = X_{2x}(0,t) \tag{1.78}$$

Our goal is to find the wave field that satisfies the wave equation at each regular point of the bar and boundary conditions, eqs. 1.78. Assuming that the primary (incident) wave propagates along the x–axis, it is obvious that in the first portion of the bar, $(x < 0)$, there are two waves – the incident and reflected waves – whereas in the second portion, $(x > 0)$, we observe the transmitted wave. Respectively, expressions for the displacements are

$$u_1(x,t) = Af[a(t - \frac{x}{c_{1l}})] + Bf[a(t + \frac{x}{c_{1l}})] \quad \text{if } x < 0 \tag{1.79}$$

$$\text{and} \qquad u_2(x,t) = C\ f\ [a(t - \frac{x}{c_{2l}})] \qquad \text{if } x > 0$$

Here A is known, while B and C have to be determined, and

$$c_{1l} = \sqrt{\frac{E_1}{\rho_1}} \qquad \text{and} \qquad c_{2l} = \sqrt{\frac{E_2}{\rho_2}} \tag{1.80}$$

Of course, functions u_1 and u_2 satisfy the wave equation at each portion of the bar. Now it is proper to make a comment. We supposed that the incident, reflected, and transmitted waves are described by the same function

$$f\ [a(t \pm \frac{x}{c_l})]$$

This is only an assumption. However, if $u_1(x,t)$ and $u_2(x,t)$ satisfy boundary conditions, then we can say that this guess is correct. Now, making use of eqs. 1.79 and Hooke's law

$$X_x = Ee_{xx} = E\frac{\partial u}{\partial x},$$

system 1.78 gives two equations with two unknowns

$$A + B = C \qquad \text{and} \qquad -AZ_1 + BZ_1 = -Z_2C_1 \tag{1.81}$$

since

$$Z = \frac{E}{c} = \rho c$$

Their solution yields the expressions

$$B = \frac{Z_1 - Z_2}{Z_1 + Z_2}A \qquad \text{and} \qquad C = \frac{2Z_1}{Z_1 + Z_2}A, \tag{1.82}$$

which represent the famous Fresnel formulas. Thus, we have demonstrated that $u_1(x,t)$ and $u_2(x,t)$ satisfy wave equations for corresponding parts of the bar and boundary conditions, provided that B and C are given by eqs. 1.82. This means that all three waves depend on time and distance in the same manner and that coefficients of reflection and transmission, eqs. 1.82, are defined by the ratio of impedances. Certainly, this important result is always valid, as long as boundary conditions are independent of time. Fresnel's coefficients can be also written in the other form:

$$B = \frac{\sqrt{\rho_1 E_1} - \sqrt{\rho_2 E_2}}{\sqrt{\rho_1 E_1} + \sqrt{\rho_2 E_2}}A = \frac{n - m}{n + m}A \qquad \text{and} \qquad C = \frac{2n}{n + m}A \tag{1.83}$$

where $n = c_1/c_2$ is the refraction coefficient and $m = \rho_2/\rho_1$ is the ratio of densities. Thus, expressions for displacement u in each part of the bar are

$$u_1(x,t) = Af[a(t - \frac{x}{c_{1l}})] + \frac{Z_1 - Z_2}{Z_1 + Z_2}A\ f[a(t + \frac{x}{c_{1l}})] \tag{1.84}$$

$$\text{and} \qquad u_2(x,t) = \frac{2Z_1}{Z_1 + Z_2}A\ f[a(t - \frac{x}{c_{2l}})]$$

In particular, at the boundary, $(x = 0)$, we have

$$u_1(o,t) = (1 + \frac{Z_1 - Z_2}{Z_1 + Z_2})A\ f(at) \quad \text{and} \quad u_2(o,t) = u_1(o,t) = \frac{2Z_1}{Z_1 + Z_2}A\ f(at) \tag{1.85}$$

Eq. 1.85 emphasizes that at the boundary, the reflected and incident waves vary synchronously, and this fact plays a fundamental role in migration (Part II). Of course, the transmitted wave displays similar behavior. As follows from the first equation of the 1.84, the strain at the boundary caused by the incident and reflected waves is

$$e_{xx} = \frac{aA}{c_{1l}}(-1 + \frac{Z_1 - Z_2}{Z_1 + Z_2})\ f'(at)$$

Thus, at the boundary the strain of the incident and reflected waves has the same sign, if $Z_1 < Z_2$. This means that either both waves are compressional or both are extensional. For instance, at the rigid boundary, $Z_2 \to \infty$, the compressional wave gives rise to the reflected compressional wave. In contrast, if $Z_1 > Z_2$, the incident and reflected waves are of different types. This is clearly illustrated in the case of a free boundary, $Z_2 = 0$, when the compressional wave causes the extensional reflected wave.

Now we illustrate wave behavior, considering several examples. In all of them it is assumed that the external force changes instantly and then remains constant for some time. Speaking strictly, such behavior is impossible, since a finite time interval is always needed to generate the constant force. For this reason we treat such wave as the limiting case when the real force arises very quickly. Behind the wave front the force remains constant and, of course, all functions – $u(x,t)$, $e_{xx}(x,t)$, and $X_x(x,t)$ – satisfy the wave equation.

Example one First, suppose that the incident wave propagates along a homogeneous bar and the stress, X_x, behaves as

$$X_x = A\ h[a(t - \frac{x}{c_l})], \tag{1.86}$$

where h is the step function.

$$h[a(t - \frac{x}{c_l})] = 0 \quad \text{if} \quad t < \frac{x}{c_l} \qquad \text{and} \qquad h[a(t - \frac{x}{c_l})] = 1 \quad \text{if} \quad t > \frac{x}{c_l}$$

Consider an elementary volume of the bar with an infinitely small extension dx. Because of this, the wave almost instantly reaches the back and front faces of the volume. Correspondingly, external forces acting on both faces have the same magnitude but opposite directions. Therefore, the total force is equal to zero, and this element moves at a constant velocity. Thus, stress, strain, and particle velocity behave as step-functions, whereas displacement is a linear function, Fig. 1.3c. In other words, functions, $\dot{u}$, e_{xx}, and X_x remain constant behind the wave front, but the distance of the particle from the original position changes linearly. We have discussed a compressional wave. Behavior of these functions in the case of the extensional wave, $e_{xx} > 0$, is similar.

Second, assume that a semi-infinite bar has a rigid end. When the incident wave reaches this end, the whole bar is compressed, $e_{xx} < 0$, and it moves at a constant velocity, $\dot{u} > 0$. At this moment the reflected compressional wave arises and, as result, the strain and stress magnitudes double behind the wave front of the reflected wave. Since the particle velocities of the incident and reflected waves have opposite directions, the resultant velocity of the bar is equal to zero between the rigid end and the front of the reflected wave, Fig. 1.3d. Thus, one portion of the bar does not move, and it is twice as deformed as the rest of the bar, which moves with the velocity $\dot{u}$. If the bar end is free, then the compressional wave causes the extensional wave. Because of this behind the wave's front the particle velocity doubles but deformation vanishes. Unlike the previous case of the rigid end, both portions of the bar move but with different velocities, and in one part both the stress and strain are equal to zero. Now we are prepared to consider several more examples.

Example two Suppose that the bar of length l is under an action of two constant forces that are applied at the same moment $t = 0$, Fig. 1.4a. At this instant, two compressional waves arise and move in opposite directions. Since they reach the center of the bar simultaneously, the strain at this point becomes equal to $2e_{xx}$, but the particle velocity is zero. Here e_{xx} is the strain caused by the single wave. At the instant $t = l/c_l$ extensional waves arise, and they again arrive at the middle point O at the same time. Correspondingly, this point is still at rest, but a deformation disappears. Because reflected waves regularly appear at both ends of the bar, we can say that the velocity at point O is always equal to zero,

$$\dot{u}(0, t) = 0, \tag{1.87}$$

and the center of mass is located at this point. At the same time strain like stress, is a periodic function, Fig. 1.4b. Of course, condition 1.87 follows from the fact that the resultant external force is equal to zero, while the wave propagation explains how it happens. The periodic function $e_{xx}(0, t)$ can be represented as a sum of sinusoidal functions and the constant. Because of attenuation, harmonic functions decay, and in equilibrium strain becomes constant. Unlike the middle point, other points of the bar experience motion, which is also described by the periodic function of time. Again due to attenuation, sinusoidal harmonics disappear and only the constant portion remains. It turns out that the latter linearly decreases in approaching the middle point O.

Example three As is well known, the essence of Hooke's law is the fact that a displacement Δl is directly proportional to the bar length $\Delta l \sim l$. For instance,

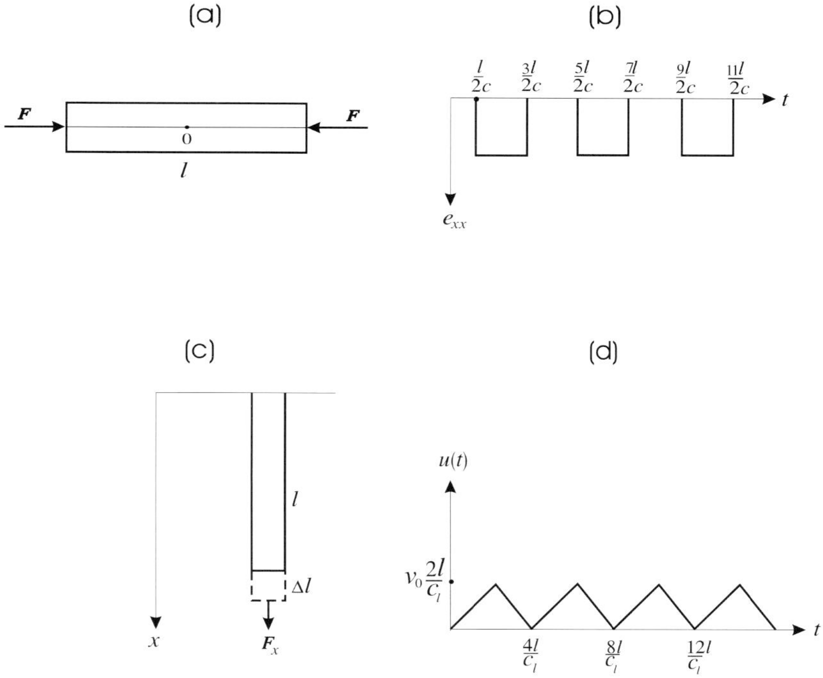

Figure 1.4: (a) Bar under the action of two forces (b) Strain at the middle point as a function of time (c) Illustration of Hooke's law (d) Displacement of the low end of the bar as a function of time

a semi-infinite bar has an infinite extension, $\Delta l \to \infty$. This behavior of Δl poses the following question. How does the low end of the bar, Fig. 1.4c, "know" where the opposite end is located? In order to find an explanation, consider the motion of the low end under action of the constant force F_x:

$$F_x = 0 \quad \text{if} \quad t < 0 \qquad \text{and} \qquad F_x = \text{const} \qquad \text{if} \quad t > 0,$$

provided that the upper end of the bar is fixed. Since the force F_x is directed downward, an extensional wave arises at the instant $t = 0$ and propagates toward the upper end. At the same time, the low end moves with constant velocity along the x–axis. Because of a reflection at the upper end, the extensional wave arises at the instant $t = l/c$ and travels downward, but the particle motion has an opposite direction. For this reason, the bar stops to move behind the wave front. At the instant $t = 2l/c$ the wave reaches the low end, and the whole bar is at rest. However, at the same moment the compressional wave appears at the low end and propagates upward. Correspondingly, the direction of the motion of the low end changes. Thus, its velocity v_0 remains constant only within the time interval

$$0 < t < \frac{2l}{c_l}$$

When the compressional wave reaches the upper fixed end at $t = 3l/2c_l$, it gives rise to another compressional wave. The new wave propagates downward and the particle velocity is directed along the x–axis. In particular, at the instant $t = 4l/c_l$ all points of the bar are at rest. At this moment, the extensional wave starts to move upward. Thus, we begin to observe the same behavior of the particle velocity as before. It is obvious that the function $v(t)$ is periodic, and its period is equal to

$$T = \frac{4l}{c_l} \tag{1.88}$$

Within each period we have

$$v(t) = v_0 \quad \text{if} \quad -\frac{T}{2} < t < 0 \qquad \text{and} \qquad v(t) = -v_0 \quad \text{if} \quad 0 < t < \frac{T}{2} \tag{1.89}$$

Therefore during the first half of the period, the displacement of the low end $u(l,t)$ linearly increases with time, Fig. 1.4d, and reaches its maximal value

$$u = \frac{2l}{c_l} v_0 \tag{1.90}$$

Then, in the other half of the period, the displacement linearly decreases and the low end returns to the original position at the instant $t = 4l/c_l$. By analogy with the previous example, it is important to represent the even periodic function $u(l,t)$ as the Fourier series

$$u(l,t) = \frac{b_0}{2} + \sum_{n=1}^{\infty} b_n \cos 2\pi n \frac{t}{T}, \tag{1.91}$$

where

$$b_n = \frac{4}{T} \int_0^{2l/c_l} u(l,t) \cos 2\pi n \frac{t}{T} dt \tag{1.92}$$

In particular, the constant b_0 is equal to

$$b_0 = \frac{c_l}{l} v_0 \int_0^{\frac{2l}{c_l}} t dt = 2\frac{v_0 l}{c_l} \qquad \text{or} \qquad \frac{b_0}{2} = \frac{v_0 l}{c_l} \tag{1.93}$$

Thus, the total displacement of the low end of the bar is a sum of the constant displacement and a system of harmonics with different frequencies. It is remarkable that the constant part of the displacement

$$\Delta l = \frac{b_0}{2} = \frac{v_0 l}{c_l} \tag{1.94}$$

is directly proportional to the original length of the bar. In other words, Δl is proportional to the time during which the wave travels from the lower to the upper end. As follows from eq. 1.94, $\Delta l/l = v_0/c_l$, and this result is not surprising, since the left and right sides of this equality characterize the strain. Since

$$\frac{v_0}{c_l} = \frac{X_x}{E},$$

we have

$$\frac{\Delta l}{l} = \frac{X_x}{E} \qquad \text{or} \qquad e_{xx} = \frac{X_x}{E}$$

This describes Hooke's law, provided that Δl is half of the maximal displacement of the low end. In this light it is appropriate to notice the following. Hooke's law is based on an experimental fact, namely, that under an action of constant force, the displacement of the bar end reaches some value and then remains constant. Certainly, such behavior is different from one prescribed by the function $u(l,t)$. This discrepancy is easily explained if we take into account an effect of attenuation. In reality, vibrations described by harmonics of Fourier's series $(n \geq 1)$ decay relatively quickly with time, since the period T is very small. Correspondingly, observations performed at times significantly exceeding T allow us to find Δl. Of course, if the external force F_x varies slowly, the influence of harmonics is strongly reduced even at earlier times. This analysis shows that the wave "informs" the low end of the bar about its original length l.

Example four: propagation of pulse and Newton's first law Suppose that at the instant $t = 0$ the narrow impulse of the force

$$N_x = F_x \delta t \tag{1.95}$$

is applied to the left end of the bar, which has a length l and the cross-section S, Fig. 1.5a. At this moment, the pulse of the compressional wave arises and moves with the velocity c_l. We assume that $c_l \delta t \ll l$ and that a change of the bar length due to either a compression or extension is extremely small. At the beginning we study a

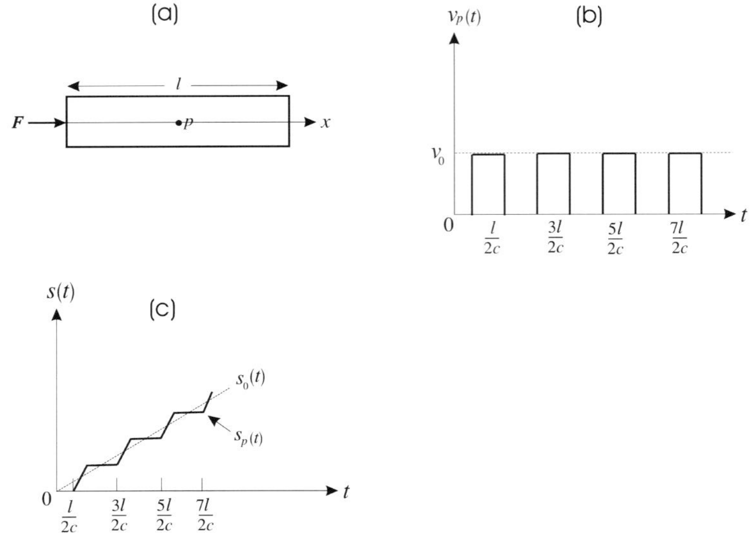

Figure 1.5: (a) The bar under action of constant force (b) Velocity of the middle point as a function of time (c) Displacements of the point p and the center of mass as functions of time

motion of the middle point p, which is the center of mass at equilibrium o. As we know, propagation of the compressional pulse is accompanied by particle motion with the velocity v_0, ($v_0 \ll c$), in the same direction. For this reason, at the instant $t = l/2c$ point p starts to move along the x–axis during the time interval δt, and then it stops. The wave pulse reaches the right free end of the bar at the instant $t = l/c_l$. At this moment the extensional reflected wave arises and starts to propagate toward the left end. When it reaches point p, this point begins to move again with velocity v_0 along the x–axis, but only during time interval δt. Finally, at moment $t = 2l/c_l$ this wave pulse approaches the left free end, and the compressional reflected wave arises. Thus, we see that the velocity $v(t)$ of the middle point is a periodic function, Fig. 1.5b, and its period is equal to

$$T = \frac{l}{c_l} \tag{1.96}$$

Here T is time of wave traveling between the bar ends. Similar behavior of the velocity $v(t)$ takes place at other points. For instance, at the bar ends, both the period becomes and the particle velocity double. The latter happens due to superposition of the incident and reflected waves. As is seen from Fig. 1.5b, the velocity of the middle point p during each period is either zero or is equal to v_0. On the other hand, in accordance with Newton's first law, the velocity of the center of mass is constant, and it is defined from the equality

$$mV = N_x = F_x \delta t, \tag{1.97}$$

where m is the bar mass $m = \rho \, lS$ and ρ is the density of the bar in equilibrium. Certainly, there is a difference between the actual velocity $v(t)$ at the middle point p, Fig. 1.5b, and the constant velocity V of the center of mass o. In order to find a relationship between them, we introduce an average value of the function $v(t)$. By definition we have

$$v^{av} = \frac{1}{T} \int_0^T v(t) dt \qquad \text{or} \qquad v^{av} = \frac{\delta t}{T} v_0 = \frac{c_l \, \delta t}{c_l \, T} v_0 \tag{1.98}$$

Thus, the coefficient of proportionality between v^{av} and v_0 is the ratio of the width of the pulse to the bar length. Multiplication by m of both sides of eq. 1.98 gives

$$m \, v^{av} = \frac{m c_l \delta t}{l} v_0 = \rho \, c_l S \, \delta t \, v_0 \tag{1.99}$$

As was demonstrated earlier

$$v_0 = \frac{c_l F_x}{ES}$$

Its substitution into eq. 1.99 yields

$$m v^{av} = F_x \delta t \tag{1.100}$$

Comparison with eq. 1.97 shows that the constant velocity V in Newton's first law represents an average velocity, $v^{av}(t)$, of the middle point p. The same is valid for all other points of the bar. Note that if the width of the wave pulse coincides with the bar length, then

$$V = v_0$$

At the same time, if $\delta t \ll T$, velocity V is much smaller than particle velocity v_0. Next, consider displacement of the middle point p. It moves relatively quickly, as the linear function, $(s_p(t) = v_0 t)$, in the presence of the pulse. Then point p is at rest until arrival of the next pulse. This jerk-like motion is shown in Fig. 1.5c. Displacement $s_0(t)$ of the center of mass is different. During the time interval δt, when the constant external force F_x is applied, displacement is parabolic. This result directly follows from Newton's second law. After it, $(t > \delta t)$, $s_0(t)$ is a linear function

$$s_0(t) = Vt \qquad \text{if} \qquad t \geq \delta t$$

So far the effect of attenuation associated with particle vibrations along the x–axis, as well as in the radial direction, has been ignored. Since $v(t)$ is the periodical function, it can be represented as the Fourier series

$$v(t) = V + \sum_{n=1}^{\infty} b_n \cos 2\pi n \frac{t}{T}$$

With an increase of time, sinusoidal functions decay due to attenuation and point p starts to move with constant velocity V. In such a case all points of the bar begin to move with the same velocity V, as if it were ideally rigid body. At the same time the middle point p coincides with the center of mass, and Newton's first law describes its motion. We can say that in limiting cases of an elementary particle, $(l \to 0)$, or an ideally rigid body, $(c_l \to \infty)$, Newton's first law describes their motion at any time. As is seen from Fig. 1.5c, the center of mass is located either in front of or behind the bar center. During each period these points coincide when the wave pulse is located in the vicinity of the bar center or near its ends. Let us notice that the maximal separation between these points is usually very small and it is approximately equal to

$$\frac{v_0 \delta t}{2} = \frac{Vl}{2c_l} \tag{1.101}$$

Example five Now consider the arising of the reflected wave at the free end of the bar, when the incident waveform is an arbitrary function, Fig. 1.6a. First of all, it is useful to represent this wave as a system of narrow pulses, following one after another, Fig. 1.6a. Each pulse causes a reflected pulse. Correspondingly, the resultant reflected wave has two important features, namely

a. At the bar end the reflected and incident waves are of different types.

b. The front of the reflected wave is caused by the front of the incident wave.

Superposition of these waves is shown in Fig. 1.6b-g. Suppose that the extensional wave approaches the free end, Fig. 1.6b. Then, due to a reflection, the compressional wave appears, Fig. 1.6c. The thin line corresponds to this wave. Superposition of both waves shows that at the beginning the resultant wave (thick line) is still extensional. It happens

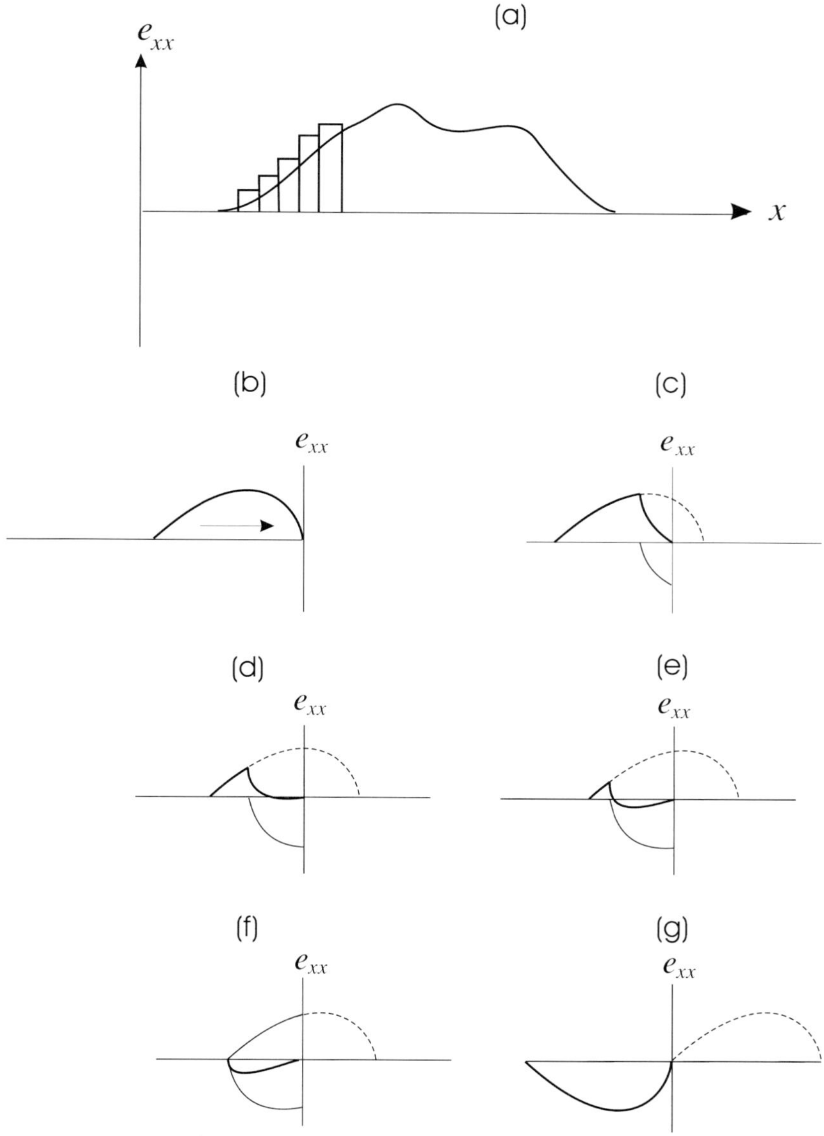

Figure 1.6: (a) The incident wave is an arbitrary function of x (b-g) A superposition of the incident and reflected waves at different instances near the free end. [After Kolsky, 1963]

because near the bar end, the magnitude of the reflected wave is smaller than that of the incident wave. With an increase of time, we start to observe an appearance of the compressional wave, which becomes more and more noticeable, Fig. 1.6e. Finally, when the back of the incident wave reaches the bar end, we see only the compressional reflected wave.

Example six: wave propagation and Newton's second law Now assume that at the instant $t = 0$ the constant force is applied to the left end of the bar, Fig. 1.5a. Correspondingly, the compressional wave arises and travels at a velocity c_l. In this case, unlike in example four, a portion of the bar between the wave front and the left end becomes deformed, and its particles move with the same velocity v_0. At the instant $t = l/c_l$, the whole bar is compressed and moves with this velocity. It is essential that the reflected extensional wave appears at the right end at this moment and propagates through the bar. It causes both an expansion of elementary volumes and their movement with the velocity v_0. Therefore, behind the front of this wave, particles move with the velocity $2v_0$, but deformation disappears. For example, at the instant $t = 2l/c_l$ the bar is not deformed, and each of its particles has the velocity $v = 2v_0$. Also at this moment the reflected wave of compression arises and propagates toward the right end of the bar. Because of this, at the instant $t = 3l/c_l$ the whole bar moves with the velocity $v = 3v_0$, and the reflected wave of tension appears at the right end. It is clear that this process of reflections repeats itself, and propagation of waves between bar ends causes an increase of velocity at any point along the body. As illustration, behavior of the velocity $v(t)$ at the middle point p is shown in Fig. 1.7a. It is evident that the function $v(t)$ is similar at other points. As follows from Newton's second law

$$F_x = ma_x \tag{1.102}$$

behavior of the velocity, V, of the center of mass is completely different, and this velocity linearly increases with time

$$V = a_x t = \frac{F_x}{m} t \tag{1.103}$$

Since

$$v_0 = \frac{c\, F_x}{ES} \qquad \text{and} \qquad m = \rho l S, \tag{1.104}$$

eq. 1.103 becomes

$$V(t) = \frac{v_0}{T} t \tag{1.105}$$

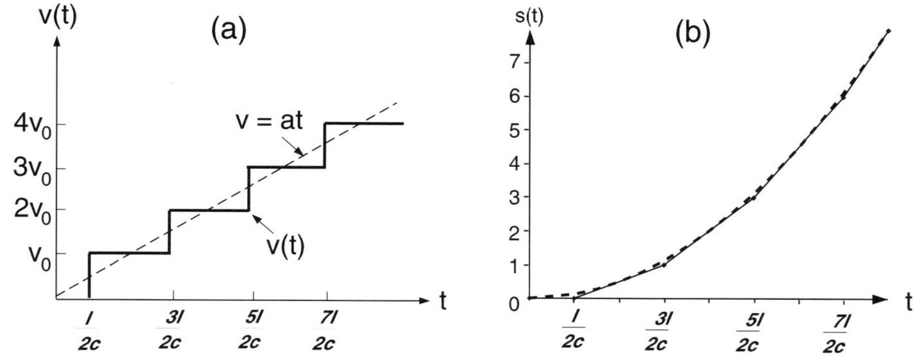

Figure 1.7: (a) Velocity of point p and the center of mass as a function of time (b) Displacement of point p (solid line) and center mass (dashed line) as a function of time

Certainly, the motion of the middle point p and the center of mass differ from each other. This is especially seen at the beginning of a motion when observation time is comparable with T. However, with an increase of time motion of point p asymptotically approaches that described by Newton's second law. In fact, if $t \gg T$, an increase of velocity by v_0 at instances

$$t_n = (2n-1)\frac{l}{2c_l}$$

becomes very small with respect to velocity $v(t_n)$. This means that a discontinuous function $v(t)$ can be practically replaced by the linear function $V(t)$, given by eq. 1.103. From the definition of acceleration and eqs. 1.103 and 1.104, we can express acceleration in three ways:

$$a_x = \frac{v_0}{T} = \frac{v_0 c_l}{l}, \qquad a_x = \frac{\partial V}{\partial t}, \qquad a_x = \frac{F_x}{m} \tag{1.106}$$

The last two formulas characterize the rate of change of the velocity of the center of mass, and in this case $V(t)$ is a continuous function. However, at all points of the bar, particularly at the middle point p, a change of velocity occurs abruptly, and the first equation of the 1.106 may serve as a measure of such behavior. Suppose that there are several bars and the same force F_x is applied to one of their ends. They may differ from each other by density, length, and cross-section. In general, their motion under an action of force F_x is also different. For instance, if an acceleration, a_x, of the center of mass of some bar is higher than for some other bar, velocity $V(t)$ grows more rapidly, and we usually say that inertia of this body is smaller. Also, proceeding from Newton's second law, it is conventional to consider mass, m, as the parameter that characterizes inertia. In other words, mass m defines a time interval during which the center of mass of the moving bar reaches a certain value of the velocity, if $F_x = \text{const}$. For instance, with an increase of mass acceleration decreases and therefore this time interval increases. It is natural to raise the following question. Why does the mass m, that is the product $lS\rho$, define inertia? To answer this question, consider in detail the influence of each factor. First, with an increase of the bar length, l, the time of the wave traveling between bar ends, T, also increases, and in accordance with eqs. 1.106,

$$a_x = \frac{v_0}{T},$$

the rate of velocity change becomes smaller. This linear dependence between length l and inertia is obvious. An influence of density ρ manifests itself in two ways. First of all, with an increase of ρ the wave velocity becomes smaller, since

$$c_l \sim \frac{1}{\sqrt{\rho}}$$

Second, as follows from eq. 1.65, the particle velocity also decreases in the same manner:

$$v_0 \sim \frac{1}{\sqrt{\rho}}$$

Thus, the effect of both factors results in a linear relationship between inertia and density. Finally, with a change of cross-section S, the stress and therefore the particle velocity v_0, changes too. For instance, with an increase of S, the stress X_x becomes smaller and the velocity v_0 decreases. Respectively, inertia shows itself to be stronger, because the time interval during which the velocity $v(t)$ reaches a certain value increases. Thus, as in the cases of length and density, inertia linearly depends on cross-section S. Now we can say that inertia, as an intrinsic property of a body, is related to the time of wave propagation between the bar ends and to particle velocity. It seems we have found out why mass m, as the product $\rho l S$, characterizes inertia and, certainly, this explanation is applied to an arbitrary body. Next, let us demonstrate that inertia is independent of the Young modulus. For instance, with an increase of E, the particle velocity decreases as, eq. 1.65,

$$v_0 \sim \frac{1}{\sqrt{E}}$$

while the wave velocity increases in the same manner. For this reason, these two effects cancel each other, and inertia does not change. In fact, from eq. 1.64 we have

$$c_l v_0 = \frac{c_l^2 X_x}{E} = \frac{X_x}{\rho}, \tag{1.107}$$

and the influence of E vanishes.

Consider, as in example four, a transition from an elastic bar to an ideally rigid one. As we already know, with an increase of the Young modulus the wave velocity increases, but the particle velocity becomes smaller. In other words, with an increase of E, both the particle velocity v_0 and the time interval $T = l/c_l$, during which the velocity $v(t)$ remains constant, decrease. At the same time the product $c_l v_0$ is preserved. Respectively, in the limiting case of an ideally rigid body

$$c_l \to \infty \qquad \text{and} \qquad v_0 \to 0$$

the velocity becomes a continuous function $V(t)$, and it describes the motion of all points of the bar. It is also instructive to study displacement, $s(t)$, of different points of an elastic bar. For instance, within the time interval

$$(2n-1)\frac{l}{c_l} < t < (2n+1)\frac{l}{c_l}$$

the velocity $v(t)$ of the middle point is constant, and its displacement $s(t)$ is a linear function of time. Besides, in each successive time interval, an increase of the slope of the line describing the displacement is the same, Fig. 1.7b. At the same time, in accordance with Newton's second law, the motion of the center of mass is described by the parabola:

$$s_0(t) = \frac{a_x t^2}{2}$$

Comparison of functions $s_p(t)$ and $s_0(t)$, Fig. 1.7b, shows that within each time interval T, the center of mass is either in front of or behind the middle point p. As in case of the impulse of the force (example four), the separation between these points is very small. Also it is clear that with an increase of time, $(t \gg T)$, these functions practically coincide. In conclusion let us make some comments:

1. Newton's first and second laws describe the motion of the center of mass, and its relative position changes with time under the action of impulsive or constant applied force.

2. Because of this, Newton's second law shows the linear change of velocity $V(t)$ of the center of mass in the case of constant external force. Meanwhile, in reality, particles move by jerks.

3. Propagation of reflected waves between bar ends explains the process of a summation of the particle velocities. In other words, it is understandable why the velocity of the bar increases with time even if the external force F_x is constant.

4. Motion of a body is accompanied by a periodic change of stress and strain at each of its points.

5. The geometric parameters of the bar and its density define a rate of change of the particle velocity $v(t)$, that is, inertia of a body.

1.3 Longitudinal sinusoidal waves in a bar

We found out earlier that the solution of the one-dimensional wave equation has the form:

$$u(x,t) = Af[a(t - \frac{x}{c_l})] + Bg[a(t + \frac{x}{c_l})] \tag{1.108}$$

where $u(x,t)$ is the particle displacement of the bar, A and B are constants, and f and g are practically arbitrary functions of distance and time. Thus, all results obtained in the previous section are completely applied to the sinusoidal waves. At the same time it is also useful to consider them separately (Parts I and II), taking into account their special role in the theory of wave phenomena and numerous practical applications. As is well known, the convenient use of sinusoidal functions is related to the following factors:

1. Linear operations, such as a summation of sinusoidal waves of the same frequency, as well as differentiation and integration, do not change the shape of the sinusoidal (harmonic) function. In other words, their frequency remains the same. This fact greatly simplifies the study of sinusoidal waves.

2. The shape of transient waves is preserved when they propagate along a bar and attenuation is absent. However, this factor causes a change in wave shape, i.e., it is impossible to describe this process by either single function $f[a(t-x/c_l]$ or $g[a(t+x/c_l]$, or by a sum of them. At the same time, even in the presence of attenuation, the sinusoidal wave as a function of time preserves the same frequency. This is the second reason why it is very convenient to study wave phenomena using sinusoidal waves even when part of an elastic energy is transformed into heat.

3. The use of Fourier's integral allows us to treat an arbitrary transient wave as superposition of sinusoidal waves (Part I).

4. Finally, in many cases, sources of waves generate sinusoidal oscillations that create harmonic waves.

As before, we will often deal with sinusoidal waves, and so let us recall the basic features of a sinusoidal wave. Suppose that at some point of the bar that coincides with the origin of coordinates, $(x = 0)$, there is a source of sinusoidal vibrations

$$u(0, t) = A \sin \omega t \tag{1.109}$$

Certainly, the sinusoidal wave is generated and propagates away from the source. In accordance with eq. 1.108 we have:

$$u(x, t) = A \sin \omega (t - \frac{x}{c_l}) \qquad \text{or} \qquad u(x, t) = A \sin(\omega t - kx) \tag{1.110}$$

Here ω is an angular frequency and

$$k = \frac{\omega}{c_l} \tag{1.111}$$

is the wave number. By definition, $\omega t - kx$ is the phase of the outgoing wave and, in the same manner, $\omega t + kx$ is the phase of the incoming wave. The period T, and the wavelength, λ, are defined as

$$T = \frac{1}{f} = \frac{2\pi}{\omega}, \qquad \lambda = c_l T = \frac{c_l}{f} = \frac{2\pi c_l}{\omega} = \frac{2\pi}{k} \tag{1.112}$$

The period T and the wavelength λ characterize the time and distance intervals during which the phase changes by 2π, and in this sense they are similar. There is an evident analogy between the angular frequency ω, and the wavenumber k, and correspondingly, the latter is often called the spatial frequency. For instance, with an increase of the wave number, the wavelength becomes smaller. As follows from eq. 1.111 both frequencies are related to each other. If the wave velocity c_l is frequency-independent, then there is a linear relationship between k and ω. In a dispersive medium, where the wave velocity is a function of ω, this relation becomes more complicated.

Next let us describe fields that accompany the sinusoidal waves. First of all, they do not have a beginning or an end. By definition, for the outgoing wave we have

$$u(x, t) = A \sin(\omega t - kx), \qquad v_x(x, t) = A\,\omega \cos(\omega t - kx), \tag{1.113}$$

$$e_{xx}(x, t) = -A\,k \cos(\omega t - kx), \qquad X_x(x, t) = -A\,kE \cos(\omega t - kx)$$

Similar formulas describe the incoming sinusoidal wave. Of course, as in a general case of nonstationary waves, eqs. 1.113 give

$$\frac{v_x(x, t)}{c_l} = -\frac{X_x(x, t)}{E} \qquad \text{or} \qquad v_x(x, t) = -\frac{X_x(x, t)}{Z}, \tag{1.114}$$

where $Z = \rho c_l$ is the bar impedance. The reflected and transmitted waves are also sinusoidal waves with the same frequency as the incident wave, and the reflection and transmission coefficients

$$\frac{Z_1 - Z_2}{Z_1 + Z_2} \qquad \text{and} \qquad \frac{2Z_1}{Z_1 + Z_2}$$

are independent of a frequency. Both of these features greatly simplify the study of wave behavior. Now let us consider one example.

Normal modes Suppose that a source is located at one end of the bar of a finite length l, and it generates a nonstationary wave. As we know, due to a reflection at both bar ends, the resultant wave consists of the system of waves traveling in opposite directions. It is convenient to discuss superposition of these waves in terms of sinusoidal waves with different frequencies. In fact, in accordance with Fourier's integral, the nonstationary wave can be represented as a superposition of sinusoidal waves with all possible frequencies, and they have infinitely small amplitudes and different phases. Each of these sinusoidal harmonics gives rise to a system of reflected waves with the same frequency. Considering their interference for each frequency we can expect that interference has either a constructive or destructive character. However, in the presence of the primary source, the resultant wave contains all frequencies. This happens because the effect of a destructive interference at some frequencies is compensated by an action of the primary source which generates waves at such frequencies. Completely different behavior is observed when this source ceases to act, since only the interference of waves moving in the opposite directions defines the frequency content of the resultant wave. In other words, the resultant oscillations are formed only by sinusoidal waves that experience the constructive interference.

Next, we discuss what determines frequencies corresponding to constructive interference, and frequencies' relationship with the bar length, wave velocity, and boundary conditions. Consider sinusoidal solutions of the wave equation, and as follows from eq. 1.108 they have the form:

$$u(x,t) = A\sin(\omega t - kx) + B\sin(\omega t + kx) + C\cos(\omega t - kx) + D\cos(\omega t + kx) \qquad (1.115)$$

It is clear that all four $\sin(\omega t \pm kx)$ and $\cos(\omega t \pm kx)$ functions obey the wave equation. Making use of known trigonometric formulas, we obtain

$$u(x,t) = a\cos kx \sin\omega t + b\sin kx \cos\omega t + c\cos kx \cos\omega t + d\sin kx \sin\omega t \qquad (1.116)$$

Each term of this sum can be interpreted as a superposition of two sinusoidal waves with equal amplitudes traveling in opposite directions. For instance,

$$\frac{a}{2}\sin(\omega t - kx) + \frac{a}{2}\sin(\omega t + kx) = a\cos kx \sin\omega t, \tag{1.117}$$

and this describes the standing wave because all points of the bar vary synchronously. Besides, there are points (nodes) at which motion is absent. Such behavior shows that each term of the sum given by eq. 1.116 describes the standing wave. It is essential that the terms represent the result of an interference of sinusoidal waves of the same frequency. It is natural to expect that in the case of destructive interference, amplitudes of these standing waves are equal to zero. To illustrate a calculation of frequencies corresponding to the constructive interference consider one example. Suppose that both ends of the bar do not move, i.e.,

$$u(0,t) = 0 \qquad \text{and} \qquad u(l,t) = 0, \tag{1.118}$$

and the origin of coordinates coincides with one of the bar ends. As follows from eq. 1.116 at all times

$$0 = a\sin\omega t + c\cos\omega t$$

Thus, $a = c = 0$. Correspondingly, eq. 1.116 is simplified and we have

$$u(x,t) = \sin kx\ (b\cos\omega t + d\sin\omega t) \tag{1.119}$$

The second boundary condition gives

$$\sin kl = 0 \qquad \text{or} \qquad kl = \frac{\omega\, l}{c_l} = \pi n, \tag{1.120}$$

i.e.,

$$k_n = \frac{\pi n}{l} \qquad \text{or} \qquad \omega_n = \frac{\pi n\ c_l}{l} \tag{1.121}$$

Thus, boundary conditions are satisfied only if frequencies (wavenumbers) obey eq. 1.121. In other words, constructive interference occurs when frequencies are related to the bar length in a certain manner. Wave numbers k_n are called eigenvalues, and the corresponding solution $u_n(x,t)$ is written as

$$u_n(x,t) = \sin k_n x\ (b_n\cos\omega_n t + d_n\sin\omega_n t) \tag{1.122}$$

It represents a sum of two standing waves, which are shifted in time by $\pi/2$. The latter can be also written in the form

$$u_n(x,t) = A_n \sin k_n x \sin(\omega_n t + \varphi) \qquad (1.123)$$

where A_n and φ_n are independent of both distance x and time t. The function $\sin k_n x$ is called the eigenfunction, and $u_n(x,t)$ represents the normal mode. As follows from eq. 1.123 there is an infinite number of normal modes, and therefore the resultant wave inside the bar is a sum of normal modes:

$$u(x,t) = \sum_{n=1}^{\infty} \sin k_n x \, (b_n \cos \omega_n t + d_n \sin \omega_n t) \qquad (1.124)$$

To determine unknown coefficients b_n and d_n we have to define the initial conditions:

$$u(x,0) = u_0(x) \qquad \text{and} \qquad \frac{\partial u(x,0)}{\partial t} = v_0(x), \qquad (1.125)$$

which describe behavior of the displacement and its velocity at some instant $t = 0$. Then, as follows from eq. 1.124,

$$u_0(x) = \sum_{n=1}^{\infty} b_n \sin k_n x \qquad \text{and} \qquad v_0(x) = \sum_{n=1}^{\infty} \omega_n d_n \sin k_n x \qquad (1.126)$$

Thus, the given functions $u_0(x)$ and $v_0(x)$ are represented as the Fourier's series and, using the known formulas for its coefficients, both sets of amplitudes, b_n and d_n, are easily determined. The same approach is used in a general case, when the wave propagates inside an elastic body of an arbitrary shape. This means that as before, the oscillations of this body result from a superposition of normal modes, which are characterized by an infinite set of eigenvalues.

1.4 Hooke's law for shear stresses and torsional waves along a bar

In order to study propagation of waves caused by a bar twist, we first consider shear stress and shear strain, as well as the relationship between them. Suppose that at some instant $t = 0$, one face of a rectangular parallelepiped inside a medium is subjected to the action of the tangential force $\mathbf{F}_y$, Fig. 1.8a. Because of this force, a wave arises and propagates toward the opposite face. Since the volume is very small, the wave reaches the back face very quickly and the same force $\mathbf{F}_y$ acts on a medium behind the volume. In accordance with Newton's third law, this medium acts on the back face with the force

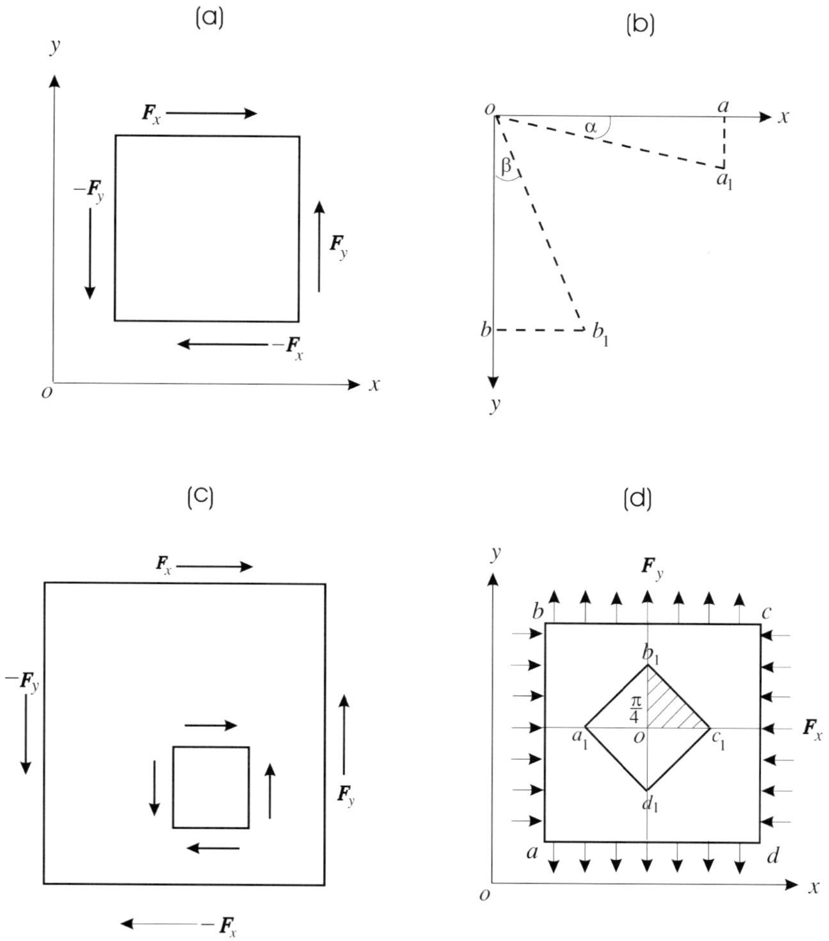

Figure 1.8: (a) Transmission of the shear force (b) Illustration of eqs. 1.128 (c) Stresses inside elementary volume (d) Forces acting on the rectangular parallelepiped $a_1b_1c_1d_1$

$-\mathbf{F}_y$. In equilibrium, both forces $\mathbf{F}_y$ and $-\mathbf{F}_y$ have equal magnitudes but opposite directions. Their resultant force is equal to zero, but they form the couple, which tries to rotate the volume counterclockwise. Such motion causes a deformation of a medium in the vicinity of the horizontal faces and, correspondingly, the tangential (shear) forces $\mathbf{F}_x$ and $-\mathbf{F}_x$ appear, which act on these faces, Fig. 1.8a. Of course, in equilibrium the total force and the resultant moment of all four forces are equal to zero. However, before equilibrium occurred, motion of the elementary volume resulted in a change of its shape. This means that the angle between intersecting faces varies, and instead of $\pi/2$ it becomes $\pi/2-\gamma$. First of all we assume that the elementary volume is cubical and that forces are uniformly distributed over the cube's faces. Because of this, at equilibrium, magnitudes of forces $\mathbf{F}_x$ and $\mathbf{F}_y$ are equal to each other. For simplicity it is also supposed that the z-component of forces is zero, $\mathbf{F}_z = 0$. Now we introduce shear stresses τ_{xy} and τ_{yx} in the following way:

$$\mathbf{F}_x = \tau_{xy} dS\, \mathbf{i} \qquad \text{and} \qquad \mathbf{F}_y = \tau_{yx} dS\, \mathbf{j} \tag{1.127}$$

Here dS is the face area and $\mathbf{i}$ and $\mathbf{j}$ are unit vectors. Since the resultant moment is equal to zero, we have

$$\tau_{xy} = \tau_{yx}$$

The notation τ_{ij} indicates that this stress characterizes the force component, directed along the $\mathbf{i}$-axis and applied to the face, which is normal to the $\mathbf{j}$-axis (Appendix C). In equilibrium, stress τ_{xy} has the same value at all four faces of the elementary volume. Next we express the angle of distortion, γ, in terms of the displacement derivatives. As is seen from Fig. 1.8b, after a deformation the angle between intersecting faces becomes

$$\frac{\pi}{2} - \gamma = \frac{\pi}{2} - (\alpha + \beta),$$

where $\alpha + \beta = \gamma$. It is obvious that

$$\tan\alpha = \frac{v}{x} \quad \text{or} \quad \alpha = \frac{\partial v}{\partial x} \qquad \text{and} \qquad \tan\beta = \frac{u}{y} \quad \text{or} \quad \beta = \frac{\partial u}{\partial y}, \tag{1.128}$$

because α and β are very small. Thus

$$\gamma = \frac{\partial u}{\partial y} + \frac{\partial v}{\partial x} \tag{1.129}$$

Making use of notations for strain (Appendix D), we have

$$e_{xy} = e_{yx} = \frac{\partial u}{\partial y} + \frac{\partial v}{\partial x} = \gamma \tag{1.130}$$

Thus, strain e_{xy} is equal to the angle of distortion γ. In general, angles α and β are different, and later we will study two special cases: $\alpha = \beta$ and $\beta = 0$. Taking into account that deformations are very small, we have, as in the case of normal stress, the linear relationship between shear strain and shear stresses

$$\tau_{yx} = \mu\, e_{xy}, \tag{1.131}$$

where the coefficient of proportionality μ is called the modulus of rigidity. In a general case, when distortion is also observed in planes XOZ and YOZ we have

$$\tau_{xz} = \mu\, e_{xz}, \qquad \tau_{yz} = \mu\, e_{yz} \tag{1.132}$$

where

$$e_{xz} = \frac{\partial u}{\partial z} + \frac{\partial w}{\partial x}, \qquad e_{yz} = \frac{\partial v}{\partial z} + \frac{\partial w}{\partial y} \tag{1.133}$$

In essence, eqs. 1.131 and 1.132 represent Hooke's law for shear stresses and shear strains. Earlier we pointed out that at equilibrium,

$$\tau_{xy} = \tau_{yx} \tag{1.134}$$

In the same manner we obtain

$$\tau_{xz} = \tau_{zx} \qquad \text{and} \qquad \tau_{yz} = \tau_{zy} \tag{1.135}$$

This clearly demonstrates that stress is a symmetrical tensor (Appendix C). Note that forces applied to the volume faces are transmitted inside of the elementary volume. Their distribution is shown in Fig. 1.8c.

In previous sections we considered deformation of an elementary volume due to an action of normal stresses and found that equilibrium can be provided by only two forces, applied to the opposite faces of the volume. As a result its shape does not change, and the angle between intersecting faces remains equal to $\pi/2$. In contrast, in the presence of shear stresses, equilibrium takes place when there are shear stresses at all four faces, Fig. 1.8a. In describing the longitudinal waves, we used two elastic parameters, Young modulus E and Poisson's ratio σ. Now we will show that the modulus of rigidity μ can be expressed in terms of E and σ. In order to demonstrate this important fact, consider two special cases, when either $\alpha = \beta$ or $\beta = 0$.

Case one: pure shear For simplicity we restrict ourselves to the two-dimensional case when stress τ_{xy} is independent of the z-coordinate. Suppose that a rectangular parallelepiped $abcd$ is simultaneously subjected to the action of forces that produce extension and compression along the $y-$ and x-axes, respectively (Fig. 1.8d). As a result, the angle between intersecting faces remains equal to $\pi/2$, but the length of sides ab and bc varies. As follows from Hooke's law and Poisson's relation, as well as from the principle of superposition, we have for strains e_{yy} and e_{xx}:

$$e_{yy} = \frac{\tau_{yy}}{E} - \frac{\sigma\tau_{xx}}{E} \quad \text{and} \quad e_{xx} = \frac{\tau_{xx}}{E} - \frac{\sigma\tau_{yy}}{E} \tag{1.136}$$

Since we assume that $\tau_{xx} = -\tau_{yy}$, eq. 1.136 gives

$$e_{yy} = \frac{1+\sigma}{E}\tau_{yy}, \qquad e_{xx} = -\frac{(1+\sigma)}{E}\tau_{yy}, \tag{1.137}$$

that is they differ by a sign only and are constant within $abcd$. Next consider forces acting on the rectangular parallelepiped $a_1b_1c_1d_1$ located inside $abcd$, Fig. 1.8d. Since this volume is at equilibrium, the resultant force, acting on each element of volume, for instance ob_1c_1, is equal to zero. Taking into account that the face ob_1 is parallel to ab, the stress at its points is equal to τ_{xx}, but at the face oc_1 it coincides with τ_{yy}. Correspondingly, forces applied to these faces are

$$\mathbf{F}_x = -\ \tau_{xx} ob_1 \mathbf{i} \quad \text{and} \quad \mathbf{F}_y = -\ \tau_{yy} oc_1 \mathbf{j}$$

because $\tau_{xx} < 0$ and $\tau_{yy} > 0$. In order to provide equilibrium, the force acting on the face b_1c_1 has to be equal to

$$\mathbf{F} = \ \tau_{xx} ob_1 \mathbf{i} + \tau_{yy} oc_1 \mathbf{j} \quad \text{or} \quad \mathbf{F} = \ \tau_{yy}(\mathbf{j}-\mathbf{i}) ob_1 \tag{1.138}$$

As is seen from Fig. 1.9a, the normal component F_n is equal to zero, but the tangential component is

$$F_t = \tau_{yy}\sqrt{2}ob_1$$

This means that the tangential strain τ_t is equal to

$$\tau_t = \frac{F_t}{b_1c_1} = \tau_{yy}\sqrt{2}\frac{ob_1}{b_1c} = \tau_{yy} \tag{1.139}$$

It is clear that there is only shear strain at all faces of volume $a_1b_1c_1d_1$ (its extension along the z-axis is implied). In essence, we applied Cauchy's formula of stress transformation for this simple case.

Because of an expansion along the y-axis and shortening along the x-axis, the parallelepiped $a_1b_1c_1d_1$ is deformed into a rhombus, Fig. 1.9b, and the angle between

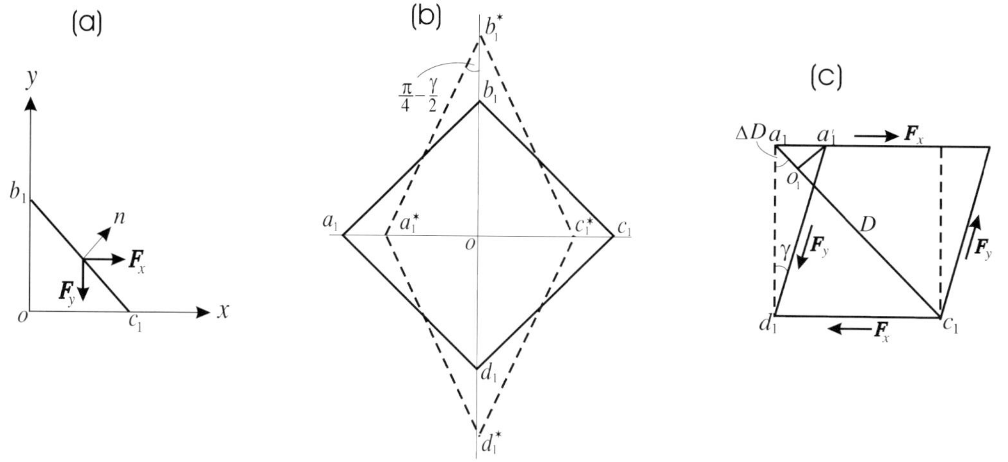

Figure 1.9: (a) Normal and tangential forces applied to face b_1c_1 (b) Deformation into rhombus (c) Illustration of eq. 1.145

intersecting faces slightly changes. The remarkable feature of this deformation is the fact that an orientation of diagonals a_1c_1 and b_1d_1 does not change, i.e., $\alpha = \beta = \gamma/2$. Such a deformation is called pure shear, and it may occur for different types of waves. As follows from eqs. 1.137

$$\text{div } \mathbf{s} = e_{xx} + e_{yy} = 0, \tag{1.140}$$

since $e_{zz} = 0$, and therefore volume does not change. In accordance with eqs. 1.128 we see that

$$\text{curl}_z\, \mathbf{s} = \frac{\partial v}{\partial x} - \frac{\partial u}{\partial y} = 0, \tag{1.141}$$

and rotation is absent.

Now we find a relationship between the shearing strain e_{xy}, i.e., the angle γ, and the stress τ_t (τ_{yy}). After deformation (Fig. 1.9b) we have

$$\frac{o\, c_1^*}{o\, b_1^*} = \tan(\frac{\pi}{4} - \frac{\gamma}{2}) = \frac{1 + e_{xx}}{1 + e_{yy}} \tag{1.142}$$

Note that for small γ

$$\tan(\frac{\pi}{4} - \frac{\gamma}{2}) = \frac{\tan \pi/4 - \tan \gamma/2}{1 + \tan \pi/4 \tan \gamma/2} \approx \frac{1 - \gamma/2}{1 + \gamma/2} \approx 1 - \gamma,$$

eq. 1.142 becomes

$$1-\gamma=\frac{1+e_{xx}}{1+e_{yy}}=1-2e_{yy},$$

since $e_{xx}=-\,e_{yy}$ (eq. 1.137), and $e_{yy}\ll 1$. Thus

$$\gamma=2e_{yy} \tag{1.143}$$

Substitution of eq. 1.143 into eq. 1.137 yields

$$\gamma=\frac{2(1+\sigma)}{E}\tau_t \qquad \text{or} \qquad e_{xy}=\frac{2\,(1+\sigma)}{E}\tau_{xy}$$

i.e., the modulus of rigidity μ is equal to

$$\mu=\frac{E}{2(1+\sigma)}, \tag{1.144}$$

and we expressed μ in terms of the Young modulus and Poisson's ratio.

Case two: simple shear Now we study deformation that accompanies propagation of shear waves, as is shown in Fig. 1.9c. In this case, face d_1c_1 does not move and the distortion angle γ coincides with α. As before, it is assumed that forces producing a deformation, are parallel to diagonals in their vicinity, and they cause either their extension or shortening. For instance, as follows from Hooke's law (eq. 1.137), the relative change of diagonal D Fig. 1.9c, is

$$\frac{\Delta D}{D}=\frac{1+\sigma}{E}\tau \tag{1.145}$$

At the same time from the triangle $a_1d_1a_1'$ we have

$$\tan\gamma=\frac{\delta}{a_1d_1}$$

Also

$$a_1d_1=D\cos\frac{\pi}{4} \qquad \text{and} \qquad \delta=a_1a_1'=\frac{\Delta D}{\cos\pi/4}$$

Since γ is small, eq. 1.145 gives

$$\gamma=2\frac{\Delta D}{D}=\frac{2(1+\sigma)}{E}\tau,$$

which gives again the known expression of μ. It is clear that deformation does not change volume, and, correspondingly:

$$\text{div }\mathbf{s}=0$$

The distortion angle γ characterizes the rate of change of the displacement component u with respect to y:

$$\gamma = \frac{\partial u}{\partial y},$$

while $\partial v/\partial x = 0$ and $w = 0$. We conclude that $\text{curl}\,\mathbf{s} \neq 0$. This suggests that simple shear is a combination of pure shear and rotation (Appendix D). From the geometric point of view, this is illustrated by the change of an orientation of the volume diagonals. Next we apply our knowledge of Hooke's law for normal and shear stresses to one special case, which will allow us to understand some features of shear waves.

Torsion of a circular bar with a constant cross-section

Suppose that one end of the bar is fixed, while the shear forces $\mathbf{T}$ are applied to the free end and their action is equivalent to that of a pair of forces with the moment $\mathbf{M}$ (directed along the z-axis), Fig. 1.10a,b. It causes a deformation (twist) of the bar, and shear stresses arise. The solution of this problem was given by Coulomb at the end of the eighteenth century and is based on two assumptions:

1. The twist does not change the distance between cross-sections, i.e., displacement along the bar axis is absent.

2. After deformation, cross-sections remain planar and the radii drawn in these planes do not bend.

Taking into account these assumptions, we investigate the distribution of the displacement at each cross-section and demonstrate that this field $\mathbf{s}$ satisfies the following conditions: (a) Particles do not move at the fixed end. (b) External forces are absent on the lateral surface of the bar (c) At the free end, the distribution of stresses is such that their action is equivalent to the given moment $\mathbf{M}$. In order to solve this problem it is convenient to first represent the bar as a system of thin coaxial cylindrical shells of thickness Δr and find stresses and strains for each shell, Fig. 1.10c.

Consider the elementary volume of a shell that has extension Δz and thickness Δr, Fig. 1.10c. Due to deformation, points of lines ab and a_1b_1 or cd and c_1d_1 remain in the same planes. At the same time, distance d_1d exceeds distance a_1a. Correspondingly, the angle between faces ab and a_1d_1 becomes equal to $\pi/2-\gamma$. It is essential that the twist of all cross-sections of the bar is characterized by the same angle γ. This deformation gives rise to shear forces acting on four faces of the volume, Fig. 1.10d. However, they are absent on the external and internal lateral faces of the shell

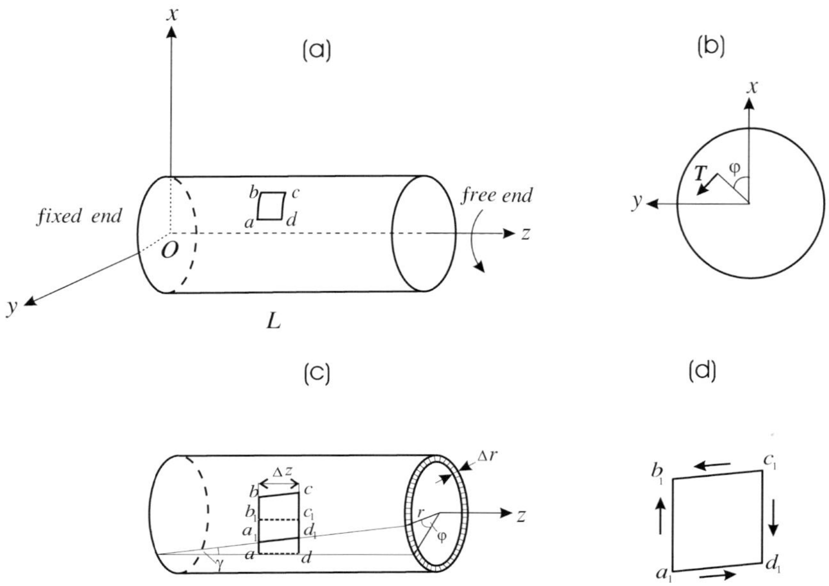

Figure 1.10: (a,b) Shear stress and moment applied to cylindrical bar (c) Deformation of the elementary shell of the bar (d) Illustration of eq. 1.160

element. In accordance with eq. 1.132 we have for shear stress

$$\tau = \mu\,\gamma, \tag{1.146}$$

where τ describes the force tangential to the cross-section of the bar. Now we relate the shear stress τ to the angular displacement φ. As is seen from Fig. 1.10c, we have:

$$\tan\gamma = \frac{r\varphi}{z} \qquad \text{or} \qquad \gamma = \frac{r\varphi}{z}, \tag{1.147}$$

since γ is small. Here z is the distance between a cross-section and the fixed end, $(z = 0)$. In particular, at the free end

$$\gamma = \frac{r\varphi}{L} \tag{1.148}$$

Substitution of eq. 1.147 into eq. 1.146 gives the relationship between the stress τ and the angle φ:

$$\tau = \mu\frac{r}{z}\varphi \tag{1.149}$$

Thus, at equilibrium the shear stress τ is directly proportional to the angle φ and the radius r of the shell, and inversely proportional to the distance between a cross-section $S(z)$ and the fixed end of the bar. Since at n equilibrium τ =const, we conclude that the angle φ is directly proportional to z and, in particular, it linearly decreases in approaching the origin, $(z = 0)$. As follows from eq. 1.149 the shear stress τ varies at points of a cross-section and disappears at the bar axis, $(r = 0)$. This is an example of an inhomogeneous deformation along r. It is instructive to compare eq. 1.149 with Hooke's law describing a longitudinal displacement $w(z)$:

$$\tau_{zz} = E\frac{w(z)}{z} = E\frac{\partial w}{\partial z} \tag{1.150}$$

Certainly there is a similarity between them, and a displacement along the arc $r\varphi(z)$ plays the same role as $w(z)$. Note that in both cases the displacements, $w(z)$ or $r\varphi$, are directly proportional to distance z. Now we will represent eq. 1.149 in a different form. Consider two cross-sections of the bar, located at distances z and $z + dz$ from the fixed end. Since

$$\frac{\varphi(L)}{L} = \frac{\varphi(z)}{z} = \frac{\varphi(z + dz)}{z + dz},$$

we have

$$z\varphi(z) + dz\ \varphi(z) = z\varphi(z + dz)$$

Applying the Taylor expansion and neglecting higher-order terms, we obtain

$$z\varphi(z) + \varphi(z)dz \approx z\varphi(z) + z\varphi'(z)dz$$

or

$$\frac{\varphi(z)}{z} = \varphi'(z)$$

Correspondingly, eq. 1.149 becomes

$$\tau = \mu\, r\frac{\partial \varphi}{\partial z} \tag{1.151}$$

or

$$\tau = \mu\frac{\partial l_\varphi}{\partial z} \tag{1.152}$$

The analogy with Hooke's law for normal stress, eq. 1.150, is obvious. As follows from eq. 1.151, at the state of equilibrium the derivative $\partial\varphi/\partial z$ is a constant along the cylindrical shell.

Displacement s and stresses in a solid bar

Now we will show that under certain conditions an approximate theory given by Coulomb represents the exact solution. Let us consider a displacement field $\mathbf{s}$ and stresses inside the solid bar. As is seen in Fig. 1.10b, at any cross-section the φ-component of the field $\mathbf{s}$ is

$$s_\varphi = r\varphi \qquad \text{or} \qquad s_\varphi = r\frac{\varphi}{z}z = \frac{\partial\varphi}{\partial z}rz, \tag{1.153}$$

and it is directly proportional to radius r and distance z from the fixed end, ($\partial\varphi/\partial z$ =const). Respectively, at the Cartesian system of coordinates we have

$$u = \frac{\partial\varphi}{\partial z}zr(-\frac{y}{r}) = -\frac{\partial\varphi}{\partial z}yz \qquad \text{and} \qquad v = \frac{\partial\varphi}{\partial z}zr\frac{x}{r} = \frac{\partial\varphi}{\partial z}xz, \tag{1.154}$$

while $w = 0$. Here

$$\mathbf{s} = u\,\mathbf{i} + v\,\mathbf{j}$$

By definition

$$e_{xx} = e_{yy} = e_{zz} = 0$$

and dilatation is equal to zero. As we know, this means that deformation does not cause a change in volume. At the same time, shear strains in this plane, (z =const), are

$$\frac{\partial u}{\partial y} = -\frac{\partial\varphi}{\partial z}z \qquad \text{and} \qquad \frac{\partial v}{\partial x} = \frac{\partial\varphi}{\partial z}z \tag{1.155}$$

Therefore

$$\text{curl}_z\,\mathbf{s} = 2\frac{\partial\varphi}{\partial z}z, \tag{1.156}$$

and an elementary volume experiences rotation about the z-axis. As follows from Fig. 1.10b and eqs. 1.154

$$\tau_{xz} = -\mu\frac{\partial\varphi}{\partial z}y, \qquad \tau_{yz} = \mu\frac{\partial\varphi}{\partial z}x \tag{1.157}$$

In order to find other components of the stress tensor, we make use of the principle of superposition, as well as Hooke's law and Poisson's relation. This gives

$$e_{xx} = \frac{\tau_{xx}}{E} - \frac{\sigma}{E}\tau_{yy} - \frac{\sigma}{E}\tau_{zz}, \quad e_{yy} = \frac{\tau_{yy}}{E} - \frac{\sigma}{E}\tau_{xx} - \frac{\sigma}{E}\tau_{zz}, \quad e_{zz} = \frac{\tau_{zz}}{E} - \frac{\sigma}{E}\tau_{xx} - \frac{\sigma}{E}\tau_{yy}$$

Since $e_{xx} = e_{yy} = e_{zz} = 0$, the system has only zero solution:

$$\tau_{xx} = \tau_{yy} = \tau_{zz} = 0, \tag{1.158}$$

that is, normal stresses are equal to zero. Taking into account also the relationships between shear stresses and strains, we have

$$\tau_{xy} = \mu(\frac{\partial u}{\partial y} + \frac{\partial v}{\partial x}) = 0$$

$$\tau_{xz} = \mu(\frac{\partial w}{\partial x} + \frac{\partial u}{\partial z}) = -\mu\frac{\partial \varphi}{\partial z}y \tag{1.159}$$

$$\tau_{yz} = \mu(\frac{\partial v}{\partial z} + \frac{\partial w}{\partial y}) = \mu\frac{\partial \varphi}{\partial z}x$$

Thus, at equilibrium all three normal components of stress are zero, and there are only two components of shear stresses. They describe forces acting on an elementary area normal to the bar axis. Note that the stress components τ_{xz} and τ_{yz} were obtained in two ways. In order to determine stresses on the lateral surface of the bar, consider an elementary volume near this surface. From the condition of equilibrium (Appendix C), we obtain

$$\tau_{xr} = \tau_{xx}l + \tau_{xy}m + \tau_{xz}n,$$

$$\tau_{yr} = \tau_{yx}l + \tau_{yy}m + \tau_{yz}n, \tag{1.160}$$

$$\tau_{zr} = \tau_{zx}l + \tau_{zy}m + \tau_{zz}n$$

Here

$$l = \frac{x}{r}, \qquad m = \frac{y}{r}, \qquad n = 0$$

Thus

$$\tau_{xr} = \tau_{yr} = \tau_{zr} = 0, \tag{1.161}$$

and the solution given by eqs. 1.154 satisfies the boundary condition at the lateral surface, if it is not subjected to an action of external forces. It is also clear that the field $\mathbf{s}$ obeys the boundary condition at the bar end, since $\mathbf{s}(0) = 0$. Now we focus on the free end, $z = L$, and determine the external forces acting on this cross-section, which create the displacement field $\mathbf{s}$, given by eqs. 1.154. Directional cosines of the plane are

$$l = 0, \qquad m = 0, \qquad n = 1$$

Substitution of eqs. 1.158 and 159 into set 1.160, which also describes stresses at any cross-section of the bar, gives

$$\tau_{xz} = -\mu\frac{\partial\varphi}{\partial z}y, \qquad \tau_{yz} = \mu\frac{\partial\varphi}{\partial z}x, \qquad \tau_{zz} = 0 \tag{1.162}$$

Therefore, on the free end there are only shear stresses, and they are distributed in the same manner as in any other cross-section of the bar. Next it is useful to find the sum of forces acting on a bar cross-section. Performing an integration, we have

$$\int_S \tau_{xz} dS = -\mu\frac{\partial\varphi}{\partial z}\int_S y\, dxdy = 0$$

and

$$\int_S \tau_{yz} dS = \mu\frac{\partial\varphi}{\partial z}\int_S x\, dxdy = 0\ ,$$

since across the surface S x and y are odd functions. At the same time, the z-component of the resultant moment differs from zero, and it is defined as

$$M_z = \int_S (\mathbf{r}\times\mathbf{T})_z\, dS$$

or

$$M_z = \int_S (x\tau_{yz} - y\tau_{xz})dS = \mu\frac{\partial\varphi}{\partial z}\int_S (x^2+y^2)dS = \mu\frac{\partial\varphi}{\partial z}\int_S r^2 dxdy \tag{1.163}$$

Thus, an action of tangential forces uniformly distributed over the cross-section, in particular on the free end, is equivalent to the torque $\mathbf{M}$, which has only the component M_z:

$$M_z = \mu\frac{\partial\varphi}{\partial z}I_0 \tag{1.164}$$

Here

$$I_0 = \int_S r^2 dS \tag{1.165}$$

is called the polar moment of inertia of the cross-section. Since $dS = 2\pi r dr$, eq. 1.165 becomes

$$I_0 = 2\pi \int_0^a r^3 dr = \frac{\pi a^4}{2} \tag{1.166}$$

and

$$M_z = \mu \frac{\partial \varphi}{\partial z} \frac{\pi a^4}{2} = \mu \frac{\pi a^4}{2} \frac{\varphi(L)}{L} \tag{1.167}$$

The coefficient of proportionality between the resultant moment and the torsional angle

$$k = \frac{M_z}{\varphi(L)} = \frac{\mu \pi a^4}{2L} \tag{1.168}$$

is called torsional stiffness. It is directly proportional to the fourth power of the bar radius a and inversely proportional to the distance from the fixed end. Note that in the case of a thin cylindrical shell, we have

$$I_0 = 2\pi r^3 \Delta r$$

Thus, we have demonstrated that Coulomb's theory correctly describes the displacement field $\mathbf{s}$ and stresses at an equilibrium, when shear stresses produce the moment at the free end and external forces are not applied to the lateral surface.

1.5 Torsional waves

Until now we have considered the bar in equilibrium, when the resultant moment M_z (eqs. 1.167)

$$M_z = \mu \frac{\pi a^4}{2} \frac{\partial \varphi}{\partial z}$$

is the same in all cross-sections. Next, suppose that at some instant $t = 0$, shear forces are applied to the free end, $z = L$. Because of deformation (twisting), the wave starts to propagate along the bar, and moment M_z becomes a function of time and a position of the cross-section z. At the same time, particles of the bar move along arcs with the radius r $(0 \le r \le a)$, in a direction perpendicular to the z-axis. Let us derive an equation of motion of an elementary volume of the bar, bounded by its lateral surface and cross-sections $S(z)$ and $S(z + dz)$, where z is the distance from the fixed end.

As we know, wave propagation is accompanied by an appearance of shear forces. They create moment $M_z(z+dz)$ at the face $S(z+dz)$, as well as moment $M_z(z)$, acting on the portion of the bar, located in front of the elementary volume. In accordance with Newton's third law, the moment applied to the face $S(z)$ is

$$M = -M_z(z)$$

Thus, the resultant moment, producing torsion of the volume, is equal to

$$M_z(z+dz) - M_z(z) = M$$

Taking into account that dz is small and M_z is a continuous function, we have

$$M = \frac{\partial M_z}{\partial z} dz \tag{1.169}$$

Earlier we demonstrated that motion of an elementary volume can be represented as a superposition of a pure shear and rotation as a rigid body. In accordance with Newton's second law, such motion (rotation) is described by the equation (Appendix A):

$$M = I\,\frac{\partial^2 \varphi}{\partial t^2}, \tag{1.170}$$

where I is the moment of inertia of the cylindrical element with the length dz, and $\partial^2\varphi/\partial t^2$ is angular acceleration. Since the polar moment of inertia I_0, eq. 1.166, characterizes the moment of inertia of the cylindrical bar with unit density $\rho = 1 kg/m^3$ and length $dz = 1\,m$, it is easy to find I. In fact, applying the principle of superposition, we obtain

$$I = \frac{\pi a^4}{2}\rho\, dz \tag{1.171}$$

Thus, as in the case of the longitudinal waves, we proceed from Hooke's and Newton's second law, applied to the elementary volume. Substitution of eq. 1.169 and eq. 1.171 into eq. 1.170 gives

$$\frac{\partial M_z}{\partial z} = \frac{\pi a^4}{2}\rho\frac{\partial^2 \varphi}{\partial t^2}$$

Finally, making use of eq. 1.164, we have

$$\mu\,\frac{\partial^2 \varphi}{\partial z^2} = \rho\,\frac{\partial^2 \varphi}{\partial t^2} \qquad \text{or} \qquad \frac{\partial^2 \varphi}{\partial z^2} = \frac{1}{c_s^2}\frac{\partial^2 \varphi}{\partial t^2} \tag{1.172}$$

Here

$$c_s = \sqrt{\frac{\mu}{\rho}} \tag{1.173}$$

is the velocity of propagation of torsional (shear) waves. It is clear that eq. 1.173 is the wave equation, which is similar to that for longitudinal waves. However, waves satisfying these equations differ from one another by direction of particle motion, type of deformation, and velocity of propagation. As follows from eq. 1.144, we have

$$c_s = \sqrt{\frac{E}{2(1+\sigma)\rho}}, \qquad \text{while} \qquad c_l = \sqrt{\frac{E}{\rho}},$$

whence

$$c_s = \frac{1}{\sqrt{2(1+\sigma)}} c_l \tag{1.174}$$

Since Poisson's ratio varies within the range

$$0 \leq \sigma \leq \frac{1}{2},$$

we always have

$$c_s < c_l, \tag{1.175}$$

that is, shear waves propagate more slowly than longitudinal waves. In particular, when $\sigma = 1/2$,

$$c_s \approx 0.6 c_l \tag{1.176}$$

Comparison of these waves shows that the modulus of rigidity μ plays the same role for shear waves as the Young modulus E plays for longitudinal waves, whereas the angular displacement φ is an analogy of the displacement w along the bar. Note that eq. 1.173 correctly describes the velocity of shear waves in any elastic medium. At the same time, the expression

$$c_l = \sqrt{\frac{E}{\rho}}$$

is valid for longitudinal waves traveling along the bar, when the lateral surface is not subjected to action by external forces.

Boundary conditions

Suppose that the bar consists of two homogeneous portions with different parameters μ and ρ. Since the wave equation cannot be applied to the interface, we replace eq. 1.172 with boundary conditions. First of all, angles φ_1 and φ_2 at both sides of the boundary have to be equal to each other; otherwise the bar would be broken. Thus, we have

$$\varphi_1(z) = \varphi_2(z) \tag{1.177}$$

Consider an elementary volume bounded by cross-sections $S(z+\Delta z/2)$ and $S(z-\Delta z/2)$, which are located at opposite sides of the interface. The resultant moment

$$M_z(z + \frac{\Delta z}{2}) - M_z(z - \frac{\Delta z}{2}),$$

acting on an infinitely thin cylinder ($\Delta z \to 0$ and $I \to 0$), has to be equal to zero. If it had a nonzero value, angular acceleration would be infinitely large, which is impossible. Correspondingly, the second boundary condition is

$$M_{1z}(z) = M_{2z}(z) \tag{1.178}$$

It is easy to show that wave equations and boundary conditions for longitudinal and shear waves are similar. Boundary problems for these waves become identical if we change the notations in the following way

$$\varphi \to w \qquad \text{and} \qquad \mu \to E$$

This allows us to use results obtained for longitudinal waves in the previous sections and represent angle φ as

$$\varphi(z,t) = Af\ [\alpha(z + c_s t)] + Bg[\alpha(z - c_s t)] \tag{1.179}$$

Also in studying reflection and transmission of shear waves, we can use the same coefficients as in the case of longitudinal waves, provided that the impedance is equal to

$$Z_s = c_s \rho = \rho\sqrt{\frac{\mu}{\rho}} = \sqrt{\mu\rho} \tag{1.180}$$

Bearing in mind the analogy with longitudinal waves and the fact that the incident wave arises at point $z = L$, we can represent this wave and a reflected wave in the form

$$\varphi(z,t) = \varphi_0 f\ [\alpha(z + c_s t)] + \frac{Z_1 - Z_2}{Z_1 + Z_2}\varphi_0 f[\alpha(z - c_s t)] \tag{1.181}$$

This shows that when the incident wave travels along the part of bar with impedance Z_1 and $Z_2 > Z_1$, the reflected wave at the boundary causes a rotation of the bar in the opposite direction. In particular, at the fixed end the angular displacements due to both waves are equal by a magnitude but have opposite signs. At the same time, the resultant moment becomes twice as large. If the second medium has smaller impedance, $Z_2 < Z_1$, both waves at the boundary cause a twist in the same direction. In the limiting case of the free end $(Z_2 = 0)$, the angles, due to the incident and reflected waves, are equal to each other

$$\varphi = \varphi_i + \varphi_r = 2\varphi_i,$$

while the resultant moment at the free boundary vanishes.

Let us also notice that frequencies of normal modes arising in the bar, having either free or fixed ends or a combination of them, are defined from the same expressions as in the case of longitudinal waves. Before we consider an example, it is useful to write down relationships between the angular and wave velocities, as well as the moment M_z. As follows from eq. 1.179, in the case of the incident wave we have

$$\varphi(z,t) = \varphi_0 f[\alpha(z + c_s t)], \qquad \frac{\partial \varphi}{\partial z}(z,t) = \alpha\, \varphi_0 f\,'[\alpha(z + c_s t)], \tag{1.182}$$

$$\omega_0(z,t) = \frac{\partial \varphi(z,t)}{\partial t} = \varphi_0\, c_s\, \alpha\, f'[\alpha(z + c_s t)], \qquad M_z(z,t) = \mu\, I_0\, \alpha\, \varphi_0\, f\,'[\alpha(z + c_s t)]$$

Therefore

$$\frac{\omega_0}{c_s} = \frac{M_z(z,t)}{I_0 \mu} = \frac{\partial \varphi}{\partial z} \tag{1.183}$$

and the expression for the reflected wave differs by a sign only. Certainly, there is a similarity with the analogous relationships for longitudinal waves.

Example: wave propagation and Hooke's law for torsion Suppose that at the instant $t = 0$, the constant moment M_z is applied to the free end of the bar, $(z = L)$, while the opposite end, $(z = 0)$, is fixed. As follows from eqs. 1.183, the free end starts to rotate with constant velocity ω_0 and the wave propagates along the bar. At the instant $t = L/c_s$, the reflected wave arises at the fixed end and twists the bar in the opposite direction. Correspondingly, behind the wave front a deformation disappears. At the moment $t = 2L/c_s$, the angular displacement of the free end is equal to

$$\varphi(L) = \omega_0 \frac{2L}{c_s}, \tag{1.184}$$

and the whole bar is not deformed. At this instant, the reflected wave arises and decreases the angle φ. Because of this, wave the bar again experiences deformation. When the wave reaches the fixed end, it gives rise to the reflected wave with rotation in the opposite direction. Therefore, behind the wave front deformation again vanishes and at the instant $t = 4L/c_s$ the whole bar is not deformed, and the angular displacement at the free end is equal to zero. In other words, this end returns to the original position. When the reflected wave arises, we observe the same motion as at the beginning, $(t = 0)$. Thus the function describing the behavior of the angle $\varphi(L,t)$ is periodic, and it coincides with the function shown in Fig. 1.4d. Expanding $\varphi(L,t)$ in Fourier's series, we see that the constant part is equal to

$$\frac{b_0}{2} = \frac{1}{2}\frac{4}{T}\int\limits_0^{T/2} \varphi(t)dt = \frac{2}{T}\omega_0 \int\limits_0^{T/2} t dt = \frac{\omega_0 T}{4} = \frac{\omega_0 L}{c_s}$$

Because of attenuation of sinusoidal harmonics, the angle φ tends to be a constant corresponding to an equilibrium:

$$\varphi_0 = \frac{\omega_0 L}{c_s} \qquad \text{or} \qquad \frac{\omega_0}{c_s} = \frac{\varphi_0}{L} = \frac{\partial \varphi}{\partial z} \tag{1.185}$$

that coincides with eq. 1.183. Note that variations of the angle $\varphi(L,t)$ take place within the range

$$0 \leq \varphi \leq \frac{2\omega_0 L}{c_s}$$

and due to attenuation, with time φ gradually approaches to the average value φ_0.

1.6 Bending of a bar at equilibrium and bending waves

Until now we have studied two relatively simple types of waves, namely longitudinal and torsional waves. Each is characterized by one kind of motion and deformation. Next we discuss bending waves, where elastic deformation is more complex and wave propagation changes an elementary volume as well as its rotation. Note that similar behavior characterizes Rayleigh and Stoneley waves. We will limit ourselves to a one-dimensional case related to the seismic responses of some constructions, i.e., dams, during strong earthquakes. It is also worthwhile to mention that the bending waves may also propagate along elastic plates, for example, ice sheets. As before, in order to derive an equation describing the bending waves, we have to establish the relationship between deformation and internal forces at equilibrium.

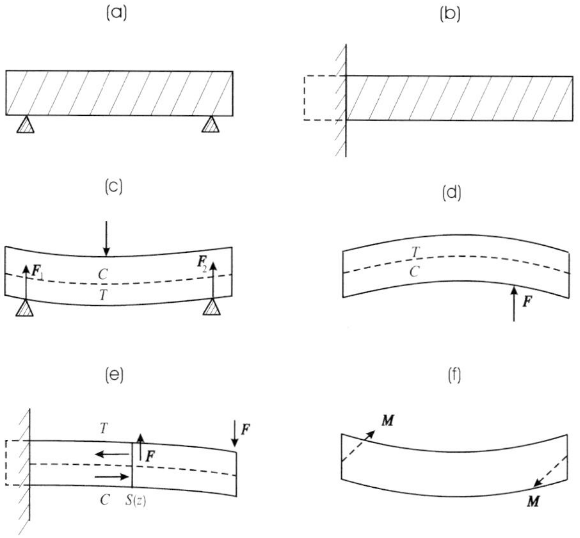

Figure 1.11: (a,b) Horizontal beams (c,d,e) Compression and tension of different portions of deformed beam (f) Deformation due to two moments $\mathbf{M}$ and $-\mathbf{M}$

As in the case of torsion, Coulomb was the first to suggest an approximate theory of bending. Under certain boundary conditions, this theory correctly describes the distribution of stress and strain inside thin bar (beam). Let us imagine a horizontal beam that is either supported at two ends or at one end, as is shown in Fig. 1.11a,b. We will neglect the effect of its weight. When vertically oriented force is applied, the bar becomes deformed, Fig. 1.11c,d,e. It is essential that there is always a line (surface) whose length remains unchanged; it is called the neutral line. Because of deformation, the bar portion located above this line, Fig. 1.11c, is shortened and in a state of a compression (C). At the same time, below the neutral line we observe a stretching of lines that are parallel to the bar axis, and correspondingly, tension (T) takes place. If the external force is directed upward, Fig. 1.11d, tension and compression occur above and beneath the neutral line, respectively. The same picture is observed in the case of the cantilever, Fig. 1.11e. Longitudinal stress is equal to zero at points on the neutral line and it increases with increased distance from this line. In order to describe qualitatively a distribution of stresses at equilibrium, let us mentally draw a cross-section $S(z)$ of the bar at some point z, Fig. 1.11e. As an example, consider the right portion of the bar, subjected to the force $\mathbf{F}$, directed downward. Since this beam element is at equilibrium, the resultant force must be equal to zero. Correspondingly, the left portion of the bar causes the shear force $-\mathbf{F}$, which is applied to the face $S(z)$ and

$$\sum \mathbf{F} = 0$$

Besides, the external force $\mathbf{F}$ causes deformation, and therefore normal stresses arise. They have different signs above and beneath the neutral line. As a result, internal forces associated with these stresses form the moment, which is trying to rotate the beam counterclockwise. This moment compensates an action of the moment due to external force $\mathbf{F}$, which has an opposite direction. Thus, the resultant moment is also equal to zero:

$$\sum \mathbf{M} = 0$$

There is one important case, when the internal shear force $\mathbf{F}$ is absent. This occurs if bending is caused by two moments applied to the bar ends, Fig. 1.11f. In other words, equilibrium takes place when only normal stresses exist. This case was studied by Coulomb.

Coulomb's theory of pure bending

As was pointed out, we assume that bending of a thin beam arises due to moments applied to its ends, and after deformation cross-sections remain plane. Also it is implied that the radius of the curvature is the same for all points of the beam, Fig. 1.12a,

$$R = \text{const} \tag{1.186}$$

By definition, the normal stress τ_{zz} is equal to

$$\tau_{zz}(x) = E\,\frac{\Delta z(x)}{z} \tag{1.187}$$

It is positive above the neutral line and has an opposite sign below. Here z is the original length of an elementary volume and $\Delta z(x)$ is its change, which varies over the cross-section. Thus, we have one more example of an inhomogeneous strain. It is easy to relate τ_{zz} to the radius of curvature R. In fact as is seen from Fig. 1.12b:

$$\frac{z}{R} = \frac{z+\Delta z}{R+x} \qquad \text{or} \qquad \frac{\Delta z}{z} = \frac{x}{R}, \tag{1.188}$$

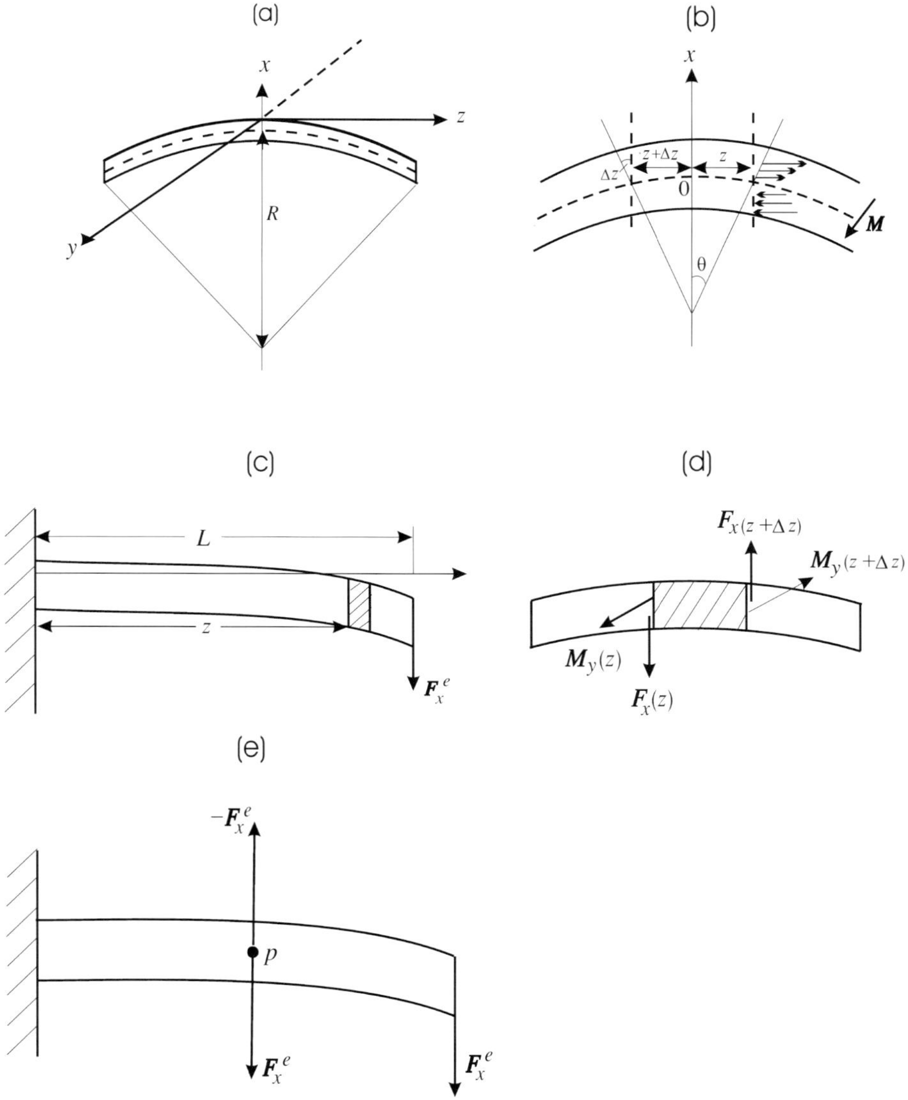

Figure 1.12: (a) Bending bar with constant radius of curvature (b) Illustration of eq. 1.188 (c) External force F_x^e applied to free end of beam (d) Illustration of eq. 1.199 (e) Replacement of force $F_x^e(l)$ by a couple and shear force

and both strain and stress linearly change with increased distance from the neutral line. Thus, in accordance with eq. 1.187 we have

$$e_{zz} = \frac{x}{R} \qquad \text{and} \qquad \tau_{zz} = \frac{Ex}{R} \tag{1.189}$$

Correspondingly, the z-component of the force associated with this stress and acting on the element dS of a cross-section is equal to

$$dF_z(x) = \frac{Ex}{R} dS \tag{1.190}$$

These forces are continuously distributed over area S, and they have opposite directions above and beneath the neutral line. It is clear that each force $dF_z(x)$ forms the moment with respect to this line:

$$dM_y = x\, dF_z,$$

and in order to find the resultant moment we have to perform an integration, which gives

$$M_y = \int_S x\, dF_z$$

Making use of eq. 1.190, we obtain

$$M_y = \frac{E}{R} \int_S x^2 dS \tag{1.191}$$

The integral describes the polar moment of inertia I:

$$I = \int_S x^2 dS \tag{1.192}$$

Thus, we express the torque caused by forces dF_z in terms of the radius of curvature and the polar moment of inertia

$$M_y = \frac{EI}{R} \qquad \text{or} \qquad R = \frac{EI}{M_y} \tag{1.193}$$

It is obvious that with an increase of R, bending decreases, and correspondingly the moment M_y decreases. As follows from eqs. 1.193 and assuming that the moment M_y is given, we see that with an increase of the product EI, the radius R increases too. This means that EI plays the role of "the bending stiffness", since with its increase the bending lessens. For instance, if more masses are placed away from the neutral line, the beam is able to sustain the moment of greater value. As is well known from calculus (Part I), the curvature $1/R$ can be approximately represented as

$$\frac{1}{R} = \frac{d^2x(z)}{dz^2}, \tag{1.194}$$

provided that bending is sufficiently small. Substitution of eq. 1.194 into eq. 1.193 yields

$$M_y = EI\frac{d^2x(z)}{dz^2} \tag{1.195}$$

Here $x(z)$ is the displacement of the neutral line with respect to its original position. As follows from eq. 1.186, at all points of the beam

$$\frac{d^2x(z)}{dz^2} = \text{const} \tag{1.196}$$

Consider an elementary volume, bounded by the lateral surface of the beam and its cross-section $S(z)$ and $S(z+\Delta z)$. In accordance with Newton's third law, moments applied to these faces have opposite directions but the same magnitude, eq. 1.193:

$$\mathbf{M}_y(z) + \mathbf{M}_y(z+\Delta z) = 0 \tag{1.197}$$

Therefore, they provide equilibrium at each element of the bar, and the shear force F_x is absent:

$$\mathbf{F}_x = 0 \tag{1.198}$$

Equilibrium of a cantilever

For comparison, it is useful to consider a more complicated case which is important for deriving the wave equation. Suppose that force F_x^e is applied to the free end of a cantilever, Fig. 1.12c. Note that the word "cantilever" describes a beam supported in such a way that both the position and a slope are fixed at one end. As usual, it is assumed that the length of the beam L is much greater than its cross-section dimensions. Unlike the previous case, the radius of curvature R varies along the bar and, correspondingly, the moment M_y becomes a function of z. Our goal is to demonstrate the presence of shear force F_x and also to find the function $x(z)$ when the beam is at equilibrium. Under action of the external force F_x^e the beam experiences a deformation and, respectively, normal stresses arise at each cross-section and they depend on the coordinate z. For this reason, the resultant moment M_y also changes along the bar axis. Therefore, in order to provide equilibrium of an elementary volume, Fig. 1.12d, we have to assume the presence of shear force F_x. Otherwise, the resultant moment about the y-axis would not be equal to zero and equilibrium would not take place. As is seen from Fig. 1.12d, the condition of equilibrium for the moments is

$$M_y(z+\Delta z) - M_y(z) + F_x \Delta z = 0 \qquad \text{or} \qquad \frac{\partial M_y}{\partial z}\Delta z + F_x \Delta z = 0 \tag{1.199}$$

that is,

$$F_x(z) = -\frac{\partial M_y}{\partial z}$$

In deriving the latter we discarded the term, proportional to $(\Delta z)^2$, since force $F_x(z+\Delta z) \simeq F_x(z) + \Delta z F'_x(z)$. As follows from eq. 1.199 the shear force, acting on the cross-section, is defined by the rate of change of the moment M_y. Earlier derived eqs. 1.193 and 1.195 for the resultant moment M_y, caused by internal forces, remain valid for any external forces applied to the free end. For instance, taking into account eqs. 1.195 and 1.199 we have

$$F_x = -EI\frac{\partial^3 x(z)}{dz^3}, \tag{1.200}$$

provided that I=const.

Before we continue, let us demonstrate that the shear force $F_x(z)$ does not change along the beam. To do this we mentally apply forces $F_x^e(z)$ and $-F_x^e(z)$ at any point p of the bar, Fig. 1.12e. Certainly, this new system of forces is equivalent to the original one, i.e., the force F_x^e is at the free end. These three forces can be treated as the couple of forces $F_x^e(L)$ and $-F_x^e(z)$ and the shear force $F_x^e(z)$ acting at point p. The moment of the couple is directly proportional to the distance between point p and the free end, whereas the shear force F_x^e remains constant. In order to provide equilibrium, a deformation has to produce the couple and the single force with the same behavior as the external force. Thus,

$$F_x = \text{const.}$$

Next, as illustration, we define function $x(z)$. Consider a cross-section $S(z)$, which can be treated as the face of the right portion of the beam. In order to provide its equilibrium, the normal stress τ_{zz} at this face has to create the moment M_y about the line $x=0$ and z =const with the same magnitude as the moment due to the external force F_x^e, but acting in the opposite direction::

$$M_y = F_x(L-z) \tag{1.201}$$

Thus, making use of eq. 1.195, we obtain

$$F_x(L-z)=EI\frac{d^2x}{dz^2} \tag{1.202}$$

Its integration by z gives

$$F_xL\,z-F_x\frac{z^2}{2}=EI\frac{dx}{dz}+C_1$$

At the fixed end of the beam we assume that

$$x(0)=0 \qquad \text{and} \qquad \frac{dx(0)}{dz}=0 \tag{1.203}$$

The second boundary condition yields:

$$F_xL\,z-F_x\frac{z^2}{2}=EI\frac{dx}{dz}$$

Integrating again, we have

$$F_xL\frac{z^2}{2}-F_x\frac{z^3}{6}=EI\;x(z)+C_2$$

The first equality of set 1.203 gives $C_2=0$ and

$$x(z)=\frac{F_x}{EI}\left(\frac{Lz^2}{2}-\frac{z^3}{6}\right) \tag{1.204}$$

It is useful to notice that displacement of the free end, $z=L$, is

$$x(L)=\frac{F_x}{EI}\frac{L^3}{3}, \tag{1.205}$$

and it increases as a cube of the distance from the fixed end.

Stress and displacement fields for pure bending

As we know equilibrium takes place after attenuation of waves and, correspondingly, distribution of displacements, stresses and strains at this second stage preserves some important features of these fields, which are carried by bending waves. Now we describe Coulomb's solution in detail and also investigate displacement of bar particles. Consider the beam with the rectangular cross-section and, as before, assume that the origin of the Cartesian system of coordinates is located at the neutral line at the middle of the beam. The z-axis is oriented along the beam, the x-axis is directed downward and the y-axis is perpendicular to the plane XOZ. We restrict ourselves to the case of pure bending, when both ends of the beam are subjected to an action of the force couples with moments

$$\mathbf{M}_y \qquad \text{and} \qquad -\mathbf{M}_y$$

They have the same magnitude but opposite directions. Coulomb assumed that stresses arising due to the bending are

$$\tau_{xx} = \tau_{yy} = \tau_{xy} = \tau_{yz} = \tau_{xz} = 0, \qquad \tau_{zz} = \frac{Ex}{R}, \tag{1.206}$$

where E is the Young modulus.

After a deformation, the $z-$line is transformed into a circle with radius R. In other words, R is constant at all points of the beam, and it represents its radius of curvature. Thus, as follows from Coulomb's solution, there is only one component of stress τ_{zz}, and it is the normal stress at elementary areas perpendicular to the beam axis. This stress is directly proportional to distance x. Our purpose is to formulate conditions under which Coulomb's approximation becomes an exact solution. Since τ_{zz} is independent of z, it is clear that stresses given by eqs. 1.206 provide equilibrium of an elementary volume, if volume forces are absent.

Next, consider stresses at the lateral surface of the beam, which is formed by four plane strips. Letting $\boldsymbol{\nu}$ be the unit vector normal to the lateral surface and making use of Cauchy's formulas (Appendix C), we have

$$\begin{aligned} \tau_{x\nu} &= \tau_{xx} l + \tau_{xy} m + \tau_{xz} n \\ \tau_{y\nu} &= \tau_{xy} l + \tau_{yy} m + \tau_{yz} n \\ \tau_{z\nu} &= \tau_{xz} l + \tau_{yz} m + \tau_{zz} n \end{aligned} \tag{1.207}$$

Since the normal to the lateral surface, $\boldsymbol{\nu}$, is perpendicular to the z-axis, $(n = 0)$, we obtain

$$\tau_{x\nu} = \tau_{y\nu} = \tau_{z\nu} = 0 \tag{1.208}$$

Thus, the solution given by eqs. 1.206 obeys the boundary conditions at the lateral surface, if the external forces are absent at its points. Cross-sections of free ends are parallel to the plane XOY and, correspondingly, directional cosines are

$$l = m = 0, \qquad n = 1 \tag{1.209}$$

Substitution of eqs. 1.206 and 1.209 into set 1.207 yields

$$\tau_{x\nu} = \tau_{y\nu} = 0 \qquad \text{and} \qquad \tau_{zz} = \pm\frac{Ex}{R} \tag{1.210}$$

Therefore, Coulomb's formulas satisfy the boundary condition at the free ends, if external forces have only the normal component, F_z, which is distributed over these ends in accordance with eq. 1.206. As was shown previously the normal stress τ_{zz} produces an extension above the neutral line but a compression beneath it. Respectively, the resultant force, acting on a cross-section of the beam, vanishes, but the total moment $\mathbf{M}_y$ differs from zero. In fact, we have

$$\int\limits_S \tau_{zz} dxdy = \frac{E}{R}\int\limits_S xdxdy = 0,$$

but the scalar component of the moment is

$$M_y = \int\limits_S \tau_{zz} xdxdy = \frac{E}{R}\int\limits_S x^2 dxdy \qquad \text{or} \qquad M_y = \frac{EI}{R}$$

and it was derived above.

The displacement field

Now we begin to study the behavior of the function $\mathbf{s}$:

$$\mathbf{s} = u\mathbf{i} + v\mathbf{j} + w\mathbf{k},$$

which characterizes displacement of beam particles. Proceeding from Hooke's law:

$$e_{xx} = \frac{1}{E}[\tau_{xx} - \sigma\left(\tau_{yy} + \tau_{zz}\right)]$$

$$e_{yy} = \frac{1}{E}[\tau_{yy} - \sigma(\tau_{xx} + \tau_{zz})]$$

$$e_{zz} = \frac{1}{E}[\tau_{zz} - \sigma(\tau_{xx} + \tau_{yy})] \tag{1.211}$$

$$e_{xy} = \frac{1}{\mu}\tau_{xy}, \qquad e_{xz} = \frac{1}{\mu}\tau_{xz}, \qquad e_{yz} = \frac{1}{\mu}\tau_{yz}$$

and making use of eqs. 1.206, we obtain

$$e_{xx} = e_{yy} = -\frac{\sigma}{E}\tau_{zz} = -\sigma\frac{x}{R}, \qquad e_{zz} = \frac{\tau_{zz}}{E} = \frac{x}{R}, \qquad e_{xy} = e_{xz} = e_{yz} = 0 \qquad (1.212)$$

Therefore, according to the definition of strains:

$$\frac{\partial u}{\partial x} = \frac{\partial v}{\partial y} = -\frac{\sigma x}{R}, \qquad \frac{\partial w}{\partial z} = \frac{x}{R} \qquad (1.213)$$

$$\frac{\partial u}{\partial y} + \frac{\partial v}{\partial x} = 0 \qquad \frac{\partial u}{\partial z} + \frac{\partial w}{\partial x} = 0 \qquad \frac{\partial v}{\partial z} + \frac{\partial w}{\partial y} = 0 \qquad (1.214)$$

The set of eqs. 1.213–1.215 permits us to determine all three components of the displacement **s**. First, integrating the second equation of set 1.213, we have

$$w = \frac{xz}{R} + \omega_0(x, y), \qquad (1.215)$$

where $\omega_0(x, y)$ is an arbitrary function of x and y. Substitution of the latter into the last two equations of set 1.214 gives:

$$\frac{\partial u}{\partial z} = -\frac{\partial w}{\partial x} = -\frac{z}{R} - \frac{\partial \omega_0}{\partial x}, \qquad \frac{\partial v}{\partial z} = -\frac{\partial w}{\partial y} = -\frac{\partial \omega_0}{\partial y} \qquad (1.216)$$

Their integration yields

$$u = -\frac{z^2}{2R} - z\frac{\partial \omega_0}{\partial x} + u_0(x, y) \quad \text{and} \quad v = -z\frac{\partial \omega_0}{\partial y} + v_0(x, y) \qquad (1.217)$$

Here $u_0(x, y)$ and $v_0(x, y)$ are arbitrary functions of x and y. In order to find these unknown functions, we substitute eqs. 1.217 into eq. 1.213, and it gives

$$-z\frac{\partial^2 \omega_0}{\partial x^2} + \frac{\partial u_0}{\partial x} = -\frac{\sigma x}{R} \quad \text{and} \quad -z\frac{\partial^2 \omega_0}{\partial y^2} + \frac{\partial v_0}{\partial y} = -\frac{\sigma x}{R} \qquad (1.218)$$

These equations are valid for any z, and this fact allows us to greatly simplify them. Correspondingly, in place of this set we have:

$$\frac{\partial^2 \omega_0}{\partial x^2} = 0, \qquad \frac{\partial^2 \omega_0}{\partial y^2} = 0 \qquad (1.219)$$

and

$$\frac{\partial u_0}{\partial x} = -\frac{\sigma x}{R}, \qquad \frac{\partial v_0}{\partial y} = -\frac{\sigma x}{R} \qquad (1.220)$$

An integration of eqs. 1.220 yields

$$u_0 = -\frac{\sigma x^2}{2R} + f_1(y) \qquad v_0 = -\frac{\sigma xy}{R} + f_2(x), \tag{1.221}$$

where $f_1(y)$ and $f_2(x)$ are arbitrary functions and each of them depends on a single argument. Next we establish a relationship among functions $f_1(y)$, $f_2(x)$, and $w_0(x,y)$. Substitution of eqs. 1.217 and 1.221 into the first equation of set 1.214 gives

$$2z\,\frac{\partial^2 \omega_0}{\partial x \partial y} - \frac{\partial f_1}{\partial y} - \frac{\partial f_2}{\partial x} + \frac{\sigma y}{R} = 0$$

Since the latter is valid for any z, it can be replaced by the set:

$$\frac{\partial^2 \omega_0}{\partial x \partial y} = 0 \qquad \text{and} \qquad \frac{\partial f_1(y)}{\partial y} + \frac{\partial f_2(x)}{\partial x} - \frac{\sigma y}{R} = 0 \tag{1.222}$$

From eqs. 1.219 and 1.222, it follows that the function $w_0(x,y)$ is linear with respect to x and y, i.e.,

$$w_0(x,y) = Ax + By + C, \tag{1.223}$$

where A, B, and C are constants. At the same time, from eq. 1.222 we have:

$$\frac{\partial f_1(y)}{\partial y} - \frac{\sigma y}{R} = -\frac{\partial f_2(x)}{\partial x} = D, \tag{1.224}$$

because functions at the left and right sides depend on different arguments. Integration of the last equality gives

$$f_2(x) = -Dx + D_1 \qquad \text{and} \qquad f_1(y) = \frac{\sigma y^2}{2R} + Dy + D_2 \tag{1.225}$$

Here D, D_1, and D_2 are constants. Substitution of eqs. 1.224 and 1.225 into eqs. 1.215 and 1.217 and the use of eq. 1.221 give

$$u = -\frac{z^2}{2R} - \sigma\frac{x^2 - y^2}{2R} - Az + Dy + D_2, \tag{1.226}$$

$$v = -\frac{\sigma xy}{R} - Bz - Dx + D_1, \qquad w = \frac{xz}{R} + Ax + By + C,$$

which contain six unknowns: A, B, C, D, D_1, and D_2. In order to determine them, we make use of the fact that the origin of coordinates is at the middle of the beam. Because of symmetry, the cross-section $z = 0$ does not move during bending, i.e.,

$$w = 0 \qquad \text{for} \qquad z = 0 \tag{1.227}$$

This gives

$$A = B = C = 0 \tag{1.228}$$

Also due symmetry at the origin $0, (x = y = z = 0)$, we have

$$u = v = w = 0$$

Therefore

$$D_1 = D_2 = 0 \tag{1.229}$$

Finally, consider a linear element of the bar, oriented along the y-axis and passing through the origin. After deformation, it preserves its orientation. This behavior can described as

$$\frac{du}{dy} = 0 \qquad \text{if} \qquad x = y = z = 0$$

Correspondingly, the first equation of set 1.226 gives

$$D = 0 \tag{1.230}$$

Thus, expressions for the displacement components are

$$u = -\frac{1}{2R}[z^2 + \sigma(x^2 - y^2)], \qquad v = -\frac{\sigma xy}{R}, \qquad w = \frac{zx}{R} \tag{1.231}$$

For instance, points of the neutral line, $(x = y = 0)$, experience displacement

$$u = -\frac{z^2}{2R}, \qquad v = w = 0,$$

that is, the straight line is transformed into a parabola. Also consider a cross-section of the beam, $z = z_0$. After deformation, its points are situated at the surface:

$$z = z_0 + w = z_0(1 + \frac{x}{R})$$

The latter also describes the plane; that is, due to pure bending, cross-sections remain plane. As follows from eqs. 1.231, the displacement component along the x-axis is directed toward the neutral line and it is proportional to the square of coordinates. It is interesting to note that the y-component, v, is independent of the coordinate z, and its magnitude linearly increases with an increase of x and y, but the sign is defined by that of these coordinates. Component w has similar behavior, but it is a function of x and z. In accordance with eq. 1.213, the divergence of the field $\mathbf{s}$ is

$$\text{div } \mathbf{s} = \frac{(1-2\sigma)\, x}{R} = \frac{(1-2\sigma)}{E}\tau_{zz}, \tag{1.232}$$

and it is independent of coordinates y and z. Thus, bending is accompanied by a change of elementary volumes, and the normal stress τ_{zz} can be treated as the source of the field $\mathbf{s}$. Now consider the second important characteristic of the vector field, namely, curl $\mathbf{s}$:

$$\text{curl}_x \mathbf{s} = \frac{\partial w}{\partial y} - \frac{\partial v}{\partial z}, \qquad \text{curl}_y \mathbf{s} = \frac{\partial u}{\partial z} - \frac{\partial w}{\partial x}, \qquad \text{curl}_z \mathbf{s} = \frac{\partial v}{\partial x} - \frac{\partial u}{\partial y}$$

The use of eq. 1.231 gives

$$\text{curl}_x \mathbf{s} = 0, \qquad \text{curl}_y \mathbf{s} = -\frac{2z}{R} \qquad \text{curl}_z \mathbf{s} = -\frac{2\sigma y}{R} \tag{1.233}$$

This means that bending also causes rotation of an elementary volume as a rigid body. For this reason, we may say that bending waves differ from both longitudinal and torsional waves.

Bending waves

Until now we have assumed that a bar is at equilibrium. Next, suppose that either the force $\mathbf{F}_x$ or the moment $\mathbf{M}_y$, or both, are applied at some place on the bar, for example, at its ends. Due to deformation, internal forces arise and the bending waves start to propagate. Unlike longitudinal and torsional (shear) waves, bending waves carry more complicated motion of elementary volumes that includes both translation and rotation. At the same time, its motion along the bar axis is absent.

To derive an equation describing bending waves, consider an element of the bar with an extension Δz, Fig. 1.13. Suppose that the wave propagates along the z-axis toward large values of z. When it reaches the back face of the element, $S(z)$, this surface becomes subjected to an action of the moment $\mathbf{M}_y\ (z)$ and the shear force $\mathbf{F}_x(z)$. Inasmuch as the volume length, Δz, is extremely small, the wave almost instantly reaches the opposite face and acts on a medium located in front of this element. Its action is characterized by the same direction of the moment and the shear force as at the back face. Then, in accordance with the Newton's third law, the front face of the volume element is subjected to the moment and the force, which have opposite directions. Thus, the resultant force and the resultant moment applied to the bar element are

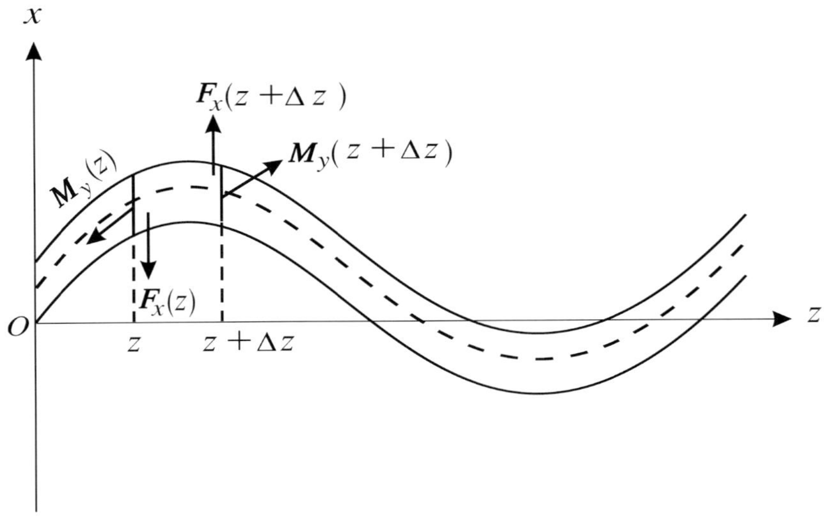

Figure 1.13: Forces and moments acting on an elementary volume

$$F_x = F_x(z + \Delta z) - F_x(z) \tag{1.234}$$

and

$$M_y = M_y(z + \Delta z) - M_y(z) \tag{1.235}$$

The moment M_y causes rotation of the element as a rigid body around an axis that is parallel to the y-axis, and it can be represented as

$$M_y = \frac{\partial M_y}{\partial z}\Delta z \tag{1.236}$$

An action of force F_x displays in two ways. First of all, it displaces the elementary volume along the x-axis, and in accordance with Newton's second law we have:

$$F_x(z + \Delta z) - F_x(z) = \rho\ \Delta z\ S\frac{\partial^2 u}{\partial t^2} \qquad \text{or} \qquad \frac{\partial F_x}{\partial z} = \rho\ S\ \frac{\partial^2 u}{\partial t^2} \tag{1.237}$$

Here S is the cross-section of the element, ρ is its density, and u is a displacement of the neutral line due to a wave. We assume that this displacement is very small, and, correspondingly, the length of this line practically remains the same, i.e., its extension is neglected. It may be proper to note that at equilibrium, the moment M_y linearly changes between faces and, therefore, forces applied to $S(z)$ and $S(z+\Delta z)$ are equal by a magnitude. In other words, they do not cause translation of the bar element. Second, shear forces give rise to the moment about the axis, which is parallel to the y-axis and passes through the center of the elementary volume. It is equal to

$$F_x \Delta z \tag{1.238}$$

Thus, the total moment causing rotation of this element is

$$(\frac{\partial M_y}{\partial z} + F_x)\Delta z \tag{1.239}$$

Applying Newton's second law for a rotation (Appendix A), we obtain:

$$\rho\, I \Delta z \frac{\partial^2 \alpha}{\partial t^2} = (\frac{\partial M_y}{\partial z} + F_x)\Delta z \qquad \text{or} \qquad \rho\, I \frac{\partial^2 \alpha}{\partial t^2} = \frac{\partial M_y}{\partial z} + F_x \tag{1.240}$$

Here α is an angle of rotation of the element, and I is the polar moment of inertia. Thus, we arrived at two equations:

$$\rho\, S \frac{\partial^2 u}{\partial t^2} = \frac{\partial F_x}{\partial z} \qquad \text{and} \qquad \rho\, I \frac{\partial^2 \alpha}{\partial t^2} = \frac{\partial M_y}{\partial z} + F_x, \tag{1.241}$$

which describe simultaneous translation and rotation of an elementary volume bounded by the lateral surface of the bar and cross-sections $S(z)$ and $S(z+\Delta z)$. This system contains several unknowns, namely, $u(z,t)$, $\alpha(z,t)$, $F_x(z,t)$, and $M_y(z,t)$, and our goal is to obtain one equation with respect to displacement $u(z,t)$. By definition, α characterizes a slope of the neutral line with respect to the z-axis at some point z and the instant t. It is obvious that for small values of α we have

$$\alpha \simeq \tan \alpha = \frac{\partial u}{\partial z}$$

Thus, set 1.241 becomes

$$\rho S \frac{\partial^2 u}{\partial t^2} = \frac{\partial F_x}{\partial z} \qquad \rho I \frac{\partial^3 u}{\partial z\, \partial t^2} = \frac{\partial M_y}{\partial z} + F_x \tag{1.242}$$

Differentiation of the last of eqs. 1.242 with respect to z and the use of the first of these equations allow us to eliminate the unknown force F_x. Thus we obtain one equation:

$$\rho I\, \frac{\partial^4 u}{\partial z^2\, \partial t^2} = \frac{\partial M_y}{\partial z^2} + \rho S \frac{\partial^2 u}{\partial t^2} \tag{1.243}$$

with two unknowns. Taking into account eq. 1.195

$$M_y = IE \frac{\partial^2 u}{\partial z^2},$$

we arrive at the equation with respect to the single unknown, $u(z,t)$:

$$\rho\, I \frac{\partial^4 u}{\partial z^2 \partial t^2} = IE \frac{\partial^4 u}{\partial z^4} + \rho\, S \frac{\partial^2 u}{\partial t^2} \tag{1.244}$$

This is the linear partial differential equation of the fourth order. Certainly, it is not one-dimensional wave equation at its conventional form

$$\frac{\partial^2 u}{\partial z^2} = \frac{1}{c^2}\frac{\partial^2 u}{\partial t^2},$$

but some solutions of eq. 1.244 describe a wave phenomenon.

Sinusoidal waves

Suppose that external forces are sinusoidal functions of time. Taking into account that eq. 1.244 is linear, we observe the sinusoidal vibrations of the bar, too. Correspondingly, the displacement $u(z,t)$ can be represented in the form

$$u(z,t) = \mathrm{Re}[\widetilde{u}(z)e^{-i\omega t}] \tag{1.245}$$

Its substitution into eq. 1.244 transforms the latter into an ordinary differential equation of the fourth order with respect to the complex amplitude $\widetilde{u}(z)$, and we obtain:

$$-\rho\, I\, \omega^2 \frac{d^2\widetilde{u}}{dz^2} = IE \frac{d^4\widetilde{u}}{dz^2} - \rho S \omega^2\, \widetilde{u} = 0$$

or

$$\frac{\partial^4 \widetilde{u}}{\partial z^4} + 2a \frac{d^2\widetilde{u}}{dz^2} - b\, \widetilde{u} = 0, \tag{1.246}$$

where

$$a = \frac{\rho\omega^2}{2E}, \qquad b = \frac{\omega^2 S\rho}{IE} \tag{1.247}$$

Respectively, a solution of eq. 1.246 has the form:

$$\widetilde{u}(z) = A_1 e^{k_1 z} + A_2 e^{k_2 z} + A_3 e^{k_3 z} + A_4 e^{k_4 z} \tag{1.248}$$

Here k_n are roots of the characteristic equation:

$$k_n^4 + 2a\, k_n^2 - b = 0 \tag{1.249}$$

So

$$k_n^2 = -a \pm \sqrt{a^2 + b}$$

or

$$k_n = b^{\frac{1}{4}} \left(-\frac{a}{\sqrt{b}} \pm \sqrt{1 + \frac{a^2}{b}} \right)^{1/2} \tag{1.250}$$

As follows from eq. 1.247,

$$\frac{a^2}{b} = \frac{\rho\, \omega^2 I}{4ES}$$

Since

$$I = r_0^2\, S \qquad \text{and} \qquad \frac{E}{\rho} = c_l^2,$$

we have

$$\frac{a^2}{b} = \frac{\pi^2 r_0^2}{c_l^2 T^2} = \frac{\pi^2 r_0^2}{\lambda^2}, \tag{1.251}$$

where λ is the wavelength of the longitudinal waves and r_0 is the so-called the radius of inertia of the cross-section. In deriving eq. 1.244, we assumed that the normal stress, τ_{zz}, linearly increases with a distance from the $z-$axis. Otherwise, the equality

$$M_y = EI\frac{\partial u^2}{\partial z^2}$$

becomes invalid. This condition implies that the wavelength has to greatly exceed the bar width. Thus, eq. 1.244 is applied, when an equality

$$r_0 \ll \lambda \tag{1.252}$$

takes place. Of course, the same relation is valid for the wavelength of bending waves. This inequality greatly simplifies eq. 1.250, and we have:

$$k_n = \left(\frac{\omega^2 S \rho}{IE} \right)^{1/4} (\pm 1)^{1/2}$$

or

$$k_1 = i \left(\frac{\rho}{E r_0^2} \right)^{1/4} \omega^{1/2}, \qquad k_2 = -i \left(\frac{\rho}{E r_0^2} \right)^{1/4} \omega^{1/2}, \tag{1.253}$$

$$k_3 = \left(\frac{\rho}{Er_0^2}\right)^{1/4} \omega^{1/2}, \qquad k_4 = -\left(\frac{\rho}{Er_0^2}\right)^{1/4} \omega^{1/2},$$

Therefore, the general solution for the displacement $u(z,t)$ is

$$u(z,t) = \mathrm{Re}\{A_1 \exp[i(kz - \omega t)] + A_2 \exp[-i(kz + \omega t)] \tag{1.254}$$

$$+A_3 \exp(kz - i\omega t) + A_4 \exp(-kz - i\omega t)\}$$

Here

$$k = \left(\frac{\rho}{Er_0^2}\right)^{1/4} \omega^{1/2} \tag{1.255}$$

is the magnitude of wavenumber of a bending wave. It can be also written in the form

$$k = \left(\frac{\omega}{c_l r_0}\right)^{1/2} \tag{1.256}$$

The first two terms in eq. 1.254 describe waves traveling along the bar in the opposite direction with the phase velocity

$$c_b(\omega) = \frac{\omega}{k} = (c_l r_0 \omega)^{1/2} \qquad \text{or} \qquad \frac{c_b(\omega)}{c_l} = \sqrt{2\pi}\left(\frac{r_0}{\lambda}\right)^{1/2} \tag{1.257}$$

The latter clearly shows that the velocity of bending waves is less than that of longitudinal waves, $(r_0 \ll \lambda)$, and the difference between them becomes more noticeable with a decrease of frequency. Unlike with longitudinal and torsional waves, the velocity of bending waves, $c_b(\omega)$, depends on a frequency, and it is directly proportional to the square root of ω. It is obvious that due to dispersion, propagation of the bending wave is accompanied by a change of its shape. As is well known, two factors usually cause a dispersion. One is a transformation of elastic energy into heat, and the other is interference of waves, propagating, for example, in a waveguide. The latter occurs when the wavelength is comparable to or smaller than the distance between interfaces. In our case, both of these factors are absent. Two more terms in a solution with coefficients A_3 and A_4, eq. 1.254, represent periodic oscillations as functions of time, which exponentially change with distance. Each term varies simultaneously at different points along the bar. Certainly, they do not characterize wave propagation and have rather a diffusion behavior. The equation of bending waves describes displacement, $u(z,t)$, of the neutral line along the x-axis. At the same time, motion of bar particles is much more complicated and results in translation and rotation of an elementary volume. Earlier, considering an equilibrium, we demonstrated that

$$\text{curl}\mathbf{s} \neq 0 \qquad \text{and} \qquad \text{div}\mathbf{s} \neq 0 \tag{1.258}$$

This means that the bending waves belong to a more general type of wave than do longitudinal and torsional waves, and in this sense they are similar to the Rayleigh waves.

Boundary conditions

Since propagation of bending waves is described by a differential equation of the fourth order, there are more boundary conditions than in the case of longitudinal and shear waves. First, suppose that the bar consists of two homogeneous portions and the cross-section $S(z)$ is the boundary between them. It is obvious that at the interface both the displacement $u(z,t)$ and its derivative $\partial u(z,t)/\partial z$ have to be continuous functions:

$$u_1(z,t) = u_2(z,t) \qquad \frac{\partial u_1(z,t)}{\partial z} = \frac{\partial u_2(z,t)}{\partial z} \tag{1.259}$$

Otherwise the bar would be broken. Also the moment M_y and the shear force F_x are continuous, that is, in accordance with eqs. 1.195 and 1.200:

$$E_1 \frac{\partial^2 u_1}{\partial z^2} = E_2 \frac{\partial^2 u_2}{\partial z^2}, \qquad E_1 \frac{\partial^3 u_1}{\partial z^3} = E_2 \frac{\partial^3 u_2}{\partial z^3} \tag{1.260}$$

Discontinuity of one of these functions leads to infinitely large angular or linear acceleration. Thus, eqs. 1.259 and 1.260 describe behavior of bending waves at the interface. Let us also consider two more boundary conditions, when waves are sinusoidal functions of time.

Case one The displacement u and its derivative with respect to z are given functions at some point z_0:

$$u(z_0,t) = a_1 \cos \omega t, \qquad \frac{\partial u(z_0,t)}{\partial z} = b_1 \sin \omega t \tag{1.261}$$

In particular, it may happen that $a_1 = b_1 = 0$. Note that $\partial u(z_0,t)/\partial z$ characterizes the bar slope at point z_0.

Case two The moment M_y and the shear force F_x are given at point z_0. This means that

$$\frac{\partial^2 u(z_0,t)}{\partial z^2} = a_2 \cos \omega t, \qquad \frac{\partial^3 u(z_0,t)}{\partial z^3} = b_2 \sin \omega t \tag{1.262}$$

For instance, at the free end without external forces:

$$a_2 = b_2 = 0$$

Of course, one can introduce different boundary conditions. To illustrate the behavior of bending waves, consider several examples and start from the simplest model of a bar.

Example one: infinite bar Suppose that bending waves are caused by a displacement at some point $z = 0$, and the boundary conditions are given by eq. 1.261. Inasmuch as the wave field has everywhere a finite value and incoming waves are absent, the solution of eq. 1.254 can be written as

$$u(z,t) = A_1 \cos(\omega t - kz) + A_4 e^{-kz} \cos \omega t \qquad \text{if } z > 0 \tag{1.263}$$

For negative values of z the displacement has a similar form. Applying boundary conditions, we obtain

$$A_1 + A_4 = a_1, \qquad kA_1 - kA_4 = b_1,$$

whence

$$A_1 = \frac{ka_1 + b_1}{2k}, \qquad A_4 = \frac{ka_1 - b_1}{2k} \tag{1.264}$$

and the displacement is

$$u(z,t) = \frac{ka_1 + b_1}{2k} \cos(\omega t - kz) + \frac{ka_1 - b_1}{2k} e^{-kz} \cos \omega t \tag{1.265}$$

Thus, $u(z,t)$ represents a superposition of the wave, traveling away from point $z = 0$, and vibrations, which exponentially decay with distance. Taking into account that

$$e^{-kz} = e^{-2\pi\, z/\lambda_b}, \tag{1.266}$$

we see that the second (diffusion) part of the wave field is noticeable only in the vicinity of point z_0, at distances that are smaller than the wavelength.

Example two: reflection from free end Suppose that the incident wave propagates toward the free end, $(z = 0)$, where the moment M_y and shear force F_x are equal to zero

$$M_y(0,t) = 0, \qquad F_x(0,t) = 0$$

or

$$\frac{\partial^2 u}{\partial z^2} = 0 \qquad \text{and} \qquad \frac{\partial^3 u}{\partial z^3} = 0 \qquad \text{at } z = 0$$

It is natural to expect an appearance of the reflected wave at the free end, as well as decaying vibrations. Therefore, the displacement can be written as a sum:

$$\widetilde{u}(z,t) = A_1 e^{-ikz} + A_2 e^{ikz} + B e^{-kz}\,, \qquad \text{if} \quad z > 0 \tag{1.267}$$

where A_1 is given. From the boundary conditions we obtain two equations with two unknowns

$$-A_2 + B = A_1 \qquad\qquad -iA_2 - B = -iA_1$$

whence

$$A_2 = \frac{i-1}{i+1}A_1, \qquad\qquad B = \frac{2\,i}{1+i}A_1 \tag{1.268}$$

and the displacement field is

$$u(z,t) = A_1[\cos(kz+\omega t) - \sin(kz-\omega t) + \sqrt{2}\,e^{-kz}\cos(\omega t - \frac{\pi}{4})] \tag{1.269}$$

In particular, at the free end

$$u(0,t) = 2A_1(\cos\omega t + \sin\omega t) = 2\sqrt{2}A_1\sin(\omega t + \frac{\pi}{4}), \tag{1.270}$$

and its amplitude is almost three times that of the incident wave.

Example three: reflection from the fixed end As we already know, at the fixed end we have

$$u(0,t) = 0 \qquad\qquad \frac{\partial u(0,t)}{\partial z} = 0$$

Then, making use of eq. 1.267, these conditions give

$$A_2 + B = -A_1 \qquad\qquad iA_2 - B = iA_1$$

Thus

$$A_2 = \frac{i-1}{i+1}A_1, \qquad\qquad B = -\frac{2i}{1+i}A_1 \tag{1.271}$$

Comparison with eqs. 1.268 shows that in both cases, the amplitude and phase of reflected waves coincide. At the same time the complex amplitudes of vibrations differ by a sign only. As follows from eq. 1.254,

$$u(z,t) = A_1[\cos(kz+\omega t) - \sin(kz-\omega t) - \sqrt{2}\cos(\omega t - \frac{\pi}{4})e^{-kz}] \tag{1.272}$$

Of course, at the boundary both $u(0,t)$ and $\partial u(0,t)/\partial z$ vanish.

Example four: normal modes By analogy with longitudinal and torsional waves, consider a formation of normal modes in a bar of finite length l. In order to solve this task, it is convenient to take the real part of the complex amplitude of the displacement and represent it in terms of the sinusoidal and hyperbolic functions:

$$\mathrm{Re}\,\tilde{u}(z) = A_1 \cos kz + A_2 \sin kz + B_1 cosh\ kz + B_2 sinh\ kz \tag{1.273}$$

For illustration, suppose that boundary conditions are

$$u(0,t) = 0, \qquad \frac{\partial^2 u(0,t)}{\partial z^2} = 0, \qquad u(l,t) = 0, \qquad \frac{\partial^2 u(l,t)}{\partial z^2} = 0$$

The first two equations give

$$A_1 + B_1 = 0, \qquad -A_1 + B_1 = 0, \qquad \text{i.e.,} \qquad A_1 = B_1 = 0$$

and

$$\tilde{u}(z) = A_2 \sin kz + B_2 \sinh kz \tag{1.274}$$

The second set of boundary conditions yields

$$A_2 \sin kl + B_2 sinh\ kl = 0 \qquad -A_2 \sin kl + B_2 \sinh kl = 0 \tag{1.275}$$

Therefore, this system has a nonzero solution when

$$B_2 = 0 \qquad \text{and} \qquad kl = \pi n \tag{1.276}$$

and

$$u(z,t) = A_2 \sin \frac{\pi n}{l} z \cos \omega t, \tag{1.277}$$

where A_2 is an arbitrary constant. As follows from eq. 1.276

$$\frac{l}{\lambda_b} = \frac{n}{2}$$

and normal modes arise, provided that the bar length, l, is equal to the integer number of $\lambda_b/2$. A similar relationship was observed for other waves. As follows from eq. 1.257, frequencies of normal modes can be represented as

$$\omega_n = c_l r_0 (\frac{\pi n}{l})^2 \tag{1.278}$$

By analogy we can determine frequencies of normal modes for different boundary conditions.

Chapter 2

Basic equations of elastic waves

In this chapter, we will derive an equation that describes the displacement field $\mathbf{s}$. Then we introduce scalar and vector potentials and derive the wave equations and boundary conditions that characterize the behavior of these functions. Finally, we will focus on the relationship between kinetic and potential energies in an elastic medium and on Poynting's vector.

2.1 Equations of motion of an elementary volume

Let us consider an elementary cube inside a medium, as is shown in Fig. 2.1a. When a wave passes through this volume, it becomes deformed, and internal forces arise. As a result, the medium surrounding the elementary volume, acts on each face of the cube. Since the faces are small, it is assumed that the forces are uniformly distributed over them. Correspondingly, it is natural to introduce the vector $\mathbf{t}$, which characterizes the force per unit area:

$$\mathbf{F} = \mathbf{t}\, dS \tag{2.1}$$

Here $dS = dxdy = dxdz = dydz$, and in the Cartesian system of coordinates

$$\mathbf{t} = t_x\mathbf{i} + t_y\mathbf{j} + t_z\mathbf{k} \tag{2.2}$$

The mutual orientation of forces applied to the cube faces is not arbitrary but obeys two rules that follow from the physical considerations. First, consider opposite faces, for example, $S(x - \Delta x/2,\, y,\, z)$ and $S(x + \Delta x/2,\, y,\, z)$, that are perpendicular to the x-axis. When the wave approaches the back face $S(x - \Delta x/2,\, y,\, z)$ at some instant t, the surrounding medium acts on this face with a force that, in general, has the normal

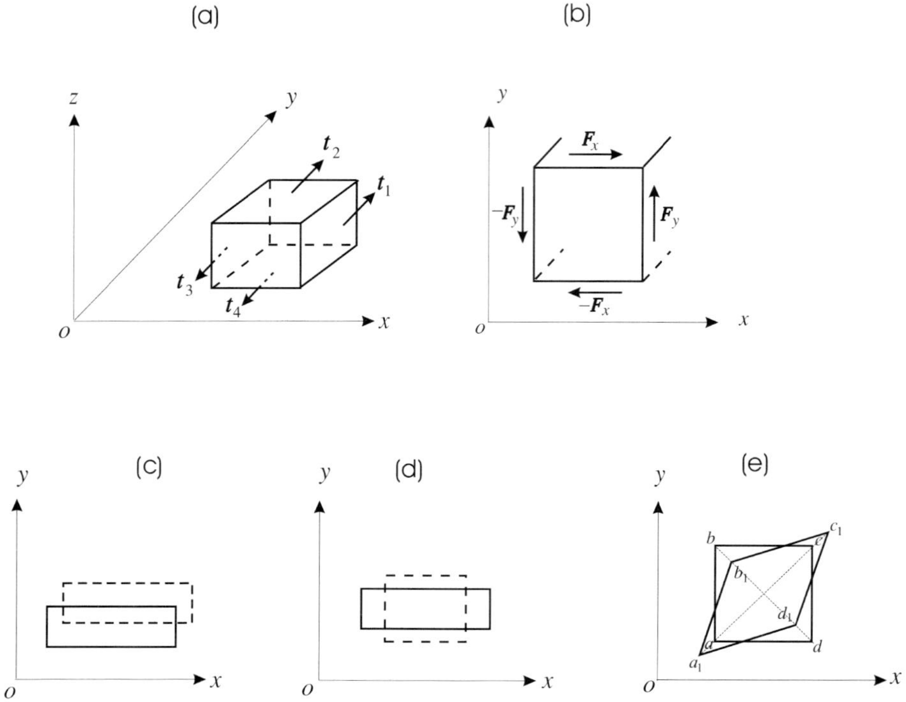

Figure 2.1: (a) Surface forces acting on elementary volume in a medium (b) Behavior of shear forces (c) Translation of elementary volume (d) Compression (expansion) of elementary volume (e) Pure shear deformation

and tangential components:

$$F_n(x - \frac{\Delta x}{2}, y, z, t) \qquad \text{and} \qquad F_t(x - \frac{\Delta x}{2}, y, z, t) \tag{2.3}$$

After a very small time interval Δt, the wave reaches face $S(x+\Delta x/2, y, z)$, and forces given by eqs. 2.3 act on the medium in front of the elementary volume. In accordance with Newton's third law, face $S(x + \Delta x/2, y, z)$ is subjected to action of forces:

$$-F_n(x + \frac{\Delta x}{2}, y, z, t) \qquad \text{and} \qquad -F_t(x + \frac{\Delta x}{2}, y, z, t) \tag{2.4}$$

Here $t_1 = t + \Delta t$. During the extremely small time interval $(\Delta t \to 0)$, forces applied to the back face may slightly change by a magnitude, but their direction remains the same. Otherwise, the rate of change of wavefields would be infinitely large. Thus, at opposite faces perpendicular to the x-axis, the normal components, as well as the tangential ones, have opposite directions. It is clear that the same behavior of forces is observed on other faces of the volume.

In order to formulate the second rule, let us discuss the action of shear forces, and, as an example, consider faces perpendicular to the $x-$ and y-axes, Fig. 2.1b. In general, magnitudes of forces applied to opposite faces are not equal to each other, and for this reason they produce both translation and rotation. In our case, the latter takes place about the z-axis, Fig. 2.1b, and, as is well known (Appendix A),

$$M_z = I\alpha_z \tag{2.5}$$

Here M_z is the z-component of torque, I is the moment of inertia, and α_z is the z-component of angular acceleration. By definition (Appendix A),

$$M_z = F_x dy \pm F_y dx = (t_x \pm t_y) dx dy dz \tag{2.6}$$

and

$$I = \frac{\rho}{12}(dx^2 + dy^2) dx dy dz \tag{2.7}$$

Thus, in place of eq. 2.5 we have

$$t_x \pm t_y = \frac{\rho}{12}(dx^2 + dy^2)\alpha_z \tag{2.8}$$

Here ρ is density and t_x and t_y are components of traction on faces perpendicular to the $y-$ and x-axes, respectively. Inasmuch as acceleration, α_z, cannot be infinitely large, we conclude that with a decrease of the volume, t_x tends to t_y and in the limit

$$t_x = t_y \tag{2.9}$$

This means that at intersecting faces shear forces are directed toward each other. The same behavior takes place on other faces of the cube.

Equation of motion

Now we will derive an equation of motion of an elementary volume. We introduce three vectors – $\mathbf{X}$, $\mathbf{Y}$, and $\mathbf{Z}$ – in the following way (Appendix C):

$$t_x = \mathbf{X} \cdot \mathbf{n}, \qquad t_y = \mathbf{Y} \cdot \mathbf{n}, \qquad t_z = \mathbf{Z} \cdot \mathbf{n}, \tag{2.10}$$

Here $\mathbf{n}$ is the unit vector normal to the cube faces, and

$$\mathbf{X} = \tau_{xx}\mathbf{i} + \tau_{xy}\mathbf{j} + \tau_{xz}\mathbf{k}, \quad \mathbf{Y} = \tau_{yx}\mathbf{i} + \tau_{yy}\mathbf{j} + \tau_{yz}\mathbf{k}, \quad \mathbf{Z} = \tau_{zx}\mathbf{i} + \tau_{zy}\mathbf{j} + \tau_{zz}\mathbf{k} \tag{2.11}$$

Each scalar component τ_{mn} is a continuous function of $x, y,$ and $z,$ and all of them form the stress tensor. It is essential to note that usually vectors **X**, **Y**, and **Z** can be arbitrarily oriented with respect to the face normal, **n**. However, as we have demonstrated, their mutual orientation at different faces obeys certain rules. Our goal is to use of Newton's second law and write equations of motion along the coordinate axes. By definition, we have:

$$m\frac{\partial^2 u}{\partial t^2} = \sum_{k=1}^{6} F_{xk}, \qquad m\frac{\partial^2 v}{\partial t^2} = \sum_{k=1}^{6} F_{yk}, \qquad m\frac{\partial^2 w}{\partial t^2} = \sum_{k=1}^{6} F_{zk} \tag{2.12}$$

where m is the mass of an elementary volume; $u, v,$ and w are scalar components of displacement **s**; $F_{xk}, F_{yk},$ and F_{zk} are components of the force applied to the k-face of the volume. First, we add the x-components of forces acting on all faces. It is obvious that the sum of forces F_{x1} and $F_{x3},$ applied to two opposite faces perpendicular to the x-axis, is

$$F_{x1} + F_{x3} = \mathbf{i} \cdot \left[\mathbf{X}(x + \frac{dx}{2}, y, z) - \mathbf{X}(x - \frac{dx}{2}, y, z)\right] dydz$$

or
$$F_{x1} + F_{x3} = \left[\tau_{xx}(x + \frac{dx}{2}, y, z) - \tau_{xx}(x - \frac{dx}{2}, y, z)\right] dydz \tag{2.13}$$

The presence of the minus sign in front of the second term is related to the fact that at the back face "3,″, the normal **n** and the unit vector **i** have opposite directions. In the same manner we obtain:

$$F_{x2} + F_{x4} = \mathbf{j} \cdot \left[\mathbf{X}(x, y + \frac{dy}{2}, z) - \mathbf{X}(x, y - \frac{dy}{2}, z)\right] dxdz \qquad \text{or}$$

$$F_{x2} + F_{x4} = \left[\tau_{xy}(x, y + \frac{dy}{2}, z) - \tau_{xy}(x, y - \frac{dy}{2}, z)\right] dxdz \tag{2.14}$$

and
$$F_{x5} + F_{x6} = \left[\tau_{xz}(x, y, z + \frac{dz}{2}) - \tau_{xz}(x, y, z - \frac{dz}{2})\right] dxdy$$

Taking into account that distances between opposite faces are very small, we may assume that stresses change linearly inside the volume, and this gives

$$F_{x1} + F_{x3} = \frac{\partial \tau_{xx}}{\partial x} dV, \qquad F_{x2} + F_{x4} = \frac{\partial \tau_{xy}}{\partial y} dV, \qquad F_{x5} + F_{x6} = \frac{\partial \tau_{xz}}{\partial z} dV \tag{2.15}$$

Substitution of eqs. 2.15 into the first equation of set 2.12 yields

$$\rho\frac{\partial^2 u}{\partial t^2}=\frac{\partial\tau_{xx}}{\partial x}+\frac{\partial\tau_{xy}}{\partial y}+\frac{\partial\tau_{xz}}{\partial z}\tag{2.16}$$

In the same manner, we have

$$\rho\frac{\partial^2 v}{\partial t^2}=\frac{\partial\tau_{yx}}{\partial x}+\frac{\partial\tau_{yy}}{\partial y}+\frac{\partial\tau_{yz}}{\partial z},\qquad \rho\frac{\partial^2 w}{\partial t^2}=\frac{\partial\tau_{zx}}{\partial x}+\frac{\partial\tau_{zy}}{\partial y}+\frac{\partial\tau_{zz}}{\partial z}\tag{2.17}$$

Thus, we have obtained a system of three equations with twelve unknowns – namely, nine components of stress tensor and three components of displacement. Now it is appropriate to make several comments.

1. Stress components, τ_{mn}, are considered to exist at faces of an elementary volume. Since the latter is small, stress components can be treated as linear functions within the volume. Correspondingly, their derivatives are constants.

2. At opposite faces of the volume, each stress component has the same sign. This reflects the fact that the whole elementary volume is either expanded or compressed. At the same time, as was already demonstrated, each component of the force applied to opposite faces has opposite signs. In equilibrium, magnitudes of the force components are equal at opposite faces.

3. As follows from eq. 2.9 and similar equalities

$$t_x=t_z,\qquad t_y=t_z,\qquad t_x=t_y,$$

and so we have

$$\tau_{xy}=\tau_{yx},\qquad \tau_{xz}=\tau_{zx},\qquad \text{and}\qquad \tau_{yz}=\tau_{zy},\tag{2.18}$$

i.e., the stress tensor is symmetrical (Appendix C).

4. Along with surface forces, an elementary volume is subjected to action of the volume force

$$\mathbf{F}=\mathbf{f}dV,\tag{2.19}$$

where $\mathbf{f}$ is the force per unit volume.

Taking into account eqs. 2.18 and 2.19, system 2.16–2.17 becomes

$$\frac{\partial\tau_{xx}}{\partial x}+\frac{\partial\tau_{xy}}{\partial y}+\frac{\partial\tau_{xz}}{\partial z}+f_x=\rho\frac{\partial^2 u}{\partial t^2}$$

$$\frac{\partial \tau_{xy}}{\partial x} + \frac{\partial \tau_{yy}}{\partial y} + \frac{\partial \tau_{yz}}{\partial z} + f_y = \rho \frac{\partial^2 v}{\partial t^2} \tag{2.20}$$

$$\frac{\partial \tau_{xz}}{\partial x} + \frac{\partial \tau_{yz}}{\partial y} + \frac{\partial \tau_{zz}}{\partial z} + f_z = \rho \frac{\partial^2 w}{\partial t^2}$$

This system of equations describes both translation and rotation of an elementary volume. In other words, the equation of rotation follows from set 2.20. This question is discussed in detail in Appendix E.

Stress in terms of strains

Our next step is to replace set 2.20 with equations that contain only components of displacement. To do this, we will use Hooke's law and the principle of superposition. In the first chapter, it was shown that normal stresses τ_{xx}, τ_{yy}, and τ_{zz} and strains $\partial u/\partial x$, $\partial v/\partial y$, and $\partial w/\partial z$ are related to each other as

$$E\frac{\partial u}{\partial x} = \tau_{xx} - \sigma\tau_{yy} - \sigma\tau_{zz} \tag{2.21}$$

$$E\frac{\partial v}{\partial y} = -\sigma\tau_{xx} + \tau_{yy} - \sigma\tau_{zz} \tag{2.22}$$

$$E\frac{\partial w}{\partial z} = -\sigma\tau_{xx} - \sigma\tau_{yy} + \tau_{zz} \tag{2.23}$$

Multiplying eq. 2.22 by σ and adding eq. 2.21, we obtain

$$E\left(\frac{\partial u}{\partial x} + \sigma\frac{\partial v}{\partial y}\right) = (1-\sigma^2)\tau_{xx} - \sigma(1+\sigma)\tau_{zz} \tag{2.24}$$

Again multiplying eq. 2.22 by σ but adding eq. 2.23, we have

$$E\left(\sigma\frac{\partial v}{\partial y} + \frac{\partial w}{\partial z}\right) = -\sigma(1+\sigma)\tau_{xx} + (1-\sigma^2)\tau_{zz} \tag{2.25}$$

Multiplication of eqs. 2.24 and 2.25 by $(1-\sigma)$ and σ, respectively, and their summation gives

$$E(1-\sigma)\frac{\partial u}{\partial x} + E\sigma\frac{\partial v}{\partial y} + E\sigma\frac{\partial w}{\partial z} = (1+\sigma)(1-2\sigma)\tau_{xx}$$

or
$$E(1-2\sigma)\frac{\partial u}{\partial x} + E\sigma\left(\frac{\partial u}{\partial x} + \frac{\partial v}{\partial y} + \frac{\partial w}{\partial z}\right) = (1+\sigma)(1-2\sigma)\tau_{xx} \tag{2.26}$$

Thus
$$\tau_{xx} = \frac{E}{(1+\sigma)}\frac{\partial u}{\partial x} + \frac{E\sigma}{(1+\sigma)(1-2\sigma)}\left(\frac{\partial u}{\partial x} + \frac{\partial v}{\partial y} + \frac{\partial w}{\partial z}\right) \tag{2.27}$$

or
$$\tau_{xx} = 2\mu\frac{\partial u}{\partial x} + \lambda\left(\frac{\partial u}{\partial x} + \frac{\partial v}{\partial y} + \frac{\partial w}{\partial z}\right) \tag{2.28}$$

Here
$$\mu = \frac{E}{2(1+\sigma)} \qquad \text{and} \qquad \lambda = \frac{E\sigma}{(1+\sigma)(1-2\sigma)} \tag{2.29}$$

are Lame constants. In the same manner, we obtain expressions for τ_{yy} and τ_{zz} in terms of strains:
$$\tau_{xx} = 2\mu\frac{\partial u}{\partial x} + \lambda \operatorname{div} \mathbf{s}, \qquad \tau_{yy} = 2\mu\frac{\partial v}{\partial y} + \lambda \operatorname{div} \mathbf{s}, \qquad \tau_{zz} = 2\mu\frac{\partial w}{\partial z} + \lambda \operatorname{div} \mathbf{s} \tag{2.30}$$

These equalities clearly show that normal stresses are functions of the diagonal elements of the strain tensor only. By definition, shear stresses and shear strains are related as
$$\tau_{xy} = \tau_{yx} = \mu\left(\frac{\partial u}{\partial y} + \frac{\partial v}{\partial x}\right), \qquad \tau_{xz} = \tau_{zx} = \mu\left(\frac{\partial w}{\partial x} + \frac{\partial u}{\partial z}\right) \tag{2.31}$$
$$\text{and} \qquad \tau_{yz} = \tau_{zy} = \mu\left(\frac{\partial v}{\partial z} + \frac{\partial w}{\partial y}\right),$$

and they do not contain diagonal elements of strain (Appendix D). Now we are ready to replace the stresses in eq. 2.20 in terms of displacement. Since
$$\frac{\partial \tau_{xx}}{\partial x} = 2\mu\frac{\partial^2 u}{\partial x^2} + \lambda\frac{\partial}{\partial x}\operatorname{div} \mathbf{s} = \mu\frac{\partial^2 u}{\partial x^2} + \mu\frac{\partial}{\partial x}\left(\frac{\partial u}{\partial x}\right) + \lambda\frac{\partial}{\partial x}\operatorname{div} \mathbf{s}$$

and
$$\frac{\partial \tau_{xy}}{\partial y} = \mu\frac{\partial^2 u}{\partial y^2} + \mu\frac{\partial}{\partial x}\left(\frac{\partial v}{\partial y}\right), \qquad \frac{\partial \tau_{xz}}{\partial z} = \mu\frac{\partial^2 u}{\partial z^2} + \mu\frac{\partial}{\partial x}\left(\frac{\partial w}{\partial z}\right),$$

the first equation of the set 2.20 becomes
$$\mu\nabla^2 u + (\lambda+\mu)\frac{\partial}{\partial x}\operatorname{div} \mathbf{s} = \rho\frac{\partial^2 u}{\partial t^2} \tag{2.32}$$

By analogy we have
$$\mu\nabla^2 v + (\lambda+\mu)\frac{\partial}{\partial y}\operatorname{div} \mathbf{s} = \rho\frac{\partial^2 v}{\partial t^2} \tag{2.33}$$

and
$$\mu\nabla^2 w + (\lambda+\mu)\frac{\partial}{\partial z}\operatorname{div} \mathbf{s} = \rho\frac{\partial^2 w}{\partial t^2} \tag{2.34}$$

Equation for displacement $\mathbf{s}$

Multiplication of eqs. 2.32–2.34 by unit vectors $\mathbf{i}$, $\mathbf{j}$, and $\mathbf{k}$, respectively, and their summation yields

$$\mu\nabla^2\mathbf{s} + (\lambda+\mu)\text{graddiv }\mathbf{s} = \rho\frac{\partial^2\mathbf{s}}{\partial t^2} \tag{2.35}$$

This equation plays a fundamental role in the theory of elastic waves. It is also useful to represent eq. 2.35 differently. Taking into account the equality

$$\text{curlcurl }\mathbf{s} = \text{graddiv }\mathbf{s} - \nabla^2\mathbf{s}, \tag{2.36}$$

in place of eq. 2.35 we have

$$\text{curlcurl }\mathbf{s} + (\lambda+2\mu)\nabla^2\mathbf{s} = \rho\frac{\partial^2\mathbf{s}}{\partial t^2} \tag{2.37}$$

Both of these forms are rather complicated equations, and, certainly, they do not correspond to the conventional form of a wave equation. At the same time, in two important cases, eqs. 2.35 and 2.37 are reduced to two wave equations that describe elastic waves propagating with different velocities. Before we discuss this subject, it is proper to notice the following. The equation for displacement was derived in the Cartesian system of coordinates. However, as is well known (Part I), spatial derivatives of the scalar and vector fields

$$\text{grad }\varphi, \qquad \text{div }\mathbf{M}, \qquad \text{curl }\mathbf{M}, \qquad \nabla^2\varphi$$

are invariants, and, therefore, eqs. 2.35 and 2.37 are valid in any orthogonal system of coordinates. It has also been demonstrated that every vector field, including $\mathbf{s}$, is characterized at regular points by a system of two equations

$$\text{curl }\mathbf{s} = \alpha\mathbf{W}(p) \qquad \text{div }\mathbf{s} = \beta\delta(p) \tag{2.38}$$

Since they are linear differential equations, their solution can be written as a sum:

$$\mathbf{s}(p) = \mathbf{s}_1(p) + \mathbf{s}_2(p) \tag{2.39}$$

In general, these different vector fields are related to each other, and they obey the systems

$$\text{curl }\mathbf{s}_1 = 0 \qquad \text{div }\mathbf{s}_1 = \beta\delta(p) \tag{2.40}$$

and

$$\text{curl } \mathbf{s}_2 = \alpha \mathbf{W} \qquad \text{div } \mathbf{s}_2 = 0 \tag{2.41}$$

Let us note that if functions $\mathbf{s}_1$ and $\mathbf{s}_2$ satisfy eqs. 2.40 and 2.41, respectively, then the sums $\mathbf{s}_1 + \mathbf{s}_0$ and $\mathbf{s}_2 + \mathbf{s}_0$ also obey these systems, provided that $\mathbf{s}_0$ is a solution of the homogeneous system:

$$\text{curl } \mathbf{s}_0 = 0 \qquad \text{and} \qquad \text{div } \mathbf{s}_0 = 0 \tag{2.42}$$

Now consider separately three important cases.

1. The wave associated with the field $\mathbf{s}_1$ Suppose that $\mathbf{s}_2 = 0$ and the displacement field is described by eqs. 2.40, i.e.,

$$\mathbf{s} = \mathbf{s}_1 \tag{2.43}$$

Since $\text{curl } \mathbf{s}_1 = 0$, eq. 2.37 is greatly simplified, and we arrive at the wave equation

$$(\lambda + 2\mu)\nabla^2 \mathbf{s} = \rho \frac{\partial^2 \mathbf{s}}{\partial t^2} \tag{2.44}$$

or

$$\nabla^2 \mathbf{s} = \frac{1}{c_l^2} \frac{\partial^2 \mathbf{s}}{\partial t^2}, \tag{2.45}$$

where

$$c_l = \sqrt{\frac{\lambda + 2\mu}{\rho}} \tag{2.46}$$

is the velocity of waves that cause a displacement field $\mathbf{s}_1$, eqs. 2.40. In accordance with eqs. 2.29 we have

$$c_l = \left[\frac{E(1-\sigma)}{\rho(1+\sigma)(1-2\sigma)} \right]^{1/2}, \tag{2.47}$$

and it depends on the density ρ, as well as the Young modulus and Poisson's ratio.

Earlier we studied propagation of waves along a thin bar with the velocity

$$c_l = \sqrt{\frac{E}{\rho}} \tag{2.48}$$

and called them longitudinal waves. In that case it was assumed that the parameter σ is equal to zero, i.e. the particle displacement in the direction perpendicular to the bar axis was neglected. It is obvious that eq. 2.48 follows from eq. 2.47, if we let $\sigma = 0$. At the same time, waves described by eq. 2.44 may cause displacement of particles in different directions if

$$\text{curl } \mathbf{s} = 0 \tag{2.49}$$

By analogy to acoustics fields, they are usually called dilatational waves. This is because, by definition, the second equation of set 2.40 characterizes a relative change of volume, i.e., the dilatation:

$$\text{div } \mathbf{s} = \Theta = \frac{\Delta V}{V} = \beta\, \delta(p) \tag{2.50}$$

This means that these waves are associated with deformations that change the volume provided that condition 2.49 is met. Now it is proper to make two comments:

a. The right side of eq. 2.50 can be represented in terms of stress. Performing a summation of eqs. 2.21–2.23, we obtain

$$\text{div } \mathbf{s} = \frac{\Delta V}{V} = \frac{1-2\sigma}{E}(\tau_{xx} + \tau_{yy} + \tau_{zz}) \tag{2.51}$$

or

$$\text{div } \mathbf{s} = \frac{1}{M}(\tau_{xx} + \tau_{yy} + \tau_{zz}), \tag{2.52}$$

where

$$M = \frac{E}{1-2\sigma} \tag{2.53}$$

is the bulk modulus. This suggests, eq. 2.52, that volume change occurs due to forces that are normal to the volume faces. Applying the conventional terminology of vector analysis, we may say that the diagonal elements of the stress tensor are sources of this field $\mathbf{s}$. Taking into account that div $\mathbf{s}$ characterizes the wavefield, it is obvious that the sum:

$$\tau_{xx} + \tau_{yy} + \tau_{zz}$$

is also an invariant with respect to the coordinate systems.

b. Dilatational waves are also called longitudinal or P waves. Sometimes the term "irrotational" is used. Now, proceeding from eq. 2.49, we will begin to study the types of motion and deformation that accompany longitudinal waves. For simplicity, we restrict ourselves to two-dimensional cases, where

$$\mathbf{s} = u(x,y)\mathbf{i} + v(x,y)\mathbf{j}, \qquad w = 0 \qquad \text{and} \qquad \frac{\partial u}{\partial z} = \frac{\partial v}{\partial z} = 0 \tag{2.54}$$

Correspondingly,

$$\text{div } \mathbf{s} = \frac{\Delta V}{V} = \frac{1}{M}(\tau_{xx} + \tau_{yy}) \qquad \text{and} \qquad \text{curl } \mathbf{s} = \begin{vmatrix} \mathbf{i} & \mathbf{j} & \mathbf{k} \\ \dfrac{\partial}{\partial x} & \dfrac{\partial}{\partial y} & \dfrac{\partial}{\partial z} \\ u & v & 0 \end{vmatrix} = 0 \tag{2.55}$$

As follows from eqs. 2.54 and 2.55,

$$\text{curl}_x\ \mathbf{s} = \text{curl}_y\ \mathbf{s} = 0,$$

and $\text{curl}_z\ \mathbf{s}$ has to be equal to zero:

$$\frac{\partial v}{\partial x} - \frac{\partial u}{\partial y} = 0 \tag{2.56}$$

As was described in Appendix A, the motion of an elementary volume as a rigid body is, in general, a superposition of translation and rotation. At the same time, deformation of this volume can be represented as a combination of compression (expansion) and pure shear, which are in general accompanied by both types of motion (Appendix D). First, consider translation, Fig. 2.1c. This motion takes place when all particles of the volume have the same displacement, i.e.,

$$u(x,y) = \text{const} \qquad \text{and} \qquad v(x,y) = \text{const} \tag{2.57}$$

Therefore, all derivatives of these components with respect to coordinates are equal to zero, and condition 2.49 is met.

Next we focus on compression (expansion) of an elementary volume, when displacement components u and v may vary only along the x and y coordinates, respectively (Fig. 2.1d). Then

$$\frac{\partial u}{\partial y} = \frac{\partial v}{\partial x} = 0$$

As follows from eq. 2.56, condition 2.49 is met again. One more deformation also obeys this equality, and it is called pure shear, Fig. 2.1e (Appendix D). Unlike with compression, the volume remains the same, i.e., $\text{div } \mathbf{s} = \mathbf{0}$, but the angle between intersecting faces slightly changes. Due to such deformation, the angles formed by two pairs of sides – a_1b_1, b_1c_1 and a_1d_1, d_1c_1 – are equal. As is seen from Fig. 2.1e, points of side ab experience different displacements along the x-axis, $u(y)$, and we have

$$\frac{u(y)}{y} = \frac{u(y+dy)}{y+dy} = \tan\gamma \simeq \gamma,$$

since the angle γ is very small. The last equality gives

$$y\,u(y) + dy\,u(y) = y\,u(y+dy) = y\,u(y) + y\,\frac{\partial u}{\partial y}\,dy \quad \text{or} \quad \frac{\partial u}{\partial y} = \frac{u(y)}{y} = \gamma \tag{2.58}$$

This derivative characterizes a distortion angle. In the same manner, considering displacement of side ad, we obtain

$$\frac{\partial v}{\partial x} = \gamma \tag{2.59}$$

Therefore, in accordance with eq. 2.56, condition 2.49 is met. We see that translation, compression (expansion), and pure shear may accompany longitudinal waves. Note that both translation and pure shear are described by the displacement field, $\mathbf{s}_0$, which satisfies the homogeneous system, eqs. 2.42. For this reason, they may be observed for both fields $\mathbf{s}_1$ and $\mathbf{s}_2$ as well as in the general case, eqs. 2.38. Thus, change in an elementary volume is a typical property of longitudinal waves. As will be shown later, the second type of motion (rotation) is absent for these waves.

2. The wave associated with the field $\mathbf{s}_2$ Next, assume that the displacement field is described by eqs. 2.41, i.e., $\mathbf{s} = \mathbf{s}_2$ and

$$\text{div}\,\mathbf{s} = 0 \tag{2.60}$$

Then, the equation for displacement, 2.35, becomes

$$\mu\nabla^2\mathbf{s} = \rho\frac{\partial^2\mathbf{s}}{\partial t^2} \quad \text{or} \quad \nabla^2\mathbf{s} = \frac{1}{c_s^2}\frac{\partial^2\mathbf{s}}{\partial t^2}, \tag{2.61}$$

where

$$c_s = \sqrt{\frac{\mu}{\rho}} \tag{2.62}$$

Thus, we have again arrived at a wave equation that characterizes propagation of waves with velocity c_s. An example of such a wave is the torsional wave in a thin bar (Chapter 1), and its velocity is still defined by eq. 2.62. These waves are accompanied by rotation of elementary volumes, and for this reason they are called rotational or shear waves. The last term is used because rotation is caused by shear forces. It is obvious that velocities of longitudinal and shear waves depend on the same elastic parameters, namely E and σ, as well as on density ρ. Taking into account eq. 2.29, we have

$$c_s = \sqrt{\frac{E}{2(1+\sigma)\rho}}, \tag{2.63}$$

and the ratio of velocity is

$$\frac{c_s}{c_l} = \sqrt{\frac{1-2\sigma}{2(1-\sigma)}} \tag{2.64}$$

or

$$\frac{c_s}{c_l} = \sqrt{\frac{\mu}{\lambda+2\mu}} = \sqrt{\frac{1}{2+\lambda/\mu}} \tag{2.65}$$

Since $\lambda > 0$ and $\mu > 0$, we see that the velocity of shear waves is always smaller than that of longitudinal waves, $c_s < c_l$. Because of this, the former are also called S (secondary) waves, and they arrive after the longitudinal P (primary) waves. As follows from eq. 2.65,

$$0 \le c_s < \frac{1}{\sqrt{2}} c_l \tag{2.66}$$

Now let us describe types of motion and deformation that can be caused by shear waves, and, therefore, obey eq. 2.60. By definition, this wave does not produce compression or expansion of an elementary volume; i.e., such deformation is impossible for shear waves. Also translation and a pure shear alone cannot describe the displacement field of these waves. However, they can contribute to the field $\mathbf{s}_2$ along with rotation of an elementary volume (Appendix D). As is clearly seen from Fig. 2.2a, the strains associated with rotation are

$$\frac{\partial u}{\partial y} = -\gamma \qquad \text{and} \qquad \frac{\partial v}{\partial x} = \gamma \tag{2.67}$$

The presence of the minus sign in the first equation is related to the fact that angle γ is positive, but $\partial u/\partial y < 0$. By definition, we have

$$\text{curl}_z\, \mathbf{s} = \frac{\partial v}{\partial x} - \frac{\partial u}{\partial y} = 2\gamma,$$

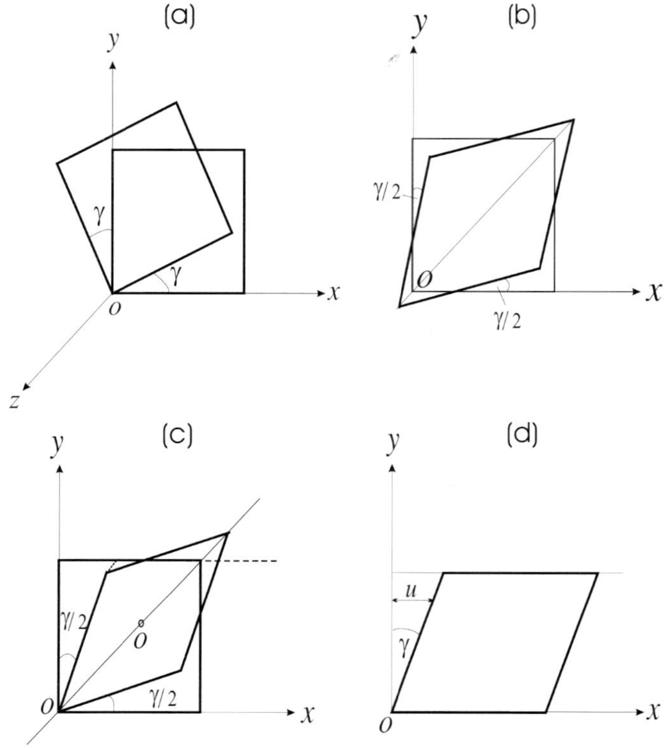

Figure 2.2: (a) Rotation of elementary volume (b,c) Pure shear and translation of elementary volume (d) Simple shear as superposition of pure shear, translation, and rotation

while dilatation is equal to zero. Thus, rotation is a characteristic of shear waves, i.e., they cannot exist without rotation of elementary volumes of a medium. By analogy with longitudinal waves it is possible to perform one generalization and represent the displacement field of shear waves, as

$$\mathbf{s} = \mathbf{s}_r + \mathbf{s}_0 \tag{2.68}$$

Here $\mathbf{s}_r$ is the displacement field due to rotation, while $\mathbf{s}_0$ is associated with translation and pure shear. Of course, one or both of them may be absent. Taking into account eqs. 2.42, we obtain:

$$\text{curl}\ \mathbf{s} = \text{curl}\ \mathbf{s}_r \tag{2.69}$$

We use this equality in order to express curl $\mathbf{s}_r$ in terms of stresses. Let us consider a so-called simple shear (Appendix D). In this case, as is illustrated in Fig. 2.2b,c,d, a shear wave involves an elementary volume in both types of motion (translation and rotation), as well as deformation of the pure shear. Since the displacement $\mathbf{s}$ has only one component $u(y)$, we have:

$$\text{curl}_z\mathbf{s} = -\frac{\partial u}{\partial y} = -\gamma \tag{2.70}$$

On the other hand, in accordance with Hooke's law

$$e_{xy} = \frac{\partial u}{\partial y} + \frac{\partial v}{\partial x} = \frac{1}{\mu}\tau_{xy} = \frac{\partial u}{\partial y}, \qquad \text{because} \quad v = 0$$

Therefore, eq. 2.70 gives

$$\text{curl}_z\mathbf{s} = -\frac{1}{\mu}\tau_{xy} \tag{2.71}$$

It is easy to generalize this result for a three-dimensional case. For instance, if there is rotation of an elementary volume around the $x-$ and y-axes, we have

$$\text{curl}_x\mathbf{s} = -\frac{1}{\mu}\tau_{yz}, \qquad \text{curl}_y\mathbf{s} = -\frac{1}{\mu}\tau_{xz} \tag{2.72}$$

Multiplication of eqs. 2.71 and 2.72 by the corresponding unit vectors and then a summation yield

$$\text{curl }\mathbf{s} = -\frac{1}{\mu}(\tau_{yz}\mathbf{i} + \tau_{xz}\mathbf{j} + \tau_{xy}\mathbf{k}) \tag{2.73}$$

The vector

$$\mathbf{W} = \tau_{yz}\mathbf{i} + \tau_{xz}\mathbf{j} + \tau_{xy}\mathbf{k} \tag{2.74}$$

is formed with the help of the nondiagonal elements of the stress tensor, and it can be treated as the density of vortexes of field $\mathbf{s}$. Thus, the system of equations of the displacement field $\mathbf{s}$, which accompanies shear waves, has the form

$$\text{curl }\mathbf{s} = -\frac{1}{\mu}\mathbf{W}, \qquad \text{div }\mathbf{s} = 0 \tag{2.75}$$

3. **General case** We have demonstrated that in a homogeneous medium, either longitudinal or shear waves may exist. The more general case is also possible, when the system of equations for the displacement field $\mathbf{s}$ is

$$\text{curl }\mathbf{s} = -\frac{1}{\mu}(\tau_{yz}\mathbf{i} + \tau_{xz}\mathbf{j} + \tau_{xy}\mathbf{k}) \tag{2.76}$$

and

$$\text{div } \mathbf{s} = \frac{1}{M}(\tau_{xx} + \tau_{yy} + \tau_{zz}) \tag{2.77}$$

Therefore propagation of a wave associated with this field $\mathbf{s}$ may cause both types of motion and deformation, namely: a. translation, b. rotation, c. pure shear, and d. compression (expansion).

It turns out that the velocity of propagation of these waves is smaller than that of shear waves. This may be related to the fact that waves produce different types of motion and deformation. The bending waves considered in Chapter 1, is an example of this type of wave, as are Rayleigh waves, which will be studied later. Previously, we represented the total field as a sum: $\mathbf{s} = \mathbf{s}_1 + \mathbf{s}_2 + \mathbf{s}_0$. In general, $\mathbf{s}_1, \mathbf{s}_2$, and $\mathbf{s}_0$ are related to each other and describe one wave, propagating with the same velocity. However, there is one obvious case in which longitudinal and shear waves exist simultaneously. Of course, these independent waves travel with different velocities, c_l and c_s. Later we will demonstrate that within some range of distances from the source, the field $\mathbf{s}_0$ may also describe some wave disturbance (Laplace motion).

Boundary conditions

Since eq. 2.35 is valid only at regular points of a medium, at interfaces between media with different elastic parameters this equation must be replaced by boundary conditions. They characterize the behavior of forces and displacement $\mathbf{s}$ at points of such an interface. For simplicity, suppose that the latter is the plane XOY, i.e., $z = 0$. Also, we imply that elastic media are welded at the surface of their contact. Then the tangential and normal components of displacement have to be continuous functions at points of the boundary:

$$\mathbf{s}^{(1)} = \mathbf{s}^{(2)} \quad \text{or}$$

$$u_1 = u_2, \qquad v_1 = v_2, \qquad w_1 = w_2 \qquad \text{at} \quad z = 0 \tag{2.78}$$

Besides, the normal and shear stresses acting on each element of the boundary are also continuous functions:

$$\tau_{zz}^{(1)} = \tau_{zz}^{(2)}, \qquad \tau_{xz}^{(1)} = \tau_{xz}^{(2)}, \qquad \tau_{yz}^{(1)} = \tau_{yz}^{(2)} \tag{2.79}$$

These equalities follow from Newton's second law describing translation and rotation. If eqs. 2.79 were invalid, either the linear or angular acceleration, or both, would be infinitely large, which, of course, is impossible. Taking into account the relationship between stress and strain, eqs. 2.79 can be also written in the form:

$$\lambda_1\Theta_1 + 2\mu_1\frac{\partial w_1}{\partial z} = \lambda_2\Theta_2 + 2\mu_2\frac{\partial w_2}{\partial z}$$

$$\mu_1\left(\frac{\partial u_1}{\partial z} + \frac{\partial w_1}{\partial x}\right) = \mu_2\left(\frac{\partial u_2}{\partial z} + \frac{\partial w_2}{\partial x}\right) \tag{2.80}$$

$$\mu_1\left(\frac{\partial v_1}{\partial z} + \frac{\partial w_1}{\partial y}\right) = \mu_2\left(\frac{\partial v_2}{\partial z} + \frac{\partial w_2}{\partial y}\right)$$

where $\Theta = \text{div}\,\mathbf{s}$ is dilatation. Thus, eqs. 2.78 and 2.80 allow us to formulate boundary conditions in terms of displacement and its derivatives. It is useful to describe boundary conditions for three special cases.

Case one: free interface In order to provide continuity of stresses at such a boundary, they have to be equal to zero at its points:

$$\tau_{xz} = 0, \qquad \tau_{yz} = 0, \qquad \tau_{zz} = 0 \qquad \text{at} \qquad z = 0 \tag{2.81}$$

Eqs. 2.81 represent boundary conditions at the free surface. However, displacement components are not defined at this surface. Note that if instead of the solid there is a fluid, eqs. 2.79 are simplified and we have

$$\tau_{zz} = 0\ , \tag{2.82}$$

since shear stresses are absent.

Case two: ideally rigid boundary By definition, particles belonging to this interface cannot move and, correspondingly, all components of the displacement are equal to zero:

$$u = 0, \qquad v = 0, \qquad w = 0 \qquad \text{at} \qquad z = 0 \tag{2.83}$$

Case three: boundary between solid and fluid A slippage may occur along such an interface and, therefore, only the normal component of displacement has to be continuous. Since shear stresses are absent in the fluid, they are equal to zero at points of the interface. The normal stress is a continuous function. This gives

$$w_{1z} = w_{2z}, \qquad \tau_{xz}^{(1)} = \tau_{yz}^{(1)} = 0, \qquad \tau_{zz}^{(1)} = \tau_{zz}^{(2)} \tag{2.84}$$

As before, index “1” corresponds to an elastic medium.

Boundary value problem

Boundary conditions, eqs. 2.78 and 2.79, can be treated as a surface analogy of the differential equation for displacement field $\mathbf{s}$. From the physical point of view, it is clear that an infinite number of functions satisfy eq. 2.35 and these conditions. In fact, changing the type of primary source and its position, we obtain different wavefields, but each of them obeys eq. 2.35 and sets 2.78 and 2.79. In order to remove this ambiguity, we need additional information, which can be derived from the theorem of uniqueness. In other words, this allows us to formulate the boundary value problem. By analogy with the case of an acoustic medium (Part I), it is possible to show that the wave fields are uniquely defined, provided that they satisfy a certain set of conditions. Let us assume that the multilayered medium with homogeneous layers is bounded by some surface S_0 and surface S_∞ at infinity. It turns out that a solution of the boundary problem is unique if it obeys

1. The initial condition.
2. Eq. 2.35 for displacement field $\mathbf{s}$ at regular points.
3. Boundary conditions at surfaces S_0 and S_∞.
4. Conditions 2.78 and 2.79 at interfaces.

The initial condition implies a knowledge of wavefields at all points of a medium at some instant $t = 0$. Usually we assume that the wavefield is absent at the initial moment. The behavior of a wave at points of surface S_∞ corresponds to the outgoing spherical wave, so that it satisfies the Sommerfeld condition (Part II). As far as surface S_0 is concerned, information about wave behavior at its points can be formulated in different ways. First, suppose that surface S_0 surrounds the primary source, and it is located at its vicinity. In such a case, we usually know stresses at points of S_0 as functions of time. Instead of this condition one can imagine that the force $\mathbf{F}(t)$ is given at some point or that several forces are specified at different points of a medium. In particular, distribution of external forces is often known at the free surface. Of course, it is possible to introduce different conditions at surface S_0 that also describe strains as well as the displacement field.

Scalar and vector potentials

In accordance with eqs. 2.39, displacement field $\mathbf{s}$ is, in general, a sum of two fields, $\mathbf{s}_1$ and $\mathbf{s}_2$, which obey eqs. 2.40 and 2.41, respectively. By analogy with the acoustic and electromagnetic fields, we introduce two functions that may often simplify wave studies.

As usual, we deal with a layered medium. In each layer, longitudinal and shear waves propagate with constant velocities. Consider again three different cases.

Case one: dilatational waves From the equation curl $\mathbf{s}_l = 0$, it follows that the vector field $\mathbf{s}_l$ can be described with the help of the scalar potential U only:

$$\mathbf{s}_l = \text{grad } U \tag{2.85}$$

In order to obtain the equation for U, we substitute eq. 2.85 into eq. 2.45. This gives

$$\text{grad}\left(\nabla^2 U - \frac{1}{c_l^2}\frac{\partial^2 U}{\partial t^2}\right) = 0 \qquad \text{or} \qquad \nabla^2 U - \frac{1}{c_l^2}\frac{\partial^2 U}{\partial t^2} = C \tag{2.86}$$

Here C is some constant. In accordance with eq. 2.85, there is an infinite number of functions U that describe the same displacement field $\mathbf{s}_l$. Taking this into account, we choose such U that the constant C is equal to zero. Then, the scalar potential U also obeys the wave equation

$$\nabla^2 U = \frac{1}{c_l^2}\frac{\partial^2 U}{\partial t^2} \tag{2.87}$$

Certainly, a transition from the vector field $\mathbf{s}_l$ to the scalar one may greatly simplify a wave study, even when displacement has a single component. Knowing the function U, we can calculate displacement $\mathbf{s}$ and then, making use of Hooke's law, determine stresses.

Case two: shear waves Taking into account that dilatation for shear waves is equal to zero (div $\mathbf{s}_2 = 0$), we have

$$\mathbf{s}_2 = \text{curl } \mathbf{A}, \tag{2.88}$$

where $\mathbf{A}$ is the vector potential. It is obvious that there is an infinite number of functions $\mathbf{A}$ that describe the same field $\mathbf{s}_2$. Substituting eq. 2.88 into eq. 2.61 and applying the same approach as in the first case, we obtain

$$\nabla^2 \mathbf{A} = \frac{1}{c_s^2}\frac{\partial^2 \mathbf{A}}{\partial t^2} \tag{2.89}$$

The use of the vector-potential $\mathbf{A}$ instead of $\mathbf{s}_2$ is not so obvious, because they are both vectors. However, in some cases it is sufficient to deal only with the single component of $\mathbf{A}$. Moreover, this component may be directed along the Cartesian axis, which brings additional simplification. In general, a reflection of the longitudinal waves causes an appearance of shear waves and vice versa. For this reason, in order to satisfy boundary conditions it is necessary to use both potentials.

Case three In general, when displacement $\mathbf{s}$ is a sum of $\mathbf{s}_1$ and $\mathbf{s}_2$, we obviously have

$$\mathbf{s} = \text{grad}\, U + \text{curl}\, \mathbf{A} \tag{2.90}$$

In essence, we have already demonstrated that potentials obey wave equations. The same result can be derived differently. To show this, we rewrite eq. 2.35 as

$$\mu\nabla^2(\mathbf{s}_1 + \mathbf{s}_2) + (\lambda + \mu)\text{grad div}(\mathbf{s}_1 + \mathbf{s}_2) = \rho\frac{\partial^2}{\partial t^2}(\mathbf{s}_1 + \mathbf{s}_2)$$

or

$$\mu\nabla^2(\mathbf{s}_1 + \mathbf{s}_2) + (\lambda + \mu)\text{grad div}\, \mathbf{s}_1 = \rho\frac{\partial^2}{\partial t^2}(\mathbf{s}_1 + \mathbf{s}_2) \tag{2.91}$$

since $\text{div}\, \mathbf{s}_2 = 0$. Taking into account that $\text{curl curl}\, \mathbf{s}_1 = \text{grad div}\, \mathbf{s}_1 - \nabla^2\mathbf{s}_1$ and $\text{curl}\, \mathbf{s}_1 = 0$, we have

$$(\lambda + 2\mu)\nabla^2\mathbf{s}_1 - \rho\frac{\partial^2\mathbf{s}_1}{\partial t^2} + \mu\nabla^2\mathbf{s}_2 - \rho\frac{\partial^2\mathbf{s}_2}{\partial t^2} = 0 \tag{2.92}$$

This result is trivial, since both fields $\mathbf{s}_1$ and $\mathbf{s}_2$ satisfy wave equations. Substitution of eq. 2.90 into eq. 2.92 gives

$$\text{grad}\left[(\lambda + 2\mu)\nabla^2 U - \rho\frac{\partial^2 U}{\partial t^2}\right] + \text{curl}\left[\mu\nabla^2\mathbf{A} - \rho\frac{\partial^2\mathbf{A}}{\partial t^2}\right] = 0 \tag{2.93}$$

Because there is an infinite number of pair of functions U and A that describe the same field $\mathbf{s}$, we can always choose the functions U and A that obey wave equations 2.45 and 2.61, respectively. It is clear that eq. 2.92 is satisfied.

2.2 Kinetic and potential energy and its flux

In studying wave propagation in acoustic media, we demonstrated that the density of kinetic and potential energy is defined as (Part I):

$$e_0 = \frac{1}{2}\rho\mathbf{v}^2 \qquad \text{and} \qquad u_0 = \frac{M}{2}\text{div}^2\mathbf{s}, \tag{2.94}$$

respectively. Here ρ is density of a medium, $\mathbf{v}$ is particle velocity, $\mathbf{v} = \partial \mathbf{s}/\partial t$, and M is the bulk modulus. Therefore, the total mechanical energy in an arbitrary volume is

$$W(t) = \frac{1}{2} \int\limits_V (\rho\, \mathbf{v}^2 + M\, \mathrm{div}^2 \mathbf{s})\, dV \tag{2.95}$$

In general, three factors cause a change in this energy, namely

1. The presence of external (primary) sources of waves inside volume V.
2. Transformation of mechanical energy W into heat Q.
3. The flux of this energy through a surface surrounding the volume.

Then, in accordance with the principle of conservation of energy, we have

$$\frac{\partial W}{\partial t} = L - Q - \oint\limits_S \mathbf{N} \cdot d\mathbf{S}, \tag{2.96}$$

where L is an amount of the kinetic and potential energy produced by external sources per unit of time and Q is an amount of the mechanical energy that is transformed into heat, also per unit of time. Finally

$$\oint\limits_S \mathbf{N} \cdot d\mathbf{S} \tag{2.97}$$

is called the energy flux, and it defines an amount of energy passing through the closed surface S during one second. Correspondingly, the vector $\mathbf{N}$ is the density of the flux, and it plays the same role as the Poynting vector for electromagnetic fields or the current density $\mathbf{j}$ for electric current. This shows that the magnitude of $\mathbf{N}$ equals the amount of energy passing through an elementary surface, dS, with unit of area per unit of time. It is essential that this surface is perpendicular to vector $\mathbf{N}$. In SI units we have

$$[\mathbf{N}] = \frac{\text{joule}}{m^2 s} = \frac{\text{watt}}{m^2}$$

Since vector $d\mathbf{S}$ in eq. 2.97 is directed away from the volume, the positive value of the flux means that energy leaves the volume. In contrast, energy increases inside the volume if the flux is negative. In general, vector $\mathbf{N}$ has a different magnitude and direction at different points of a closed surface S. For instance, at some points it can be directed inward, whereas at other points it is directed outward or is parallel to the surface. As was shown in Part I, in an acoustic medium

$$\mathbf{N} = P\, \mathbf{v}, \tag{2.98}$$

where P is pressure caused by a wave. Thus, vector $\mathbf{N}$ is equal to the product of the additional pressure and the particle velocity. For example, if pressure P is positive (compression), vectors $\mathbf{v}$ and $\mathbf{N}$ have the same direction. In contrast, at points where P is negative (expansion), these vectors have opposite directions. Vector $\mathbf{N}$ always shows the direction in which the wave moves. We can say that vector $\mathbf{N}$ allows us to visualize wave propagation as transmission of energy. For instance, assuming that the primary sources are absent inside the volume, $L=0$, and the process of propagation is adiabatic, $Q=0$, eq. 2.96 becomes

$$\frac{\partial W}{\partial t} = -\oint_S \mathbf{N}\cdot\mathbf{dS} \tag{2.99}$$

In this case, any change of energy W can only be caused by its flux.

Elastic potential

By analogy with eq. 2.98, we derive an expression for vector $\mathbf{N}$ in an elastic medium. Since a distribution of forces in this medium is characterized by the stress tensor $\mathbf{T}_*$, it is natural to expect that in eq. 2.98 the additional pressure P has to be replaced by $\mathbf{T}_*$, which gives

$$\mathbf{N} = \mathbf{T}_*\mathbf{v}$$

In order to prove this relationship we assume that surface S surrounding volume V is subjected to action of surface forces. For simplicity, the influence of volume forces is neglected. Forces applied to surface S can be treated as external, and the work produced by them is equal to

$$A = \oint_S \mathbf{t}(q)\cdot\mathbf{s}(q)dS \tag{2.100}$$

Here q is the point on surface S and $\mathbf{s}(q)$ is its displacement:

$$\mathbf{s}(q) = u(q)\,\mathbf{i} + v(q)\,\mathbf{j} + w(q)\,\mathbf{k} \tag{2.101}$$

By definition, the traction vector $\mathbf{t}(q)$ is the force per unit of area and

$$\mathbf{t}(q) = t_x\mathbf{i} + t_y\mathbf{j} + t_z\mathbf{k} \tag{2.102}$$

In the previous section we applied the relations

$$t_x = \mathbf{X}\cdot\mathbf{n}, \qquad t_y = \mathbf{Y}\cdot\mathbf{n}, \qquad t_z = \mathbf{Z}\cdot\mathbf{n}, \tag{2.103}$$

where $\mathbf{n}$ is the unit vector normal to the surface element,

$$\mathbf{n} = \cos(n,x)\ \mathbf{i} + \cos(n,y)\ \mathbf{j} + \cos(n,z)\ \mathbf{k}, \tag{2.104}$$

and scalar components of vectors $\mathbf{X}$, $\mathbf{Y}$, and $\mathbf{Z}$ are elements of the stress tensor:

$$\mathbf{X} = \tau_{xx}\mathbf{i}+\tau_{xy}\mathbf{j}+\tau_{xz}\mathbf{k}, \qquad \mathbf{Y} = \tau_{yx}\mathbf{i}+\tau_{yy}\mathbf{j}+\tau_{yz}\mathbf{k}, \qquad \mathbf{Z} = \tau_{zx}\mathbf{i}+\tau_{zy}\mathbf{j}+\tau_{zz}\mathbf{k} \tag{2.105}$$

Taking into account eq. 2.103, we have

$$A = \oint_S (\mathbf{X}\ u + \mathbf{Y}\ v + \mathbf{Z}\ w) \cdot \mathbf{dS} \tag{2.106}$$

Therefore, the rate at which the surface forces perform the work is equal to

$$\frac{\partial A}{\partial t} = \oint_S \left(\mathbf{X}\frac{\partial u}{\partial t} + \mathbf{Y}\frac{\partial v}{\partial t} + \mathbf{Z}\frac{\partial w}{\partial t}\right) \cdot \mathbf{dS} \tag{2.107}$$

Here it is appropriate to make two comments:

1. Taking derivatives with respect to time, we neglected higher-order terms. For instance

$$\frac{\partial}{\partial t}(\mathbf{X}u) = \mathbf{X}\ \frac{\partial u}{\partial t} + \frac{\partial \mathbf{X}}{\partial t}\ u$$

Since $u \ll 1$ and $\partial \mathbf{X}/\partial t \ll 1$, their product is neglected.

2. It is obvious that $\partial A/\partial t$ can be treated as a change of the work per unit time, because

$$dA = \frac{\partial A}{\partial t}\delta t \qquad \text{if} \qquad \delta t = 1 s$$

This work produces a change in kinetic and potential energy inside the volume during the same time, $(\delta t = 1\ s)$. The rate of change of both types of energy can be represented in terms of volume integrals:

$$\frac{\partial E}{\partial t} = \int_V \frac{\partial e_0}{\partial t} dV \qquad \text{and} \qquad \frac{\partial U}{\partial t} = \int_v \frac{\partial u_0}{\partial t} dV \tag{2.108}$$

Here $e_0(p)$ and $u_0(p)$ are densities of kinetic and potential energy at any point p inside an elastic medium. The function $e_0(p)$ is known, because every elementary volume moves like a rigid body and, therefore,

$$e_0 = \frac{1}{2}\rho \mathbf{v}^2 = \frac{\rho}{2}\left[\left(\frac{\partial u}{\partial t}\right)^2 + \left(\frac{\partial v}{\partial t}\right)^2 + \left(\frac{\partial w}{\partial t}\right)^2\right] \tag{2.109}$$

To derive an expression for density u_0, which characterizes a deformation, we proceed from eq. 2.107, which can be written as

$$dA = dE + dU \tag{2.110}$$

Note that the work of surface forces produces mechanical energy, which arrives inside the volume as the flux of vector **Y**. This means that

$$dA = -\oint_S \mathbf{Y}\cdot d\mathbf{S} \qquad \text{if} \qquad \delta t = 1s \tag{2.111}$$

As follows from eqs. 2.108 and 2.109,

$$dE = \int_V \rho(\dot{u}\ddot{u} + \dot{v}\ddot{v} + \dot{w}\ddot{w})dV, \tag{2.112}$$

where

$$\dot{u} = \frac{\partial u}{\partial t} \qquad \text{and} \qquad \ddot{u} = \frac{\partial^2 u}{\partial t^2}$$

Next we also represent the work dA in terms of the volume integral. Taking into account eq. 2.106 and the Gauss formula, we have

$$dA = \int_V \left[\text{div}\left(\mathbf{X}\frac{\partial u}{\partial t}\right) + \text{div}\left(\mathbf{Y}\frac{\partial v}{\partial t}\right) + \text{div}\left(\mathbf{Z}\frac{\partial w}{\partial t}\right)\right] dV \quad \text{if} \ \delta t = 1\ s\,. \tag{2.113}$$

The use of equality

$$\text{div}\ (\varphi \mathbf{a}) = \varphi\ \text{div}\ \mathbf{a} + \mathbf{a}\cdot \text{grad}\ \varphi \tag{2.114}$$

gives

$$dA = \int_V \left(\frac{\partial u}{\partial t}\ \text{div}\ \mathbf{X} + \frac{\partial v}{\partial t}\ \text{div}\ \mathbf{Y} + \frac{\partial w}{\partial t}\ \text{div}\ \mathbf{Z}\right) dV + \tag{2.115}$$

$$\int_V \left(\mathbf{X}\cdot \text{grad}\ \frac{\partial u}{\partial t} + \mathbf{Y}\cdot \text{grad}\ \frac{\partial v}{\partial t} + \mathbf{Z}\cdot \text{grad}\ \frac{\partial w}{\partial t}\right) dV$$

Thus, we expressed the work of surface forces per unit of time as a sum of two volume integrals. It turns out that the first integral in eq. 2.114 characterizes a change of kinetic energy during the same time δt. In fact, using equations of motion derived in the previous section,

$$\text{div } \mathbf{X} = \rho \ddot{u}, \qquad \text{div } \mathbf{Y} = \rho \ddot{v}, \qquad \text{div } \mathbf{Z} = \rho \ddot{w},$$

and eq. 2.112, we find that

$$dE = \int\limits_V \left(\dot{u} \ \text{div } \mathbf{X} + \dot{v} \ \text{div } \mathbf{Y} + \dot{w} \ \text{div } \mathbf{Z} \right) dV$$

Thus, eq. 2.115 can be rewritten as

$$dA - dE = \int\limits_V \left(\mathbf{X} \cdot \nabla \dot{u} + \mathbf{Y} \cdot \nabla \dot{v} + \mathbf{Z} \cdot \nabla \dot{w} \right) dV \tag{2.116}$$

Comparison with eq. 2.110 shows that the integrand in this equality can be treated as a change of potential energy density per unit of time:

$$du_0 = \frac{\partial u_0}{\partial t} dt \qquad \text{and} \qquad dt = 1 \ s$$

Correspondingly, we have

$$du_0 = \left(\mathbf{X} \cdot \nabla \dot{u} + \mathbf{Y} \cdot \nabla \dot{v} + \mathbf{Z} \cdot \nabla \dot{w} \right) dt = \left[\tau_{xx} \frac{\partial \dot{u}}{\partial x} + \tau_{xy} \frac{\partial \dot{u}}{\partial y} + \tau_{xz} \frac{\partial \dot{u}}{\partial z} \right.$$

$$\left. + \tau_{yx} \frac{\partial \dot{v}}{\partial x} + \tau_{yy} \frac{\partial \dot{v}}{\partial y} + \tau_{yz} \frac{\partial \dot{v}}{\partial z} + \tau_{zx} \frac{\partial \dot{w}}{\partial x} + \tau_{zy} \frac{\partial \dot{w}}{\partial y} + \tau_{zz} \frac{\partial \dot{w}}{\partial z} \right] dt$$

Since the stress tensor is symmetrical, we have

$$du_0 = \tau_{xx} de_{xx} + \tau_{yy} de_{yy} + \tau_{zz} de_{zz} + \tau_{yx} de_{yx} + \tau_{yz} de_{yz} + \tau_{xz} de_{xz} \tag{2.117}$$

This equation demonstrates that the density of potential energy u_0 is a function of strains, and it permits us to represent the right side of eq. 2.117 as

$$du_0 = \frac{\partial u_0}{\partial e_{xx}} de_{xx} + \frac{\partial u_0}{\partial e_{yy}} de_{yy} + \frac{\partial u_0}{\partial e_{zz}} de_{zz} + \frac{\partial u_0}{\partial e_{xy}} de_{xy} + \frac{\partial u_0}{\partial e_{yz}} de_{yz} + \frac{\partial u_0}{\partial e_{xz}} de_{xz} \tag{2.118}$$

From the last two equations, we obtain

$$\tau_{xx} = \frac{\partial u_0}{\partial e_{xx}}, \qquad \tau_{yy} = \frac{\partial u_0}{\partial e_{yy}}, \qquad \tau_{zz} = \frac{\partial u_0}{\partial e_{zz}}, \tag{2.119}$$

$$\tau_{xy} = \frac{\partial u_0}{\partial e_{xy}}, \quad \tau_{yz} = \frac{\partial u_0}{\partial e_{yz}}, \quad \tau_{xz} = \frac{\partial u_0}{\partial e_{xz}}$$

Thus, the stresses are expressed in terms of derivatives of the function u_0, which is often called the elastic potential (Appendix E). At the beginning we neglected volume forces. It is easy to show that our results remain valid in a more general case when these forces are taken into account.

Flux density Y

As follows from eqs. 2.107 and 2.111, we have

$$\oint_S \left(\mathbf{X}\dot{u} + \mathbf{Y}\dot{v} + \mathbf{Z}\dot{w}\right) \cdot \mathbf{dS} = -\int_S \mathbf{N} \cdot \mathbf{dS}$$

Therefore, it is natural to assume that

$$\mathbf{N} = -\left(\mathbf{X}\dot{u} + \ \mathbf{Y}\dot{v} + \ \mathbf{Z}\dot{w}\right) = N_x\mathbf{i} + N_y\mathbf{j} + N_z\mathbf{k}, \tag{2.120}$$

where

$$N_x = -\tau_{xx}\dot{u} - \tau_{yx}\dot{v} - \tau_{zx}\dot{w}, \qquad N_y = -\tau_{yx}\dot{u} - \tau_{yy}\dot{v} - \tau_{yz}\dot{w} \tag{2.121}$$

$$N_z = -\tau_{zx}\dot{u} - \tau_{yz}\dot{v} - \tau_{zz}\dot{w}$$

The latter represents the product of the stress tensor and the particle velocity

$$\mathbf{N} = -\mathbf{T}_*\mathbf{v} \tag{2.122}$$

For illustration, consider three cases. First, suppose that a wave propagates through a homogeneous fluid. Then

$$\tau_{xx} = \tau_{yy} = \tau_{zz} = -P \qquad \text{and} \qquad \tau_{xy} = \tau_{xz} = \tau_{yz} = 0$$

Correspondingly

$$N_x = P\dot{u}, \qquad N_y = P\dot{v},, \qquad N_z = P\dot{w} \qquad \text{and} \qquad \mathbf{N} = P\mathbf{v},$$

which coincides with eq. 2.98. Second, assume that shear stresses are absent. This gives

$$N_x = -\tau_{xx}\dot{u}, \qquad N_y = -\tau_{yy}\dot{v}, \qquad N_z = -\tau_{zz}\dot{w}$$

and

$$N_x = -\tau_{xx}\dot{u}\ \mathbf{i} - \tau_{yy}\dot{v}\ \mathbf{j} - \tau_{zz}\dot{w}\ \mathbf{k}, \tag{2.123}$$

which describes the flux of energy in the case of longitudinal waves.
Finally, if $\tau_{xx} = \tau_{yy} = \tau_{zz}$, we have

$$N_x = -\tau_{yx}\dot{v} - \tau_{xz}\dot{w}, \qquad N_y = -\tau_{yx}\dot{v} - \tau_{yz}\dot{w}, \qquad N_z = -\tau_{xz}\dot{u} - \tau_{yz}\dot{v}$$

For instance, when torsional waves propagate along the bar, we have $\tau_{yx} = 0$ and $\dot{w} = 0$. Then

$$N_x = N_y = 0, \qquad \text{but} \qquad N_z = -\tau_{xz}\dot{u} - \tau_{yz}\dot{v},$$

and vector $\mathbf{N}$ is directed along the bar axis z.

2.3 Strain, stress, and Hooke's law in the curvilinear orthogonal system of coordinates

As we know, forces applied to an elastic body cause a change in its shape and size, and in order to determine the changes it is necessary to find the length of every line after deformation. Our goal is to evaluate the distance between two points located close to each other and find its expression in different systems of coordinates.

Cartesian coordinates

Consider two arbitrary points in a medium,

$$P\ (x, y, z) \qquad \text{and} \qquad Q\ (x_1, y_1, z_1),$$

and suppose that the distance between them, r, is very small, Fig. 2.3. Let us introduce directional cosines, l, m, and n of the line PQ. By definition we have

$$\frac{x_1 - x}{r} = l, \qquad \frac{y_1 - y}{r} = m, \qquad \frac{z_1 - z}{r} = n \tag{2.124}$$

Correspondingly, coordinates of point Q are

$$x_1 = x + lr, \qquad y_1 = y + mr, \qquad z_1 = z + nr \tag{2.125}$$

After deformation, the particle that was at point P comes to point $P_1(x', y', z')$, where

$$x' = x + u(x, y, z), \qquad y' = y + v(x, y, z) \qquad z' = z + w(x, y, z) \tag{2.126}$$

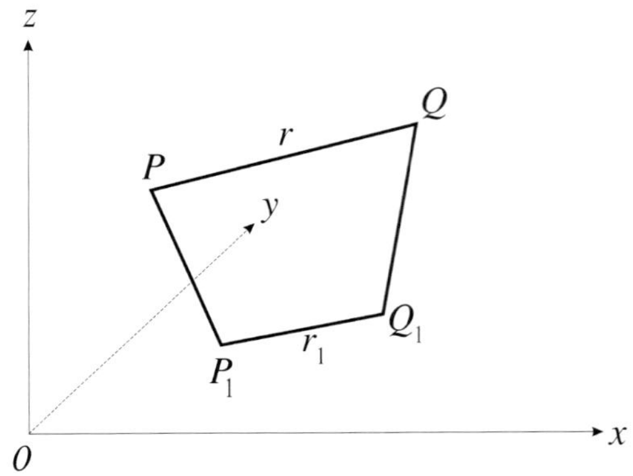

Figure 2.3: Relative change of line length

and $u, v,$ and w are scalar components of displacement. At the same time the particle at point Q moves at point $Q_1(x_1^{'}, y_1^{'}, z_1^{'})$. Here

$$x_1^{'} = x + lr + u_1, \qquad y_1^{'} = y + mr + v_1 \qquad z_1^{'} = z + nr + w_1 \tag{2.127}$$

Taking into account that displacement is a continuous function of coordinates, we expand each scalar component, u_1, v_1, and w_1, in the Taylor series. Discarding all of its terms except the linear ones, we get

$$u_1 = u + r\left(\frac{\partial u}{\partial x}l + \frac{\partial u}{\partial y}m + \frac{\partial u}{\partial z}n\right) = u + \mathbf{r}\cdot \text{grad } u$$

$$v_1 = v + r\left(\frac{\partial v}{\partial x}l + \frac{\partial v}{\partial y}m + \frac{\partial v}{\partial z}n\right) = v + \mathbf{r}\cdot \text{grad } v \tag{2.128}$$

$$w_1 = w + r\left(\frac{\partial w}{\partial x}l + \frac{\partial w}{\partial y}m + \frac{\partial w}{\partial z}n\right) = w + \mathbf{r}\cdot \text{grad } w$$

In deriving this system of equations, we used the fact that coordinates of points P and Q differ by $lr, mr,$ and $nr,$ respectively. Thus, coordinates of point Q are

$$x_1^{'} = x + lr + u + r\left(\frac{\partial u}{\partial x}l + \frac{\partial u}{\partial y}m + \frac{\partial u}{\partial z}n\right)$$

$$y_1^{'} = y + mr + v + r\left(\frac{\partial v}{\partial x}l + \frac{\partial v}{\partial y}m + \frac{\partial v}{\partial z}n\right) \tag{2.129}$$

$$z_1^{'} = z + nr + w + r\left(\frac{\partial w}{\partial x}l + \frac{\partial w}{\partial y}m + \frac{\partial w}{\partial z}n\right)$$

For the distance r_1 between points P_1 and Q_1, we have

$$r_1 = r(1+e) = [(x_1^{'} - x^{'})^2 + (y_1^{'} - y^{'})^2 + (z_1^{'} - z^{'})^2]^{1/2} \tag{2.130}$$

Here e is a relative change in the distance between points P and Q. Then substitution of 2.129 into 2.130 gives

$$r_1 = r\left\{\left[l\left(1+\frac{\partial u}{\partial x}\right) + m\frac{\partial u}{\partial y} + n\frac{\partial u}{\partial z}\right]^2 + \left[l\frac{\partial v}{\partial x} + m\left(1+\frac{\partial v}{\partial y}\right) + n\frac{\partial v}{\partial z}\right]^2\right. \tag{2.131}$$

$$\left. + \left[l\frac{\partial w}{\partial x} + m\frac{\partial w}{\partial y} + n\left(1+\frac{\partial w}{\partial z}\right)\right]^2\right\}^{1/2}$$

Since $l^2 + m^2 + n^2 = 1$ and the terms that contain the product of derivatives are small, we obtain:

$$r_1 = r\,(1 + e_{xx}l^2 + e_{yy}m^2 + e_{zz}n^2 + e_{yz}mn + e_{zx}nl + e_{xy}lm), \tag{2.132}$$

where

$$e_{xx} = \frac{\partial u}{\partial x}, \qquad e_{yy} = \frac{\partial v}{\partial y}, \qquad e_{zz} = \frac{\partial w}{\partial z},$$

$$e_{yz} = \frac{\partial v}{\partial z} + \frac{\partial w}{\partial y}, \qquad e_{zx} = \frac{\partial w}{\partial x} + \frac{\partial u}{\partial z}, \qquad e_{xy} = \frac{\partial u}{\partial y} + \frac{\partial v}{\partial x}$$

are components of the strain tensor (Appendix D). In general, they vary from point to point. As follows from eq. 2.132, a relative change in the distance of the short line is

$$e = e_{xx}l^2 + e_{yy}m^2 + e_{zz}n^2 + e_{xz}nl + e_{yz}mn + e_{xy}lm \tag{2.133}$$

and it is defined by the strain components. Also, we can treat strains as coefficients of terms in eq. 2.132 that contain the square of directional cosines or their product. Taking this fact into account, we find a relationship between r_1 and r at different systems of coordinates. This will permit us to determine expressions for strains.

Curvilinear orthogonal system of coordinates

To facilitate our study of strains, it is useful to recall the main features of these systems (Part I). We start from the simplest Cartesian coordinates, where the position of any point is defined as

$$x = C_1, \qquad y = C_2, \qquad z = C_3, \tag{2.134}$$

where C_1, C_2, and C_3 are constants. From the geometric point of view, each of these equalities describes a plane perpendicular to the corresponding coordinate axis. For instance, $x = C_1$ means a plane that is parallel to plane YOZ. Varying C_1 we arrive at a family of planes that are all perpendicular to the x-axis. Similarly, equations $y = C_2$ and $z = C_3$ characterize families of planes that are parallel to planes XOZ and XOY, respectively. These three families represent coordinate planes that are perpendicular to each other. By definition, at every coordinate plane one coordinate $(x, y,$ or $z)$ is constant, while other two vary. Note that (a) there are always three coordinate planes that pass through the same point, and their intersection determines the position of this point; and (b) planes YOZ, XOZ, and XOY belong to corresponding families of the coordinate planes.

There is another approach that allows us to determine the point's location. Consider first planes $x = \alpha$ and $y = \beta$. Their intersection gives a straight line parallel to the z-axis. Changing α or β or both of them, we obtain a family of straight lines that are parallel to each other. In the same manner, the intersection of planes $x = \alpha$ and $z = \gamma$, as well as of $y = \beta$ and $z = \gamma$, produces two other families of lines, which are called coordinate lines. It is obvious that three coordinate lines pass through each point and form direct angles with each other. Along each line, two coordinates – for instance, y and z – remain constant, but one changes. The axes of coordinates passing through origin O are examples of these lines. Thus, at each point three coordinate lines, x, y, and z, are normal to the corresponding coordinate planes, and the intersection of such lines defines the position of the point. Because coordinate lines are straight, expressions of grad, div, curl, and laplacian in the Cartesian system are greatly simplified. First, elementary displacement along coordinate lines is equal to a change of coordinates:

$$dl_x = dx, \qquad dl_y = dy, \qquad dl_z = dz \tag{2.135}$$

Respectively, elementary areas of coordinate planes that are formed by coordinate lines are

$$dS_x = dydz, \qquad dS_y = dxdz, \qquad dS_z = dxdy \tag{2.136}$$

and the elementary volume surrounded by elements of coordinate planes is

$$dV = dx\, dy\, dz \tag{2.137}$$

Also, if the elementary vector $\mathbf{ds}$ is arbitrarily oriented, then its magnitude is equal to

$$|\mathbf{ds}| = (dx^2 + dy^2 + dz^2)^{1/2},$$

where dx, dy, and dz are its scalar projections on the coordinate lines. As follows from the vector analysis (Part I),

$$\nabla\varphi = \frac{1}{\Delta V}\oint_S \varphi\, \mathbf{dS}, \qquad \nabla\mathbf{M} = \frac{1}{\Delta V}\int_S \mathbf{M}\cdot\mathbf{dS}, \qquad \nabla\times\,\mathbf{M} = \frac{\mathbf{n}}{\Delta S}\oint_S \mathbf{M}\cdot\mathbf{dl} \tag{2.138}$$

Here $\mathbf{n}$ is the unit vector showing a direction of vortexes. Then, replacing the integration in eqs. 2.138 by a differentiation and using eqs. 2.135–2.137, we obtain

$$\text{grad}\ \varphi = \frac{\partial\varphi}{\partial x}\mathbf{i} + \frac{\partial\varphi}{\partial y}\mathbf{j} + \frac{\partial\varphi}{\partial z}\mathbf{k} \qquad\qquad \text{div}\ \mathbf{M} = \frac{\partial M_x}{\partial x} + \frac{\partial M_y}{\partial y} + \frac{\partial M_x}{\partial z}$$

$$\text{curl}\ \mathbf{M} = \begin{vmatrix} \mathbf{i} & \mathbf{j} & \mathbf{k} \\ \dfrac{\partial}{\partial x} & \dfrac{\partial}{\partial y} & \dfrac{\partial}{\partial z} \\ M_x & M_y & M_z \end{vmatrix}, \tag{2.139}$$

$$\nabla^2\varphi = \frac{\partial^2\varphi}{\partial x^2} + \frac{\partial^2\varphi}{\partial y^2} + \frac{\partial^2\varphi}{\partial z^2} = \text{grad div}\ \varphi$$

Next, we discuss a more general case in which coordinate surfaces are not usually planes, and so the coordinate lines can be curvilinear. However, they are still perpendicular to each other. Suppose that α, β, and γ are coordinates of a point and their relationships with Cartesian coordinates are

$$f_1(x, y, z) = \alpha, \qquad f_2(x, y, z) = \beta, \qquad f_3(x, y, z) = \gamma \tag{2.140}$$

For instance, the function $f_1(x, y, z) = \alpha$ describes the coordinate surface, where one coordinate, α, is constant, but two others, β and γ, vary. In accordance with eqs. 2.140, we have three families of coordinate surfaces, and their intersection forms three families of coordinate lines: l_α, l_β, and l_γ. Unlike in the Cartesian system, these lines are in general curvilinear, but at each point they form direct angles. As before, along each coordinate line only one coordinate varies, but two others are constant. This fact is obvious, since every line is the result of an intersection of two coordinate surfaces – for instance, $\beta =$ const and $\gamma =$ const. It is also convenient to introduce three unit vectors, $\mathbf{i}_\alpha$, $,\mathbf{i}_\beta$, and $\mathbf{i}_\gamma$, each of which defines the direction of the coordinate line. Thus, we have three families of coordinate surfaces, as well as three families of coordinate lines, and both of these groups equally characterize the position of a point. By definition, every coordinate surface can be treated as the surface of equal values of the corresponding coordinate. Therefore, the gradient of the latter is perpendicular to this surface, i.e., it is directed along the coordinate line. Thus means that

$$\nabla\alpha = |\nabla\alpha|\ \mathbf{i}_\alpha = \frac{d\alpha}{dl_\alpha}\mathbf{i}_\alpha, \qquad \nabla\beta = |\nabla\beta|\ \mathbf{i}_\beta = \frac{d\beta}{dl_\beta}\mathbf{i}_\beta, \qquad \nabla\gamma = |\nabla\gamma|\ \mathbf{i}_\gamma = \frac{d\gamma}{dl_\gamma}\mathbf{i}_\gamma, \tag{2.141}$$

where dl_α, dl_β, and dl_γ are elementary displacements along coordinate lines. Taking into account that coordinate lines are usually curvilinear, these displacements do not coincide with a change of coordinates, and unlike in the Cartesian system, we have

$$dl_\alpha = h_\alpha d\alpha, \qquad dl_\beta = h_\beta d\beta, \qquad dl_\gamma = h_\gamma d\gamma \tag{2.142}$$

Here h_α, h_β, and h_γ are called the metric coefficients and, as a rule, they are functions of coordinates of the point. From eqs. 2.141 and 2.142, we have

$$\frac{1}{h_\alpha^2} = \left(\frac{\partial\alpha}{\partial x}\right)^2 + \left(\frac{\partial\alpha}{\partial y}\right)^2 + \left(\frac{\partial\alpha}{\partial z}\right)^2, \tag{2.143}$$

$$\frac{1}{h_\beta^2} = \left(\frac{\partial\beta}{\partial x}\right)^2 + \left(\frac{\partial\beta}{\partial y}\right)^2 + \left(\frac{\partial\beta}{\partial z}\right)^2, \qquad \frac{1}{h_\gamma^2} = \left(\frac{\partial\gamma}{\partial x}\right)^2 + \left(\frac{\partial\gamma}{\partial y}\right)^2 + \left(\frac{\partial\gamma}{\partial z}\right)^2$$

It is assumed that within elementary displacements, eqs. 2.142, metric coefficients: h_α, h_β, and h_γ are constants. Since the angles between unit vectors are equal to $\pi/2$, eqs. 2.141 give

$$\frac{\partial\alpha}{\partial x}\frac{\partial\beta}{\partial x} + \frac{\partial\alpha}{\partial y}\frac{\partial\beta}{\partial y} + \frac{\partial\alpha}{\partial z}\frac{\partial\beta}{\partial z} = 0,$$

$$\frac{\partial\alpha}{\partial x}\frac{\partial\gamma}{\partial x} + \frac{\partial\alpha}{\partial y}\frac{\partial\gamma}{\partial y} + \frac{\partial\alpha}{\partial z}\frac{\partial\gamma}{\partial z} = 0, \qquad \frac{\partial\beta}{\partial x}\frac{\partial\gamma}{\partial x} + \frac{\partial\beta}{\partial y}\frac{\partial\gamma}{\partial y} + \frac{\partial\beta}{\partial z}\frac{\partial\gamma}{\partial z} = 0$$

In accordance with eqs. 2.142, elementary areas of coordinate surfaces are

$$dS_\alpha = dl_\beta\, dl_\gamma = h_\beta\, h_\gamma\, d\beta\, d\gamma, \qquad dS_\beta = dl_\alpha\, dl_\gamma = h_\alpha\, h_\gamma\, d\alpha\, d\gamma, \tag{2.144}$$

$$dS_\gamma = dl_\alpha\, dl_\beta = h_\alpha\, h_\beta\, d\alpha\, d\beta,$$

and the elementary volume formed by coordinate surfaces is equal to

$$dV = h_\alpha\, h_\beta\, h_\gamma\, d\alpha\, d\beta\, d\gamma \tag{2.145}$$

It is obvious that the length of an elementary displacement, ds, arbitrarily oriented with respect to coordinate lines, is defined as

$$ds^2 = dl_\alpha^2 + dl_\beta^2 + dl_\gamma^2 \tag{2.146}$$

As was demonstrated in Part I, we have:

$$\nabla\varphi = \frac{1}{h_\alpha}\frac{\partial\varphi}{\partial\alpha}\mathbf{i}_\alpha + \frac{1}{h_\beta}\frac{\partial\varphi}{\partial\beta}\mathbf{i}_\beta + \frac{1}{h_\gamma}\frac{\partial\varphi}{\partial\gamma}\mathbf{i}_\gamma$$

$$\text{div}\,\mathbf{M} = \frac{1}{h_\alpha h_\beta h_\gamma}\left[\frac{\partial}{\partial\alpha}(h_\beta h_\gamma M_\alpha) + \frac{\partial}{\partial\beta}(h_\alpha h_\gamma M_\beta) + \frac{\partial}{\partial\gamma}(h_\alpha h_\beta M_\gamma)\right]$$

$$\text{curl}\,\mathbf{M} = \frac{1}{h_\alpha h_\beta h_\gamma}\begin{vmatrix} h_\alpha\,\mathbf{i}_\alpha & h_\beta\,\mathbf{i}_\beta & h_\gamma\,\mathbf{i}_\gamma \\ \dfrac{\partial}{\partial\alpha} & \dfrac{\partial}{\partial\beta} & \dfrac{\partial}{\partial\gamma} \\ h_\alpha M_\alpha & h_\beta M_\beta & h_\gamma M_\gamma \end{vmatrix} \quad \text{and}$$

$$\nabla^2\varphi = \frac{1}{h_\alpha h_\beta h_\gamma}\left[\frac{\partial}{\partial\alpha}\left(\frac{h_\beta h_\gamma}{h_\alpha}\frac{\partial\varphi}{\partial\alpha}\right) + \frac{\partial}{\partial\beta}\left(\frac{h_\alpha h_\gamma}{h_\beta}\frac{\partial\varphi}{\partial\beta}\right) + \frac{\partial}{\partial\gamma}\left(\frac{h_\alpha h_\beta}{h_\gamma}\frac{\partial\varphi}{\partial\gamma}\right)\right] \tag{2.147}$$

Usually metric coefficients are derived from the geometry of the coordinate system. For instance, in the cylindrical system of coordinates r, φ, and z, they are

$$h_r = 1, \qquad h_\varphi = r \qquad \text{and} \qquad h_z = 1, \tag{2.148}$$

whereas in the spherical system R, θ, and φ we have

$$h_R = 1, \qquad h_\theta = R \quad \text{and} \quad h_\varphi = R\sin\theta \tag{2.149}$$

Strain in the curvilinear orthogonal system of coordinates: α, β, **and** γ

As was pointed out earlier, in order to obtain expressions for strains we have to derive eq. 2.132 in the system of coordinates α, β, and γ. Consider two neighboring points, $P(\alpha, \beta, \gamma)$ and $Q\ (\alpha+a,\ \beta+b,\ \gamma+c)$, situated at a small distance $r = PQ$ from each other. Suppose that l, m, and n are directional cosines of vector **PQ** at point P, i.e.,

$$l = \frac{dl_\alpha}{r}, \qquad m = \frac{dl_\beta}{r}, \qquad n = \frac{dl_\gamma}{r} \tag{2.150}$$

Using eqs. 2.142, we also have

$$l = \frac{h_\alpha a}{r}, \qquad m = \frac{h_\beta b}{r}, \qquad n = \frac{h_\gamma c}{r}, \tag{2.151}$$

since

$$d\alpha = a, \qquad d\beta = b, \qquad d\gamma = c,$$

whence

$$a = \frac{l\ r}{h_\alpha}, \qquad b = \frac{m\ r}{h_\beta}, \qquad c = \frac{n\ r}{h_\gamma} \tag{2.152}$$

The equations above allow us to represent a change of coordinates of point P in terms of metric coefficients, directional cosines and distance r. Because of deformation, points P and Q move at points P_1 and Q_1, respectively, Fig. 2.3. Let us assume that coordinates of point P_1 are

$$\alpha + \zeta, \qquad \beta + \eta, \qquad \gamma + \xi$$

In order to determine coordinates of point Q_1, we take into account the fact that ζ, η, and ξ are functions of α, β, and γ. Correspondingly, in the linear approximation, coordinates of point Q_1 are equal to

$$\alpha + a + \zeta + a\frac{\partial \zeta}{\partial \alpha} + b\frac{\partial \zeta}{\partial \beta} + c\frac{\partial \zeta}{\partial \gamma}, \tag{2.153}$$

$$\beta + b + \eta + a\frac{\partial \eta}{\partial \alpha} + b\frac{\partial \eta}{\partial \beta} + c\frac{\partial \eta}{\partial \gamma}, \qquad \gamma + c + \xi + a\frac{\partial \xi}{\partial \alpha} + b\frac{\partial \xi}{\partial \beta} + c\frac{\partial \xi}{\partial \gamma}$$

Therefore, the difference in corresponding coordinates of points P_1 and Q_1 is

$$\zeta_1 = a\left(1 + \frac{\partial \zeta}{\partial \alpha}\right) + b\frac{\partial \zeta}{\partial \beta} + c\frac{\partial \zeta}{\partial \gamma}, \tag{2.154}$$

$$\eta_1 = a\frac{\partial \eta}{\partial \alpha} + b\left(1 + \frac{\partial \eta}{\partial \beta}\right) + c\frac{\partial \eta}{\partial \gamma}, \qquad \xi_1 = a\frac{\partial \xi}{\partial \alpha} + b\frac{\partial \xi}{\partial \beta} + c\left(1 + \frac{\partial \xi}{\partial \gamma}\right)$$

To evaluate the distance $r_1 = P_1 Q_1$, we have to find metric coefficients h'_α, h'_β, and h'_γ at point P_1. Again in the linear approximation we have:

$$\begin{aligned} h'_\alpha &= h_\alpha + \zeta\frac{\partial h_\alpha}{\partial \zeta} + \eta\frac{\partial h_\alpha}{\partial \eta} + \xi\frac{\partial h_\alpha}{\partial \xi}, \\ h'_\beta &= h_\beta + \zeta\frac{\partial h_\beta}{\partial \zeta} + \eta\frac{\partial h_\beta}{\partial \eta} + \xi\frac{\partial h_\beta}{\partial \xi}, \\ h'_\gamma &= h_\gamma + \zeta\frac{\partial h_\gamma}{\partial \zeta} + \eta\frac{\partial h_\gamma}{\partial \eta} + \xi\frac{\partial h\,\gamma}{\partial \xi} \end{aligned} \tag{2.155}$$

By definition, projections of vector $\mathbf{P}_1\,\mathbf{Q}_1$ on the coordinate lines at point P_1 are

$$(r_1)_\alpha = h'_\alpha\,\zeta_1, \qquad (r_1)_\beta = h'_\beta\,\eta_1, \qquad (r_1)_\gamma = h'_\gamma\,\xi_1 \tag{2.156}$$

Performing multiplications on the right side of this set, we have to discard from our approximation terms of the second and higher orders with respect to ζ, η, and ξ, as well as their derivatives. For instance, component $(r_1)_\alpha$ becomes

$$(r_1)_\alpha = \left(h_\alpha + \zeta\frac{\partial h_\alpha}{\partial \zeta} + \eta\frac{\partial h_\alpha}{\partial \eta} + \xi\frac{\partial h_\alpha}{\partial \xi}\right)\left[a\left(1 + \frac{\partial \zeta}{\partial \alpha}\right) + b\frac{\partial \zeta}{\partial \beta} + c\frac{\partial \zeta}{\partial \gamma}\right] \tag{2.157}$$

$$= a h_\alpha\left(1 + \frac{\partial \zeta}{\partial \alpha}\right) + b\,h_\alpha\frac{\partial \zeta}{\partial \beta} + c\,h_\alpha\frac{\partial \zeta}{\partial \gamma} + a\zeta\frac{\partial h_\alpha}{\partial \alpha} + a\eta\frac{\partial h_\alpha}{\partial \beta} + a\xi\frac{\partial h_\alpha}{\partial \gamma}$$

Taking into account that components of vector $\mathbf{PP}_1$,

$$\mathbf{PP}_1 = u_\alpha \mathbf{i}_\alpha + u_\beta \mathbf{i}_\beta + u_\gamma \mathbf{i}_\gamma,$$

at point P are related to ζ, η, and ξ as

$$u_\alpha = h_\alpha \zeta, \qquad u_\beta = h_\beta \eta, \qquad u_\gamma = h_\gamma \xi,$$

and making use of eqs. 2.152, we have

$$(r_1)_\alpha = r\left[l\left(1 + \frac{\partial \zeta}{\partial \alpha} + \frac{\zeta}{h_\alpha}\frac{\partial h_\alpha}{\partial \alpha} + \frac{\eta}{h_\alpha}\frac{\partial h_\alpha}{\partial \beta} + \frac{\xi}{h_\alpha}\frac{\partial h_\alpha}{\partial \gamma}\right) + m\frac{h_\alpha}{h_\beta}\frac{\partial \zeta}{\partial \beta} + n\frac{h_\alpha}{h_\gamma}\frac{\partial \zeta}{\partial \gamma}\right]$$

$$= r\left[l\left(1+\frac{1}{h_\alpha}\frac{\partial u_\alpha}{\partial \alpha}+u_\alpha\frac{\partial h_\alpha^{-1}}{\partial \alpha}-h_\alpha^{-2}u_\alpha\frac{\partial h_\alpha^{-1}}{\partial \alpha}h_\alpha^2+u_\beta h_\alpha^{-1}h_\beta^{-1}\frac{\partial h_\alpha}{\partial \beta}\right.\right. \tag{2.158}$$

$$\left.\left.+\,u_\gamma h_\alpha^{-1}h_\gamma^{-1}\frac{h_\alpha}{h_\gamma}\right)+m\,\frac{h_\alpha}{h_\beta}\frac{\partial}{\partial \beta}(h_\alpha^{-1}u_\alpha)+n\,\frac{h_\alpha}{h_\gamma}\frac{\partial}{\partial \gamma}(h_\alpha^{-1}u_\alpha)\right]$$

Finally, we obtain

$$(r_1)_\alpha = r\left[l\left(1+h_\alpha^{-1}\frac{\partial u_\alpha}{\partial \alpha}+h_\alpha^{-1}h_\beta^{-1}u_\beta\frac{\partial h_\alpha}{\partial \beta}+h_\alpha^{-1}h_\gamma^{-1}u_\gamma\frac{\partial h_\alpha}{h_\gamma}\right)\right. \tag{2.159}$$

$$\left.+\,m\frac{h_\alpha}{h_\beta}\frac{\partial}{\partial \beta}(h_\alpha^{-1}u_\alpha)+n\frac{h_\alpha}{h_\gamma}\frac{\partial}{\partial \gamma}(h_\alpha^{-1}u_\alpha)\right]$$

By analogy,

$$(r_1)_\beta = r\left[m\left(1+h_\beta^{-1}\frac{\partial u_\beta}{\partial \beta}+h_\beta^{-1}h_\gamma^{-1}u_\gamma\frac{\partial h_\beta}{\partial \gamma}+h_\alpha^{-1}h_\beta^{-1}u_\alpha\frac{\partial h_\beta}{h_\alpha}\right)\right.$$

$$\left.+\,n\frac{h_\beta}{h_\gamma}\frac{\partial}{\partial \gamma}(h_\beta^{-1}u_\beta)+l\,\frac{h_\beta}{h_\alpha}\frac{\partial}{\partial \alpha}(h_\beta^{-1}u_\beta)\right] \tag{2.160}$$

and $$(r_1)_\gamma = r\left[n\left(1+h_\gamma^{-1}\frac{\partial u_\gamma}{\partial \gamma}+h_\alpha^{-1}h_\gamma^{-1}u_\alpha\frac{\partial h_\gamma}{\partial \alpha}+h_\beta^{-1}h_\gamma^{-1}u_\beta\frac{\partial h_\gamma}{h_\beta}\right)\right.$$

$$\left.+\,l\,\frac{h_\gamma}{h_\alpha}\frac{\partial}{\partial \alpha}(h_\gamma^{-1}u_\gamma)+m\frac{h_\gamma}{h_\beta}\frac{\partial}{\partial \beta}(h_\gamma^{-1}u_\gamma)\right]$$

Let e be a relative extension of the linear element PQ, i.e.,

$$r_1^2 = r^2(1+e)^2 = (r_1)_\alpha^2+(r_1)_\beta^2+(r_1)_\gamma^2 \tag{2.161}$$

Substituting eqs. 2.159 and 2.160 into eq. 2.161 and discarding squares and products of u_α, u_β, and u_γ, we obtain

$$(1+e)^2 = l^2+m^2+n^2+2l^2e_{\alpha\alpha}+2m^2e_{\beta\beta}+2n^2e_{\gamma\gamma}+ \tag{2.162}$$

$$2mne_{\beta\gamma}+2nle_{\alpha\gamma}+2lme_{\alpha\beta}$$

Here

$$e_{\alpha\alpha}=\frac{1}{h_\alpha}\frac{\partial u_\alpha}{\partial\alpha}+\frac{u_\beta}{h_\alpha h_\beta}\frac{\partial h_\alpha}{\partial\beta}+\frac{u_\gamma}{h_\alpha h_\gamma}\frac{\partial h_\alpha}{\partial\gamma},\qquad e_{\beta\beta}=\frac{1}{h_\beta}\frac{\partial u_\beta}{\partial\beta}+\frac{u_\gamma}{h_\beta h_\gamma}\frac{\partial h_\beta}{\partial\gamma}+\frac{u_\alpha}{h_\alpha h_\beta}\frac{\partial h_\beta}{\partial\alpha},$$

$$e_{\gamma\gamma}=\frac{1}{h_\gamma}\frac{\partial u_\gamma}{\partial\gamma}+\frac{u_\alpha}{h_\alpha h_\gamma}\frac{\partial h_\gamma}{\partial\alpha}+\frac{u_\beta}{h_\beta h_\gamma}\frac{\partial h_\gamma}{\partial\beta},\qquad e_{\beta\gamma}=\frac{h_\gamma}{h_\beta}\frac{\partial}{\partial\beta}\left(\frac{u_\gamma}{h_\gamma}\right)+\frac{h_\beta}{h_\gamma}\frac{\partial}{\partial\gamma}\left(\frac{u_\beta}{h_\beta}\right),\tag{2.163}$$

$$e_{\gamma\alpha}=\frac{h_\alpha}{h_\gamma}\frac{\partial}{\partial\gamma}\left(\frac{u_\alpha}{h_\alpha}\right)+\frac{h_\gamma}{h_\alpha}\frac{\partial}{\partial\alpha}\left(\frac{u_\gamma}{h_\gamma}\right),\qquad e_{\alpha\beta}=\frac{h_\beta}{h_\alpha}\frac{\partial}{\partial\alpha}\left(\frac{u_\beta}{h_\beta}\right)+\frac{h_\alpha}{h_\beta}\frac{\partial}{\partial\beta}\left(\frac{u_\alpha}{h_\alpha}\right)$$

Since $l^2+m^2+n^2=1$, after taking a square root eq. 2.162 becomes

$$e=e_{\alpha\alpha}\,l^2+e_{\beta\beta}\,m^2+e_{\gamma\gamma}\,n^2+e_{\beta\gamma}\,mn+e_{\gamma\alpha}\,nl+e_{\alpha\beta}\,ml,\tag{2.164}$$

which coincides with eq. 2.133. Coefficients

$$e_{\alpha\alpha},\quad e_{\beta\beta},\quad e_{\gamma\gamma},\quad e_{\beta\gamma},\quad e_{\gamma\alpha},\quad e_{\alpha\beta}$$

are six components of the strain tensor in the curvilinear orthogonal system of coordinates, and they were first derived by Lame. As an illustration, consider the cylindrical and spherical system of coordinates. As we already know, in the first system

$$\alpha=r,\quad \beta=\varphi,\quad \text{and}\quad \gamma=z,\qquad \text{while}\qquad h_\alpha=h_\gamma=1\quad\text{and}\quad h_\beta=r$$

Therefore

$$e_{rr}=\frac{\partial u_r}{\partial r},\qquad e_{\varphi\varphi}=\frac{1}{r}\frac{\partial u_\varphi}{\partial\varphi}+\frac{u_r}{r},\qquad e_{zz}=\frac{\partial u_z}{\partial z},\tag{2.165}$$

$$e_{\varphi z}=\frac{1}{r}\frac{\partial u_z}{\partial\varphi}+\frac{\partial u_\varphi}{\partial z},\qquad e_{rz}=\frac{\partial u_r}{\partial z}+\frac{\partial u_z}{\partial r},\qquad e_{r\varphi}=\frac{\partial u_\varphi}{\partial r}-\frac{u_\varphi}{r}+\frac{1}{r}\frac{\partial u_r}{\partial\varphi}$$

In the spherical system R,θ,φ

$$h_\alpha=1,\qquad h_\beta=R,\qquad h_\gamma=R\sin\theta,$$

we have

$$e_{RR}=\frac{\partial u_R}{\partial R},\qquad e_{\theta\theta}=\frac{1}{R}\frac{\partial u_\theta}{\partial\theta}+\frac{u_R}{R},\qquad e_{\varphi\varphi}=\frac{1}{R\sin\theta}\frac{\partial u_\varphi}{\partial\varphi}+\frac{u_\theta}{R}\cot\theta+\frac{u_R}{R},$$

$$e_{\varphi\theta}=\frac{1}{R}\left(\frac{\partial u_\varphi}{\partial\theta}-u_\varphi\cot\theta\right)+\frac{1}{R\sin\theta}\frac{\partial u_\theta}{\partial\varphi},\tag{2.166}$$

$$e_{\varphi R}=\frac{1}{R\sin\theta}\frac{\partial u_R}{\partial\varphi}+\frac{\partial u_\varphi}{\partial R}-\frac{u_\varphi}{R},\qquad e_{R\theta}=\frac{\partial u_\theta}{\partial R}-\frac{u_\theta}{R}+\frac{1}{R}\frac{\partial u_R}{\partial\theta}$$

Hooke's law

In the curvilinear orthogonal system of coordinates, angles between faces of an elementary volume are equal to $\pi/2$, and they can be treated as planar faces. Correspondingly, relationships between stresses and strains (Hooke's law), derived in the Cartesian system of coordinates, remain valid in the general case. Note that a difference of face areas of the elementary volume is an order of magnitude smaller and in the limit it can be neglected. As was shown earlier,

$$\begin{aligned} \tau_{xx} &= \lambda\,\Theta + 2\mu e_{xx} \qquad & \tau_{yz} = \mu e_{yz} \\ \tau_{yy} &= \lambda\,\Theta + 2\mu e_{yy} \qquad & \tau_{zx} = \mu e_{zx} \\ \tau_{zz} &= \lambda\,\Theta + 2\mu e_{zz} \qquad & \tau_{xy} = \mu e_{xy} \end{aligned}$$

Here $\Theta =$ div $\mathbf{s}$ is invariant. Therefore, we have

$$\tau_{\alpha\alpha} = \lambda\,\Theta + 2\mu e_{\alpha\alpha}, \qquad \tau_{\beta\beta} = \lambda\,\Theta + 2\mu e_{\beta\beta}, \qquad \tau_{\gamma\gamma} = \lambda\,\Theta + 2\mu\, e_{\gamma\gamma} \tag{2.167}$$

$$\tau_{\gamma\alpha} = \mu\, e_{\gamma\alpha}, \qquad \tau_{\beta\gamma} = \mu e_{\beta\gamma}, \qquad \tau_{\alpha\beta} = \mu\, e_{\alpha\beta}$$

Chapter 3

Elastic waves in a homogeneous medium

We will first consider longitudinal and shear waves caused by spherical sources, and next we will study waves caused by the action of a single force $\mathbf{F}(t)$. Finally, we will discuss propagation of plane waves in a homogeneous medium.

3.1 Longitudinal spherical waves

First suppose that a longitudinal wave is caused by vibrations of a spherical shell having a very small radius. Also, it is assumed that each point of the shell surface is subjected to an action of force that has only the radial component and the same magnitude:

$$\mathbf{F}(t) = \tau_0 \; S \; \mathbf{i}_r \tag{3.1}$$

Here S is the surface area, τ_0 is the normal stress, and $\mathbf{i}_r$ is the unit vector normal to S. Because of stress variations caused by very small changes in the shell radius, a wave arises and propagates through a homogeneous medium. It is obvious that a distribution of wavefields that accompany the wave – namely, stresses, strains, displacement and velocity of particles – possesses spherical symmetry. Taking this into account, we choose a spherical system of coordinates with the origin at the source center and assume that all fields depend on the coordinate R only and on time t.

Scalar potential

To determine wavefields, we use the scalar potential U related to displacement $\mathbf{s}$ (Chapter 2):

$$\mathbf{s} = \text{grad}\, U \tag{3.2}$$

U is also a function of R and t:

$$U = U(R, t) \tag{3.3}$$

At regular points, U satisfies the wave equation

$$\nabla^2 U = \frac{1}{c_l^2} \frac{\partial^2 U}{\partial t^2}, \tag{3.4}$$

where

$$c_l = \sqrt{\frac{\lambda + 2\mu}{\rho}} \tag{3.5}$$

is the velocity of propagation of longitudinal waves. In addition, we have to describe the behavior of the potential on the shell surface and formulate initial conditions. Vibrations of the source generate the outgoing wave, the magnitude of which decreases with distance R and in the limit:

$$U(R, t) \to 0 \qquad \text{if} \qquad R \to \infty \tag{3.6}$$

This simply means that sources of waves are absent at infinity. The same behavior follows if we imagine a spherical surface with a radius so large that the wave does not reach it before or during the time of observation. Suppose that displacement of the shell surface is the given function of time

$$s_R(t) = \begin{cases} 0 & t < 0 \\ s_0\, f(at) & t \geq 0 \end{cases}$$

Then, as follows from eq. 3.2, the boundary condition for the potential at points of the shell surface can be represented as

$$\frac{\partial U}{\partial R} = \begin{cases} 0 & t < 0 \\ s_0\, f(at) & t \geq 0 \end{cases} \tag{3.7}$$

Assume that R_0 is the shell radius at rest. The wave was absent at each point of the medium until the instant $t = 0$, when the source was turned on, that is,

$$s_R(R, 0) = 0 \qquad \text{or} \qquad \frac{\partial U}{\partial R} = 0 \qquad \text{for} \qquad t < 0 \tag{3.8}$$

In accordance with Hooke's law, the second derivative, $\partial^2 U(R, 0)/\partial R^2$, also vanishes. Note that the initial condition contains information about waves at infinity. Thus, the solution of the boundary value problem in terms of scalar potential should satisfy the following conditions.

1. At regular points of the medium

$$\nabla^2 U = \frac{1}{c_l^2}\frac{\partial^2 U}{\partial t^2}$$

2. At the surface of the source

$$\frac{\partial U}{\partial R} = \begin{cases} 0 & t < 0 \\ s_0\ f(at) & t \geq 0 \end{cases} \qquad \text{if} \quad R = R_0 + s_R(t)$$

3. At the initial moment, the wavefield is absent inside the medium

$$\frac{\partial U(R,0)}{\partial R} = 0 \qquad \text{if} \qquad R > R_0, \qquad t = 0$$

First, we solve the wave equation, which, in the spherical system of coordinates (Chapter 2), is

$$\frac{1}{R^2}\frac{\partial}{\partial R}\left(R^2\frac{\partial U}{\partial R}\right) = \frac{1}{c_l^2}\frac{\partial^2 U}{\partial t^2} \tag{3.9}$$

since the potential depends on coordinate R only. To simplify its solution, we introduce a new function: $W = R\,U$. Differentiation and multiplication by R^2 give

$$R^2\frac{\partial U}{\partial R} = -W + R\frac{\partial W}{\partial R}$$

Therefore, eq. 3.9 becomes

$$\frac{\partial^2 W}{\partial R^2} = \frac{1}{c_l^2}\frac{\partial^2 W}{\partial t^2}$$

This equation was already derived in studying the propagation of waves along the bar (Chapter 1). Its solution consists of two independent functions,

$$f_1\left[a_1\left(t - \frac{R}{c_l}\right)\right] \qquad \text{and} \qquad g_1\left[a_1\left(t - \frac{R}{c_l}\right)\right]$$

Correspondingly, for scalar potential we have

$$U(R,t) = \frac{A}{R}f_1\left[a_1\left(t - \frac{R}{c_l}\right)\right] + \frac{B}{R}g_1\left[a_1\left(t + \frac{R}{c_l}\right)\right], \tag{3.10}$$

where A and B are constants. The second term at the right side of this equation does not satisfy the initial condition, since it describes the wave traveling from infinity to the source. For this reason, we let $B = 0$, which gives

$$U(R,t) = \frac{A}{R}\, f_1\left[a_1\left(t - \frac{R}{c_l}\right)\right] \tag{3.11}$$

It is clear that $U(R,t)$ describes the propagation of a wave away from the source at the origin with the phase velocity c_l and satisfies the wave equation and eq. 3.6. In order to determine the unknown coefficient A and function f_1, we use condition 3.7, which is valid at points of the source surface. It follows from eq. 3.2 and eq. 3.11 that

$$s_R(R,t) = \frac{\partial U}{\partial R} = -\frac{A}{R^2}\, f_1\left[a_1\left(t - \frac{R}{c_l}\right)\right] - \frac{Aa_1}{c_l R}\, f_1'\left[a_1\left(t - \frac{R}{c_l}\right)\right],$$

where $f_1'\left[a_1\left(t - R/c_l\right)\right]$ is the first derivative of the function with respect to its argument: $a_1(t - R/c_l)$. Now we make three assumptions. First, suppose that pulsations of the source are characterized by very small displacement:

1. $|s_R(t)| \ll R_0$

For this reason, in satisfying boundary condition 3.7, we approximately define the position of the shell by the constant coordinate R_0. Correspondingly, displacement of the surface point becomes:

$$s_0 f(at) = -\frac{A}{R_0^2}\, f_1\left[a_1\left(t - \frac{R_0}{c_l}\right)\right] - \frac{Aa_1}{c_l R_0}\, f_1'\left[a_1\left(t - \frac{R_0}{c_l}\right)\right] \tag{3.12}$$

Thus, the determination of unknowns is related to the solution of the differential equation. To simplify this procedure, we also suppose that the second term on the right side of this equation is relatively small,

2. $\left|\frac{1}{R_0^2}\, f_1\left[a_1\left(t - \frac{R_0}{c_l}\right)\right]\right| \gg \left|\frac{a_1}{c_l R_0}\, f_1'\left[a_1\left(t - \frac{R_0}{c_l}\right)\right]\right|$

and that

3. $t \gg \frac{R_0}{c_l}$

Then, instead of eq. 3.12, we obtain

$$s_0 f(at) = -\frac{Af_1(a_1 t)}{R_0^2} \tag{3.13}$$

The latter equality takes place if

$$a_1 = a, \qquad f_1(a_1 t) = f(at), \qquad \text{and} \qquad A = -R_0^2 \, s_0$$

Substitution of eq. 3.13 into eq. 3.11 gives an approximate expression for scalar potential:

$$U(R,t) = -\frac{R_0^2 \, s_0}{R} \, f \left[a \left(t - \frac{R}{c_l} \right) \right] \tag{3.14}$$

It is obvious that the function $U(R,t)$ satisfies the wave equation and the condition near the source. Since the function $f\,[a\,(t - R/c_l)]$ and its derivatives are equal to zero when the argument $a(t - R/c_l)$ is negative, scalar potential U also obeys the initial condition. Therefore, we have solved the boundary value problem. Eq. 3.14 describes scalar potential in a homogeneous medium, provided that the spherical source has a very small radius.

Equations for displacement, velocity of particles, and dilatation

Taking into account eq. 3.2, we have for the radial component of displacement s_R and for the velocity of radial displacement $v_R = \dot{s}_R(R,t)$ the following expressions:

$$s_R(R,t) = \frac{R_0^2}{R^2} \, s_0 \, f \left[a \left(t - \frac{R}{c_l} \right) \right] + \frac{R_0^2 \, s_0}{c_l R} \, a f^{\,\prime} \left[a \left(t - \frac{R}{c_l} \right) \right] \quad \text{and} \tag{3.15}$$

$$v_R(R,t) = \frac{R_0^2}{R^2} \, s_0 a \, f^{\,\prime} \left[a \left(t - \frac{R}{c_l} \right) \right] + \frac{R_0^2 \, s_0 \, a^2}{c_l R} \, f^{\,\prime\prime} \left[a \left(t - \frac{R}{c_l} \right) \right],$$

while

$$s_\theta = s_\varphi = 0 \qquad \text{and} \qquad v_\theta = v_\varphi = 0$$

By definition, dilatation is equal to

$$\theta = \text{div } \mathbf{s} = \frac{1}{R^2} \frac{\partial}{\partial R} (R^2 s_R) \tag{3.16}$$

Performing a differentiation, we have

$$\theta = -\frac{R_0^2 \, a^2}{c_l^2 R} \, s_0 \, f^{\,\prime\prime} \left[a \left(t - \frac{R}{c_l} \right) \right] \tag{3.17}$$

At the same time,

$$\text{curl } \mathbf{s} = \text{curl grad } U \equiv 0,$$

and, as we know, longitudinal waves do not cause rotation of an elementary volume, regardless of its orientation.

Spherical wave

Let us emphasize again the fact that if the argument $a(t - R/c_l)$ is negative, the function f and its derivatives are equal to zero. Also, it is clear that distribution of wavefields in space is defined by the single coordinate R. In other words, these functions remain constant on the spherical surface with the center at the source. This wave possesses spherical symmetry and, for this reason, it is called a spherical wave. The displacement and velocity of particles have only the radial component perpendicular to these surfaces, $(R = \text{ const})$, and this fact reflects one of the features of longitudinal (dilatational), waves. Suppose that the source surface changes its position during time interval Δt:

$$0 \leq t \leq \Delta t,$$

and we study the wave at some point located at distance R from the source. As follows from eqs. 3.15, motion of particles is absent if

$$t < \frac{R}{c_l},$$

since the argument is negative. Then, at instant $t = R/c_l$, the wave arrives simultaneously at all points having the same coordinate R, and we observe the wavefront. Within the time interval

$$\frac{R}{c_l} \leq t \leq \frac{R}{c_l} + \Delta t,$$

particles of a medium are involved in motion, and elementary volumes become deformed. At instant

$$t = \frac{R}{c_l} + \Delta t,$$

the rear of the wave passes the point of observation, and the medium is then again at equilibrium. Thus, regardless of the distance, the duration time of of the wave is equal to Δt, and the arrival time of the wave increases with R. As an illustration, consider wave distribution as a function of distance R. First of all, the wave may exist only at distances R that satisfy the condition

$$R < c_l t$$

Let us distinguish two cases, namely

$$1. \quad \Delta t > t \qquad \text{and} \qquad 2. \quad \Delta t < t$$

In the first case, the source still vibrates at instant t, and therefore the wave occupies the spherical volume having a radius equal to $R_c = c_l t$. In the second case, the source was turned off before the wave arrived at the observation point. Respectively, at instant $t = R/c_l$, the rear of the wave reaches points with the coordinate

$$R = c_l(t - \Delta t)$$

Thus, we observe the wave within spherical layer having the thickness $\Delta R = c_l \Delta t$.

Wavefields

Now we outline some features of wavefields s_R and v_R. The work of external forces of the source is transformed into the mechanical energy of the wave, and its amount remains the same since the influence of dissipation is neglected. As we know, when time elapses, the volume of the spherical layer occupied by the wave becomes large, and the energy density decreases. This shows that with an increase of distance R, wavefields become smaller, which of course follows from eqs. 3.15. In accordance with eq. 3.17, we see that behavior of dilatation θ, regardless of the distance from the source, is defined by the second derivative of the function $f\ [a\,(t - R/c_l)]$. For instance, in the vicinity of the source, the first term of the expression for the displacement

$$R_0^2\ s_0\ f\ \left[a\left(t - \frac{R}{c_l}\right)\right]$$

plays the dominant role, but dilatation essentially depends on the derivative of the second term

$$\frac{R_0^2\ s_0 a}{c_l R}\ f\,'\ \left[a\left(t - \frac{R}{c_l}\right)\right]$$

The other important feature of function Θ is that its dependence on distance remains the same everywhere. Finally, as follows from eq. 3.17, dilatation of the spherical wave decreases relatively slowly, $(1/R)$, and it is inversely proportional to phase velocity c_l. This suggests that a finite value of this velocity is a vital factor in producing a deformation. For instance, if the wave propagates instantly, $(c_l \to \infty)$, then the displacement, $s_R(R,t)$, becomes equal to

$$s_R(R,t) = \frac{R_0^2\ s_0}{R^2} f\ (at),$$

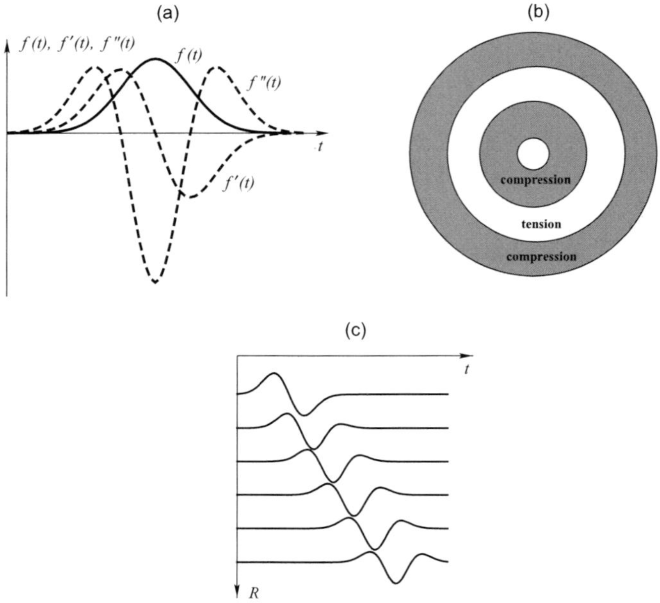

Figure 3.1: (a) The function $f(t)$ and its derivatives (b) Zones of compression and tension (c) Distribution of s_R as a function of time

and, therefore, dilatation Θ, characterizing a change in an elementary volume, vanishes. This result is not accidental, and it is true for any wave.

To demonstrate the wavefields, suppose that displacement of the shell surface, $(R = R_0)$, begins to increase at instant $t = 0$, then reaches the maximum, and then gradually approaches zero. In addition, we assume that the first and second derivatives of $f(at)$ are also equal to zero at the beginning $t = 0$ and at instant Δt. Behavior of those functions is shown in Fig. 3.1a. As we see, the function $f''(at)$ changes its sign twice. Let us suppose again that we observe the wave at some point with the coordinate R. Until the moment $t = R/c_l$, the wave is absent. Then, during the time interval

$$\frac{R}{c_l} < t < \frac{R}{c_l} + \Delta t,$$

there are two subintervals during which dilatation is negative, but between the intervals it is positive. We assumed that displacement of the source shell is characterized by one maximum only. In a more general case, the number of time intervals with positive and negative values of dilatation would increase. The zones of compression and expansion carried by the wave always accompany each other within the moving spherical layer, Fig. 3.1b. In order to prove this fact, consider the integral from dilatation

$$\int_{t_1}^{t_2} \Theta \, dt = -\frac{R_0^2 \, a \, s_0}{c_l^2 R} I, \tag{3.18}$$

where

$$t_1 = \frac{R}{c_l} - \varepsilon, \qquad t_2 = \frac{R}{c_l} + \Delta t + \varepsilon, \qquad \varepsilon \ll 1,$$

Δt is the duration of the source action, and

$$I = a \int_{t_1}^{t_2} f'' \left[a \left(t - \frac{R}{c_l} \right) \right] dt \tag{3.19}$$

Since at instant $t = t_1$ the wave has not yet arrived at the observation point, the scalar potential U and its derivatives are zero. Therefore, eq. 3.19 becomes

$$I = f' \left[a \left(t_2 - \frac{R}{c_l} \right) \right]$$

If we assume that the function $f'[a(t_2 - R/c_l)]$ differs from zero, then in accordance with eqs. 3.15 there must be particle motion behind the wave rear. Because this is impossible, we have to set

$$f' \left[a \left(t_2 - \frac{R}{c_l} \right) \right] = 0, \qquad \text{i.e.,} \qquad \int_{t_1}^{t_2} \Theta \, dt = 0 \tag{3.20}$$

The latter shows that if the source generates vibrations during a finite time interval, then for any function $f\,(at)$ there are zones of compression and expansion within the spherical wave. Moreover, distribution of dilatation between the front and rear of the wave is such that condition 3.20 is met.

In accordance with eqs. 3.15, the displacement of particles and their velocity are described by the sum of two terms, and each term has a different dependence on distance R. For this reason it is natural, as in the case of acoustic waves (Part I), to distinguish three intervals of distances – the near, intermediate, and far zones.

1. Near zone If distances from the source are relatively small, the first term of expressions for s_R and v_R is dominant, eqs. 3.15, and we have

$$s_R(R,t) = \frac{R_0^2 \, s_0}{R^2} f \left[a \left(t - \frac{R}{c_l} \right) \right], \qquad v_R(R,t) = \frac{R_0^2 \, s_0 \, a}{R^2} f' \left[a \left(t - \frac{R}{c_l} \right) \right] \tag{3.21}$$

In this zone, functions s_R and v_R decrease relatively quickly with distance from the source, and their dependence on time is the same as that for corresponding functions on the source surface. However, dilatation cannot be derived from the approximate equation for displacement, eq. 3.21. As was pointed out earlier, in order to obtain the correct expression of Θ it is necessary to take into account the second term of the function s_R (R,t), eqs. 3.15.

2. Intermediate zone With an increase of distance from the source, the influence of the second term at the right side of eqs. 3.15 becomes more noticeable. Because of this, with a change of distance R, the shape of wavefields $s_R\ (R,t)$ and $v_R\ (R,t)$ as functions of time also varies. An example of such behavior of displacement is shown in Fig. 3.1c.

3. Far zone As follows from eqs. 3.15, at sufficiently large distance we have

$$s_R(R,t) = \frac{R_0^2\, s_0}{c_l R}\, a\, f'\left[a\left(t - \frac{R}{c_l}\right)\right], \qquad v_R(R,t) = \frac{R_0^2\, s_0}{c_l R}\, a^2\, f''\left[a\left(t - \frac{R}{c_l}\right)\right] \tag{3.22}$$

In this zone, unlike in the previous one, all wavefields decrease with distance relatively slowly. As in the near zone, wave behavior as a function of time is independent of distance from the source. With an increase of R, the curvature of the wave surfaces becomes smaller. This leads to an increase of an area, where the wave can be treated as the plane one with almost the same magnitude.

If we assume that the wave is sinusoidal, then in accordance with eqs. 3.15 we have

$$s_R(R,t) = \frac{R_0^2}{R^2} s_0 \sin\omega\left(t - \frac{R}{c_l}\right) + \frac{R_0^2\, s_0\, \omega}{c_l R} \cos\omega\left(t - \frac{R}{c_l}\right)$$

Correspondingly, the far zone is observed when the distance exceeds the wave length, λ_l:

$$R > \lambda_l \tag{3.23}$$

It is obvious that behavior of particle displacement $\mathbf{s}$ associated with the longitudinal wave is the same an elastic and acoustic media (Part I).

Strain and stress

Now we focus our attention on forces acting on faces of an elementary volume bounded by coordinate surfaces. As was shown in Chapter 2

$$e_{RR} = \frac{\partial s_R}{\partial R}, \qquad e_{\theta\theta} = e_{\varphi\varphi} = \frac{s_R}{R} \tag{3.24}$$

and

$$e_{\theta\varphi} = e_{\varphi R} = e_{R\theta} = 0, \qquad \text{since} \qquad s_\theta = s_\varphi = 0 \tag{3.25}$$

Taking into account eqs. 3.15, we have

$$e_{RR} = -\frac{2R_0^2\, s_0}{R^3}\, f\left[a\left(t-\frac{R}{c_l}\right)\right] - \frac{2R_0^2\, s_0\, a}{c_l R_2}\, f'\left[a\left(t-\frac{R}{c_l}\right)\right] \tag{3.26}$$

$$-\frac{R_0^2\, s_0\, a^2}{c_l^2 R}\, f''\left[a\left(t-\frac{R}{c_l}\right)\right]$$

and

$$e_{\theta\theta} = \frac{R_0^2}{R^3}\, s_0\, f\left[a\left(t-\frac{R}{c_l}\right)\right] + \frac{R_0^2\, s_0\, a}{c_l R^2}\, f'\left[a\left(t-\frac{R}{c_l}\right)\right] = e_{\varphi\varphi}$$

We can again distinguish the near, intermediate, and far zones. In the first two zones, both strains e_{RR} and $e_{\theta\theta}$ ($e_{\varphi\varphi}$) can be comparable, but in the far zone the radial strain becomes dominant:

$$e_{RR} > e_{\theta\theta} \qquad \text{and} \qquad e_{RR} > e_{\varphi\varphi} \tag{3.27}$$

Note that the existence of strains does not require the presence of the corresponding component of the displacement. For instance, particles do not move along the θ and φ coordinate lines ($s_\theta = s_\varphi = 0$), but strains $e_{\theta\theta}$ and $e_{\varphi\varphi}$ differ from zero. In accordance with Hooke's law:

$$\tau_{RR} = \lambda\,\Theta + 2\mu\, e_{RR}, \qquad \tau_{\theta\theta} = \lambda\,\Theta + 2\mu\, e_{\theta\theta}, \qquad \tau_{\varphi\varphi} = \lambda\,\Theta + 2\mu\, e_{\varphi\varphi} \tag{3.28}$$

and

$$\tau_{R\theta} = \mu\, e_{R\theta}, \qquad \tau_{R\varphi} = \mu\, e_{R\varphi}, \qquad \tau_{\theta\varphi} = \mu\, e_{\theta\varphi} \tag{3.29}$$

Since nondiagonal elements of the strain tensor are equal to zero, we have

$$\tau_{R\theta} = \tau_{R\varphi} = \tau_{\theta\varphi} = 0, \tag{3.30}$$

and the elementary volume surrounded by coordinate surfaces is subjected to the action of normal stresses only. As in the case of strains, the behavior of stresses varies depending on distance from the source. For instance, taking into account eqs. 3.17 and 3.26, stresses in the wave zone are

$$\tau_{RR} = \lambda\,\Theta + 2\mu\, e_{RR} = -\frac{R_0^2\, a^2\, s_0\, \rho}{R}\, f''\left[a\left(t-\frac{R}{c_l}\right)\right], \tag{3.31}$$

while

$$\tau_{\theta\theta} = \tau_{\varphi\varphi} = \lambda\Theta \qquad \text{or} \qquad \tau_{\theta\theta} = -\frac{R_0^2\, a^2 \lambda\, s_0}{c_l^2\, R}\, f''\left[a\left(t - \frac{R}{c_l}\right)\right] \tag{3.32}$$

Thus, in the wave zone, stresses decay with distance in the same manner, and the ratio between them is

$$\frac{\tau_{\theta\theta}}{\tau_{RR}} = \frac{\lambda}{c_l^2 \rho} = \frac{\lambda}{\lambda + 2\mu} = \frac{\tau_{\varphi\varphi}}{\tau_{RR}} \tag{3.33}$$

Certainly, unlike the acoustic wave, forces acting on faces of the elementary volume differ from each other. In accordance with eqs. 3.15, the velocity of particles in the wave zone is

$$v_R(R,t) = \frac{R_0^2\, s_0\, a^2}{c_l R}\, f''\left[a\left(t - \frac{R}{c_l}\right)\right] \tag{3.34}$$

and, therefore,

$$\frac{v_R}{\tau_{RR}} = -\frac{1}{c_l \rho} = -\frac{1}{Z}\,, \qquad \frac{v_R}{\tau_{\theta\theta}} = \frac{c_l}{\lambda}\,, \tag{3.35}$$

where $Z = c_l \rho$ is the impedance of a medium.

Spherical source with finite radius

Until now we have implied that the radius of the source, R_0, is very small, and observations are performed at times essentially exceeding R_0/c_l. Next we remove these restrictions and consider sinusoidal waves. Of course, as before we can define the displacement of the source. However, let us approach it differently and assume that the stress

$$\tau_{RR}(R_0, t) = \mathrm{Re}\,\tilde{\tau}_0 e^{-i\omega t} \tag{3.36}$$

is the same at all points of the source surface. As follows from eqs. 3.28,

$$\tau_{RR}(R,t) = \frac{\lambda}{R^2}\frac{\partial}{\partial R}(R^2 s_R) + 2\mu \frac{\partial s_R}{\partial R}$$

or

$$\tau_{RR}(R,t) = (\lambda + 2\mu)\frac{\partial s_R}{\partial R} + \frac{2\lambda}{R}\, s_R \tag{3.37}$$

Since

$$s_R = \frac{\partial U}{\partial R},$$

we have in the vicinity of the source

$$\tau_{RR}(R_0,t) = (\lambda+2\mu)\frac{\partial^2 U}{\partial R^2} + \frac{2\lambda}{R_0}\frac{\partial U}{\partial R} \tag{3.38}$$

This allows us to determine the unknown coefficient A in the expression for the potential

$$U(R,t) = \text{Re}\,(\mathcal{U}\, e^{-i\omega t}), \tag{3.39}$$

where

$$\mathcal{U} = \frac{A}{R}\, e^{ikR} \tag{3.40}$$

is the complex amplitude of the potential and $k = \omega/c_l$ is the wavenumber. Substitution of eq. 3.36 and eq. 3.40 into eq. 3.38 gives

$$\tilde{\tau}_0 = A\left[(\lambda+2\mu)\frac{\partial^2}{\partial R^2}\left(\frac{e^{ikR}}{R}\right) + \frac{2\lambda}{R_0}\frac{\partial}{\partial R}\left(\frac{e^{ikR}}{R}\right)\right]$$

Performing differentiations, we obtain

$$A(\omega) = \frac{\tilde{\tau}_0 R_0 e^{-i\,\omega R_0/c_l}}{\left(\frac{4\mu}{R_0^2} - \rho\,\omega^2\right) - i\,\omega\,\frac{4\mu}{R_0 c_l}} \tag{3.41}$$

Also, it is useful to represent A as:

$$A(\omega) = \frac{\tilde{\tau}_0 R_0 e^{-i\omega R_0/c_l}}{\rho(\omega_{0p}^2 - \omega^2 - 2\,i\,h_p\omega)} \tag{3.42}$$

Here

$$\omega_{0p} = \frac{2c_s}{R_0}, \qquad h_p = \omega_{0p}\frac{c_s}{c_l}, \qquad c_s = \sqrt{\frac{\mu}{\rho}} \tag{3.43}$$

where c_s is the velocity of shear waves. Now it is easy to derive an expression for the potential of a transient wave. Suppose that $\tau_{RR}(R_0,t)$ is an arbitrary function of time and $\tau(\omega)$ is its spectrum. Then, applying Fourier's integral, we obtain

$$U(R,t) = \frac{1}{2\pi R}\int_{-\infty}^{\infty} A\,(\omega)\, e^{-ik(R_0-R)}\, e^{-i\omega t}\, d\omega \tag{3.44}$$

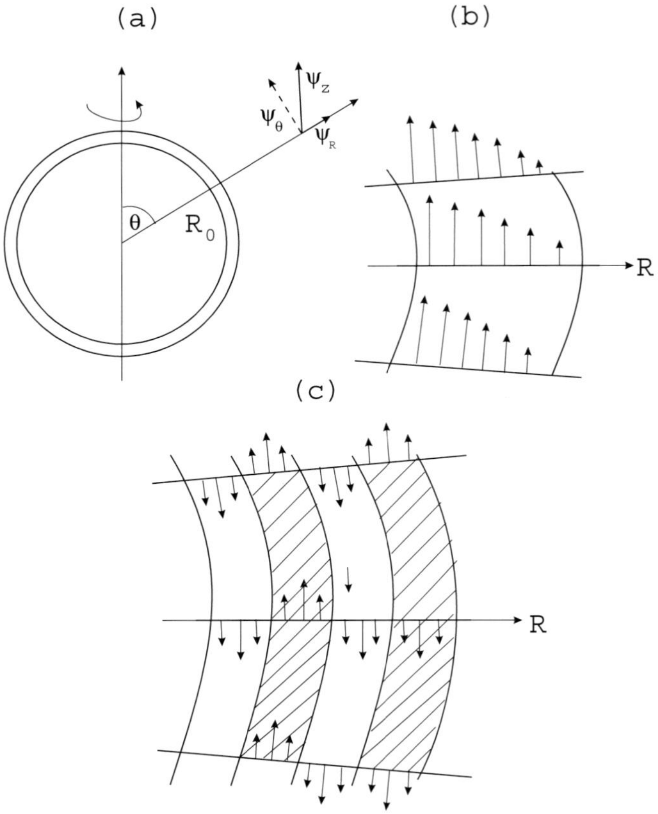

Figure 3.2: (a) Source of shear wave (b) Displacement s_φ in the near zone (c) Zones of a rotation with different directions

3.2 Spherical shear wave in a homogeneous medium

Suppose that a spherical shell or a solid sphere is placed in a homogeneous medium and that it vibrates around the z-axis, Fig. 3.2a. In order to determine wavefields, it is convenient to introduce, as before, the spherical system of coordinates originating at the source center. It is clear that all points of the spherical shell move around the z-axis at the same angle φ, which in general varies with time. Correspondingly, the source displacement has only the φ-component. In particular, for points of its external surface, $R = R_0$, we have:

$$s_0(R_0, t) = R_0 \sin\theta \; f(at) \tag{3.45}$$

Thus, maximal displacement occurs at the equatorial plane, $(\theta = \pi/2)$, while points located at the z-axis remain at rest. Such motion of the source causes in its vicinity deformation of a medium, and therefore the wave arises. To simplify derivations, we make several assumptions. First, assume that as well as the source, particles of a medium have only the φ-component of displacement, but

$$s_R = s_\theta = 0 \tag{3.46}$$

Then, taking into account the symmetry, $(\partial s_\varphi / \partial \varphi = 0)$, the expression for divergence of the field $\mathbf{s}$ has the form

$$\text{div }\mathbf{s} = \frac{1}{R^2 \sin\theta}\frac{\partial}{\partial\varphi}(Rs_\varphi) = \frac{1}{R\sin\theta}\frac{\partial s_\varphi}{\partial\varphi}, \qquad \text{i.e.,} \qquad \text{div }\mathbf{s} = 0 \tag{3.47}$$

This means that propagation of such a wave is not accompanied by a change in the volume, and it suggests that we are dealing with a shear wave.

The potential of the spherical wave

As was demonstrated in the previous chapter, we can represent displacement associated with a shear wave in terms of the vector potential $\boldsymbol{\psi}$:

$$\mathbf{s} = \text{curl }\boldsymbol{\psi}, \tag{3.48}$$

which satisfies the wave equation

$$\nabla^2 \boldsymbol{\psi} = \frac{1}{c_s^2}\frac{\partial^2 \boldsymbol{\psi}}{\partial t^2} \tag{3.49}$$

Here

$$c_s = \sqrt{\frac{\mu}{\rho}} \tag{3.50}$$

is the velocity of propagation of shear waves. Note that function $\mathbf{s}$ also obeys the same wave equation. In principle, we can solve this equation, assuming that vector $\mathbf{s}_\varphi$ depends on the azimuthal angle θ in the same manner as that of the source, eq. 3.45. However, it is simpler to proceed from the vector potential $\boldsymbol{\psi}$. In accordance with eq. 3.48, an infinite number of vectors $\boldsymbol{\psi}$ describe the same field $\mathbf{s}$. This clearly shows that, usually, functions $\boldsymbol{\psi}$ do not have any physical meaning. However, it does not exclude a case in which some vector $\boldsymbol{\psi}$ characterizes a certain physical quantity. Bearing in mind an ambiguity in choosing $\boldsymbol{\psi}$, let us attempt to solve the boundary value problem, provided that $\boldsymbol{\psi}$ has a single component along the axis of rotation, Fig. 3.2a, i.e.,

$$\boldsymbol{\psi} = \psi\,\mathbf{k}, \tag{3.51}$$

and the scalar function ψ depends on time t and coordinate R only. Here $\mathbf{k}$ is the unit vector. Since it has the same direction at all points of a medium, eq. 3.49 is simplified and we have

$$\nabla^2 \psi = \frac{1}{c_s^2} \frac{\partial^2 \psi}{\partial t^2} \tag{3.52}$$

Taking into account our assumptions

$$\frac{\partial \psi}{\partial \theta} = \frac{\partial \psi}{\partial \varphi} = 0,$$

eq. 3.52 in the spherical system of coordinates becomes

$$\frac{1}{R^2} \frac{\partial}{\partial R} \left(R^2 \frac{\partial \psi}{\partial R} \right) = \frac{1}{c_s^2} \frac{\partial^2 \psi}{\partial t^2} \tag{3.53}$$

Eq. 3.53 has exactly the same form as the wave equation for scalar potential that describes the longitudinal spherical wave. For this reason, again applying the substitution $W = R\,\psi$, we arrive at the equation

$$\frac{\partial^2 W}{\partial R^2} = \frac{1}{c_s^2} \frac{\partial^2 W}{\partial t^2}$$

Therefore, the function ψ characterizing the outgoing shear wave is

$$\psi(R,t) = \frac{A_0}{R} f_1 \left[a_1 (t - \frac{R}{c_s}) \right] \tag{3.54}$$

The simplicity of eq. 3.54 is due to several assumptions that require justification. Because of this, our first goal is to demonstrate that the wavefields described by eq. 3.54 satisfy boundary conditions, provided that constants A_0 and a_1, as well as the function

$$f_1 \left[a_1 (t - \frac{R}{c_s}) \right],$$

are properly chosen. First, we will find an expression for displacement. From eq. 3.48, it follows that

$$\mathbf{s} = \frac{1}{R^2 \sin\theta} \begin{vmatrix} \mathbf{i}_R & R\,\mathbf{i}_\theta & R\sin\theta\,\mathbf{i}_\varphi \\ \dfrac{\partial}{\partial R} & \dfrac{\partial}{\partial \theta} & \dfrac{\partial}{\partial \varphi} \\ \psi_R & R\,\psi_\theta & R\sin\theta\,\psi_\varphi \end{vmatrix}, \tag{3.55}$$

where $\mathbf{i}_R$, $\mathbf{i}_\theta$, and $\mathbf{i}_\varphi$ are unit vectors oriented along coordinate lines. As is seen from Fig. 3.2a

$$\psi_R = \psi\cos\theta, \qquad \psi_\theta = -\psi\sin\theta, \qquad \psi_\varphi = 0$$

or

$$\psi_R = \frac{A_0}{R} f_1\left[a_1(t - \frac{R}{c_s})\right]\cos\theta, \qquad \psi_\theta = -\frac{A_0}{R} f_1\left[a_1(t - \frac{R}{c_s})\right]\sin\theta, \qquad \psi_\varphi = 0 \tag{3.56}$$

Substitution of eq. 3.56 into eq. 3.55 gives

$$s_R = 0 \qquad \text{and} \qquad s_\theta = 0, \tag{3.57}$$

which agrees with our assumption about field geometry. Moreover,

$$s_\varphi = \frac{1}{R}\left[\frac{\partial}{\partial R}(R\psi_\theta) - \frac{\partial\psi_R}{\partial\theta}\right]$$

Performing differentiation, we obtain

$$s_\varphi(R,t) = A_0\left\{\frac{1}{R^2} f_1\left[a_1(t - \frac{R}{c_s})\right] + \frac{a_1}{R\,c_s}\, f_1{}'\left[a_1(t - \frac{R}{c_s})\right]\right\}\sin\theta \tag{3.58}$$

It is clear from eq. 3.58 that particle displacement has the same dependence on angle θ as the source. As we did with longitudinal waves, we will determine unknowns from the condition near the source.

Small spherical source

First, consider a transient wave caused by the spherical source with a relatively small radius, R_0 $(R_0 \to 0)$. Then, as follows from eq. 3.58, near the source, displacement is approximately equal to

$$s_\varphi(R,t) = \frac{A_0}{R^2}\, f_1\left[a_1(t - \frac{R}{c_s})\right]\sin\theta \tag{3.59}$$

We suppose that the time of observation greatly exceeds the ratio R/c_s:

$$t \gg \frac{R}{c_s}, \qquad \text{so that} \qquad s_\varphi(R,t) = \frac{A_0\, f_1(a_1 t)}{R^2}\sin\theta, \qquad \text{if } R \sim R_0 \tag{3.60}$$

In the vicinity of the source, particles of a medium move in the same way as the surface of the source, and therefore

$$s_0\, f(at) = \frac{A_0}{R_0^2}\, f_1(a_1 t)$$

It is obvious that this expression is true at any time, if

$$a_1 = a, \qquad f_1(a_1 t) = f(a\, t), \qquad \text{and} \qquad A_0 = R_0^2\, s_0 \tag{3.61}$$

Thus, the expression the displacement becomes

$$s_\varphi(R,t) = R_0^2\, s_0 \left\{ \frac{1}{R^2} f\left[a(t - \frac{R}{c_s})\right] + \frac{a}{R\, c_s}\, f'\left[a(t - \frac{R}{c_s})\right]\right\} \sin\theta \tag{3.62}$$

It is essential that $s_\varphi(R,t)$ obeys all conditions of the boundary value problem and, therefore, in accordance with the theorem of uniqueness, describes the wavefield caused by the given source. In fact, direct substitution shows that vector $\mathbf{s} = s_\varphi \mathbf{i}_\varphi$ satisfies the wave equation

$$\nabla^2 \mathbf{s} = \frac{1}{c_s^2}\frac{\partial^2 \mathbf{s}}{\partial t^2}$$

at regular points. Also, $s_\varphi(R,t)$ vanishes everywhere when $t = 0$. It tends to zero with an increase of distance R and coincides with particle displacement of the source surface if $R = R_0$. This means that all assumptions were correct and the shear wave is accompanied by the field $s_\varphi(R,t)$, eq. 3.62. Taking a derivative from $s_\varphi(R,t)$ with respect to time, we obtain for particle velocity

$$v_\varphi(R,t) = R_0^2\, s_0 \left\{ \frac{a}{R^2} f'\left[a(t - \frac{R}{c_s})\right] + \frac{a^2}{R\, c_s}\, f''\left[a(t - \frac{R}{c_s})\right]\right\} \sin\theta \tag{3.63}$$

Comparison of eqs. 3.62 and 3.63 with eqs. 3.15 in the previous section shows that expressions for displacement and velocity of particles caused by longitudinal and shear waves, respectively, coincide if $c_s = c_l$.

The field s

Now we are ready to describe behavior of the spherical shear wave. For illustration, suppose that motion of the source is

$$s_\varphi(at) = s_0 \begin{cases} 0 & t \le 0 \\ f(at) & 0 \le t \le \tau \\ 0 & t > \tau \end{cases} \tag{3.64}$$

Because of source vibrations, the shear wave appears and moves away from the source. Its phase surface

$$t - \frac{R}{c_s} = \text{const}$$

is spherical, and the motion of particles is tangential to this surface. Unlike the longitudinal wave caused by the spherical source, displacement s_φ varies on the phase surface of the shear wave as $\sin\theta$, and it reaches maximal value at the equatorial plane, $\theta = \pi/2$. Again, as with longitudinal waves, it is natural to distinguish the near, intermediate, and far zones. In the first zone, particle displacement varies synchronously with the source and decays relatively quickly with distance. As an example, distribution of the function s_φ in the near zone between the wavefront and its rear is shown in Fig. 3.2b. In the intermediate zone, the field s_φ is defined by both terms in eq. 3.62, which differently depend on R. Correspondingly, a wave shape varies with distance. Finally, in the far zone the field changes rather slowly, and its behavior is controlled by the derivative of the function $f\,[a(t-R/c_s)]$. Suppose that the shear wave is sinusoidal. As follows from eq. 3.62, we have

$$s_\varphi = s_0 R_0^2 \left[\frac{1}{R^2}\sin\omega(t-\frac{R}{c_s}) + \frac{\omega}{Rc_s}\cos\omega\,(t-\frac{R}{c_s})\right]\sin\theta$$

This clearly shows that as with the longitudinal wave, the far zone exists at distances R exceeding the wavelength λ_s:

$$R > \lambda_s = \frac{c_s}{f}$$

Next consider motion and deformation of an elementary volume of a medium. First of all, since $\text{div}\ \mathbf{s} = 0$, the shear wave does not produce a change in the volume. The elementary volume can experience rotation as well as pure shear and translation (Appendix E). Because $\text{curl}\ \mathbf{s}$ is sensitive only to rotation, it is useful to find its components. By definition we have

$$\text{curl}\ \mathbf{s} = \frac{1}{R^2\sin\theta}\begin{vmatrix} \mathbf{i}_R & R\,\mathbf{i}_\theta & R\sin\theta\,\mathbf{i}_\varphi \\ \dfrac{\partial}{\partial R} & \dfrac{\partial}{\partial\theta} & \dfrac{\partial}{\partial\varphi} \\ 0 & 0 & R\sin\theta\, s_\varphi \end{vmatrix}$$

So

$$\text{curl}_R\ \mathbf{s} = \frac{1}{R\sin\theta}\frac{\partial}{\partial\theta}(\sin\theta\ s_\varphi), \quad \text{curl}_\theta\ \mathbf{s} = -\frac{1}{R}\frac{\partial}{\partial R}(R\ s_\varphi), \quad \text{curl}_\varphi\ \mathbf{s} = 0 \tag{3.65}$$

Making use of eq. 3.62, we obtain

$$\text{curl}_R\ \mathbf{s} = 2R_0^2\ s_0\left\{\frac{1}{R^3}\ f\ \left[a(t-\frac{R}{c_s})\right] + \frac{a}{R^2c_s}f'\left[a(t-\frac{R}{c_s})\right]\right\}\cos\theta \quad \text{and} \tag{3.66}$$

$$\text{curl}_\theta\ \mathbf{s} = R_0^2\ s_0 \left\{ \frac{1}{R^3} f\left[a(t - \frac{R}{c_s})\right] + \frac{a}{R^2 c_s} f'\left[a(t - \frac{R}{c_s})\right] + \frac{a^2}{c_s^2 R} f''\left[a(t - \frac{R}{c_s})\right] \right\} \sin\theta$$

Thus, the axis of rotation of an elementary volume is located in the plane $\varphi = \text{const}$, and its orientation changes from point to point. As follows from eqs. 3.66, the θ-component of curl $\mathbf{s}$ prevails in the far zone, and we have

$$\text{curl}_\theta\ \mathbf{s} = \frac{R_0^2\ s_0\ a^2}{c_s^2 R} \sin\theta\ f''\left[a(t - \frac{R}{c_s})\right] \tag{3.67}$$

Consider the integral

$$I = \int_{t_1-\varepsilon}^{t_2+\varepsilon} \text{curl}_\theta\ \mathbf{s}\ dt$$

Here t_1 and t_2 are arrival times of the front and rear of the wave, respectively, and ε is a very small number. By analogy with the similar integral for dilatation carried out by longitudinal waves, we conclude that

$$I = 0 \tag{3.68}$$

This means that within the spherical wave, there are intervals of distance R with opposite directions of rotation, Fig. 3.2c. However, equality 3.68 is an approximate one, and it becomes more accurate with an increase of R. Taking into account that displacement related to translation obeys the homogeneous system

$$\text{curl}\ \mathbf{s} = 0 \quad \text{and} \quad \text{div}\ \mathbf{s} = 0,$$

we can say that an elementary volume is in general involved in both types of motion: rotation and translation.

Strain and stress

Deformation related to a shear wave may cause only pure shear; otherwise, div $\mathbf{s}$ would not be equal to zero. Proceeding from equations for strain derived in Chapter 2, we have

$$e_{RR} = 0, \quad e_{\theta\theta} = 0, \qquad e_{\varphi\varphi} = 0, \qquad e_{\varphi\theta} = \frac{1}{R}\left(\frac{\partial s_\varphi}{\partial\theta} - s_\varphi \cot\theta\right), \tag{3.69}$$

$$e_{\varphi R} = \frac{\partial s_\varphi}{\partial R} - \frac{s_\varphi}{R}, \qquad e_{R\,\theta} = 0$$

Let us make some comments:

1. The absence of diagonal elements of the strain tensor indicates that a volume does not change.

2. Since neither type of motion causes deformation, shear strains characterize a pure shear, i.e., a change of the angle between neighboring faces of the volume. 3. As we know (Appendix D), a superposition of two types of motion and pure shear is equivalent to simple shear. Substitution of eq. 3.62 into set 3.69 yields

$$e_{\varphi\theta} = 0 \tag{3.70}$$

and

$$e_{\varphi R} = -R_0^2\, s_0 \left\{ \frac{3}{R^3} f\left[a(t - \frac{R}{c_s})\right] + \frac{3a}{R^2 c_s} f'\left[a(t - \frac{R}{c_s})\right] \right. \tag{3.71}$$

$$\left. + \frac{a^2}{Rc_s^2} f''\left[a(t - \frac{R}{c_s})\right] \right\} \sin\theta$$

Thus, only one shear strain, $e_{\varphi R}$, differs from zero. By definition, this means that the angle between coordinate lines R and φ changes, whereas angles formed by lines R and θ, as well as φ and θ, remain equal to $\pi/2$. Of course, if an elementary volume is arbitrarily oriented, then all angles between faces can be distorted. In accordance with Hooke's law, normal stresses are equal to zero, since

$$e_{RR} = e_{\theta\theta} = e_{\varphi\varphi} = 0 \qquad \text{and} \qquad \text{div}\, \mathbf{s} = 0$$

Besides

$$\tau_{R\theta} = \tau_{\varphi\theta} = 0$$

There is only shear stress acting on faces of the volume perpendicular to the coordinate line R. It defines the surface force oriented along the φ-line. From eq. 3.71 and the relationship $\tau_{\varphi R} = \mu\, e_{\varphi R}$, we have

$$\tau_{\varphi R} = -R_0^2\, s_0 \mu \left\{ \frac{3}{R^3} f\left[a(t - \frac{R}{c_s})\right] + \frac{3a}{R^2 c_s} f'\left[a(t - \frac{R}{c_s})\right] \right. \tag{3.72}$$

$$\left. + \frac{a^2}{R\, c_s^2} f''\left[a(t - \frac{R}{c_s})\right] \right\} \sin\theta,$$

and its behavior varies with distance R. In particular, in the far zone,

$$\tau_{\varphi R} = -\frac{R_0^2 \, s_0 \, \mu \, a^2}{R \, c_s^2} f'' \left[a(t - \frac{R}{c_s}) \right] \sin\theta \tag{3.73}$$

As follows from eq. 3.63, the velocity in this zone is

$$v_\varphi(R,t) = \frac{R_0^2 \, s_0 \, a^2}{R \, c_s} f'' \left[a(t - \frac{R}{c_s}) \right] \sin\theta \tag{3.74}$$

Therefore, we have

$$\frac{v_\varphi(R,t)}{\tau_{\varphi R}(R,t)} = -\frac{c_s}{\mu} = -\frac{1}{Z_s} \tag{3.75}$$

Here

$$Z_s = \frac{\mu}{c_s} = \frac{\rho \, c_s^2}{c_s} = \rho \, c_s \tag{3.76}$$

is the impedance of the shear wave in the far zone. Earlier we demonstrated that the Poynting vector is defined as

$$\mathbf{Y} = \tilde{\tau}\mathbf{v}$$

where $\tilde{\tau}$ is the stress tensor, and in our case

$$\mathbf{Y} = \begin{pmatrix} 0 & 0 & \tau_{R\varphi} \\ 0 & 0 & 0 \\ \tau_{R\varphi} & 0 & 0 \end{pmatrix} \begin{pmatrix} 0 \\ 0 \\ v_\varphi \end{pmatrix}$$

Hence

$$Y_R = \tau_{R\varphi} \, v_\varphi, \qquad Y_\theta = 0, \qquad Y_\varphi = 0 \tag{3.77}$$

and, as we can expect, the energy travels only in the radial direction away from the source.

The unknown A_0 In order to determine wavefields for an arbitrarily radius R_0, we assume at the beginning a sinusoidal dependence on time. Then, as with longitudinal waves, the complex amplitude of the z-component of the potential ψ can be written in the form

$$\tilde{\psi} = A_0 \frac{e^{ik_s R}}{R}, \qquad \text{where} \qquad k_s = \frac{\omega}{c_s} \tag{3.78}$$

In accordance with eq. 3.62, the complex amplitude of displacement s_φ is equal to

$$\tilde{s}_\varphi(R,k_s) = A_0 \left(\frac{e^{ik_sR}}{R^2} + \frac{ik_s}{R}\, e^{ik_sR} \right) \sin\theta \tag{3.79}$$

At the surface of the source we must have an equality:

$$\tilde{s}_0 = A_0 \frac{e^{ik_sR_0}}{R_0^2}(i + ik_sR_0)$$

Therefore

$$A_0 = \frac{s_0\ R_0^2}{1 + ik_sR_0} e^{-ik_sR_0} \tag{3.80}$$

Next, suppose that instead of displacement, we know shear stress, $\tau_{\varphi R}$, acting at points of the source surface. As follows from eq. 3.71 and Hooke's law, the function $\tau_{\varphi R}(R_0, t)$ varies as $\sin\theta$, i.e.,

$$\tau_{\varphi R}(R_0, t) = \text{Re}\ \tau_0 e^{-i\omega t} \sin\theta \tag{3.81}$$

Taking into account eq. 3.72, the complex amplitude of $\tau_{\varphi R}$ is

$$\tilde{\tau}_{\varphi R} = -B_0\ \mu \left(\frac{3}{R^3} + \frac{3ik_s}{R^2} - \frac{k_s^2}{R} \right)\ e^{-ik_sR} \sin\theta$$

At the boundary $R = R_0$ it has to coincide with $\tau_{\varphi R}(R_0, t)$, given by eq. 3.81, whence we obtain

$$B_0 = -\frac{\tau_0\ R_0^3\ e^{-ik_sR_0}}{\mu(3 + 3\ ik_sR_0 - k_s^2\ R_0^2)} \tag{3.82}$$

Now, knowing A_0 or B_0 and applying Fourier's integral, we can find the field associated with transient waves.

3.3 The displacement field s caused by the point force

Prior to a study of wave propagation, it is useful from the mathematical and physical points of view to consider the case of equilibrium that follows the dynamic stage. Our goal is to establish a relationship between the displacement field **s** and the given volume force **F**, causing deformation of a medium, i.e., we have to solve the boundary value problem. In order to solve this task, we use (a) the Helmholtz formula, allowing one to represent the vector field as a sum of the source and vortex fields; (b) the solution of the Poisson equation; (c) the condition of equilibrium; and (d) formulas of vector analysis that relate volume and surface integrals.

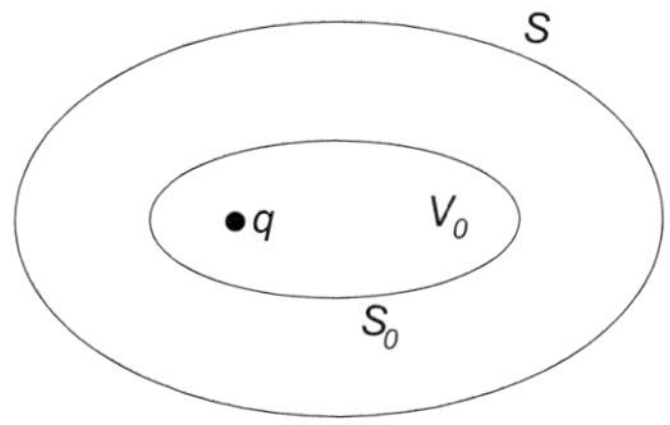

Figure 3.3: Illustration of eq. 3.83

Scalar and vector potentials of the force field F

Suppose that within some portion, V_0, of a homogeneous medium, Fig. 3.3, constant forces are applied and their distribution is known:

$$\mathbf{F}(q) = \rho\, \mathbf{f}(q)\, dV \tag{3.83}$$

Here q is the point of volume V_0 and $\mathbf{f}(q)$ is the force per unit mass

$$\mathbf{f}(q) = f_x(q)\mathbf{i} + f_y(q)\mathbf{j} + f_z(q)\mathbf{k}, \tag{3.84}$$

and $\mathbf{i}$, $\mathbf{j}$, and $\mathbf{k}$ are unit vectors along coordinate axes. As follows from the theory of fields, two equations,

$$\text{curl}\ \mathbf{f} = \mathbf{W} \qquad \text{and} \qquad \text{div}\ \mathbf{f} = \Theta, \tag{3.85}$$

characterize behavior of the field $\mathbf{f}(q)$. By analogy with the field of displacement, we represent the force $\mathbf{f}(q)$ as a sum,

$$\mathbf{f}(q) = \mathbf{f}_1(q) + \mathbf{f}_2(q), \tag{3.86}$$

where

$$\text{curl}\ \mathbf{f}_1 = 0 \qquad \text{div}\ \mathbf{f}_1 = \Theta(q) \tag{3.87}$$

and

$$\text{curl}\ \mathbf{f}_2 = \mathbf{W}(q) \qquad \text{div}\ \mathbf{f}_2 = 0 \tag{3.88}$$

From the first equation of set 3.87, we have

$$\mathbf{f}_1(q) = \text{grad}\ U(q) \tag{3.89}$$

The second equation of the system of eq. 3.88 gives

$$\mathbf{f}_2(q) = \text{curl } \mathbf{A}(q) \tag{3.90}$$

Thus for the total field $\mathbf{f}(q)$, we have

$$\mathbf{f}(q) = \text{grad } U(q) + \text{curl } \mathbf{A}(q) \tag{3.91}$$

Here $U(q)$ and $\mathbf{A}(q)$ are the scalar and vector potentials of the force field $\mathbf{f}(q)$. Certainly there is complete similarity with the displacement field $\mathbf{s}$, which was earlier represented as

$$\mathbf{s} = \text{grad } \varphi + \text{curl } \boldsymbol{\psi} \tag{3.92}$$

Next we arrive at equations describing functions U and A. This procedure was discussed earlier, and substituting eq. 3.89 into the second equation of the 3.87 we obtain

$$\text{div grad } U = \Theta(q) \quad \text{or} \quad \nabla^2 U = \Theta\,(q) \tag{3.93}$$

From the first equation of system 3.88 and eq. 3.90, we have

$$\text{curl curl } \mathbf{A} = \mathbf{W}(q) \tag{3.94}$$

The latter can be presented in a different form. Taking into account the known equality

$$\text{curl curl } \mathbf{A} = \text{grad div } \mathbf{A} - \nabla^2 \ \mathbf{A}, \tag{3.95}$$

eq. 3.94 becomes

$$\text{grad div } \mathbf{A} - \nabla^2 \ \mathbf{A} = \mathbf{W}(q) \tag{3.96}$$

Since

$$\text{curl } \mathbf{A} = \text{curl } (\mathbf{A} + \ \text{grad } T)$$

where $T(q)$ is an arbitrary scalar function, we conclude that an infinite number of vector potentials $\mathbf{A}(q)$ describe the same force field $\mathbf{f}_2(q)$. Among them we choose those that greatly simplify eq. 3.96. To do this, let us assume that

$$\text{div } \mathbf{A} = 0, \tag{3.97}$$

which is usually called the gauge condition. Correspondingly, eq. 3.96 becomes

$$\nabla^2 \mathbf{A} = -\mathbf{W} \tag{3.98}$$

Note that there is also an infinite number of scalar potentials $U(q)$, but they characterize the same force field $\mathbf{f}_1(q)$ and differ from each other by a constant. Thus, we have arrived at Poisson's equations for both potentials

$$\nabla^2 U = \Theta(q) \qquad \text{and} \qquad \nabla^2 \mathbf{A} = -\mathbf{W}(q) \tag{3.99}$$

Here functions $\Theta(q)$ and $\mathbf{W}(q)$ are known, and they are

$$\Theta(q) = \text{div}\,\mathbf{f} = \frac{\partial f_x}{\partial x} + \frac{\partial f_y}{\partial y} + \frac{\partial f_z}{\partial z} \tag{3.100}$$

$$\text{and} \qquad \mathbf{W}(q) = \text{curl}\,\mathbf{f} = \begin{vmatrix} \mathbf{i} & \mathbf{j} & \mathbf{k} \\ \dfrac{\partial}{\partial x} & \dfrac{\partial}{\partial y} & \dfrac{\partial}{\partial z} \\ f_x & f_y & f_z \end{vmatrix}$$

Here $f_x(q)$, $f_y(q)$, and $f_z(q)$ are scalar components of the known field $\mathbf{f}(q)$, which differs from zero in volume V_0. As forces $\mathbf{f}(q)$ are absent outside volume V_0, both potentials obey Laplace equations

$$\nabla^2 U(q) = 0 \quad \text{and} \quad \nabla^2 \mathbf{A}(q) = 0 \qquad \text{if} \qquad V \neq V_0 \tag{3.101}$$

The Poisson and Laplace equations for scalar potential play a fundamental role in the theory of potential fields, such as gravitational and electric fields. For example, in the case of gravitational field $\mathbf{g}(q)$, we have

$$\text{curl}\,\mathbf{g} = 0 \qquad \text{and} \qquad \text{div}\,\mathbf{g} = -4\,\pi\gamma\rho$$

Here γ is the gravitational constant and $\rho(q)$ is the volume density of masses. Correspondingly, the equation for potential, ($\mathbf{g} = \text{grad}\,U$), is

$$\nabla^2 U = -4\,\pi\gamma\rho \tag{3.102}$$

Its fundamental solution is

$$U(p) = \gamma \int\limits_{V_0} \frac{\rho(q)}{R} dV, \tag{3.103}$$

where V_0 is volume occupied by masses, R is the distance between points q and p where p is a point of observation. Substituting eq. 3.103 into eq. 3.102 and performing differentiation and volume integration, it is possible to prove that $U(p)$ obeys the Poisson equation. The same result can be obtained differently. From Newton's law of mass attraction, it follows that potential due to elementary mass is equal to

$$dU(p) = \gamma \frac{\rho(q)dV}{R}$$

Then, applying the principle of superposition, we obtain eq. 3.103. Since this equation describes potential of the gravitational field, it has to satisfy eq. 3.102. Note that point p can be located everywhere – inside volume V_0, at its surface or outside. In the last case, function $U(p)$ obeys the Laplace equation. At the surface, surrounding V_0, the Poisson equation is not defined. In the same manner as the gravitational field, potential $U(p)$ differs from zero outside volume V_0, and this fact is hardly surprising.

When point p is situated inside V_0, the denominator R in eq. 3.103 can tend to zero. However, this singularity is easily removed because the elementary volume near point p decreases more rapidly. These results are entirely applied to potential U of the force field. Comparison of eqs. 3.93 and 3.102 shows that

$$U(p) = -\frac{1}{4\pi} \int\limits_{V_0} \frac{\Theta(q)\, dV}{R} \tag{3.104}$$

Because scalar components of vector potential satisfy the same equations as U:

$$\nabla^2 A_x = -W_x(q), \qquad \nabla^2 A_y = -W_y(q), \qquad \nabla^2 A_z = -W_z(q),$$

we have

$$A_x(p) = \frac{1}{4\pi} \int\limits_{V_0} \frac{W_x(q)\, dV}{R}, \quad A_y(p) = \frac{1}{4\pi} \int\limits_{V_0} \frac{W_y(q)}{R} dV, \quad A_z(p) = \frac{1}{4\pi} \int\limits_{V_0} \frac{W_z(q)}{R} dV$$

or

$$\mathbf{A}(p) = \frac{1}{4\pi} \int\limits_{V_0} \frac{\mathbf{W}(q)}{R} dV \tag{3.105}$$

Thus, we have expressed both potentials in terms of the given force $\mathbf{f}\ (q)$ and, in principle, our first task is solved. However, functions Θ and W contain derivatives from scalar components of $\mathbf{f}$, eq. 3.100, and this is certainly a shortcoming of eqs. 3.104

and 3.105. It is much more attractive to express integrands in these equations in terms of f_x, f_y, and f_z themselves, and this is our next task.

Relationship between potentials and components of volume force f

We use two known formulas of vector analysis (Part I):

$$\frac{1}{R}\overset{q}{\operatorname{div}}\,\mathbf{f} = \overset{q}{\operatorname{div}}\frac{\mathbf{f}}{R} - \mathbf{f}\cdot\overset{q}{\operatorname{grad}}\frac{1}{R} \tag{3.106}$$

and

$$\frac{1}{R}\overset{q}{\operatorname{curl}}\,\mathbf{f} = \overset{q}{\operatorname{curl}}\frac{\mathbf{f}}{R} + \mathbf{f}\times\overset{q}{\operatorname{grad}}\frac{1}{R} \tag{3.107}$$

Here index q means that differentiation is performed with respect to coordinates of point q. Substitution of eq. 3.106 into the integrand of eq. 3.104 gives

$$U(p) = -\frac{1}{4\pi}\int\limits_{V_0}\overset{q}{\operatorname{div}}\left(\frac{\mathbf{f}}{R}\right)dV + \frac{1}{4\pi}\int\limits_{V_0}\mathbf{f}(q)\cdot\overset{q}{\operatorname{grad}}\frac{1}{R}dV \tag{3.108}$$

Beyond volume V_0, force $\mathbf{f}$ is equal to zero. Accounting for this, we can rewrite the first integral as

$$\int\limits_{V_0}\operatorname{div}\frac{\mathbf{f}(q)}{R}dV = \int\limits_{V}\overset{q}{\operatorname{div}}\frac{\mathbf{f}(q)}{R}dV, \tag{3.109}$$

where V_0 is a portion of V.

Now, applying the Gauss formula (Part I), we have

$$\int\limits_{V}\overset{q}{\operatorname{div}}\frac{\mathbf{f}(q)}{R}dV = \oint\limits_{S}\frac{\mathbf{f}(q)\cdot\mathbf{dS}}{R} \tag{3.110}$$

Here S is the surface surrounding volume V and $\mathbf{dS} = dS\mathbf{n}$. The unit vector $\mathbf{n}$ is directed outside volume V, and q is any point of the volume. Taking into account that force $\mathbf{f}(q)$ is equal to zero at points of the surface S, Fig 3.3, we conclude that

$$\int\limits_{V_0}\overset{q}{\operatorname{div}}\frac{\mathbf{f}(q)}{R}dV = 0 \tag{3.111}$$

and therefore

$$U(p) = \frac{1}{4\pi}\int\limits_{V_0}\mathbf{f}(q)\cdot\overset{q}{\nabla}\frac{1}{R}\,dV \tag{3.112}$$

It is also convenient to represent this expression in the form

$$U(p) = -\frac{1}{4\pi} \int\limits_{V_0} \mathbf{f}(q) \cdot \overset{p}{\nabla} \frac{1}{R}\, dV, \tag{3.113}$$

since

$$\overset{q}{\nabla} \frac{1}{R} = -\overset{q}{\nabla} \frac{1}{R} \tag{3.114}$$

and integration and differentiation are performed with respect to different points. By definition we have

$$U(p) = -\frac{1}{4\pi} \int\limits_{V_0} \left[f_x(q) \frac{\partial}{\partial x} \frac{1}{R} + f_y(q) \frac{\partial}{\partial y} \frac{1}{R} + f_z(q) \frac{\partial}{\partial z} \frac{1}{R} \right] dV, \tag{3.115}$$

and it represents the relationship between scalar potential U and components of force $\mathbf{f}(q)$. Here x, y, and z are coordinates of point p.

Next we derive similar expressions for vector potential $\mathbf{A}$. Substitution of eq. 3.107 into eq. 3.105 yields

$$\mathbf{A}(p) = \frac{1}{4\pi} \int\limits_{V_0} \text{curl}\, \frac{\mathbf{f}(q)}{R}\, dV + \frac{1}{4\pi} \int\limits_{V_0} \mathbf{f}(q) \times \overset{q}{\nabla} \frac{1}{R}\, dV \tag{3.116}$$

Applying Stokes' formula of vector analysis (Part I), the first integral of eq. 3.116 becomes

$$\int\limits_{V} \overset{q}{\nabla} \times \frac{\mathbf{f}(q)}{R}\, dV = \oint\limits_{S} \left(\mathbf{n} \times \frac{\mathbf{f}}{R} \right) dS \tag{3.117}$$

As before, S surrounds volume V, and V_0 is its portion, Fig 3.3a, whence

$$\int\limits_{V_0} \overset{q}{\nabla} \times \frac{\mathbf{f}(q)}{R}\, dV = 0$$

We obtain

$$\mathbf{A}(p) = \frac{1}{4\pi} \int\limits_{V_0} \mathbf{f}(q) \times \overset{q}{\nabla} \frac{1}{R}\, dV$$

or

$$\mathbf{A}(p) = \frac{1}{4\pi} \int\limits_{V_0} \overset{p}{\nabla} \frac{1}{R} \times \mathbf{f}(q)\, dV \tag{3.118}$$

This gives

$$A_x(p) = \frac{1}{4\pi} \int\limits_{V_0} \left(f_z \frac{\partial}{\partial y} \frac{1}{R} - f_y \frac{\partial}{\partial z} \frac{1}{R} \right) \, dV, \tag{3.119}$$

$$A_y(p) = \frac{1}{4\pi} \int\limits_{V_0} \left(f_x \frac{\partial}{\partial z} \frac{1}{R} - f_z \frac{\partial}{\partial x} \frac{1}{R} \right) \, dV,$$

$$A_z(p) = \frac{1}{4\pi} \int\limits_{V_0} \left(f_y \frac{\partial}{\partial x} \frac{1}{R} - f_x \frac{\partial}{\partial y} \frac{1}{R} \right) \, dV$$

After performing differentiation with respect to coordinates of point p, it is easy to see that potential $\mathbf{A}(p)$ obeys the gauge condition, eq. 3.97. Both sets of equations – 3.104 and 3.105 and 3.115 and 3.118 allow us to determine scalar and vector potentials. However, the second set is more suitable for our purpose, because it does not require a knowledge of derivatives of force components. Besides, eqs. 3.115 and 3.118 permit us to study one limiting but important case, when the force is applied at the point.

Point force

Again we start from an analogy with the gravitational field and the known expression of the potential:

$$U(p) = \gamma \int\limits_{V_0} \frac{\rho(q) \, dV}{R(q,p)} \tag{3.120}$$

Suppose that the field is considered at distances R, which greatly exceed linear dimensions of the volume occupied by masses, V_0. This means that the denominator R is almost constant, and eq. 3.120 becomes

$$U(p) = \frac{\gamma}{R} \int\limits_{V_0} \rho(q) \, dV = \frac{\gamma \, m}{R}, \tag{3.121}$$

where R is the distance from observation point p to any point q of volume V_0, and m is its mass. Certainly, with an increase of R, eq. 3.121 gives a more accurate value of the potential. It is essential that in such an approximation $U(p)$ depends on mass m, but it is independent of distribution of density $\rho(q)$ and of the size and shape of the volume, provided that m remains the same. For instance, a decrease of volume V_0 and an increase of density $\rho(q)$ do not change the potential, as long as $m =$ const.

Performing this procedure, $(V_0 \to 0)$ in the limit we arrive at the point mass, $(V_0 \to 0,$ $\rho \to \infty)$. Of course, such a mass does not exist, but this concept greatly simplifies calculations of the field when distances from the real mass are sufficiently large. Exactly the same approach is used when we introduce the notion of the point force. Suppose that the force $\mathbf{f}(q)$ is applied in volume V_0 and its magnitude and direction are, in general, functions of point q. Also assume that observation point p is located at a great distance from V_0. Then vector $\overset{p}{\nabla} 1/R$ in eqs. 3.113 and 3.118 is practically constant, and we have

$$U(p) = -\frac{1}{4\pi} \int\limits_{V_0} \mathbf{f}(q)\, dV \cdot \overset{p}{\nabla} \frac{1}{R} \tag{3.122}$$

and
$$\mathbf{A}(p) = -\frac{1}{4\pi} \int\limits_{V_0} \mathbf{f}(q)\, dV \times \overset{p}{\nabla} \frac{1}{R}$$

Here R is the distance between point p and any point q of volume V_0. We see that in such a case, potentials U and $\mathbf{A}$ are independent of a distribution of $\mathbf{f}(q)$, but they are defined by the resultant force $\mathbf{f}^0$, which is equal to

$$\mathbf{f}^{\,0} = \int\limits_{V} \mathbf{f}(q)\, dV \tag{3.123}$$

Similarly to the gravitational field we can imagine that volume V_0 tends to zero, but $\mathbf{f}(q)$ unlimitedly increases, so that $\mathbf{f}^0$ remains the same. In the limit we obtain the point force. Respectively, expressions for potentials, eqs. 3.122 are written in the form:

$$U(p) = -\frac{1}{4\pi} \mathbf{f}^0 \cdot \overset{p}{\nabla} \frac{1}{R} \tag{3.124}$$

and

$$\mathbf{A}(p) = -\frac{1}{4\pi} \mathbf{f}^0 \times \overset{p}{\nabla} \frac{1}{R} \tag{3.125}$$

or

$$U(p) = -\frac{1}{4\pi} \left(f_x^0 \frac{\partial}{\partial x} \frac{1}{R} + f_y^{\,0} \frac{\partial}{\partial y} \frac{1}{R} + f_z^0 \frac{\partial}{\partial z} \frac{1}{R} \right) \tag{3.126}$$

and

$$A_x(p) = \frac{1}{4\pi} \left(f_z^0 \frac{\partial}{\partial y} \frac{1}{R} - f_y^0 \frac{\partial}{\partial z} \frac{1}{R} \right), \quad A_y(p) = \frac{1}{4\pi} \left(f_x^0 \frac{\partial}{\partial z} \frac{1}{R} - f_z^0 \frac{\partial}{\partial x} \frac{1}{R} \right), \tag{3.127}$$

$$A_z(p) = \frac{1}{4\pi}\left(f_y^0 \frac{\partial}{\partial x}\frac{1}{R} - f_x^0 \frac{\partial}{\partial y}\frac{1}{R}\right)$$

For illustration, consider several cases.

Case one Assume that the resultant force $\mathbf{f}^0$ is directed along the x-axis: $\mathbf{f}^0 = (f_x^0,\ 0,\ 0)$. Then eqs. 3.124 and 3.125 are greatly simplified and it gives

$$U(p) = -\frac{f_x^0}{4\pi}\frac{\partial}{\partial x}\frac{1}{R}, \quad A_x = 0, \quad A_y(p) = \frac{f_x^0}{4\pi}\frac{\partial}{\partial z}\frac{1}{R}, \quad A_z(p) = -\frac{f_x^0}{4\pi}\frac{\partial}{\partial y}\frac{1}{R} \tag{3.128}$$

It is clear that

$$\frac{\partial}{\partial x}\frac{1}{R} = -\frac{x - x_q}{R^3}, \qquad \frac{\partial}{\partial y}\frac{1}{R} = -\frac{y - y_q}{R^3}, \qquad \text{and} \qquad \frac{\partial}{\partial z}\frac{1}{R} = -\frac{z - z_q}{R^3}$$

Case two If the direction of $\mathbf{f}^0$ coincides with the y-axis, $\mathbf{f}^0 = (0,\ f_y^0,\ 0)$, we have

$$U(p) = -\frac{f_y^0}{4\pi}\frac{\partial}{\partial y}\frac{1}{R}, \quad A_x(p) = -\frac{f_y^0}{4\pi}\frac{\partial}{\partial z}\frac{1}{R}, \quad A_y(p) = 0, \quad A_z(p) = \frac{f_y^0}{4\pi}\frac{\partial}{\partial x}\frac{1}{R} \tag{3.129}$$

Case three When the force is oriented along the z-axis, $\mathbf{f}^0 = (0,\ 0,\ f_z{}^0)$, we obtain

$$U(p) = -\frac{f_z{}^0}{4\pi}\frac{\partial}{\partial z}\frac{1}{R}, \quad A_x(p) = \frac{f_z{}^0}{4\pi}\frac{\partial}{\partial y}\frac{1}{R}, \quad A_y(p) = -\frac{f_z{}^0}{4\pi}\frac{\partial}{\partial x}\frac{1}{R}, \quad A_z^0(p) = 0 \tag{3.130}$$

Case four Suppose that the point force is applied at the origin of the spherical system of coordinates, $(R,\ \Theta,\ \varphi)$. Taking into account that

$$\overset{p}{\nabla}\frac{1}{R} = -\frac{\mathbf{R}}{R^3}, \tag{3.131}$$

eqs. 3.124 and 3.125 give

$$U(p) = \frac{\mathbf{f}_0 \cdot \mathbf{R}}{4\pi\ R^3}, \qquad \mathbf{A}(p) = \frac{\mathbf{f}_0 \times \mathbf{R}}{4\pi\ R^3} \tag{3.132}$$

For instance, if the force has only the radial component f_R, we have

$$U(p) = \frac{f_R}{4\pi\ R^2} \qquad \text{but} \qquad \mathbf{A}(p) = 0 \tag{3.133}$$

On the contrary, when $\mathbf{f}^0 = f_\varphi{}^0\ \mathbf{i}_\varphi$, we have

$$U(p) = 0, \qquad A_R = A_\varphi = 0, \qquad \text{but} \qquad A_\theta = \frac{f_\varphi{}^0}{4\pi\ R^2} \tag{3.134}$$

The condition of an equilibrium and relationships between potentials of the force and displacement fields

Earlier we found the linkage between potentials U and $\mathbf{A}$ and force components, eqs. 3.115 and 3.119. In order to determine the field of displacement caused by volume force, we have to establish a relationship between potentials of fields $\mathbf{f}$ and $\mathbf{s}$. To do this, we use the condition of equilibrium (Chapter 2):

$$\mu\nabla^2 \mathbf{s} + (\mu + \lambda) \text{ grad div } \mathbf{s} + \rho\, \mathbf{f} = 0 \tag{3.135}$$

This means that the resultant force, which consists of the surface and volume forces, is equal to zero. Substitution of eqs. 3.91 and 3.92 into eq. 3.135 yields

$$\mu\nabla^2(\text{grad } \varphi + \text{curl } \boldsymbol{\psi}) + (\mu + \lambda) \text{ grad div}(\text{grad } \varphi + \text{curl } \boldsymbol{\psi}) + \rho\, (\text{grad } U + \text{curl } \mathbf{A}) = 0$$

Since we can change an order of differentiation and

$$\text{div curl } \boldsymbol{\psi} \equiv \mathbf{0},$$

we have

$$\mu \text{ grad } \nabla^2\, \varphi + \mu \text{ curl } \nabla^2\, \boldsymbol{\psi} + (\mu + \lambda) \text{ grad } \nabla^2\, \varphi + \rho\, (\text{grad } U + \text{curl } \mathbf{A}) = 0$$

or

$$\text{grad } \left[(\lambda + 2\mu)\, \nabla^2\varphi + \rho\, U\right] + \text{ curl } \left[\mu\nabla^2\boldsymbol{\psi} + \rho\mathbf{A}\right] = 0 \tag{3.136}$$

This equality takes place if

$$(\lambda + 2\mu)\, \nabla^2\varphi + \rho\, U = 0 \qquad \text{and} \qquad \mu\, \nabla^2\boldsymbol{\psi} + \rho\, \mathbf{A} = 0 \tag{3.137}$$

or

$$\nabla^2\varphi = -\frac{1}{c_l^2}U, \qquad \nabla^2\boldsymbol{\psi} = -\frac{1}{c_s^2}\mathbf{A} \tag{3.138}$$

Here c_l and c_s are the velocity of propagation of longitudinal and shear waves, respectively. This fact clearly shows that equilibrium occurs as a result of propagation of both types of waves. Note that scalar and vector potentials of displacement field $\mathbf{s}$ are defined only by corresponding potentials of the force field. Finally, Poisson eqs. 3.138 allow us to find functions φ and $\boldsymbol{\psi}$, since U and $\mathbf{A}$ are known and

$$\varphi\,(p) = \frac{1}{4\pi c_l^2}\int\limits_V \frac{U\,(q)}{R}dV, \qquad \boldsymbol{\psi}\,(p) = \frac{1}{4\pi c_s^2}\int\limits_V \frac{\mathbf{A}\,(q)}{R}dV \tag{3.139}$$

Here integration is performed over the whole space because potentials $U(q)$ and $\mathbf{A}(q)$ differ from zero beyond volume V_0. In accordance with eqs. 3.115–3.118, determination of U and $\mathbf{A}$ requires integration over volume V_0. Correspondingly, calculation of potentials of the field $\mathbf{s}$ is in general related to double integration, a rather cumbersome task. This is the main reason why we will pay attention only to the case of the point force.

Displacement $\mathbf{s}$ due to the point force $\mathbf{f}^0$ To begin with, we find expressions for the potentials $\varphi(p)$ and $\boldsymbol{\psi}(p)$ and then derive formulas for the displacement field. Let us assume that force is applied at the origin of coordinates. Taking into account eqs. 3.139 and 3.124–3.125, we have:

$$\varphi(p) = -\frac{1}{16\pi^2 c_l^2}\,\mathbf{f}^0\cdot\int\limits_V \frac{1}{R}\overset{p}{\nabla}\frac{1}{R}dV, \qquad \boldsymbol{\psi}(p) = -\frac{1}{16\pi^2 c_s^2}\,\mathbf{f}^0\times\int\limits_V \frac{1}{R}\overset{p}{\nabla}\frac{1}{R}dV \tag{3.140}$$

Certainly, eqs. 3.140 are much simpler than those in a general case, when it is necessary to perform a double integration. However, we still need to integrate the vector function over a whole space, and it is rather a tedious procedure. Because of this, we will solve our task differently and suppose that force $\mathbf{f}^0$ is directed along the $x-$axis: $\mathbf{f} = (f_x, 0, 0)$. This approach was suggested by Stokes, and it greatly simpifies derivations. Then, as follows from eqs. 3.128 and 3.138,

$$\nabla^2\varphi = \frac{f_x^0}{4\pi c_l^2}\frac{\partial}{\partial x}\frac{1}{R} \tag{3.141}$$

Now we use the relationship, which can be easily checked by differentiation:

$$\nabla^2(\text{grad } R) = 2\text{ grad }\frac{1}{R} \tag{3.142}$$

or

$$\nabla^2\frac{\partial R}{\partial x} = 2\frac{\partial}{\partial x}\frac{1}{R}, \qquad \nabla^2\frac{\partial R}{\partial y} = 2\frac{\partial}{\partial y}\frac{1}{R}, \qquad \nabla^2\frac{\partial R}{\partial z} = 2\frac{\partial}{\partial z}\frac{1}{R}$$

Respectively, eq. 3.141 becomes

$$\nabla^2\varphi = \frac{f_x^0}{8\pi c_l^2}\nabla^2\frac{\partial R}{\partial x}$$

Therefore,

$$\varphi(p) = \frac{f_x^0}{8\pi c_l^2}\frac{\partial R}{\partial x} \tag{3.143}$$

As follows from the same eqs. 3.128 and 3.138:

$$\nabla^2\psi_x = 0, \qquad \nabla^2\psi_y = -\frac{A_y}{c_s^2}, \qquad \nabla^2\psi_z = -\frac{A_z}{c_s^2}$$

or

$$\nabla^2\psi_x = 0, \qquad \nabla^2\psi_y = -\frac{f_x^0}{4\pi c_s^2}\frac{\partial}{\partial z}\frac{1}{R}, \qquad \nabla^2\psi_z = \frac{f_x^0}{4\pi c_s^2}\frac{\partial}{\partial y}\frac{1}{R}$$

whence

$$\psi_x = 0, \qquad \psi_y(p) = -\frac{f_x^0}{8\pi c_s^2}\frac{\partial R}{\partial z}, \qquad \psi_z(p) = \frac{f_x^0}{8\pi c_s^2}\frac{\partial R}{\partial y} \tag{3.144}$$

Now we are finally ready to determine field **s**. From eq. 3.92, we have:

$$\mathbf{s} = \text{grad}\,\varphi + \text{curl}\,\boldsymbol{\psi},$$

i.e.,

$$u(p) = \frac{\partial\varphi}{\partial x} + \frac{\partial\psi_z}{\partial y} - \frac{\partial\psi_y}{\partial z}, \qquad v(p) = \frac{\partial\varphi}{\partial y} - \frac{\partial\psi_z}{\partial x}, \qquad w(p) = \frac{\partial\varphi}{\partial z} + \frac{\partial\psi_y}{\partial x} \tag{3.145}$$

Substitution of eqs. 3.143 and 3.144 into eq. 3.145 yields

$$u(p) = \frac{f_x^0}{8\pi}\left(\frac{1}{c_l^2} - \frac{1}{c_s^2}\right)\frac{\partial^2 R}{\partial x^2} + \frac{f_x^0}{8\pi c_s^2}\nabla^2 R$$

Since

$$\nabla^2 R = \frac{2}{R},$$

we have

$$u(p) = \frac{f_x^0}{8\pi}\left(\frac{1}{c_l^2} - \frac{1}{c_s^2}\right)\frac{\partial^2 R}{\partial x^2} + \frac{f_x^0}{4\pi c_s^2}\frac{1}{R}, \tag{3.146}$$

also

$$v(p) = \frac{f_x^0}{8\pi}\left(\frac{1}{c_l^2} - \frac{1}{c_s^2}\right)\frac{\partial^2 R}{\partial x\partial y}, \tag{3.147}$$

and

$$w(p) = \frac{f_x^0}{8\pi}\left(\frac{1}{c_l^2} - \frac{1}{c_s^2}\right)\frac{\partial^2 R}{\partial x\partial z} \tag{3.148}$$

We see that the point force f_x^0 causes displacement of particles in all directions, and, as was pointed out earlier, this field arises as a result of superposition of the longitudinal and shear waves. It is also instructive to determine the dilatation and curl of field $\mathbf{s}$. Performing differentiations, we obtain

$$\text{div } \mathbf{s} = \frac{f_x^0}{4\pi c_l^2} \frac{\partial}{\partial x} \frac{1}{R}$$

or

$$\text{div } \mathbf{s} = -\frac{f_x^0}{4\pi c_l^2} \frac{x}{R^3} \tag{3.149}$$

and

$$\text{curl}_x \mathbf{s} = 0, \qquad \text{curl}_y \mathbf{s} = \frac{f_x^0}{4\pi c_s^2} \frac{\partial}{\partial z} \frac{1}{R}, \qquad \text{curl}_z \mathbf{s} = -\frac{f_x^0}{4\pi c_s^2} \frac{\partial}{\partial y} \frac{1}{R}$$

or

$$\text{curl}_x \mathbf{s} = 0, \qquad \text{curl}_y \mathbf{s} = -\frac{f_x^0}{4\pi c_s^2} \frac{z}{R^3}, \qquad \text{curl}_z \mathbf{s} = \frac{f_x^0}{4\pi c_s^2} \frac{y}{R^3} \tag{3.150}$$

Thus, in general, the point force produces both deformation of an elementary volume and its rotation. It is easy to derive by analogy expressions for field $\mathbf{s}$ when the point force is directed along either the $y-$ or $z-$axis.

3.4 Propagation of waves caused by the point force

As in the previous section, we assume that force $\mathbf{F}$ is applied in the vicinity of the coordinate origin, and it has the x-component only:

$$\mathbf{F} = (\mathbf{F}_x, \ 0, \ 0) \tag{3.151}$$

Here $\mathbf{F}_x$ is an arbitrary function of time. To begin with, we will use results derived from studying the displacement field in equilibrium. First of all, let us introduce the potentials of the body force $\mathbf{f}$ acting on unit mass,

$$\mathbf{f} = \text{grad } U + \text{curl } \mathbf{A}, \tag{3.152}$$

where all three functions, $\mathbf{f}$, U, and $\mathbf{A}$, depend on time and coordinates of a point. Taking a divergence and after it curl from both sides of eq. 3.152, we obtain Poisson's equations for both potentials:

$$\nabla^2 U = \text{div } \mathbf{f}, \qquad \nabla^2 \mathbf{A} = -\text{curl } \mathbf{f} \tag{3.153}$$

The last equality is valid, provided that

$$\text{div}\ \mathbf{A} = 0$$

These equations are exactly the same as in the case of equilibrium, and, correspondingly, for the point force f_x^0 at the origin, eq. 3.128, we have

$$U(q,t) = -\frac{f_x{}^0(t)}{4\pi}\frac{\partial r_0^{-1}}{\partial x'}, \qquad A_x = 0, \qquad A_y(q,t) = \frac{f_x{}^0(t)}{4\pi}\frac{\partial r_0^{-1}}{\partial z'}, \tag{3.154}$$

$$A_z(q,t) = -\frac{f_x{}^0(t)}{4\pi}\frac{\partial r_0^{-1}}{\partial y}$$

Here r_0 is the distance from the origin to an arbitrary point $q(x',\ y',\ z')$. Note that potentials $U(q,t)$ and $\mathbf{A}(q,t)$ synchronously change with the force applied at the origin and this happens regardless of the position of point q. This fact vividly demonstrates the auxiliary character of functions U and $\mathbf{A}$.

Scalar and vector potentials of field s

The relationship between the displacement $\mathbf{s}$ and its potentials,

$$\mathbf{s} = \text{grad}\ \varphi + \text{curl}\ \boldsymbol{\psi}, \tag{3.155}$$

is always valid, since it follows from the system of equations

$$\text{curl}\ \mathbf{s} = \mathbf{W} \qquad \text{and} \qquad \text{div}\ \mathbf{s} = \Theta$$

Now, as in the case of equilibrium, we establish a linkage between the potential of fields $\mathbf{f}(q,t)$ and $\mathbf{s}(q,t)$. With this purpose in mind, we use the equation of motion (Chapter 2):

$$(\lambda+\mu)\text{grad div}\ \mathbf{s} + \mu\nabla^2\mathbf{s} + \rho\,\mathbf{f} = \rho\frac{\partial^2\mathbf{s}}{\partial t^2} \tag{3.156}$$

Substitution of eqs. 3.155 and 3.152 into eq. 3.156 yields

$$(\lambda+\mu)\text{grad div}(\text{grad}\varphi + \text{curl}\ \boldsymbol{\psi}) + \mu\nabla^2(\text{grad}\varphi + \text{curl}\ \boldsymbol{\psi})$$

$$+\rho(\text{grad}\ U + \text{curl}\ \mathbf{A}) = \rho\frac{\partial^2}{\partial t^2}(\text{grad}\varphi + \text{curl}\ \boldsymbol{\psi})$$

or

$$(\lambda+2\mu)\mathrm{grad}\nabla^2\varphi+\rho\ \mathrm{grad}\ U+\mu\ \mathrm{curl}\ \nabla^2\boldsymbol{\psi}+\rho\ \mathrm{curl}\ \mathbf{A}=\rho\ \mathrm{grad}\frac{\partial^2\varphi}{\partial t^2}+\rho\ \mathrm{curl}\ \frac{\partial^2\boldsymbol{\psi}}{\partial t^2}$$

This means that

$$(\lambda+2\mu)\nabla^2\varphi+\rho\, U=\rho\frac{\partial^2\varphi}{\partial t^2}, \qquad \mu\nabla^2\boldsymbol{\psi}+\rho\,\mathbf{A}=\rho\frac{\partial^2\boldsymbol{\psi}}{\partial t^2},$$

that is,

$$\nabla^2\varphi-\frac{1}{c_l^2}\frac{\partial^2\varphi}{\partial t^2}=-\frac{1}{c_l^2}U \tag{3.157}$$

and

$$\nabla^2\boldsymbol{\psi}-\frac{1}{c_s^2}\frac{\partial^2\boldsymbol{\psi}}{\partial t^2}=-\frac{1}{c_s^2}\mathbf{A} \tag{3.158}$$

These are inhomogeneous wave equations, and their right sides are represented by potentials of the body force. Earlier, we learned that longitudinal and shear waves caused by the point spherical source obey homogeneous wave equations at regular points. As we know, the solutions for scalar potential φ and a scalar component of $\boldsymbol{\psi}$ have the forms

$$\frac{C_1}{R}f\left(t-\frac{R}{c_l}\right) \quad \text{and} \quad \frac{C_2}{R}f\left(t-\frac{R}{c_s}\right),$$

respectively, whereas coefficients C_1 and C_2 are determined from the condition near the source. This procedure is equivalent to solving an inhomogeneous wave equation with the given right side. Then, applying the principle of superposition, we see that functions

$$\varphi(p,t)=\frac{1}{4\pi c_l^2}\int\limits_V\frac{1}{r}U\left(t-\frac{r}{c_l}\right)\,dx'\,dy'\,dz' \tag{3.159}$$

and

$$\boldsymbol{\psi}(p,t)=\frac{1}{4\pi c_s^2}\int\limits_V\frac{1}{r}\mathbf{A}\left(t-\frac{r}{c_l}\right)\,dx'\,dy'\,dz' \tag{3.160}$$

obey their corresponding wave equations (eqs. 3.157 and 3.158). Here r is the distance between points $q\,(x',y',z')$ and $p\,(x,y,z)$, Fig. 3.4a. First we focus on scalar potential. In order to perform an integration, we imagine a medium as a system of thin spherical

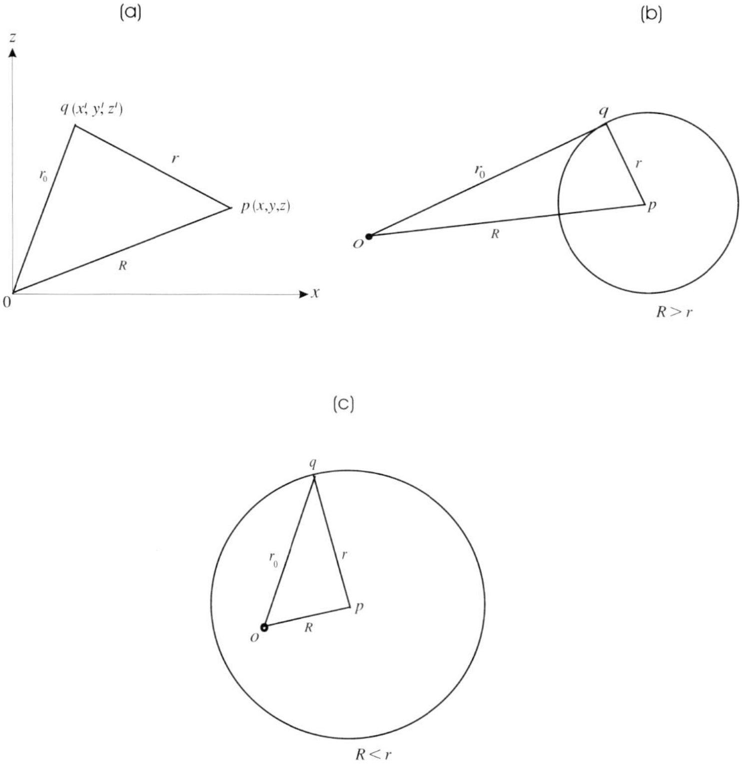

Figure 3.4: Mutual position of points p and q and origin 0

shells having the point p as the center, Fig. 3.4b. Their radius, r, changes from zero to infinity: $0 \leq r < \infty$.

Let r be the radius of one such shell and dr its thickness, while dS is the elementary surface. Taking into account eqs. 3.154 and bearing in mind that the product

$$\frac{1}{r} f_x \left(t - \frac{r}{c_l} \right)$$

is constant at points of the shell surface, we can write

$$\varphi(p,t) = -\frac{1}{16\pi^2 c_l^2} \int_0^\infty \frac{1}{r} f_x \left(t - \frac{r}{c_l} \right) dr \oint_S \frac{\partial r_0^{-1}}{\partial x'} dS \tag{3.161}$$

Calculation of the integral I

First, consider the surface integral

$$I = \oint\limits_S \frac{\partial r_0^{-1}}{\partial x'} dS, \tag{3.162}$$

where

$$r_0 = (x'^{\,2} + y'^{\,2} + z'^{\,2})^{1/2},$$

as is shown in Fig. 3.4a. The integrand can be written as

$$\frac{\partial r_0^{-1}}{\partial x'} = -\frac{x'}{r_0^3} = -\frac{\cos(\mathbf{r}_0, \mathbf{i})}{r_0^2} \tag{3.163}$$

Here

$$\frac{x'}{r_0} = \cos(\mathbf{r}_0, \mathbf{i}) \tag{3.164}$$

is the cosine of the angle between the unit vector $\mathbf{r}_0$ and the x-axis. Therefore

$$I = -\oint\limits_S \frac{x'}{r_0^3} dS = -\oint\limits_S \frac{\cos(\mathbf{r}_0, \mathbf{i})}{r_0^2} dS \tag{3.165}$$

To calculate this integral, we recall the behavior of the gravitational field when a mass is uniformly distributed over a spherical surface. In accordance with Newton's law of attraction and the principle of superposition, the field at the origin of coordinates due to such a mass is

$$\mathbf{g}(0) = -\gamma\rho \oint\limits_S \frac{\mathbf{r}_0(q)}{r_0^2(q)} dS \tag{3.166}$$

Here $\mathbf{r}_0$ is the unit vector directed from the origin to a point of the surface, q. As follows from eq. 3.166, the x-component of the field the point 0 is

$$q_x(0) = -\gamma\rho \oint\limits_S \frac{\mathbf{r}_0(q) \cdot \mathbf{i}}{r_0^2} dS = -\gamma\rho \oint\limits_S \frac{\cos(\mathbf{r}_0 \cdot \mathbf{i})}{r_0^2}\, dS$$

or

$$q_x(0) = \gamma\rho\, I \tag{3.167}$$

The integral I characterizes the x-component of the field caused by the spherical mass, provided that the surface density ρ is constant.

It is clear that with an increase of the shell radius r, the origin of coordinates can be located either outside or inside a shell. Both cases are shown in Fig. 3.4b,c. As has been pointed out, it is not easy directly find an analytical expression for I. However, equality 3.167 allows us to solve this task in a very simple manner. In fact, from symmetry it follows that in the spherical system of coordinates with the origin at point $p(x, y, z)$, the field has only the radial component

$$\mathbf{g} = (q_r, 0,\ 0)$$

This means that field $\mathbf{g}$ is perpendicular at the coordinate surface $r =$ const. Then, use of Gauss's formula

$$\oint_S \mathbf{g} \cdot \mathbf{dS} = -4\ \pi\gamma\ m,$$

where m is mass inside S and $\mathbf{dS} = dS\ \mathbf{n}$, shows that at each point inside the shell the field vanishes, since $m = 0$, i.e.,

$$I = 0 \qquad \text{if} \qquad R < r \tag{3.168}$$

At the same time, outside the shell, the field coincides with that of elementary mass located at point p. In particular, at the origin of the Cartesian system we have

$$\mathbf{g}(0) = \gamma \frac{m}{R^2}\ \mathbf{R}_0 \tag{3.169}$$

Here $\mathbf{R}_0$ is the unit vector directed from the origin to point p. Since $m = 4\pi r^2 \rho$, we have

$$\mathbf{g}(0) = 4\pi r^2 \gamma\ \rho \frac{\mathbf{R}_0}{R^2}$$

Correspondingly,

$$q_x(0) = \mathbf{q}(0) \cdot \mathbf{i} = 4\pi\ r^2\ \gamma\rho\ \frac{\mathbf{R}_0 \cdot \mathbf{i}}{R^2}$$

or

$$q_x(0) = 4\pi\gamma\rho\ r^2 \frac{\cos(\mathbf{R}_0 \cdot \mathbf{i})}{R^2} = -4\pi\ \gamma\rho\ r^2\ \frac{\partial}{\partial x} R_0^{-1} \tag{3.170}$$

Comparison with eq. 3.167 gives

$$I = 4\pi r^2\ \frac{\partial R^{-1}}{\partial x} \qquad \text{if} \qquad r < R \tag{3.171}$$

Substituting eqs. 3.168 and 3.171 into eq. 3.161 and introducing a new variable, $\tau = R/c_l$, we obtain

$$\varphi(p,t) = -\frac{1}{4\pi}\frac{\partial R^{-1}}{\partial x}\int_0^{R/c_l} \tau f_x^{\,0}(t-\tau)\,d\tau \tag{3.172}$$

In the same manner, components of the vector potential can be represented as

$$\psi_x = 0, \qquad \psi_y(p,t) = \frac{1}{4\pi}\frac{\partial R^{-1}}{\partial z}\int_0^{R/c_s} \tau f_x^0(t-\tau)\,d\tau, \tag{3.173}$$

and

$$\psi_z(p,t) = -\frac{1}{4\pi}\frac{\partial R^{-1}}{\partial y}\int_0^{r/c_l} \tau f_x^0(t-\tau)\,d\tau$$

Displacement components

In deriving expressions for displacement components, we proceed from eqs. 3.155 and 3.172–3.173, as well as from the equality

$$\frac{\partial}{\partial x}\int_0^{R/c} \tau\, f_x^{\,0}(t-\tau)\,d\tau = \frac{R}{c} f_x^{\,0}(t-\frac{R}{c})\,\frac{1}{c}\frac{\partial R}{\partial x} \tag{3.174}$$

This equality is obtained using the rule of differentiation of integrals with respect to the upper limit. Since determination of **s** requires some special effort, let us for illustration consider the component u, which is related to potentials as

$$u = \frac{\partial\varphi}{\partial x} + \frac{\partial\psi_z}{\partial y} - \frac{\partial\psi_y}{\partial z}$$

Performing differentiations, we have

$$\frac{\partial\varphi}{\partial x} = -\frac{1}{4\pi}\frac{\partial^2 R^{-1}}{\partial x^2}\int_0^{R/c_l} \tau f_x^{\,0}(t-\tau)\,d\tau - \frac{1}{4\pi}\frac{\partial R^{-1}}{\partial x}\frac{R}{c_l^2}\, f_x^{\,0}(t-\frac{R}{c_l})\,\frac{\partial R}{\partial x},$$

$$\frac{\partial\psi_z}{\partial y} = -\frac{1}{4\pi}\frac{\partial^2 R^{-1}}{\partial y^2}\int_0^{R/c_s} \tau f_x^{\,0}(t-\tau)\,d\tau - \frac{1}{4\pi}\frac{\partial R^{-1}}{\partial y}\frac{R}{c_s^2}\, f_x^{\,0}(t-\frac{R}{c_s})\,\frac{\partial R}{\partial y}, \tag{3.175}$$

$$\frac{\partial \psi_y}{\partial z} = \frac{1}{4\pi}\frac{\partial^2 R^{-1}}{\partial z^2}\int\limits_0^{R/c_s} \tau f_x^{\,0}(t-\tau)\, d\tau + \frac{1}{4\pi}\frac{\partial R^{-1}}{\partial z}\frac{R}{c_s^2}\, f_x^{\,0}(t-\frac{R}{c_s})\,\frac{\partial R}{\partial z}$$

Because

$$\int\limits_0^{R/c_s} \tau f_x^{\,0}(t-\tau)\, d\tau = \int\limits_0^{R/c_l} \tau f_x^{\,0}(t-\tau)\, d\tau + \int\limits_{R/c_l}^{R/c_s} \tau f_x^{\,0}(t-\tau)\, d\tau, \tag{3.176}$$

the sum of terms containing the integral

$$\int\limits_0^{R/c_l} \tau f_x^{\,0}(t-\tau)\, d\tau$$

is

$$-\frac{1}{4\pi}\left(\frac{\partial^2 R^{-1}}{\partial x^2} + \frac{\partial^2 R^{-1}}{\partial y^2} + \frac{\partial^2 R^{-1}}{\partial z^2}\right)\int\limits_0^{R/c_l} \tau f_x^{\,0}(t-\tau)\, d\tau = \tag{3.177}$$

$$-\frac{1}{4\pi}\left(\nabla^2 \frac{1}{R}\right)\int\limits_0^{R/c_l} \tau f_x^{\,0}(t-\tau)\, d\tau = 0,$$

since

$$\nabla^2 1/R = 0 \tag{3.178}$$

Then, substitution of eq. 3.176 into last two derivatives of set 3.175 yields:

$$u(p,t) = \frac{1}{4\pi R\, c_l^2}\left(\frac{\partial R}{\partial x}\right)^2 f_x^0\left(t-\frac{R}{c_l}\right) + \frac{1}{4\pi R\, c_s^2}\left(\frac{\partial R}{\partial y}\right)^2 f_x^0\left(t-\frac{R}{c_s}\right) +$$

$$\frac{1}{4\pi R\, c_s^2}\left(\frac{\partial R}{\partial z}\right)^2 f_x^0\left(t-\frac{R}{c_s}\right) - \frac{1}{4\pi}\left(\frac{\partial^2 R^{-1}}{\partial y^2} + \frac{\partial^2 R^{-1}}{\partial z^2}\right)\int\limits_{R/c_l}^{R/c_s} \tau f_x^0(t-\tau)\, d\tau \tag{3.179}$$

Taking into account eq. 3.178 and the equality

$$\left(\frac{\partial R}{\partial x}\right)^2 + \left(\frac{\partial R}{\partial y}\right)^2 + \left(\frac{\partial R}{\partial z}\right)^2 = 1,$$

we finally have

$$u(p,t) = \frac{1}{4\pi R\, c_s^2}\, f_x^{\,0}\left(t - \frac{R}{c_s}\right) + \frac{1}{4\pi R}\left(\frac{\partial R}{\partial x}\right)^2 \cdot \tag{3.180}$$

$$\left[\frac{1}{c_l^2} f_x^0\left(t - \frac{R}{c_l}\right) - \frac{1}{c_s^2} f_x^0\left(t - \frac{R}{c_s}\right)\right] + \frac{1}{4\pi}\frac{\partial^2 R^{-1}}{\partial x^2} \int\limits_{R/c_l}^{R/c_s} \tau f_x^{\,0}(t-\tau)\, d\tau$$

In a similar manner, we obtain formulas for two other components of displacement:

$$v(p,t) = \frac{1}{4\pi R}\frac{\partial R}{\partial x}\,\frac{\partial R}{\partial y}\left[\frac{1}{c_l^2} f_x^0\left(t - \frac{R}{c_l}\right) - \frac{1}{c_s^2} f_x^0\left(t - \frac{R}{c_s}\right)\right]$$

$$+ \frac{1}{4\pi}\frac{\partial^2 R^{-1}}{\partial x \partial y} \int\limits_{R/c_l}^{R/c_s} \tau f_x^{\,0}(t-\tau)\, d\tau \qquad \text{and} \tag{3.181}$$

$$w(p,t) = \frac{1}{4\pi R}\frac{\partial R}{\partial x}\,\frac{\partial R}{\partial z}\cdot\left[\frac{1}{c_l^2} f_x^0\left(t - \frac{R}{c_l}\right) - \frac{1}{c_s^2} f_x^0\left(t - \frac{R}{c_s}\right)\right] + \frac{1}{4\pi}\frac{\partial^2 R^{-1}}{\partial x \partial z} \int\limits_{R/c_l}^{R/c_s} \tau f_x^0(t-\tau) d\tau$$

Thus, we have found the field of displacement $\mathbf{s}$ caused by the point force at any point of a homogeneous medium. Correspondingly, it is possible to determine stresses and strains, as well as the divergence and curl of field $\mathbf{s}$. Now, making use of eqs. 3.180 and 3.181, we represent field $\mathbf{s}$ as a sum of three terms,

$$\mathbf{s} = \mathbf{s}_1 + \mathbf{s}_2 + \mathbf{s}_3, \tag{3.182}$$

where $\mathbf{s}_1$ is displacement along the radius-vector $\mathbf{R}$, $\mathbf{s}_2$ is located in the plane normal to $\mathbf{R}$, and $\mathbf{s}_3$ is the vector that can be arbitrarily oriented with respect to $\mathbf{R}$. In order to find these terms, we use relations between components of any vector $\mathbf{M}$ in the spherical and Cartesian systems of coordinates:

$$M_R = M_x \sin\theta\cos\varphi + M_y \sin\theta\sin\varphi + M_z\cos\theta$$

$$M_\theta = M_x \cos\theta\cos\varphi + M_y \cos\theta\sin\varphi - M_z\sin\theta \tag{3.183}$$

$$M_\varphi = -M_x \sin\varphi + M_y \cos\varphi$$

We also need to express derivatives $\partial R/\partial x$, $\partial R/\partial y$, and $\partial R/\partial z$ in terms of θ and φ. Since

$$x = R\sin\theta\cos\varphi, \qquad y = R\sin\theta\sin\varphi, \qquad \text{and} \qquad z = R\cos\theta,$$

we have

$$\frac{\partial R}{\partial x} = \frac{x}{R} = \sin\theta\cos\varphi, \qquad \frac{\partial R}{\partial y} = \frac{y}{R} = \sin\theta\sin\varphi, \qquad \frac{\partial R}{\partial z} = \frac{z}{R} = \cos\theta \tag{3.184}$$

First, collect terms in eqs. 3.180 and 3.181 that are proportional to

$$f_x^0\left(t - \frac{R}{c_l}\right)$$

This gives

$$\mathbf{s}_1(p,t) = \frac{1}{4\pi R\, c_l^2}\, f_x^{\,0}\left(t - \frac{R}{c_l}\right)\frac{\partial R}{\partial x}\left(\frac{\partial R}{\partial x}\,\mathbf{i} + \frac{\partial R}{\partial y}\,\mathbf{j} + \frac{\partial R}{\partial z}\,\mathbf{k}\right)$$

It is clear that $\partial R/\partial x$, $\partial R/\partial y$, and $\partial R/\partial z$ are directional cosines of unit vector $\mathbf{i}_R$ directed along the radius-vector $\mathbf{R}$. Thus

$$\mathbf{s}_1(p,t) = \frac{1}{4\pi R\, c_l^2}\,\mathbf{N}_1\left(t - \frac{R}{c_l}\right) \tag{3.185}$$

Here

$$\mathbf{N}_1\left(t - \frac{R}{c_l}\right) = f_x^{\,0}\left(t - \frac{R}{c_l}\right)\frac{\partial R}{\partial x}\,\mathbf{i}_R \tag{3.186}$$

is the vector component of the point force in the radial direction.

Next, we find the part of the displacement that is proportional to the function

$$f_x^0\left(t - \frac{R}{c_s}\right)$$

In accordance with eqs. 3.180 and 3.181, we have

$$\begin{aligned}\mathbf{s}_2(p,t) &= \frac{1}{4\pi R\, c_s^2}\left\{\left[1 - \left(\frac{\partial R}{\partial x}\right)^2\right]\mathbf{i} - \frac{\partial R}{\partial x}\frac{\partial R}{\partial y}\,\mathbf{j} - \frac{\partial R}{\partial x}\frac{\partial R}{\partial z}\,\mathbf{k}\right\} f_x^0\left(t - \frac{R}{c_s}\right)\\ &\quad \frac{1}{4\pi R\, c_s^2}\left[\mathbf{i} - \frac{\partial R}{\partial x}\left(\frac{\partial R}{\partial x}\,\mathbf{i} + \frac{\partial R}{\partial y}\,\mathbf{j} + \frac{\partial R}{\partial z}\,\mathbf{k}\right)\right] f_x^0\left(t - \frac{R}{c_s}\right)\end{aligned}$$

or

$$\mathbf{s}_2(p,t) = \frac{1}{4\pi R\, c_s^2}\left(\mathbf{i} - \frac{\partial R}{\partial x}\mathbf{i}_R\right)\, f_x^0\left(t - \frac{R}{c_s}\right) \tag{3.187}$$

As follows from eqs. 3.183, unit vector $\mathbf{i}$ in the spherical system of coordinates is

$$\mathbf{i} = \sin\theta\cos\varphi\;\mathbf{i}_R + \cos\theta\cos\varphi\;\mathbf{i}_\theta - \sin\varphi\;\mathbf{i}_\varphi$$

Therefore, eq. 3.187 becomes

$$\mathbf{s}_2(p,t) = \frac{1}{4\pi R\, c_s^2}\mathbf{N}_2\left(t - \frac{R}{c_s}\right), \tag{3.188}$$

where

$$\mathbf{N}_2\left(t - \frac{R}{c_s}\right) = f_x^0\left(t - \frac{R}{c_s}\right)(\cos\theta\cos\varphi\;\mathbf{i}_\theta - \sin\varphi\;\mathbf{i}_\varphi) \tag{3.189}$$

is the vector component of the force in the direction perpendicular to the radius-vector $\mathbf{R}$. Finally, the sum of terms in eqs. 3.180 and 3.181 containing the integral is

$$\mathbf{s}_3(p,t) = \frac{1}{4\pi}\left(\frac{\partial^2 R^{-1}}{\partial x^2}\mathbf{i} + \frac{\partial^2 R^{-1}}{\partial x \partial y}\mathbf{j} + \frac{\partial^2 R^{-1}}{\partial x \partial z}\mathbf{k}\right)\int\limits_{R/c_l}^{R/c_s} \tau f_x{}^0(t-\tau)\, d\tau$$

Performing differentiation, we obtain

$$\left(\frac{\partial^2 R^{-1}}{\partial x^2}\,\mathbf{i} + \frac{\partial^2 R^{-1}}{\partial x\;\partial y}\,\mathbf{j} + \frac{\partial^2 R^{-1}}{\partial x\;\partial z}\,\mathbf{k}\right) = \frac{\partial}{\partial x}\left(\frac{\partial R^{-1}}{\partial x}\,\mathbf{i} + \frac{\partial R^{-1}}{\partial y}\,\mathbf{j} + \frac{\partial R^{-1}}{\partial z}\,\mathbf{k}\right)$$

$$= -\frac{\partial}{\partial x}\frac{1}{R^3}\,(x\;\mathbf{i} + y\;\mathbf{j} + z\;\mathbf{k}) = -\frac{1}{R^3}\,\mathbf{i} + \frac{3x\;\mathbf{i}_R}{R^4} = -\frac{1}{R^3}\left(\mathbf{i} - 3\frac{x}{R}\;\mathbf{i}_R\right) \tag{3.190}$$

$$= \frac{1}{R^3}\left[2\;\frac{x}{R}\;\mathbf{i}_R - \left(\mathbf{i} - \frac{x}{R}\;\mathbf{i}_R\right)\right]$$

Taking into account eqs. 3.186–3.189, we can represent $\mathbf{s}_3(p,t)$ as

$$\mathbf{s}_3(p,t) = \frac{1}{4\pi R^3}\int\limits_{R/c_l}^{R/c_s} \tau\left[2\mathbf{N}_1(t-\tau) - \mathbf{N}_2(t-\tau)\right]\, d\tau \tag{3.191}$$

Behavior of a wavefield associated with displacement

We assume that each term in eq. 3.182 can be interpreted as a wave and consider its main features. This approach may create an impression that such waves may exist independently from each other. However, as will be shown later, wavefields associated with either displacement $\mathbf{s}_1$ or displacement $\mathbf{s}_3$ are always accompanied by wave $\mathbf{s}_2$. Because of this, we will also investigate the behavior of the resultant wave in the near, intermediate, and far zones. Note that sometimes the total wave is represented by the last term, $\mathbf{s}_3$, of eq. 3.182 only.

Wave field $\mathbf{s}_1$ By definition, eqs. 3.185 and 3.186, we have

$$s_{1R}(p,t) = \frac{1}{4\pi c_l^2 R}\, N_1\left(t - \frac{R}{c_l}\right) \tag{3.192}$$

Here

$$N_1\left(t - \frac{R}{c_l}\right) = f_x^0\left(t - \frac{R}{c_l}\right)\sin\theta\cos\varphi \tag{3.193}$$

It is clear that the function $s_{1R}(p,t)$ describes a spherical wave propagating from the point force with the velocity of a longitudinal wave, c_l. As follows from eq. 3.192, the amplitude of this wave decreases as $1/R$, regardless of the distance between the force and an observation point. In other words, we cannot distinguish the near, intermediate, and far zones. At each point of the wave surface, $R =$ const, displacement has only the radial component s_{1R}, which coincides with the direction of propagation. In this sense, this wave is similar to a longitudinal wave generated by a pulsating sphere. Both waves travel with the same velocity c_l. As follows from eq. 3.192, displacement $s_{1R}(p,t)$ is proportional to the radial component of force f_x^0 at instant $t - R/c_l$, which depends on angles θ and φ. The factor $\sin\theta\cos\varphi$ in eq. 3.193 defines the radiation pattern of the point source. The magnitude and direction of vector $\mathbf{s}_1$ change from point to point of the wave surface, Fig. 3.5a. As we see, this vector has a very peculiar behavior. For instance, its magnitude reaches a maximum at the x-axis, and it is equal to zero at the plane $x = 0$. Displacement $\mathbf{s}_1$ and the radius-vector $\mathbf{R}$ have the same direction if $x > 0$, and they have opposite directions when $x < 0$. We can say that the wave behaves either as a compressional $(x > 0)$ or a tensional $(x < 0)$ wave. In order to determine the type of this wave, eq. 3.191, we evaluate both the divergence and curl of vector $\mathbf{s}_1$. In the spherical system of coordinates, we have

$$\text{div}\ \mathbf{s} = \frac{1}{R^2\sin\theta}\left[\sin\theta\frac{\partial(R^2 s_R)}{\partial R} + R\frac{\partial(\sin\theta\ s_\theta)}{\partial\theta} + R\frac{\partial s_\varphi}{\partial\varphi}\right] \tag{3.194}$$

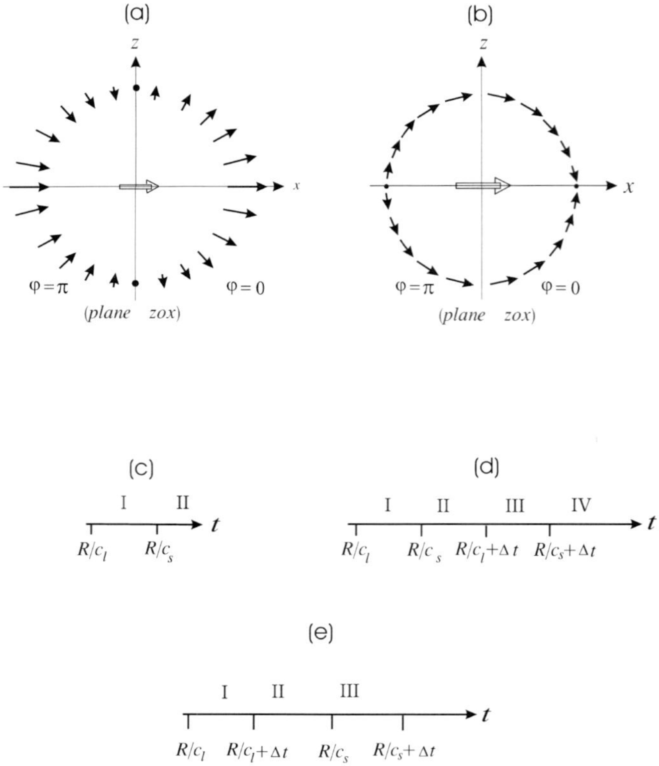

Figure 3.5: (a) Orientation of s_{1R} on the wave surface (b) Orientation of $s_{2\theta}$ on the wave surface (c) Illustration of case one (d) Four time intervals in case two (e) Time intervals in case three

and $$\text{curl}\,\mathbf{s} = \frac{1}{R^2\sin\theta}\begin{vmatrix} \mathbf{i}_R & R\,\mathbf{i}_\theta & R\sin\theta\,\mathbf{i}_\varphi \\ \dfrac{\partial}{\partial R} & \dfrac{\partial}{\partial\theta} & \dfrac{\partial}{\partial\varphi} \\ s_R & Rs_\theta & R\sin\theta\; s_\varphi \end{vmatrix}$$

Taking into account that $s_{1\theta} = s_{1\varphi} = 0$ and performing differentiation, we obtain

$$\text{div}\,\mathbf{s}_1 = \frac{\sin\theta\cos\varphi}{4\pi c_l^2}\left[\frac{1}{R^2}\,f_x^0\left(t-\frac{R}{c_l}\right) - \frac{1}{R\,c_l}\,f_x^{0'}\left(t-\frac{R}{c_l}\right)\right] \tag{3.195}$$

Also

$$\text{curl}_R\mathbf{s}_1 = 0, \qquad \text{curl}_\theta\mathbf{s}_1 = \frac{1}{R\sin\theta}\frac{\partial\, s_{1R}}{\partial\varphi}, \qquad \text{curl}_\varphi\mathbf{s}_1 = -\frac{1}{R}\frac{\partial\, s_{1R}}{\partial\,\theta}$$

Hence

$$\text{curl}_\theta \mathbf{s}_1 = -\frac{1}{4\pi c_l^2 \, R^2} \, f_x^0 \left(t - \frac{R}{c_l}\right) \sin\varphi, \qquad \text{curl}_\varphi \mathbf{s}_1 = -\frac{1}{4\pi c_l^2 \, R^2} f_x^0 \left(t - \frac{R}{c_l}\right) \cos\theta \cos\varphi$$

i.e.,

$$\text{curl } \mathbf{s}_1 = -\frac{1}{4\pi c_l^2 \, R^2} \, f_x^0 \left(t - \frac{R}{c_l}\right) [\sin\varphi \; \mathbf{i}_\theta + \cos\theta \cos\varphi \; \mathbf{i}_\varphi] \tag{3.196}$$

As follows from eqs. 3.183, in the spherical system of coordinates the force components are

$$f_R^0 = f_x^0 \sin\theta \cos\varphi, \qquad f_\theta^0 = f_x^0 \cos\theta \cos\varphi, \qquad f_\varphi^0 = -f_x^0 \sin\varphi \tag{3.197}$$

Correspondingly, eq. 3.196 can be represented in the form

$$\text{curl } \mathbf{s}_1 = -\frac{1}{4\pi c_l^2 R^3} \, \mathbf{M} \left(t - \frac{R}{c_l}\right), \tag{3.198}$$

where $\mathbf{M}$ is the moment of force $\mathbf{f}_x^0 \, (t - R/c_l)$:

$$\mathbf{M} = \mathbf{R} \times \mathbf{f}_x^0 \tag{3.199}$$

In accordance with eqs. 3.195 and 3.198, this wave is neither dilatational nor rotational, because both $\text{div } \mathbf{s}_1$ and $\text{curl } \mathbf{s}_1$ differ from zero. It is essential that this feature displays itself regardless of the distance from the point force. However, in reality this wave is always accompanied by field $\mathbf{s}_3$, and these fields should be interpreted only in combination.

Wavefield $\mathbf{s}_2$ From eqs. 3.188 and 3.189, we have for the vector component of field $\mathbf{s}_2$

$$\mathbf{s}_2(p,t) = \frac{1}{4\pi c_s^2 \, R} \, f_x{}^0 \left(t - \frac{R}{c_s}\right) \mathbf{i}_s(p) \tag{3.200}$$

Here

$$\mathbf{i}_s(p) = \cos\theta \cos\varphi \; \mathbf{i}_\theta - \sin\varphi \; \mathbf{i}_\varphi \tag{3.201}$$

Again, as in the case of field $\mathbf{s}_1$, we assume that the function $\mathbf{s}_2(p,t)$ describes a spherical wave that is moving away from the origin with the velocity of a rotational (shear) wave, c_s. At points of the phase surface, $R =$ const, displacement $\mathbf{s}_2$ has only the tangential component, and in this sense it is similar to a shear wave generated by a rotating sphere. Note that the orientation of vector $\mathbf{s}_2$ does not usually coincide with the orientation of coordinate axes. However, there are two exceptions, corresponding to the directions $\varphi = 0$ or $\varphi = \pi$ and $\theta = \pi/2$. The first is shown in Fig. 3.5b for $f_x^0 > 0$. In the same manner as for $\mathbf{s}_1$, field $\mathbf{s}_2$ decays with distance as $1/R$, provided that the argument $t - R/c_s$ remains constant. Correspondingly, we again cannot distinguish the near, intermediate, and far zones. From eq. 3.200, it follows that field $\mathbf{s}_2$ is directly proportional to projection of the point force on the direction defined by unit vector $\mathbf{i}_s$, which is perpendicular to the radius-vector $\mathbf{R}$.

Now, using eq. 3.200, we determine div $\mathbf{s}_2$ and curl $\mathbf{s}_2$. Since

$$s_{2R} = 0, \qquad s_{2\theta} = \frac{1}{4\pi c_s^2 R} f_x^0 \left(t - \frac{R}{c_s}\right) \cos\theta \cos\varphi, \tag{3.202}$$

and
$$s_{2\varphi} = -\frac{1}{4\pi c_s^2 R} f_x^0 \left(t - \frac{R}{c_s}\right) \sin\varphi,$$

we have

$$\text{div } \mathbf{s}_2 = \frac{1}{4\pi c_s^2 R^2 \sin\theta} f_x^0 \left(t - \frac{R}{c_s}\right) (\cos 2\theta \cos\varphi - \cos\varphi) \tag{3.203}$$

$$= -\frac{1}{2\pi c_s^2 R^2 \sin\theta} f_x^0 \left(t - \frac{R}{c_s}\right) \sin\theta \cos\varphi$$

or

$$\text{div } \mathbf{s}_2 = -\frac{1}{2\pi c_s^2 R^2} N_1 \left(t - \frac{R}{c_s}\right), \tag{3.204}$$

where $N_1 (t - R/c_s)$ is given by eq. 3.186. By definition

$$\text{curl}_R \mathbf{s}_2 = \frac{1}{R\sin\theta}\left[\frac{\partial(\sin\theta \ s_{2\varphi})}{\partial\theta} - \frac{\partial \ s_{2\theta}}{\partial\varphi}\right], \quad \text{curl}_\theta \mathbf{s}_2 = -\frac{1}{R^2}\frac{\partial(R \ s_{2\varphi})}{\partial R},$$

$$\text{curl}_\varphi \mathbf{s}_2 = \frac{1}{R}\frac{\partial(R \ s_{2\theta})}{\partial R}$$

Performing differentiation, we obtain

$$\text{curl}_R \mathbf{s}_2 = 0$$

$$\text{curl}_\theta \mathbf{s}_2 = -\frac{1}{4\pi c_s^3 R} f_x^{0'}\left(t - \frac{R}{c_s}\right) \sin\varphi \tag{3.205}$$

and $$\text{curl}_\varphi \mathbf{s}_2 = -\frac{1}{4\pi c_s^3 R} f_x^{0'}\left(t - \frac{R}{c_s}\right) \cos\theta\cos\varphi$$

Correspondingly,

$$\text{curl}\ \mathbf{s}_2 = -\frac{1}{4\pi c_s^3\ R} f_x^{0'}\left(t - \frac{R}{c_s}\right) \mathbf{i}_t, \tag{3.206}$$

where

$$\mathbf{i}_t = \sin\varphi\ \mathbf{i}_\theta + \cos\theta\cos\varphi\ \mathbf{i}_\varphi \tag{3.207}$$

From eq. 3.201, it follows that

$$\mathbf{i}_t \cdot \mathbf{i}_s = 0,$$

and, therefore, vector $\mathbf{i}_t$ is located in the plane perpendicular to the radius-vector $\mathbf{R}$, and it forms angle $\pi/2$ with $\mathbf{i}_s$. Similarly to $\mathbf{s}_1$, the wave associated with vector $\mathbf{s}_2$ is neither dilatational nor rotational.

Wave field $\mathbf{s}_3$ In accordance with eqs. 3.186 and 3.189, in place of eq. 3.191 we have

$$\mathbf{s}_3(p,t) = \frac{2\sin\theta\cos\varphi\ \mathbf{i}_R - \ \mathbf{i}_s}{4\pi R^3} \int\limits_{R/c_l}^{R/c_s} \tau f_x^0(t-\tau)d\tau, \tag{3.208}$$

where $\mathbf{i}_s$ is defined by eq. 3.201. Thus, field $\mathbf{s}_3$ has, in general, all three components, and both its orientation and the magnitude depend on point position. In the same manner as $\mathbf{s}_1$ and $\mathbf{s}_2$, field $\mathbf{s}_3(p,t)$ can be treated as a wave. In fact, by definition, the function $f_x^0(t-\tau)$ is equal to zero when the argument $t-\tau$ is negative. Since the smallest value of τ is R/c_l, the integral vanishes, until $t = R/c_l$. This time delay increases with an increase of distance between the origin and an observation point. This suggests that $\mathbf{s}_3(p,t)$ describes a wave propagating away from the point force. If the time of observation, t, satisfies the condition

$$\frac{R}{c_l} < t < \frac{R}{c_s},$$

then eq. 3.208 can be written as

$$\mathbf{s}_3(p,t) = \frac{2\sin\theta\cos\varphi\,\mathbf{i}_R - \mathbf{i}_s}{4\pi R^3} \int\limits_{R/c_l}^{t} \tau f_x^{\,0}(t-\tau)\,d\tau \tag{3.209}$$

Such replacement is correct, because for larger values of τ the argument of the function $f_x^{\,0}(t-\tau)$ becomes negative. In such a case, τ varies from R/c_l to t, and this allows us to treat wave $\mathbf{s}_3$ as a superposition of wave impulses propagating with different velocities. In particular, when $t < R/c_s$, these velocities are in the range between c_l and c_s. The coefficient in front of the integral, eq. 3.191, is inversely proportional to R^3, but dependence of the integral on R is defined by several factors. Next, we will derive expressions of $\operatorname{div}\mathbf{s}_3$ and $\operatorname{curl}\mathbf{s}_3$ for three different cases.

Case one Suppose that the point force arises at instant $t = 0$ and acts at all times. Then, for any observational point there are two distinct time intervals after a wave arrival, Fig. 3.5c:

$$\frac{R}{c_l} \le t \le \frac{R}{c_s} \qquad \text{and} \qquad t > \frac{R}{c_s}$$

During the first one there are two wavefields, $\mathbf{s}_1$ and $\mathbf{s}_3$, but during the second interval all three waves, $\mathbf{s}_1$, $\mathbf{s}_2$, and $\mathbf{s}_3$, are present.

Case two Assume that the point force acts only during time interval Δt:

$$f_x^0(t) = \begin{cases} 0 & t < 0 \\ f_x^0(t) & 0 \le t \le \Delta t \\ 0 & t > \Delta t \end{cases} \tag{3.210}$$

and

$$\Delta t > \frac{R}{c_s} - \frac{R}{c_l} \tag{3.211}$$

Consider four time intervals, Fig. 3.5d. During them, displacement is formed as

$$\mathbf{s} = \mathbf{s}_1 + \mathbf{s}_3, \qquad \mathbf{s} = \mathbf{s}_1 + \mathbf{s}_2 + \mathbf{s}_3, \qquad \mathbf{s} = \mathbf{s}_2 + \mathbf{s}_3, \qquad \text{and} \qquad \mathbf{s} = \mathbf{0}$$

Case three Suppose that

$$\Delta t < \frac{R}{c_s} - \frac{R}{c_l}$$

Then the time interval between the arrival of wave $\mathbf{s}_1$ and that of the rear of wave $\mathbf{s}_2$ is naturally divided into three subintervals (Fig. 3.5e), which are

$$\frac{R}{c_l} < t < \frac{R}{c_l} + \Delta t, \qquad \frac{R}{c_l} + \Delta t < t < \frac{R}{c_s}, \qquad \text{and} \qquad \frac{R}{c_s} < t < \frac{R}{c_s} + \Delta t$$

Corresponding wavefields are

$$\mathbf{s} = \mathbf{s}_1 + \mathbf{s}_3, \qquad \mathbf{s} = \mathbf{s}_3, \qquad \text{and} \qquad \mathbf{s} = \mathbf{s}_2 + \mathbf{s}_3$$

We pay attention only to the last case and derive expressions for divergence and curl of field $\mathbf{s}_3$. As will be demonstrated, these functions change with the time interval. For instance, it turns out that they are equal to zero when fields $\mathbf{s}_1$ and $\mathbf{s}_2$ are absent, $(R/c_l + \Delta t < t < R/c_s)$. Before we perform differentiation, it is convenient to represent function $\mathbf{s}_3(p,t)$ in the form

$$\mathbf{s}_3(p,t) = s_{3R}\,\mathbf{i}_R + s_{3\theta}\,\mathbf{i}_\theta + s_{3\varphi}\,\mathbf{i}_\varphi, \tag{3.212}$$

where

$$s_{3R} = \frac{1}{2\pi R^3} \int\limits_{R/c_l}^{R/c_s} f_x^0(t-\tau)\;\tau\; d\tau \sin\theta\cos\varphi\;, \tag{3.213}$$

$$s_{3\theta} = -\frac{1}{4\pi R^3} \int\limits_{R/c_l}^{R/c_s} f_x^0(t-\tau)\;\tau\; d\tau \cos\theta\cos\varphi\;,$$

$$s_{3\varphi} = \frac{1}{4\pi R^3} \int\limits_{R/c_l}^{R/c_s} f_x^0(t-\tau)\;\tau\; d\tau \sin\varphi$$

To perform differentiations, we use equality 3.174:

$$\frac{\partial}{\partial R} \int\limits_{R/c_l}^{R/c_s} \tau\; f_x^0(t-\tau) d\tau = R\left[\frac{1}{c_s^2}\; f_x{}^0(t-\frac{R}{c_s}) - \frac{1}{c_l^2}\; f_x^0(t-\frac{R}{c_l})\right] \tag{3.214}$$

From eqs. 3.194, we obtain

$$\text{div}\;\mathbf{s}_3 = \frac{\sin\theta\cos\varphi}{2\pi R^2}\left[\frac{1}{c_s^2}\; f_x^0(t-\frac{R}{c_s}) - \frac{1}{c_l^2}\; f_x^0(t-\frac{R}{c_l})\right] \tag{3.215}$$

Taking into account eq. 3.210, we find three different expressions for divergence:

$$\text{div } \mathbf{s}_3 = -\frac{\sin\theta\cos\varphi}{2\pi c_l^2\, R^2}\, f_x^0\,(t-\frac{R}{c_l}) \qquad \text{if} \quad \frac{R}{c_l} < t < \frac{R}{c_l} + \Delta t \tag{3.216}$$

$$\text{div } \mathbf{s}_3 = 0 \qquad \text{if} \quad \frac{R}{c_l} + \Delta t < t < \frac{R}{c_s} \tag{3.217}$$

and

$$\text{div } \mathbf{s}_3 = \frac{\sin\theta\cos\varphi}{2\pi c_s^2\, R^2}\, f_x^0\,(t-\frac{R}{c_s}) \qquad \text{if} \quad t > \frac{R}{c_s} \tag{3.218}$$

In accordance with eqs. 3.216 and 3.217, divergence is directly proportional to the radial component of force $f_x^{\,0}$, taken at either instant $t - R/c_l$ or $t - R/c_s$. As follows from eqs. 3.194, we have:

$$\text{curl}_R \mathbf{s}_3 = 0$$

$$\text{curl}_\theta \mathbf{s}_3 = \frac{1}{4\pi R^2}\left[\frac{1}{c_l^2}\, f_x^0(t-\frac{R}{c_l}) - \frac{1}{c_s^2}\, f_x^0(t-\frac{R}{c_s})\right]\sin\varphi \tag{3.219}$$

$$\text{curl}_\varphi \mathbf{s}_3 = \frac{1}{4\pi R^2}\left[\frac{1}{c_l^2}\, f_x^0(t-\frac{R}{c_l}) - \frac{1}{c_s^2}\, f_x^0(t-\frac{R}{c_s})\right]\cos\theta\cos\varphi$$

Respectively, we have

$$\text{curl}_R \mathbf{s}_3 = 0, \qquad \text{curl}_\theta \mathbf{s}_3 = \frac{1}{4\pi R^2\, c_l^2}\, f_x^0(t-\frac{R}{c_l})\sin\varphi, \tag{3.220}$$

$$\text{curl}_\varphi \mathbf{s}_3 = \frac{1}{4\pi R^2\, c_l^2}\, f_x^0(t-\frac{R}{c_l})\cos\theta\cos\varphi, \qquad \text{if} \quad \frac{R}{c_l} < t < \frac{R}{c_l} + \Delta t$$

and

$$\text{curl } \mathbf{s}_3 = 0, \qquad \text{if} \quad \frac{R}{c_l} + \Delta t < t < \frac{R}{c_s} \tag{3.221}$$

while

$$\text{curl}_R \mathbf{s}_3 = 0, \qquad \text{curl}_\theta \mathbf{s}_3 = -\frac{1}{4\pi R^2\, c_s^2}\, f_x^0(t-\frac{R}{c_s})\sin\varphi, \tag{3.222}$$

$$\text{curl}_\varphi \mathbf{s}_3 = -\frac{1}{4\pi R^2\, c_s^2}\, f_x^0(t-\frac{R}{c_s}) \cos\theta \cos\varphi \qquad \text{if} \quad \frac{R}{c_s} < t < \frac{R}{c_s} + \Delta t$$

Thus, in the first and third time intervals wavefield $\mathbf{s}_3$ carries out both dilatation and rotation and overlaps with waves associated with $\mathbf{s}_1$ and $\mathbf{s}_2$. Note that wave $\mathbf{s}_3$ is spherical, and vector $\mathbf{s}_3$ is arbitrarily oriented with respect to its phase surface, $(R = \text{const})$. The times of arrival of its front and rear are equal to

$$t = \frac{R}{c_l} \qquad \text{and} \qquad t = \frac{R}{c_s} + \Delta t,$$

respectively. Now we are prepared to study field $\mathbf{s}(p,t)$ as a function of time and distance from the point force.

The resultant wave as a function of time

We continue to assume that force $f_x^{\,0}$ differs from zero within some time interval only.

$$0 \le t \le \Delta t$$

If an observational point is located relatively close to the origin, then an equality

$$\Delta t > \frac{R}{c_s} - \frac{R}{c_l}$$

takes place, and "waves" $\mathbf{s}_1$ and $\mathbf{s}_2$ can overlap in time. We, however, suppose that the offset R is sufficiently large, $(\Delta t < R/c_s - R/c_l)$, and that such superposition is absent. Then, as we know, motion in a receiver point can be split into three intervals:

1. Between the front and rear of wave $\mathbf{s}_1$,
2. Between the rear of wave $\mathbf{s}_1$ and the front of wave $\mathbf{s}_2$, and
3. Between the front and rear of wave $\mathbf{s}_2$.

The first interval $(R/c_l < t < R/c_l + \Delta t)$ For this time interval, we have

$$\mathbf{s}(p,t) = \mathbf{s}_1(p,t) + \mathbf{s}_3(p,t),$$

and in accordance with eqs. 3.185 and 3.209

$$\mathbf{s}(p,t) = \frac{\sin\theta\cos\varphi}{4\pi c_l^2\, R}\, f_x^0(t-\frac{R}{c_l})\mathbf{i}_R + \frac{2\sin\theta\cos\varphi\,\mathbf{i}_R - \mathbf{i}_s}{4\pi R^3} \int\limits_{R/c_l}^{t} \tau\, f_x^0(t-\tau)\, d\tau \tag{3.223}$$

This expression describes a spherical wave traveling away from the point force with displacement vector $\mathbf{s}$ being arbitrarily oriented with respect to the phase surface.

From eqs. 3.195 and 3.196 and eqs. 3.216 and 3.220, we have

$$\text{div } \mathbf{s}_1 = \frac{\sin\theta\cos\varphi}{4\pi c_l^2}\left[\frac{1}{R^2}\, f_x^0(t-\frac{R}{c_l}) - \frac{1}{R\, c_l}\, f_x^{0\prime}(t-\frac{R}{c_l})\right],$$

$$\text{curl } \mathbf{s}_1 = -\frac{1}{4\pi c_l^2\, R^2}\, f_x^0(t-\frac{R}{c_l})\,(\sin\varphi\;\mathbf{i}_\theta + \cos\theta\cos\varphi\;\mathbf{i}_\varphi)$$

and

$$\text{div } \mathbf{s}_3 = -\frac{\sin\theta\cos\varphi}{2\pi c_l^2\, R^2}\, f_x^0(t-\frac{R}{c_l}),$$

$$\text{curl } \mathbf{s}_3 = \frac{1}{4\pi c_l^2\, R^2}\, f_x^0(t-\frac{R}{c_l})\,(\sin\varphi\;\mathbf{i}_\theta + \cos\theta\cos\varphi\;\mathbf{i}_\varphi)$$

This gives

$$\text{div } \mathbf{s} = -\frac{\sin\theta\cos\varphi}{4\pi c_l^2}\left[\frac{1}{R^2}\, f_x^0(t-\frac{R}{c_l}) + \frac{1}{c_l\, R}\, f_x^{0\prime}(t-\frac{R}{c_l})\right] \tag{3.224}$$

$$\text{and} \qquad \text{curl } \mathbf{s} = \mathbf{0}$$

Thus, during the first interval of a motion, we deal with the longitudinal wave, (curl $\mathbf{s} = \mathbf{0}$), which propagates with velocity c_l. Elementary volumes of an elastic medium experience deformation, but rotation is absent. As was mentioned before, the direction of vector $\mathbf{s}$ does not usually coincide with that of wave propagation. When we considered wavefields $\mathbf{s}_1$ and $\mathbf{s}_3$ separately, we found that each is accompanied by rotation. However, physical meaning has only the resultant wave, in which this type of motion is absent.

The second interval $(R/c_l + \Delta t < t < R/c_s)$ Unlike with the first and last intervals, the duration of this interval depends on distance R. This interval is absent near the origin and appears when $\Delta t = R(1/c_s - 1/c_l)$. Then it becomes wider with an increase of R. In the limit, $(R \to \infty)$ the time interval also tends to infinity. In the second interval only, field $\mathbf{s}_3$ is present, and, correspondingly, it describes the real wave. As follows from eqs. 3.209,

$$\mathbf{s}(p,t) = \frac{1}{4\pi R^3}\int_{R/c_l}^{R/c_s} \tau\; f_x^0(t-\tau)\; d\tau\,(\sin\theta\cos\varphi\;\mathbf{i}_R - \mathbf{i}_s)\,, \tag{3.225}$$

while

$$\text{curl } \mathbf{s} = 0 \qquad \text{and} \qquad \text{div } \mathbf{s} = 0 \tag{3.226}$$

Since both divergence and curl are equal to zero, motion in the second phase is not accompanied by a change of volume and its rotation. Taking curl from both sides of the first equation in set 3.226, we obtain:

$$\text{curl curl } \mathbf{s} = \text{grad div } \mathbf{s} - \nabla^2 \mathbf{s} = 0 \qquad \text{or} \qquad \nabla^2 \mathbf{s} = 0 \tag{3.227}$$

Thus, displacement $\mathbf{s}$ obeys the Laplace equation; this is why the motion of particles during the second interval is called Laplace motion. Substituting eqs. 3.226 and 3.227 into the known differential equation of elastic waves:

$$\rho \frac{\partial^2 \mathbf{s}}{\partial t^2} = (\lambda + \mu) \text{ grad div } \mathbf{s} - \mu \nabla^2 \mathbf{s},$$

we see that acceleration is equal to zero, and particles move with constant velocity. In this sense, Laplace motion is similar to the potential motion of the ideal noncompressed field. Let us also consider the integral in eq. 3.225:

$$I_1 = \int\limits_{R/c_l}^{R/c_s} \tau \, f_x^0(t-\tau) \, d\tau \tag{3.228}$$

Introducing new variable t_1:

$$\tau = t - t_1 \qquad \text{and} \qquad d\tau = -dt_1$$

we obtain

$$I_1 = t \int\limits_{t-R/c_s}^{t-R/c_l} f_x^0(t_1) \, dt_1 - \int\limits_{t-R/c_s}^{t-R/c_l} t_1 \, f_x^0(t_1) \, dt_1 \tag{3.229}$$

By definition, $R/c_l + \Delta t < t < R/c_s$, which shows that the upper limit of the integral is not smaller than Δt, but the lower limit is not positive. Thus, integration is always performed over the time interval during which f_x^0 differs from zero. In such a case, integrals in eqs. 3.228 are independent of time and of distance R. They are defined by the behavior of the force oriented along the x-axis. Applying notations

$$A_1 = \int\limits_{t-R/c_s}^{t-R/c_l} f_x^0(t_1) \, dt_1 = \int\limits_0^{\Delta t} f_x^0(t) \, dt \tag{3.230}$$

and $$A_2 = \int_{t-R/c_s}^{t-R/c_l} t_1\, f_x^{\,0}(t_1)\, dt_1 = \int_0^{\Delta t} t\, f_x^0(t)\; dt\;,$$

we obtain for the displacement of Laplace motion

$$\mathbf{s}(p,t) = \frac{1}{4\pi R^3}(t\, A_1 - A_2)\; (\sin\theta\cos\varphi\; \mathbf{i}_R - \mathbf{i}_s) \tag{3.231}$$

Therefore, during the second phase, particles move along a straight line at a constant velocity. The value of velocity essentially depends on the function $f_x^{\,0}(t)$. In particular, if velocity changes in such a way that an area beneath function $f_x^0(t)$ is equal to zero, then the velocity of motion is equal to zero, $A_1 = 0$. At the same time, constant displacement with respect to equilibrium can be observed, and it is defined by A_2.

The third interval $(R/c_s < t < R/c_s + \Delta t)$ During this last interval, total displacement is a sum

$$\mathbf{s} = \mathbf{s}_2 + \mathbf{s}_3,$$

and in accordance with eqs. 3.200 and 3.209, we have

$$\mathbf{s}(p,t) = \frac{f_x^0(t - R/c_s)}{4\pi c_s^2\, R}\,\mathbf{i}_s + \frac{2\sin\theta\cos\varphi\;\mathbf{i}_R - \mathbf{i}_s}{4\pi R^3}\int_{R/c_l}^{R/c_s} \tau\; f_x^0(t-\tau)\; d\tau \tag{3.232}$$

Thus, as before, the wave is spherical and it propagates with velocity c_s. At the same time, vector $\mathbf{s}$ is usually arbitrarily oriented with respect to the phase surface. As follows from eqs. 3.204, 3.206, and 3.224:

$$\text{div}\;\mathbf{s}_2 = -\frac{1}{2\pi c_s^2\; R^2} f_x^0(t - \frac{R}{c_s})\sin\theta\cos\varphi,$$

$$\text{curl}\;\mathbf{s}_2 = -\frac{1}{2\pi c_s^2\; R^2} f_x^{0\prime}(t - \frac{R}{c_s})\;(\sin\varphi\;\mathbf{i}_\theta + \cos\theta\cos\varphi\;\mathbf{i}_\varphi)$$

and

$$\text{div}\;\mathbf{s}_3 = \frac{\sin\theta\cos\varphi}{2\pi c_s^2\; R^2}\; f_x^0(t - \frac{R}{c_s}),$$

$$\text{curl}\;\mathbf{s}_3 = -\frac{1}{4\pi c_s^2\; R^2}\; f_x^0(t - \frac{R}{c_s})\;(\sin\varphi\;\mathbf{i}_\theta + \cos\theta\cos\varphi\;\mathbf{i}_\varphi)$$

Performing a summation, we obtain

$$\text{div } \mathbf{s} = 0 \qquad \text{and} \tag{3.233}$$

$$\text{curl } \mathbf{s} = -\frac{1}{4\pi c_s^2}\left[\frac{1}{R^2} f_x^0(t-\frac{R}{c_s}) + \frac{1}{R}\, f_x^{0\prime}(t-\frac{R}{c_s})\right](\sin\varphi\ \mathbf{i}_\theta + \cos\theta\cos\varphi\ \mathbf{i}_\varphi)$$

Thus, due to a superposition of fields $\mathbf{s}_2$ and $\mathbf{s}_3$, dilatation vanishes and the rotational wave is formed. This wave produces rotation of elementary volumes of a medium, as well as pure shear and translation. Since the divergence of total field $\mathbf{s}$ is equal to zero, dilatation of fields $\mathbf{s}_2$ and $\mathbf{s}_3$ has no physical meaning.

Thus, considering propagation of wave impulses of dilatational and rotational waves, we can distinguish the near, intermediate, and far zones. Within the first two zones the field of both waves changes relatively quickly with distance, and it is described by all three components, s_R, s_θ, and s_φ. With an increase of distance, the behavior of displacement becomes much simpler. In the far zone, dilatational and rotational waves decay practically inversely proportionally to distance R, and either the radial or tangential component plays the dominant role. This is the reason they are also called the longitudinal and shear waves, respectively. In conclusion, let us point out again that representation of displacement as a sum of three terms – $\mathbf{s} = \mathbf{s}_1 + \mathbf{s}_2 + \mathbf{s}_3$ – is a result of a solution of the boundary value problem. However, only dilatational and rotational waves, as well as Laplace motion, have physical meaning.

3.5 Longitudinal and shear plane waves

As in the case of acoustic waves (Part II), we pay special attention to the behavior of plane waves in an elastic medium. By definition, the phase surface of such waves is planar, and at the beginning we assume that at its points the magnitude and direction of particle displacement $\mathbf{s}$ do not change. Of course, the same is valid for other characteristics of wavefields, and this means that the plane wave is homogeneous. Later we will study the more complicated inhomogeneous wave, in which displacement, as well as strain and stress, may change very rapidly along its phase surface. As our main goal is to describe reflection and transmission of plane waves at the planar interface, let us introduce the Cartesian system of coordinates, so that the $y-$axis is parallel to the wave surface. Later (in the next chapter) we will assume that the boundary between two elastic media is situated in the plane XOY, Fig. 3.6a. It is obvious that at each plane perpendicular to the $y-$axis, wave behavior is identical. This allows us to study a wave in one such plane,

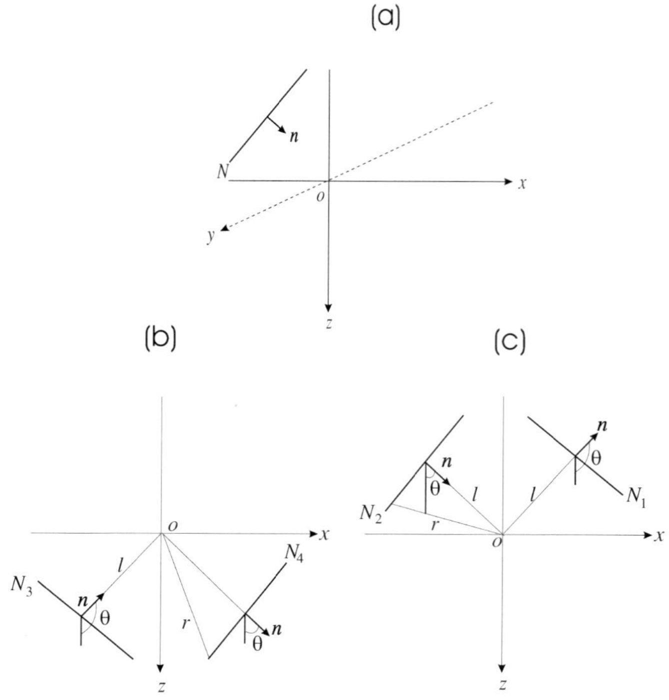

Figure 3.6: (a) Wave front N of plane wave (b,c) Different orientations of unit vector $\mathbf{n}$

for instance XOZ, which is usually called the plane of incidence. We distinguish three possible types of homogeneous plane waves: (1) longitudinal wave P, in which particles move in the direction of propagation, that is, perpendicular to the phase surface; (2) shear wave SV, in which particle motion is tangential to the phase surface and occurs in the plane of incidence; and (3) shear wave SH, in which the vector of displacement $\mathbf{s}$ is tangential to the phase surface but perpendicular to the plane of incidence. In other words, it is parallel to the $y-$axis. Now we will demonstrate that plane wave P is compressional and propagates with the velocity

$$c_l = \sqrt{\frac{\lambda + 2\mu}{\rho}}$$

At the same time, plane waves SV and SH are shear waves, and they propagate with the velocity

$$c_s = \sqrt{\frac{\mu}{\rho}}$$

We will use the system of coordinates x_1, y_1, and z_1 with unit vectors $\mathbf{i}_1$, $\mathbf{j}_1$, and $\mathbf{k}_1$, Fig. 3.6a, where $\mathbf{j} = \mathbf{j}_1$ and $\mathbf{k}_1 = \mathbf{n}$. By definition, for wave P we have

$$\mathbf{s} = w_1\, \mathbf{k}_1, \qquad u_1 = v_1 = 0$$

and

$$\frac{\partial w_1}{\partial x_1} = \frac{\partial w_1}{\partial y_1} = 0$$

Therefore,

$$\text{div } \mathbf{s} = \frac{\partial w_1}{\partial z_1} \qquad \text{but} \qquad \text{curl } \mathbf{s} = 0 \tag{3.234}$$

This means that wave P is compressional, and it advances with velocity c_l. The displacement components of wave SV in the new system of coordinates are

$$\mathbf{s} = u_1\, \mathbf{i}_1, \qquad \text{since} \qquad v_1 = w_1 = 0$$

Also

$$\frac{\partial u_1}{\partial x_1} = \frac{\partial u_1}{\partial y_1} = 0$$

It follows from these equalities that

$$\text{div } \mathbf{s} = 0 \qquad \text{and} \qquad \text{curl } \mathbf{s} \neq 0 \tag{3.235}$$

Thus, wave SV is a shear wave and propagates with velocity c_s. Finally, in the case of wave SH:

$$\mathbf{s} = v_1\, \mathbf{j}_1, \qquad u_1 = w_1 = 0$$

and

$$\frac{\partial v_1}{\partial x_1} = \frac{\partial v_1}{\partial y_1} = 0$$

Again, this gives div $\mathbf{s} = 0$ and curl $\mathbf{s} \neq \mathbf{0}$ and, correspondingly, SH is also a shear wave. Because divergence and curl are invariants with respect to a change of a system of coordinates, we have proved, that both wave SV and waves SH are rotational, whereas wave P is compressional. They can be described by vector and scalar potentials that satisfy the wave equations

$$\nabla^2 \varphi = \frac{1}{c_l^2} \frac{\partial^2 \varphi}{\partial t^2} \qquad \text{and} \qquad \nabla^2 \boldsymbol{\psi} = \frac{1}{c_s^2} \frac{\partial^2 \boldsymbol{\psi}}{\partial t^2}, \tag{3.236}$$

where

$$\nabla^2 = \frac{\partial^2}{\partial x^2} + \frac{\partial^2}{\partial z^2},$$

since the wavefields are independent on the $y-$coordinate. Note that an infinite number of functions φ and $\boldsymbol{\psi}$ define the same field $\mathbf{s}$, and this fact will be used in choosing components of vector $\boldsymbol{\psi}$. For any compressional wave, we have

$$\mathbf{s} = \text{grad } \varphi \tag{3.237}$$

In particular, in the case of the plane wave, eq. 3.237 yields

$$u = \frac{\partial \varphi}{\partial x} \qquad \text{and} \qquad w = \frac{\partial \varphi}{\partial z} \tag{3.238}$$

Thus, one scalar function φ characterizes both components of displacement vector $\mathbf{s}$. For rotational (shear) waves, we have

$$\mathbf{s} = \text{curl } \boldsymbol{\psi} = \begin{vmatrix} \mathbf{i} & \mathbf{j} & \mathbf{k} \\ \dfrac{\partial}{\partial x} & \dfrac{\partial}{\partial y} & \dfrac{\partial}{\partial z} \\ \psi_x & \psi_y & \psi_z \end{vmatrix},$$

and in the case of plane waves we obtain

$$u = -\frac{\partial \psi_y}{\partial z}, \qquad v = \frac{\partial \psi_x}{\partial z} - \frac{\partial \psi_z}{\partial x}, \qquad w = \frac{\partial \psi_y}{\partial x}, \tag{3.239}$$

because derivatives with respect to y vanish. Let us assume that for wave SV field $\mathbf{s}$ is described by the single component ψ_y. Correspondingly, in place of eq. 3.239 we obtain

$$u = -\frac{\partial \psi_y}{\partial z}, \qquad v = 0, \qquad w = \frac{\partial \psi_y}{\partial x} \tag{3.240}$$

Suppose that the vector potential for wave SH has the component ψ_x only, and, as a result, eqs. 3.239 become

$$u = 0, \qquad v = \frac{\partial \psi_x}{\partial z}, \qquad w = 0 \tag{3.241}$$

Thus, due to our assumptions, potentials for all three waves obey wave equations with respect to scalar functions φ, ψ_y, and ψ_x, which greatly simplifies the determination of wavefields. Note that the validity of these assumptions will be confirmed in solving the boundary value problems.

Expression for potentials In accordance with eqs. 3.236, functions φ, ψ_x, and ψ_y are solutions of ae wave equation that has the form

$$\frac{\partial^2 U}{\partial x^2}+\frac{\partial^2 U}{\partial z^2}=\frac{1}{c^2}\frac{\partial^2 U}{\partial t^2} \tag{3.242}$$

In order to find the function U, we assume that the plane wave is sinusoidal. Then, U can be represented as

$$U(x,z,t)=\text{Re}\ \mathcal{U}\ (\omega,x,z)\ e^{-i\omega t} \tag{3.243}$$

Its substitution into eq. 3.242 yields

$$\frac{\partial^2 \mathcal{U}}{\partial x^2}+\frac{\partial^2 \mathcal{U}}{\partial z^2}+k^2\mathcal{U}=0 \tag{3.244}$$

Here $\mathcal{U}\ (\omega,x,z)$ is the complex amplitude of U, and

$$k=\frac{\omega}{c} \tag{3.245}$$

is the wavenumber for either dilatational or rotational waves. Now, applying the method of separation of variables, we represent the function $\mathcal{U}\ (\omega,x,z)$ as the product of two functions:

$$\mathcal{U}\ (x,z,\omega)=X(x,\omega)\ Z(z,\omega) \tag{3.246}$$

Then, eq. 3.244 becomes

$$Z\ \frac{d^2X}{dx^2}+X\ \frac{d^2Z}{dz^2}+k^2XZ=0 \qquad \text{or} \qquad \frac{1}{X}\ \frac{d^2X}{dx^2}+\frac{1}{Z}\frac{d^2Z}{dz^2}+k^2=0 \tag{3.247}$$

This equality takes place if the first two terms are constants, and it gives

$$\frac{1}{X}\frac{d^2X}{dx^2}=-k^2\ n_x^2 \qquad \text{and} \qquad \frac{1}{Z}\frac{d^2Z}{dz^2}=-k^2n_z^2, \tag{3.248}$$

where n_x and n_z are quantities that are related to each other (Part II). In fact, from eqs. 3.247 and 3.248, we obtain an important relationship:

$$n_x^2+n_z^2=1 \tag{3.249}$$

It is obvious that functions

$$\exp\left(\pm\ i\ k\ n_x x\right) \qquad \text{and} \qquad \exp\left(\pm\ i\ k\ n_z z\right)$$

satisfy eqs. 3.248, respectively. Taking into account eq. 3.246, the function $\mathcal{U}$ and, therefore, the complex amplitudes of potentials $\tilde{\varphi}$, $\tilde{\psi}_x$, and $\tilde{\psi}_y$ can be written in the form

$$\mathcal{U}\,(x,z,\omega) = A \exp\left[\pm\, i\, k\ \left(n_x x + n_z z\right)\right] \tag{3.250}$$

Here A is a constant. It is also convenient to use a slightly different form:

$$\mathcal{U}\,(x,z,\omega) = \frac{A}{i\,k} \exp\left[\pm\, i\, k\ \left(n_x x + n_z z\right)\right] \tag{3.251}$$

Since determination of displacement components is related to differentiation with respect to x and z, the latter form is often more preferable. In general, n_x and n_z can be arbitrary numbers, satisfying condition 3.249, but we are interested in two cases only, namely

$n_x < 1, \qquad n_z < 1$ and

$n_x > 1, \qquad n_z = i\, b_z,$

where b_z is a real number.

As follows from eq. 3.251, these cases describe homogeneous and inhomogeneous plane waves, respectively. In fact, in the first case, the argument

$$\pm\, i\, k\ \left(n_x x + n_z z\right)$$

and $1/i$ characterize the phase of the wave, and since

$$\left|\exp\left[\pm\, i\, k\ \left(n_x x + n_z z\right)\right]\right| = 1,$$

the amplitude, (A/k), remains constant at the phase surface.

We have considered the homogeneous plane wave. The picture is completely different in the case of the inhomogeneous plane wave, where eq. 3.251 can be written as

$$\mathcal{U}\,(x,z,\omega) = \frac{A}{ik} e^{\pm b_z z} \exp\left(\pm\, i\, k\ n_x x\right) \tag{3.252}$$

It is clear that the phase surfaces are perpendicular to the $x-$axis, and the wave amplitude

$$\frac{A}{k}\, e^{\pm b_z z}$$

varies at their points. This shows that we are dealing with an inhomogeneous or evanescent plane wave. As was demonstrated in Part II, they may arise in the vicinity of an interface. There are several types of such waves, some of which are called surface waves.

In the case of a homogeneous plane wave, numbers n_x and n_z are directional cosines of unit vector $\mathbf{n}$, normal to the wavefront. Since the wave travels along line l, Fig. 3.6b,c, the complex amplitude $\mathcal{U}$ can be written in the form

$$\mathcal{U} = \frac{A}{ik}\, e^{ikl}, \tag{3.253}$$

where $|l|$ is the distance between the origin of coordinates O and phase surface N. Indeed, from eq. 3.243 it follows that

$$U(x,z,t) = \frac{A}{k}\,\mathrm{Re}\,\frac{1}{i}\, e^{-i\,(\omega t - kl)} \qquad \text{or} \qquad U(x,z,t) = -\frac{A}{k}\sin(\omega t - kl)$$

or

$$U(x,z,t) = \frac{A}{k}\sin(kl - \omega t) \tag{3.254}$$

This equation describes a homogeneous plane wave with amplitude A/k propagating along the $l-$line. If l is positive, then we are dealing with the outgoing wave, which moves away from the origin. On the contrary, when l is negative, the wave is incoming, approaching origin O. We express parameter l in terms of coordinates of a point, (x,z), located on the wave surface. From Fig. 3.6b,c, it follows that

$$\mathbf{n} = \sin\theta\;\mathbf{i} + \cos\theta\;\mathbf{k}, \tag{3.255}$$

where θ is the angle formed by the normal $\mathbf{n}$ and the $z-$axis, and

$$0 \le \theta \le \pi$$

Inasmuch as l can be treated as the scalar component of the radius-vector

$$\mathbf{r} = x\;\mathbf{i} + z\;\mathbf{k}$$

of any point of the wave plane, Fig. 3.6b,c, we have

$$l = \mathbf{r}\cdot\mathbf{n} = x\sin\theta + z\cos\theta \tag{3.256}$$

Therefore, eq. 3.257 is written as

$$\mathcal{U}\;(x,z,\omega) = \frac{A}{i\;k}\exp ik\,(x\sin\theta + z\cos\theta) \tag{3.257}$$

Comparison with eq. 3.255 shows that

$$n_x = \sin\theta \qquad \text{and} \qquad n_z = \cos\theta\;, \tag{3.258}$$

and we have solved our task. Of course, condition 3.249 is met. To illustrate eq. 3.257, consider several examples, Fig. 3.6b,c. In the first quadrant, we have

$$x > 0, \qquad z < 0, \qquad \text{and} \qquad \theta > \frac{\pi}{2}$$

Correspondingly, l is positive and the wave (N_1) is outgoing. In the case of N_2 (the second quadrant):

$$x < 0, \qquad z < 0, \qquad \text{and} \qquad \theta < \frac{\pi}{2}$$

Thus, l is negative and the wave is incoming. In the third quadrant (N_3), we have

$$x < 0, \qquad z > 0, \qquad \text{and} \qquad \theta > \frac{\pi}{2},$$

and, correspondingly, $l < 0$, i.e., the wave is incoming, as is indicated by the direction of the normal $\mathbf{n}$. Finally, for the wave in the fourth quadrant (N_4), we have

$$x > 0, \qquad z > 0, \qquad \text{and} \qquad \theta < \frac{\pi}{2}$$

Hence $l > 0$ and the wave is outgoing. We demonstrated that the same argument of complex amplitude, eq. 3.257, characterizes all possible directions of the plane wave. However, later we will use angles that are always smaller than or equal to $\pi/2$, and for this reason the sign in front of z will be changed in some cases. Now we are prepared to study reflection and transmission of plane waves in an elastic medium.

Chapter 4

Plane waves in a layered medium

The main subject of this chapter is the study of reflection and transmission of the longitudinal and shear plane waves at the planar interface. To solve this boundary value problem we need to understand how surface forces act in the vicinity of the boundary of two elastic media. We will begin by discussing the behavior of these forces.

4.1 Strain and stress in plane waves

As was demonstrated in Chapter 2, Hooke's law in the Cartesian system of coordinates is

$$\tau_{xx} = \lambda\, \Theta + 2\mu\, e_{xx}, \qquad \tau_{yy} = \lambda\, \Theta + 2\mu\, e_{yy}, \qquad \tau_{zz} = \lambda\, \Theta + 2\mu\, e_{zz} \tag{4.1}$$

and

$$\tau_{yz} = \mu\, e_{yz}, \qquad \tau_{xz} = \mu\, e_{xz}, \qquad \tau_{xy} = \mu\, e_{xy} \tag{4.2}$$

Here

$$e_{xx} = \frac{\partial u}{\partial x}, \qquad e_{yy} = \frac{\partial v}{\partial y}, \qquad e_{zz} = \frac{\partial w}{\partial z}, \qquad \Theta = \frac{\partial u}{\partial x} + \frac{\partial v}{\partial y} + \frac{\partial w}{\partial z} \tag{4.3}$$

$$\text{and} \quad e_{xy} = \frac{\partial u}{\partial y} + \frac{\partial v}{\partial x}, \qquad e_{xz} = \frac{\partial u}{\partial z} + \frac{\partial w}{\partial x}, \qquad e_{yz} = \frac{\partial v}{\partial z} + \frac{\partial w}{\partial y}$$

Since we are considering wavefields that are independent of the y-axis, eqs. 4.1–4.3 may be slightly simplified, and we have:

$$e_{xx} = \frac{\partial u}{\partial x}, \qquad e_{yy} = 0, \qquad e_{zz} = \frac{\partial w}{\partial z}, \tag{4.4}$$

$$e_{yz} = \frac{\partial v}{\partial z}, \qquad e_{xz} = \frac{\partial u}{\partial z} + \frac{\partial w}{\partial x}, \qquad e_{xy} = \frac{\partial v}{\partial x}$$

and

$$\Theta = \frac{\partial u}{\partial x} + \frac{\partial w}{\partial z} \tag{4.5}$$

Also

$$\tau_{xx} = \lambda\,\Theta + 2\mu\, e_{xx}, \qquad \tau_{yy} = \lambda\,\Theta, \qquad \tau_{zz} = \lambda\,\Theta + 2\mu\, e_{zz}, \tag{4.6}$$

and, as in the general case,

$$\tau_{yz} = \mu\, e_{yz}, \qquad \tau_{xz} = \mu\, e_{xz}, \qquad \tau_{xy} = \mu\, e_{xy} \tag{4.7}$$

Note that strain e_{yy} is absent, but the normal stress, τ_{yy}, characterizing the surface force along the y-axis has a nonzero value if $\Theta \neq 0$.

Next we will study strains and stresses for each type of homogeneous plane wave.

Incident P wave

In the case of the P wave, $v = 0$ and therefore

$$e_{xx} = \frac{\partial u}{\partial x}, \quad e_{yy} = 0, \quad e_{zz} = \frac{\partial w}{\partial z}, \quad e_{yz} = 0, \quad e_{xy} = 0, \quad e_{xz} = \frac{\partial u}{\partial z} + \frac{\partial w}{\partial x} \tag{4.8}$$

Thus, there are two diagonal elements of the strain tensor and one shear strain that describes a change of the angle in any plane parallel to XOZ. Respectively, stresses are

$$\tau_{xx} = \lambda\,\Theta + 2\mu\, e_{xx}, \qquad \tau_{yy} = \lambda\,\Theta, \qquad \tau_{zz} = \lambda\,\Theta + 2\mu\, e_{zz}, \tag{4.9}$$

and

$$\tau_{yz} = \tau_{xy} = 0, \qquad \tau_{xz} = \mu\, e_{xz}$$

We see that an elementary volume surrounded by coordinate surfaces is subjected to an action of three normal stresses and one shear stress. The shear stress characterizes the force oriented along the x-axis and applied to the face of the volume perpendicular to the z-axis. All of these forces produce deformation, but rotation is absent, since the P wave is compressional. In particular, if this wave moves along the z-axis, we have

$$u = v = 0$$

and

$$e_{xx} = e_{yy} = 0 \qquad \text{and} \qquad e_{zz} = \frac{\partial w}{\partial z}, \tag{4.10}$$

whereas

$$e_{yz} = e_{xy} = 0 \qquad \text{and} \qquad e_{xz} = 0, \tag{4.11}$$

because the displacement component w does not change on the wave surface $(z = \text{const})$. Therefore, there is only one strain element, e_{zz}, and respectively

$$\tau_{xx} = \tau_{yy} = \lambda \, e_{zz} \qquad \text{and} \qquad \tau_{zz} = (\lambda + 2\mu) \; e_{zz}, \tag{4.12}$$

but

$$\tau_{yz} = \tau_{xz} = \tau_{xy} = 0 \tag{4.13}$$

We see that all shear stresses disappear, and propagation of the longitudinal wave along the z-axis is accompanied by forces acting in the perpendicular direction, $(\tau_{xx} = \tau_{yy} \neq 0)$. This is the reason why the velocity of the P wave depends on the rigid modulus μ.

Incident SV wave

Since the displacement is situated in the plane of incidence $(v = 0)$ and is tangential to the wave surface, it has in general two components, u and w. This means that eqs. 4.8 and 4.9 describe strains and stresses for the SV wave, too. However, the action of surface forces in this case is completely different. Because SV is a rotational plane wave, these forces do not cause deformation of an elementary volume but only produce its rotation. Suppose that SV wave moves along the z-axis. As follows from eqs. 4.8,

$$e_{xx} = e_{yy} = e_{zz} = 0 \qquad \text{and} \qquad e_{yz} = e_{xy} = 0, \qquad e_{xz} = \frac{\partial u}{\partial z} \tag{4.14}$$

There is one nondiagonal element of the strain tensor that describes a distortion of angles in the plane of incidence. This strain also characterizes the rate of change of displacement u along the z-axis. For stresses, eqs. 4.9, we have:

$$\tau_{xx} = \tau_{yy} = \tau_{zz} = 0, \qquad \tau_{yz} = \tau_{xy} = 0, \qquad \text{and} \qquad \tau_{xz} = \tau_{zx} = \mu \, e_{xz} \tag{4.15}$$

Thus, normal stresses are absent, and tangential surface forces directed along the $x-$ and z-axes are applied to the corresponding faces of the volume.

Incident SH wave

Because displacement is oriented along the y-axis, i.e.,

$$u = w = 0,$$

eqs. 4.4 yield

$$e_{xx} = e_{yy} = e_{zz} = 0 \qquad \text{and} \qquad e_{yz} = \frac{\partial v}{\partial z}, \qquad e_{xz} = e_{xy} = 0 \tag{4.16}$$

For stresses, we have (eqs. 4.6 and 4.7)

$$\tau_{xx} = \tau_{yy} = \tau_{zz} = 0 \qquad \text{and} \qquad \tau_{yz} = \mu\frac{\partial v}{\partial z}, \qquad \tau_{xz} = \tau_{xy} = 0 \tag{4.17}$$

The simplicity of such a rotational wave is obvious and, regardless of the orientation of the wavefront, to describe the strain and stress we need only e_{yz} and τ_{yz}.

4.2 Reflection from the free surface (normal incidence)

Assume that the sinusoidal plane wave moves along the z-axis and approaches the plane interface between two media having different elastic parameters. Our goal is to describe the reflected and transmitted waves arising at the boundary. We start from the simplest case of the free surface (Fig. 4.1a). We assume that elastic parameters of the upper medium are:

$$\lambda = \mu = 0 \qquad \text{if} \qquad z < 0$$

This means that waves are absent in that part of the portion of space and that stresses are equal to zero. Then, taking into account continuity of stresses, we conclude that at the free boundary, stresses vanish:

$$\tau_{zz} = \tau_{xz} = \tau_{yz} = 0 \qquad \text{on} \qquad z = 0 \tag{4.18}$$

Respectively, the normal and tangential components of the force applied to any element of the interface are equal to zero. First, consider the case of the incident P wave.

Incident P wave

As was shown in Chapter 3, the complex amplitude of scalar potential for the incident P wave can be written in the form

$$\widetilde{\varphi}_i\,(z, \omega) = \frac{A_i}{i\,k_l}\exp\left(-i\,k_l\,z\right), \qquad \text{if} \qquad z > 0 \tag{4.19}$$

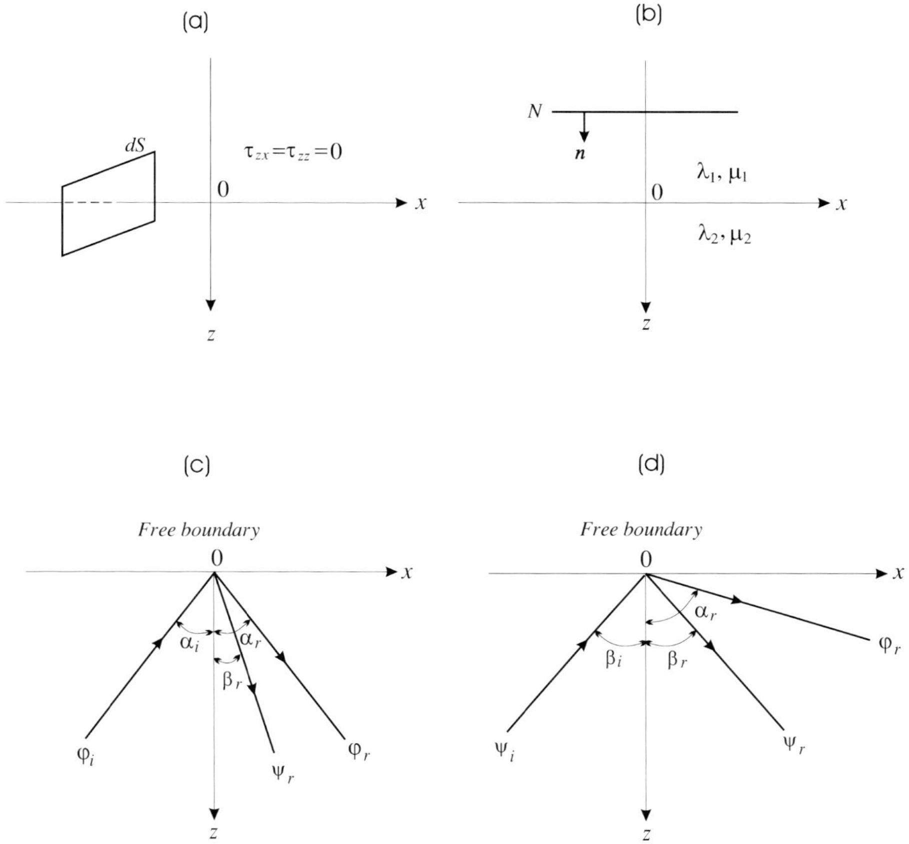

Figure 4.1: (a) Free interface (b) Boundary between elastic media (c) Reflection of P wave from free boundary (d) Reflection of SV wave from free boundary

Here A_i is known and $k_l = \omega / c_l$. Suppose that the reflected wave is also a P wave, and correspondingly, its potential is

$$\widetilde{\varphi}_r(z,\omega) = \frac{A_r}{i\,k_l}\exp(i\,k_l\,z) \tag{4.20}$$

We have changed the sign of the argument because the complex amplitude, $\widetilde{\varphi}_r$, describes the outgoing wave. Thus, the wavefield is a superposition of two P waves propagating in opposite directions:

$$\widetilde{\varphi}(z,\omega) = \frac{1}{i\,k_l}\left[A_i \exp(-i\,k_l\,z) + A_r \exp(i\,k_l\,z)\right] \tag{4.21}$$

Here A_r is unknown. In the previous section, we demonstrated that on the face of an elementary volume that is normal to the z-axis, there is only one component of stress, $\tilde{\tau}_{zz}$. As follows from Hooke's law, eq. 4.12:

$$\tilde{\tau}_{zz} = (\lambda + 2\mu)\,\frac{\partial \tilde{w}}{\partial z} \qquad \text{or} \qquad \tilde{\tau}_{zz} = (\lambda + 2\mu)\,\frac{\partial^2 \tilde{\varphi}}{\partial z^2} \tag{4.22}$$

since

$$\mathbf{s} = \text{grad}\ \varphi \qquad \text{or} \qquad w = \frac{\partial \varphi}{\partial z}$$

Thus, the boundary condition in terms of potential is

$$\frac{\partial^2 \tilde{\varphi}}{\partial\ z^2} = 0 \qquad \text{on} \qquad z = 0 \tag{4.23}$$

Substitution of eq. 4.21 into eq. 4.23 gives

$$A_r = -A_i, \tag{4.24}$$

and the boundary condition is met, if both waves are longitudinal and amplitudes of their potentials are the same. However, their phases differ by π, since $-1 = e^{i\pi}$. In essence, we have confirmed our assumption and proved that the normal incidence of the P wave does not cause shear plane waves. Note that in the vicinity of the boundary, two other stresses, τ_{xx} and τ_{yy}, are also equal to zero. In fact, as follows from eqs. 4.12,

$$\tilde{\tau}_{xx} = \tilde{\tau}_{yy} = \lambda\,\frac{\partial \tilde{w}}{\partial\ z} = \lambda\,\frac{\partial^2 \tilde{\varphi}}{\partial\ z^2},$$

and, therefore, the stresses vanish on $z = 0$. From eqs. 4.21 and 4.24, we have for the complex amplitude of the resultant wave

$$\tilde{\varphi}\,(z, \omega) = \frac{A_i}{i\ k_l}\,[\exp\,(-i\ k_l\ z) - \exp\,(i\ k_l\ z)] \tag{4.25}$$

Because scalar potential is an auxiliary function, we focus on the physically meaningful wavefield properties. For instance, displacement has only the z-component, and $\tilde{w} = \partial\tilde{\varphi}/\partial z$. For the complex amplitude of w, it gives

$$\tilde{w} = -A_i\,[\exp\,(-ik_l\ z) + \exp\,(ik_l z)] \tag{4.26}$$

$$\text{or} \qquad w(z,t) = w_i(z,t) + w_r(z,t) = -A_i\,[\cos\,(\omega t + k_l\ z) + \cos\,(\omega t - k_l z)]$$

and, by definition, the reflection coefficient for displacements is equal to

$$\mathcal{R}_{PP} = \frac{w_r(0,t)}{w_i(0,t)} = 1 \tag{4.27}$$

It is convenient to represent eq. 4.26 in the form

$$w(z,t) = -2A_i \cos k_l z \cos \omega t, \tag{4.28}$$

and this representation clearly demonstrates that the resultant wave, $w(z,t)$, is the standing wave with nodes at points

$$k_l\, z_n = \frac{\pi}{2}\,(2n+1) \qquad \text{or} \qquad \frac{z_n}{\lambda_l} = \frac{1}{4}\,(2n+1)\,, \tag{4.29}$$

where $n = 0, 1, 2, ...$ It is obvious that at the free surface, as in the case of acoustic waves, displacement is doubled:

$$w(0,t) = -2A_i \cos \omega t \tag{4.30}$$

With an increase of distance z, displacement behaves as a sinusoidal function, which happens because there is, in general, a phase shift between the two waves. As a result of their interference, a standing wave is formed. In accordance with eqs. 4.12 and 4.28, we have

$$\tau_{zz} = 2\,(\lambda + 2\mu)\, k_l\; A_i \sin k_l\; z\;\; \cos \omega t \tag{4.31}$$

$$\text{and} \qquad \tau_{xx} = \tau_{yy} = 2\lambda\; k_l\; A_i \sin k_l\; z\;\; \cos \omega t$$

Thus, the resultant stresses are also described by the standing wave with nodes at points

$$k_l\; z = \pi n \qquad \text{or} \qquad \frac{z_n}{\lambda_l} = \frac{1}{2} n \tag{4.32}$$

We see that at all points of a medium, except nodes

$$\tau_{zz} > \tau_{xx} = \tau_{yy} \tag{4.33}$$

Now let us consider incident and reflected waves separately. For instance, particle velocity and stress τ_{zz} associated with the incident wave are

$$\dot{w}(z,t) = \frac{\partial w}{\partial t} = A_i\; \omega \sin(\omega t + k_l z) \tag{4.34}$$

$$\text{and} \qquad \tau_{zz}(z,t) = A_i\, k_l\,(\lambda + 2\mu)\sin(\omega t + k_l z)$$

In studying acoustic waves (Part II), we introduced the concept of acoustic impedance,

$$Z = \rho\, c = \sqrt{\rho\, M},$$

where M is the bulk modulus, and it relates particle velocity $\dot{w}$ and the pressure P:

$$\dot{w} = \frac{P}{Z}$$

From eqs. 4.34, we obtain a similar relationship for the elastic P wave:

$$\frac{\tau_{zz}}{\dot{w}} = \frac{k_l\,(\lambda + 2\mu)}{\omega} = \frac{\lambda + 2\mu}{c_l} = \sqrt{(\lambda + 2\mu)\,\rho} \tag{4.35}$$

Thus, the impedance of a medium for the plane longitudinal wave is

$$Z_l = \sqrt{(\lambda + 2\mu)\,\rho} \tag{4.36}$$

Z_l characterizes resistance to motion caused by P wave.

Incident SV wave

Next, suppose that the plane wave SV propagates along the z-axis toward the free interface and that XOZ is the plane of incidence. Then, as we already know, $u \neq 0$ but $v = w = 0$, and the wavefield is described by the y-component of vector potential ψ_y. Omitting subscript y, the complex amplitude of potential can be written as

$$\tilde{\psi}_i(z,\omega) = \frac{B_i}{i\,k_s}\exp(-ik_s z) \tag{4.37}$$

Here B_i is known, $k_s = \omega/c_s$, and c_s is the velocity of propagation of the shear wave. Let us assume that when the incident wave reaches the interface, the reflected wave of the same type, SV arises, and we have:

$$\tilde{\psi}_r = \frac{B_r}{i\,k_s}\exp(ik_s z) \tag{4.38}$$

Then the resultant wave is

$$\tilde{\psi}(z,\omega) = \frac{1}{i\,k_s}\left[B_i \exp(-ik_s z) + B_r \exp(ik_s z)\right] \tag{4.39}$$

Propagation of these waves is accompanied by the stress $\tau_{xz}(z,t)$ as well as τ_{zx}, and, in accordance with eq. 4.15

$$\tilde{\tau}_{xz} = \mu \frac{\partial \tilde{u}}{\partial z} \tag{4.40}$$

In order to determine the unknown B_r, we use the boundary condition

$$\tilde{\tau}_{xz} = 0 \qquad \text{or} \qquad \frac{\partial \tilde{u}}{\partial z} = 0 \tag{4.41}$$

Taking into account that

$$\tilde{u} = -\frac{\partial \tilde{\psi}}{\partial z}, \tag{4.42}$$

eq. 4.41 becomes

$$\frac{\partial^2 \tilde{\psi}}{\partial z^2} = 0, \qquad \text{on} \qquad z = 0, \tag{4.43}$$

and an analogy with the case of P wave is obvious. Substitution of eq. 4.39 into eq. 4.43 gives

$$B_r = -B_i, \tag{4.44}$$

and, correspondingly,

$$\tilde{\psi}(z,\omega) = \frac{B_i}{i\, k_s} \left[\exp\left(-ik_s z\right) - \exp\left(ik_s z\right)\right] \tag{4.45}$$

Performing a differentiation with respect to z (eq. 4.42), we obtain for the complex amplitude of displacement

$$\tilde{u}(z,\omega) = B_i \left[\exp\left(-ik_s z\right) + \exp\left(ik_s z\right)\right],$$

or

$$u(z,\omega) = u_i(z,\omega) + u_r(z,\omega) = B_i \left[\cos\left(\omega t + k_s z\right) + \cos\left(\omega t - k_s z\right)\right]$$

Therefore, the coefficient of reflection is

$$\mathcal{R}_{SS} = \frac{u_r(0,t)}{u_i(0,t)} = 1 \tag{4.46}$$

Moreover,

$$u(z,t) = 2B_i \cos k_s z \cos \omega t, \tag{4.47}$$

which practically coincides with eq. 4.28. As before we observe a standing wave, but the position of nodes is different, since $k_s \neq k_l$. At the free boundary, the tangential component of displacement, u, is equal to

$$u(0,t) = 2B_i \cos \omega t, \tag{4.48}$$

i.e., it is twice as big as displacement carried by the incident wave. From eq. 4.40, we have for shear stress

$$\tau_{xz}(z,t) = -2\mu\ k_s B_i \sin k_s z \cos \omega t, \tag{4.49}$$

and it is also describes a standing wave. Of course, at the free interface τ_{xz} and τ_{zx} vanish. Since the function $\psi(z,t)$ is a solution of the boundary value problem, our assumption about the reflected wave is correct. In accordance with eq. 4.37, particle velocity and stress, caused by the SV incident wave are

$$\dot{u}(z,t) = -B_i\ \omega \sin(\omega t + k_s z)\,, \qquad \tau_{xz}(z,t) = -B_i\ k_s \mu \sin(\omega t + k_s z)$$

Hence

$$Z_s = \frac{\tau_{xz}}{\dot{u}} = \frac{k_s \mu}{\omega} = \frac{\mu}{c_s} = \sqrt{\mu\rho} \tag{4.50}$$

Comparison with eq. 4.36 shows that $Z_l > Z_s$. This means that if $\tau_{zz} = \tau_{xz}$, the shear wave causes higher particle velocity. The same result is obtained if displacement has only the component along the y-axis (SH wave): $u = w = 0$ and $\mathbf{s} = v\mathbf{j}$.

4.3 Reflection and transmission at the plane boundary of two elastic media (normal incidence)

In this case, unlike in the previous one, waves exist in both half-spaces, Fig. 4.1b. That is, at the interface, the incident wave gives rise to reflected and transmitted waves. As before, we start with longitudinal waves.

Incident P wave

Let us assume that if the incident P wave moves along the z-axis, the reflected and transmitted waves are of the same type and also propagate along the z-axis. Correspondingly, the scalar potential for all of these waves can be represented as

$$\widetilde{\varphi}_i(z,\omega) = \frac{A_i}{i\,k_{1l}} \exp\left(i\,k_{1l}\,z\right), \qquad \widetilde{\varphi}_r(z,\omega) = \frac{A_r}{i\,k_{1l}} \exp\left(-i\,k_{1l}\,z\right), \tag{4.51}$$

$$\widetilde{\varphi}_2(z,\omega) = \frac{A_2}{i\,k_{2l}} \exp\left(i\,k_{2l}\,z\right)$$

Here $k_{1l} = \omega/c_{1l}$, $k_{2l} = \omega/c_{2l}$ and c_{1l}, c_{2l} are velocities of propagation of the P waves in each medium. Now we will attempt to satisfy the boundary conditions, provided that the shear waves SV and SH are absent. As is well known, at the interface, displacement and stresses are continuous functions. Because the P wave has only the component $w(z,t)$ and the stress $\tau_{zz}(z,t)$ differs from zero (normal incidence), we have:

$$w_z^{(1)}(0,t) = w_z^{(2)}(0,t) \qquad \text{and} \qquad \tau_{zz}^{(1)}(0,t) = \tau_{zz}^{(2)}(0,t) \tag{4.52}$$

In terms of the complex amplitude of the potential, eqs. 4.52 become

$$\frac{\partial \widetilde{\varphi}_1}{\partial z} = \frac{\partial \widetilde{\varphi}_2}{\partial z} \qquad \text{and} \qquad (\lambda_1 + 2\,\mu_1)\,\frac{\partial^2 \widetilde{\varphi}_1}{\partial z^2} = (\lambda_2 + 2\,\mu_2)\,\frac{\partial^2 \widetilde{\varphi}_2}{\partial z^2}, \tag{4.53}$$

where

$$\widetilde{\varphi}_1(\omega,z) = \widetilde{\varphi}_i(\omega,z) + \widetilde{\varphi}_r(\omega,z)$$

Substitution of eqs. 4.51 into eqs. 4.53 gives the system for determining of unknowns A_r and A_2:

$$A_i - A_r = A_2$$

$$(\lambda_1 + 2\,\mu_1)\; k_{1l}\,(A_i + A_r) = (\lambda_2 + 2\,\mu_2)\; k_{2l}\,A_2$$

or

$$A_i - A_r = A_2 \qquad \text{and} \qquad Z_{1l}\,(A_i + A_r) = Z_{2l}\,A_2 \tag{4.54}$$

Here Z_{1l} and Z_{2l} are impedances for the P waves, hence

$$A_r = \frac{Z_{2l} - Z_{1l}}{Z_{2l} + Z_{1l}} A_i \tag{4.55}$$

and

$$A_2 = \frac{2\, Z_{1l}}{Z_{1l} + Z_{2l}} A_i \tag{4.56}$$

These formulas do not differ from those that describe reflection and transmission of acoustic waves. It is not surprising, since the latter are also compressional waves (Part II). From eqs. 4.51, 4.55, and 4.56, we have

$$\tilde{\varphi}_1(z,\omega) = \frac{A_i}{ik_{1l}} \left[\exp(i\, k_{1l}\, z) + \frac{Z_{2l} - Z_{1l}}{Z_{2l} + Z_{1l}} \exp(-i\, k_{1l}\, z)\right] \tag{4.57}$$

and $$\tilde{\varphi}_2(z,\omega) = \frac{2\, Z_{1l}}{Z_{1l} + Z_{2l}} \frac{A_i}{ik_{2l}} \exp(i\, k_{2l}\, z)$$

For displacement, we obtain

$$\tilde{w}_1(z,\omega) = \tilde{w}_i(z,\omega) + \tilde{w}_r(z,\omega) = A_i \left[\exp(i\, k_{1l}\, z) - \frac{Z_{2l} - Z_{1l}}{Z_{2l} + Z_{1l}} \exp(-i\, k_{1l}\, z)\right] \tag{4.58}$$

and $$\tilde{w}_2(z,\omega) = \frac{2\, Z_{1l}}{Z_{1l} + Z_{2l}} A_i \exp(i\, k_{2l}\, z)$$

Correspondingly, coefficients of reflection and transmission are

$$\mathcal{R}_{PP} = \frac{A_r}{A_i}, \qquad \mathcal{T}_{PP} = \frac{A_2}{A_i}$$

Note that these coefficients change in the following ranges:

$$-1 \leq \mathcal{R}_{PP} \leq 1 \qquad \text{and} \qquad 0 \leq \mathcal{T}_{PP} \leq 2$$

It is easy to derive formulas for limiting cases from eq. 4.58. For instance, when the lower medium is a free space, we arrive at the known expression

$$\tilde{w}_1(z,\omega) = A_i \left[\exp(i\, k_{1l}\, z) + \exp(-i\, k_{1l}\, z)\right],$$

since $Z_{2l} = 0$. On the contrary, if the second medium is ideally rigid, $Z_{2l} \to \infty$, then

$$\tilde{w}_1(z,\omega) = A_i \left[\exp(i\, k_{1l}\, z) - \exp(-i\, k_{1l}\, z)\right], \tag{4.59}$$

while

$$\tilde{w}_2(z, \omega) = 0$$

This is obvious, because particles in an ideally rigid medium cannot move. As follows from eq. 4.59, at the boundary

$$\tilde{w}_1(0, \omega) = 0,$$

and continuity of displacement takes place. Of course, eqs. 4.58 remain valid when we consider the boundary between an elastic medium and a fluid. In such a case, impedance Z_l is replaced by the acoustic impedance Z. For instance, if the upper medium is a fluid, we write $\mu_1 = 0$ and have

$$Z_{1l} = \sqrt{\lambda_1 \rho}$$

Respectively, the parameter λ_1 plays the role of the bulk modulus M, i.e.,

$$Z_{1l} = Z = \sqrt{M \rho} \tag{4.60}$$

Note that superposition of the incident and reflected waves, eqs. 4.58, does not form the standing wave in the upper half-space. This is because amplitudes of these waves are different.

Incident SV wave

Suppose that the incident wave is an SV plane wave advancing along the z-axis, and that reflected and transmitted waves are of the same type. Then, the wavefields are described by the $y-$component of the vector potential. The complex amplitude of this component at each part of a medium is written in the form

$$\tilde{\psi}_1(z, \omega) = \frac{1}{i\, k_{1s}} \left[B_i \exp\left(i\, k_{1s}\, z\right) + B_r \exp\left(-i\, k_{1s}\, z\right) \right] \tag{4.61}$$

and $$\tilde{\psi}_2(z, \omega) = \frac{1}{i\, k_{2s}} B_2 \exp\left(i\, k_{2s}\, z\right)$$

At the boundary, the component of displacement u and stress τ_{xz} are continuous functions, and in terms of complex amplitudes, we have

$$\frac{\partial \tilde{\psi}_1}{\partial z} = \frac{\partial \tilde{\psi}_2}{\partial z} \quad \text{and} \quad \mu_1 \frac{\partial^2 \tilde{\psi}_1}{\partial z^2} = \mu_2 \frac{\partial^2 \tilde{\psi}_2}{\partial z^2} \qquad \text{on} \quad z = 0 \tag{4.62}$$

The similarity of eqs. 4.53 and 4.62 is obvious. From eqs. 4.61 and 4.62 we obtain

$$B_i - B_r = B_2 \qquad \text{and} \qquad Z_{1s}\,(B_i + B_r) = Z_{2s} B_2 \tag{4.63}$$

Hence

$$B_r = \frac{Z_{2s} - Z_{1s}}{Z_{2s} + Z_{1s}}\,B_i \qquad \text{and} \qquad B_2 = \frac{2\,Z_{1s}}{Z_{2s} + Z_{1s}}\,B_i \tag{4.64}$$

where $Z_s = \sqrt{\mu\,\rho}$ is the impedance of a medium to the shear waves. Formulas 4.64 coincide with corresponding expressions of set 4.55–4.56, if B_i is replaced by A_i and Z_s by Z_l. Inasmuch as

$$\tilde{u}(z,\omega) = -\frac{\partial\tilde{\psi}}{\partial z} \qquad \text{and} \qquad \tilde{\tau}_{xz} = -\mu\frac{\partial^2\tilde{\psi}}{\partial z^2},$$

we obtain

$$\tilde{u}_1(z,\omega) = -B_i\left[\exp\left(i\,k_{1s}\,z\right) - \frac{Z_{2s} - Z_{1s}}{Z_{2s} + Z_{1s}}\exp\left(-i\,k_{1s}\,z\right)\right]$$

$$\text{and} \quad \tilde{u}_2(z,\omega) = -B_i\frac{2\,Z_{1s}}{Z_{2s} + Z_{1s}}\exp\left(i\,k_{2s}\,z\right) \tag{4.65}$$

Similarly to the previous case of the incident P wave, the coefficients of reflection and transmission are

$$\mathcal{R}_{SS} = \frac{B_r}{B_i}, \qquad \mathcal{T}_{SS} = \frac{B_2}{B_1}$$

Moreover,

$$\tilde{\tau}^{(1)}_{xz} = -ik_{1s}\,\mu_1 B_i\left[\exp\left(i\,k_{1s}\,z\right) + \frac{Z_{2s} - Z_{1s}}{Z_{2s} + Z_{1s}}\exp\left(-i\,k_{1s}\,z\right)\right] \tag{4.66}$$

$$\text{and} \qquad \tilde{\tau}^{(2)}_{xz} = -ik_{2s}\,\mu_2\,B_i\frac{2\,Z_{1s}}{Z_{2s} + Z_{1s}}\exp\left(i\,k_{2s}\,z\right)$$

For illustration, consider two special cases. First of all, if the lower medium is an ideally rigid one, $Z_{2s} \to \infty$, we have

$$\tilde{u}_1(z,\omega) = -B_i\left[\exp\left(i\,k_{1s}\,z\right) - \exp\left(-i\,k_{1s}\,z\right)\right] \tag{4.67}$$

and $\overset{\sim(1)}{\tau_{xz}}(z,\omega) = -ik_{1s}\,\mu_1 B_i\,[\exp(i\,k_{1s}\,z) + \exp(-i\,k_{1s}\,z)]$

In particular, at the boundary,

$$\widetilde{u}_1(0,\omega) = 0, \qquad \overset{\sim(1)}{\tau_{xz}} = -2\,i\,k_{1s}\mu_1 B_i \tag{4.68}$$

Thus, as in the case of the P wave, stress is doubled at the interface, but displacement is equal to zero.

Next, suppose that the lower medium is a fluid, i.e.,

$$\mu_2 = 0 \qquad \text{or} \qquad Z_{2s} = 0$$

Correspondingly, eq. 4.65 becomes

$$\widetilde{u}_1(z,\omega) = -B_i\,[\exp(i\,k_{1s}\,z) + \exp(-i\,k_{1s}\,z)]\,,$$

while $\widetilde{u}_2$ is not defined since $k_{2s} \to \infty$. This indicates that the incident wave does not cause motion of fluid particles. As concerns stresses, we have

$$\overset{\sim(1)}{\tau_{xz}} = -ik_{1s}\mu_1 B_i\,[\exp(i\,k_{1s}\,z) - \exp(-i\,k_{1s}\,z)] \qquad \text{and} \qquad \overset{\sim(2)}{\tau_{xz}} = 0,$$

which is obvious, since shear stresses are absent in fluid.

The case of the incident SH wave is completely analogous to the case of the SV wave.

Summing up, we state the following:

1. The incident plane wave (P, SV, or SH) generates secondary waves of the same type.

2. Expressions describing coefficients of reflection (transmission) of secondary waves are similar for any type of incident wave.

3. Since coefficients for sinusoidal waves are independent of a frequency, they are the same for transient waves.

4.4 Reflection from the free surface (oblique incidence)

Now we will study a more general case of oblique incidence of the plane wave at the free surface, Fig. 4.1c, beginning with the incident longitudinal wave.

Incident P wave

As was shown in Chapter 3, scalar potential of the incident P wave can be written as

$$\widetilde{\varphi}_i(x, z, \omega) = \frac{A_i}{i\,k_l} \exp\left[ik_l\left(x \sin\alpha_i - z\cos\alpha_i\right)\right] \tag{4.69}$$

Here α_i is the angle of incidence formed by the ray and the z-axis:

$$0 \le \alpha_i \le \frac{\pi}{2}$$

First assume, as before, that a reflected wave of the same type arises at the interface only. Correspondingly, the complex amplitude of scalar potential is

$$\widetilde{\varphi}_r(x, z, \omega) = \frac{A_r}{i\,k_l} \exp\left[i\,k_l\left(x \sin\alpha_r + z\cos\alpha_r\right)\right] \tag{4.70}$$

where α_r is the angle of reflection formed by the z-axis and the ray of the reflected wave, Fig. 4.1c. Thus, the resultant potential is

$$\widetilde{\varphi}_1(x, z, \omega) = \widetilde{\varphi}_i(x, z, \omega) + \widetilde{\varphi}_r(x, z, \omega) \tag{4.71}$$

or

$$\widetilde{\varphi}_1 = \frac{1}{i\,k_l} A_i \exp\left[i\,k_l\left(x \sin\alpha_i - z\cos\alpha_i\right)\right] + \tag{4.72}$$

$$\frac{1}{i\,k_l} A_r \exp\left[i\,k_l\left(x \sin\alpha_r + z\cos\alpha_r\right)\right]$$

Our goal is to find A_r and α_r, that cause all stresses to disappear at the free boundary

$$\widetilde{\tau}_{zz}(x, 0, \omega) = 0, \qquad \widetilde{\tau}_{xz}(x, 0, \omega) = 0, \qquad \widetilde{\tau}_{yz}(x, 0, \omega) = 0 \tag{4.73}$$

Since the displacement component $v = 0$ and the wavefields do not vary along the y-axis, the last equality is satisfied regardless of the values of A_r and α_r. Taking into account Hooke's law, eqs. 4.73 can be written as

$$(\lambda + 2\mu)\frac{\partial w}{\partial z} + \lambda\frac{\partial u}{\partial x} = 0 \quad \text{and} \quad \frac{\partial u}{\partial z} + \frac{\partial w}{\partial x} = 0, \quad \text{on} \quad z = 0 \tag{4.74}$$

or, in terms of the complex amplitude of the potential, we have

$$(\lambda + 2\mu)\,\frac{\partial^2\widetilde{\varphi}}{\partial z^2} + \lambda\frac{\partial^2\widetilde{\varphi}}{\partial x^2} = 0 \quad \text{and} \quad \frac{\partial^2\widetilde{\varphi}}{\partial x\partial z} = 0 \quad \text{on} \quad z = 0 \tag{4.75}$$

Substitution of eq. 4.72 into eqs. 4.75 gives two equations with respect to A_r and α_r:

$$ik_l A_i \left[(\lambda + 2\mu)\cos^2\alpha_i + \lambda \sin^2\alpha_i\right] \exp(ik_l \sin\alpha_i)$$

$$+ ik_l\ A_r \left[(\lambda + 2\mu)\cos^2\alpha_r + \lambda \sin^2\alpha_r\right] \exp(ik_l \sin\alpha_r) = 0 \tag{4.76}$$

and $\quad - ik_l \sin\alpha_i \cos\alpha_i\ A_i \exp(ik_l \sin\alpha_i) + ik_l \sin\alpha_r \cos\alpha_r\ A_r \exp(ik_l \sin\alpha_r) = 0$

Since these equalities are valid for any x, we conclude that

$$k_l \sin\alpha_i = k_l \sin\alpha_r,$$

$$\text{i.e.,} \qquad \alpha_i = \alpha_r, \tag{4.77}$$

and in place of eqs. 4.76, we obtain

$$A_i + A_r = 0 \qquad \text{and} \qquad -A_i + A_r = 0 \tag{4.78}$$

This system does not have a solution, and, therefore, our assumption that the reflected wave consists only of the P wave is incorrect. This suggests that both longitudinal and shear reflected waves are generated, and we attempt to satisfy the boundary conditions with the help of the P and SV waves. In such a case, the total displacement is the sum of displacements caused by each wave. Therefore, we have:

$$\widetilde{\mathbf{s}} = \text{grad}\ \widetilde{\varphi} + \text{ curl}\ \widetilde{\boldsymbol{\psi}}$$

or

$$\widetilde{u} = \frac{\partial\widetilde{\varphi}}{\partial x} - \frac{\partial\widetilde{\psi}}{\partial z}, \quad \widetilde{v} = 0, \quad \widetilde{w} = \frac{\partial\widetilde{\varphi}}{\partial z} + \frac{\partial\widetilde{\psi}}{\partial x}, \tag{4.79}$$

because

$$\boldsymbol{\psi} = \psi_y \mathbf{j}$$

and wavefields are independent of the $y-$coordinate. This gives for the complex amplitude of dilatation, $\widetilde{\Theta}$:

$$\widetilde{\Theta} = \frac{\partial\widetilde{u}}{\partial x} + \frac{\partial\widetilde{w}}{\partial z} = \frac{\partial^2\widetilde{\varphi}}{\partial x^2} + \frac{\partial^2\widetilde{\varphi}}{\partial z^2} = \nabla^2\widetilde{\varphi} = -\ k_l^2\widetilde{\varphi} \tag{4.80}$$

Also,

$$e_{zz} = \frac{\partial \widetilde{w}}{\partial z} = \frac{\partial^2 \widetilde{\varphi}}{\partial z^2} + \frac{\partial^2 \widetilde{\psi}}{\partial x \partial z}, \qquad e_{xz} = \frac{\partial u}{\partial z} + \frac{\partial w}{\partial x} = 2\frac{\partial^2 \widetilde{\varphi}}{\partial x \partial z} + \frac{\partial^2 \widetilde{\psi}}{\partial x^2} - \frac{\partial^2 \widetilde{\psi}}{\partial z^2}$$

Therefore, in place of eqs. 4.74, we have

$$-\lambda\, k_l^2 \widetilde{\varphi} + 2\mu \left(\frac{\partial^2 \widetilde{\varphi}}{\partial z^2} + \frac{\partial^2 \widetilde{\psi}}{\partial x\, \partial z} \right) = 0 \quad \text{and} \quad 2\frac{\partial^2 \widetilde{\varphi}}{\partial x\, \partial z} + \frac{\partial^2 \widetilde{\psi}}{\partial x^2} - \frac{\partial^2 \widetilde{\psi}}{\partial z^2} = 0 \tag{4.81}$$

Expressions for potentials are

$$\widetilde{\varphi}_1 = \frac{A_i}{ik_l} \exp\left[ik_l\left(x \sin\alpha_i - z\cos\alpha_i\right)\right] + \frac{A_r}{ik_l} \exp\left[ik_l\left(x \sin\alpha_r + z\cos\alpha_r\right)\right] \tag{4.82}$$

and $$\widetilde{\psi}_1 = \frac{B_r}{ik_s} \exp\left[ik_s\left(x\sin\beta_r + z\cos\beta_r\right)\right],$$

where β_r is an angle of reflection of the SV wave. Substituting eqs. 4.82 into set 4.81, we arrive at a system of equations with respect to A_r, B_r, α_r, and β_r. First of all, proceeding from the boundary conditions, it is easy to derive Snell's law for an elastic medium. In fact, every term in eqs. 4.81 contains either $\exp(ik_l \sin\alpha_i)$ or $\exp(ik_l \sin\alpha_r)$ or $\exp(ik_s \sin\beta_r)$. Since the boundary conditions take place regardless of x, we conclude that all three arguments are equal to each other:

$$k_l \sin\alpha_i = k_l \sin\alpha_r = k_s \sin\beta_r \tag{4.83}$$

or

$$\alpha_i = \alpha_r \quad \text{and} \quad \frac{\sin\beta_r}{c_s} = \frac{\sin\alpha_i}{c_l} \tag{4.84}$$

This represents Snell's law of reflection in an elastic medium. Eq. 4.83 shows that the apparent velocity of three waves along the x-axis is the same. Making use of eqs. 4.83 and performing differentiations, eqs. 4.81 give

$$k_l\left(A_i + A_r\right)\left(\lambda + 2\mu\cos^2\alpha_i\right) + 2\mu\, k_s\, B_r \sin\beta_r\, \cos\beta_r = 0$$

and $$2k_l \sin\alpha_i \cos\alpha_i \left(A_r - A_i\right) + k_s\, B_r\left(\sin^2\beta_r - \cos^2\beta_r\right) = 0$$

or

$$(A_i + A_r)\left[(\lambda + 2\mu)\cos^2\alpha_i + \lambda \sin^2\alpha_i\right] + 2\mu m B_r \sin\beta_r \cos\beta_r = 0 \tag{4.85}$$

and $$2\sin\alpha_i \cos\alpha_i\,(A_r - A_i) + m\;B_r\left(\sin^2\beta_r - \cos^2\beta_r\right) = 0$$

Here

$$m = \frac{c_l}{c_s}$$

It is convenient to express the left side of these equations in terms of $\cot\alpha_i$ and $\cot\beta_r$. This yields

$$(A_i + A_r)\left[(\lambda + 2\mu)\cot^2\alpha_i + \lambda\right] m + 2\mu\;B_r \cot\beta_r = 0 \tag{4.86}$$

and $$2m\cot\alpha_i\,(A_r - A_i) + B_r\left(1 - \cot^2\beta_r\right) = 0$$

Whence

$$B_r = \frac{2m\cot\alpha_i\,(A_i - A_r)}{1 - \cot^2\beta_r} \tag{4.87}$$

Substitution of eq. 4.87 into the first equation of set 4.86 gives

$$(A_i + A_r)\left[(\lambda + 2\mu)\cot^2\alpha_i + \lambda\right]\left(1 - \cot^2\beta_r\right) + 4\mu\cot\alpha_i\cot\beta_r\,(A_i - A_r) = 0$$

Thus,

$$A_r = \frac{4\mu\cot\alpha_i\cot\beta_r - [(\lambda + 2\mu)\cot^2\alpha_i + \lambda]\,(\cot^2\beta_r - 1)}{D_1}\,A_i \tag{4.88}$$

and $$B_r = \frac{-4m\cot\alpha_i\,[(\lambda + 2\mu)\cot^2\alpha_i + \lambda]}{D_1}\,A_i,$$

where

$$D_1 = 4\mu\cot\alpha_i\cot\beta_r + \left[(\lambda + 2\mu)\cot^2\alpha_i + \lambda\right]\left(\cot^2\beta_r - 1\right)$$

It is easy to find the relationship between $\cot^2 \beta_r$ and $\cot^2 \alpha_i$, taking into account that the apparent velocity, c, along the x-axis is the same. As can be shown

$$\cot^2 \alpha_i = \frac{c^2 - c_l^2}{c_l^2} \qquad \text{or} \qquad c^2 = c_l^2 \left(1 + \cot^2 \alpha_i\right)$$

and $$\cot^2 \beta_r = \frac{c^2 - c_s^2}{c_s^2} \qquad \text{or} \qquad c^2 = c_s^2 \left(1 + \cot^2 \beta_r\right)$$

Whence

$$c_l^2 \left(1 + \cot^2 \alpha_i\right) = c_s^2 \left(1 + \cot^2 \beta_r\right),$$

i.e.

$$\cot^2 \beta_r = m^2 \left(1 + \cot^2 \alpha_i\right) - 1 \tag{4.89}$$

Thus, we have demonstrated that scalar and vector potentials given by eqs. 4.82 and 4.88 are solutions of the Helmholtz equations

$$\nabla^2 \tilde{\varphi} + k_l^2 \tilde{\varphi} = 0 \qquad \text{and} \qquad \nabla^2 \tilde{\psi} + k_s^2 \tilde{\psi} = 0,$$

correspondingly, and they obey the boundary conditions. In other words, our assumptions were correct, and these functions, $\tilde{\varphi}$ and $\tilde{\psi}$, describe the incident and reflected waves in the presence of the free plane boundary. Besides, we have demonstrated that the SH wave is absent. As we can see, the incident wave P gives rise to two reflections, namely longitudinal and shear waves. This fundamental feature of wave behavior is not observed in a fluid medium (Part II). As follows from eqs. 4.88, in such a case $(\mu = 0)$, we have

$$A_r = -A_i \qquad \text{and} \qquad B_r = -\frac{4m \cot \alpha_i}{\cot^2 \beta_r - 1} A_i$$

or, taking into account eq. 4.89,

$$B_r = -\frac{4m \cot \alpha_i}{m^2 \left(1 + \cot^2 \alpha_i\right) - 2} A_i$$

Since $c_s \to 0$ and $m \to \infty, B_r \to 0$, and the SV wave vanishes. Note that coefficient A_r, eqs. 4.88, depends on the elastic parameters (λ, μ), incident angle α_i, and amplitude A_i, whereas in fluid, amplitudes of incident and reflected waves are equal to each other. At the same time, in both media, the geometry of waves obeys Snell's law.

In accordance with eqs. 4.84, the angles of incidence and reflection of the P waves are equal to each other, and angle β_r is defined from the relation

$$\alpha_r = \alpha_i, \qquad \sin\beta_r = \frac{c_s}{c_l} \sin\alpha_i, \qquad \text{i.e.,} \qquad \beta_r < \alpha_i, \tag{4.90}$$

as is shown in Fig. 4.1c. Snell's law indicates that, regardless of the the value of angle α_i, both secondary waves, P and SV, remain homogeneous.

Reflection coefficients As follows from eqs. 4.79 and 4.82, the displacement vectors of P waves at the boundary are

$$\mathbf{s}_i = A_i \cos(k_l x \sin\alpha_i - \omega t)\ \mathbf{n}_i \qquad \text{and} \qquad \mathbf{s}_r = A_r \cos(k_l x \sin\alpha_i - \omega t)\ \mathbf{n}_r,$$

where $\mathbf{n}_i = \sin\alpha_i\ \mathbf{i} - \cos\alpha_i\ \mathbf{k}$ and $\mathbf{n}_r = \sin\alpha_i\ \mathbf{i} + \cos\alpha_i\ \mathbf{k}$ are the unit vectors of rays. The reflection coefficient $\mathcal{R}_{PP} = A_r/A_i$, and if $\mathcal{R}_{PP}$ is positive, the displacement vector is directed downward along the reflected P ray. The indices PP and PS mean that P and S waves are caused by the P incident wave. The displacement vector of the reflected SV wave is

$$\mathbf{s}_r = B_r(-\cos\beta_r\ \mathbf{i} + \sin\beta_r\ \mathbf{k}) \cos(k_s x \sin\beta_r - \omega t),$$

and the unit vector of the reflected S ray is $\mathbf{n}_r = \sin\beta_i\ \mathbf{i} + \cos\beta_i\ \mathbf{k}$. It is easy to see that $\mathbf{n}_r \cdot \mathbf{s}_r = 0$, i.e.,the displacement carried by the reflected SV wave is orthogonal to the SV ray. The reflection coefficient $\mathcal{R}_{PS} = B_r/A_i$, and if $\mathcal{R}_{PS}$ is positive, the displacement is directed toward the z-axis.

Case $\lambda = \mu$ Formulas 4.88 are essentially simplified when Poisson's ratio is $\sigma = 0.25$ or $\lambda = \mu$, which is often a rather good approximation. Then $c_l^2 = 3c_s^2$, and instead of eq. 4.89 we have

$$\cot^2\beta_r = 3\cot^2\alpha_i + 2$$

Its substitution into set 4.88 gives

$$A_r = \frac{4\cot\alpha_i \cot\beta_r - (1 + 3\cot^2\alpha_i)^2}{4\cot\alpha_i \cot\beta_r + (1 + 3\cot^2\alpha_i)^2} A_i \tag{4.91}$$

$$\text{and} \qquad B_r = -\frac{4m\cot\alpha_i\,(1 + 3\cot^2\alpha_i)}{4\cot\alpha_i \cot\beta_r + (1 + 3\cot^2\alpha_i)^2} A_i$$

Here $m = \sqrt{3}$. Simplicity of eq. 4.91 allows us to see the following features of waves. The reflected P wave vanishes if

$$4 \cot \alpha_i \cot \beta_r = \left(1 + 3 \cot^2 \alpha_i\right)^2$$

or

$$4 \cot \alpha_i \left(3 \cot^2 \alpha_i + 2\right)^{1/2} = \left(1 + 3 \cot^2 \alpha_i\right)^2 \tag{4.92}$$

This equation has two roots:

$$\alpha_i = 60^\circ \qquad \text{and} \qquad \alpha_i = 77^\circ 13'$$

There is some similarity of this case with the reflection at the boundary between two acoustic media (Brewster's angle). In both cases, the wave disappears due to the destructive interference of elementary spherical waves that arise at the boundary (Huygen's principle). Behavior of the P-wave amplitude is similar for different values of m and σ (Fig. 4.2a). First of all, with an increase of α_i, amplitude $|\mathcal{R}_{PP}|$ decreases because of the destructive interference of elementary waves. If m and σ are relatively large, $|\mathcal{R}_{PP}(\alpha_i)|$ passes a minimum value and then begins to increase, approaching 1. In such a case, the reflected wave exists for all values of the incident angle. With a decrease of these parameters, we observe two values of angle, as in the case $\lambda = \mu$, where the P reflected wave vanishes.

Behavior of the function $|\mathcal{R}_{PS}(\alpha_i)|$ is completely different. For normal incidence the amplitude of reflected SV-wave is equal to zero, but with an increase of α_i due to constructive interference, it becomes larger. Depending on the value of m, it reaches maximum in the range between 35° and 65°. Then it decreases, and for the grazing angle $\mathcal{R}_{PS}$ becomes equal to zero.

Incident SV wave

Next suppose that the shear incident wave, SV, approaches the free surface and generates there two plane waves, SV and P. Correspondingly, the complex amplitudes of potentials are

$$\tilde{\varphi}_r(x, z, \omega) = \frac{A_r}{i\, k_l} \exp\left[i\, k_l \left(x \sin \alpha_r + z \cos \alpha_r\right)\right] \quad \text{and} \tag{4.93}$$

$$\tilde{\psi}_1(x, z, \omega) = \frac{B_i}{i\, k_s} \exp\left[i\, k_s \left(x \sin \beta_i - z \cos \beta_i\right)\right] + \frac{B_r}{i\, k_s} \exp\left[i\, k_s \left(x \sin \beta_r + z \cos \beta_r\right)\right]$$

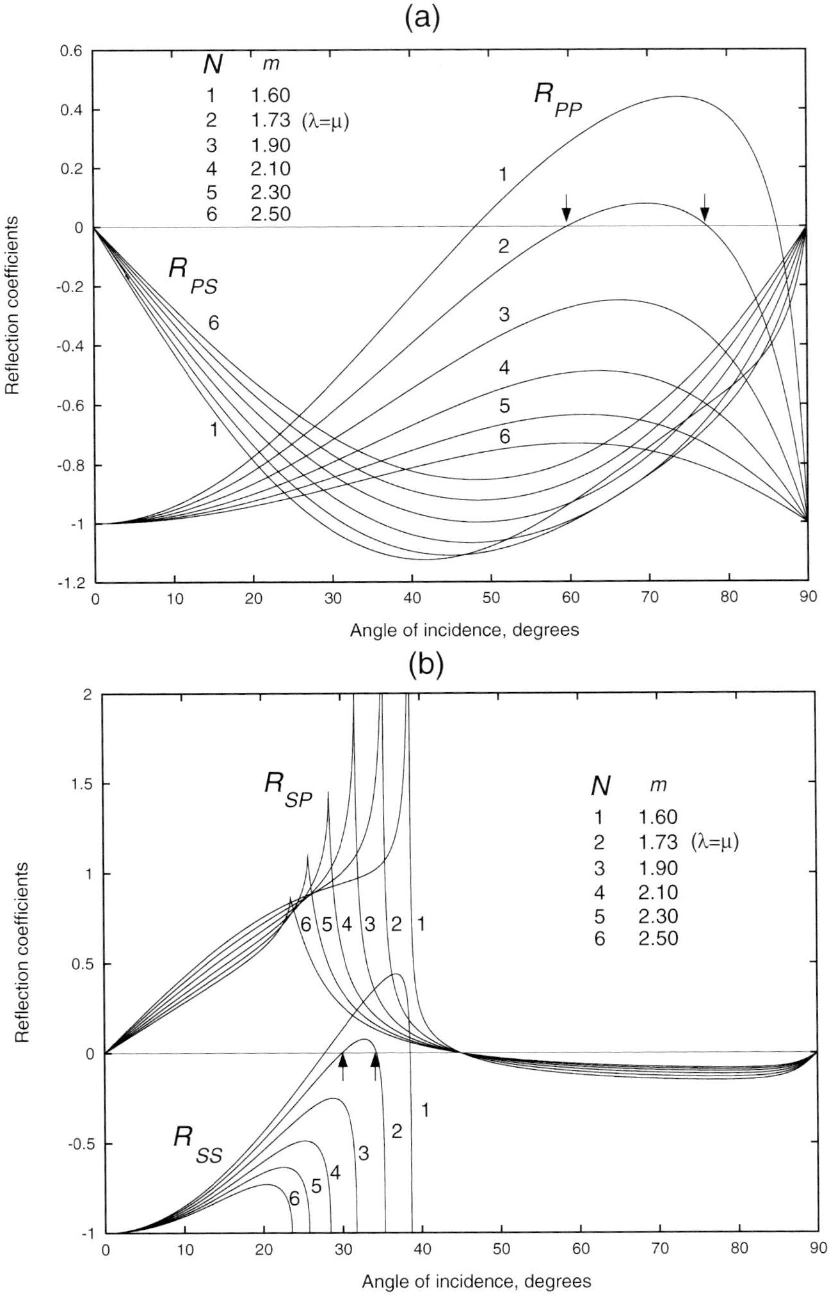

Figure 4.2: Reflection from the free surface for different values of $m = c_l/c_s$: (a) incident P wave (b) incident SV wave. Numbers near curves correspond to different values of m; arrows show Brewster's angles for $\lambda = \mu$ $(\sigma = 0.25)$.

To determine the unknowns A_r, B_r, α_r, and β_r, we again use the boundary conditions, eqs. 4.81. As we know, they can be satisfied if all arguments in eqs. 4.93 are equal to each other at the free surface:

$$k_s\, x \sin\beta_i = k_s\, x \sin\beta_r = k_l x \sin\alpha_r$$

or

$$\frac{\sin\beta_i}{c_s} = \frac{\sin\beta_r}{c_s} = \frac{\sin\alpha_r}{c_l} \qquad (4.94)$$

This shows that the incident and reflection angles of the SV wave are equal:

$$\beta_r = \beta_i \qquad (4.95)$$

At the same time, we have

$$\frac{\sin\beta_i}{c_s} = \frac{\sin\alpha_r}{c_l} \qquad (4.96)$$

Eqs. 4.95 and 4.96 represent Snell's law for the incident SV wave. As follows from eq. 4.96,

$$\sin\alpha_r = m\sin\beta_i, \qquad (4.97)$$

Since $m > 1$ (Fig. 4.1d), $\alpha_r > \beta_i$. If

$$\sin\beta_i = \frac{c_s}{c_l},$$

then the angle α_r becomes equal to $\pi/2$, and the reflected wave P slides along the free surface. By analogy with acoustic waves, angle β_c is called the critical angle. When $\beta_i > \beta_c$, the P wave becomes inhomogeneous.

In order to determine the unknowns, A_r and B_r, we substitute eqs. 4.93 into set 4.81 and obtain

$$\left(\lambda + 2\mu\cos^2\alpha_r\right) k_l A_r + 2\mu\; k_s\,(B_r - B_i)\sin\beta_i\cos\beta_i = 0$$

and

$$2k_l\sin\alpha_r\cos\alpha_r\; A_r + k_s\,(B_i + B_r)\left(\sin^2\beta_i - \cos^2\beta_i\right) = 0$$

or

$$\left(\lambda + 2\mu\cos^2\alpha_r\right)\; A_r + 2\mu\; m\,(B_r - B_i)\sin\beta_i\cos\beta_i = 0 \qquad (4.98)$$

and $$2\sin\alpha_r\cos\alpha_r A_r + m\,(B_i + B_r)\left(\sin^2\beta_i - \cos^2\beta_i\right) = 0$$

Using Snell's law, we obtain

$$B_r = \frac{4\mu\cot\alpha_r\cot\beta_i - [\lambda + (\lambda+2\mu)\cot^2\alpha_r](\cot^2\beta_i - 1)}{D_2}\,B_i \tag{4.99}$$

and $$A_r = \frac{1}{m}\frac{4\mu\cot\beta_i\,(\cot^2\beta_i - 1)}{D_2}\,B_i$$

where

$$D_2 = 4\mu\cot\alpha_r\cot\beta_i + \left[\lambda + (\lambda+2\mu)\cot^2\alpha_r\right]\left(\cot^2\beta_i - 1\right) \tag{4.100}$$

Reflection coefficients The displacement vectors of the incident and reflected SV waves at the boundary are

$$\mathbf{s}_i = B_i(\cos\beta_i\,\mathbf{i} + \sin\beta_i\,\mathbf{k})\cos(k_s x\sin\beta_i - \omega t)$$

and $$\mathbf{s}_r = B_r(-\cos\beta_i\,\mathbf{i} + \sin\beta_i\,\mathbf{k})\cos(k_s x\sin\beta_i - \omega t),$$

whereas $\mathbf{n}_i = \sin\beta_i\,\mathbf{i} - \cos\beta_i\,\mathbf{k}$ and $\mathbf{n}_r = \sin\beta_i\,\mathbf{i} + \cos\beta_i\,\mathbf{k}$ are the unit vectors of rays. It is evident that $\mathbf{n}_i\cdot\mathbf{s}_i = \mathbf{n}_r\cdot\mathbf{s}_r = 0$. The reflection coefficient of wave SV is $\mathcal{R}_{SS} = B_r/B_i$. The indices SS and SP mean that P and S waves are caused by the S incident wave. We see that displacement carried by the reflected SV wave is orthogonal to the ray, and its sign depends on the sign of $\mathcal{R}_{SS}$. If $\mathcal{R}_{SS}$ is positive, the displacement is directed toward the z-axis, like the displacement in the incident wave. At the same time $\mathcal{R}_{SP} = A_r/B_i$, and if $\mathcal{R}_{SP}$ is positive, the displacement vector for the P wave is directed downward along the reflected P ray. [1]

Case $\lambda = \mu$ Then eqs. 4.99 and 4.100 are simplified, and we have

$$B_r = \frac{4\cot\alpha_r\cot\beta_i - (1 + 3\cot^2\alpha_r)^2}{4\cot\alpha_r\cot\beta_i + (1 + 3\cot^2\alpha_r)^2}\,B_i \tag{4.101}$$

and $$A_r = \frac{4}{m}\,\frac{\cot\beta_i\,(1 + 3\cot^2\alpha_r)}{4\cot\alpha_r\cot\beta_i + (1 + 3\cot^2\alpha_r)^2}\,B_i$$

---

[1] It is possible, of course, to present coefficients R_{PP}, R_{PS}, R_{SS}, *and* R_{SP} in a slightly different way, using equalities $\lambda + 2\mu = c_l^2\rho$, $\mu = c_s^2\rho$, and Snell's law (see, for example, Aki and Richards, 1980).

The expression for B_r shows that the reflected wave SV vanishes when angles obey the equation

$$4 \cot \alpha_r \cot \beta_i = \left(1 + 3 \cot^2 \alpha_r\right)^2,$$

which has two roots:

$$\beta_i = 34°76' \qquad \text{and} \qquad \beta_i = 30°$$

Behavior of the reflected waves is shown in Fig. 4.2b. For relatively large values of m and σ the amplitude of the reflected SV wave becomes smaller with an increase of β_i and reaches a minimum near $\beta_i = 32°$. Then it starts to increase. For smaller values of m and σ, this wave disappears at two angles between $25°$ and $40°$. Amplitudes of the reflected P wave are equal to zero values for normal incidence, smoothly increase with increased angle and pass sharp peaks near the critical angle.

Reflected waves beyond the critical angle As we already know, if $\beta_i > \beta_c$, Snell's law for the reflected P wave becomes invalid ($\sin \alpha_r > 1$). Correspondingly,

$$\cos \alpha_r = \sqrt{1 - \frac{c_l^2}{c_s^2} \sin^2 \beta_i} = i\, b_z \qquad \text{and} \qquad b_z = \sqrt{\frac{c_l^2}{c_s^2} \sin^2 \beta_i - 1}$$

Therefore, for the potential of the reflected P wave, eqs. 4.93, we have

$$\tilde{\varphi}_r (x, z, \omega) = \frac{A_r}{ik_l} \exp(-k_l b_z z + ik_l x \sin \alpha_r), \tag{4.102}$$

where A_r is now complex. We see that an evanescent P wave propagates along the free surface with the velocity $c = c_s / \sin \beta_i$, which varies with the angle of incidence β_i within the range $c_s < c \le c_l$. The amplitude of this wave, $A_r \exp(-k_l b_z z)$, decays exponentially with an increase of depth z. The rate of its change depends on the angle of incidence. In particular, at the critical angle, the parameter b_z is equal to zero, and the reflected P wave becomes homogeneous. With an increase of β_i, the evanescent wave decays more rapidly. Since $k_l = \omega / c_l$, the exponential term $\exp(-k_l b_z z)$ depends on a frequency. Correspondingly, the high frequency harmonics concentrate near a surface. As follows from equations for displacement components, during each period particles of a medium move along an ellipse whose parameters change with depth z. The evanescent (inhomogeneous) P wave is always accompanied by the reflected homogeneous SV wave, which moves away from the free surface through an elastic medium. Its apparent velocity along the boundary is the same as that of the inhomogeneous P wave. As is seen from eqs. 4.99 and 4.100, the reflection coefficient $\mathcal{R}_{SS}$ is now complex and may be presented as a ratio,

$$\mathcal{R}_{SS} = \frac{iE - F}{iE + F}, \tag{4.103}$$

where

$$E = 4\mu b_z \frac{c}{c_l} \cot \beta_i \quad \text{and} \quad F = \left[\lambda + (\lambda + 2\mu) \left(\frac{b_z c}{c_l} \right)^2 \right] \left(\cot^2 \beta_i - 1 \right) \tag{4.104}$$

It is evident that in this case

$$\mathcal{R}_{SS} = \exp\left(-i\Psi\right), \quad \Psi = 2 \tan^{-1} \frac{E}{F} \tag{4.105}$$

This means that the reflected SV wave has the same amplitude as the incident SV wave, but it is shifted in phase. Thus, the resulting total reflected wave consists of two parts, namely, the evanescent P wave and the homogeneous SV wave. For an observer at the surface, it is impossible to distinguish them. For instance, the displacement field of the reflected waves can be represented as

$$\mathbf{s} = \text{grad}\ \varphi + \text{curl}\ \boldsymbol{\psi},$$

and simultaneously we observe different types of motion and deformation that are typical for compressional and shear waves. Because a portion of the energy of this reflected wave moves away from the boundary, with an increase of depth z the reflected wave SV becomes dominant and, therefore, polarization becomes linear. This analysis shows that only the superposition of the incident wave and both reflected waves satisfies the boundary conditions – that is, the normal and shear stresses are equal to zero:

$$\tau_{zz} = 0, \qquad \tau_{xz} = 0$$

This means that at each point of the boundary, the wavefield is a result of the superposition of all waves. Of course, the same is correct when the angle of incidence is smaller than the critical angle. Thus, we have demonstrated that a plane wave cannot create a surface wave that satisfies the boundary conditions. The same is true if the boundary is ideally rigid. Also note that there is a phase shift between the incident and reflected waves at points of the boundary. Because of this, at every instant part of the energy of the incident wave is transformed into energy of the reflected SV wave, but the other part is transformed into energy of the evanescent P wave. As was mentioned earlier, Snell's law for the reflected SV wave is valid for all values of β_i.

Incident SH wave

Finally, assume that the SH wave is incident and that displacement has the v-component only.

$$\tilde{v}_i\,(x,z,\omega) = C_i \exp\left[i\;k_s\,(x\sin\gamma_i + z\cos\gamma_i)\right] \tag{4.106}$$

Since this wave is accompanied by the single stress component τ_{yz}, let us suppose that the reflected wave is also an SH wave, and therefore

$$\tilde{v}\,(x,z,\omega) = C_i \exp\left[i\;k_s\,(x\sin\gamma_i + z\cos\gamma_i)\right] + C_r \exp\left[i\;k_s\,(x\sin\gamma_r - z\cos\gamma_r)\right] \tag{4.107}$$

Here γ_i and γ_r are angles of incidence and reflection, respectively. At the boundary, we have

$$\tilde{\tau}_{yz} = \mu\,\frac{\partial\tilde{v}}{\partial z} = 0$$

By analogy with the two previous cases, we conclude that this equality is satisfied, provided that

$$\sin\gamma_i = \sin\gamma_r \qquad \text{or} \qquad \gamma_i = \gamma_r, \tag{4.108}$$

and we again arrive at Snell's law. In the same way as before, we obtain

$$C_r = C_i, \quad \text{and} \quad \mathcal{R}_{SS} = 1 \tag{4.109}$$

Hence

$$\tilde{v}\,(x,z,\omega) = C_i\{\exp\left[i\;k_s\,(x\sin\gamma_i + z\cos\gamma_i)\right] + \exp\left[i\;k_s\,(x\sin\gamma_i - z\cos\gamma_i)\right]\} \tag{4.110}$$

We see that displacement at the free surface is doubled, which happens for any angle of incidence. The similarity with the behavior of acoustic waves is obvious (Part II). Since

$$v\,(x,z,t) = \text{Re}\left[\,\tilde{v}\,(x,z,\omega)\;e^{-i\omega t}\right],$$

we have

$$v\,(x,z,t) = C_i\{\cos\left[k_s\,(x\sin\gamma_i + z\cos\gamma_i) - \omega t\right] + \cos\left[k_s\,(x\sin\gamma_i - z\cos\gamma_i) - \omega t\right]\}$$

or

$$v\,(x,z,t) = 2C_i\cos\left(k_s z\cos\gamma_i\right)\cos\left(\omega t - k_s x\sin\gamma_i\right) \tag{4.111}$$

Thus, superposition of the incident and reflected waves produces a standing wave along the z-axis and a wave propagating along the x-axis with the velocity

$$c = \frac{c_s}{\sin \gamma_i} \tag{4.112}$$

In conclusion note the following. The energy of the incident wave is transformed into that of the reflected waves. For instance, in the case of the incident SV wave, the energy is distributed between the reflected SV and P waves. Imagine three elementary vector tubes with a common point at the interface. Then the amount of energy moving through the tube of the incident wave is equal to the sum of energies flowing through the two other tubes during the same time interval. If the angle of incidence exceeds the critical angle $\beta_i > \beta_c$, the mean values of the energy of the incident and reflected SV waves are equal to each other.

4.5 Reflection from the rigid surface (oblique incidence)

Suppose now that plane $z = 0$ is the interface between an elastic medium and an ideally rigid medium. To illustrate reflection, consider the incidence of an SV wave and assume that the reflected P and SV waves arise. Correspondingly, the complex amplitudes of the scalar potential and the y-component of the vector potential are

$$\tilde{\varphi}_r(x, z, \omega) = \frac{A_r}{i\, k_l} \exp\left[i\, k_l\, (x \sin \alpha_r + z \cos \alpha_r)\right] \tag{4.113}$$

and $$\tilde{\psi}(x, z, \omega) = \frac{B_i}{i\, k_s} \exp\left[i\, k_s\, (x \sin \beta_i - z \cos \beta_i)\right] + \frac{B_r}{i\, k_s} \exp\left[i\, k_s\, (x \sin \beta_r + z \cos \beta_r)\right]$$

By definition, at the boundary all three components of displacement are equal to zero, i.e.,

$$\tilde{u}\,(x, 0, \omega) = 0, \qquad \tilde{v}\,(x, 0, \omega) = 0, \qquad \tilde{w}\,(x, 0, \omega) = 0 \tag{4.114}$$

Our assumptions about waves imply that the component v is absent everywhere. In terms of potentials we have

$$\frac{\partial \tilde{\varphi}}{\partial x} - \frac{\partial \tilde{\psi}}{\partial z} = 0 \qquad \text{and} \qquad \frac{\partial \tilde{\varphi}}{\partial z} + \frac{\partial \tilde{\psi}}{\partial x} = 0 \quad \text{at} \quad z = 0 \tag{4.115}$$

Substitution of eqs. 4.113 into set 4.115 allows us to determine the unknowns A_r and B_r, and to establish again Snell's law:

$$\beta_r = \beta_i \qquad \text{and} \qquad \frac{\sin\alpha_r}{c_l} = \frac{\sin\beta_i}{c_s}$$

The system of equations with respect to A_r and B_r is

$$\sin\alpha_r\ A_r + (B_i - B_r)\cos\beta_i = 0, \tag{4.116}$$

$$\cos\alpha_r\ A_r + (B_i + B_r)\sin\beta_i = 0$$

Hence

$$B_r = \frac{\cos\alpha_r\ \cos\beta_i - \sin\alpha_r\ \sin\beta_i}{\cos\alpha_r\ \cos\beta_i + \sin\alpha_r\ \sin\beta_i}\, B_i \tag{4.117}$$

$$\text{and} \qquad A_r = \frac{-2\sin\beta_i\cos\beta_i}{\cos\alpha_r\ \cos\beta_i + \sin\alpha_r\ \sin\beta_i}\, B_i$$

or

$$B_r = \frac{\cos(\alpha_r + \beta_i)}{\cos(\alpha_r - \beta_i)}\, B_i \qquad \text{and} \qquad A_r = -\frac{\sin 2\beta_i}{\cos(\alpha_r - \beta_i)}\, B_i \tag{4.118}$$

First consider the behavior of reflected waves when the angle of incidence is smaller than the critical angle:

$$\beta_i < \beta_c = \sin^{-1}\frac{c_s}{c_l}$$

For the normal incidence, $\beta_i = 0$, we obtain

$$B_r = B_i \qquad \text{and} \qquad A_r = 0,$$

i.e., only the reflected SV wave arises. With an increase of β_i, the amplitude of the reflected SV wave becomes smaller and the P wave appears. Beyond the critical angle, $(\beta > \beta_c)$, the P wave becomes evanescent and exponentially decays with distance from the boundary. At the boundary, there is a phase shift between the incident wave and the reflected SV wave, and their amplitudes are equal to each other: $|B_r| = B_i$.

4.6 Reflection and transmission at the boundary between a fluid and an elastic medium

As is well known (Appendix D), in the vicinity of the boundary, $(z = 0)$, particles of a fluid medium and an elastic medium are not rigidly connected to each other. Correspondingly, tangential components of displacement may have different values on each side of the interface. In other words, in general, they are discontinuous functions on $z = 0$. At the same time, the normal component of displacement is a continuous function. Otherwise we would observe either a gap between the two media or their overlapping. We also take into account that all stresses are continuous functions, and shear stresses are absent in a fluid. This means that in the vicinity of the boundary, shear stresses in an elastic medium are also equal to zero. Thus, the boundary conditions are

$$w = w_1 \quad \text{and} \quad \tau_{zz} = \tau_{zz}^{(1)}, \quad \tau_{xz}^{(1)} = 0, \quad \tau_{yz}^{(1)} = 0 \quad \text{on} \quad z = 0 \tag{4.119}$$

Here the index "1" shows that displacement and stresses, as well as other wave characteristics, are considered in an elastic medium. Recalling that in a fluid $\mu = 0$ and using Hooke's law, set 4.119 can be represented as

$$w = w_1, \qquad \lambda \text{ div } \mathbf{s} = \lambda_1 \text{ div } \mathbf{s}_1 + 2\mu_1 \frac{\partial w_1}{\partial z} \tag{4.120}$$

$$\text{and} \qquad \frac{\partial u_1}{\partial z} + \frac{\partial w_1}{\partial x} = 0, \qquad \frac{\partial v_1}{\partial z} + \frac{\partial w_1}{\partial y} = 0 \qquad \text{on} \quad z = 0$$

Here u_1, v_1, and w_1 are components of displacement in an elastic medium, while u, v, and w describe vector $\mathbf{s}$ in a fluid. Parameters λ_1 and μ_1 along with density ρ_1 define the velocity of propagation of longitudinal and shear waves:

$$\lambda_1 + 2\mu_1 = \rho_1 c_l^2, \qquad \mu_1 = \rho_1 c_s^2 \tag{4.121}$$

Parameter λ plays the role of the bulk modulus of the fluid: $\lambda = \rho\, c^2$. In studying reflection, our main attention is paid to the case in which the incident plane wave propagates through a fluid and its phase surface is parallel to the y-axis, Fig. 4.3a. In other words, wavefields are independent of the y-coordinate. In particular,

$$\frac{\partial u}{\partial y} = \frac{\partial v}{\partial y} = \frac{\partial w}{\partial y} = 0 \tag{4.122}$$

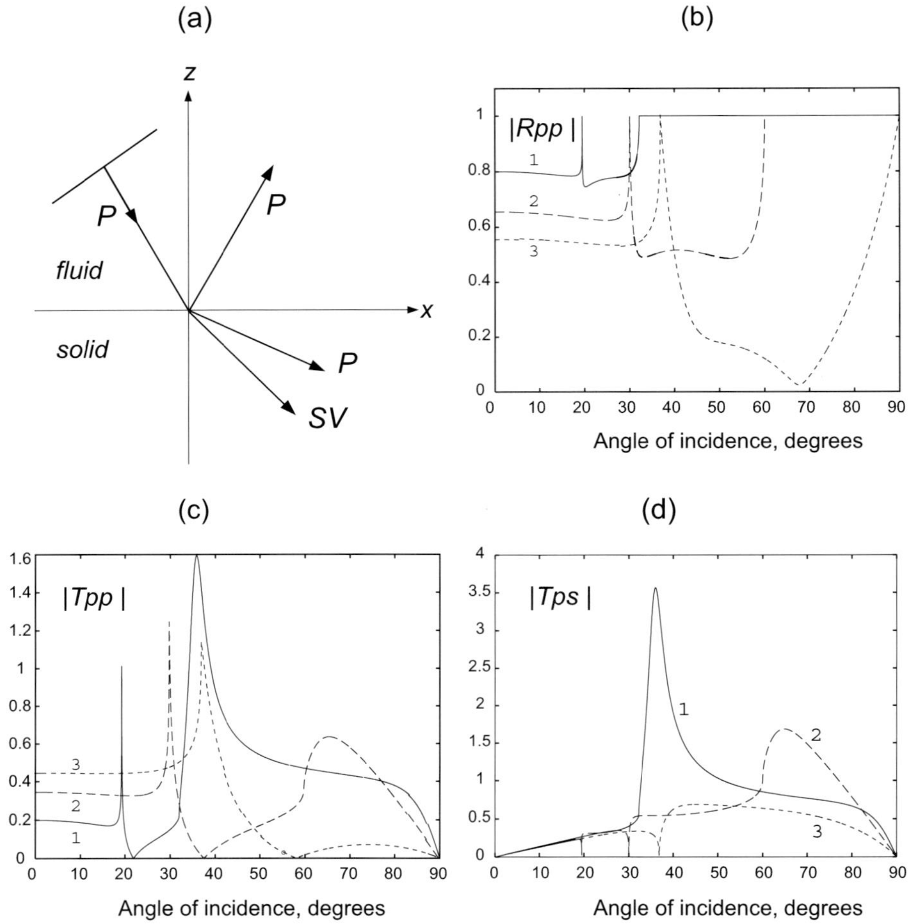

Figure 4.3: Reflection and transmission at the boundary between fluid and solid media. The incident P wave in fluid: (a) Ray scheme (b) Reflection coefficient $|Rpp|$ as a function af the angle of incidence and parameters of media (c) Transmission coefficient $|Tpp|$ (d) Transmission coefficient $|Tps|$. Numbers near the curves correspond to different solid media: 1. $c_l = 4.5$ km/s, $c_s = 2.81$ km/s, $\rho_1 = 3.0$ g/cm^3; 2. $c_l = 3.0$ km/s, $c_s = 1.73$ km/s, $\rho_1 = 2.4$ g/cm^3; 3. $c_l = 2.5$ km/s, $c_s = 1.25$ km/s, $\rho_1 = 2.1$ g/cm^3. In fluid , $c = 1.5$ km/s, $\rho = 1.0$ g/cm^3.

At the boundary, the incident wave gives rise to secondary waves, and we assume that the plane wave P appears in a fluid, whereas in an elastic medium, the plane waves P and SV arise. This implies that the displacement component along the y-axis is absent: $v = 0$. As in previous sections, it is convenient to introduce scalar and vector potentials:

$$\mathbf{s} = \text{grad}\ \varphi \qquad \text{if} \qquad z < 0, \tag{4.123}$$

and

$$\mathbf{s}_1 = \text{grad}\ \varphi_1 + \text{curl}\ \boldsymbol{\psi}_1, \qquad \text{where} \qquad \boldsymbol{\psi}_1 = \psi_1 \mathbf{j} \tag{4.124}$$

Then, eqs. 4.122–4.124 give

$$u = \frac{\partial \varphi}{\partial x}, \qquad v = 0, \qquad w = \frac{\partial \varphi}{\partial z} \tag{4.125}$$

$$\text{and} \qquad u_1 = \frac{\partial \varphi_1}{\partial x} - \frac{\partial \psi_1}{\partial z}, \qquad v_1 = 0, \qquad w_1 = \frac{\partial \varphi_1}{\partial z} + \frac{\partial \psi_1}{\partial x}$$

Because we are considering sinusoidal waves, we can also apply eqs. 4.119–4.125 to the complex amplitudes of displacement and potentials. Then, substituting eqs. 4.125 into set 4.120 and using equalities

$$\text{div}\ \widetilde{\mathbf{s}} = \text{div grad}\ \widetilde{\varphi} = \nabla^2 \widetilde{\varphi} = \frac{\partial \widetilde{u}}{\partial x} + \frac{\partial \widetilde{w}}{\partial z}$$

$$\text{and} \qquad \text{div}\ \widetilde{\mathbf{s}}_1 = \text{div grad}\ \widetilde{\varphi}_1 = \nabla^2 \widetilde{\varphi}_1 = \frac{\partial \widetilde{u}_1}{\partial x} + \frac{\partial \widetilde{w}_1}{\partial z},$$

we obtain the boundary conditions in terms of potentials

$$\begin{gathered}
\frac{\partial \widetilde{\varphi}}{\partial z} = \frac{\partial \widetilde{\varphi}_1}{\partial z} + \frac{\partial \widetilde{\psi}_1}{\partial x} \\
\lambda\ \nabla^2 \widetilde{\varphi} = \lambda_1\ \nabla^2 \widetilde{\varphi}_1 + 2\mu_1 \left(\frac{\partial^2 \widetilde{\varphi}_1}{\partial z^2} + \frac{\partial^2 \widetilde{\psi}_1}{\partial x \partial z} \right) \\
2\, \frac{\partial^2 \widetilde{\varphi}_1}{\partial x \partial z} + \frac{\partial^2 \widetilde{\psi}_1}{\partial x^2} - \frac{\partial^2 \widetilde{\psi}_1}{\partial z^2} = 0 \qquad \text{on} \quad z = 0
\end{gathered} \tag{4.126}$$

Note that the last boundary condition of set 4.119, ($\tau_{yz}=0$), is met for any field **s**, as soon as eqs. 4.122 and 4.125 are valid. As was demonstrated earlier, potentials can be represented in the form

$$\widetilde{\varphi}(x,y,\omega)=\frac{A_i}{ik}\exp\left[ik\left(x\sin\alpha_i+z\cos\alpha_i\right)\right]+\frac{A_r}{ik}\exp\left[ik\left(x\sin\alpha_r-z\cos\alpha_r\right)\right],$$

$$\widetilde{\varphi}_1(x,z,\omega)=\frac{A_2}{i\,k_l}\exp\left[ik_l\ \left(x\sin\alpha_2+z\cos\alpha_2\right)\right], \tag{4.127}$$

and $$\widetilde{\psi}_1(x,z,\omega)=\frac{B_2}{i\,k_s}\exp\left[ik_s\left(x\sin\beta_2+z\cos\beta_2\right)\right]$$

First of all, substitution of eqs. 4.127 into the boundary conditions, eqs. 4.126, leads us again to Snell's laws of reflection and refraction:

$$\alpha_r=\alpha_i \qquad \text{and} \qquad \frac{\sin\alpha_2}{c_l}=\frac{\sin\alpha_i}{c}, \qquad \frac{\sin\beta_2}{c_s}=\frac{\sin\alpha_i}{c}, \tag{4.128}$$

or

$$\frac{\sin\alpha_i}{c}=\frac{\sin\alpha_2}{c_l}=\frac{\sin\beta_2}{c_s}$$

Correspondingly, the system of equations for determining of unknowns A_r, A_2, and B_2, eqs. 4.126, is greatly simplified, and we obtain

$$\cos\alpha_i\left(A_i-A_r\right)=\cos\alpha_2\ A_2+\sin\beta_2\ B_2$$

$$\lambda\ k\left(A_i+A_r\right)=\lambda\ k_lA_2+2\mu_1\left(k_l\cos^2\alpha_2\ A_2+k_s\sin\beta_2\cos\beta_2\ B_2\right) \tag{4.129}$$

$$2\ k_l\sin\alpha_2\cos\alpha_2\ A_2+k_s\ B_2\left(\sin^2\beta_2-\cos^2\beta_2\right)=0$$

Introducing notations (Part II)

$$Z^*=\frac{\rho\ c}{\cos\alpha_i}, \qquad Z_l^*=\frac{\rho_1\ c_l}{\cos\alpha_2}, \qquad Z_s^*=\frac{\rho_1\ c_s}{\cos\beta_2}$$

and making use of an equality

$$k_l\left(\lambda_1+2\mu_1\cos^2\alpha_2\right)=k_l\left(\lambda_1+2\mu_1\right)-k_l\ 2\mu_1\sin^2\alpha_2=$$

$$k_l\rho_1 c_l^2 - 2k_l\rho_1\, c_s^2\, \frac{c_l^2}{c_s^2}\, \sin^2\beta_2 = k_l\rho_1 c_l^2\, \cos 2\beta_2,$$

set 4.129 becomes

$$\cos\alpha_i\,(A_i - A_r) = \cos\alpha_2\, A_2 + \sin\beta_2\, B_2$$

$$Z^*\cos\alpha_i\,(A_i + A_r) = Z_l^*\cos\alpha_2\cos 2\beta_2\, A_2 + Z_s^*\sin 2\beta_2\cos\beta_2 B_2 \tag{4.130}$$

$$Z_s^*\sin 2\alpha_2\cos\beta_2\, A_2 - Z_l^*\cos 2\beta_2\cos\alpha_2 B_2 = 0$$

The first two equations give

$$2Z^*\cos\alpha_i A_i = (Z^* + Z_l^*\cos 2\beta_2)\cos\alpha_2\, A_2 + \tag{4.131}$$

$$\left(Z^* + 2Z_s^*\cos^2\beta_2\right)\sin\beta_2 B_2 = 0$$

From the last equation of set 4.130, we have

$$B_2 = \frac{Z_s^*\sin 2\alpha_2\ \cos\beta_2}{Z_l^*\cos 2\beta_2\ \cos\alpha_2}A_2 \tag{4.132}$$

Its substitution into eq. 4.131 gives

$$2Z^*Z_l^*\cos\alpha_i\cos\alpha_2\ \cos 2\beta_2\ A_i = D\ A_2, \tag{4.133}$$

where

$$D = (Z^* + Z_l^*\cos 2\beta_2)\, Z_l^*\cos^2\alpha_2\cos 2\beta_2 + \left(Z^* + 2Z_s^*\cos^2\beta_2\right)\cdot$$

$$Z_s^*\sin 2\alpha_2\ \sin\beta_2\cos\beta_2 = Z^*\left(Z_l^*\cos 2\beta_2\ \cos^2\alpha_2\ + Z_s^*\sin 2\alpha_2\ \sin\beta_2\cos\beta_2\right) +$$

$$Z_l^{*2}\cos^2 2\beta_2\cos^2\alpha_2\ + Z_s^{*2}\cos^2\beta_2\sin 2\beta_2\sin 2\alpha_2 =$$

$$Z^*Z_l^*\left(\cos 2\beta_2\cos^2\alpha_2\ + \frac{\cos\alpha_2\sin\beta_2}{\cos\beta_2\sin\alpha_2}\sin 2\alpha_2\ \sin\beta_2\cos\beta_2\right) +$$

$$Z_l^*(Z_l^* \cos^2 2\beta_2 \cos^2 \alpha_2 + Z_s^* \frac{\sin \beta_2 \cos \alpha_2}{\sin \alpha_2 \cos \beta_2} \cos^2 \beta_2 \sin 2\beta_2 \sin 2\alpha_2) =$$

$$Z^* Z_l^* \cos^2 \alpha_2 + Z_l^* \cos^2 \alpha_2 \left(Z_l^* \cos^2 2\beta_2 + Z_s^* \sin^2 2\beta_2 \right)$$

Whence

$$A_2 = \frac{2Z^* \cos \alpha_i \cos 2\beta_2 \; A_i}{\cos \alpha_2 \left(Z_l^* \cos^2 2\beta_2 + Z_s^* \sin^2 2\beta_2 + Z^* \right)} \tag{4.134}$$

or

$$A_2 = \frac{1}{m \; n_l} \frac{2Z_l^* \cos 2\beta_2 \; A_i}{Z_l^* \cos^2 2\beta_2 + Z_s^* \sin^2 2\beta_2 + Z^*}$$

Here

$$m = \frac{\rho_1}{\rho} \qquad \text{and} \qquad n_l = \frac{c_l}{c}$$

From eq. 4.132, we have

$$B_2 = \frac{1}{m \; n_s} \frac{2Z_s^* \sin 2\beta_2 \; A_i}{Z_l^* \cos^2 2\beta_2 + Z_s^* \sin^2 2\beta_2 + Z^*} \tag{4.135}$$

and

$$n_s = \frac{c_s}{c}$$

Substitution of eqs. 4.134 and 4.135 into the first equation of set 4.130 yields

$$A_i - A_r = \frac{2Z^* A_i}{Z_l^* \cos^2 2\beta_2 + Z_s^* \sin^2 2\beta_2 + Z^*}$$

Thus

$$A_r = \frac{Z_l^* \cos^2 2\beta_2 + Z_s^* \sin^2 2\beta_2 - Z^*}{Z_l^* \cos^2 2\beta_2 + Z_s^* \sin^2 2\beta_2 + Z^*} A_i \tag{4.136}$$

We have demonstrated that if coefficients A_r, A_2, and B_2 are given by eqs. 4.134–4.136, boundary conditions are satisfied and, therefore, our assumptions about reflected and transmitted waves are correct.

Wave behavior

As follows from Snell's law, in the simplest case of normal incidence $(\alpha_i = 0)$, we have

$$\alpha_2 = \beta_2 = 0,$$

and, correspondingly:

$$A_r = \frac{Z_l - Z}{Z_l + Z}\,A_i, \qquad A_2 = \frac{2Z}{Z_l + Z}\,A_i, \qquad B_2 = 0,$$

where

$$Z = \rho\, c, \qquad Z_l = \rho_1 c_l, \qquad Z_s = \rho_1 c_s$$

are impedances for acoustic, longitudinal, and shear waves, respectively. In such a case a shear wave does not arise, and we again arrive at formulas that correspond to an acoustic medium.

Also, it is interesting to note that there is an angle of incidence α_i when the P wave is absent in an elastic medium (Fig. 4.3c). In fact, from eq. 4.134 it follows that

$$A_2 = 0 \qquad \text{if} \qquad \beta_2 = \pi/4$$

Since

$$\frac{\sin\beta_2}{c_s} = \frac{\sin\alpha_i}{c},$$

we have

$$\alpha_i = \sin^{-1}\frac{c}{c_s\sqrt{2}} \tag{4.137}$$

Certainly, this happens because of the destructive interference of elementary spherical waves of the P type. As a result, the only wave in an elastic medium is the SV wave, and eqs. 4.134–4.136 give

$$A_2 = 0\,, \qquad B_2 = \frac{1}{m\,n_s}\,\frac{2Z_s^*\,A_i}{Z_s^* + Z^*}, \qquad \text{and} \qquad A_r = \frac{Z_s^* - Z^*}{Z_s^* + Z^*}\,A_i \tag{4.138}$$

In analyzing wave behavior, suppose that at the beginning

1. $c_s < c < c_l$

From Snell's law,

$$\sin\alpha_2 = \frac{c_l}{c}\sin\alpha_i, \qquad \sin\beta_2 = \frac{c_s}{c}\sin\alpha_i,$$

it is clear that for all angles of incidence an equality $\sin\beta_2 < 1$ takes place and, therefore, the plane wave SV is homogeneous, regardless of α_i. However, the P wave in an elastic medium is homogeneous only if

$$\sin\alpha_i \leq \frac{c}{c_l}$$

At greater angles, it becomes inhomogeneous, and it exponentially decays with distance from the boundary. In previous sections, we have shown that beyond the critical angle α_{c1} $\left(\alpha_{c1} = \sin^{-1}\, c/c_l\right)$,

$$\cos\alpha_2 = \sqrt{1-\sin^2\alpha_2} = i\ b_z$$

Therefore, $Z_l^* = -i\ |Z_l^*|$, and the reflection coefficient A_r becomes

$$A_r = \frac{Z_s^*\sin^2 2\beta_2 - Z^* - i\ |Z_l^*|\cos^2 2\beta_2}{Z_s^*\sin^2 2\beta_2 + Z^* - i\ |Z_l^*|\cos^2 2\beta_2}\, A_i \tag{4.139}$$

Coefficient A_r is a complex number and, unlike in the case of total internal reflection in an acoustic medium (Part II), its magnitude is smaller than unity. This fact is easily explained, since the part of the energy of the incident wave is transformed into the energy of the shear wave. Thus, if the angle of incidence does not exceed the critical angle α_c, both transmitted waves are homogeneous, and at the boundary the incident and reflected waves are in phase. When $\alpha_i > \alpha_{c1}$, the shear wave is still homogeneous, but the P wave in an elastic medium becomes evanescent and propagates along the interface. At the same time, there is a phase shift between the incident and reflected waves at the interface points.

2. $c < c_s < c_l$

In the second case, it is convenient to distinguish several ranges of the angle of incidence α_i. If $0 < \sin\alpha_i \leq c/c_l$, then both transmitted waves are homogeneous, and the coefficient of reflection is real, $(A_r < A_i)$. In the second range

$$\frac{c}{c_l} < \sin\alpha_i < \frac{c}{c_s},$$

angle β_2 is real but angle α_2 is complex. In other words, the P wave is evanescent and moves along the boundary, whereas the SV wave remains homogeneous. Finally, when

$$\sin \alpha_i > \frac{c}{c_s},$$

angles α_2 and β_2 are both complex. This means that the transmitted P and SV waves are inhomogeneous, and they travel with the same velocity along the boundary. Correspondingly, Z_l^* and Z_s^* are purely imaginary numbers, and the reflected coefficient is equal to

$$A_r = \frac{i\ \left(|Z_l^*|\cos^2 2\beta_2 + \ |Z_s^*|\sin^2 2\beta_2\right) + Z^*}{i\ \left(|Z_l^*|\cos^2 2\beta_2 + |Z_s^*|\sin^2 2\beta_2\right) - Z^*}\, A_i \tag{4.140}$$

Since $|A_r| = |A_i|$, total internal reflection occurs – that is, during a half-period the energy of the incident wave is transferred to the energies of both evanescent waves, and then it returns to the fluid. Results of calculation of the secondary fields, eqs. 4.134–4.136, as functions of the incident angle α_i are shown in Fig. 4.3b–d. As we know, this parameter strongly influences on interference of elementary spherical waves, which creates reflected and refracted waves. For instance, in approaching the critical angle $\alpha_{c1} = \sin^{-1}(c/c_l)$, the amplitude of the reflected P wave rapidly increases, and this is understandable, since total internal reflection for the P wave is observed. At the same time, for slightly larger values of α_i, we see a decrease of the amplitude of this wave accompanied by a sharp increase of the SV wave. The latter indicates the strong constructive interference of the SV elementary waves. Finally, near the critical angle $\alpha_{c2} = \sin^{-1}(c/c_s)$, the amplitude of the SV wave tends to zero, and total internal reflection takes place. Derivation of formulas describing reflection and transmission when the incident wave propagates through an elastic medium is similar to the previous case.

In conclusion, let us note the following: When the incident SH wave propagates through an elastic medium, it is easy to demonstrate that the reflected wave is of the same type. The sum of these two wavefields satisfies the boundary condition

$$\tau_{yz} = 0 \qquad \text{on} \quad z = 0$$

This means that an incident wave does not generate waves in a fluid. Of course, the geometry of waves obeys Snell's law

$$\gamma_r = \gamma_i,$$

and at the boundary for the y-component of displacement we have

$$\tilde{v}\left(x,0,\omega\right)=2\tilde{v}_i\left(x,0,\omega\right),$$

whereas other components (u and w) are absent.

4.7 Reflection and transmission at the boundary of two elastic media

Incident P wave

By analogy with the previous case, first suppose that the incident P wave propagates through the upper medium, Fig. 4.4a, and at the boundary the reflected and transmitted P and SV waves arise. Thus, the secondary waves include two P and two SV waves, but the SH wave is absent. Correspondingly, our assumption implies that the y-component of displacement is equal to zero:

$$v=0 \tag{4.141}$$

Since the wavefields are independent on y-coordinate, stress τ_{yz} vanishes. At the boundary of two elastic media, both components of displacement, u and w, as well as stresses, are continuous functions:

$$u_1=u_2, \qquad w_1=w_2 \tag{4.142}$$

$$\tau_{zz}^{(1)}=\tau_{zz}^{(2)} \qquad \tau_{xz}^{(1)}=\tau_{xz}^{(2)} \qquad \text{on} \qquad z=0$$

or, using Hooke's law, we have

$$u_1=u_2, \qquad w_1=w_2$$

$$\lambda_1 \operatorname{div} \mathbf{s}_1+2\mu_1\frac{\partial w_1}{\partial z}=\lambda_2 \operatorname{div} \mathbf{s}_2+2\mu_2\frac{\partial w_2}{\partial z}, \tag{4.143}$$

$$\mu_1\left(\frac{\partial u_1}{\partial z}+\frac{\partial w_1}{\partial x}\right)=\mu_2\left(\frac{\partial u_2}{\partial z}+\frac{\partial w_2}{\partial x}\right), \qquad \text{on} \qquad z=0$$

where

$$\operatorname{div} \mathbf{s}=\frac{\partial u}{\partial x}+\frac{\partial w}{\partial z}$$

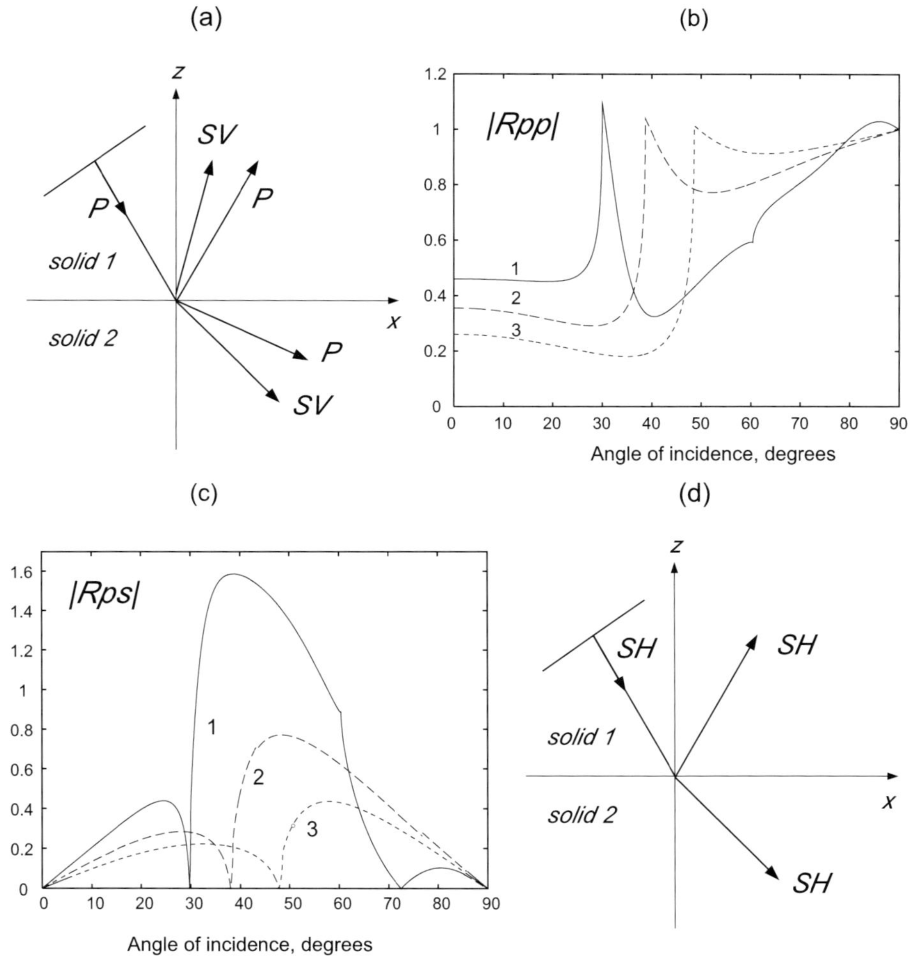

Figure 4.4: Reflection and transmission at the boundary between two solid media. (a) Ray scheme for $P - SV$ waves. (b) Reflection coefficient $|Rpp|$ as a function of the angle of incidence and parameters of media. (c) Reflection coefficient $|Rps|$. Numbers near the curves correspond to different solid media: 1. $c_{1l}/c_{2l} = 0.5$, $\rho_1/\rho_2 = 0.74$. 2. $c_{1l}/c_{2l} = 0.62$, $\rho_1/\rho_2 = 0.76$. 3. $c_{1l}/c_{2l} = 0.75$, $\rho_1/\rho_2 = 0.78$. Poisson' ratio is 0.25 for all cases. (d) Ray scheme for SH waves.

Once again we introduce, but in a slightly different form, the complex amplitudes of potentials

$$\widetilde{\varphi}_1 = A_i \exp\left[ik_{1l}\left(x \sin\alpha_i - z\cos\alpha_i\right)\right] + A_r \exp\left[ik_{1l}\left(x\sin\alpha_r + z\cos\alpha_r\right)\right]$$

$$\widetilde{\psi}_1 = B_r \exp\left[ik_{1s}\left(x\sin\beta_r + z\cos\beta_r\right)\right]$$

$$\widetilde{\varphi}_2 = A_2 \exp\left[ik_{2l}\left(x\sin\alpha_2 - z\cos\alpha_2\right)\right] \tag{4.144}$$

$$\widetilde{\psi}_2 = B_2 \exp\left[ik_{2s}\left(x\sin\beta_2 - z\cos\beta_2\right)\right],$$

From the equality

$$\mathbf{s} = \text{grad } \varphi + \text{curl } \boldsymbol{\psi},$$

where

$$\boldsymbol{\psi} = \psi \mathbf{j},$$

we have

$$u = \frac{\partial\varphi}{\partial x} - \frac{\partial\psi}{\partial z}, \qquad w = \frac{\partial\varphi}{\partial z} + \frac{\partial\psi}{\partial x}$$

and

$$\tau_{zz} = \lambda\nabla^2\varphi + 2\mu\left(\frac{\partial^2\varphi}{\partial z^2} + \frac{\partial^2\psi}{\partial x\,\partial z}\right)$$

and

$$\tau_{xz} = \mu\left(2\frac{\partial^2\varphi}{\partial x\partial z} + \frac{\partial^2\psi}{\partial x^2} - \frac{\partial^2\psi}{\partial z^2}\right)$$

Correspondingly, the boundary conditions in terms of complex amplitudes of potentials are

$$\frac{\partial\widetilde{\varphi}_1}{\partial x} - \frac{\partial\widetilde{\psi}_1}{\partial z} = \frac{\partial\widetilde{\varphi}_2}{\partial x} - \frac{\partial\widetilde{\psi}_2}{\partial z}, \qquad \frac{\partial\widetilde{\varphi}_1}{\partial z} + \frac{\partial\widetilde{\psi}_1}{\partial x} = \frac{\partial\widetilde{\varphi}_2}{\partial z} + \frac{\partial\widetilde{\psi}_2}{\partial x},$$

$$-\lambda_1\, k_{1l}^2\widetilde{\varphi}_1 + 2\mu_1\left(\frac{\partial^2\widetilde{\varphi}_1}{\partial z^2} + \frac{\partial^2\widetilde{\psi}_1}{\partial x\,\partial z}\right) = -\lambda_2\, k_{2l}^2\widetilde{\varphi}_2 + 2\mu_2\left(\frac{\partial^2\widetilde{\varphi}_2}{\partial z^2} + \frac{\partial^2\widetilde{\psi}_2}{\partial x\,\partial z}\right), \tag{4.145}$$

$$\mu_1 \left(2\,\frac{\partial^2 \widetilde{\varphi}_1}{\partial x\,\partial z} + \frac{\partial^2 \widetilde{\psi}_1}{\partial x^2} - \frac{\partial^2 \widetilde{\psi}_1}{\partial z^2} \right) = \mu_2 \left(2\frac{\partial^2 \widetilde{\varphi}_2}{\partial x\,\partial z} + \frac{\partial^2 \widetilde{\psi}_2}{\partial x^2} - \frac{\partial^2 \widetilde{\psi}_2}{\partial z^2} \right) \qquad \text{on} \quad z = 0$$

In order to satisfy this system of equations, arguments of all potentials have to be equal, and we again arrive at Snell's law:

$$\frac{\sin \alpha_i}{c_{1l}} = \frac{\sin \alpha_r}{c_{1l}} = \frac{\sin \beta_r}{c_{1s}} = \frac{\sin \alpha_2}{c_{2l}} = \frac{\sin \beta_2}{c_{2s}} \tag{4.146}$$

or

$$\alpha_r = \alpha_i$$

and

$$\frac{\sin \alpha_r}{c_{1l}} = \frac{\sin \beta_r}{c_{1s}} = \frac{\sin \alpha_2}{c_{2l}} = \frac{\sin \beta_2}{c_{2s}} \tag{4.147}$$

Directions of rays corresponding to the incident and secondary waves are shown in Fig. 4.4a. Following Snell's law, substitution of eqs. 4.144 into set 4.145 gives

$$k_{1l} \sin \alpha_i \;(A_i + A_r) - k_{1s} \cos \beta_r \; B_r = k_{2l} \sin \alpha_2 \; A_2 + k_{2s} \cos \beta_2 \; B_2$$

$$k_{1l} \cos \alpha_i \;(A_r - A_i) + k_{1s} \sin \beta_r \; B_r = -k_{2l} \cos \alpha_2 \; A_2 + k_{2s} \sin \beta_2 \; B_2 \tag{4.148}$$

$$-\lambda_1 k_{1l}^2 \;(A_i + A_r) + 2\mu_1 \left[k_{1l}^2 \cos^2 \alpha_i \,(A_i + A_r) + k_{1s}^2 \sin \beta_r \cos \beta_r \; B_r\right]$$

$$= -\lambda_2 \; k_{2l}^2 \; A_2 + 2\mu_2 \;[k_{2l}^2 \cos^2 \alpha_2 \; A_2 - k_{2s}^2 \sin \beta_2 \cos \beta_2 \; B_2]$$

$$\mu_1[2k_{1l}^2 \sin \alpha_i \cos \alpha_i \,(A_r - A_i) + k_{1s}^2 \left(\sin^2 \beta_r - \cos^2 \beta_r\right) B_r]$$

$$= \mu_2 \;\left[2k_{2l}^2 \sin \alpha_2 \cos \alpha_2 \; A_2 + k_{2s}^2 \left(\sin^2 \beta_2 - \cos^2 \beta_2\right) B_2\right]$$

Thus, we have obtained a system of four linear equations with four unknowns (A_r, B_r, A_2, and B_2). A numerical solution of this system allows us to find all wavefields at any point in an elastic medium. Reflection coefficients $\mathcal{R}_{PP}$ and $\mathcal{R}_{PS}$ are

$$\mathcal{R}_{PP} = \frac{A_r}{A_i}, \qquad \mathcal{R}_{PS} = \frac{B_r}{A_i}, \tag{4.149}$$

and transmission coefficients $\mathcal{T}_{PP}$ and $\mathcal{T}_{PS}$ are

$$\mathcal{T}_{PP} = \frac{A_2}{A_i}, \qquad \mathcal{T}_{PS} = \frac{B_2}{A_i} \tag{4.150}$$

As an illustration, the behavior of coefficients characterizing the reflected P and SV waves is shown in Fig. 4.4b,c, when the P wave is the incident wave. The case of the incident SV wave can be treated in a similar way.

Incident SH wave

Next we assume that the incident SH wave moves through the upper medium, Fig. 4.4d, and that reflected and transmitted SH waves arise at the interface. Since

$$\mathbf{s} = \text{curl}\, \boldsymbol{\psi} \qquad \text{and} \qquad \boldsymbol{\psi} = \psi\, \mathbf{i},$$

we have

$$u = w = 0 \qquad \text{and} \qquad v = \frac{\partial \psi}{\partial z} \tag{4.151}$$

Taking into account that $\text{div}\, \mathbf{s} = 0$ and that the field is independent of the y-coordinate, the stresses are

$$\tau_{zz} = 0, \qquad \tau_{xz} = 0, \qquad \tau_{yz} = \mu \frac{\partial v}{\partial z} \tag{4.152}$$

Therefore the boundary conditions have the form

$$v_1 = v_2 \qquad \text{and} \qquad \mu_1 \frac{\partial v_1}{\partial z} = \mu_2 \frac{\partial v_2}{\partial z} \qquad \text{on} \quad z = 0 \tag{4.153}$$

Since displacement is described by the single component that obeys the wave equation, we solve the boundary value problem with respect to the complex amplitude, $\widetilde{v}(x, z, \omega)$, where

$$\widetilde{v}_1(x, z, \omega) = C_i \exp\left[i\ k_{1s}\left(x \sin \gamma_i - z \cos \gamma_i\right)\right] +$$

$$C_r \exp\left[i\ k_{1s}\left(x \sin \gamma_r + z \cos \gamma_r\right)\right] \tag{4.154}$$

$$\text{and} \qquad \widetilde{v}_2(x, z, \omega) = C_2 \exp\left[i\ k_{2s}\left(x \sin \gamma_2 - z \cos \gamma_2\right)\right]$$

Substitution of eqs. 4.154 into set 4.153 leads us, first of all, to the Snell's law:

$$\gamma_r = \gamma_i \qquad \text{and} \qquad \frac{\sin \gamma_i}{c_{1s}} = \frac{\sin \gamma_2}{c_{2s}} \tag{4.155}$$

Also, we arrive at a system of two simple equations:

$$C_i + C_r = C_2, \qquad \mu_1\ k_{1s} \cos \gamma_i \left(C_i - C_r\right) = \mu_2\ k_{2s} \cos \gamma_2\ C_2 \tag{4.156}$$

Solution of eqs. 4.156 gives

$$C_r = \frac{Z_{1s} \cos \gamma_i - Z_{2s} \cos \gamma_2}{Z_{1s} \cos \gamma_i + Z_{2s} \cos \gamma_2}\, C_i \tag{4.157}$$

$$\text{and} \qquad C_2 = \frac{2\ Z_{1s}\cos\gamma_i}{Z_{1s}\cos\gamma_i + Z_{2s}\cos\gamma_2}\ C_i$$

The simplicity of these formulas and their resemblance to formulas for the reflection and transmission of acoustic plane waves is obvious (Part II). In particular, the reflected wave vanishes due to destructive interference for some value of angle of incidence γ_i. Also, beyond the critical angle $(c_{2s} > c_{1s})$, an evanescent wave in a medium with higher velocity is formed and propagates along the boundary (total internal reflection). In this case, we have

$$\widetilde{v}_2(x,z,w) = C_2 \exp(-k_{1s}b_z z)\exp(-ik_{2s}x\sin\gamma_i),$$

From eq. 4.156 it follows that

$$C_r = \frac{Z_{1s}\cos\gamma_i - iZ_{2s}b_z}{Z_{1s}\cos\gamma_i + iZ_{2s}b_z}\ C_i, \qquad C_2 = \frac{2\ Z_{1s}\cos\gamma_i}{Z_{1s}\cos\gamma_i + iZ_{2s}b_z}\ C_i$$

where

$$b_z = \sqrt{\frac{c_{2s}^2}{c_{1s}^2}\sin\gamma_i - 1}$$

It is easy to see that $|C_r| = C_i$ and the reflection coefficient $\mathcal{R}_{SS} = C_r/C_i = \exp(-i\Psi)$, where

$$\Psi = \arg C_r = 2\tan^{-1}\frac{Z_{2s}b_z}{Z_{1s}\cos\gamma_i} \tag{4.158}$$

In conclusion, let us note the following. If the angles of incidence of the P, SV, or SH plane waves do not exceed the critical angle, coefficients of reflection and transmission derived for sinusoidal waves are also valid for arbitrary transient waves. Beyond the critical angle, a phase shift between the incident and reflected waves occurs at the boundary. This shift is independent of frequency. Because of this, use of the known coefficients of reflection and transmission and the Hilbert transform allow us to find in a relatively simple way a transient reflected wave (Part II). Since the complex amplitude of evanescent waves depends on frequency, nonstationary wavefields are defined by Fourier's transform.

4.8 Ray tubes and flux of energy

Suppose that a plane wave propagates through a homogeneous medium, and choose the Cartesian system of coordinates, x_1, y_1, z_1, so that the wavefront coincides with the plane $z_1 =$ const. By definition, a lateral surface of any elementary ray tube is formed by rays, which are straight lines, and the area of the cross-section of the tube is constant. First we will demonstrate that the flux of elastic energy travels along these tubes. We will consider this subject separately for compressional and shear plane waves. As was derived in Appendix D, the vector of the density of flux energy, $\mathbf{Y}$, is the product of the symmetrical tensor of stress and particle velocity, $\dot{\mathbf{s}}$:

$$\mathbf{Y} = -\begin{pmatrix} \tau_{x_1x_1} & \tau_{x_1y_1} & \tau_{x_1z_1} \\ \tau_{y_1x_1} & \tau_{y_1y_1} & \tau_{y_1z_1} \\ \tau_{z_1x_1} & \tau_{z_1y_1} & \tau_{z_1z_1} \end{pmatrix} \begin{pmatrix} \dot{u}_1 \\ \dot{v}_1 \\ \dot{w}_1 \end{pmatrix}, \tag{4.159}$$

where $\dot{u}_1, \dot{v}_1, \dot{w}_1$ are components of particle velocity in the coordinate system x_1, y_1, z_1. Now we take into account that derivatives with respect to x_1 and y_1 are equal to zero and find expressions of $\mathbf{Y}$ for each type of plane wave.

Incident P wave

Since displacement $\mathbf{s}$ has only one component w_1 that is directed along a ray we have in accordance with Hooke's law

$$\tau_{x_1x_1} = \lambda\frac{\partial w_1}{\partial z_1}, \qquad \tau_{y_1y_1} = \lambda\frac{\partial w_1}{\partial z_1}, \qquad \tau_{z_1z_1} = (\lambda + 2\mu)\frac{\partial w_1}{\partial z_1} \tag{4.160}$$

$$\text{and} \qquad \tau_{x_1y_1} = \tau_{x_1z_1} = \tau_{y_1z_1} = 0$$

Therefore

$$\mathbf{Y} = -\begin{pmatrix} \tau_{x_1x_1} & 0 & 0 \\ 0 & \tau_{y_1y_1} & 0 \\ 0 & 0 & \tau_{z_1z_1} \end{pmatrix} \begin{pmatrix} 0 \\ 0 \\ \dot{w}_1 \end{pmatrix}$$

or

$$Y_{x_1} = 0, \qquad Y_{y_1} = 0, \qquad Y_{z_1} = -\dot{w}_1\tau_{z_1z_1} \tag{4.161}$$

or

$$Y_{z_1} = -(\lambda + 2\mu)\,\dot{w}_1\frac{\partial w_1}{\partial z_1} \tag{4.162}$$

We see that elastic energy moves along a ray tube, and the flux through its lateral surface is equal to zero.

Incident *SV* wave

In the case of the SV wave the single component of displacement, u_1, is tangential to the wavefront, and since

$$\operatorname{div}\mathbf{s} = 0,$$

the stresses are

$$\tau_{x_1x_1} = 2\mu\,\frac{\partial u_1}{\partial x_1} = 0, \qquad \tau_{y_1y_1} = 2\mu\,\frac{\partial v_1}{\partial y_1} = 0, \qquad \tau_{z_1z_1} = 2\mu\,\frac{\partial w_1}{\partial z_1} = 0 \tag{4.163}$$

$$\text{and} \qquad \tau_{x_1y_1} = \mu\,\frac{\partial u_1}{\partial y_1} = 0, \qquad \tau_{x_1z_1} = \mu\,\frac{\partial u_1}{\partial z_1}, \qquad \tau_{y_1z_1} = 0$$

Then

$$\mathbf{Y} = -\begin{pmatrix} 0 & 0 & \tau_{x_1z_1} \\ 0 & 0 & 0 \\ \tau_{x_1z_1} & 0 & 0 \end{pmatrix}\begin{pmatrix} \dot{u}_1 \\ 0 \\ 0 \end{pmatrix}, \tag{4.164}$$

that is,

$$Y_{x_1} = 0, \qquad Y_{y_1} = 0, \qquad Y_{z_1} = -\dot{u}_1\tau_{x_1z_1} \tag{4.165}$$

or

$$Y_{z_1} = -\mu\dot{u}_1\frac{\partial u_1}{\partial z}, \tag{4.166}$$

and again energy flux advances along the ray tube, in spite of the fact that particles in the medium are moving in a perpendicular direction.

Incident *SH* wave

In the case of the SH wave, since displacement is oriented along the y_1-axis and, as before, $\operatorname{div}\mathbf{s} = 0$, we obtain

$$\tau_{x_1x_1} = \tau_{y_1y_1} = \tau_{z_1z_1} = 0 \qquad \text{and} \qquad \tau_{x_1y_1} = \tau_{x_1z_1} = 0, \tag{4.167}$$

$$\text{but} \qquad \tau_{y_1z_1} = \mu\,\frac{\partial v_1}{\partial z_1}$$

Thus

$$\mathbf{Y} = -\begin{pmatrix} 0 & 0 & 0 \\ 0 & 0 & \tau_{y_1 z_1} \\ 0 & \tau_{y_1 z_1} & 0 \end{pmatrix} \begin{pmatrix} 0 \\ \dot{v} \\ 0 \end{pmatrix}, \tag{4.168}$$

that is,

$$Y_{x_1} = Y_{y_1} = 0 \qquad \text{and} \qquad Y_{z_1} = -\dot{v}\, \tau_{y_1 z_1}$$

or

$$Y_{z_1} = -\mu \dot{v}_1 \frac{\partial v_1}{\partial z} \tag{4.169}$$

It is clear that regardless of the direction of particle motion, energy flux moves along ray tubes. As an illustration, consider sinusoidal waves. In the case of the P wave, for example, displacement w_1 is

$$w_1(z_1, t) = A \sin(\omega t - k_l z_1) \tag{4.170}$$

Correspondingly, velocity and stress are equal to

$$\dot{w}_1(z_1, t) = A\, \omega \cos(\omega t - k_l z_1) \tag{4.171}$$

$$\text{and} \qquad \tau_{z_1 z_1}(z_1, t) = -(\lambda + 2\mu)\, k_l A \cos(\omega t - k_l z_1)$$

Whence

$$Y_{z_1} = (\lambda + 2\mu)\, \omega\, k_l\, A^2 \cos^2(\omega t - k_l z_1) \tag{4.172}$$

As is well known, the mean value of Poynting's vector is defined from the relationship

$$Y_{z_1}^{av} = \frac{1}{T} \int_0^T Y_{z_1}(t)\, dt,$$

and the latter gives

$$Y_{z_1}^{av} = \frac{1}{2} (\lambda + 2\mu)\, \omega\, k_l\, A^2 \tag{4.173}$$

or

$$Y_{z_1}^{av} = \frac{\omega^2}{2} Z_l\, A^2 \tag{4.174}$$

Here Z_l is the impedance for the P wave. In a similar manner, for SV and SH waves we have

$$Y_{z_1} = \frac{\omega^2}{2} Z_s\, A^2, \tag{4.175}$$

and $Z_s = \rho\, c_s$ is the impedance of shear waves. These results are very useful for understanding the nature of the high-frequency spectrum of elastic waves in an inhomogeneous medium, where the velocity of propagation (c_l or c_s) varies. By analogy with acoustic waves (Part II), in such a case we can treat each elastic wave as a plane wave, and its amplitude, phase, and direction all depend on the observation point. Correspondingly, wave amplitude can be represented as the asymptotic series with respect to inverse powers of ω (Debye expansion). Its zero approximation describes the wave amplitude when energy flux propagates along the ray tubes. Inasmuch as the flux inside a tube remains the same, eqs. 4.174 and 4.175, let us write the equality

$$Z(p_1)\, A^2(p_1)\, S(p_1) = Z(p_2)\, A^2(p_2)\, S(p_2) \tag{4.176}$$

Here Z is the impedance of either the compressional or the shear wave, and $S(p_1)$ and $S(p_2)$ are two cross-sections of the ray tube. From eq. 4.176 we have

$$A(p_2) = A(p_1)\sqrt{\frac{S(p_1)\, Z(p_1)}{S(p_2)\, Z(p_2)}} \tag{4.177}$$

or

$$A(p_2) = \frac{1}{\sqrt{F}}\, A(p_1), \tag{4.178}$$

where

$$F = \frac{Z(p_2)\, S(p_2)}{Z(p_1)\, S(p_1)} \tag{4.179}$$

is the spreading factor (Part II). As in the case of acoustic waves, eq. 4.178 permits us to determine the change of the displacement amplitude. The direction of rays is defined by Snell's law.

The previously outlined general features of reflection and transmission of plane longitudinal and shear waves are important for many applications in exploration seismology. These coefficients are used in the ray theory of seismic waves in inhomogeneous media as approximations to the coefficients of reflection and refraction of nonplane waves. The AVO (amplitude versus offset) technique for determining the lithological properties of reflectors is based on knowledge about the behavior of these coefficients as functions of the angle of incidence and elastic parameters. Because potentials of these waves obey the wave equation, the principles of migration developed for acoustic waves can be applied for each elastic wave. In particular, at points of a reflector, the incident and reflected waves are in phase, provided that the angle of incidence does not exceed the critical angle. However, there is one important difference – namely, in calculating amplitudes, it is necessary to take into account the appearance of the P and SV waves at the boundary, except in the case of the incident SH wave.

4.9 Reflection and transmission of plane waves in a multilayered elastic medium

Let us consider propagation of stationary plane waves in an elastic medium consisting of n homogeneous layers between two homogeneous half-spaces (Fig. 4.5). All layers are supposed to be in the welded contact, i.e., displacements and stresses are continuous across boundaries of layers. Our goal is to find reflection and transmission coefficients for plane waves P, SV, and SH incident on the upper boundary of the n-layered "sandwich". To do this, we need to construct a recurrent formalism linking displacements and stresses at boundaries of this medium.

P-SV case

The chosen Cartesian system of coordinates is shown in Fig. 4.5. Let us consider an arbitrary layer m bounded by the planes $z = z_m$ and $z = z_{m+1}$, of thickness H_m, with elastic parameters λ_m and μ_m, density ρ_m, and compressional and shear speeds $c_{lm} = \sqrt{(\lambda_m + 2\mu_m)/\rho_m}$ and $c_{sm} = \sqrt{\mu_m/\rho_m}$.

Expressions for potentials of stationary P and SV plane waves of frequency ω propagating in this layer in positive and negative directions away from the z-axis can be written in the form

$$\varphi_m = \frac{1}{ik_{lm}} \left(A_m e^{i\,\alpha_m\, z} + B_m e^{-i\,\alpha_m\, z} \right) e^{i(px\,-\,\omega t)} \tag{4.180}$$

and
$$\psi_m = \frac{1}{ik_{sm}} \left(C_m e^{i\beta_m\, z} + D_m e^{-i\beta_m z} \right)\, e^{i(px\,-\,\omega t)}$$

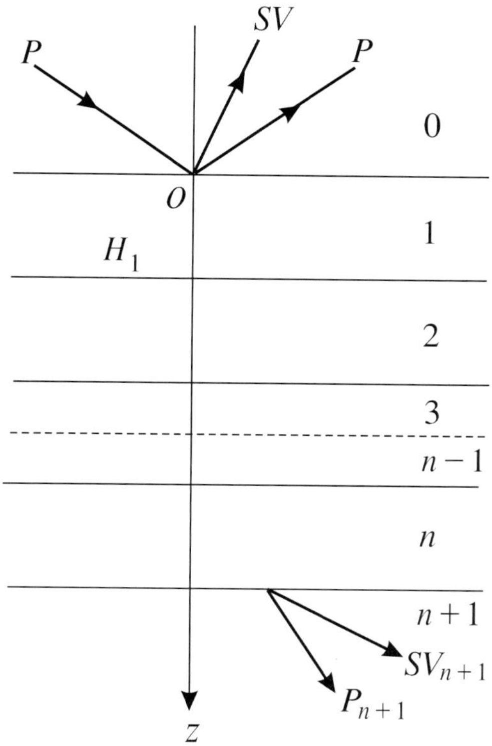

Figure 4.5: Plane wave in n-layered medium

Here

$$k_{lm} = \omega/c_{lm}, \quad k_{sm} = \omega/c_{sm}, \quad \alpha_m = \sqrt{k_{lm}^2 - p^2}, \quad \beta_m = \sqrt{k_{sm}^2 - p^2}, \tag{4.181}$$

where $p = \omega/c$ and c is the phase velocity of the waves along x-direction. As follows from Snell's law, this velocity is the same for all waves at each boundary. Values of α_m, β_m and the coefficients A_m, B_m, C_m, D_m vary from one layer to another. Now we will use the known expressions for displacements,

$$u^{(m)} = \frac{\partial \varphi_m}{\partial x} - \frac{\partial \psi_m}{\partial z}, \qquad w^{(m)} = \frac{\partial \varphi_m}{\partial z} + \frac{\partial \psi_m}{\partial x}; \tag{4.182}$$

velocities of displacements,

$$\dot{u}^{(m)} = -i\omega\, u^{(m)}, \quad \dot{w}^{(m)} = -i\omega\, w^{(m)}; \tag{4.183}$$

and stresses,

$$\tau_{xz}^{(m)} = \mu_m \left(\frac{\partial u^{(m)}}{\partial z} + \frac{\partial w^{(m)}}{\partial x} \right), \qquad \tau_{zz}^{(m)} = \lambda_m \frac{\partial u^{(m)}}{\partial x} + (\lambda_m + 2\mu_m) \frac{\partial w^{(m)}}{\partial z} \tag{4.184}$$

Let us now introduce two vectors,

$$\mathbf{X}_m(z) = \begin{pmatrix} \dot{\tilde{u}}^{(m)} \\ \dot{\tilde{w}}^{(m)} \\ \tilde{\tau}_{xz}^{(m)} \\ \tilde{\tau}_{zz}^{(m)} \end{pmatrix} \quad \text{and} \quad \mathbf{N}_m = \begin{pmatrix} A_m \\ B_m \\ C_m \\ D_m \end{pmatrix}, \tag{4.185}$$

where components of $\mathbf{X}_m$ are complex amplitudes of displacement velocities and stresses. We also introduce the matrix $Q_m(z)$:

$$Q_m(z) = \begin{pmatrix} e^{i\,\alpha_m\, z} & 0 & 0 & 0 \\ 0 & e^{-i\,\alpha_m\, z} & 0 & 0 \\ 0 & 0 & e^{i\,\beta_m\, z} & 0 \\ 0 & 0 & 0 & e^{-i\,\beta_m\, z} \end{pmatrix} \tag{4.186}$$

Using eqs. 4.185 and 4.186 and applying the rules of matrix algebra (Appendix B), we can relate displacement velocities and stresses in the mth layer with coefficients A_m, B_m, C_m, D_m:

$$\mathbf{X}_m(z) = L_m Q_m(z) \mathbf{N}_m \tag{4.187}$$

Elements m_{ij} of the 4×4 matrix L_m depend on elastic characteristics of the mth layer and phase velocity c.

$$L_m = -ip \begin{pmatrix} c_{lm} & c_{lm} & -q_m & q_m \\ g_m & -g_m & c_{sm} & c_{sm} \\ -2\mu_m \dfrac{g_m}{c} & 2\mu_m \dfrac{g_m}{c} & 2\mu_m \dfrac{b_m}{c_s} & 2\mu_m \dfrac{b_m}{c_s} \\ d_m & d_m & -2\mu_m \dfrac{q_m}{c_s} & 2\mu_m \dfrac{q_m}{c_s} \end{pmatrix} \tag{4.188}$$

Here

$$q_m = \sqrt{c^2 - c_{sm}^2}, \qquad g_m = \sqrt{c^2 - c_{lm}^2}, \qquad b_m = c^2 - 2c_s^2,$$

$$d_m = -\lambda_m \frac{c_l}{c} - \frac{2\mu_m g_m^2}{c c_{lm}}$$

From eq. 4.187 it follows that

$$\mathbf{N}_m = Q_m^{-1}(z_m) L_m^{-1} \mathbf{X}_m(z_m) = Q_m^{-1}(z_{m-1}) L_m^{-1} \mathbf{X}_m(z_{m-1})$$

where Q_m^{-1} and L_m^{-1} are inverse matrices of Q_m and L_m . Now it is possible to connect expressions for X_m at the top and bottom of the mth layer as

$$\mathbf{X}_m(z_m) = L_m Q_m(z) Q_m^{-1}(z_{m-1}) L_m^{-1}(z_{m-1}) \mathbf{X}_m(z_{m-1}) \tag{4.189}$$

$$= L_m Q_m(H_m) L_m^{-1} \mathbf{X}_m(z_{m-1}) \ = \ M_m \mathbf{X}_m(z_{m-1})$$

where $M_m = L_m Q_m(H_m) L_m^{-1}$. The elements of matrix M_m depend on phase velocity c, frequency ω, elastic characteristics, and the thickness H_m of the layer. The conditions of the welded contact allow us to link vectors $\mathbf{X}_m(z_m)$ and $\mathbf{X}_m(z_{m+1})$ at two adjacent layers at the same boundary $z = z_m$:

$$\mathbf{X}_m(z_m) = \mathbf{X}_{m+1}(z_m) \tag{4.190}$$

As a result we obtain the relation between vectors $\mathbf{X}_{m+1}(z_m)$ and $\mathbf{X}_m(z_{m-1})$

$$\mathbf{X}_{m+1}(z_m) = M_m \ \mathbf{X}_m(z_{m-1}) \tag{4.191}$$

Using eq. 4.191, we can connect vector $\mathbf{X}_1(0)$ at the top of the layered "sandwich" with vector $\mathbf{X}_n(z_n)$ at its bottom:

$$\mathbf{X}_n(z_n) = M_n \mathbf{X}_{n-1}(z_n) = M_n M_{n-1} \mathbf{X}_{n-2}(z_{n-1}) = M_n M_{n-1} ... M_1 Y_1(0) = S Y_1(0) \tag{4.192}$$

Here the matrix

$$S = \prod_{m=1}^{n} M_m \tag{4.193}$$

is called a propagator, as it "propagates" the wavefield from the top to the bottom of the layered media. The P and SV waves transmitted in the lower half-space and propagating in the positive direction of the z-axis may be described by potentials

$$\varphi_{n+1} = \frac{1}{i k_{l(n+1)}} A_{n+1} \, e^{i \ \alpha_{n+1} \ z} e^{i(px - \omega t)} \tag{4.194}$$

and $$\psi_{n+1} = \frac{1}{ik_{s(n+1)}} C_{n+1}\, e^{i\,\beta_{n+1}\,z} e^{i(px-\omega t)}$$

The corresponding vector $\mathbf{X}_{n+1}(z_n)$ in the lower half-space is defined by the expression

$$\mathbf{X}_{n+1}(z) = L_{n+1}Q_{n+1}(z) \begin{pmatrix} A_{n+1} \\ 0 \\ C_{n+1} \\ 0 \end{pmatrix} \tag{4.195}$$

Incident P wave If the incident wave in the upper half-space is a P wave, the displacement velocity–stress vector in this part of the medium can be presented as

$$\mathbf{X}_0(z) = L_0Q_0(z) \begin{pmatrix} A_0 \\ B_0 \\ 0 \\ D_0 \end{pmatrix} \tag{4.196}$$

Here coefficients A_0, B_0 correspond to the incident and reflected P waves, and coefficient D_0 corresponds to the reflected SV wave. Taking into account equalities

$$\mathbf{X}_0(0) = \mathbf{X}_1(0)\,, \qquad Q_0(0) = E, \qquad \text{and} \qquad \mathbf{X}_n(z_n) = \mathbf{X}_{n+1}(z_n) \tag{4.197}$$

where E is a unit matrix, we arrive at a system of linear algebraic equations,

$$\begin{pmatrix} A_{n+1} \\ 0 \\ C_{n+1} \\ 0 \end{pmatrix} = Q^{-1}_{n+1}(z_n)L^{-1}_{n+1}S\ L_0(0) \begin{pmatrix} A_0 \\ B_0 \\ 0 \\ D_0 \end{pmatrix} = G \begin{pmatrix} A_0 \\ B_0 \\ 0 \\ D_0 \end{pmatrix}, \tag{4.198}$$

for unknown coefficients $B_0, D_0, A_{n+1}, C_{n+1}$. Solving this equation, we find the following expressions for reflection coefficients,

$$\mathcal{R}^{(0)}_{PP} = \frac{B_0}{A_0} = \frac{g_{24}\ g_{41} - g_{21}\ g_{44}}{g_{22}\ g_{44} - g_{24}\ g_{42}}, \qquad \mathcal{R}^{(0)}_{PS} = \frac{D_0}{A_0} = -\frac{g_{21} + g_{22}R^{(0)}_{PP}}{g_{24}}, \tag{4.199}$$

and transmission coefficients,

$$\mathcal{T}^{(n+1)}_{PP} = \frac{A_{n+1}}{A_0} = \left(g_{11} + g_{12}R^{(0)}_{PP} + g_{14}R^{(0)}_{PS}\right), \tag{4.200}$$

$$\mathcal{T}^{(n+1)}_{PS} = \frac{C_{n+1}}{A_0} = \left(g_{31} + g_{32}R^{(0)}_{PP} + g_{34}R^{(0)}_{PS}\right),$$

where g_{ij} are elements of matrix G.

Incident SV **wave** The single difference with the previous case (incident P wave) is in expression for the vector $\mathbf{N}_0$:

$$\mathbf{N}_0 = \begin{pmatrix} 0 \\ B_0 \\ C_0 \\ D_0 \end{pmatrix} \tag{4.201}$$

Substituting this vector into eq. 4.198 and solving it we obtain recursive expressions for reflection and transmission coefficients in the case of the incident SV wave:

$$\mathcal{R}_{SS}^{(0)} = \frac{B_0}{C_0} = \frac{g_{23}\ g_{44} - g_{24}\ g_{43}}{g_{22}\ g_{44} - g_{24}\ g_{42}}, \qquad \mathcal{R}_{SP}^{(0)} = \frac{D_0}{C_0} = -\frac{g_{23} + g_{22}R_{SS}^{(0)}}{g_{24}} \tag{4.202}$$

and

$$\mathcal{T}_{SP}^{(n+1)} = \frac{A_{n+1}}{C_0} = \left(g_{13} + g_{12}R_{PP}^{(0)} + g_{14}R_{PS}^{(0)}\right), \tag{4.203}$$

$$\mathcal{T}_{SS}^{(n+1)} = \frac{C_{n+1}}{C_0} = \left(g_{33} + g_{32}R_{PP}^{(0)} + g_{34}R_{PS}^{(0)}\right)$$

SH case

If the incident wave in the upper half-space is an SH wave, the single nonzero component of displacement in the mth layer may be presented as

$$v_m = \left(A_m e^{i\ \beta_m\ z} + B_m e^{-i\ \beta_m\ z}\right) e^{i(px\ -\ \omega t)} \tag{4.204}$$

and the stress component as

$$\tau_{yz}^{(m)} = \mu_m \frac{dv^{(m)}}{dz} \tag{4.205}$$

The displacement velocity–stress vector has only two components:

$$\mathbf{X}_m(z) = \begin{pmatrix} \dot{\tilde{v}}^{(m)} \\ \tilde{\tau}_{yz}^{(m)} \end{pmatrix} \tag{4.206}$$

The corresponding matrices Q_m, L_m, and M_m are now

$$Q_m = \begin{pmatrix} e^{i\ \beta_m\ z} & 0 \\ 0 & e^{-i\ \beta_m\ z} \end{pmatrix}, \quad L_m = \begin{pmatrix} -i\omega & -iw \\ i\mu_m\beta_m & -i\mu_m\beta_m \end{pmatrix}, \tag{4.207}$$

$$M_m = L_m Q_m L_m^{-1}$$

Vector $\mathbf{X}_{n+1}$ in the lower half-space and vector $\mathbf{X}_0$ in the upper half-space are defined as

$$\mathbf{X}_{n+1}(z) = L_{n+1} Q_{n+1}(z) \begin{pmatrix} A_{n+1} \\ 0 \end{pmatrix} \quad \text{and} \quad \mathbf{X}_0(z) = L_0 \begin{pmatrix} A_0 \\ B_0 \end{pmatrix} \tag{4.208}$$

The resulting equation defining unknown coefficients B_0, A_{n+1} is

$$\begin{pmatrix} A_{n+1} \\ 0 \end{pmatrix} = G \begin{pmatrix} A_0 \\ B_0 \end{pmatrix} \tag{4.209}$$

where, again, 2×2 matrix $G = Q_{n+1}^{-1}(z_n) L_{n+1} \prod_{m=1}^{n} M_n L_0(0)$. Correspondingly, coefficients of reflection and transmission are

$$\mathcal{R}_{SS}^{(0)} = \frac{B_0}{A_0} = -\frac{g_{21}}{g_{22}}, \qquad \mathcal{T}_{SS}^{(n+1)} = \frac{A_{n+1}}{A_0} = g_{11} + g_{12} R_{SS}^{(0)} \tag{4.210}$$

In conclusion, we should remember that coefficients defining reflected and transmitted waves in all considered cases are frequency-dependent. Moreover, they may be real (if the angles of incidence inside all layers are less than critical angles for all involved waves) or complex (if even one of these angles is above the critical angle). Thus, reflection and transmission of nonstationary (transient) waves in such a medium should be treated using Fourier's transformation.

Chapter 5

Surface waves in an elastic medium

The purpose of this chapter is to describe so-called surface or boundary waves. The energy carried by these waves concentrates near some surface, such as a free surface or a boundary between different media. Waves of this type are different from the evanescent waves discussed in the previous chapter because they are not generated by homogeneous plane waves coming to this boundary. Two classical examples of such waves are the Rayleigh wave in a homogeneous half-space with a free surface and the Stoneley wave at the boundary between two elastic half-spaces or between a fluid and an elastic half-space. These waves are composed of two evanescent waves of different types propagating along the boundary with the same speed – one that is less than the intrinsic speeds of body waves (compressional or shear) in the medium. The speed does not depend on frequency, i.e., these waves are not dispersive.

We will also consider in this chapter waves of a more complex nature that arise in layered media as a result of the constructive interference of multiply reflected body waves. They still propagate horizontally without leakage of energy in a vertical direction. As examples of interferential waves, we will analyze Love waves in a homogeneous elastic half-space overlaid by a homogeneous elastic layer and Rayleigh waves in a homogeneous elastic half-space overlaid by a homogeneous liquid layer. The speed of these waves is frequency-dependent, i.e., the waves are dispersive. They are presented as a suite of modes, and each mode is characterized by its dispersion curve and depth-depending distribution of energy. Analogs of such waves propagating in a layered fluid were considered in Part II. Waves of this kind exist, of course, in more complicated vertically or radially inhomogeneous media. Depending on the problem at hand, they are considered as a source of noise to be suppressed or as carriers of important information about the structure under study.

5.1 Rayleigh wave in a homogeneous half-space with a free boundary

Earlier we demonstrated that the SV incident wave may generate an evanescent P wave as well as the homogeneous reflected SV wave at the free surface. This means that the sum of these three waves satisfies boundary conditions, and stresses vanish at points of the free surface. In other words, the evanescent wave cannot exist alone. Moreover, since all three waves move with the same apparent velocity, it is impossible to distinguish them.

Now we pose the following question. Is there a surface wave that is similar to evanescent plane waves but that alone obeys boundary conditions at the free surface? This would imply that such a wave propagates along the boundary and exponentially decays with the coordinate z. Also, we assume that this wave would not depend on the y-coordinate, Fig. 5.1a. Let us recall that we have already studied a surface wave in a fluid, and it displayed all of these features. However, a water wave is caused by the gravitational field, whereas in the suggested scenario the influence of deformation of a fluid can be neglected (Part I).

Rayleigh wave velocity

In accordance with our assumptions about a surface wave, the complex amplitudes of the scalar potential and the y-component of the vector potential are

$$\tilde{\varphi}(x,z,\omega) = A\, e^{-k\, b_l\, z} e^{i\, k\, x} \quad \text{and} \quad \tilde{\psi}(x,z,\omega) = B\, e^{-k\, b_s\, z} e^{i\, k\, x} \tag{5.1}$$

Here $k = \omega/c$ and c is the velocity of propagation of this wave. A, B, b_l, and b_s are constants. It is clear that these equations describe a wave in which particle displacement has only components u and w, with $v = 0$. If such a surface wave exists, it has to be a solution of the Helmholtz equations

$$\frac{\partial^2 \tilde{\varphi}}{\partial x^2} + \frac{\partial^2 \tilde{\varphi}}{\partial z^2} + k_l^2 \tilde{\varphi} = 0 \quad \text{and} \quad \frac{\partial^2 \tilde{\psi}}{\partial x^2} + \frac{\partial^2 \tilde{\psi}}{\partial z^2} + k_s^2 \tilde{\psi} = 0 \tag{5.2}$$

This condition allows us to find a relationship between c and parameters b_l and b_s. Substitution of eqs. 5.1 into eqs. 5.2 gives

$$-k^2 + b_l\, k^2 + k_l^2 = 0, \qquad -k^2 + b_s\, k^2 + k_s^2 = 0$$

or

$$b_l = \sqrt{1 - \frac{c^2}{c_l^2}} \quad \text{and} \quad b_s = \sqrt{1 - \frac{c^2}{c_s^2}} \tag{5.3}$$

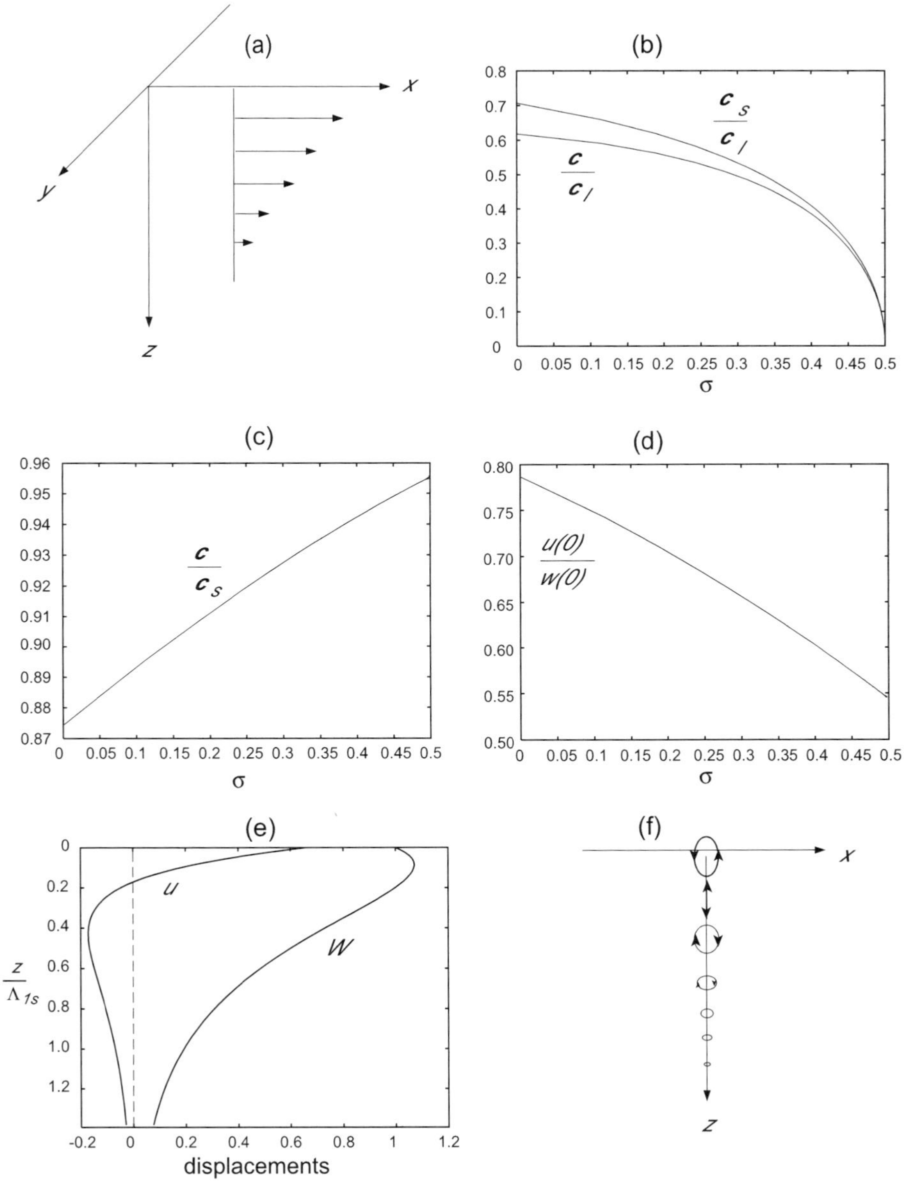

Figure 5.1: Rayleigh wave in a homogeneous half-space: (a) Scheme of wave propagation (b) Dependence of c/c_l and c_s/c_l on Poisson's ratio (c) Dependence of c/c_s on Poisson's ratio (d) Dependence of $u(0)/w(0)$ on Poisson's ratio (e) Dependence of amplitudes of horizontal and vertical components on z/Λ_{1s} (f) Particle motion at different depths

The latter shows that the velocity of propagation c of this wave has to be smaller than c_s, i.e.,

$$c < c_s < c_l \tag{5.4}$$

Otherwise, we would not observe the exponential decay of potentials with an increase of z. Now we have only three unknowns; A, B, and c. Our goal is to prove that such a surface wave may exist and that its velocity c obeys inequality 5.4. Since we are not considering how this wave is generated, we are not able to determine both constants A and B separately; we will leave that for the next chapter. Correspondingly, we focus on calculating velocity c. To find c, let us consider boundary conditions at the free surface:

$$\widetilde{\tau}_{zz} = 0 \qquad \text{and} \qquad \widetilde{\tau}_{xz} = 0 \tag{5.5}$$

This allows us to derive an equation with respect to c. Since

$$\widetilde{\tau}_{zz} = \lambda \operatorname{div} \widetilde{\mathbf{s}} + 2\mu \frac{\partial \widetilde{w}}{\partial z}, \qquad \widetilde{\tau}_{xz} = \mu \left(\frac{\partial \widetilde{u}}{\partial z} + \frac{\partial \widetilde{w}}{\partial x} \right)$$

and

$$\widetilde{\mathbf{s}} = \operatorname{grad} \widetilde{\varphi} + \operatorname{curl} \widetilde{\boldsymbol{\psi}},$$

eqs. 5.5 become

$$-\lambda\, k_l^2 \widetilde{\varphi} + 2\mu \left(\frac{\partial^2 \widetilde{\varphi}}{\partial z^2} + \frac{\partial^2 \widetilde{\psi}}{\partial x\, \partial z} \right) = 0 \tag{5.6}$$

$$\text{and} \qquad 2\, \frac{\partial^2 \widetilde{\varphi}}{\partial x\, \partial z} + \frac{\partial^2 \widetilde{\psi}}{\partial x^2} - \frac{\partial^2 \widetilde{\psi}}{\partial z^2} = 0 \qquad \text{at} \qquad z = 0$$

Now, substituting eqs. 5.1 into eqs. 5.6, we obtain two equations with three unknowns:

$$-\lambda\, k_l^2 A + 2\mu\, b_l^2\, k^2 A - i k^2 b_s\, 2\mu\, B = 0 \tag{5.7}$$

$$\text{and} \qquad -\, 2ik^2\, b_l\, A - k^2 B - k^2 b_s^2\, B = 0$$

or

$$\left(2-\frac{c^2}{c_s^2}\right)A-2i\left(1-\frac{c^2}{c_s^2}\right)^{1/2}B=0 \tag{5.8}$$

and $$2i\left(1-\frac{c^2}{c_l^2}\right)^{1/2}A+\left(2-\frac{c^2}{c_s^2}\right)B=0$$

If the surface wave exists, coefficients A and B differ from zero. Therefore, the determinant of the homogeneous system 5.8 has to be equal to zero. This gives

$$\left(2-\frac{c^2}{c_s^2}\right)^2=4\left(1-\frac{c^2}{c_s^2}\right)^{1/2}\left(1-\frac{c^2}{c_l^2}\right)^{1/2}, \tag{5.9}$$

and we obtain an equation with respect to c that is of great importance. First of all, eq. 5.9 follows from the boundary conditions. This means that if its solution does not satisfy inequality 5.4, the surface wave is absent. Before we find the roots of eq. 5.9, let us note that although system 5.8 does not allow us to determine coefficients A and B, it establishes a relation between them:

$$B=-\frac{i}{2}\left(2-\frac{c^2}{c_s^2}\right)\left(1-\frac{c^2}{c_s^2}\right)^{-1/2}A \tag{5.10}$$

In order to calculate velocity c, we square both parts of eq. 5.9, which gives

$$\left(2-\frac{c^2}{c_s^2}\right)^4=16\left(1-\frac{c^2}{c_l^2}\right)\left(1-\frac{c^2}{c_s^2}\right) \qquad \text{or} \tag{5.11}$$

$$\frac{c^2}{c_s^2}\left[\frac{c^6}{c_s^6}-8\frac{c^4}{c_s^4}+c^2\left(\frac{24}{c_s^2}-\frac{16}{c_l^2}\right)-16\left(1-\frac{c_s^2}{c_l^2}\right)\right]=0 \tag{5.12}$$

Eq. 5.12 implies that either

$$r=\left(\frac{c}{c_s}\right)^2=0 \tag{5.13}$$

or

$$r^3-8\,r^2+8\left(3-2\frac{c_s^2}{c_l^2}\right)r-16\left(1-\frac{c_s^2}{c_l^2}\right)=0 \tag{5.14}$$

where $r=(c/c_s)^2$. The root of eq. 5.13 is $c=0$, and, correspondingly, the wave is absent. Moreover, from eqs. 5.8 we have $A-iB=0$, $iA+B=0$, and its solution is

$A = B = 0$. That is, particles of a medium do not move. Therefore, the degenerative case $c = 0$ is not of interest.

Eq. 5.14 is a cubic equation relative to r. Its left side is negative when $r = 0$, $(c = 0)$ and it is positive if $r = 1$, $(c = c_s)$. Thus, eq. 5.14 has a positive real root within the interval $0 < r < 1$. In other words, velocity c obeys inequality 5.4, which means that the surface wave described by eqs. 5.1 can exist in a homogeneous half-space with a free boundary. This wave is a combination of two inhomogeneous plane waves, compressional and shear, propagating along the free surface with the same velocity c, which does not depend on frequency. Such a surface wave was predicted and theoretically investigated by Rayleigh, and for this reason it is called the Rayleigh wave. In general, the roots of eq. 5.14 can be found analytically as solutions of a cubic algebraic equation by means of the Cardano formula. Also, it is useful to consider two special cases when determination of roots is rather simple.

Case one Suppose that a medium is not compressible, i.e., deformation is absent. Then the velocity of the longitudinal waves c_l tends to infinity, and in place of eq. 5.14 we have

$$r^3 - 8\,r^2 + 24\,r - 16 = 0 \tag{5.15}$$

This cubic equation has one real root:

$$r \approx 0.91275$$

Therefore, the velocity of the surface wave is approximately equal to

$$c \approx 0.9553\; c_s \tag{5.16}$$

The other two roots of eq. 5.15 are complex and do not represent a surface wave.

Case two Next, we assume that Poisson's ratio σ is equal to $1/4$, i.e., $\lambda = \mu$ and

$$c_l = \sqrt{3}\; c_s \tag{5.17}$$

Respectively, eq. 5.14 becomes

$$r^3 - 8\,r^2 + \frac{56}{3}\,r - \frac{32}{3} = 0 \tag{5.18}$$

Its roots are

$$r_1 = 4, \qquad r_2 = 2 + \frac{2}{\sqrt{3}}, \qquad r_3 = 2 - \frac{2}{\sqrt{3}} \tag{5.19}$$

Unlike the last root,

$$r_3 \approx 0.8453, \tag{5.20}$$

the first two roots do not satisfy condition 5.4, and from eq. 5.20 we have

$$c \approx 0.9194\ c_s \tag{5.21}$$

Thus, in the case $(\lambda = \mu)$, the Rayleigh wave moves slightly slower than the shear wave, but almost twice as slow as the longitudinal wave:

$$c \ \approx 0.5309\ c_l$$

Results of calculation of functions c_s/c_l, c/c_l, and c/c_s for different values of Poisson's ratio are shown in Fig. 5.1b,c. By definition,

$$\lambda = \frac{E\,\sigma}{(1+\sigma)\,(1-2\sigma)} \qquad \text{and} \qquad \mu = \frac{E}{2\,(1+\sigma)}$$

Hence

$$\lambda + 2\mu = \frac{(1-\sigma)\,E}{(1+\sigma)\,(1-2\sigma)} \qquad \text{and} \qquad \frac{c_s}{c_l} = \sqrt{\frac{1-2\sigma}{2\,(1-\sigma)}}, \tag{5.22}$$

and the function c_s/c_l tends to 0.707, when $\sigma \to 0$. On the contrary, with an increase of σ this ratio approaches zero. The behavior of functions c_s/c_l and c/c_l is similar; in particular, they are equal to zero when an elastic medium becomes a fluid $(\sigma = 0.5)$. At the same time, the ratio c/c_s gradually increases with σ when approaching the limit at $\sigma = 0.5$.

The field of displacement carried by the Rayleigh wave

To illustrate the distribution of horizontal and vertical components of displacement, suppose, as before, that $\sigma = 1/4$. Then

$$\frac{c^2}{c_s^2} \approx 0.8453 \qquad \text{and} \qquad \frac{c_s^2}{c_l^2} = \frac{1}{3},$$

and we have

$$\sqrt{1-\frac{c^2}{c_l^2}} = \sqrt{1-\frac{c^2}{c_s^2}\,\frac{c_s^2}{c_l^2}} \approx 0.8475 \qquad \text{and} \qquad \sqrt{1-\frac{c^2}{c_s^2}} \approx 0.3933,$$

while

$$\widetilde{\varphi}(x, z, \omega) = A \ \exp\left(-k\sqrt{1 - \frac{c^2}{c_l^2}}z + \ i \ k \ x\right), \tag{5.23}$$

$$\widetilde{\psi}(x, z, \omega) = B \ \exp\left(-k\sqrt{1 - \frac{c^2}{c_s^2}}z + \ i \ k \ x\right)$$

From eq. 5.10, we have

$$B \approx -i \ 1.4679 \ A$$

This gives

$$\varphi(x, z, t) \approx \ e^{-0.8475 \ kz} A \cos\left(kx - \omega t\right) \tag{5.24}$$

and $\quad \psi(x, z, t) \approx 1.4679 \ A \ e^{-0.3933 \ kz} \sin\left(kx - \omega t\right)$

Substitution of eqs. 5.24 into the relationships

$$u = \frac{\partial \varphi}{\partial x} - \frac{\partial \psi}{\partial z}, \qquad w = \frac{\partial \varphi}{\partial z} + \frac{\partial \psi}{\partial x}$$

gives

$$u(x, z, t) \approx -k \ A \ \left(e^{-0.8475 \ kz} - 0.5773 \ e^{-0.3933 \ kz}\right) \sin\left(kx - \omega t\right) \tag{5.25}$$

and $\quad w(x, z, t) \approx -k \ A \ \left(0.8475 \ e^{-0.8475 \ kz} - 1.4679 \ e^{-0.3933 \ kz}\right) \cos\left(kx - \omega t\right)$

Since there is a phase shift $\pi/2$ between these components, and they differ by an amplitude, particle motion is elliptical. At the free surface, the vertical component w is about 1.5 times of the horizontal one (Fig. 5.1d). Note that at depth $z \approx 0.192 \ \Lambda$, where $\Lambda = 2\pi/k$ is the wavelength, the horizontal component vanishes. Below this point it changes sign. The distribution of both components with depth is shown in Fig. 5.1e. How the shapes of the ellipses are changing with depth is demonstrated in Fig. 5.1f. The major axis of ellipses is directed along the z-axis. Unlike in the case of the water wave (Part I), motion of particles at the free surface is counterclockwise. The exponential decay in eqs. 5.1 is directly proportional to frequency and, correspondingly, higher-frequency oscillations decrease more rapidly with z. In contrast, the lower is the frequency, the deeper the Rayleigh wave penetrates.

Now let us make several comments:

1. Unlike evanescent waves, the Rayleigh wave can exist alone, and it propagates with velocity c, which is less than shear-wave velocity.

2. Later we will show that a real source gives rise to a nonplane Rayleigh wave, which displays the same features as a plane wave. In other words, its velocity c is still defined by eq. 5.14, wave potentials exponentially decay with depth, and particles move along ellipses.

3. Two other roots of eq. 5.14 correspond to different wavefields that also obey the Helmholtz equations and boundary conditions at the free surface but do not describe surface wave motion.

Now let us raise two questions.

Could the Rayleigh wave exist in a homogeneous half-space with an ideally rigid boundary?

To find an answer, we will proceed from eqs. 5.1 and boundary conditions

$$u = 0 \qquad \text{and} \qquad w = 0$$

or

$$\frac{\partial \tilde{\varphi}}{\partial x} = \frac{\partial \tilde{\psi}}{\partial z} \qquad \text{and} \qquad \frac{\partial \tilde{\varphi}}{\partial z} = -\frac{\partial \tilde{\psi}}{\partial x} \tag{5.26}$$

Substitution of eqs. 5.1 into eqs. 5.26 yields

$$i\,A + \sqrt{1 - \frac{c^2}{c_s^2}}\;B = 0 \qquad \text{and} \qquad \sqrt{1 - \frac{c^2}{c_l^2}}\;A - i\,B = 0 \tag{5.27}$$

Excluding constants A and B, we obtain

$$\left(1 - \frac{c^2}{c_s^2}\right)^{1/2} \left(1 - \frac{c^2}{c_l^2}\right)^{1/2} = 1 \tag{5.28}$$

It is obvious that roots of eq. 5.28 do not satisfy inequality 5.4 and, therefore, the surface wave is absent.

Could the SH surface wave exist in a homogeneous half-space with a free boundary?

Now we will demonstrate that the SH surface wave cannot exist at the free boundary of such a medium. By definition, the complex amplitude of the x-component of the vector potential is

$$\tilde{\psi} = C \exp\left(-k \sqrt{1 - \frac{c^2}{c_s^2}} z + i\, kx\right), \qquad (5.29)$$

and at the boundary stress τ_{yz} vanishes,

$$\tilde{\tau}_{yz} = \mu \frac{\partial \tilde{v}}{\partial z} = 0 \qquad \text{or} \qquad \frac{\partial^2 \tilde{\psi}}{\partial z^2} = 0, \qquad \text{on} \qquad z = 0 \qquad (5.30)$$

Substitution of eq. 5.29 into 5.30 gives

$$\left(1 - \frac{c^2}{c_s^2}\right) C = 0$$

Since c/c_s should be less than 1, we conclude that $C = 0$ and that, indeed, this wave cannot exist.

5.2 Stoneley wave

We have shown that the Rayleigh wave may propagate along the free surface of a homogeneous half-space, but it is absent if the boundary is ideally rigid. Now let us consider a more general case and demonstrate that under certain conditions, a boundary wave similar to the Rayleigh wave moves along an interface between fluid and elastic media or between two elastic media. This wave is usually called the Stoneley wave.

Boundary between fluid and elastic half-spaces

Suppose there is a wave that propagates along the boundary (i.e., the wavefront is perpendicular to the x-axis), and its amplitude exponentially decays with increased distance from the boundary, $|z|$. As in the case of Rayleigh waves, this happens due to the destructive interference of elementary waves that arise at the interface. Because in a fluid shear waves are absent and the y-component of displacement v is equal to zero, expressions for the complex amplitudes of potentials are

$$\tilde{\varphi}_1 = A_1 e^{-k\, b_1\, z}\, e^{i\, k\, x} \qquad \text{if } z > 0 \qquad (5.31)$$

$$\text{and} \quad \tilde{\varphi}_2 = A_2\, e^{k\, b_{2l}\, z}\, e^{i\, k\, x}, \qquad \tilde{\psi}_2 = B_2\, e^{k\, b_{2s}\, z}\, e^{i\, k\, x} \qquad \text{if } z < 0$$

Here

$$b_1 = \left(1 - \frac{c^2}{c_1^2}\right)^{1/2}, \qquad b_{2l} = \left(1 - \frac{c^2}{c_{2l}^2}\right)^{1/2}, \tag{5.32}$$

$$b_{2s} = \left(1 - \frac{c^2}{c_{2s}^2}\right)^{1/2}, \quad k_1 = \frac{\omega}{c_1}, \quad \text{and} \quad k = \frac{\omega}{c},$$

where c is the velocity of propagation of the boundary wave. It is clear that exponential decay of potentials in both directions from the boundary takes place if velocity c obeys inequality

$$c < \min(c_1, c_{2s}) \tag{5.33}$$

Let us note that due to eqs. 5.32, functions $\tilde{\varphi}_1$, $\tilde{\varphi}_2$, $\tilde{\psi}_2$ are solutions of corresponding Hemholtz equations. Applying exactly the same approach we used in studying the Rayleigh wave, we find such values of c that eq. 5.33 would be met. To do this, we use known boundary conditions. At the interface between an elastic medium and a fluid, shear stresses vanish, whereas normal stress and the normal component of displacement are continuous:

$$\tau_{xz}^{(2)} = 0, \qquad \tau_{zz}^{(1)} = \tau_{zz}^{(2)}, \qquad w_1 = w_2$$

or

$$\frac{\partial u_2}{\partial z} + \frac{\partial w_2}{\partial x} = 0, \qquad \lambda_1 \operatorname{div} \mathbf{s}_1 = \lambda_2 \operatorname{div} \mathbf{s}_2 + 2\mu_2 \frac{\partial w_2}{\partial z}, \qquad w_1 = w_2 \tag{5.34}$$

Since component v is absent and the fields are independent of the y-coordinate, the second shear stress, $\tau_{yz}^{(2)}$, is also equal to zero:

$$\tau_{yz}^{(2)} = 0 \tag{5.35}$$

In terms of the complex amplitudes of potentials eqs. 5.34, become

$$2\,\frac{\partial^2 \tilde{\varphi}_2}{\partial x \partial z} + \frac{\partial^2 \tilde{\psi}_2}{\partial x^2} - \frac{\partial^2 \tilde{\psi}_2}{\partial z^2} = 0$$

$$-\lambda_1 k_1^2\, \tilde{\varphi}_1 = -\lambda_2\, k_{2l}^2\, \tilde{\varphi}_2 + 2\mu_2 \left(\frac{\partial^2 \tilde{\varphi}_2}{\partial z^2} + \frac{\partial^2 \tilde{\psi}_2}{\partial x \partial z}\right) \tag{5.36}$$

$$\frac{\partial \widetilde{\varphi}_1}{\partial z} = \frac{\partial \widetilde{\varphi}_2}{\partial z} + \frac{\partial \widetilde{\psi}_2}{\partial x} \qquad \text{on} \qquad z = 0$$

Substitution of eqs. 5.31 into eqs. 5.36 gives a homogeneous system of equations with respect to $A_1, \; A_2, \; B_2$:

$$2i\; k^2\; b_{2l}\; A_2 - k^2 B_2 - k^2\; b_{2s}^2 B_2 = 0$$

$$-\lambda_1\; k_1^2\; A_1 = -\lambda_2\; k_{2l}^2\; A_2 + 2\mu_2\; k^2\left(b_{2l}^2\; A_2 + i\; b_{2s} B_2\right) \tag{5.37}$$

$$-k\; b_1\; A_1 = k\; b_{2l}\; A_2 + i\; k\; B_2$$

Taking into account eqs. 5.32 and equalities

$$\mu_2 = \rho_2\; c_{2s}^2, \qquad \lambda_2 + 2\mu_2 = \rho_2\; c_{2l}^2, \qquad \lambda_1 = \rho_1\; c_1^2,$$

in place of eq. 5.37 we obtain

$$2i\left(1 - \frac{c^2}{c_{2l}^2}\right)^{1/2} A_2 - \left(2 - \frac{c^2}{c_{2s}^2}\right) B_2 = 0$$

$$-\frac{c^2}{c_{2s}^2}\; A_1 = \frac{\rho_2}{\rho_1}\left(2 - \frac{c^2}{c_{2s}^2}\right)\; A_2 + 2i\; \frac{\rho_2}{\rho_1}\left(1 - \frac{c^2}{c_{2s}^2}\right)^{1/2} B_2 \tag{5.38}$$

$$-\left(1 - \frac{c^2}{c_1^2}\right)^{1/2} A_1 = \left(1 - \frac{c^2}{c_{2l}^2}\right)^{1/2} A_2 + iB_2$$

As in the case of the Rayleigh wave, the wave propagating along the boundary may exist if system 5.38 has a nonzero solution. This means that determinant of system 5.38 is equal to zero:

$$\begin{vmatrix} 0 & 2i\left(1 - \frac{c^2}{c_{2l}^2}\right)^{1/2} & -\left(2 - \frac{c^2}{c_{2s}^2}\right) \\ \frac{c^2}{c_{2s}^2} & \frac{\rho_2}{\rho_1}\left(2 - \frac{c^2}{c_{2s}^2}\right) & 2i\frac{\rho_2}{\rho_1}\left(1 - \frac{c^2}{c_{2s}}\right)^{1/2} \\ \left(1 - \frac{c^2}{c_1^2}\right)^{1/2} & \left(1 - \frac{c^2}{c_{2l}^2}\right)^{1/2} & i \end{vmatrix} = 0 \tag{5.39}$$

Here ρ_1 and ρ_2 are densities of a fluid and an elastic medium, respectively. We obtain an algebraic equation with respect to the velocity of Stoneley waves, and it has the form

$$\left(\frac{\rho_1}{\rho_2}\, b_{2l} + b_1\right) r^4 - 4\, b_1 r^2 - 4\, b_1\,(b_{2l}b_{2s} - 1) = 0 \tag{5.40}$$

Here $r = c/c_{2s}$. Letting $\rho_1 = 0$, we again obtain the equation determining the velocity of Rayleigh waves. Numerical analysis of eq. 5.40 shows that for any set of parameters c_1/c_{2s}, c_{2l}/c_{2s}, ρ_2/ρ_1, there is a real root that obeys inequality 5.33 and does not depend on frequency. In other words, the wave can propagate along the boundary between a fluid and an elastic medium, and its potentials exponentially decay with increased distance from this interface. This special type of Stoneley wave is often called a Scholte wave. It is interesting to note that the velocity of this wave is smaller than that of the Rayleigh wave in an elastic half-space (Fig. 5.2a). Since displacement $\mathbf{s}$ in a fluid is described by one potential only, both displacement components decrease exponentially at the same rate. In an elastic medium, due to the existence of two potentials, the displacement components have a different dependence on z. The exponential decrease of both components with z is observed only when z exceeds some value. This value is frequency-dependent and decreases with increased frequency.

Boundary between two elastic media

In this general case, the wavefields in both media are described by two potentials, and we have

$$\tilde{\varphi}_1 = A_1 e^{-k\, b_{1l}\, z}\, e^{i\, k\, x}, \qquad \tilde{\psi}_1 = B_1 e^{-k\, b_{1s}\, z}\, e^{i\, k\, x} \quad \text{if} \quad z > 0 \tag{5.41}$$

and

$$\tilde{\varphi}_2 = A_2\, e^{k\, b_{2l}\, z}\, e^{i\, k\, x}, \qquad \tilde{\psi}_2 = B_2 e^{k\, b_{2s}\, z}\, e^{i\, k\, x} \quad \text{if} \quad z < 0 \tag{5.42}$$

Here

$$b_{nl} = \left(1 - \frac{c^2}{c_{nl}^2}\right)^{1/2}, \qquad b_{ns} = \left(1 - \frac{c^2}{c_{ns}^2}\right)^{1/2}, \quad \text{and} \quad n = 1, 2 \tag{5.43}$$

At points of the boundary $z = 0$, stresses and displacement components are continuous functions. By analogy with eqs. 5.36, we have:

$$\mu_1\left(2\frac{\partial^2\tilde{\varphi}_1}{\partial x\partial z} + \frac{\partial^2\tilde{\psi}_1}{\partial x^2} - \frac{\partial^2\tilde{\psi}_1}{\partial z^2}\right) = \mu_2\left(2\frac{\partial^2\tilde{\varphi}_2}{\partial x\partial z} + \frac{\partial^2\tilde{\psi}_2}{\partial x^2} - \frac{\partial^2\tilde{\psi}_2}{\partial z^2}\right)$$

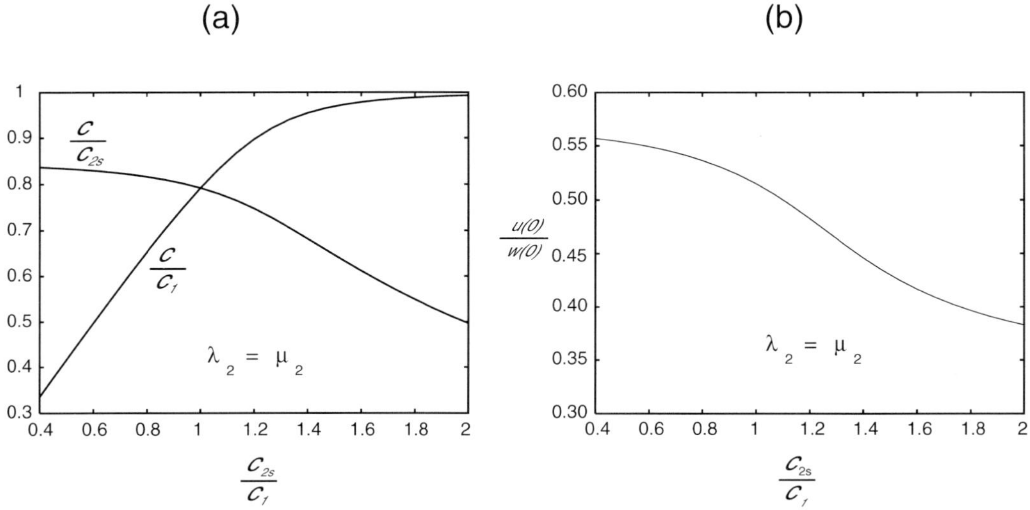

Figure 5.2: Stoneley wave along fluid:solid boundary: (a) Velocity as a function of ratio c_{2s}/c_1 (b) Ratio of amplitudes of horizontal and vertical components at the boundary as a function of c_{2s}/c_1

$$-\lambda_1\ k_{1l}^2\widetilde{\varphi}_1+2\mu_1\left(\frac{\partial^2\widetilde{\varphi}_1}{\partial z^2}+\frac{\partial^2\widetilde{\psi}_1}{\partial x\partial z}\right)=-\lambda_2\ k_{2l}^2\widetilde{\varphi}_2+2\mu_2\left(\frac{\partial^2\widetilde{\varphi}_2}{\partial z^2}+\frac{\partial^2\widetilde{\psi}_2}{\partial x\partial z}\right), \tag{5.44}$$

$$\frac{\partial\widetilde{\varphi}_1}{\partial z}+\frac{\partial\widetilde{\psi}_1}{\partial x}=\frac{\partial\widetilde{\varphi}_2}{\partial z}+\frac{\partial\widetilde{\psi}_2}{\partial x},\quad \text{and}\quad \frac{\partial\widetilde{\varphi}_1}{\partial x}-\frac{\partial\widetilde{\psi}_1}{\partial z}=\frac{\partial\widetilde{\varphi}_2}{\partial x}-\frac{\partial\widetilde{\psi}_2}{\partial z}$$

Substitution of eqs. 5.41 and 5.42 into set 5.44 gives a homogeneous system of four equations with respect to $A_1, A_2, B_1,$ and B_2. In order to obtain a nonzero solution, the determinant of this system has to be equal to zero, and we obtain:

$$\begin{vmatrix} 2\rho_1c_{1s}^2b_{1l} & 2\rho_2c_{2s}^2b_{2l} & \rho_1c_{1s}^2\left(2-\dfrac{c^2}{c_{1s}^2}\right) & \rho_2c_{2s}^2\left(2-\dfrac{c^2}{c_{2s}^2}\right) \\ \rho_1\left(c^2-2c_{1s}^2\right) & -\rho_2\left(c^2-2c_{2s}^2\right) & -2\rho_1c_{1s}^2b_{1s} & 2\rho_2c_{2s}^2b_{2s} \\ 1 & -1 & b_{1s} & -b_{2s} \\ b_{1l} & b_{2l} & 1 & 1 \end{vmatrix}=0 \tag{5.45}$$

Performing multiplication, we arrive at the equation that allows us to determine the velocity of s Stoneley wave:

$$r^4\left[\left(\frac{\rho_1}{\rho_2}-1\right)^2-\left(\frac{\rho_1}{\rho_2}\,b_{2l}+b_{1l}\right)\left(\frac{\rho_1}{\rho_2}\,b_{2s}+b_{1s}\right)\right] \tag{5.46}$$

$$+\,4r^2\left(\frac{\rho_1}{\rho_2}\,\frac{c_{1s}^2}{c_{2s}^2}-1\right)\left(\frac{\rho_1}{\rho_2}\,b_{2l}b_{2s}-b_{1l}\,b_{1s}-\frac{\rho_1}{\rho_2}+1\right)$$

$$+4\,\left(\frac{\rho_1}{\rho_2}\,\frac{c_{1s}^2}{c_{2s}^2}-1\right)^2(b_{1l}\,b_{1s}-1)\,(b_{2l}\,b_{2s}-1)=0$$

where $r=c/c_{2s}$. Assuming that $\rho_1=0$, we again obtain the equation for Rayleigh waves. In fact, eq. 5.46 becomes

$$r^4\,(1-b_{1l}\,b_{1s})-4r^2\,(1-b_{1l}\,b_{1s})-4\,(1-b_{1l}\,b_{1s})\,(b_{2l}\,b_{2s}-1)=0$$

or

$$r^4-4r^2+4=4b_{2l}\,b_{2s},$$

that is,

$$\left(\frac{c^2}{c_{2s}^2}-2\right)^2=4\left(1-\frac{c^2}{c_{1l}^2}\right)^{1/2}\left(1-\frac{c^2}{c_{1s}^2}\right)$$

Next suppose that the upper medium is a fluid. Therefore

$$c_{1s}=0 \qquad \text{and} \qquad b_{1s}\to i\,\infty,$$

and in place of eq. 5.46 we have

$$r^4\left(\frac{\rho_1}{\rho_2}\,b_{2l}+b_{1l}\right)-4r^2b_{1l}-4b_{1l}\,(b_{2l}\,b_{2s}-1)=0,$$

which coincides with eq. 5.40. Study of eq. 5.46 shows that its solution satisfies the condition

$$c<\min\left(c_{1s},c_{2s}\right), \tag{5.47}$$

if shear velocities c_{1s} and c_{2s} differ only slightly. In illustration, Fig. 5.3 shows two shaded zones where eq. 5.64 has real roots, obeying inequality 5.47. Outside of these

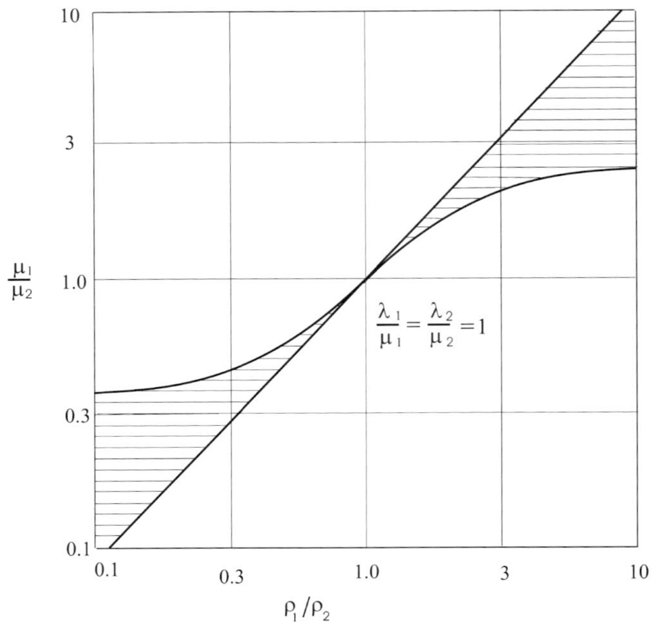

Figure 5.3: Zones of existence of Stoneley wave at the solid:solid boundary. [After Grant & West, 1965]

zones, the Stoneley wave cannot exist. It turns out that the velocity of the Stoneley wave usually falls between the velocities of Rayleigh waves and shear waves in the medium with greater density. It is important that the velocity of the Stoneley wave, as well as that of the Rayleigh wave, does not depend on frequency.

Could the boundary SH wave exist?

Finally, we demonstrate that the SH boundary wave is absent at the boundary of two elastic media. Taking into account that components u and w are equal to zero and that component v can depend on x and z only, we have for the x-component of the potential

$$\widetilde{\psi}_1 = C_1\, e^{-k\, b_{1s}\, z}\, e^{i\, k\, x}, \qquad \widetilde{\psi}_2 = C_2\, e^{k\, b_{2s}\, z}\, e^{i\, k\, x} \tag{5.48}$$

Boundary conditions are

$$\widetilde{v}_1 = \widetilde{v}_2 \quad \text{and} \quad \widetilde{\tau}_{1yz} = \widetilde{\tau}_{2yz} \quad \text{on} \quad z = 0 \tag{5.49}$$

or

$$\frac{\partial\ \tilde{\psi}_1}{\partial z} = \frac{\partial\ \tilde{\psi}_2}{\partial z} \quad \text{and} \quad \frac{\partial^2\ \tilde{\psi}_1}{\partial z^2} = \frac{\partial^2\ \tilde{\psi}_2}{\partial z^2} \quad \text{on} \quad z = 0 \tag{5.50}$$

Substitution of eqs. 5.48 into eqs. 5.50 gives

$$b_{1s}\ C_1 + b_{2s}\ C_2 = 0 \quad \text{and} \quad b_{1s}^2\ C_1 - b_{2s}^2\ C_2 = 0 \tag{5.51}$$

Since the determinant of this system differs from zero, we conclude that

$$C_1 = C_2 = 0$$

and that, therefore, the boundary SH wave cannot exist in this case.

5.3 Love waves

Now we will study propagation of the simplest interferential waves, named Love waves for the physicist who proved their existence. To explain how the interference of elementary plane body waves produces this type of surface wave, we will consider propagation of the interferential plane SH wave along the x-axis in a two-layered medium with a free surface, in which $c_{1s} < c_{2s}$ (Fig. 5.4a). We will suppose that $c_{1s} < c_{2s}$ and this wave exponentially decays in the half-space with depth z. The motion of particles is characterized by linear polarization because displacement has the single component v. In order to understand the nature of such a surface wave, imagine that the elementary plane SH wave moves downward in the layer and

$$c_{1s} < c_{2s} \tag{5.52}$$

At boundary $z = 0$, this wave will generate reflected and transmitted SH waves. In accordance with Snell's law, the critical angle is defined from the equation

$$\gamma_c = \sin^{-1} \frac{c_{1s}}{c_{2s}} \tag{5.53}$$

If the angle of incidence $\gamma_i < \gamma_c$, then both the reflected and transmitted SH waves that arise at the bottom of the layer are homogeneous. The reflected wave has a smaller amplitude than the incident wave because part of its energy has leaked into the half-space. The reflected wave propagates upward, and is reflected downward at the free surface without the loss of energy. Correspondingly, at interface $z = 0$ we again observe the reflected wave, and so on. Thus, there are two systems of SH plane waves moving

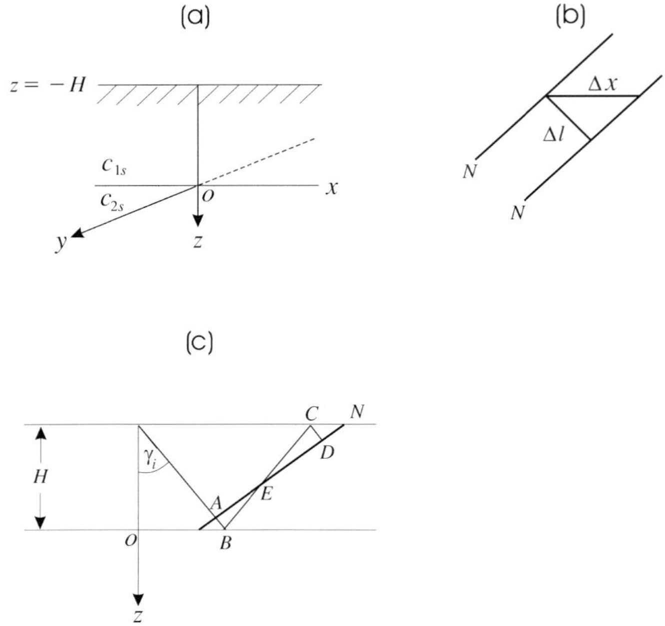

Figure 5.4: Love waves: (a) Two-layered elastic medium (b) Illustration of eq. 5.53 (c) Ray paths of the downgoing and upgoing waves

either downward or upward with the same apparent velocity along the surface. Since after each reflection the amplitudes of these waves decrease, these systems of waves rapidly decay as x increases.

The situation drastically changes if the incident angle exceeds the critical angle, $\gamma_i > \gamma_c$. In this case, homogeneous transmitted waves are absent, and total internal reflections take place. Respectively, amplitudes of the incident and reflected waves are equal to each other at both interfaces, and elastic energy remains unchanged inside the layer. However, this fact alone does not guarantee the existence of a wave propagating inside the layer without attenuation. As was pointed out, the wavefield is a superposition of two systems of homogeneous plane waves, and their fronts form the same angle with boundaries. If interference between these two sets of waves is constructive, then the resultant (interferential) wave moves along the layer without attenuation. On the contrary, when destructive interference occurs, this wave rapidly decreases with distance along the layer and finally disappears. It is obvious that as in the case of acoustic waves, the condition for constructive interference depends on frequency.

Condition for constructive interference and dispersion equation

Since we assume that $\gamma_i > \gamma_c$, propagation of the interferential wave inside the layer is accompanied by an evanescent wave in the half-space $(z > 0)$. This wave exponentially decays with depth. Wavefields inside and outside the layer constitute one wave, traveling along the x-axis. Its velocity c can be easily evaluated from Fig. 5.4b, where the position of the wavefront is shown at two instances. It is clear that

$$c = \frac{\Delta x}{\Delta t} = \frac{\Delta l}{\Delta t \, \sin \gamma_i} = \frac{c_{1s}}{\sin \gamma_i} \tag{5.54}$$

Here γ_i is the angle of incidence of the homogeneous plane wave, and it varies as

$$\gamma_c \leq \gamma_i \leq \frac{\pi}{2}$$

Respectively, velocity c changes in the following way:

$$c_{1s} \leq c \leq c_{2s} \tag{5.55}$$

Let us note that the presence of an evanescent wave in the half-space does not decrease elastic energy inside the layer. The loss of energy does not happen because during each half-period, the total energy leaving the layer is equal to zero (Part II). The wavefield inside and outside the layer is called the Love wave, and c is its velocity. It is another example of the surface wave propagating in a horizontal direction. However, unlike the Rayleigh and Stoneley waves, the Love waves require as a necessary condition of existence the presence of a layer of finite thickness in which constructive interference must take place. Correspondingly, when the layer thickness tends to zero, the Love waves vanish, as was proved earlier. The same is correct if the layer thickness becomes infinitely large, because the set of upgoing plane waves disappears. Also, it is clear that if the half-space has a lower velocity $(c_{1s} > c_{2s})$, there are always transmitted homogeneous waves, and the Love waves cannot exist.

By analogy with acoustic waves (Part II), it is useful to demonstrate that constructive interference takes place for a certain set of frequencies. Let us consider phase surface N of the downgoing wave, Fig. 5.4c. As we know, the reflected upgoing and downgoing waves appear as a result of the action of wave N. If the frequencies are such that the phase difference between the incident and twice reflected waves is equal to $2\pi n$, their superposition is constructive. Waves with other frequencies have a phase shift different from $2\pi n$, and they destructively interfere. As a result, they cancel each other out after a relatively small number of reflections. Note that the phase difference of two upgoing and downgoing waves is defined by the following factors:

1. The phase shift between the displacements carried out by incident and reflected waves at the free surface $(z = -H)$ is equal to 0.

2. The phase shift Ψ of the reflected wave at the layer bottom depends on angle γ_i as well as on elastic impedances of both media. This shift is defined by eq. 4.158 in the previous chapter.

3. The phase delay due to the extra path of the twice reflected wave relative to the incident wave is defined by the length of this path $ABECD$ and wavelength Λ_{1s} of the shear wave in the layer. Taking into account that the argument of the sinusoidal wave is

$$\omega t - k_{1s} l = \omega t - \frac{2\pi}{\Lambda_{1s}}\, l,$$

the condition for constructive interference can be written as

$$k_{1s}\,|ABECD| + \Psi = 2\pi n$$

or

$$\frac{2\pi}{\Lambda_{1s}}\,(|AB| + |BC| + |CD|) = 2n\pi - \Psi \tag{5.56}$$

Here n is some integer number. By definition $\Lambda_{1s} = c_{1s}\,/\,f$, and certainly the result of interference depends on frequency. As is seen from Fig. 5.4c, $AB = BE\cos 2\gamma_i$, $CD = CEcos2\gamma_i$, and $BC = H/\cos\gamma_i$. Substitution of these terms into eq. 5.56 gives

$$\frac{2\pi H}{\Lambda_{1s}}\left(\frac{\cos 2\gamma_i}{\cos\gamma_i} + \frac{1}{\cos\gamma_i}\right) \equiv \frac{4\pi H\cos\gamma_i}{\Lambda_{1s}} = 2n\pi - \Psi$$

or

$$\frac{\omega H\cos\gamma_i}{c_{1s}} = \frac{2n\pi - \Psi}{2} \tag{5.57}$$

In particular, when the underlying medium is ideally rigid, $\Psi = 0$, in place of eq. 5.57 we have

$$\frac{\omega_n H\cos\gamma_i}{c_{1s}} = n\pi \tag{5.58}$$

In such a case, the evanescent wave is absent, and the Love wave is confined inside the elastic layer.

Taking into account eq. 5.54, it is easy to eliminate $\cos\gamma_i$ from eq. 5.57

$$\cos\gamma_i = \left(1 - \frac{c_{1s}^2}{c^2}\right)^{1/2} = \frac{c_{1s}}{c}\left(\frac{c^2}{c_{1s}^2} - 1\right)^{1/2} = \frac{c_{1s}\tilde{b}_{1s}}{c}$$

and find the relationship between the velocity of the Love wave and frequency. This gives

$$\frac{\omega_n H}{c} = \frac{2n\pi - \Psi}{2} \tag{5.59}$$

However, this equation contains an unknown phase shift at the layer bottom. This shift can be found from the expression for the reflection coefficient $\mathcal{R}_{SH}$ at the boundary between elastic media obtained in Chapter 4. When the angle of incidence γ_i exceeds the critical angle γ_c, this coefficient is complex:

$$\mathcal{R}_{SH} = \exp(-i\Psi) \quad \text{and} \quad \Psi = -2\tan^{-1}\left(\frac{\mu_2 \tilde{b}_{2s}}{\mu_1 \tilde{b}_{1s}}\right)$$

where

$$\tilde{b}_{2s} = (1 - \frac{c^2}{c_{2s}^2})^{1/2}$$

As a result, the dispersion equation for Love waves takes the form

$$\tan\left(\frac{\omega H}{c}\tilde{b}_{1s}\right) = \frac{\mu_2 \tilde{b}_{2s}}{\mu_1 \tilde{b}_{1s}} \tag{5.60}$$

This equation allows us to find the velocity of the Love wave as a function of frequency. But before we start to investigate solutions of eq. 5.60, we will take a different approach and derive the same equation from the boundary value problem.

The boundary value problem for Love waves

We represent the complex amplitude of the v-component of displacement as

$$\tilde{v}^{(1)} = \left(C_1 e^{i\,k\,\tilde{b}_{1s}\,z} + C_2 e^{-i\,k\,\tilde{b}_{1s}\,z}\right) e^{i\,k\,x} \quad \text{if} \quad z \le 0$$

$$\text{and} \quad \tilde{v}^{(2)} = C_3\, e^{-k\,\tilde{b}_{2s}\,z}\, e^{i\,k\,x} \quad \text{if } z \ge 0 \tag{5.61}$$

Here again

$$\tilde{b}_{1s} = \left(\frac{c^2}{c_{1s}} - 1\right)^{1/2}, \quad \tilde{b}_{2s} = \left(1 - \frac{c^2}{c_{2s}^2}\right)^{1/2}$$

Now we use the boundary conditions and find the equation that relates the velocity of Love waves to ther frequency. As the normal stress at the free surface is equal to 0, we have

$$\tilde{\tau}_y^{(1)} z = \mu_1 \frac{\partial \tilde{v}^{(1)}}{\partial z} = 0 \qquad \text{on} \qquad z = -H$$

From continuity of displacements and stresses at the lower boundary of the layer, it follows that

$$\tilde{v}^{(1)} = \tilde{v}^{(2)} \quad \text{and} \quad \mu_1 \frac{\partial \tilde{v}^{(1)}}{\partial z} = \mu_2 \frac{\partial \tilde{v}^{(2)}}{\partial z} \quad \text{on} \quad z = 0$$

These conditions yield:

$$\begin{aligned} & C_1 e^{-i\,k\,\tilde{b}_{1s} H} - C_2 e^{i\,k\,\tilde{b}_{1s} H} = 0 \\ & C_1 + C_2 = C_3 \\ & i\,\mu_1\,\tilde{b}_{1s}\,(C_1 - C_2) = -\,\mu_2\,\tilde{b}_{2s}\,C_3 \end{aligned} \tag{5.62}$$

System 5.62 has a nonzero solution for C_1, C_2, and C_3 if the determinant is equal to zero:

$$\begin{vmatrix} e^{-i\,k\,\tilde{b}_{1s} H} & -e^{i\,k\,\tilde{b}_{1s} H} & 0 \\ 1 & 1 & -1 \\ i\,\mu_1\,(\tilde{b}_{1s})^2 & -i\,\mu_1\,(\tilde{b}_{1s})^2 & \mu_2\,(\tilde{b}_{2s})^2 \end{vmatrix} = 0 \tag{5.63}$$

Performing simple algebra, we can obtain the dispersion equation in a more explicit form. Of course, the same result directly follows from system 5.62. Excluding C_3 in its last two equations, we have:

$$C_2 = \frac{\mu_1\,\tilde{b}_{1s} - i\mu_2\,\tilde{b}_{2s}}{\mu_1\,\tilde{b}_{1s} + i\mu_2\,\tilde{b}_{2s}}\, C_1 = \exp\left(-2i\tan^{-1}\frac{\mu_2\,\tilde{b}_{2s}}{\mu_1\,\tilde{b}_{1s}}\right) \tag{5.64}$$

Substituting C_2 into the first equation of set 5.62, after simple transformation using Euler's formulas, we obtain

$$\tan(k\tilde{b}_{1s}H) = \frac{\mu_2\,\tilde{b}_{2s}}{\mu_1\,\tilde{b}_{1s}} \qquad \text{or} \qquad \tan(\frac{\omega}{c}\,\tilde{b}_{1s}H) = \frac{\mu_2\,\tilde{b}_{2s}}{\mu_1\,\tilde{b}_{1s}} \tag{5.65}$$

This dispersion equation is exactly the same as eq. 5.60 obtained using the condition of constructive interference. Eq. 5.65 establishes the relationship between the velocity of the Love waves c and frequency ω, as well as between this velocity and parameters of the medium. We can rewrite eq. 5.65 in a slightly different form:

$$\tan\left[kH\left(\frac{c^2}{c_{1s}^2}-1\right)^{1/2}\right]=\frac{\mu_2}{\mu_1}\left(\frac{c^2}{c_{1s}^2}-1\right)^{-1/2}\left(1-\frac{c^2}{c_{2s}^2}\right)^{1/2} \tag{5.66}$$

Since the left side of this equation is a periodic function, for any fixed value of c inside the range prescribed by inequality 5.55 there is an infinite number of roots:

$$\omega_0(c),\ \omega_1(c),\ \omega_2(c),\ldots..,\ \omega_n(c),\ldots$$

Each value of ω_n characterizes the frequency of the nth mode of the Love waves propagating with velocity c. The inverse functions $c_n(\omega)$ describe phase velocity dispersion curves of Love waves.

Since the left side of eq. 5.66 is a periodic function, it is convenient to present this equation in the form

$$kH\left(\frac{c^2}{c_{1s}^2}-1\right)^{1/2}=\pi n+\tan^{-1}\left[\frac{\mu_2}{\mu_1}\frac{\left(1-\frac{c^2}{c_{2s}^2}\right)^{1/2}}{\left(\frac{c^2}{c_{1s}^2}-1\right)^{1/2}}\right] \tag{5.67}$$

Here n is an integer number.

Now, using eq. 5.67, we will confirm inequality 5.55, beginning with the case when $n=0$. Then eq. 5.67 becomes

$$kH\left(\frac{c^2}{c_{1s}^2}-1\right)^{1/2}=\tan^{-1}\frac{\mu_2}{\mu_1}\frac{\left(1-\frac{c^2}{c_{2s}^2}\right)^{1/2}}{\left(\frac{c^2}{c_{1s}^2}-1\right)^{1/2}} \tag{5.68}$$

Suppose, first, that the velocity of the Love wave, c, approaches c_{2s}:

$$c\to c_{2s}$$

Since the right side of eq. 5.68 vanishes, we have

$$kH\left(\frac{c^2}{c_{1s}^2}-1\right)^{1/2}\to 0 \quad \text{or} \quad k\to 0, \qquad \text{i.e.,} \qquad \omega\to\infty$$

On the contrary, when $c \to c_{1s}$, the right side of eq. 5.71 tends to $\pi/2$, while

$$\left(\frac{c^2}{c_{1s}^2} - 1\right)^{1/2} \to 0$$

To preserve equality 5.68, the wave number has to be infinitely large, that is $\omega \to \infty$. Thus, in agreement with eq. 5.55, when the ratio Λ_{1s}/H varies from zero to infinity

$$0 < \frac{\Lambda_{1s}}{H} < \infty,$$

the phase velocity of the Love wave changes as $c_{1s} \leq c \leq c_{2s}$. It is easy to show that this conclusion remains valid for any mode. However, there is a difference which concerns the range of ω. In fact, as follows from eq. 5.70, if $c = c_{2s}$ and $n \neq 0$, we have

$$kH\left(\frac{c_{2s}^2}{c_{1s}^2} - 1\right)^{1/2} = \pi n$$

or

$$\frac{\Lambda_{1s}}{H} = \frac{2}{n}\left(1 - \frac{c_{1s}^2}{c_{2s}^2}\right)^{1/2} \tag{5.69}$$

In the opposite case, when $c = c_{1s}$, as before $\Lambda_{1s}/H \to 0$. Thus, for any n a change in the velocity of the Love wave (eq. 5.55) occurs within the frequency range

$$0 \leq \frac{\Lambda_{1s}}{H} \leq \frac{2}{n}\left(1 - \frac{c_{1s}^2}{c_{2s}^2}\right)^{1/2} \tag{5.70}$$

We see that with an increase of the order n, the range of frequencies (wavelengths) where the mode exists, narrows. For instance, if

$$\frac{\Lambda_{1s}}{H} > 2\left(1 - \frac{c_{1s}^2}{c_{2s}^2}\right)^{1/2}$$

there is only the fundamental mode, $n = 0$. However, within the interval

$$\left(1 - \frac{c_{1s}^2}{c_{2s}^2}\right)^{1/2} \leq \frac{\Lambda_{1s}}{H} < 2\left(1 - \frac{c_{1s}^2}{c_{2s}^2}\right)^{1/2},$$

two modes can be observed, $n = 0$ and $n = 1$. Within the interval

$$\frac{2}{3}\left(1 - \frac{c_{1s}^2}{c_{2s}^2}\right)^{1/2} \leq \frac{\Lambda_{1s}}{H} < \left(1 - \frac{c_{1s}^2}{c_{2s}^2}\right)^{1/2},$$

three modes exist, and so on. The width of such intervals decreases with an increase of n, and it is equal to

$$\frac{2}{n\,(n-1)}\left(1-\frac{c_{1s}^2}{c_{2s}^2}\right)^{1/2} \tag{5.71}$$

Dispersion curves for the first five modes are shown in Fig. 5.5a.

Displacement $v_n\,(z)$

Next consider the displacement of the nth mode as a function of z. From eqs. 5.64 and 5.65, we have

$$\frac{C_2}{C_1}=\exp\left(-i2k_n\tilde{b}_{1sn}H\right),$$

$$\tilde{v}_n=C_1\left(e^{ik_n\,\tilde{b}_{1sn}\,z}+e^{-2ik_n\,\tilde{b}_{1sn}H}\,e^{-ik_n\,\tilde{b}_{1sn}z}\right)\,e^{i\,k_n\,x}$$

$$=C_n\cos\left[k_n\tilde{b}_{1sn}(z+H)\right]\;e^{ik_nx}\quad \text{if}\quad -H<z\le 0 \tag{5.72}$$

$$\text{and}\quad \tilde{v}_n=C_n\;\cos(k_n\tilde{b}_{1sn}H)\;e^{-k_n\tilde{b}_{2sn}z}\;e^{i\,k_n\,x}\qquad \text{if } z\ge 0.$$

Here

$$\tilde{b}_{1sn}=\left(\frac{c_n^2}{c_{1s}^2}-1\right)^{1/2},\quad \tilde{b}_{2sn}=\left(1-\frac{c_n^2}{c_{2s}^2}\right)^{1/2},\quad \text{and}\quad C_n=2C_1\exp\left(-ik_n\,\tilde{b}_{1sn}H\right)$$

Both $\tilde{b}_{1sn}$ and $\tilde{b}_{2sn}$ depend on the mode order. By definition, displacement v_n is

$$v_n=D_n\cos\left[k_n\tilde{b}_{1sn}(z+H)\right]\cos\left(\omega t-k_nx\right)\qquad \text{if}\;\; -H<z\le 0 \tag{5.73}$$

$$\text{and}\qquad v_n=D_n\cos\left(k_n\tilde{b}_{1sn}H\right)\;e^{-k_n\,\tilde{b}_{2sn}\,z}\cos\left(\omega t-k_nx\right)\qquad \text{if}\quad z\ge 0$$

As in the case of Rayleigh or Stoneley waves, coefficient D_n remains unknowns, since the primary source is not taken into account. From eqs. 5.73 we see that the Love wave is oscillating wave along the z-axis inside the layer but exponentially decays outside of it. For instance, if $n=0$, displacement gradually decreases with depth. For the $n=1$,

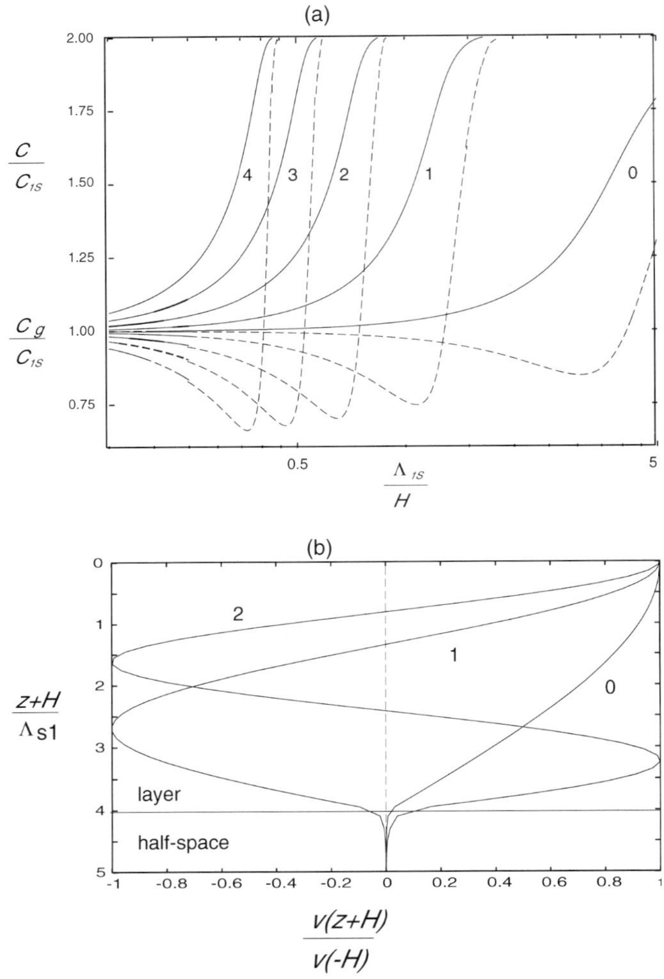

Figure 5.5: Love waves in a two-layered elastic medium: (a) Dispersion curves of phase velocities (solid lines) and group velocities (dashed lines) for the first five modes $(n = 0, 1, ..., 4)$ (b) Normalized displacement v as function of z/Λ_{1s}. Here $c_{2s}/c_{1s} = 2$

we observe a change of sign, and correspondingly, there is a nodal plane where $v_{1n} = 0$. With an increase of n, the number of such planes also increases, as is seen in Fig. 5.5b.

Transient modes of Love waves

In the case of a nonstationary source, constructive and destructive interference of sinusoidal harmonics that form each mode occurs. As a result of constructive interference, transient modes of Love waves are observed. Their waveforms are defined by the corresponding group velocity curves (Parts I and II). Behavior of the group velocities c_{gn} as functions of Λ_{1s}/H for the first five Love modes in a two-layered model is shown in Fig. 5.5a (dashed lines).

Let us briefly discuss the behavior of any mode presented in a considered frequency range. Suppose that the observation point has an offset x. Then, as follows from behavior of function $c_{gn}(\omega)$, during time interval $x/c_{2s} < t < x/c_{1s}$, we observe a wave group with relatively low apparent frequencies that increase with time. Then, at moment $t = x/c_{1s}$, a high-frequency group arrives and interferes with the first group. The apparent frequency of this group decreases with time. As time increases, the frequency contents of the two groups become very close, and they finally merge into one quasi-stationary wavetrain called the Airy phase. The resemblance to propagation along a waveguide in an acoustic medium is obvious (Part II).

5.4 Rayleigh waves in a two-layered medium

Earlier we demonstrated that in a homogeneous half-space, the velocity of propagation of Rayleigh waves is independent of frequency. We also found out that in a two-layered medium, different modes of Love waves may exist. They demonstrate a dispersive behavior – that is, their phase velocities are functions of frequency. Now we will show that Rayleigh waves in a layered medium may display similar features. For illustration, we will consider the case in which a layer of fluid overlays a homogeneous elastic half-space (Fig. 5.6a).

Fluid layer over an elastic half-space

Suppose that the surface wave propagates along the x-axis with phase velocity c, and, in an elastic medium it decays exponentially with depth z. Unlike in the case of Love waves, displacement $\mathbf{s}$ is characterized by two components, u and w, while $v = 0$. Correspondingly, expressions of complex amplitudes of potentials are

$$\tilde{\varphi}_1 = \left(A_1\, e^{k\, b_1\, z} + B_1 e^{-k\, b_1\, z}\right)\, e^{i\, k\, x}, \qquad \tilde{\varphi}_2 = A_2\, e^{-k\, b_{2l}\, z}\, e^{i\, k\, x}, \tag{5.74}$$

$$\text{and} \quad \tilde{\psi}_2 = B_2 \, e^{-k \, b_{2s} \, z} \, e^{i \, k \, x}$$

Here

$$b_1 = \left(1 - \frac{c^2}{c_1^2}\right)^{1/2}, \quad b_{2l} = \left(1 - \frac{c^2}{c_{2l}^2}\right)^{1/2}, \quad b_{2s} = \left(1 - \frac{c^2}{c_{2s}^2}\right)^{1/2}, \tag{5.75}$$

$$k_1 = \frac{\omega}{c_1}, \qquad k_{2l} = \frac{\omega}{c_{2l}}, \qquad k_{2s} = \frac{\omega}{c_{2s}}$$

Boundary conditions at the interface between the elastic medium and a fluid and at the free surface are

$$w_1 = w_2, \qquad \tau_{zz}^{(1)} = \tau_{zz}^{(2)}, \quad \tau_{xz}^{(2)} = 0 \qquad \text{on} \qquad z = 0 \tag{5.76}$$

$$\text{and} \qquad \tau_{zz}^{(1)} = 0 \qquad \text{on} \qquad z = -H$$

Taking into account that

$$\mathbf{s} = \text{grad} \; \varphi + \text{curl} \; \boldsymbol{\psi}$$

and Hooke's law, set 5.76 can be represented in the form

$$\frac{\partial \tilde{\varphi}_1}{\partial z} = \frac{\partial \tilde{\varphi}_2}{\partial z} + \frac{\partial \tilde{\psi}_2}{\partial x}, \qquad -\lambda_1 \, k_1^2 \tilde{\varphi}_1 = -\lambda_2 \, k_{2l}^2 \tilde{\varphi}_2 + 2\mu_2 \left(\frac{\partial^2 \tilde{\varphi}_2}{\partial z^2} + \frac{\partial^2 \tilde{\psi}_2}{\partial x \, \partial z}\right),$$

$$2 \frac{\partial^2 \, \tilde{\varphi}_2}{\partial x \, \partial z} + \frac{\partial^2 \tilde{\psi}_2}{\partial x^2} - \frac{\partial^2 \tilde{\psi}_2}{\partial z^2} = 0 \qquad \text{on} \qquad z = 0 \tag{5.77}$$

$$\text{and} \qquad \tilde{\varphi}_1 = 0, \qquad \text{on} \qquad z = -H$$

since

$$\tilde{\tau}_{zz}^{(1)} = \lambda_1 \; \text{div} \; \mathbf{s}_1 = \lambda_1 \nabla^2 \tilde{\varphi}_1 = -\lambda_1 \, k_1^2 \tilde{\varphi}_1$$

Substituting eqs. 5.74 into eqs. 5.77, we obtain a system of equations with respect to unknowns A_1, B_1, A_2, and B_2:

$$b_1 \; (A_1 - B_1) = -b_{2l} \; A_2 + i B_2$$

$$-\lambda_1\, k_1^2\, (A_1 + B_1) = -\lambda_2\, k_{2l}^2\, A_2 + 2\mu_2\, k^2\, \left(b_{2l}^2\, A_2 - i\, b_{2s} B_2\right) \tag{5.78}$$

$$-2i\, b_{2l}\, A_2 - (1 + b_{2s}^2)\, B_2 = 0, \qquad A_1\, e^{-k\, b_1 H} + B_1\, e^{k\, b_1 H} = 0$$

The amplitudes of a surface wave should decrease exponentially in the solid half-space ($z > 0$). This implies that b_{2s} is real, i.e., $c < c_{2s}$. The existence of this surface wave means that unknowns A_1, B_1, A_2, and B_2 differ from zero; that is, the determinant of system 5.78 has to be equal to zero.

$$\begin{vmatrix} b_1 & -b_1 & b_{2l} & 1 \\ -\rho_1 c^2 & -\rho_1 c^2 & \rho_2\,(c^2 - 2c_{2s}^2) & 2\rho_2 c_{2s}^2 b_{2s} \\ 0 & 0 & b_{2l} & \dfrac{c^2}{c_{2l}^2} - 2 \\ e^{-k\, b_1 H} & e^{k\, b_1 H} & 0 & 0 \end{vmatrix} = 0 \tag{5.79}$$

From this we obtain a dispersion (periodic) equation in the following form:

$$\tanh\left(kH\sqrt{1 - \frac{c^2}{c_1^2}}\right) = \eta = \frac{\rho_2\, c_{2s}^4\, b_1}{\rho_1\, c^4\, b_2 l}\left[4\, b_{2l} b_{2s} - \left(2 - \frac{c^2}{c_{2s}^2}\right)^2\right] \tag{5.80}$$

Let us consider two different cases.

"Hard bottom" $c_1 < c_{2s}$

Then within the range of possible values of c: $c_1 < c < c_{2s}$ b_1 becomes imaginary, and instead of eq. 5.80 we have

$$\tan\left(kH\sqrt{\frac{c^2}{c_1^2} - 1}\right) = \eta_1 = \frac{\rho_2\, c_{2s}^4\, |b_1|}{\rho_1\, c^4\, b_2 l}\left[4\, b_{2l} b_{2s} - \left(2 - \frac{c^2}{c_{2s}^2}\right)^2\right] \tag{5.81}$$

This transcendent equation relates phase velocity $c\,(\omega)$ to frequency and to the parameters of a medium. It is convenient to rewrite eq. 5.81 as

$$k_n H\, \sqrt{\frac{c_n^2}{c_1^2} - 1} = \pi n + \tan^{-1} \eta_1, \tag{5.82}$$

where $c_n(\omega)$ is phase velocity of the nth mode and $k_n = \omega / c_n$.
Letting $n = 0$, eq. 5.82 becomes

$$2\pi p\, \sqrt{\frac{c^2}{c_1^2} - 1} = \tan^{-1} \eta_1 \tag{5.83}$$

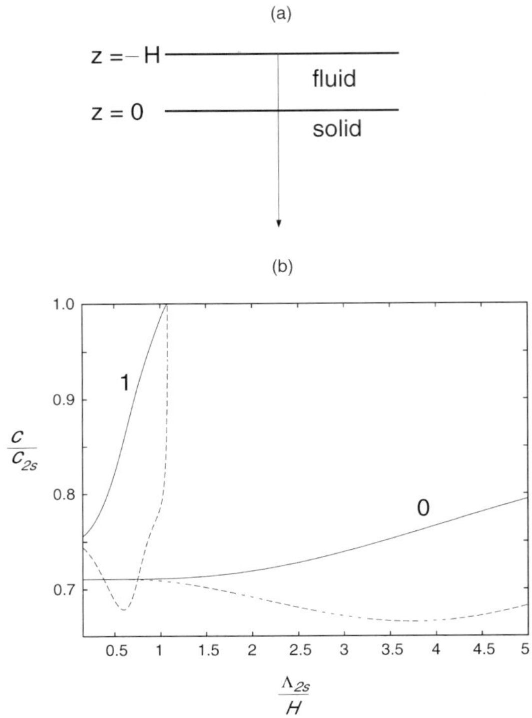

Figure 5.6: Rayleigh waves in a fluid layer overlaying an elastic half-space: (a) Model of a medium (b) Dispersion curves of the phase velocities (solid lines) and group velocities (dashed lines) for the first two modes. $\rho_2/\rho_1 = 2.2$, $c_{2l}/c_{2s} = \sqrt{3}$, $c_{2s}/c_1 = 1.333$

Here $p = H/\Lambda_{1l}$ and Λ_{1l} is the wavelength in a fluid. With an increase of Λ_{1l} the left side of eq. 5.83 tends to zero, and we have

$$\eta_1 = 4\, b_{2l} b_{2s} - \left(2 - \frac{c^2}{c_{2s}^2}\right)^2 = 0$$

This coincides with the dispersion equation for Rayleigh waves in a homogeneous half-space. Respectively, the phase velocity of the fundamental mode approaches that of the Rayleigh wave c_R when $\omega \to 0$. In other words, the influence of the upper layer vanishes. As is seen from Fig. 5.6b, the phase velocity of the fundamental mode gradually decreases with an increase of frequency, and its range of change is

$$c_{St} \le c(\omega) \le c_R$$

where c_{St} is the velocity of the Stoneley wave at the fluid/solid boundary. Also, this numerical analysis shows that the fundamental mode of the Rayleigh waves exists at any frequency. As in the case of Love waves, we observe the dependence of the phase velocity on frequency, i.e., dispersion. In particular, the fundamental mode at the lower frequencies propagates more quickly. By definition

$$\tilde{u} = \frac{\partial \tilde{\varphi}}{\partial x} - \frac{\partial \tilde{\psi}}{\partial z}, \qquad \tilde{w} = \frac{\partial \tilde{\varphi}}{\partial z} + \frac{\partial \tilde{\psi}}{\partial x},$$

and, in accordance with eqs. 5.74, it is easy to see that propagation of the fundamental mode, as well as of others modes, is accompanied by elliptical polarization of the particle motion.

Letting now $n = 1$ we define the dispersion equation of the second mode as

$$\frac{\omega}{c} H \sqrt{\frac{c^2}{c_1^2} - 1} = \pi + \tan^{-1} \eta_1 \tag{5.84}$$

As before, at high frequencies, phase velocity approaches c_1:

$$c(\omega) \to c_1 \qquad \text{if} \quad p \to \infty$$

To determine the low frequency limit, suppose that $c(\omega) = c_{2s}$. Then eq. 5.82 becomes

$$\frac{\omega}{c_{2s}} H \sqrt{\frac{c_{2s}^2}{c_1^2} - 1} = \pi \qquad \text{or} \qquad 2\pi p \sqrt{1 - \frac{c_1^2}{c_{2s}^2}} = 1 \tag{5.85}$$

It turns out that the latter defines the cut-off frequency of this mode, since at lower frequencies the wavenumber becomes complex. This means that part of the energy moves from the layer into the elastic medium $(z > 0)$, and the wave rapidly decays. Thus, the range of change of the phase velocity is

$$c_1 \le c(\omega) \le c_{2s}, \tag{5.86}$$

and below the cut-off frequency this mode is absent. Of course, there are also modes of the higher order, but we will restrict ourselves to the first two modes.

"Soft bottom" $c_1 > c_{2s}$

In this case b_1 is always real as phase velocity c is less than c_{2s}. It is easy to show that eq. 5.80 has only one root. The range of change of phase velocity $c(\omega)$ is the same as for fundamental mode $n = 0$ in a previous case.

Chapter 6

Waves in a layered medium caused by linear and point sources

We start from a relatively simple two-dimensional case in which the primary source is linear and stretched – for example, along the y-axis, Fig. 6.1a. This means that a wavefield remains the same in any plane, parallel to the coordinate plane XOZ, and deformation in the y-direction is absent. There is an evident analogy with plane waves (Chapter 4), and it is not accidental that waves generated by such a source can be represented as a superposition of plane waves. One can say that the study of waves due to a linear source is a logical intermediate step between the the study of plane waves and of a more general three-dimensional case. Of course, the linear source is hardly practical, but all derivations are significantly simpler in this case. It is essential that linear and point sources give rise to the same types of waves. For instance, in a homogeneous half-space with a free surface, we can observe longitudinal and shear direct and reflected waves, as well as Rayleigh waves. The appearance of these waves due to either a linear or the point source can be expected. In fact, their appearance follows from the theory of plane and surface waves, the ray theory, and the reflection of plane waves. Our goal is not to discover new waves, but rather to study how these waves are generated and propagate depending on source type and location as well as on elastic parameters of a medium.

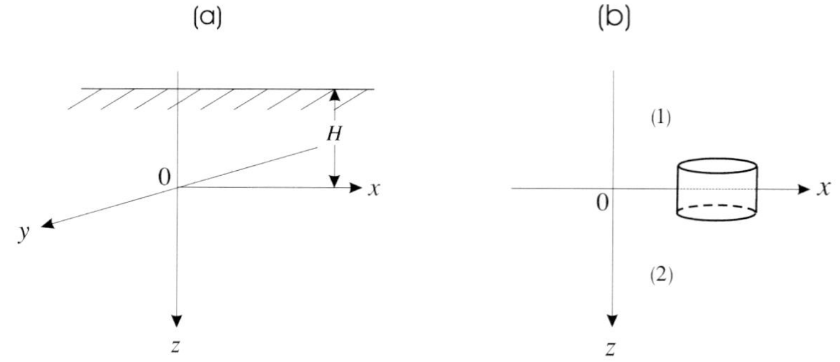

Figure 6.1: (a) Linear source beneath the surface (b) Boundary condition for stresses at the plane, $z = 0$

6.1 Linear source of P and SV waves in a homogeneous half-space with a free boundary

In order to formulate the boundary value problem in the presence of a horizontal interface, it is necessary first of all to derive expressions for potentials that describe the direct longitudinal or shear waves.

Linear source of the P wave in a homogeneous medium

Suppose that an infinitely long cylinder with a very small radius r_0 is oriented along the y-axis, and it experiences sinusoidal vibrations in the radial direction:

$$r(t) = r_0 + s_r(t) \tag{6.1}$$

Here

$$s_r(t) = s_0 \cos \omega t \tag{6.2}$$

and

$$\frac{s_0}{r_0} \ll 1 \tag{6.3}$$

It is clear that such a source generates a direct wave that has only a radial component of displacement. In the cylindrical system of coordinates

$$\mathbf{s} = s_r\, \mathbf{i}_r, \qquad \text{i.e.,} \qquad s_\theta = s_y = 0, \tag{6.4}$$

since the source is situated on the y-axis. By definition,

$$\text{div } \mathbf{s} = \frac{1}{r}\frac{\partial}{\partial r}(r s_r) \tag{6.5}$$

$$\text{and} \quad \text{curl}\ \mathbf{s} = \frac{1}{r}\begin{vmatrix} \mathbf{i}_r & r\mathbf{i}_\theta & \mathbf{i}_y \\ \frac{\partial}{\partial r} & \frac{\partial}{\partial \theta} & \frac{\partial}{\partial y} \\ s_r & 0 & 0 \end{vmatrix}$$

Because of axial symmetry and the independence of the field on the y-coordinate, we obtain

$$\text{curl}\ \mathbf{s} = 0 \tag{6.6}$$

Therefore, such a wave is longitudinal and, correspondingly, it is described by scalar potential only. Its complex amplitude is a solution of the Helmholtz equation

$$\nabla^2 \widetilde{\varphi}_0 + k_l^2 \widetilde{\varphi}_0 = 0$$

or, in the cylindrical system of coordinates,

$$\frac{d^2\widetilde{\varphi}_0}{dr^2} + \frac{1}{r}\frac{d\widetilde{\varphi}_0}{dr} + k_l^2 \widetilde{\varphi}_0 = 0 \tag{6.7}$$

This is the Bessel equation, and its solutions are Bessel functions of the first and second types,

$$J_0\left(k_l r\right) \qquad \text{and} \qquad Y_0\left(k_l r\right),$$

as well as some combinations of them. Taking into account that the source causes the outgoing wave and the function $e^{-i\omega t}$ describes time dependence, we choose the Hankel function,

$$H_0^{(1)}\left(k_l r\right) = J_0\left(k_l r\right) + i\, Y_0\left(k_l r\right)$$

as a solution of eq. 6.7. This choice is related to the fact that at large distances from the source, its asymptotic expression is

$$H_0^{(1)}\left(k_l r\right) \simeq \sqrt{\frac{2}{\pi k_l r}} e^{i\left(k_l\, r - \pi/4\right)} \tag{6.8}$$

This allows us to say that the function

$$\widetilde{\varphi}_0\left(k_l r\right) = A_1 H_0\left(k_l\, r\right) \tag{6.9}$$

characterizes the wave generated by the linear source. Here A_1 is unknown, and in order to determine we assume that the stress τ_{rr} is uniformly distributed on the source surface and depends on time t as

$$\tau_{rr}\left(r_0\right) = \tau^0_{rr} \cos \omega t \tag{6.10}$$

From Hooke's law we have

$$\tilde{\tau}_{rr} = \lambda \text{ div } \tilde{\mathbf{s}} + 2\mu \frac{\partial \tilde{s}_r}{\partial r}$$

or

$$\tilde{\tau}_{rr} = -\lambda\, k_l^2 \tilde{\varphi}_0 + 2\mu \frac{\partial^2 \tilde{\varphi}_0}{\partial r^2} \tag{6.11}$$

Since

$$H_0^{(1)}\left(k_l r\right) \to \frac{2\, i}{\pi} \ln k_l\, r \qquad \text{if} \qquad r \to 0, \tag{6.12}$$

the second term in eq. 6.11 prevails near the source, and we obtain

$$\tilde{\tau}_{rr} = -\frac{2\mu\, i\, A_1}{\pi r_0^2}$$

or

$$A_1 = \frac{\pi r_0^2\, i\, \tau^0_{rr}}{2\mu}, \tag{6.13}$$

because $\tilde{\tau}_{rr} = \tau^0_{rr}$ at points along the source. Thus,

$$\tilde{\varphi}_0 = \frac{\pi r_0^2\, i}{2\mu} \tau^0_{rr} H_0^{(1)}\left(k_l r\right), \tag{6.14}$$

and our first task is solved, since this function satisfies all conditions of the boundary problem.

Also it is interesting to consider the strains and stresses that accompany this cylindrical wave. As follows from Appendix E,

$$e_{rr} = \frac{\partial s_r}{\partial r}, \qquad e_{\theta\theta} = \frac{s_r}{r}, \qquad e_{yy} = 0$$

and

$$e_{\theta y} = e_{ry} = e_{r\theta} = 0$$

Hooke's law gives

$$\tau_{rr} = \lambda \text{ div } \mathbf{s} + 2\mu \frac{\partial s_r}{\partial r}, \qquad \tau_{yy} = \tau_{\theta\theta} = \lambda \text{ div } \mathbf{s}, \qquad \text{and} \qquad \tau_{\theta y} = \tau_{ry} = \tau_{r\theta} = 0$$

Thus, in the cylindrical system of coordinates there are three normal stresses. Now let us imagine a horizontal plane, $\theta = 0$, on which the source is located. Then at points on this surface the normal stress, $\tau_{\theta\theta}$, differs from zero.

Further we focus our attention on wavefields in a medium with a horizontal interface. For this reason it is necessary, as in the case of acoustic waves (Part II), to represent the potential of the direct wave, $\widetilde{\varphi}_0$, in terms of functions that depend on coordinates x and z. There are several integral forms of the Hankel functions; one of them

$$H_0^{(1)}(k_l r) = \frac{1}{\pi i} \int\limits_{-\infty}^{\infty} \frac{1}{m_l} e^{-m_l |z|} e^{i m x} dm, \tag{6.15}$$

where

$$m_l = \sqrt{m^2 - k_l^2} \tag{6.16}$$

Correspondingly,

$$\widetilde{\varphi}_0(k_l r) = \frac{\tau_{rr}^0 r_0^2}{2\mu} \int\limits_{-\infty}^{\infty} \frac{e^{-m_l |z|}}{m_l} e^{i m x} dm \tag{6.17}$$

or

$$\widetilde{\varphi}_0(k_l r) = \frac{\tau_{rr}^0 r_0^2}{2\mu} \int\limits_{-\infty}^{\infty} \frac{e^{-m_l |z|}}{m_l} \cos mx \, dm, \tag{6.18}$$

since $\sin mx$ is the odd function.

Homogeneous half-space with a free boundary

Next we assume that the linear source of longitudinal waves is located in a homogeneous half-space beneath the free surface, Fig. 6.1a. When the direct wave reaches the boundary, we can expect the appearance of longitudinal and shear reflected waves, as well as surface waves. The sinusoidal wavefields associated with these waves are described by complex amplitudes of scalar and vector potentials

$$\widetilde{\varphi} = \widetilde{\varphi}_0 + \widetilde{\varphi}_s \qquad \text{and} \qquad \widetilde{\psi}, \tag{6.19}$$

where $\widetilde{\varphi}_s$ is the complex amplitude of potential characterizing the secondary waves and $\widetilde{\psi}$ is the y-component of the vector potential

$$\widetilde{\psi} = \widetilde{\psi}_y$$

To determine these unknown functions we have to formulate a boundary value problem that imposes the following conditions on $\widetilde{\varphi}$ and $\widetilde{\psi}$.

1. First of all, they must satisfy the Helmholtz equations

$$\nabla^2 \widetilde{\varphi} + k_l^2 \widetilde{\varphi} = 0, \qquad \nabla^2 \widetilde{\psi} + k_s^2 \widetilde{\psi} = 0 \tag{6.20}$$

Here

$$k_l = \frac{\omega}{c_l} \qquad \text{and} \qquad k_s = \frac{\omega}{c_s}$$

2. Near the source, the scalar potential must tend to potential of the direct wave

$$\widetilde{\varphi} \to \widetilde{\varphi}_0, \qquad \text{if} \qquad r \to 0 \tag{6.21}$$

3. At infinity

$$\widetilde{\varphi} \to 0 \qquad \text{and} \qquad \widetilde{\psi} \to 0, \qquad \text{if} \qquad r \to \infty \tag{6.22}$$

and the Sommerfeld condition of radiation has to be met (Part II).

4. Finally, at the free surface, normal and shear stresses must vanish

$$\widetilde{\tau}_{zz} = 0 \qquad \text{and} \qquad \widetilde{\tau}_{xz} = 0, \qquad \text{if} \quad z = -H \tag{6.23}$$

We assume that in the Cartesian system of coordinates, displacement $\mathbf{s}$:

$$\widetilde{\mathbf{s}} = \text{grad}\, \widetilde{\varphi} + \text{curl}\, \widetilde{\boldsymbol{\psi}} \tag{6.24}$$

is characterized by only two components, u and w,

$$\mathbf{s} = u\, \mathbf{i} + w\, \mathbf{k} \tag{6.25}$$

For this reason we have chosen the y-component of vector potential ψ_y. Correspondingly, eq. 6.24 gives

$$\widetilde{u} = \frac{\partial \widetilde{\varphi}}{\partial x} - \frac{\partial \widetilde{\psi}}{\partial z}, \qquad \widetilde{w} = \frac{\partial \widetilde{\varphi}}{\partial z} + \frac{\partial \widetilde{\psi}}{\partial x} \tag{6.26}$$

Then, taking into account Hooke's law:

$$\tilde{\tau}_{zz} = \lambda \, \mathrm{div}\tilde{\mathbf{s}} + 2\mu \, \frac{\partial \tilde{w}}{\partial z}, \qquad \tilde{\tau}_{xz} = \mu \left(\frac{\partial \tilde{u}}{\partial z} + \frac{\partial \tilde{w}}{\partial x} \right), \tag{6.27}$$

eqs. 6.23 become

$$-\lambda \, k_l^2 \tilde{\varphi} + 2\mu \left(\frac{\partial^2 \tilde{\varphi}}{\partial z^2} + \frac{\partial^2 \tilde{\psi}}{\partial x \, \partial z} \right) = 0, \qquad 2\frac{\partial^2 \tilde{\varphi}}{\partial x \, \partial z} + \frac{\partial^2 \tilde{\psi}}{\partial x^2} - \frac{\partial^2 \tilde{\psi}}{\partial z^2} = 0 \tag{6.28}$$

Note that we have already used these boundary conditions studying Rayleigh waves. Thus, we have formulated the boundary value problem, and in accordance with the theorem of uniqueness, only one wavefield obeys all four conditions. First, taking into account that the free surface is a plane, $z = -H$, it is natural to find a solution of the Helmholtz equations in the Cartesian system of coordinates. For instance, in the case of scalar potential, we have

$$\frac{\partial^2 \tilde{\varphi}}{\partial x^2} + \frac{\partial^2 \tilde{\varphi}}{\partial z^2} + k_l^2 \tilde{\varphi} = 0, \tag{6.29}$$

which describes the potential at regular points. Applying the method of separation of variables (Part II), we represent the function $\tilde{\varphi}$ as

$$\tilde{\varphi} = X(x) \, Z(z) \tag{6.30}$$

Substitution of eq. 6.30 into eq. 6.29 gives

$$Z \frac{d^2 X}{dx^2} + X \frac{d^2 Z}{dz^2} + k_l^2 \, X \, Z = 0$$

or

$$\frac{1}{X} \frac{d^2 X}{dx^2} + \frac{1}{Z} \frac{d^2 Z}{dz^2} + k_l^2 = 0 \tag{6.31}$$

The latter is equivalent to one of two sets of the ordinary differential equations

$$\frac{d^2 X}{dx^2} + m^2 X = 0, \qquad \frac{d^2 Z}{dz^2} - \left(m^2 - k_l^2 \right) Z = 0 \tag{6.32}$$

or

$$\frac{d^2 X}{dx^2} - m^2 X = 0, \qquad \frac{d^2 Z}{dz^2} + \left(m^2 - k_l^2 \right) Z = 0$$

We choose the first system, eqs. 6.32, because one of its solutions corresponds to the symmetry of the wavefields with respect to plane $x = 0$ and because of the behavior of this solution at infinity, when $z \to \infty$. In fact, solutions of this system are

$$e^{\pm imx} \qquad \text{and} \qquad e^{\pm m_l z} \tag{6.33}$$

Taking into account the boundary condition at infinity, the general solution for the potential $\tilde{\varphi}_s$ can be written in the form

$$\tilde{\varphi}_s = \int_{-\infty}^{\infty} B_m e^{-m_l z} e^{i m x} dm \tag{6.34}$$

Exactly the same approach to the equation

$$\frac{\partial^2 \tilde{\psi}}{\partial x^2} + \frac{\partial^2 \tilde{\psi}}{\partial z^2} + k_s^2 \tilde{\psi} = 0 \tag{6.35}$$

yields

$$\tilde{\psi} = \int_{-\infty}^{\infty} C_m e^{-m_s z} e^{i m x} dm \tag{6.36}$$

Here

$$m_s = \sqrt{m^2 - k_s^2} \tag{6.37}$$

Thus, functions $\tilde{\varphi}_s$ and $\tilde{\psi}$ obey the Helmholtz equations and the condition at infinity. Since

$$e^{i m x} = \cos mx + i \sin mx,$$

eqs. 6.34 and 6.36 can be written as

$$\varphi_s = \int_{-\infty}^{\infty} B_m e^{-m_l z} \cos mx \, dm \qquad \text{and} \qquad \psi = \int_{-\infty}^{\infty} C_m e^{-m_s z} \cos mx \, dm \tag{6.38}$$

i.e., these functions are even with respect to x, which corresponds to the symmetry of wavefields. Next, making use of eq. 6.17, we obtain the following expressions for potentials of waves:

$$\tilde{\varphi} = C_0 \left(\int_{-\infty}^{\infty} \frac{1}{m_l} e^{-m_l |z|} e^{i m x} dm + \int_{-\infty}^{\infty} B_m e^{-m_l z} e^{i m x} dm \right) \tag{6.39}$$

and

$$\tilde{\psi} = C_0 \int_{-\infty}^{\infty} C_m \, e^{-m_s \, z} \, e^{imx} dm \tag{6.40}$$

Here

$$C_0 = \frac{\tau_{rr} \, r_0^2}{2\mu} \tag{6.41}$$

In order to determine the unknown coefficients B_m and C_m, we apply the boundary conditions. Substitution of eqs. 6.39 and 6.40 into eqs. 6.28 gives two equations with two unknowns for every value of m:

$$-\lambda \, k_l^2 \left(\frac{e^{-m_l \, H}}{m_l} + B_m \, e^{m_l \, H} \right) \tag{6.42}$$

$$+2\mu(m_l e^{-m_l \, H} + m_l^2 e^{m_l \, H} B_m - i \, m \, m_s C_m \, e^{m_s \, H}) = 0$$

and $$2\left(i \, m \, e^{-m_l \, H} - i \, m \, m_l B_m \, e^{m_l \, H}\right) - m^2 C_m e^{m_s \, H} - m_s^2 \, C_m \, e^{m_s \, H} = 0$$

After doing a simple algebra, we obtain

$$\left(2m^2 - k_s^2\right) e^{m_l \, H} B_m - 2i \, m \, m_s C_m e^{m_s \, H} = -\frac{e^{-m_l \, H}}{m_l} \left(2m^2 - k_s^2\right) \tag{6.43}$$

and $$2i \, m \, m_l \, e^{m_l \, H} B_m + \left(2m^2 - k_s^2\right) \, e^{m_s \, H} C_m = 2i \, m \, e^{-m_l \, H}$$

Note that the boundary conditions are written for integrands in eqs. 6.39 and 6.40. This great simplification is based on the main property of Fourier's transform (Part I). Solution of system 6.43 gives

$$B_m = -\frac{1}{m_l} \frac{\left(2m^2 - k_s^2\right)^2 + 4m^2 m_l \, m_s}{\left(2m^2 - k_s^2\right)^2 - 4m^2 m_l \, m_s} \, e^{-2m_l \, H} \tag{6.44}$$

and $$C_m = \frac{4im \left(2m^2 - k_s^2\right) e^{-m_l \, H} \, e^{-m_s \, H}}{\left(2m^2 - k_s^2\right)^2 - 4m^2 m_l m_s}$$

As follows from eqs. 6.44, the functions $\widetilde{\varphi}_s$ and $\widetilde{\psi}$ do not have singularities. Therefore, the condition near the source, eq. 6.21, is also met. This means that we have found potentials caused by the linear source of the compressional waves. In particular, if the source is located on the free surface we obtain

$$\widetilde{\varphi} = -8C_0 \int\limits_{-\infty}^{\infty} m^2 m_s \frac{e^{-m_l\, z}\, e^{i\, m\, x} dm}{(2m^2 - k_s^2)^2 - 4m^2 m_l\, m_s} \tag{6.45}$$

and $$\widetilde{\psi} = 4i\, C_0 \int\limits_{-\infty}^{\infty} \frac{m\,(2m^2 - k_s^2)\; e^{-m_s\, z}\, e^{i\, m\, x} dm}{(2m^2 - k_s^2)^2 - 4m^2 m_l\, m_s}$$

Linear source of SV waves in a homogeneous medium

Assume that the linear source experiences a small rotation about the y-axis that is caused by shear stress:

$$\tau_{\theta r} = \tau^0_{\theta r} \cos \omega t \qquad \text{at} \qquad r = r_0 \tag{6.46}$$

This motion gives rise to a shear wave, and particle displacement is characterized by the single component s_θ:

$$\mathbf{s} = s_\theta \mathbf{i}_\theta \qquad \text{but} \qquad s_r = s_y = 0 \tag{6.47}$$

Since the wavefield is independent of θ, we have

$$\text{div}\ \mathbf{s} = \frac{1}{r}\frac{\partial}{\partial \theta}(r s_\theta) = 0,$$

while

$$\text{curl}_y\ \mathbf{s} = \frac{1}{\partial r}(r s_\theta) \neq 0$$

Correspondingly, the wavefield can be described by the single component of the vector potential $\boldsymbol{\psi}_0$:

$$\boldsymbol{\psi}_0 = \psi_0\, \mathbf{j}, \tag{6.48}$$

which is oriented along the y-axis, and

$$\mathbf{s} = \text{curl}\ \boldsymbol{\psi}_0 \qquad \text{or} \qquad s_\theta = -\frac{\partial \psi_0}{\partial r} \tag{6.49}$$

Because of axial symmetry and the independence of the coordinate y, the complex amplitude $\tilde{\psi}_0$ satisfies eq. 6.35, and its solution has the same form as in the case of the P-waves:

$$\tilde{\psi}_0 = A_2 H_0^{(1)}(k_s r) \tag{6.50}$$

To determine the unknown A_2, we use Hooke's law:

$$\tilde{\tau}_{r\theta} = \mu \left(\frac{\partial \tilde{s}_\theta}{\partial r} - \frac{\tilde{s}_\theta}{r} \right) = -\mu \left(\frac{\partial^2 \tilde{\psi}}{\partial r^2} - \frac{1}{r} \frac{\partial \tilde{\psi}}{\partial r} \right) \tag{6.51}$$

As follows from eq. 6.12, near the source

$$H_0^{(1)}(k_s r) \to \frac{2i}{\pi} \ln k_s r, \tag{6.52}$$

and substitution of eqs. 6.50 and 6.52 into eq. 6.51 gives

$$A_2 = -\frac{\pi r_0^2 \ i \ \tau_{r\theta}^0}{4\mu} \tag{6.53}$$

Therefore,

$$\tilde{\psi}_0 = -\frac{\pi r_0^2 \ i \ \tau_{r\theta}^0}{4 \ \mu} H_0^{(1)}(k_s r) \tag{6.54}$$

or

$$\tilde{\psi}_0 = C_1 \int_{-\infty}^{\infty} \frac{1}{m_s} \ e^{-m_s \ |z|} \ e^{i \ m \ x} \ dm \tag{6.55}$$

Here

$$C_1 = -\frac{r_0^2 \ \tau_{r\theta}^0}{4 \ \mu} \tag{6.56}$$

Considering elementary volume in the cylindrical system of coordinates, we see that

$$\tau_{rr} = 0, \qquad \tau_{\theta\theta} = 2\mu \frac{\partial s_\theta}{\partial \theta}, \qquad \tau_{yy} = 0$$

and

$$\tau_{r\theta} = \mu \left(\frac{\partial s_\theta}{\partial r} - \frac{s_\theta}{r} \right), \qquad \tau_{ry} = 0, \qquad \tau_{\theta y} = 0$$

Homogeneous half-space with a free boundary

Now we assume that the linear source of shear waves is located at distance H from the free surface, Fig. 6.1a. As before, we may expect that at the boundary both longitudinal and shear reflected waves arise, as well as a surface wave. For this reason, the wavefields are described by scalar and vector potentials, and their complex amplitudes can be written in the form

$$\widetilde{\varphi} = C_1 \int_{-\infty}^{\infty} A_m \, e^{-m_l \, z} \, e^{i \, m \, x} \, dm \tag{6.57}$$

and $$\widetilde{\psi} = C_1 \left[\int_{-\infty}^{\infty} \frac{e^{-m_s \, |z|}}{m_s} \, e^{i \, m \, x} \, dm + \int_{-\infty}^{\infty} B_m \, e^{-m_s \, z} \, e^{i \, m \, x} \, dm \right]$$

By definition, at regular points functions $\widetilde{\varphi}$ and $\widetilde{\psi}$ are solutions of Helmholtz equations, and they satisfy the condition at infinity $(z \to \infty)$. Substitution of eqs. 6.57 into the boundary conditions described by eqs. 6.28 gives the system for determining unknowns A_m and B_m:

$$-\lambda k_l^2 A_m e^{m_l \, H} + 2\mu \left[m_l^2 \, A_m \, e^{m_l \, H} + i \, m \left(e^{-m_s \, H} - m_s \, B_m \, e^{m_s \, H} \right) \right] = 0 \, , \tag{6.58}$$

$$-2i \, m \, m_l \, A_m \, e^{m_l \, H} - m^2 \left(\frac{e^{-m_s \, H}}{m_s} + B_m \, e^{m_s \, H} \right) - \left(m_s^2 \, B_m \, e^{m_s \, H} + m_s \, e^{-m_s \, H} \right) = 0$$

or

$$\left(2m^2 - k_s^2\right) \, A_m \, e^{m_l \, H} - 2i \, m \, m_s \, e^{m_s \, H} \, B_m = -2i \, m \, e^{-m_s \, H} \tag{6.59}$$

and $$2i \, m \, m_l \, e^{m_l \, H} \, A_m + \left(2m^2 - k_s^2\right) \, e^{m_s \, H} \, B_m = -\frac{\left(2m^2 - k_s^2\right)}{m_s} e^{-m_s \, H}$$

Solution of this system gives

$$A_m = \frac{-4i \, m \, (2m^2 - k_s^2) \, e^{-m_l \, H} \, e^{-m_s \, H}}{\left(2m^2 - k_s^2\right)^2 - 4m^2 m_l m_s} \tag{6.60}$$

and $$B_m = -\frac{1}{m_s}\,\frac{\left(2m^2 - k_s^2\right)^2 + 4m^2 m_l m_s}{\left(2m^2 - k_s^2\right)^2 - 4m^2 m_l m_s} e^{-2m_s\, H}$$

Thus, the scalar and vector potentials are

$$\tilde{\varphi} = -C_1 \int_{-\infty}^{\infty} \frac{4i\, m\, \left(2m^2 - k_s^2\right)\, e^{-m_l\, H}\, e^{-m_s\, H}}{\left(2m^2 - k_s^2\right)^2 - 4m^2 m_l m_s}\, e^{-m_l\, z}\, e^{i\, m\, x}\, dm \tag{6.61}$$

and $$\tilde{\psi}_1 = -C_1 \int_{-\infty}^{\infty} \frac{1}{m_s}\, \frac{\left(2m^2 - k_s^2\right)^2 + 4m^2 m_l m_s}{\left(2m^2 - k_s^2\right)^2 - 4m^2 m_l m_s}\, e^{-2m_s\, H}\, e^{-m_s\, z}\, e^{i\, m\, x}\, dm$$

Here $\tilde{\psi} = \tilde{\psi}_0 + \tilde{\psi}_1$, and $\tilde{\psi}_1$ describes the secondary wavefields.

Linear source of SH wave in a homogeneous half-space

In such a case, we assume that the secondary wave is also an SH wave. Correspondingly, the field is described by the vector potential only:

$$\tilde{\psi} = C_1 \int_{-\infty}^{\infty} \left(\frac{1}{m_s}\, e^{-m_s\, |z|} + B_m e^{-m_s\, z}\right) e^{i\, m\, x}\, dm$$

Since displacement has a single component v, the boundary condition at the free surface is

$$\tau_{yz} = 0 \qquad \text{or} \qquad \frac{\partial v}{\partial z} = 0 \qquad \text{if} \qquad z = -H$$

This can be represented as

$$\frac{\partial^2 \psi}{\partial z^2} = 0 \qquad \text{if} \qquad z = -H,$$

$$B_m = -\frac{1}{m_s}\, e^{-2m_s\, H}$$

The vector potential of the secondary field is

$$\tilde{\psi}_s = -C_1 \int_{-\infty}^{\infty} \frac{1}{m_s}\, e^{-m_s\, (z + 2H)}\, e^{i\, m\, x} dm$$

This vector potential characterizes the shear wave caused by a fictitious source located at the point having coordinates, $(0, -2H)$. That is, it is a mirror reflection of the real source with respect to the free surface. Note that the same behavior was observed in an acoustic medium (Part II).

6.2 Waves in a homogeneous medium caused by force $\mathbf{F}^e$ (two-dimensional case)

Assume that the external force

$$\mathbf{F}^e_x = \mathbf{F}_x(x,0)\cos\omega\, t \qquad \text{or} \qquad \mathbf{F}^e_z = \mathbf{F}_z(x,0)\cos\omega\, t \tag{6.62}$$

is oriented either along the x-axis or the z-axis, and it acts on a very thin layer coincident with plane $z=0$, Fig. 6.1b. In general, $\mathbf{F}_x$ and $\mathbf{F}_z$ can be arbitrary functions of x. We can expect that such sources generate a direct wave, which causes deformation and the rotation of elementary volumes of a medium. The wavefields are described by both potentials, and our first goal is to find expressions for the complex amplitudes of these potentials. It is clear that the boundary value problem for the direct wave requires the following:

1. The complex amplitudes of the scalar and vector potentials must obey the Helmholtz equations at regular points.

2. At plane $z=0$, both components of displacement must be continuous functions:

$$\widetilde{u}_1 = \widetilde{u}_2 \qquad \text{and} \qquad \widetilde{w}_1 = \widetilde{w}_2 \tag{6.63}$$

Here $\widetilde{u}=\widetilde{u}_1,\quad \widetilde{w}=\widetilde{w}_1 \quad z\le 0;\quad \widetilde{u}=\widetilde{u}_2,\quad \widetilde{w}=\widetilde{w}_2 \quad z\ge 0;$ while $\widetilde{v}\equiv 0$.

Also, stresses $\widetilde{\tau}_{zz}$ and $\widetilde{\tau}_{xz}$ at both sides of plane $z=0$ are related to the external force $\mathbf{F}^e$.

3. Finally, the direct wave is outgoing, and the wavefields disappear when $|z|$ tends to infinity.

Relationship between the external force and stresses

As we see, in order to solve the boundary value problem it is necessary to relate the external force to stresses in the vicinity of plane $z=0$. Let us consider an elementary cylinder, shown in Fig. 6.1b. It is obvious that a sum of forces acting on the opposite faces of the volume has to be equal to zero, that is,

$$\left(\mathbf{F}^e + \mathbf{t}^{(1)} + \mathbf{t}^{(2)}\right)\, dS = 0 \tag{6.64}$$

Otherwise, acceleration of volume would be infinitely large when $\Delta V\to 0$. Here $\mathbf{F}^e$ is the external force per unit area of the upper face, dS, of the elementary cylinder, while $\mathbf{t}^{(1)}$ and $\mathbf{t}^{(2)}$ are the vectors of traction, applied to faces. By definition (Appendix D), the components of the vector $\mathbf{t}$ are

$$t_x = \tau_{xx}n_x + \tau_{xy}n_y + \tau_{xz}n_z, \qquad t_y = \tau_{yx}n_x + \tau_{yy}n_y + \tau_{yz}n_z \tag{6.65}$$

and $$t_z = \tau_{zx} n_x + \tau_{zy} n_y + \tau_{zz} n_z$$

where n_x, n_y, n_z are directional cosines of normals to the cylinder faces and are directed outward. Since $n_x = n_y = 0$, we obtain

$$t_x^{(1)} = -\tau_{xz}^{(1)}, \qquad t_x^{(2)} = \tau_{xz}^{(2)} \tag{6.66}$$

and

$$t_z^{(1)} = -\tau_{zz}^{(1)}, \qquad t_z^{(2)} = \tau_{zz}^{(2)} \tag{6.67}$$

Correspondingly, eq. 6.64 becomes

$$F_z^e + \tau_{xz}^{(2)} - \tau_{xz}^{(1)} = 0 \tag{6.68}$$

and

$$F_z^e + \tau_{zz}^{(2)} - \tau_{zz}^{(1)} = 0 \tag{6.69}$$

In particular, if the external force is directed either along the x-axis or the z-axis we have

$$\tau_{xz}^{(2)} - \tau_{xz}^{(1)} = -F_x^e, \quad \tau_{zz}^{(1)} = \tau_{zz}^{(2)} \quad \text{or} \quad \tau_{xz}^{(2)} = \tau_{xz}^{(1)}, \quad \tau_{zz}^{(2)} - \tau_{zz}^{(1)} = -F_z^e \tag{6.70}$$

These formulas along with eqs. 6.63 allow us to find expressions for potentials of the direct wave.

Spatial spectrum of the external force

As was demonstrated earlier, the complex amplitudes of potentials satisfying the Helmholtz equations can be represented in the form

$$\tilde{\varphi}_1 = \int\limits_{-\infty}^{\infty} A_m\, e^{m_l\, z} e^{i\, m\, x} dm, \quad \tilde{\psi}_1 = \int\limits_{-\infty}^{\infty} B_m\, e^{m_s\, z} e^{i\, m\, x} dm, \quad \text{if} \quad z < 0 \tag{6.71}$$

and

$$\tilde{\varphi}_2 = \int\limits_{-\infty}^{\infty} C_m\, e^{-m_l\, z} e^{i\, m\, x} dm, \quad \tilde{\psi}_2 = \int\limits_{-\infty}^{\infty} D_m\, e^{-m_s\, z} e^{i\, m\, x} dm, \quad \text{if} \quad z > 0 \tag{6.72}$$

Here $\tilde{\psi}$ is the y-component of the vector potential and

$$m_l = \sqrt{m^2 - k_l^2}, \qquad m_s = \sqrt{m^2 - k_s^2}$$

It is clear that eqs. 6.71 and 6.72 can be treated as Fourier's transform of the complex amplitudes of potentials. In order to determine unknowns in eqs. 6.71 and 6.72, we use conditions at plane $z = 0$ (eqs. 6.63, 6.70), and also express the force amplitude, $F_x(x, 0)$ and $F_z(z, 0)$, as the Fourier's integrals:

$$F_x(x,0) = \int_{-\infty}^{\infty} X(m)\, e^{i\,m\,x} dm, \qquad F_z(x,0) = \int_{-\infty}^{\infty} Z(m)\, e^{i\,m\,x} dm \tag{6.73}$$

Here

$$X(m) = \frac{1}{2\pi}\int_{-\infty}^{\infty} F_x(x,0)\; e^{-i\,m\,x} dx, \quad Z(m) = \frac{1}{2\pi}\int_{-\infty}^{\infty} F_z(z,0)\; e^{-i\,m\,x} dx, \tag{6.74}$$

Eqs. 6.74 allow us to calculate a spatial spectrum for an arbitrary external force. In illustration, suppose that the force amplitude behaves as the delta function:

$$F_x(x,0) = F_x\; \delta(x), \qquad F_z(x,0) = F_z\; \delta(x), \tag{6.75}$$

where F_x and F_z are constants but $\delta(x)$ is the delta function:

$$\int_{-\infty}^{\infty} \delta(x - x_0)\, dx = 1$$

This means that the sinusoidal force is applied to points of an infinitely thin strip of the plane $z = 0$, which is oriented along the y-axis. Then, as follows from eqs. 6.74, the spatial spectrum of these amplitudes is constant and equals

$$X(m) = \frac{F_x}{2\pi}, \qquad Z(m) = \frac{F_z}{2\pi} \tag{6.76}$$

Therefore, in place of eq. 6.73 we have

$$F_x(x,0) = \frac{F_x}{2\pi}\int_{-\infty}^{\infty} e^{i\,m\,x} dm \quad \text{and} \quad F_z(x,0) = \frac{F_z}{2\pi}\int_{-\infty}^{\infty} e^{i\,m\,x} dm \tag{6.77}$$

Potentials of the direct wave

Now we can find the unknowns in eqs. 6.71 and 6.72. First assume that the external force is horizontal.

Case one: $(F_z = 0)$ At plane $z = 0$, the complex amplitudes of wavefields obey the conditions

$$\widetilde{u}_1 = \widetilde{u}_2, \quad \widetilde{w}_1 = \widetilde{w}_2, \quad \widetilde{\tau}^{(1)}_{zz} = \widetilde{\tau}^{(2)}_{zz}, \quad \widetilde{\tau}^{(2)}_{xz} - \widetilde{\tau}^{(1)}_{xz} = -F_x(x,0) \tag{6.78}$$

Taking into account that

$$\widetilde{s} = \text{grad}\ \widetilde{\varphi} + \text{curl}\ \widetilde{\boldsymbol{\psi}}$$

and

$$\widetilde{\tau}_{zz} = \lambda\ \text{div}\widetilde{s} + 2\mu\frac{\partial\widetilde{w}}{\partial z}, \qquad \widetilde{\tau}_{xz} = \mu\left(\frac{\partial\widetilde{u}}{\partial z} + \frac{\partial\widetilde{w}}{\partial x}\right), \tag{6.79}$$

eqs. 6.78 can be rewritten as

$$\frac{\partial\widetilde{\varphi}_1}{\partial x} - \frac{\partial\widetilde{\psi}_1}{\partial z} = \frac{\partial\widetilde{\varphi}_2}{\partial x} - \frac{\partial\widetilde{\psi}_2}{\partial z}, \qquad \frac{\partial\widetilde{\varphi}_1}{\partial z} + \frac{\partial\widetilde{\psi}_1}{\partial x} = \frac{\partial\widetilde{\varphi}_2}{\partial z} + \frac{\partial\widetilde{\psi}_2}{\partial x},$$

$$-\lambda k_l^2\widetilde{\varphi}_1 + 2\mu\left(\frac{\partial^2\widetilde{\varphi}_1}{\partial z^2} + \frac{\partial^2\widetilde{\psi}_1}{\partial x\ \partial z}\right) = -\lambda k_l^2\widetilde{\varphi}_2 + 2\mu\left(\frac{\partial^2\widetilde{\varphi}_2}{\partial z^2} + \frac{\partial^2\widetilde{\psi}_2}{\partial x\ \partial z}\right), \tag{6.80}$$

$$\left(2\frac{\partial^2\widetilde{\varphi}_2}{\partial x\ \partial z} + \frac{\partial^2\widetilde{\psi}_2}{\partial x^2} - \frac{\partial^2\widetilde{\psi}_2}{\partial z^2}\right) - \left(2\frac{\partial^2\widetilde{\varphi}_1}{\partial x\ \partial z} + \frac{\partial^2\widetilde{\psi}_1}{\partial x^2} - \frac{\partial^2\widetilde{\psi}_1}{\partial z^2}\right) = -\frac{F_x(x,0)}{2\pi\ \mu} \quad \text{for} \quad z = 0$$

Substitution of eqs. 6.71, 6.72 and 6.77 into set 6.80 gives

$$i\ m(A_m - C_m) - m_s(B_m + D_m) = 0, \quad m_l(A_m + C_m) + im(B_m - D_m) = 0,$$

$$\left(2m^2 - k_s^2\right)(A_m - C_m) + 2i\ m\ m_s(B_m + D_m) = 0, \tag{6.81}$$

$$\text{and} \quad 2i\ m\ m_l(A_m + C_m) - \left(2m^2 - k_s^2\right)(B_m - D_m) = \frac{F_x}{2\pi\mu} \qquad \text{and}$$

From the first and third equations in set 6.81 we obtain

$$A_m = C_m \qquad \text{and} \qquad B_m = -D_m \tag{6.82}$$

Substitution of the latter into two other equations of set 6.81 gives

$$A_m = -\frac{im}{m_l}\frac{F_x}{2\pi\mu k_s^2}, \qquad B_m = \frac{F_x}{2\pi\mu k_s^2} \tag{6.83}$$

and $$C_m = -\frac{im}{m_l}\frac{F_x}{2\pi\mu k_s^2}, \qquad D_m = \frac{F_x}{2\pi\mu k_s^2}$$

Therefore, potentials of the direct wave are

$$\tilde{\varphi}_0 = -i\, C_1 \int\limits_{-\infty}^{\infty} \frac{m}{m_l} e^{-m_l\,|z|} e^{i\,m\,x}\, dm \tag{6.84}$$

and

$$\tilde{\psi}_0 = C_1 \int\limits_{-\infty}^{\infty} e^{m_s z} e^{i\,m\,x}\, dm, \qquad \text{if} \qquad z < 0, \tag{6.85}$$

$$\tilde{\psi}_0 = -C_1 \int\limits_{-\infty}^{\infty} e^{-m_s z} e^{imx}\, dm, \qquad \text{if} \qquad z > 0$$

where

$$C_1 = \frac{F_x}{2\pi\ \mu\ k_s^2} \tag{6.86}$$

In accordance with eq. 6.15,

$$H_0^{(1)}(k_l r) = \frac{1}{\pi i} \int\limits_{-\infty}^{\infty} \frac{1}{m_l}\, e^{-m_l\,|z|} e^{i\,m\,x}\, dm,$$

and, correspondingly, scalar potential can be written as

$$\tilde{\varphi}_0 = -i\,\pi\, C_1 \frac{\partial}{\partial x} H_0^{(1)}(k_l r) \tag{6.87}$$

Case two: $(F_x = 0)$ Suppose that the external force is directed along the z-axis. Then, proceeding from eqs. 6.63 and 6.70, we obtain the following system, which describes the behavior of displacement components and stresses on the $z = 0$.

$$i\,m(A_m - C_m) - m_s(B_m + D_m) = 0, \qquad m_l(A_m + C_m) + i\,m(B_m - D_m) = 0,$$

$$\left(2m^2 - k_s^2\right)(A_m - C_m) + 2i\ m\ m_s(B_m + D_m) = \frac{F_z}{2\pi\mu}, \tag{6.88}$$

$$2i\ m\ m_l(A_m + C_m) - \left(2m^2 - k_s^2\right)(B_m - D_m) = 0$$

The second and fourth equations give

$$A_m = -C_m \qquad \text{and} \qquad B_m = D_m, \tag{6.89}$$

and therefore

$$i\ m\ A_m - m_s B_m = 0 \qquad \left(2m^2 - k_s^2\right) A_m + 2i\ m\ m_s B_m = \frac{F_z}{2\pi\mu k_s^2} \tag{6.90}$$

The solution of system 6.90 is

$$A_m = -\frac{F_z}{2\pi k_s^2\ \mu}, \qquad B_m = -\frac{imF_z}{2\pi k_s^2\ \mu\ m_s}, \tag{6.91}$$

$$C_m = \frac{F_z}{2\pi k_s^2\ \mu}, \qquad D_m = -\frac{imF_z}{2\pi k_s^2\ \mu\ m_s}$$

Thus,

$$\tilde{\varphi}_0 = -C_2 \int\limits_{-\infty}^{\infty} e^{m_l\ z} e^{i\ m\ x} dm \qquad \text{if} \quad z < 0 \tag{6.92}$$

$$\tilde{\varphi}_0 = C_2 \int\limits_{-\infty}^{\infty} e^{-m_l\ z} e^{i\ m\ x} dm \qquad \text{if} \quad z > 0,$$

$$\text{and} \quad \tilde{\psi}_0 = -i\ C_2 \int\limits_{-\infty}^{\infty} \frac{m}{m_s}\ e^{-m_s\ |z|} e^{i\ m\ x} dm \tag{6.93}$$

Here

$$C_2 = \frac{F_z}{2\pi\mu\ k_s^2} \tag{6.94}$$

Note that the last integral is expressed in terms of the Hankel function.

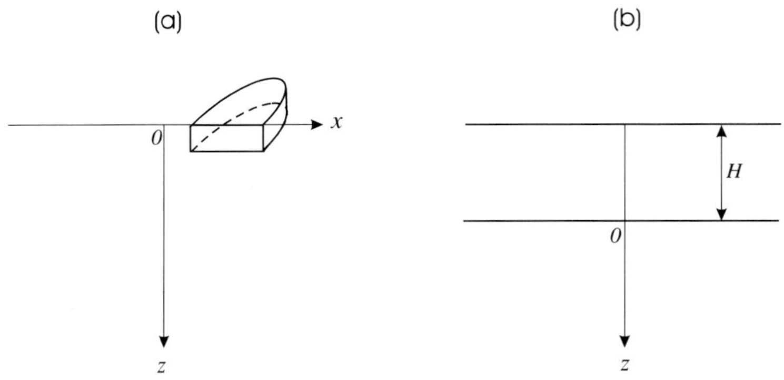

Figure 6.2: (a) External force acts on free surface (b) External force is applied beneath free surface, $z = 0$

6.3 Waves in a homogeneous half-space caused by force $\mathbf{F}^e$

Next we will derive formulas of potentials in a homogeneous half-space with a free boundary and distinguish two positions of the plane where the external force $\mathbf{F}^e$ is applied. As before, it is assumed that the force behaves as the delta function of x, and therefore

$$F_x(x,0) = \frac{F_x}{2\pi} \int\limits_{-\infty}^{\infty} e^{i\,m\,x} dm \qquad \text{and} \qquad F_z(x,0) = \frac{F_z}{2\pi} \int\limits_{-\infty}^{\infty} e^{i\,m\,x} dm \tag{6.95}$$

Force $\mathbf{F}^e$ acts at points of the free surface, Fig. 6.2a Since the z-axis is directed downward and $z > 0$, the complex amplitudes of potentials characterizing these waves can be represented as

$$\tilde{\varphi} = \int\limits_{-\infty}^{\infty} A_m\, e^{-m_l\, z} e^{i\,m\,x} dm \qquad \text{and} \qquad \tilde{\psi} = \int\limits_{-\infty}^{\infty} B_m\, e^{-m_s\, z} e^{i\,m\,x} dm \tag{6.96}$$

Eqs. 6.96 describe the resultant waves, including direct and surface waves. It is obvious that functions $\tilde{\varphi}$ and $\tilde{\psi}$ obey the corresponding Helmholtz equations and the condition at infinity. Our task is to find such coefficients A_m and B_m that boundary conditions at the free surface are also met. Consider an elementary cylinder, shown in Fig. 6.2a, and suppose that its upper face $z = 0$ is subjected to action by external force $\mathbf{F}^e$. In the limit, when the cylinder height dz tends to zero, the condition of equilibrium is written as

$$(\mathbf{F}^e + \mathbf{t})\; dS = 0 \tag{6.97}$$

In particular, when force $\mathbf{F}^e$ has either x or z components, eq. 6.97 becomes

$$\tilde{\tau}_{zz} = 0, \quad \tilde{\tau}_{xz} = -F_x(x,0) \qquad \text{or} \qquad \tilde{\tau}_{zz} = -F_z(x,0)\,, \quad \tilde{\tau}_{xz} = 0 \tag{6.98}$$

Taking into account eqs. 6.95 and 6.96, we obtain two system of equations; either

$$\left(2m^2 - k_s^2\right) A_m - 2i\; m\; m_s\; B_m = 0, \qquad 2im\; m_l\; A_m + \left(2m^2 - k_s^2\right)\; B_m = \frac{F_x}{2\pi\mu} \tag{6.99}$$

or

$$\left(2m^2 - k_s^2\right) A_m - 2i\; m\; m_s B_m = -\frac{F_z}{2\pi\mu}, \qquad 2i\; m\; m_l A_m + \left(2m^2 - k_s^2\right) B_m = 0, \tag{6.100}$$

and they have the same determinant. Thus, in the case of horizontal force, we obtain

$$A_m = \frac{2i\; m\; m_s}{2\pi\mu\; D} F_x \qquad \text{and} \qquad B_m = \frac{\left(2m^2 - k_s^2\right) F_x}{2\pi\mu\; D} \tag{6.101}$$

Here

$$D = \left(2m^2 - k_s^2\right)^2 - 4m^2 m_l\; m_s, \tag{6.102}$$

and it represents the left side of the equation, which defines velocity of Rayleigh waves. When the external force is oriented along the z-axis, eqs. 6.100, coefficients A_m and B_m are

$$A_m = -\frac{\left(2m^2 - k_s^2\right)}{2\pi\mu\; D} F_z \qquad \text{and} \qquad B_m = \frac{2i\; m\; m_l F_z}{2\pi\mu\; D} \tag{6.103}$$

Thus, expressions for the potentials are

Case one: $F_z = 0$

$$\tilde{\varphi} = \frac{iF_x}{2\pi\mu} \int\limits_{-\infty}^{\infty} \frac{2m\; m_s}{D} e^{-m_l\; z} e^{i\; m\; x} dm \tag{6.104}$$

and $$\tilde{\psi} = \frac{F_x}{2\pi\mu} \int\limits_{-\infty}^{\infty} \frac{\left(2m^2 - k_s^2\right)}{D} e^{-m_s\; z} e^{i\; m\; x} dm$$

Case two: $F_x = 0$

$$\tilde{\varphi} = -\frac{F_z}{2\pi\mu}\int_{-\infty}^{\infty} \frac{(2m^2 - k_s^2)}{D} e^{-m_l\, z} e^{i\, m\, x} dm \tag{6.105}$$

and $$\tilde{\psi} = \frac{F_x}{2\pi\mu} i \int_{-\infty}^{\infty} \frac{2m\, m_l}{D} e^{-m_s\, z} e^{i\, m\, x} dm$$

Force $\mathbf{F}^e$ acts at points of a plane beneath the free boundary, Fig. 6.2b

Suppose that the origin of coordinates 0 is situated at the plane where force $\mathbf{F}^e$ is applied, and consider again two cases: $\mathbf{F}^e = F_x(x,0)\,\mathbf{i}$ and $\mathbf{F}^e = F_z(x,0)\,\mathbf{k}$. Since the wavefields are formed by direct waves arising at plane $z = 0$ and secondary waves that appear at the free boundary $z = -H$, it is natural to represent the complex amplitude of potentials of the total wave as a sum:

$$\tilde{\varphi} = \tilde{\varphi}_0 + \tilde{\varphi}_s \quad \text{and} \quad \tilde{\psi} = \tilde{\psi}_0 + \tilde{\psi}_s \tag{6.106}$$

Here $\tilde{\varphi}_0$ and $\tilde{\psi}_0$ are potentials of direct waves given by eqs. 6.84–6.85 and 6.92–6.93. At the same time, the complex amplitudes of potentials of the secondary waves, which obey the Helmholtz equation and the condition at an infinity, can be written as

$$\tilde{\varphi}_s = \int_{-\infty}^{\infty} A'_m e^{-m_l\, z} e^{imx} dm \quad \text{and} \quad \tilde{\psi}_s = \int_{-\infty}^{\infty} B'_m e^{-m_s\, z} e^{imx} dm \tag{6.107}$$

Since at the free boundary stresses vanish, the complex amplitudes of potentials must satisfy two known equations:

$$-\lambda\, k_l^2\, \tilde{\varphi} + 2\mu \left(\frac{\partial^2 \tilde{\varphi}}{\partial z^2} + \frac{\partial^2 \tilde{\psi}}{\partial x\, \partial z} \right) = 0 \quad \text{and} \quad 2\, \frac{\partial^2 \tilde{\varphi}}{\partial x\, \partial z} + \frac{\partial^2 \tilde{\psi}}{\partial x^2} - \frac{\partial^2 \tilde{\psi}}{\partial z^2} = 0 \tag{6.108}$$

Next we obtain expressions of potentials for two orientations of force $\mathbf{F}^e$.

Horizontal external force $F_x(x,0)$ In accordance with eqs. 6.84 and 6.85, we have

$$\tilde{\varphi} = -i\, C_1 \int_{-\infty}^{\infty} \left[\frac{m}{m_l} e^{-m_l\, |z|} + A_m e^{-m_l\, z} \right] e^{i\, m\, x} dm \tag{6.109}$$

and $$\tilde{\psi} = C_1 \int_{-\infty}^{\infty} \left[e^{m_s\, z} + B_m e^{-m_s\, z}\right]\, e^{i\, m\, x} dm \qquad \text{if} \quad z < 0$$

Here

$$C_1 = \frac{F_x}{2\pi\mu\, k_s^2} \tag{6.110}$$

Substitution of eqs. 6.109 into set 6.108 yields

$$\left(2m^2 - k_s^2\right) e^{m_l H} A_m + 2m\; m_s\; e^{m_s H} B_m = L_x \tag{6.111}$$

and $$2m\; m_l e^{m_l H} A_m + \left(2m^2 - k_s^2\right)\; e^{m_s H} B_m = M_x$$

Here

$$L_x = -\left(2m^2 - k_s^2\right) \frac{m}{m_l} e^{-m_l\; H} + 2m\; m_s e^{-m_s\; H} \tag{6.112}$$

and $$M_x = 2m^2 e^{-m_l H} - \left(2m^2 - k_s^2\right)\; e^{-m_s H}$$

The solution of system 6.111 is

$$A_m = \frac{(2m^2 - k_s^2)\, L_x - 2m\; m_s\; M_x}{D} e^{m_s H} \tag{6.113}$$

and $$B_m = \frac{(2m^2 - k_s^2)\;\; M_x - 2m\; m_l L_x}{D} e^{m_l H}$$

where D is given by eq. 6.102.

Vertical external force F_z As follows from eqs. 6.92–6.93

$$\tilde{\varphi} = -C_2 \int_{-\infty}^{\infty} \left(e^{m_l\; z} + A_m e^{-m_l\; z}\right)\; e^{i\; m\; x} dm \tag{6.114}$$

and $$\tilde{\psi} = -i\; C_2 \int_{-\infty}^{\infty} \left(\frac{m}{m_s} e^{m_s\; z} + B_m e^{-m_s\; z}\right)\; e^{i\; m\; x} dm \qquad \text{if} \quad z < 0$$

Here

$$C_2 = \frac{F_z}{2\pi\mu\, k_s^2} \tag{6.115}$$

After substitution of eqs. 6.114 into set 6.108, we obtain:

$$\left(2m^2 - k_s^2\right)\, e^{m_l H} A_m + 2m\, m_s e^{m_s H} B_m = L_z, \tag{6.116}$$

$$2m\, m_l e^{m_l\, H} A_m + \left(2m^2 - k_s^2\right)\, e^{m_s\, H} B_m = M_z$$

where

$$L_z = -\left(2m^2 - k_s^2\right)\, e^{-m_l H} + 2m^2 e^{-m_s\, H} \tag{6.117}$$

and $$M_z = -2m\, m_l e^{-m_l H} + \left(2m^2 - k_s^2\right)\, e^{-m_s\, H}$$

From eqs. 6.116 we obtain

$$A_m = \frac{(2m^2 - k_s^2)\, L_z - 2m\, m_s\, M_z}{D}\, e^{m_s H} \tag{6.118}$$

and $$B_m = \frac{(2m^2 - k_s^2)\; M_z - 2m\, m_l L_z}{D}\, e^{m_l H}$$

Note the following:

1. Function D is independent of the position and orientation of force F^e. This is understandable, because the equation $D = 0$ defines the velocity of the Rayleigh wave, which is a function of elastic parameters and density only.

2. In comparison with the previous case – force applied at the free surface – it is proper to point out that coefficients A_m and B_m in eqs. 6.113 and 6.118 characterize secondary waves only.

3. As we know, the direct wave carries out both dilatational and rotational motions, and each of them gives rise to longitudinal and shear reflected waves. This directly follows from an analysis of functions L and M. Thus, there are four reflected waves and, as will be discussed later, in the far zone their propagation obeys Snell's law. Besides, the denominator of coefficients A_m and B_m is described by function D, which indicates the presence of the Rayleigh wave.

6.4 Wavefields in the far zone (linear source at the free surface)

As follows from eqs. 6.104 and 6.105, the complex amplitudes of scalar potential and the y-component of the vector potential describing waves in a homogeneous half-space are

$$\tilde{\varphi} = \frac{iF_x}{2\pi\mu}\int_{-\infty}^{\infty}\frac{2m\,m_s}{D}e^{-m_l\,z}e^{i\,m\,x}dm, \tag{6.119}$$

$$\tilde{\psi} = \frac{F_x}{2\pi\mu}\int_{-\infty}^{\infty}\frac{2m^2-k_s^2}{D}e^{-m_s z}e^{i\,m\,x}dm \qquad \text{if} \qquad F_z=0$$

and

$$\tilde{\varphi} = -\frac{F_z}{2\pi\mu}\int_{-\infty}^{\infty}\frac{2m^2-k_s^2}{D}\,e^{-m_l\,z}e^{i\,m\,x}dm, \tag{6.120}$$

$$\tilde{\psi} = \frac{iF_z}{2\pi\mu}\int_{-\infty}^{\infty}\frac{2m\,m_l}{D}\,e^{-m_s z}e^{i\,m\,x}dm \qquad \text{if} \qquad F_x=0$$

Here

$$D=\left(2m^2-k_s^2\right)^2-4m^2m_lm_s \tag{6.121}$$

These formulas were derived provided that elastic energy is not transformed into heat and, respectively, wavenumbers k_l and k_s are real. It is convenient and more realistic to assume that attenuation of waves is very small. Therefore, equations for potentials have the form

$$\nabla^2\varphi=\frac{1}{c_l^2}\frac{\partial^2\varphi}{\partial t^2}+\varepsilon_l\frac{\partial\varphi}{\partial t} \qquad \text{and} \qquad \nabla^2\boldsymbol{\psi}=\frac{1}{c_s^2}\frac{\partial^2\boldsymbol{\psi}}{\partial t^2}+\varepsilon_s\frac{\partial\boldsymbol{\psi}}{\partial t} \tag{6.122}$$

Here ε_l and ε_s are small positive numbers characterizing dissipation of mechanical energy. Taking into account the time dependence of waves, $e^{-i\omega t}$, we again arrive at the Helmholtz equations for complex amplitudes of potentials,

$$\nabla^2\tilde{\varphi}+k_l^2\,\tilde{\varphi}=0 \qquad \text{and} \qquad \nabla^2\tilde{\boldsymbol{\psi}}+k_s^2\tilde{\boldsymbol{\psi}}=0,$$

where

$$k_l^2=\frac{\omega^2}{c_l^2}+i\frac{\omega\,\varepsilon_l}{c_l}, \qquad k_s^2=\frac{\omega^2}{c_s^2}+i\frac{\omega\,\varepsilon_s}{c_s} \tag{6.123}$$

Of course, eqs. 6.119–6.121 describe sinusoidal waves even in the presence of attenuation. Note also that if the wavenumbers are real, integrands in eqs. 6.119 and 6.120 have singularities when $D(m) = 0$, but integrals exist. Taking a square root in eqs. 6.123, we obtain

$$k_l = \frac{\omega}{c_l} + i\varepsilon_1, \qquad k_l = -\frac{\omega}{c_l} - i\varepsilon_1, \tag{6.124}$$

$$\text{and} \qquad k_s = \frac{\omega}{c_s} + i\varepsilon_2, \qquad k_s = -\frac{\omega}{c_s} - i\varepsilon_2,$$

where ε_1 and ε_2 are very small positive numbers. It is essential that these roots are located either slightly above or beneath the real axis of m, Fig. 6.3a. As was demonstrated in the previous chapter, if $\text{Im}\ k_l = \text{Im}\ k_s = 0$ function $D(m)$ has at least one real root, and it corresponds to wavenumber k_R of the Rayleigh waves. It is natural to expect that if wavenumbers are complex (eqs. 6.124), roots of function $D(m)$ are not real, which may greatly simplify numerical integration along the m-axis $(-\infty < m < \infty)$.

Proceeding from eqs. 6.119 and 6.120, it is easy to derive formulas for particle displacement as well as for normal and shear stresses. Then, carrying out integration, we can study these wavefields at any point of a homogeneous half-space. However, we will restrict ourselves to the wave (far) zone, where distance from the primary source greatly exceeds the wavelength of elastic waves. We pay special attention to this range because the wave zone in which it is possible to observe different types of waves is of great practical interest. To derive asymptotic formulas for the far zone, we are going to apply the methods of stationary phase and of contour integration, based on the Cauchy theorem (Part II). For instance, use of the first method allowed us to obtain the approximate formulas for reflected and transmitted acoustic waves in the far zone, including some evanescent waves (Part II). To begin with, we apply this method for studying longitudinal and shear waves in an elastic half-space. To do this, it is necessary first to discuss behavior of radicals, m_l and m_s in eqs. 6.119 and 6.120, as functions of m.

Choice of sign of radicals m_l and m_s

From the physical point of view, it is obvious that a wavefield is a single-valued function of the coordinates of an observation point and a frequency. Correspondingly, integration in eqs. 6.119 and 6.120 has to give one value for scalar as well as vector potentials, since they are solutions of Dirichlet's boundary value problem. However, the integrands contain radicals

$$m_l = \sqrt{m^2 - k_l^2} \qquad \text{and} \qquad m_s = \sqrt{m^2 - k_s^2},$$

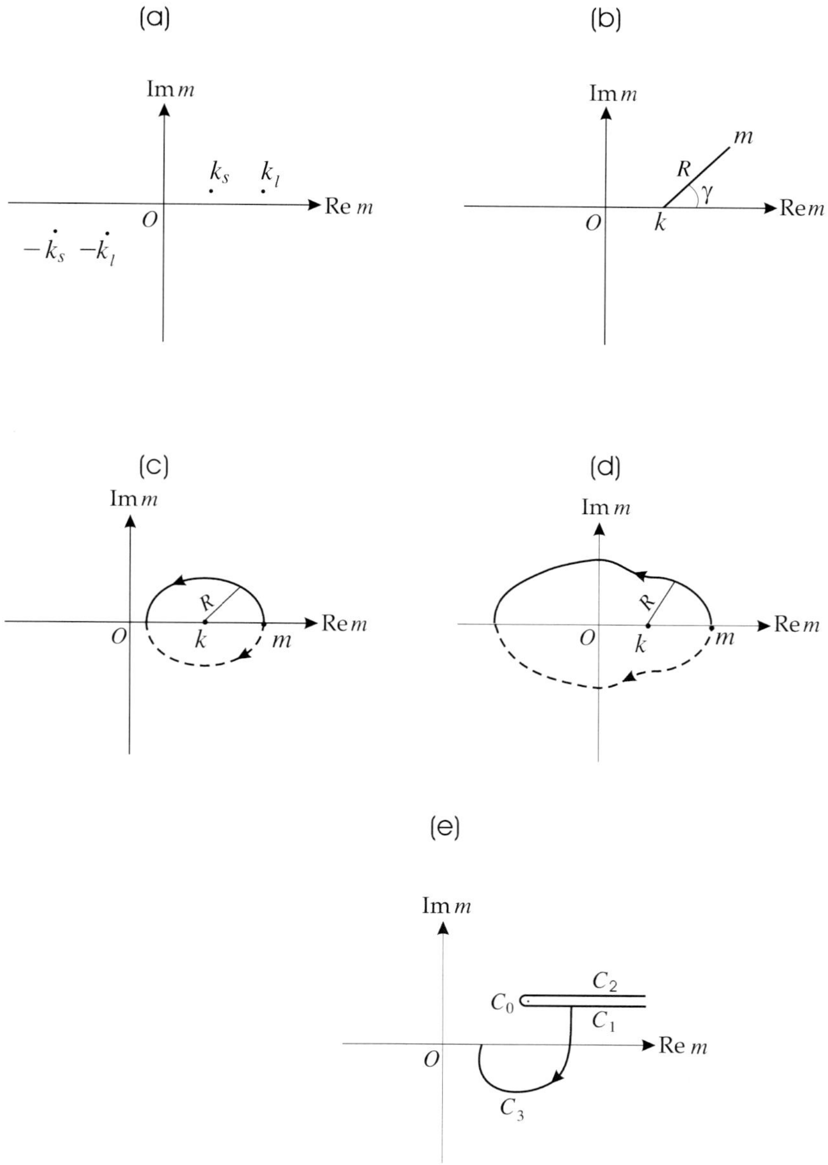

Figure 6.3: (a) Location of wavenumbers on complex plane m (b) Trigonometric form of complex number (c, d) Influence of direction of argument change (e) Branch lines and argument jumps

and each of them has two different values (Part II). Because of this, the integrals and related wavefields become multivalued functions. In order to avoid this problem, let us first recall the concepts of the branch point and branch cut, and consider several functions of a complex variable.

Example 1 For comparison, we start from the simplest single-valued function

$$w(m) = m - k, \tag{6.125}$$

where k is a real number but m is, in general, complex. In the trigonometric form, we have

$$w(m) = R\, e^{i\gamma} \tag{6.126}$$

Here

$$R = |m-k| = |k-m|\,,$$

and the positive direction along which angle γ increases from 0 to 2π is counter-clockwise, Fig. 6.3b. It is clear that this function is continuous and at each point has a derivative. In particular, in the vicinity of positive values of m (Im $m = 0$), we have the same value of w when angle γ is either 0 or 2π, since

$$R\, e^{i\,0} = R\, e^{i\,2\pi}$$

Moreover, this function is analytical everywhere on complex plane m except infinity (Part II). This feature of functions of complex variables allows us to use the Cauchy theorem and perform a deformation of the contour of integration. This procedure is the foundation of the most powerful methods of deriving asymptotic formulas of wavefields in the far zone.

Example 2 Suppose that

$$w = \sqrt{m-k} \qquad \text{or} \qquad w = \sqrt{R\, e^{i\gamma}} \tag{6.127}$$

Taking a square root:

$$w_n = \sqrt{R}\, \exp\left(i\frac{\gamma + 2\pi n}{2}\right), \qquad n = 0, 1$$

we arrive at two functions, which are called branches of function w:

$$w_1 = \sqrt{R}\, e^{i\,\gamma/2} \qquad \text{and} \qquad w_2 = -\sqrt{R}\, e^{i\,\gamma/2} \tag{6.128}$$

By definition

$$\sqrt{R} = \sqrt{|m - k|}$$

is positive, and point k is the branch point. When an expression describing a physical quantity contains a function such as $w(m)$, we always need additional information that will permit us to choose one of these branches. For instance, consider the integral

$$I(k, z) = \int_0^\infty f(m)\, e^{-\sqrt{m-k}\, z}\, dm, \tag{6.129}$$

where z is positive. As soon as $m > k$, the function

$$w = \sqrt{m - k} \tag{6.130}$$

becomes real, and in order to provide convergence we have to assume that the radical is positive. This condition is the additional information we need to choose the necessary branch at points of the m-axis, when

$$k < m < \infty$$

Our next step is to determine this branch within the integral

$$0 < m < k$$

It turns out that such a procedure can be ambiguous. In fact, as is seen from Fig. 6.3c, a change of angle γ by π either counterclockwise or clockwise gives two different values at the same point, eqs. 6.128:

$$w_1 = i\,|k - m|^{1/2} \qquad \text{or} \qquad w_2 = -i\,|k - m|^{1/2}$$

In other words, we obtain two branches of function w.

Note that in the first example, the same procedure yields only one value of the function. It is essential that such uncertainty in the last case takes place for any line C on the complex plane, Fig. 6.3d. In order to overcome this ambiguity and select one branch of w, we first assume that point k is located slightly above the m-axis and then draw a contour, shown in Fig. 6.3e. It consists of a very small circle C_0 around branch point k and two semi-infinite lines (branch cuts) located close to each other. Since the imaginary part of k is extremely small, we can treat the low line C_1 as the m-axis when $m > k$. Next we introduce the rule that a change of complex number m can take place only along paths that do not intersect branch cuts. For instance, moving along path C_0, the argument $m-k$ of the function, eq. 6.130, changes by -2π. Therefore, at the upper branch line C_2

$$w=-R^{1/2}$$

We cannot move in the opposite direction, because the path would intersect the branch. Correspondingly, in order to find the proper branch of function w within the interval

$$0\le m\le k,$$

we move a point along path C_3, and it gives

$$w=-iR^{1/2}=-i\sqrt{k-m}$$

Example 3 Consider the integral

$$I=\int\limits_{-\infty}^{\infty} f(m)\, e^{-\sqrt{m^2-k^2}z}\, dm, \tag{6.131}$$

where $z>0$. Because

$$w=\sqrt{m^2-k^2}=\sqrt{(m-k)(m+k)},$$

there are two branch points $(k,\ -k)$ as well as two branch cuts, shown in Fig. 6.4a. In the trigonometric form we have

$$m-k=R_1e^{i\,\gamma_1},\qquad m+k=R_2e^{i\,\gamma_2} \tag{6.132}$$

Because z is positive and the integrand in eq. 6.131 has to decrease with an increase of m, we conclude that at the real axis of m, when $m>k$ (branch line C_1)

$$\gamma_1=\gamma_2=0, \tag{6.133}$$

and

$$w=\sqrt{m^2-k^2}=\sqrt{R_1R_2}>0$$

A rotation clockwise about branch point k gives for the upper branch line C_2

$$\gamma_1=-2\pi,\qquad \gamma_2=0 \tag{6.134}$$

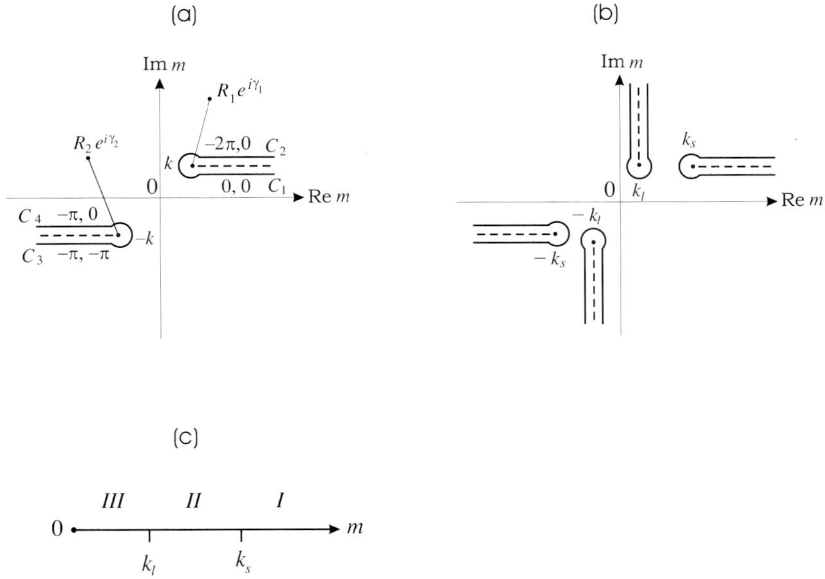

Figure 6.4: (a) Illustration of example 3 (b) Branch lines of integrands in eqs. 6.119 and 6.120 (c) Three intervals of integration

Since this procedure is not accompanied by rotation around point $-k$, it is natural that at both sides of the branch line (C_1 and C_2), angle γ_2 is the same. At the same time, a movement counterclockwise from C_2 to C_4 increases both angles by π, and at points of C_4 we have

$$\gamma_1 = -\pi, \qquad \gamma_2 = \pi \tag{6.135}$$

This result is important because branch line C_4 coincides with the interval of integration when $m < -k$. In this case

$$w = \sqrt{R_1 e^{-i\pi} R_2 e^{i\pi}} = \sqrt{R_1 R_2} > 0,$$

and the integrand becomes smaller with an increase of $|m|$. Performing a rotation around branch point $-k$, we see that γ_1 does not change, but γ_2 becomes equal to $-\pi$, which is at branch line C_3:

$$\gamma_1 = \gamma_2 = -\pi \tag{6.136}$$

Finally, movement from points of line C_3 counterclockwise to points of interval $-k < m < k$ gives

$$\gamma_1 = -\pi, \qquad \gamma_2 = 0 \tag{6.137}$$

Hence

$$w = \sqrt{R_1 R_2 e^{-i\pi}} = -i\sqrt{R_1 R_2}$$

or

$$w = -i\sqrt{k^2 - m^2} \tag{6.138}$$

This expression will be used often in deriving asymptotic formulas.

Example 4 In accordance with eqs. 6.119 and 6.120, complex amplitudes of potentials are described by two integrals:

$$\int_{-\infty}^{\infty} A_m\,(m, m_l, m_s)\, e^{-m_l\, z} e^{i\, m\, x} dm \tag{6.139}$$

and $$\int_{-\infty}^{\infty} B_m\,(m, m_l, m_s)\, e^{-m_s z} e^{i\, m\, x} dm$$

Respectively, there are four branch points and four branch lines, Fig. 6.4b. Again, assuming that $z > 0$, it is appropriate to distinguish three intervals of integration:

$$\begin{aligned}
&m > k_s, && \sqrt{m^2 - k_l^2} > 0, && \sqrt{m^2 - k_s^2} > 0 \\
&k_s > m > k_l, && \sqrt{m^2 - k_l^2} > 0, && \sqrt{m^2 - k_s^2} = -i\sqrt{k_s^2 - m^2} \\
&0 < m < k_l, && \sqrt{m^2 - k_l^2} = -i\sqrt{k_l^2 - m^2}, && \sqrt{m^2 - k_s^2} = -i\sqrt{k_s^2 - m^2}
\end{aligned} \tag{6.140}$$

Equalities are similar for negative values of m.

Wavefields as a superposition of plane waves

As follows from eqs. 6.119 and 6.120, at each point of a medium, scalar and vector potentials are represented as an infinite sum (integral) of homogeneous and inhomogeneous plane waves. Correspondingly, the complex amplitudes of potentials describing these elementary waves are

$$\tilde{\varphi}_m = \frac{iF_x}{2\pi\mu}\frac{2m\, m_s}{D(m)}\, e^{-m_l\, z} e^{i\, m\, x} dm, \tag{6.141}$$

$$\tilde{\psi}_m = \frac{F_x}{2\pi\mu}\frac{2m^2 - k_s^2}{D(m)} e^{-m_s z} e^{imx} dm, \qquad \text{if} \qquad F_z = 0$$

and

$$\tilde{\varphi}_m = -\frac{F_z}{2\pi\mu}\frac{2m^2 - k_s^2}{D(m)} e^{-m_l z} e^{i m x} dm, \tag{6.142}$$

$$\tilde{\psi}_m = \frac{iF_z}{2\pi\mu}\frac{2m\, m_l}{D(m)} e^{-m_s z} e^{i m x} dm \qquad \text{if} \qquad F_x = 0$$

It is clear that amplitudes of these waves depend on the variable of integration m, and due to the presence of the term dm, they are infinitely small. Also, we see that with an increase of m harmonics, $\tilde{\varphi}_m$ and $\tilde{\psi}_m$ as functions of x and z, change more rapidly in both horizontal and vertical directions. In contrast, they vary relatively slowly when m becomes small. This allows us to expect that at sufficiently large distances from the source, harmonics with small m values play a dominant role. In other words, they mainly form wavefields in the far zone. Respectively, in deriving asymptotic formulas, it is natural to focus on the initial interval of integration in eqs. 6.119 and 6.120. At the same time, near the source and at some moderate distance from it, harmonics with rather large values of m can make a significant contribution. It may be useful to note that integration in eqs. 6.119 and 6.120 can be carried out for positive values of m only. Indeed, taking into account Euler's formula

$$e^{i\gamma} = \cos\gamma + i\sin\gamma,$$

we have

$$\tilde{\varphi} = -\frac{F_x}{\pi\mu}\int_0^\infty \frac{2m\, m_s}{D(m)} e^{-m_l z} \sin mx\, dm \tag{6.143}$$

and $$\tilde{\psi} = \frac{F_x}{\pi\mu}\int_0^\infty \frac{2m^2 - k_s^2}{D(m)} e^{-m_s z} \cos mx\, dm \qquad \text{if} \qquad F_z = 0$$

In the same manner, we can represent complex amplitudes when the external force is directed along the z-axis. The variable of integration m, which has the dimension m^{-1}, is usually treated as the wavenumber of elementary plane waves, $\tilde{\varphi}_m$ and $\tilde{\psi}_m$. By definition, it characterizes the rate of change of these waves in the x and z directions.

It is useful to distinguish harmonics that describe either inhomogeneous or homogeneous elementary plane waves. In this light, consider three intervals of integration, Fig. 6.4c.

The first interval: $(m > k_s)$ Suppose that m corresponds to the interval in which $m > k_s$ – that is, m_l and m_s are real and positive numbers:

$$m_l = \sqrt{m^2 - k_l^2} > 0 \qquad \text{and} \qquad m_s = \sqrt{m^2 - k_s^2} > 0 \tag{6.144}$$

Then harmonics of potentials $\tilde{\varphi}_m$ and $\tilde{\psi}_m$, which satisfy this condition (eqs. 6.144), exponentially decay along the z-axis, and they characterize waves advancing in the horizontal direction. Certainly, t is clear that we are dealing with inhomogeneous plane waves, and their phase velocity

$$c(m) = \frac{\omega}{m} \tag{6.145}$$

is defined from the wave argument

$$-\omega t + mx$$

With an increase of wavenumber m, velocity $c(m)$ rapidly decreases and tends to zero, so that in the first interval

$$c_s > c(m) > 0 \tag{6.146}$$

As was demonstrated in Chapter 5, the velocity of the Rayleigh wave, c_R, is slightly smaller than that of the shear wave $(c_R < c_s)$. For this reason, we may think that the harmonics of the initial portion of this integral $(m \sim k_s)$ form the Rayleigh wave and at a great distance from the source they play the dominant role. In fact, as follows from eqs. 6.143, with an increase of m $(m \gg k_s)$, the amplitudes of harmonics $\tilde{\varphi}_m$ and $\tilde{\psi}_m$ rapidly decrease. Correspondingly, even at the free surface $(z = 0)$, the influence of inhomogeneous plane waves with large wavenumbers becomes negligible. On the other hand, when $m \sim |k_s|$, function $D\,(m)$, eq. 6.121, has a root that is located a little above the real axis of m. Therefore, amplitudes of these harmonics can be quite large.

The approach used here allows us to see again one interesting feature of the surface wavefield. As follows from eqs. 6.119 and 6.120, potentials contain terms

$$e^{-m_l\,z} \qquad \text{and} \qquad e^{-m_s\,z},$$

and at the beginning of the first interval we can approximately write

$$\tilde{\varphi}_m \sim e^{-\sqrt{k_R^2 - k_l^2}\,z}, \qquad \tilde{\psi}_m \sim e^{-\sqrt{k_R^2 - k_s^2}\,z}$$

Since

$$\sqrt{k_R^2 - k_l^2} > \sqrt{k_R^2 - k_s^2},$$

with an increase of distance from the free surface, the dilatational part of wave motion decreases more rapidly. Thus, it turns out that superposition of elementary inhomogeneous waves with wavenumbers close to k_s, but slightly exceeding it, produces the Rayleigh wave.

The second interval In this interval of integration, the wavenumber m varies within the range

$$k_l \le m \le k_s,$$

and therefore

$$m_l = \sqrt{m^2 - k_l^2} > 0 \qquad \text{and} \qquad m_s = -i\sqrt{k_s^2 - m^2} \tag{6.147}$$

As with the first interval, the harmonics $\tilde{\varphi}_m$ of the scalar potential describe inhomogeneous plane waves. Correspondingly, they may also contribute to the longitudinal evanescent wave in the far zone. At the same time, the product of the exponential terms of harmonics $\tilde{\psi}_m$ can be written in the form

$$e^{i\left(\sqrt{k_s^2 - m^2}z + m\,x\right)} \qquad \text{if} \qquad m < |k_s| \tag{6.148}$$

or

$$e^{i\,k_s\,(z\cos\theta_s + x\sin\theta_s)}$$

Here

$$\cos\theta_s = \left(1 - \frac{m^2}{k_s^2}\right)^{1/2}, \qquad \sin\theta_s = \frac{m}{k_s} \tag{6.149}$$

Eqs. 6.148 and 6.149 show that harmonics $\tilde{\psi}_m$ describe the homogeneous plane wave of the shear type, which propagates with the velocity of shear waves. The angle of incidence varies within the range

$$\sin^{-1}\frac{c_s}{c_l} \le \theta_s \le \frac{\pi}{2} \tag{6.150}$$

All of these waves make a strong contribution to the shear wave as well as to the conical wave in the far zone.

The third interval Finally, in the initial part of integration

$$0 \leq m \leq k_l$$

both radicals m_l and m_s are purely imaginary:

$$m_l = -i\sqrt{k_l^2 - m^2} \qquad \text{and} \qquad m_s = -i\sqrt{k_s^2 - m^2}$$

Therefore, the harmonics of potentials $\widetilde{\varphi}_m$ and $\widetilde{\psi}_m$ are homogeneous plane waves propagating with phase velocities c_l and c_s, respectively. The angle of incidence of elementary longitudinal waves changes as

$$0 \leq \theta_l \leq \frac{\pi}{2},$$

whereas for shear harmonics we have

$$0 \leq \theta_s \leq \sin^{-1} \frac{c_s}{c_l} \tag{6.151}$$

This interval influences both longitudinal and shear waves in the far zone and the conical shear wave.

To summarize this discussion, note the following:

1. At any point of a medium, wavefields can be represented as an infinite sum of elementary homogeneous and inhomogeneous plane waves.

2. Within the range integration $m > k_s$, all of these waves are inhomogeneous, and they move along the boundary with different phase velocities that are smaller than the velocity of the shear wave. With an increase of m, these harmonics decay more rapidly with z.

3. In the interval

$$0 \leq m \leq k_s$$

the harmonics of potential $\widetilde{\psi}_m$ are homogeneous plane waves that propagate in different directions with the velocity of shear wave c_s. The angle of incidence of these waves changes from 0 to $\pi/2$.

4. The harmonics of potential $\widetilde{\varphi}_m$ are also homogeneous plane waves, if $m < k_l$, and they move in all directions $(0 \leq \theta_l \leq \pi/2)$ with the velocity of longitudinal wave c_l. At the same time, elementary longitudinal waves in the interval

$$k_l < m < k_s$$

are inhomogeneous. They mainly form the evanescent P wave in the far zone that accompanies the shear wave.

5. Because of exponential decay along the z-axis and rapid oscillations in the horizontal direction, wavefields at the far zone are mainly due to harmonics with relatively small wavenumbers, m. For instance, the longitudinal wave is a superposition of harmonics $\tilde{\varphi}_m$ when $m < k_l$. The same takes place with shear-wave harmonics, provided that $m < k_s$.

6. The Rayleigh wave is the result of superposition of elementary waves of the dilatational and rotational types, and its wavenumber is almost minimal among inhomogeneous plane waves.

7. In accordance with eqs. 6.119 and 6.120, regardless of distance from the source, we can assume that wavefields consist of longitudinal and shear homogeneous waves as well as different inhomogeneous plane waves including the Rayleigh wave. However, at relatively small distances the effect of propagation is masked by the presence of inhomogeneous waves with large values of m. Since their contribution diminishes with increased distance, it becomes possible to observe the wave phenomenon.

Contour of integration

Now we begin to derive asymptotic formulas for wavefields using an approach based on the Cauchy theorem (Part II). As follows from eqs. 6.119 and 6.120, the complex amplitudes of potentials $\tilde{\varphi}$ and $\tilde{\psi}$ are represented as integrals along the real axis of m. Applying the Cauchy theorem, it is possible to choose different paths of integration, some of which are very useful in deriving asymptotic formulas. This procedure implies that the transition from old to new paths is not accompanied by intersection of singularities of the integrand. These singularities include the branch points $\pm k_l$ and $\pm k_s$, as well as poles that are determined from the equation

$$D = \left(2m^2 - k_s^2\right)^2 - 4m^2 m_l\, m_s = 0 \tag{6.152}$$

By analogy with the case of acoustic waves (Part II), we consider integrals along the closed path C, shown in Fig. 6.5a. Let us denote $f(m)$ as integrands in eqs. 6.119 and 6.120. Then, taking into account that its singularities are located outside the area path C surrounds, we can write:

$$\int\limits_{-\infty}^{\infty} f(m)dm + \sum_{i=1}^{4} \int\limits_{C_i} f(m)dm + \sum \oint (m)dm + \int\limits_{C_R} f(m)dm = 0 \tag{6.153}$$

The first term of the sum is the original integral along the real axis of m. The second is the sum of integrals along four branch cuts: C_1, C_2, C_3, and C_4. The third term

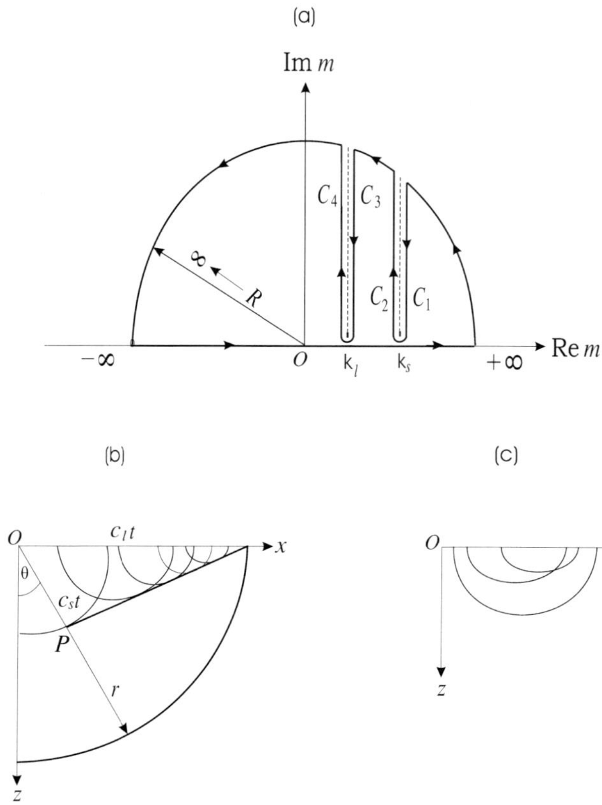

Figure 6.5: (a) Integration along closed path in complex plane m (b) Formation of shear conical wave (c) Destructive interference of longitudinal evanescent wave

is a sum of integrals around poles in the clockwise direction. Finally, the last term is the integral along the semicircle of an infinitely large radius R. Applying Jordan's lemma (Part II), it is easy to see that this integral can be neglected if $R \to \infty$. The last equality (eq. 6.153), allows us to replace integration along the real axis of m by integration along the branch lines and near poles. In general, the advantages of this procedure for numerical integration are not always obvious. However, as was already mentioned, this deformation of the integration path permits us to obtain approximate formulas for the far zone. Further, it is assumed that external force has only the vertical component. The other case $(F_x = 0)$ can be treated similarly. First, we derive asymptotic formulas for displacement at points of the free boundary.

Components of displacement $(z=0)$

As follows from the definition:

$$\mathbf{s} = \text{grad } \varphi + \text{curl } \boldsymbol{\psi},$$

we have for horizontal and vertical components of displacement

$$\widetilde{u} = \frac{\partial \widetilde{\varphi}}{\partial x} - \frac{\partial \widetilde{\psi}}{\partial z} \qquad \text{and} \qquad \widetilde{w} = \frac{\partial \widetilde{\varphi}}{\partial z} + \frac{\partial \widetilde{\psi}}{\partial x} \tag{6.154}$$

Substituting eqs. 6.120 into eqs. 6.154 and letting $z=0$, we obtain

$$\widetilde{u} = -\frac{iF_z}{2\pi\mu} \int_{-\infty}^{\infty} \frac{m\,(2m^2 - k_s^2 - 2m_l\, m_s)}{D} e^{i\,m\,x} dm \tag{6.155}$$

$$\text{and} \qquad \widetilde{w} = -\frac{F_z}{2\pi\mu} \int_{-\infty}^{\infty} \frac{k_s^2 m_l}{D} e^{i\,m\,x} dm$$

We first obtain the asymptotic formula for the tangential component of displacement $\widetilde{u}_b$ associated with the branch points. As follows from eq. 6.153, this part of the displacement can be written in the form

$$\widetilde{u}_b = -\frac{iF_z}{2\pi\mu}\,(I_s + I_l) \tag{6.156}$$

Here I_s and I_l are integrals along corresponding branch lines.

Contribution of branch lines around point k_s, , integral I_s

Movement around branch point k_s changes the sign of the radical m_s, and integration along branch cuts C_1 and C_2 is made in the opposite directions. Therefore we have

$$I_s = \int_{k_s}^{k_s + i\infty} m \left[\frac{2m^2 - k_s^2 - 2m_l\, m_s}{D\,(m_l, m_s)} - \frac{2m^2 - k_s^2 + 2m_l\, m_s}{D\,(m_l, -m_s)} \right] e^{i\,m\,x} dm \tag{6.157}$$

Note that the branch lines are parallel to the imaginary axis, Im m, – that is, the real part of the wavenumber, m, remains constant along these lines. From eq. 6.121 we have

$$I_s = \int_{k_s}^{k_s + i\infty} m\,(2m^2 - k_s^2) \left[\frac{1}{(2m^2 - k_s^2)^2 - 4m^2 m_l m_s} - \frac{1}{(2m^2 - k_s^2)^2 + 4m^2 m_l m_s} \right] e^{imx} dm \tag{6.158}$$

$$-2\int_{k_s}^{k_s+i\infty} m\, m_l\, m_s \left[\frac{1}{(2m^2-k_s^2)^2-4m^2m_l\, m_s}+\frac{1}{(2m^2-k_s^2)^2+4m^2m_l\, m_s}\right] e^{imx}dm$$

or

$$I_s=\int_{k_s}^{k_s+i\infty}\frac{8m^3m_l\, m_s\,(2m^2-k_s^2)-4m\, m_l\, m_s\,(2m^2-k_s^2)^2}{(2m^2-k_s^2)^4-16m^4\, m_l^2\, m_s^2}\, e^{i\, m\, x}dm \tag{6.159}$$

To evaluate this integral, we introduce a new variable, t:

$$m=k_s+it \tag{6.160}$$

Correspondingly, when integration is made along the branch lines, t varies as

$$0\le t<\infty$$

It is clear that

$$dm=i\, dt, \tag{6.161}$$

and in the vicinity of branch point k_s, where t is small, we have

$$m\sim k_s \tag{6.162}$$

Also

$$m_s=\sqrt{m^2-k_s^2}=\sqrt{(k_s+it)^2-k_s^2}\approx\sqrt{k_s}\sqrt{2t}\, e^{i\pi/4} \tag{6.163}$$

$$\text{and}\qquad m_l=\sqrt{m^2-k_l^2}\approx\sqrt{k_s^2-k_l^2}=k_s\left(1-\frac{c_s^2}{c_l^2}\right)^{1/2}\qquad \text{if}\quad t\to 0$$

or

$$m_l=k_s a,\qquad a=\left(1-\frac{c_s^2}{c_l^2}\right)^{1/2} \tag{6.164}$$

The oscillating term of the integrand becomes

$$e^{i\, m\, x}=e^{i\,(k_s+it)\, x}=e^{ik_sx}e^{-xt} \tag{6.165}$$

The presence of the multiplier $\exp(-xt)$ shows that after replacement of a variable, the integrand in eq. 6.159 starts to decrease rapidly even at sufficiently small values of t, if $x \gg 1$. This means that under certain conditions, I_s is mainly defined by the initial part of an integration located near branch point k_s, and this fact is used to derive the asymptotic formulas. From eqs. 6.162–6.164, it follows that

$$2m^2 - k_s^2 \approx k_s^2, \qquad m\ m_l\ m_s \approx k_s^{5/2} a\sqrt{2t}\ e^{i\pi/4}, \tag{6.166}$$

$$m^3 m_l\ m_s \approx k_s^{9/2} a\sqrt{2t}\ e^{i\pi/4}, \qquad 16m^4 m_l^2\ m_s^2 \quad {}_{\mathrm{R}} \approx 16k_s^7 a^2 2t\ i$$

Substitution of eqs. 6.161 and 6.166 into eq. 6.159 gives

$$I_s = \frac{4a\ \sqrt{2}\ i\ e^{i\pi/4}}{k_s^{3/2}}\ e^{ik_s x} \int\limits_0^\infty t^{1/2} e^{-xt} dt \tag{6.167}$$

The last integral is tabular:

$$\int\limits_0^\infty t^{1/2} e^{-xt} dt = \frac{\sqrt{\pi}}{2\ x^{3/2}} \tag{6.168}$$

Therefore

$$I_s = \frac{2a\sqrt{2\pi} i\ e^{i\pi/4}}{(k_s\ x)^{3/2}}\ e^{ik_s x}, \tag{6.169}$$

and, making use of eq. 6.156, displacement component $\tilde{u}_s$, related to branch point k_s is

$$\tilde{u}_s = \frac{F_z}{\mu}\sqrt{\frac{2}{\pi}}\left(1 - \frac{k_l^2}{k_s^2}\right)^{1/2} \frac{e^{i\,(k_s x + \pi/4)}}{(k_s\ x)^{3/2}} \tag{6.170}$$

Contribution of branch lines around point k_l, integral I_l

Since the radical m_l has opposite signs at lines C_3 and C_4, Fig. 6.5a, we have

$$I_l = 4k_s^2 \int\limits_{k_l}^{k_l + i\infty} \frac{m\ m_l\ m_s\,(2m^2 - k_s^2)\ e^{i\ m\ x}}{(2m^2 - k_s^2)^4 - 16m^4\ m_l^2\ m_s^2}\, dm$$

Introducing a variable

$$m = k_l + it \tag{6.171}$$

we have near branch point k_l

$$m \approx k_l, \quad m_s = \left(k_l^2 - k_s^2\right)^{1/2} = -i\left(k_s^2 - k_l^2\right)^{1/2} \quad \text{and} \quad m_l = k_l^{1/2}\sqrt{2t}\, e^{i\pi/4} \tag{6.172}$$

Also

$$4m\; m_l\; m_s = 4\sqrt{2}\; k_l^{3/2} e^{i\pi/4}\left(k_l^2 - k_s^2\right)^{1/2}\sqrt{t}\;\left(2m^2 - k_s^2\right) = k_l^2\left(2 - \frac{k_s^2}{k_l^2}\right)$$

Then, substitution of eqs. 6.172 into I_l yields

$$I_l = \frac{2i\sqrt{2\pi}\; e^{i\pi/4}\; k_l^3\; k_s^2\left(k_l^2 - k_s^2\right)^{1/2}}{\left(2k_l^2 - k_s^2\right)^3\left(k_l x\right)^{3/2}} \tag{6.173}$$

Therefore, displacement associated with branch point k_l is

$$\widetilde{u}_l = -\frac{F_z}{\mu}\sqrt{\frac{2}{\pi}}\frac{k_l^3\; k_s^2\;\left(k_s^2 - k_l^2\right)^{1/2} i\; e^{i\left(k_l x + \pi/4\right)}}{\left(k_s^2 - 2k_l^2\right)^3\;\left(k_l x\right)^{3/2}}, \tag{6.174}$$

since

$$\left(k_l^2 - k_s^2\right)^{1/2} = -i\left(k_s^2 - k_l^2\right)^{1/2}$$

Next consider the vertical component of displacement, $\widetilde{w}_b$, which can be written as, eqs. 6.155,

$$\widetilde{w}_b = -\frac{k_s^2 F_z}{2\pi\mu}\left(L_s + L_l\right) \tag{6.175}$$

Contribution of branch lines around point k_s, integral L_s

As follows from eqs. 6.155, we have

$$L_s = \int_{k_s}^{k_s + i\infty} m_l\left[\frac{1}{D\left(m_l, m_s\right)} - \frac{1}{D\left(m_l, -m_s\right)}\right] e^{i\,m\,x}\; dm$$

or

$$L_s = \int_{k_s}^{k_s + i\infty} \frac{8m^2\; m_l^2\; m_s\; e^{i\,m\,x}}{\left(2m^2 - k_s^2\right)^4 - 16m^4\; m_l^2\; m_s^2}\; dm \tag{6.176}$$

Introducing again the variable

$$m = k_s + it,$$

we obtain an approximate expression of L_s:

$$L_s = \frac{8\, i\, k_s^2\, (k_s^2 - k_l^2)\, \sqrt{2}\, e^{i\pi/4}\, k_s^{1/2}\, e^{ik_s x}}{k_s^8} \int_0^\infty t^{1/2} e^{-xt} dt$$

or

$$k_s^2\, L_s = \frac{4\, i \left(1 - \dfrac{k_l^2}{k_s^2}\right) \sqrt{2\pi}\, e^{i\pi/4}\, e^{ik_s x}}{(k_s x)^{3/2}} \tag{6.177}$$

Correspondingly, vertical component $\widetilde{w}_s$ associated with branch point k_s is

$$\widetilde{w}_s = -\frac{2F_z}{\mu} \sqrt{\frac{2}{\pi}}\, \frac{i \left(1 - \dfrac{k_l^2}{k_s^2}\right)}{(k_s x)^{3/2}}\, e^{i\,(k_s x + \pi/4)} \tag{6.178}$$

In the same manner, integral L_l around branch point k_l is

$$L_l = \int_{k_l}^{k_l + i\infty} m_l \left[\frac{1}{D\,(m_l, m_s)} + \frac{1}{D\,(-m_l, m_s)}\right] e^{i\,m\,x} dm$$

or

$$L_l = \int_{k_l}^{k_l + i\infty} \frac{2m_l\, (2m^2 - k_s^2)^2\, e^{i\,m\,x} dm}{(2m^2 - k_s^2)^4 - 4m^4\, m_l^2\, m_s^2}$$

After a change of the variable, $m = k_l + it$, we obtain

$$k_s^2\, L_l = \frac{2k_s^2\, \sqrt{2}\, e^{i\pi/4}\, i\, e^{ik_l x} \sqrt{\pi}\, k_l^{1/2}}{(2k_l^2 - k_s^2)^2\, 2x^{3/2}} = \frac{\sqrt{2\pi}\, i\, k_s^2\, k_l^2\, e^{i\pi/4}\, e^{ik_l x}}{(2k_l^2 - k_s^2)^2\, (k_l x)^{3/2}} \tag{6.179}$$

Therefore

$$\widetilde{w}_l = -\frac{1}{2} \sqrt{\frac{2}{\pi}}\, \frac{F_z\, i}{\mu}\, \frac{k_l^2\, k_s^2}{(2k_l^2 - k_s^2)^2}\, \frac{e^{i\,(k_l x + \pi/4)}}{(k_l x)^{3/2}} \tag{6.180}$$

Displacement field related to branch points

By definition we have

$$\widetilde{u}_b = \widetilde{u}_s + \widetilde{u}_l \qquad \text{and} \qquad \widetilde{w}_b = \widetilde{w}_s + \widetilde{w}_l,$$

and in accordance with eqs. 6.170, 6.174, 6.178, and 6.180, complex amplitudes $\tilde{u}_b$ and $\tilde{w}_b$ are

$$\tilde{u}_b = \frac{F_z}{\mu}\sqrt{\frac{2}{\pi}}\left(1-\frac{k_l^2}{k_s^2}\right)^{1/2}\frac{e^{i\,(k_s x+\pi/4)}}{(k_s x)^{3/2}} \tag{6.181}$$

$$-\frac{F_z}{\mu}\sqrt{\frac{2}{\pi}}\frac{k_l^{3/2}\,k_s^2\,i}{(k_s^2-2k_l^2)^3}\left(k_s^2-k_l^2\right)^{1/2}\frac{e^{i\,(k_l x+\pi/4)}}{(k_l\,x)^{3/2}}+\ldots$$

and

$$\tilde{w}_b = -\frac{2F_z}{\mu}\sqrt{\frac{2}{\pi}}\,i\left(1-\frac{k_l^2}{k_s^2}\right)\frac{e^{i\,(k_s x+\pi/4)}}{(k_s x)^{3/2}} \tag{6.182}$$

$$-\frac{i}{2}\frac{F_z}{\mu}\sqrt{\frac{2}{\pi}}\frac{k_l^2\,k_s^2}{(2k_l^2-k_s^2)^2}\frac{e^{i\,(k_l x+\pi/4)}}{(k_l x)^{3/2}}+\ldots$$

These formulas represent a sum of two terms, and each of them describes a sort of fringe of the cylindrical elastic wave, propagating through a medium. The first one moves along the free boundary with the velocity of the shear wave, but the other advances in the same direction with the velocity of the longitudinal wave, c_l. This fact may create the impression that these terms characterize the shear and longitudinal waves, respectively. However, as will be shown later, it turns out that each of them is a combination of both types of waves. Because the tangential and normal components of displacement for each wave, eqs. 6.181 and 6.182, have different amplitudes and phases, an orbit of the vector $\mathbf{s}$, $(\mathbf{s} = u\,\mathbf{i} + w\,\mathbf{k})$, is an ellipse, i.e., elliptical polarization takes place. The ratios of vertical and horizontal diameters of the ellipse for waves moving with velocities c_s and c_l are

$$2\sqrt{1-\frac{k_l^2}{k_s^2}} \qquad \text{and} \qquad \frac{1}{2}\left(\frac{k_s}{k_l}-\frac{2k_l}{k_s}\right)\left(1-\frac{k_l^2}{k_s^2}\right)^{-1/2} \tag{6.183}$$

In particular, if $\lambda = \mu$, these ratios are 1.33 and 0.35, respectively. We see that in the case of the wave propagating with velocity c_s along the boundary, the ellipse is extended in the vertical direction. For the second wave, the major axis of the ellipse is oriented along the x-axis.

Until now, we have discussed the contribution of integrals along branch lines. Next we will evaluate integrals around poles. As was demonstrated in Part II, integration around a simple pole gives

$$\oint f(m)dm = 2\pi i \, \frac{\varphi_1(m_p)}{\varphi_2'(m_p)} \tag{6.184}$$

where

$$f(m) = \frac{\varphi_1(m)}{\varphi_2(m)} \tag{6.185}$$

and m_p is the pole of the integrand. The poles are roots of eq. 6.152

$$D = \left(2m^2 - k_s^2\right)^2 - 4m^2 m_l \, m_s = 0 \tag{6.186}$$

and, introducing function D_-,

$$D_- = \left(2m^2 - k_s^2\right)^2 + 4m^2 \, m_l \, m_s, \tag{6.187}$$

we obtain

$$DD_- = \left(2m^2 - k_s^2\right)^4 - 16m^4 m_l^2 m_s^2 \tag{6.188}$$

$$= k_s^8 \left[1 - \frac{8m^2}{k_s^2} + \left(24 - 16\,\frac{k_l^2}{k_s^2}\right)\frac{m^4}{k_s^4} - 16\left(1 - \frac{k_l^2}{k_s^2}\right)\frac{m^6}{k_s^6}\right] = 0$$

This is the cubic equation with respect to m^2/k_s^2. It was derived when we described the Rayleigh wave (Chapter 5). For instance, since $k_s^2 > k_l^2$, there is a real root within the interval

$$1 < \frac{m^2}{k_s^2} < \infty$$

This means that radicals m_l and m_s real and positive and, correspondingly, function $D_- \neq 0$. In other words, m/k_s is the root of eq. 6.186, $(D = 0)$. Also it is not difficult to prove that the remaining roots, if they are real, are located within the interval

$$0 < \frac{m^2}{k_s^2} < \frac{k_l^2}{k_s^2}$$

If m/k_s belong to this interval m_l and m_s are negative and imaginary and cannot be, therefore, roots of eq. 6.186. For instance, assuming that $\lambda = \mu$, eq. 6.188 gives three real roots $\dfrac{m^2}{k_s^2}$:

$$\frac{1}{4}, \qquad \frac{1}{4}\left(3 - \sqrt{3}\right), \qquad \frac{1}{4}\left(3 + \sqrt{3}\right)$$

It is clear that in such a case, there is only one root of eq. 6.186,

$$k_R = \frac{1}{2}\sqrt{3+\sqrt{3}k_s} \simeq 1.088\ k_s \tag{6.189}$$

and it is the wavenumber of the Rayleigh wave.

Now we are ready to find expressions for components of displacement associated with the Rayleigh wave. Taking into account eqs. 6.155 and 6.184, we have

$$u_R = -\frac{F_z}{\mu} H e^{ik_R x} \qquad \text{and} \qquad w_R = -\frac{iF_z}{\mu} K\ e^{ik_R x} \tag{6.190}$$

Here

$$H = \frac{k_R \left(2k_R^2 - k_s^2 - 2\sqrt{k_R^2 - k_l^2}\ \sqrt{k_R^2 - k_s^2}\right)}{-D'\left(k_R, k_l, k_s\right)} \qquad \text{and} \qquad K = \frac{k_s^2\ \sqrt{k_R^2 - k_l^2}}{D'\left(k_R, k_l, k_s\right)} \tag{6.191}$$

As follows from eqs. 6.190, the Rayleigh wave, unlike the two others, does not decrease with distance. Correspondingly, this surface wave plays the dominant role in the far zone. In accordance with eqs. 6.191,

$$H = 0.125, \qquad K = 0.183 \qquad \text{if} \quad \lambda = \mu \tag{6.192}$$

Asymptotic behavior of waves beneath the free surface

By analogy with case $z = 0$, we represent complex amplitudes of potentials in the form

$$\tilde{\varphi} = -\frac{F_z}{2\pi\mu} I = -\frac{F_z}{2\pi\mu}\left(I_s + I_l + I_p\right) \tag{6.193}$$

$$\text{and} \qquad \tilde{\psi} = \frac{F_z\ i}{2\pi\mu} M = \frac{iF_z}{2\pi\mu}\left(M_s + M_l + M_p\right),$$

where

$$I = \int\limits_{-\infty}^{\infty} \frac{2m^2 - k_s^2}{D} e^{-m_l\ z} e^{i\ m\ x} dm \qquad \text{and} \qquad M = \int\limits_{-\infty}^{\infty} \frac{2mm_l}{D} e^{-m_s\ z} e^{i\ m\ x} dm \tag{6.194}$$

I_s, M_s, I_l, M_l, and I_p, M_p are integrals along the branch lines and poles, respectively. To begin with, we will derive asymptotic formulas for scalar potential $\tilde{\varphi}$ associated with branch points.

Contribution of branch cuts around point k_s, integral I_s

From eqs. 6.194 we have

$$I_s = \int_{k_s}^{k_s+i\infty} \left(2m^2 - k_s^2\right) \left[\frac{1}{D\left(m_l, m_s\right)} - \frac{1}{D\left(m_l, -m_s\right)}\right] e^{-m_l\, z} e^{i\, m\, x} dm$$

or, taking into account eq. 6.121,

$$I_s = \int_{k_s}^{k_s+i\infty} \frac{8\left(2m^2 - k_s^2\right)\, m^2 m_l\, m_s\, e^{-m_l\, z} e^{i\, m\, x}}{\left(2m^2 - k_s^2\right)^4 - 16m^4\, m_l^2\, m_s^2} dm \tag{6.195}$$

Here integration is done along branch cuts C_1 and C_2, as is shown in Fig. 6.5a. To simplify the evaluation of I_s, we introduce, as before, the variable

$$m = k_s + it$$

In the vicinity of branch point k_s,

$$m_l = \left(k_s^2 - k_l^2\right)^{1/2}, \qquad m_s = \sqrt{2} e^{i\pi/4}\, k_s^{1/2}\, t^{1/2}, \qquad \left(2m^2 - k_s^2\right) = k_s^2$$

and $\quad dm = idt, \qquad 8m^2 m_l\, m_s = 8\, k_s^2\, \left(k_s^2 - k_l^2\right)^{1/2} \sqrt{2}\, e^{i\pi/4}\, k_s^{1/2}\, t^{1/2}$

Thus, integral I_s becomes

$$I_s = \frac{8i\,\sqrt{2}\, e^{i\pi/4} \left(1 - \dfrac{k_l^2}{k_s^2}\right)^{1/2} e^{ik_s x}\, e^{-\sqrt{k_s^2 - k_l^2}\, z}}{k_s k_s^{3/2}} L_1 \tag{6.196}$$

where

$$L_1 = \int_0^\infty t^{1/2} e^{-xt} dt = \frac{\sqrt{\pi}}{2x^{3/2}}$$

Therefore

$$I_s = \frac{4\, i\, \sqrt{2\pi}\, e^{i\pi/4} \left(1 - \dfrac{k_l^2}{k_s^2}\right)^{1/2} e^{ik_s x}\, e^{-\sqrt{k_s^2 - k_l^2}\, z}}{k_s \left(k_s x\right)^{3/2}} \tag{6.197}$$

Thus, the complex amplitude of scalar potential associated with point k_s is equal to

$$\widetilde{\varphi}_s = -\frac{2\,i\,F_z}{\mu}\sqrt{\frac{2}{\pi}}\,\frac{\left(1-\dfrac{k_l^2}{k_s^2}\right)^{1/2}}{k_s}\,e^{-\sqrt{k_s^2-k_l^2}\,z}\,\frac{e^{i\,(k_s x+\pi/4)}}{(k_s x)^{3/2}} \tag{6.198}$$

It is clear that eq. 6.198 describes the wave propagating along the free boundary with the velocity of shear waves, c_s. This wave exponentially decays with depth z, since

$$\sqrt{k_s^2-k_l^2} = k_s a > 1, \qquad\qquad a = \left(1-\frac{k_l^2}{k_s^2}\right)^{1/2}$$

Oscillations with higher frequencies decrease more rapidly. Taking into account that displacement $\widetilde{\mathbf{s}}$ is related to the scalar potential as

$$\widetilde{\mathbf{s}} = \text{grad}\,\widetilde{\varphi},$$

we conclude that

$$\text{curl}\,\widetilde{\mathbf{s}} = 0$$

This means that propagation of this wave does not cause rotation of elementary volumes of a medium, also they can experience deformation and translation. Thus, function $\widetilde{\varphi}_s$ characterizes the evanescent longitudinal wave, moving with the shear-wave velocity, c_s. This wave has another important feature – namely, and unlike the Rayleigh wave, it cannot exist alone. To demonstrate such dependence on other waves, it is sufficient to show that stresses of the wave, $\widetilde{\varphi}_s$, do not vanish at the free surface. As an example, consider the normal stress, $\widetilde{\tau}_{zz}$:

$$\widetilde{\tau}_{zz} = -\lambda\,k_l^2\widetilde{\varphi}_s + 2\mu\,\frac{\partial^2\widetilde{\varphi}_s}{\partial z^2} \qquad\qquad \text{if}\ \ z=0 \tag{6.199}$$

Substitution of eq. 6.198 gives

$$\left(-\lambda\,k_l^2+2\mu\,k_s^2a^2\right)\widetilde{\varphi}_s = \left(-\lambda\,k_l^2+2\mu\,k_s^2-2\mu\,k_l^2\right)\widetilde{\varphi}_s$$

$$= -\left[-\left(\lambda+2\mu\right)k_l^2+2\mu\,k_s^2\right]\,\widetilde{\varphi}_s = \left(-\rho\,\omega^2+2\rho\,\omega^2\right)\widetilde{\varphi}_s \neq 0,$$

$$\text{since} \qquad \lambda+2\mu = \rho\,c_l^2, \qquad\qquad \mu = \rho\,c_s^2$$

It turns out that this wave, $\widetilde{\varphi}_s$, accompanies the shear wave.

Contribution of branch lines around branch point k_l, integral I_l

Since radical m_l has different signs at lines C_3 and C_4, Fig 6.5a, the integral along them can be written as

$$I_l = \int_{k_l}^{k_l+i\infty} \left(2m^2-k_s^2\right)^2 \left[\frac{e^{-m_l\, z}}{\left(2m^2-k_s^2\right)^2 - 4m^2 m_l\, m_s}\right. \tag{6.200}$$

$$\left. - \frac{e^{m_l\, z}}{\left(2m^2-k_l^2\right)^2 + 4m^2 m_l\, m_s}\right] e^{i\,m\,x} dm$$

or

$$I_l = 2\int_{k_l}^{k_l+i\infty} \frac{\left(2m^2-k_s^2\right)\left[4m^2 m_l\, m_s \cosh m_l z - \left(2m^2-k_s^2\right)^2 \sinh m_l z\right]}{\left(2m^2-k_s^2\right)^4 - 16m^4 m_l^2\, m_s^2}\, e^{i\,m\,x} dm \tag{6.201}$$

Introducing again variable $m = k_l + i\,t$, we have $dm = i\,d\,t$, and in the vicinity of branch point k_l

$$m = k_l, \qquad m^2 = k_l^2, \quad m_s = \left(k_l^2-k_s^2\right)^{1/2} = -i\, \left(k_s^2-k_l^2\right)^{1/2},$$

$$m_l = \sqrt{2}\, e^{i\pi/4}\, k_l^{1/2}\, t^{1/2}, \qquad 2m^2-k_s^2 = 2k_l^2-k_s^2,$$

$$4m^2 m_l\, m_s = -4i\,\sqrt{2}\, e^{i\pi/4}\, k_l^2\, k_l^{1/2} \left(k_s^2-k_l^2\right)^{1/2} t^{1/2}, \quad e^{i\,mx} = e^{i\,k_l x}\, e^{-xt}$$

Therefore

$$I_l = \frac{8\sqrt{2}\left(\dfrac{k_s^2}{k_l^2}-1\right)^{1/2} e^{i\,\pi/4}}{\left(2-\dfrac{k_s^2}{k_l^2}\right)^3 k_l^{5/2}}\, e^{i\,k_l x} L_1 - \frac{2i}{\left(2-\dfrac{k_s^2}{k_l^2}\right) k_l^2}\, e^{i\,k_l\,x}\, L_2 \tag{6.202}$$

Here

$$L_1 = \int_0^\infty t^{1/2} \cosh\left(\alpha_l \sqrt{t}\right)\, e^{-x\,t} dt \tag{6.203}$$

and

$$L_2 = \int_0^\infty \sinh\left(\alpha_l \sqrt{t}\right)\ e^{-x\ t} dt, \tag{6.204}$$

$$\alpha_l = \sqrt{2k_l}\ z\ e^{i\ \pi/4} \tag{6.205}$$

Both integrals are expressed in terms of elementary functions. For instance, introducing variable $v = \sqrt{t}$, we have

$$dv = \frac{dt}{2v}$$

and

$$L_2 = 2\int_0^\infty v \sinh\left(\alpha_l v\right)\ e^{-x\ v^2} dv$$

or

$$L_2 = 2\frac{\partial}{\partial \alpha_l}\int_0^\infty \cosh\left(\alpha_l v\right)\ e^{-x\ v^2} dv$$

The last integral is tabular:

$$\int_0^\infty \cosh\ \left(\alpha_l v\right)\ e^{-x\ v^2} dv = \frac{1}{2}\sqrt{\frac{\pi}{x}} \exp\left(\frac{\alpha_l^2}{4x}\right) \tag{6.206}$$

and

$$L_2 = \sqrt{\frac{\pi}{x}}\ \frac{\partial}{\partial \alpha_l}\ e^{\alpha_l^2/4x} \tag{6.207}$$

As follows from eqs. 6.203 and 6.204,

$$L_1 = \frac{\partial L_2}{\partial \alpha_l} \quad \text{or} \quad L_1 = \sqrt{\frac{\pi}{x}}\ \frac{\partial^2}{\partial \alpha_l^2}\ e^{\alpha_l^2/4x} \tag{6.208}$$

Now, taking into account eqs. 6.193 and 6.203, we obtain for scalar potential $\tilde{\varphi}_l$ associated with branch point k_l

$$\tilde{\varphi}_l = -\frac{F_z}{\sqrt{2\pi x}\mu}\ \frac{\partial}{\partial \alpha_l}\left[\frac{8\left(\dfrac{k_s^2}{k_l^2}-1\right)^{1/2} e^{i\pi/4}}{\left(2-\dfrac{k_s^2}{k_l^2}\right)^3 k_l^{5/2}}\ \frac{\partial}{\partial \alpha_l}\ e^{\alpha_l^2/4x} - \frac{\sqrt{2}\ i\ \ e^{\alpha_l^2/4x}}{\left(2-\dfrac{k_s^2}{k_l^2}\right)k_l^2}\right] e^{ik_l x} \tag{6.209}$$

Each term in brackets contains the exponent, $\exp(\alpha_l^2/4x)$. Correspondingly, at the right side of eq. 6.209, we have the common multiplier

$$\exp\left(ix+\frac{\alpha_l^2}{4x}\right), \tag{6.210}$$

or, using eq. 6.205,

$$\exp i\,k_l\left(x+\frac{1}{2}\frac{z^2}{x}\right) \tag{6.211}$$

If an observation point is located relatively close to the free surface $(z \ll x)$, then distance r from the source can be presented as

$$r=\left(x^2+z^2\right)^{1/2}=x\left(1+\frac{z^2}{x^2}\right)^{1/2}\approx x+\frac{1}{2}\frac{z^2}{x} \qquad \text{if} \qquad z\ll x \tag{6.212}$$

Thus, in place of eq. 6.211 we have

$$\exp i\,k_l\left(x+\frac{1}{2}\frac{z^2}{x}\right)\simeq e^{i\,k_l\,r} \tag{6.213}$$

This means that we are dealing with a longitudinal wave, $\widetilde{\varphi}_l$, that propagates through a medium with velocity c_l. At the boundary $z=0$, the wave's fringe is described by eq. 6.209. Now we make use of the vector potential, which has the y-component only.

Contribution of branch lines around branch point k_s, integral M_s

As follows from eq. 6.193, the integral along C_1 and C_2 is

$$M_s=\int\limits_{k_s}^{k_s+i\infty} 2m\,m_l\left[\frac{e^{-m_s z}}{D(m_l,m_s)}-\frac{e^{m_s z}}{D(m_l,-m_s)}\right]e^{i\,m\,x}dm$$

or

$$M_s=\int\limits_{k_s}^{k_s+i\infty}\frac{16m^3\,m_l^2\,m_s\cosh m_s\,z-4m\,m_l\,\left(2m^2-k_s^2\right)^2\sinh m_s\,z}{\left(2m^2-k_s^2\right)^4-16m^4\,m_l^2\,m_s^2}\,e^{i\,m\,x}dm \tag{6.214}$$

Applying the known replacement of the variable

$$m=k_s+it,$$

we obtain an approximate expression of M_s

$$M_s = \frac{16\sqrt{2}\, i\left(1-\frac{k_l^2}{k_s^2}\right) e^{i\,\pi/4}}{k_s^{5/2}} e^{i\,k_s x}\, M_1 - \frac{4\, i\left(1-\frac{k_l^2}{k_s^2}\right)^{1/2}}{k_s^2} e^{i\,k_s x}\, M_2, \tag{6.215}$$

where M_1 and M_2 are given by eqs. 6.203 and 6.204 and

$$M_1 = \sqrt{\frac{\pi}{x}}\,\frac{\partial^2}{\partial\alpha_s^2}\, e^{\alpha_s^2/4x}, \qquad M_2 = \sqrt{\frac{\pi}{x}}\,\frac{\partial}{\partial\alpha_s}\, e^{\alpha_s^2/4x} \tag{6.216}$$

and

$$\alpha_s = \sqrt{2\,k_s}\; z\; e^{i\,\pi/4} \tag{6.217}$$

From eqs. 6.194 and 6.214–6.216, we have for potential $\tilde{\psi}_s$:

$$\tilde{\psi}_s = -\frac{2\,F_z\left(1-\dfrac{k_l^2}{k_s^2}\right)^{1/2} e^{i\,k_s x}}{\sqrt{\pi}\;\mu\;\sqrt{x}\;k_s^2} \tag{6.218}$$

$$\times\frac{\partial}{\partial\alpha_s}\left[\frac{4\sqrt{2}\left(1-\dfrac{k_l^2}{k_s^2}\right)^{1/2} e^{i\,\pi/4}}{\sqrt{k_s}}\,\frac{\partial}{\partial\alpha_s}\left(e^{\alpha_s^2/4x}\right) - e^{\alpha_s^2/4x}\right]$$

By analogy with eq. 6.209, we can say that function $\tilde{\psi}_s$ describes the rotational (shear) wave propagating through a medium with velocity c_s, and eq. 6.218 characterizes its behavior near the free surface $(z \ll r)$.

Contribution of branch lines around branch point k_l, integral M_l

Finally, accomplishing integration along the closed path C, we have for the integral along branch cuts C_3 and C_4

$$M_l = \int\limits_{k_l}^{k_l+i\infty} 2m\;m_l\left[\frac{1}{D\;(m_l, m_s)} + \frac{1}{D\;(-m_l, m_s)}\right]\; e^{-m_s\;z} e^{i\;m\;x}\, dm \tag{6.219}$$

or

$$M_l = 4\int\limits_{k_l}^{k_l+i\infty} \frac{m\;m_l\,(2m^2-k_s^2)^2\,e^{-m_s\;z} e^{i\;m\;x}}{(2m^2-k_s^2)^4 - 16\;m^4 m_l^2, m_s^2}\; dm \tag{6.220}$$

Introducing again the variable $m = k_l + it$ and taking into account that

$$\sqrt{k_l^2 - k_s^2} = -i\,\sqrt{k_s^2 - k_l^2},$$

we obtain

$$M_l = \frac{4\,i\sqrt{2\pi}\;e^{i\,\pi/4}\;e^{i\left(k_l x + \sqrt{k_s^2 - k_l^2}\,z\right)}}{k_s^{5/2}\left(2 - \dfrac{k_s^2}{k_l^2}\right)^2} \int\limits_0^\infty t^{1/2} e^{-xt}\,dt$$

or

$$M_l = \frac{2\,i\,\sqrt{2\pi}\;e^{i\,\pi/4}\;e^{i\left(k_l x + \sqrt{k_s^2 - k_l^2}\,z\right)}}{k_l^{5/2}\left(2 - \dfrac{k_s^2}{k_l^2}\right)^2 x^{3/2}} \tag{6.221}$$

Therefore,

$$\tilde{\psi}_l = -\frac{F_z}{\mu}\sqrt{\frac{2}{\pi}}\,\frac{1}{k_l}\,\frac{1}{\left(2 - \dfrac{k_s^2}{k_l^2}\right)^2}\,\frac{e^{i\left(k_l x + \sqrt{k_s^2 - k_l^2}\,z + \pi/4\right)}}{(k_l\,x)^{3/2}} \tag{6.222}$$

It is obvious that the argument of $\tilde{\psi}_l$ remains constant at points of the line

$$k_l\,x + \sqrt{k_s^2 - k_l^2}\;z = r\,\left(k_l \sin\theta + \sqrt{k_s^2 - k_l^2}\cos\theta\right) = \text{const}, \tag{6.223}$$

which lies in plane $X0Z$, Fig 6.5b. Therefore, the phase surface of the wave, eq. 6.222, is formed by two planes located symmetrically with respect to coordinate plane $x = 0$. The two planes represent the lateral surface of the two-dimensional cone, and this is why this wave is called the conical wave. At points of the line, eq. 6.223, both cylindrical coordinates r and θ vary, but the argument (phase) remains the same. Let us choose angle θ so that

$$k_l = k_s\,\sin\theta \qquad \text{and} \qquad \sqrt{k_s^2 - k_l^2} = k_s\,\cos\theta \tag{6.224}$$

Substitution of the latter into eq. 6.223 yields

$$k_l\,x + \sqrt{k_s^2 - k_l^2}\;z = k_s\,r = \text{const} \tag{6.225}$$

This shows that eq. 6.222 describes a conical shear wave propagating through a medium with velocity c_s, and its direction is defined by the ratio

$$\sin\theta = \frac{c_s}{c_l} \tag{6.226}$$

It is essential that its apparent velocity along the boundary $(z = 0)$ is equal to the velocity of the longitudinal wave, c_l, and this coincidence is not occasional. The appearance of this conical wave can be explained in the following way. When the longitudinal wave, $\tilde{\varphi}_l$, moves through a medium, each point of the wavefront can be treated as the source of an elementary cylindrical wave of the same type (Huygen's principle). Superposition of these waves produces the resultant P wave, moving away from the origin, where the external force, F_z, is applied. In addition, as soon as the longitudinal (P) wave reaches some point of the free surface, it also becomes a source of an elementary shear wave advancing with velocity c_s (Huygen's principle). Certainly, at any instant t the radius of the elementary shear wave that arises at the coordinate origin is $r = c_s t$. However, with an increase of distance x, this radius becomes smaller and is defined as

$$r(x) = c_s\left(t - \frac{x}{c_l}\right)$$

We may say that the longitudinal wave plays the role of a moving source of this shear wave. In particular, the radius of the wavefront of the elementary wave with origin $x = c_l t$ is equal to zero, since the P wave has just arrived at this point. Now again applying Huygen's principle, we see that the envelope of elementary waves is the plane that forms angle θ with the free boundary, eq. 6.226. This approach also shows that the wavefront of the conical wave, $\tilde{\psi}_l$, must be tangential to that of the shear wave, Fig. 6.5b. In fact, at the initial moment $t = 0$, both P and S waves arise at origin $x = 0$, and the former causes the elementary shear wave. Therefore, shear and conical waves arrive simultaneously at point P, Fig. 6.5b, where their wavefronts are tangential to each other. There is some evident similarity between conical and head waves.

In the same manner it is possible to explain the appearance of the evanescent longitudinal wave $\tilde{\varphi}_s$ that moves along the free surface with velocity c_s. When the shear wave propagates, each point of the boundary $(z = 0)$ becomes the source of a longitudinal elementary wave. The superposition of these elementary waves gives rise to the resultant P wave, which advances along the free surface with velocity c_s. In other words, the shear wave is the moving source of this P wave. However, unlike in the previous case, $\tilde{\psi}_l$, the interference of elementary waves has a destructive character and causes exponential decay of the wave with depth z. Such behavior also follows from Huygen's principle, since an envelope of elementary waves is absent, Fig. 6.5c. This happens because their velocity, c_l, exceeds the velocity of their source moving along the free surface.

Wavefields in the far zone

In evaluating integrals in eqs. 6.193 along branch lines, we were able to distinguish four waves, namely

1. The longitudinal wave P.
2. The shear wave S.
3. The conical shear wave that accompanies the P wave at the free boundary.
4. The evanescent longitudinal wave that accompanies the S wave at surface $z = 0$.

Integration around the pole, $m_p = k_R$ allows us to obtain an expression for the Rayleigh wave in a homogeneous half-space. Unlike points of the free surface, the integrands in eqs. 6.194 contain terms $e^{-m_l\, z}$ and $e^{-m_s\, z}$. Correspondingly, the complex amplitudes of scalar and vector potentials of this wave exponentially decay with depth, and their decrease is proportional to $e^{-\sqrt{k_R^2 - k_l^2}\, z}$ and $e^{-\sqrt{k_R^2 - k_s^2}\, z}$, respectively. The Rayleigh wave has already been described in detail, so let us focus on the first four waves, which form two groups.

1. Longitudinal and conical waves Both of these waves, $\tilde{\varphi}_l$ and $\tilde{\psi}_l$, move with the same velocity c_l along the free surface. Correspondingly, they cannot be distinguished at these points $(z = 0)$ if time of arrival is measured. However, if we observe quantities characterizing dilatation

$$\frac{\partial u}{\partial x} + \frac{\partial w}{\partial z}$$

and rotation

$$\frac{\partial u}{\partial z} - \frac{\partial w}{\partial x},$$

these waves can be separated from each other. As is seen from Fig. 6.5b, the conical wave arrives after the longitudinal wave at points beneath the free surface. This means that by measuring the arrival time at these points, each wave can be observed.

Now let us evaluate displacement caused by these waves at the boundary. Performing differentiations in eq. 6.209, we obtain :

$$\tilde{\varphi}_l = -\frac{F_z}{4\mu\,\sqrt{\pi}\; x^{3/2}}\;(A - \alpha\, B)\; e^{\alpha^2/4x} e^{i\, k_l x} \tag{6.227}$$

Here

$$A = \frac{8\sqrt{2}\left(\dfrac{k_s^2}{k_l^2}-1\right)^{1/2} e^{i\,\pi/4}}{\left(2-\dfrac{k_s^2}{k_l^2}\right)^3 k_l^{3/2}} \qquad \text{and} \qquad B = \frac{2\,i}{\left(2-\dfrac{k_s^2}{k_l^2}\right) k_l^2}, \qquad \alpha_l = \sqrt{2}\,k_l^{1/2} e^{i\,\pi/4} z$$

Because displacement is studied at the boundary $(z=0)$, the term in eq. 6.227 proportional to z^2 is discarded, and therefore

$$\frac{\partial\,\tilde{\varphi}_l}{\partial x} = -\frac{F_z\, i\, k_l\, A}{4\,\mu\,\sqrt{\pi}\,x^{3/2}}$$

or

$$\frac{\partial\,\tilde{\varphi}_l}{\partial x} = -2\frac{F_z}{\mu}\sqrt{\frac{2}{\pi}}\,\frac{k_l^5\,(k_s^2-k_l^2)^{1/2}}{(2k_l^2-k_s^2)^3}\,i\,\frac{e^{i\,(k_l x+\pi/4)}}{(k_l\,x)^{3/2}} \tag{6.228}$$

Also

$$\frac{\partial\,\tilde{\varphi}_l}{\partial z} = \frac{F_z\,k_l^{1/2}\,\sqrt{2}\,e^{i\,\pi/4}}{4\mu\,\sqrt{\pi}\,x^{3/2}}\,B$$

or

$$\frac{\partial\,\tilde{\varphi}_l}{\partial z} = \frac{F_z}{2\mu}\,i\,\sqrt{\frac{2}{\pi}}\,\frac{k_l^2}{(2k_l^2-k_s^2)}\,\frac{e^{i\,(k_l x+\pi/4)}}{(k_l x)^{3/2}} \tag{6.229}$$

From eq. 6.222 we have

$$\frac{\partial\,\tilde{\psi}_l}{\partial x} = -\frac{F_z}{\mu}\sqrt{\frac{2}{\pi}}\,i\,\frac{k_l^4}{(2k_l^2-k_s^2)^2}\,\frac{e^{i\,(k_l x+\pi/4)}}{(k_l x)^{3/2}} \tag{6.230}$$

and

$$\frac{\partial\tilde{\psi}_l}{\partial z} = -\frac{F_z}{\mu}\,i\,\sqrt{\frac{2}{\pi}}\,\frac{\sqrt{k_s^2-k_l^2}\,k_l^3}{(2k_l^2-k_s^2)^2}\,\frac{e^{i\,(k_l x+\pi/4)}}{(k_l x)^{3/2}} \tag{6.231}$$

Correspondingly, displacement components advancing along the boundary with velocity c_l are

$$\tilde{u}_l = \frac{\partial\tilde{\varphi}_l}{\partial x} - \frac{\partial\tilde{\psi}_l}{\partial z} \qquad \text{and} \qquad \tilde{w}_l = \frac{\partial\tilde{\varphi}_l}{\partial x} + \frac{\partial\tilde{\psi}_l}{\partial z} \tag{6.232}$$

Their summation gives the expressions derived earlier. As follows from eqs. 6.228–6.231, the ratio of displacements caused by longitudinal and conical waves is independent of distance x and frequency ω:

$$\left|\frac{\partial\tilde{\varphi}_l}{\partial x}\right| = \left|\frac{\partial\tilde{\psi}_l}{\partial z}\right| \frac{2}{\left(2-\dfrac{k_s^2}{k_l^2}\right)} \quad \text{and} \quad \left|\frac{\partial\tilde{\varphi}_l}{\partial z}\right| = \left|\frac{\partial\tilde{\psi}_l}{\partial x}\right| \frac{\left(2-\dfrac{k_s^2}{k_l^2}\right)}{2}$$

2. Shear and inhomogeneous longitudinal waves These two waves move along the free boundary with the same velocity c_s. As in the first case, they cannot be separated from each other if we observe time of arrival of the wave combination. However, measuring dilatation as well as rotation, it is possible to distinguish them. By analogy with the first group, it is interesting to evaluate displacement at the boundary caused by each wave. As follows from eq. 6.198

$$\frac{\partial\tilde{\varphi}_s}{\partial x} = \frac{2F_z}{\mu}\sqrt{\frac{2}{\pi}}\left(1-\frac{k_l^2}{k_s^2}\right)^{1/2} \frac{e^{i\ (k_s x+\pi/4)}}{(k_s\ x)^{3/2}}, \tag{6.233}$$

$$\frac{\partial\tilde{\varphi}_s}{\partial z} = \frac{2iF_z}{\mu}\sqrt{\frac{2}{\pi}}\left(1-\frac{k_l^2}{k_s^2}\right) \frac{e^{i\ (k_s x+\pi/4)}}{(k_s\ x)^{3/2}},$$

and they characterize displacement due to the inhomogeneous longitudinal wave. Performing a differentiation of eq. 6.218 and discarding the term proportional to α_s^2, we obtain

$$\tilde{\psi}_s = -\frac{F_z\left(1-\dfrac{k_l^2}{k_s^2}\right)^{1/2} e^{ik_s x}}{\sqrt{\pi}\mu\ x^{3/2}k_s^2}\left[\frac{4\sqrt{2}\left(1-\dfrac{k_l^2}{k_s^2}\right)^{1/2} e^{i\ \pi/4}}{k_s^{1/2}} - \sqrt{2}\ k_s^{1/2}e^{i\pi/4}z\right],$$

whence

$$\frac{\partial\tilde{\psi}_s}{\partial x} = -\frac{F_z 4\ i}{\mu}\sqrt{\frac{2}{\pi}}\left(1-\frac{k_l^2}{k_s^2}\right) \frac{e^{i\ (k_s x+\pi/4)}}{(k_s\ x)^{3/2}} \tag{6.234}$$

and

$$\frac{\partial\tilde{\psi}_s}{\partial z} = \frac{F_z}{\mu}\sqrt{\frac{2}{\pi}}\frac{1}{(k_s\ x)^{3/2}}\left(1-\frac{k_l^2}{k_s^2}\right)^{1/2} \frac{e^{i\ (k_s x+\pi/4)}}{(k_s\ x)^{3/2}}$$

6.5 Buried linear source

In the previous section we described behavior of waves in the far zone caused by the linear source located at the boundary. This explains why the direct and secondary waves were inseparable and only the resultant wave was studied. Next assume that the source of either the P or the SV wave is placed inside the half-space at depth H from the boundary, Fig. 6.6a. Our goal is to derive asymptotic formulas that describe waves in the far zone. Because the linear source is situated beneath the boundary, it is natural to expect the appearance of different waves – first of all, reflected waves. In studying acoustic waves (Part II), we demonstrated that the stationary-phase method is often useful in deriving formulas that characterize reflected waves. At the same time, the contour deformation allows us to obtain expressions for the other waves. Correspondingly, we will investigate the asymptotic behavior of these two groups of waves separately, starting with the reflected waves, which obey Snell's law.

The stationary-phase method

First, let us recall the main features of this method (Part II) and consider the integral

$$I = \int_{-\infty}^{\infty} f(m)\, e^{i\,\alpha\, h(m)}\, dm \tag{6.235}$$

Its integrand is the product of two terms: $f(m)$ and $\exp\left[i\,\alpha\, h(m)\right]$. The first term usually changes relatively slowly with m, whereas the exponential term varies rapidly when parameter α is large. Therefore, in this case the oscillating nature of the integrand is due to the second term. It may be instructive to treat the integrand as a sinusoidal wave with wavenumber m. Its amplitude and initial phase, i.e., the magnitude and argument of $f(m)$, are, in general, functions of m. From this point of view, integral $I\,(\alpha)$ describes the superposition of these waves. As is well known (Part I), constructive interference takes place within the range of wavenumbers where the phase (phase function)

$$\alpha\, h(m) \tag{6.236}$$

varies only slightly. Respectively, the sum of these sinusoids may become rather large. In contrast, outside this range the phase $\alpha\, h(m)$ can change very quickly, and for this reason sinusoids cancel each other out (destructive interference). As a result, the sum of such sinusoids is relatively small. Correspondingly, integral $I\,(\alpha)$ is mainly defined by

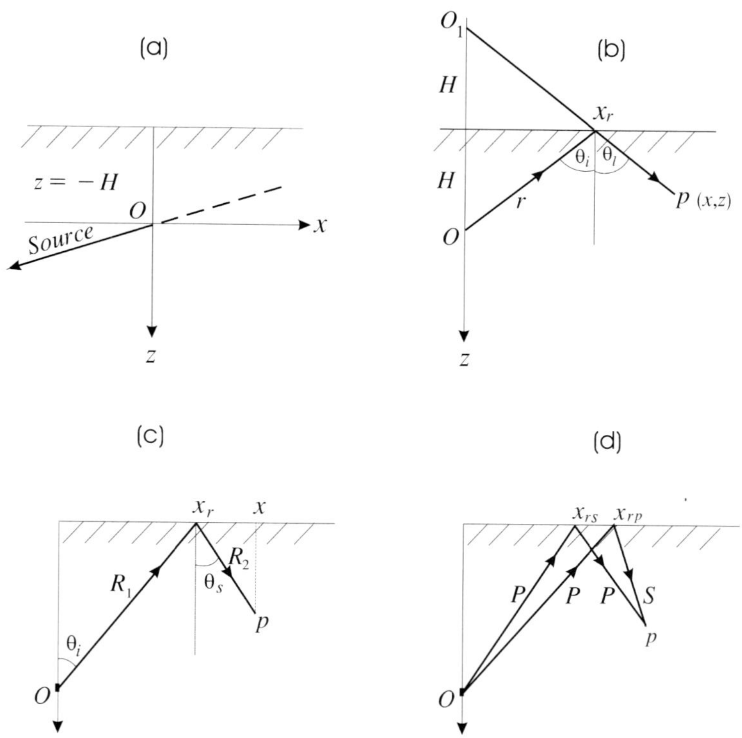

Figure 6.6: (a) Linear source beneath free surface (b) Reflection of P wave at the surface (c) S reflected wave (d) P and S reflected waves

the first range. Of course, it is impossible to draw the exact boundary between these two intervals, since in reality there is always a transition range.

This analysis clearly shows that our purpose is to determine the position of the first range and then evaluate the integral over this interval m. In essence, the stationary-phase method consists of these two steps. It is natural to characterize the position of the first interval with the help of point m_0, at which the phase change is small. Its location can be determined from the equation

$$\frac{\partial\, h(m)}{\partial m} = 0 \tag{6.237}$$

Point m_0, satisfying eq. 6.237, is called the stationary point, and it corresponds to either a maximum or minimum of function $h(m)$. Certainly, the difference

$$\alpha\ [h(m) - h(m_0)]$$

in the vicinity of m_0 strongly depends on the value of parameter α. If it is relatively small, then even sufficiently large deviations from stationary point m_0 may cause a weak change of phase. On the contrary, when parameter α is rather large, an insignificant difference, Δm, may produce a great change of phase. In other words, in this case the width of the interval at which the phase is almost constant becomes small, and outside of this interval the phase varies very rapidly.

These two important features of phase function $\alpha\, h(m)$ allow us to greatly simplify the integral in eq. 6.235. First, function $f(m)$ is replaced by its value at the stationary point,

$$f(m) = f(m_0), \tag{6.238}$$

since its change within the narrow interval is usually rather small. Because the change in $h(m)$ is also small, we can expand this function in Taylor's series and restrict ourselves to the first three terms,

$$h(m) = h(m_0) + h'(m_0)(m - m_0) + \frac{h''(m_0)}{2}(m - m_0)^2 + \ldots.., \tag{6.239}$$

because $\Delta m \to 0$. Taking into account that $h'(m_0) = 0$, we have

$$h(m) = h(m_0) + \frac{h''(m_0)}{2}(m - m_0)^2, \tag{6.240}$$

i.e., the phase behaves like a parabolic function inside the interval. Substitution of eqs. 6.238 and 6.240 into eq. 6.235 yields

$$I = f(m_0)\, e^{i\,\alpha\, h\,(m_0)} \int\limits_{-\infty}^{\infty} \exp\left[i\alpha \frac{h''(m_0)}{2}\,(m - m_0)^2\right]\, dm \tag{6.241}$$

Of course, this replacement is based on the assumption that integral $I(\alpha)$ is practically defined by the interval of integration around stationary point m_0. Now let us introduce a new variable,

$$v^2 = \frac{\alpha\,|\, h''(m_0)|}{2}\,(m - m_0)^2$$

Hence

$$v = \sqrt{\frac{\alpha\,|\, h''(m_0)|}{2}}\,(m - m_0) \quad \text{and} \quad dv = \sqrt{\frac{\alpha\,|\, h''(m_0)|}{2}}\, dm,$$

that is,

$$dm = \sqrt{\frac{2}{\alpha \,|\, h''(m_0)|}} \; dv$$

Therefore, in place of eq. 6.241 we obtain

$$I\,(\alpha) = \frac{\sqrt{2}\;\; e^{i\,\alpha\, h\,(m_0)}}{\sqrt{\alpha \,|\, h''(m_0)|}} \; f\,(m_0) \int\limits_{-\infty}^{\infty} e^{\pm\, i\, v^2} dv \tag{6.242}$$

The appearance of different signs in the exponent is related to the fact that

$$h''(m_0) = |\, h''(m_0)| \qquad \text{if} \qquad h''(m_0) > 0$$

and

$$h''(m_0) = -\,|\, h''(m_0)| \qquad \text{if} \qquad h''(m_0) < 0$$

Because

$$\int\limits_{-\infty}^{\infty} e^{\pm\, i\, v^2} dv = \sqrt{\frac{\pi}{2}} \;\; (1 \pm i)\,,$$

we have

$$I\,(\alpha) = f\,(m_0)\,(1 \pm i) \; \sqrt{\frac{\pi}{\alpha \,|\, h''(m_0)|}} \;\; e^{i\,\alpha\, h\,(m_0)}$$

or

$$I\,(\alpha) = f\,(m_0) \sqrt{\frac{2\,\pi}{\alpha \,|\, h''(m_0)|}} \;\; e^{i\,[\alpha\, h\,(m_0)\;\pm\;\pi/4]}, \tag{6.243}$$

and the sign in front of $\pi/4$ corresponds to that of the second derivative $h''(m_0)$. Thus, in place of the exact expression, eq. 6.235, we have arrived at the approximate one, and its accuracy increases with an increase of parameter α. Note again that in deriving eq. 6.243, it is assumed that only the interval around stationary point m_0 makes the main contribution. Also, we suppose that the change of magnitude and argument of function $f\,(m)$ is very small inside this interval.

1. Direct wave P We are now prepared to use this method to derive asymptotic formulas for reflected waves when the primary source generates a longitudinal wave. In accordance with formulas obtained in section 6.3, the complex amplitudes of scalar and vector potentials are

$$\tilde{\varphi} = \tilde{\varphi}_0 + \tilde{\varphi}_1 \qquad \text{and} \qquad \tilde{\psi} = \tilde{\psi}_1 \tag{6.244}$$

where

$$\tilde{\varphi}_0 = C_0 \, H_0^{(1)} \, (k_l \, r) \tag{6.245}$$

is the scalar potential of the longitudinal direct wave and

$$C_0 = \frac{\pi r_0^2 \, i}{2 \, \mu} \, \tau_{rr}^{(0)} \tag{6.246}$$

Here $\tau_{rr}^{(0)}$ is the amplitude of normal stress on the surface of a cylindrical source with a very small radius, r_0. As is well known, Hankel's function $H_0^{(1)} \, (k_l r)$ can be represented as the asymptotic series

$$H_0^{(1)} \, (k_l r) = \sqrt{\frac{2}{\pi k_l \, r}} \; e^{i \, (k_l \, r \; - \; \pi/4)} \sum_{n=0} \frac{a_n}{(k_l \, r)^n}, \tag{6.247}$$

provided that $k_l r \gg 1$ and that a_n represents given numbers. In essence, eq. 6.247 is an example of Debye's series, which characterizes high-frequency wavefields (Part II). In this case these wavefields result from interference of waves caused by all elements of an infinitely long source. From eqs. 6.39, 6.40, and 6.44, we have for potentials of secondary fields that arise due to the presence of the free boundary

$$\tilde{\varphi}_1 = C_0 \int_{-\infty}^{\infty} B_m \, e^{-m_l \, z} \, e^{i \, m \, x} dm \quad \text{and} \quad \tilde{\psi} \; = \; \tilde{\psi}_1 = C_0 \int_{-\infty}^{\infty} C_m e^{-m_s z} e^{imx} dm, \tag{6.248}$$

where

$$B_m = -\frac{1}{m_l} \frac{(2m^2 - k_s^2)^2 + 4m^2 m_l \, m_s}{(2m^2 - k_s^2)^2 - 4m^2 m_l \, m_s} e^{-m_l \, H} \tag{6.249}$$

and

$$C_m = \frac{4 \, i \, m \; (2m^2 - k_s^2) \, e^{-m_l \, H} e^{-m_s \, H}}{(2m^2 - k_s^2)^2 - 4m^2 m_l \, m_s} \tag{6.250}$$

Before we apply the stationary-phase method it may be useful to note the following. When an direct wave reaches the interface, at each point P and S elementary reflected waves arise, and at any point of a medium we observe the result of interference of elementary waves of the same type. In other words, every point of the free surface is the source of secondary waves. At the same time, in the far zone superposition of these waves – except for elementary waves that arise in the vicinity of point x_r, Fig. 6.6b – has the destructive character. Correspondingly, at an observation point, these waves experience constructive interference. This means that a reflected wave of any type is mainly caused by secondary sources around point x_r, which is different for P and S waves. It is essential that the angles of incidence and reflection at this point $(x_r, -H)$ obey Snell's law. In general, with a change of the observation point, coordinate x_r also varies. In this light we can say that the stationary-phase method allows us to evaluate both the dominant spatial frequency of elementary waves and the result of their constructive interference.

Reflected longitudinal wave From eqs. 6.248 and 6.249 we have

$$\tilde{\varphi}_1 = -C_0 \int\limits_{-\infty}^{\infty} \frac{1}{m_l} \frac{(2m^2 - k_s^2)^2 + 4m^2 m_l\, m_s}{(2m^2 - k_s^2)^2 - 4m^2 m_l\, m_s} e^{-m_l\, z_1} e^{i\, m\, x} dm \tag{6.251}$$

Here

$$z_1 = z + 2H$$

First, we find the position of stationary points located somewhere inside the interval of integration. To start with, consider the case when $m > 0$. It is natural to distinguish two intervals:

$$m > k_l \qquad \text{and} \qquad m < k_l$$

If $m > k_l$ the phase, eq. 6.236, is

$$\alpha\, h(m) = xm,$$

where

$$\alpha = x \qquad \text{and} \qquad h(m) = m$$

Correspondingly,

$$\frac{\partial h(m)}{\partial m} = 1 \neq 0$$

That is, within the interval of integration

$$k_l \leq m < \infty,$$

stationary points are absent. In the second interval, the radical m_l can be represented as

$$m_l = -i\,\sqrt{k_l^2 - m^2},$$

and therefore the phase is

$$mx + \sqrt{k_l^2 - m^2}\; z_1 \tag{6.252}$$

This means that

$$h(m) = m + \sqrt{k_l^2 - m^2}\;\frac{z_1}{x}, \tag{6.253}$$

and its differentiation gives

$$\frac{\partial h(m)}{\partial m} = 1 - \frac{m}{\sqrt{k_l^2 - m^2}}\,\frac{z_1}{x} \tag{6.254}$$

The stationary point m_{0l} is defined from the equality

$$1 = \frac{m_{0l}}{\sqrt{k_l^2 - m_{0l}^2}}\,\frac{z_1}{x}$$

or

$$m_{0l} = \frac{k_l\,x}{\left(x^2 + z_1^2\right)^{1/2}} = \frac{k_l\,x}{r_1} = k_l \sin\theta_i \tag{6.255}$$

Here

$$r_1 = \left[x^2 + (z + 2H)^2\right]^{1/2} \tag{6.256}$$

is the distance between the observation point and point O_1, which is a mirror reflection of the origin O with respect to the boundary. Thus, the stationary point is located inside the interval

$$0 \leq m \leq k_l,$$

and with an increase of x it approaches k_l. In particular, if $z_1/x \ll 1$, we have

$$m_{0l} \approx k_l\left(1 - \frac{1}{2}\,\frac{z_1^2}{x^2}\right)$$

As follows from eq. 6.254, if $m < 0$, the first derivative, $h'(m)$, differs from zero, and therefore only one stationary point is given by eq. 6.254. In accordance with eq. 6.253,

$$h(m) = k_l \sin\theta_i + k_l \cos\theta_i \cot\theta_i = \frac{k_l}{\sin\theta_i}, \tag{6.257}$$

while the phase function is

$$x\, h\left(m_{0l}\right) = \frac{k_l\, x}{\sin\theta_i} = k_l\, r_1 \tag{6.258}$$

We see that a geometric approach allows us to determine the point of the free surface x_r that gives rise to the reflected longitudinal wave observed at point (x, z). Next we find functions $h''\left(m_{0l}\right)$ and $f\left(m_{0l}\right)$, eq. 6.243. It is clear that

$$\begin{aligned} m_l &= \sqrt{k_l^2 \sin^2\theta_i - k_l^2} = -i\, k_l \cos\theta_i, \\ m_s &= \sqrt{k_l^2 \sin^2\theta_i - k_s^2} = -i\, k_l \left(\frac{k_s^2}{k_l^2} - \sin^2\theta_i\right)^{1/2} \end{aligned}$$

From eq. 6.251 we have,

$$f\ \left(m_{0l}\right) = \frac{-i\ C_0}{k_l \cos\theta_i} \frac{\left(2\sin^2\theta_i - \dfrac{c_l^2}{c_s^2}\right)^2 - 4\sin^2\theta_i\cos\theta_i\left(\dfrac{c_l^2}{c_s^2} - \sin^2\theta_i\right)^{1/2}}{\left(2\sin^2\theta_i - \dfrac{c_l^2}{c_s^2}\right)^2 + 4\sin^2\theta_i\cos\theta_i\left(\dfrac{c_l^2}{c_s^2} - \sin^2\theta_i\right)^{1/2}} \tag{6.259}$$

Differentiation of eq. 6.254 yields

$$h''\ (m) = -\frac{z_1}{x} \frac{k_l^2}{\left(k_l^2 - m^2\right)^{3/2}} \tag{6.260}$$

Respectively

$$h''\left(m_{0l}\right) = -\frac{z_1}{x\ k_l \cos^3\theta_i}$$

or

$$h''\left(m_{0l}\right) = -\frac{1}{k_l \sin\theta_i \cos^2\theta_i} \tag{6.261}$$

Since $h''\left(m_{0l}\right) < 0$, the sign in front of $\pi/4$, eq. 6.242, is negative. Substitution of eqs. 6.258, 6.259, and 6.261 into eq. 6.243 gives

$$\tilde{\varphi}_1 = -\frac{-i\ C_0\sqrt{2\pi}}{\left(k_l\ r_1\right)^{1/2}} \tag{6.262}$$

$$\times \frac{\left(2\sin^2\theta_i - \frac{c_l^2}{c_s^2}\right)^2 - 4\sin^2\theta_i\cos\theta_i\left(\frac{c_l^2}{c_s^2} - \sin^2\theta_i\right)^{1/2}}{\left(2\sin^2\theta_i - \frac{c_l^2}{c_s^2}\right)^2 + 4\sin^2\theta_i\cos\theta_i\left(\frac{c_l^2}{c_s^2} - \sin^2\theta_i\right)^{1/2}}\, e^{i\,(k_l\,r_1\,-\,\pi/4)}, \quad \text{since} \quad r_1 = \frac{x}{\sin\theta_i}$$

Thus, we have found an expression for the complex amplitude of scalar potential of the reflected longitudinal wave, and its amplitude is inversely proportional to the product $(k_l\, r_1)^{1/2}$. This is the cylindrical wave, propagating with velocity c_l, and its fictitious source is located at he point O_1, Fig. 6.6b. Speaking strictly, at each point of the wavefront normal and tangential components of displacement, s_{r_1} and s_θ, are shifted in phase with respect to each other. As we know, this indicates that the vector of displacement has an elliptical orbit. However, with an increase of distance r_1 from O_1, the normal component s_{r_1} becomes dominant, and we observe nearly linear polarization. Comparison with eq. 6.241 shows that the stationary-phase method allows us to obtain the leading term of Debye's series, which is zero approximation. In this case elastic energy moves along the elementary ray tube and, correspondingly, the flux of the Poynting vector through tube's lateral surface is absent. This allows us to derive the same expression for the field differently. For instance, we can calculate the amplitude of the direct wave at point x_r, where the angles of incidence and reflection are equal to each other (Snell's law):

$$\theta_l = \theta_i \tag{6.263}$$

Then, multiplying this amplitude by the reflection coefficient of the P plane wave and taking into account a change of the cross-sections of the tube at points $(x_r, -H)$ and (x, z), we again find the amplitude of the reflected wave at the observation point.

Reflected shear waves As follows from eqs. 6.248 and 6.250,

$$\tilde{\psi} = C_0 \int\limits_{-\infty}^{\infty} \frac{4\,i\,m\,(2m^2 - k_s^2)\,e^{-m_l\,H}e^{-m_s\,z_2}}{(2m^2 - k_s^2)^2 - 4m^2 m_l\, m_s}\, e^{i\,m\,x} dm \tag{6.264}$$

Here

$$z_2 = z + H \tag{6.265}$$

As before, it is clear that the stationary point is absent, if $m > k_s$. In fact, in this case

$$h(m) = m \qquad \text{and} \qquad h'(m) \neq 0$$

Next suppose that stationary point m_{0s} is situated within the interval

$$k_l \leq m_{0s} < k_s$$

Then the phase function is equal to

$$m\,x + \sqrt{k_s^2 - m^2}\;z_2 \qquad \text{and} \qquad h(m) = m + \sqrt{k_s^2 - m^2}\;\frac{z_2}{x}$$

Taking the first derivative, we obtain

$$h'(m) = 1 - \frac{m_{0s}}{\sqrt{k_s^2 - m_{0s}^2}}\;\frac{z_2}{x}, \tag{6.266}$$

and the condition

$$h'(m_{0s}) = 0 \qquad \text{gives} \qquad m_{os} = \frac{k_s}{\sqrt{1 + \left(\frac{z_2}{x}\right)^2}}$$

As may be seen from Fig. 6.6c,

$$\frac{z_2}{x} = \tan\theta_0,$$

where θ_0 is the angle between the x-axis and radius r_2:

$$r_2 = \sqrt{x^2 + (z+H)^2} = \sqrt{x^2 + z_2^2} \tag{6.267}$$

Respectively, we have

$$m_{0s} = k_s \cos\;\theta_0 \tag{6.268}$$

Because the minimal value of the stationary point is k_l, angle θ_0 varies within the range

$$0 \leq \theta_0 < \cos^{-1}\frac{c_s}{c_l} \tag{6.269}$$

For the phase we have

$$\alpha\,h(m_{0s}) = x\;k_s \cos\;\theta_0 + \;k_s\;z_2 \sin\;\theta_0$$

or

$$\alpha\,h(m_{0s}) = \;k_s r_2 \tag{6.270}$$

As follows from eq. 6.266,

$$h''(m) = -\frac{z_2}{x} \frac{k_s^2}{(k_s^2 - m^2)^{3/2}}$$

and

$$h''(m_{0s}) = -\frac{z_2}{x} \frac{1}{k_s \sin\, \theta_0} = -\frac{1}{k_s \cos\, \theta_0} \tag{6.271}$$

For function $f(m_{0s})$ we have

$$f(m_0) = \frac{4\, i\, C_0 \cos\, \theta_0\, (2\cos^2\theta_0 - 1)\; e^{-\sqrt{k_s^2 \cos^2\theta_0 - k_l^2}H}}{k_s \left[(2\cos^2\theta_0 - 1)^2 + 2\, i \cos\theta_0 \sin 2\theta_0 \left(\cos^2\, \theta_0 - \frac{c_s^2}{c_l^2} \right)^{1/2} \right]} \tag{6.272}$$

Therefore

$$\tilde{\psi}_{1s} = \frac{4\, i\, \sqrt{2\pi} C_0 \cos\theta_0\, (2\cos^2\theta_0 - 1)\; e^{-\sqrt{k_s^2 \cos^2\theta_0 - k_l^2}H}\; e^{i(k_s r_2 - \pi/4)}}{(k_s r_2)^{1/2} \left[(2\cos^2\theta_0 - 1)^2 + 2\, i \cos\, \theta_0 \sin 2\theta_0 \left(\cos^2\, \theta_0 - \frac{c_s^2}{c_l^2} \right)^{1/2} \right]} \tag{6.273}$$

The presence of the factor $\exp\left(-\sqrt{k_s^2\cos^2\theta_0 - k_l^2}\; H\right)$ and $\exp(ik_s r_2)$ may suggest that function $\tilde{\psi}_1$ describes a shear wave that propagates through a medium with velocity c_s and is confined to the portion of the medium described by eq. 6.269. This wave has two more features, namely, it exponentially decays with an increase of distance between the real source and the boundary, and it behaves as if its fictitious source were located at point $(0, -H)$. The appearance of this wave can be imagined in the following way. When elementary longitudinal and inhomogeneous waves of the primary source reach the vicinity of point $(0, -H)$, they give rise to homogeneous plane waves of the S type. Their constructive interference produces the shear wave $\tilde{\psi}_{1s}$. With an increase of H, the amplitude of this wave rapidly becomes smaller.

Next we will demonstrate the presence of the second stationary point inside the interval

$$0 \le m \le k_l$$

and study the wave related to this point. If $m < k_l$, the phase function is

$$\alpha\; h(m) = m\; x + m_s\; z_2 + m_l\; H \tag{6.274}$$

This means that

$$h(m) = m + \sqrt{k_l^2 - m^2}\frac{z_2}{x} + \sqrt{k_s^2 - m^2}\frac{H}{x} \tag{6.275}$$

Its differentiation gives

$$\frac{\partial h(m)}{\partial m} = 1 - \frac{m}{\sqrt{k_s^2 - m^2}}\frac{z_2}{x} - \frac{m}{\sqrt{k_l^2 - m^2}}\frac{H}{x} \tag{6.276}$$

The stationary point m_{0s} is then defined from the equality

$$x = \frac{m_{0s}}{\sqrt{k_s^2 - m_{0s}^2}}\, z_2 + \frac{m_{0s}}{\sqrt{k_l^2 - m_{0s}^2}} H \tag{6.277}$$

In order to solve this equation, m_{0s}, we apply the geometric approach. As is seen from Fig. 6.6c,

$$x = H \tan\theta_i + z_2 \tan\theta_s \tag{6.278}$$

Here θ_s is the reflected angle of the shear wave. Comparison of eqs. 6.277 and 6.278 gives

$$\frac{m_{0s}}{\sqrt{k_s^2 - m_{0s}^2}} = \tan\theta_s \qquad \text{and} \qquad \frac{m_{0s}}{\sqrt{k_l^2 - m_{0s}^2}} = \tan\theta_i \tag{6.279}$$

From these equations we arrive at Snell's law for plane waves. In fact,

$$m_{0s} = k_s \sin\theta_s \qquad \text{or} \qquad m_{0s} = k_l \sin\theta_i, \tag{6.280}$$

and therefore

$$\frac{\sin\theta_i}{c_l} = \frac{\sin\theta_s}{c_s} \tag{6.281}$$

As a result, the phase function is

$$xh\left(m_{0s}\right) = \left(m_{0s}x_r + \sqrt{k_l^2 - m_{0s}^2}\, H\right) \tag{6.282}$$

$$+ \left[m_{0s}\left(x - x_r\right) + m_{0s}\, z_2\right] = k_l R_1 + k_s R_2,$$

where x_r, R_1, and R_2 are indicated in Fig. 6.6c. Differentiation of eq. 6.276 gives

$$h''(m) = -\frac{z_2}{x}\frac{k_s^2}{\left(k_s^2 - m^2\right)^{3/2}} - \frac{H}{x}\frac{k_l^2}{\left(k_l^2 - m^2\right)^{3/2}},$$

i.e.,

$$x\,h''(m_{0s}) = -\frac{z_2}{k_s\cos^3\,\theta_s} - \frac{H}{k_l\cos^3\,\theta_i} \tag{6.283}$$

or

$$x\,h''(m_{0s}) = -\frac{R_2}{k_s\cos^2\,\theta_s} - \frac{R_1}{k_l\cos^2\,\theta_i} \tag{6.284}$$

For function $f\,(m_{0s})$ we have

$$f\,(m_0) = \frac{4\,i\,C_0\sin\theta_s\left(2\sin^2\theta_s - 1\right)}{k_s\left[\left(2\sin^2\theta_s - 1\right)^2 + 4\,i\sin^2\theta_s\cos\theta_s\left(\sin^2\theta_s - \frac{c_s^2}{c_l^2}\right)^{1/2}\right]} \tag{6.285}$$

Therefore

$$\tilde{\psi} = \frac{4\,i\,C_0\sqrt{2\pi}\sin\theta_s\left(2\sin^2\theta_s - 1\right)}{\left[\left(2\sin^2\theta_s - 1\right)^2 + 4i\sin^2\theta_s\cos\,\theta_s\left(\sin^2\theta_s - \frac{c_s^2}{c_l^2}\right)^{1/2}\right]} \tag{6.286}$$

$$\times\frac{e^{i(k_lR_1 + k_sR_2 - \pi/4)}}{(k_sR_2)^{1/2}\left[\frac{1}{\cos^2\theta_s} + \frac{c_l}{c_s}\frac{R_1}{R_2\cos^2\theta_i}\right]^{1/2}}$$

It is obvious that the vector potential $\tilde{\psi}$ describes a cylindrical shear wave propagating along the radius vector $\mathbf{R}_2$ with velocity c_s.

Thus, applying the stationary-phase method we have distinguished two reflected waves – namely, longitudinal and shear waves – that obey Snell's law of reflection. Of course, they arise at different points of the free surface, but they arrive at the same observation point, Fig. 6.6d. Earlier we found out that when the source is located at the free boundary, various inhomogeneous waves, including the Rayleigh wave, appear. However, this method does not permit us to describe them, and for this reason we will apply the second approach.

Contour-integration method

As before, we use the path along branch cuts and around poles, Fig. 6.5a, and start from the scalar potential. In accordance with eq. 6.248, we have

$$\tilde{\varphi} = -C_0\,(I_s + I_l + I_p)\,, \tag{6.287}$$

where I_s, I_l and I_p are integrals around branch cuts and poles, respectively.

Integral I_s Integration along path C_1 and C_2, Fig. 6.5a, gives

$$I_s = \int_{k_s}^{k_s+i\infty} \frac{e^{-m_l z_1 + imx}}{m_l} \left[\frac{(2m^2-k_s^2)^2 + 4m^2 m_l\, m_s}{(2m^2-k_s^2)^2 - 4m^2 m_l\, m_s} - \frac{(2m^2-k_s^2)^2 - 4m^2 m_l\, m_s}{(2m^2-k_s^2)^2 + 4m^2 m_l\, m_s} \right] dm \tag{6.288}$$

since the radical m_s has opposite signs on lines C_1 and C_2. Thus,

$$I_s = 16 \int_{k_s}^{k_s+i\infty} \frac{m^2 m_s\,(2m^2-k_s^2)^2\, e^{-m_l\, z_1} e^{imx}}{(2m^2-k_s^2)^4 - 16m^4\, m_l^2\, m_s^2}\, dm \tag{6.289}$$

Replacement of the variable of integration

$$m = k_s + it$$

gives near branch point k_s

$$m = k_s, \quad (2m^2-k_s^2)^2 = k_s^4, \quad m_l = \sqrt{k_s^2 - k_l^2}, \quad \text{and} \quad m_s = \sqrt{2}\, e^{i\pi/4}\, k_s^{1/2}\, t^{1/2}$$

Because $dm = i\, dt$, we obtain an approximate expression of I_s :

$$I_s = \frac{16\sqrt{2}\, i\, e^{i\pi/4} e^{-\sqrt{k_s^2-k_l^2}\; z_1}}{k_s^{3/2}}\, e^{ik_s x} \int_0^\infty t^{1/2} e^{-xt} dt$$

Since

$$\int_0^\infty t^{1/2} e^{-x\,t} dt = \frac{\sqrt{\pi}}{2x^{3/2}},$$

we have

$$I_s = \frac{8\sqrt{2\pi}\; i\; e^{-\sqrt{k_s^2-k_l^2}\;\; z_1}}{(k_s x)^{3/2}}\, e^{i\,(k_s x + \pi/4)} \tag{6.290}$$

and

$$\widetilde{\varphi}_s = -C_0 \frac{8\sqrt{2\pi}\; i\; e^{-\sqrt{k_s^2-k_l^2}\;\; z_1}}{(k_s\; x)^{3/2}}\, e^{i\,(k_s x + \pi/4)} \tag{6.291}$$

This function represents an inhomogeneous longitudinal wave that propagates along the boundary with the velocity of the shear wave, c_s, and that exponentially decreases with parameter z_1. We considered this wave in the previous section, where we showed that its appearance results from the destructive interference of elementary longitudinal waves that arise at the free surface due to the shear wave. In other words, the shear wave plays the role of the moving source of this evanescent wave, $\widetilde{\varphi}_s$.

Integral I_l Integration along paths C_3 and C_4 can be written as

$$I_l = 2 \int\limits_{k_l}^{k_l+i\infty} \frac{1}{m_l} \frac{\left[(2m^2-k_s^2)^4 + 16m^4 m_l^2\, m_s^2\right] \cosh m_l\, z_1 e^{i\,m\,x}}{(2m^2-k_s^2)^4 - 16m^4m_l^2m_s^2}\, dm \tag{6.292}$$

$$-16 \int\limits_{k_l}^{k_l+i\infty} \frac{m^2 m_s\,(2m^2-k_s^2)^2 \sinh m_l\, z_1}{(2m^2-k_s^2)^4 - 16m^4m_l^2\, m_s^2}\, e^{i\,m\,x} dm,$$

because m_l has different signs on C_3 and C_4. After change of variable

$$m = k_l + it,$$

we obtain a sum of three integrals,

$$I_l = I_{1l} + I_{2l} + I_{3l}, \tag{6.293}$$

where

$$I_{1l} = \frac{i\sqrt{2}\, e^{-i\pi/4} e^{-\sqrt{k_s^2-k_l^2}z_1}}{k_l^{1/2}}\, e^{ik_l x} \int\limits_0^\infty \frac{\cosh(\alpha_l\sqrt{t})}{t^{1/2}}\, e^{-x\,t} dt,$$

$$I_{2l} = \frac{64\, i\sqrt{2}\, e^{i\pi/4}\left(1-\dfrac{k_s^2}{k_l^2}\right)}{k_l^{3/2}\left(2-\dfrac{k_s^2}{k_l^2}\right)^4}\, e^{i\,k_l\,x} \int\limits_0^\infty t^{1/2} \cosh\left(\alpha_l\sqrt{t}\right) e^{-x\,t} dt, \tag{6.294}$$

and $$I_{3l} = \frac{16\, i\left(1-\dfrac{k_s^2}{k_l^2}\right)^{1/2}}{k_l\left(2-\dfrac{k_s^2}{k_l^2}\right)^2}\, e^{i\,k_l\,x} \int\limits_0^\infty \sinh\left(\alpha_l\sqrt{t}\right) e^{-x\,t} dt$$

Here

$$\alpha_l = \sqrt{2k_l} e^{i\pi/4} z_1 \tag{6.295}$$

Since

$$\int_0^\infty \frac{\cosh \alpha_l \sqrt{t}}{t^{1/2}} \, e^{-x\,t} dt = \sqrt{\frac{\pi}{x}} e^{\alpha_l^2/4x}, \tag{6.296}$$

we have

$$I_{1l} = \frac{i\sqrt{2\pi}\, e^{-i\pi/4}}{(k_l x)^{1/2}} \, e^{-i\pi/4} \, e^{i\,k_l\,x} \, e^{\alpha_l^2/4x} \tag{6.297}$$

and

$$\widetilde{\varphi}_{1l} = -C_0 \, i \frac{\sqrt{2\pi}}{(k_l x)^{1/2}} \, e^{-i\pi/4} \, e^{i\,k_l\,x} \, e^{\alpha_l^2/4x} \tag{6.298}$$

The same result follows from eq. 6.262 when $\theta_i \to \pi/2$ or $z_1/x \to 0$, since $e^{i\,k_l\,x} e^{\alpha_l^2/4x} = e^{i\,k_l\,x} \left[1 + 1/2\,(z_1^2/x^2)\right] \approx e^{i\,k_l\,r_1}$. This means that function $\widetilde{\varphi}_{1l}$ characterizes the longitudinal wave reflected from the boundary, and its fictitious source is located at point O_1, Fig. 6.6b. However, unlike with the stationary-phase method, eq. 6.262, we have obtained an expression of $\widetilde{\varphi}$ that is valid only very far away from the source and close to the free surface. Differentiation of eq. 6.296 with respect to α_l gives

$$\int_0^\infty \sinh\left(\alpha_l \sqrt{t}\right) e^{-xt} dt = \sqrt{\frac{\pi}{x}} \frac{\partial}{\partial \alpha_l} e^{\alpha_l^2/4x}$$

or

$$\int_0^\infty \sinh\left(\alpha_l \sqrt{t}\right) e^{-xt} dt = \frac{1}{2} \frac{\sqrt{\pi}}{x^{3/2}} \, \alpha_l \, e^{\alpha_l^2/4x} \tag{6.299}$$

and

$$\int_0^\infty t^{1/2} \cosh\left(\alpha_l \sqrt{t}\right) e^{-xt} dt = \frac{\partial}{\partial \alpha_l} \int_0^\infty \sinh\left(\alpha_l \sqrt{t}\right) \, e^{-xt} dt$$

or

$$\int_0^\infty t^{1/2} \cosh\left(\alpha_l \sqrt{t}\right) e^{-xt} dt = \left(\frac{1}{2} \frac{\sqrt{\pi}}{x^{3/2}} + \frac{\sqrt{\pi}}{4\,x^{5/2}} \, \alpha_l^2\right) \, e^{\alpha_l^2/4x} \tag{6.300}$$

Substitution of eqs. 6.299 and 6.300 into eq. 6.293 shows that functions $\widetilde{\varphi}_{2l}$ and $\widetilde{\varphi}_{3l}$, as well as $\widetilde{\varphi}_{1l}$, describe the same reflected wave, including the next term of Debye's series, which is proportional to $1/(k_l\, x)^{3/2}$. This part of the field does not obey Snell's law, and elastic energy moves through the lateral surface of the ray tube.

Now we will derive asymptotic formulas for vector potential $\widetilde{\psi}$, which can be also written as a sum:

$$\widetilde{\psi} = C_0\,(M_s + M_l + M_p) \tag{6.301}$$

First, consider the integral along branch cuts C_1 and C_2.

Integral M_s As follows from eqs. 6.248 and 6.250,

$$M_s = \int\limits_{k_s}^{k_s+i\infty} 4im\,(2m^2 - k_s^2)\, e^{-m_l H}$$

$$\times \left[\frac{e^{-m_s\, z_2}}{(2m^2 - k_s^2)^2 - 4m^2 m_l\, m_s} - \frac{e^{m_s\, z_2}}{(2m^2 - k_s^2)^2 + 4m^2\, m_l\, m_s} \right] e^{i\,m\,x} dm$$

or

$$M_s = i \int\limits_{k_s}^{k_s+i\infty} \frac{32\, m^3 m_l\, m_s\,(2m^2 - k_s^2)\cosh m_l\, z_2 - 8m\,(2m^2 - k_s^2)^3 \sinh m_l\, z_2}{(2m^2 - k_s^2)^4 - 16m^4 m_l^2 m_s^2} \tag{6.302}$$

$$\times\, e^{-m_l\, H} e^{i\,m\,x} dm$$

Here

$$z_2 = z + H \tag{6.303}$$

After regular change of variables, we obtain an approximate expression of M_s that is valid in the far zone:

$$M_s = \left[-\frac{32\,\sqrt{2}\, e^{i\,\pi/4} \left(1 - \frac{k_l^2}{k_s^2}\right)^{1/2}}{k_s^{3/2}} \int\limits_0^\infty t^{1/2} e^{-xt} \cosh\left(\alpha_s \sqrt{t}\right) dt \right. \tag{6.304}$$

$$+ \frac{8}{k_s} \int_0^\infty \sinh\left(\alpha_s \sqrt{t}\right) dt \Bigg] \; e^{-\sqrt{k_s^2 - k_l^2}H} \; e^{i\, k_s x}$$

Here

$$\alpha_s = \sqrt{2}\; e^{i\pi/4}\; k_s^{1/2}\; z_2 \tag{6.305}$$

Taking into account eqs. 6.299 and 6.300, we obtain

$$\tilde{\psi}_s = \tilde{\psi}_{s1} + \tilde{\psi}_{s2},$$

where

$$\tilde{\psi}_{s1} = -C_0 \frac{16\sqrt{2\pi}\left(1 - \dfrac{k_l^2}{k_s^2}\right)^{1/2}}{(k_s x)^{3/2}} \left(1 + \frac{\alpha_s^2}{2x}\right) \; e^{-\sqrt{k_s^2 - k_l^2}H} \; e^{i\,(k_s x + \pi/4)} \; e^{\alpha_s^2/4x} \tag{6.306}$$

and $$\tilde{\psi}_{s2} = \frac{4\sqrt{2\pi}C_0}{(k_s x)^{3/2}} \frac{z_2}{x} \; e^{-\sqrt{k_s^2 - k_l^2}H} \; e^{i\,(k_s x + \pi/4)} \; e^{\alpha_s^2/4x}$$

This shows that the sum

$$\tilde{\psi}_{s1} + \tilde{\psi}_{s2}$$

describes the shear wave that may arise due to inhomogeneous elementary waves radiated by the primary source, as was discussed earlier.

Integral M_l By analogy with eq. 6.302, integration along paths C_3 and C_4 gives

$$M_l = i \int_{k_l}^{k_l + i\infty} \frac{32m^3\; m_l\; m_s\,(2m^2 - k_s^2)\cosh m_l H - 8m\,(2m^2 - k_s^2)^3 \sinh m_l H}{(2m^2 - k_s^2)^4 - 16m^4\; m_l^2\; m_s^2} \tag{6.307}$$

$$\times\, e^{-m_s\, z_2}\; e^{i\, m\, x} dm$$

Introducing variable $m = k_l + i\, t$, we have

$$M_l = \left[\frac{32\, \sqrt{2}\; e^{i\pi/4} \left(\dfrac{k_s^2}{k_l^2} - 1\right)^{1/2}}{k_l^{3/2}\left(2 - \dfrac{k_s^2}{k_l^2}\right)^3} \int_0^\infty t^{1/2} \cosh\left(\alpha_l H\right)\; e^{-x\, t} dt \right. \tag{6.308}$$

$$+\frac{8}{k_l\left(2-\dfrac{k_s^2}{k_l^2}\right)}\int\limits_0^\infty \sinh\left(\alpha_l H\right)\, e^{-xt}dt\Bigg]\, e^{i\sqrt{k_s^2-k_l^2}z_2}\, e^{i\,k_l x}$$

Here

$$\alpha_l=\sqrt{2}\, e^{i\pi/4}k_l^{1/2}H \tag{6.309}$$

Substitution of eqs. 6.299 and 6.300 gives

$$\tilde{\psi}_l = C_0 4\sqrt{2\pi}\left[\frac{4\left(\dfrac{k_s^2}{k_l^2}-1\right)^{1/2}}{\left(2-\dfrac{k_s^2}{k_l^2}\right)^3}\left(1+\frac{\alpha_l^2}{2x}\right)+\frac{k_l H}{\left(2-\dfrac{k_s^2}{k_l^2}\right)}\right] \tag{6.310}$$

$$\times\ \frac{e^{i\left(k_l x+\sqrt{k_s^2-k_l^2}\ z_2+\pi/4\right)}}{\left(k_l x\right)^{3/2}}\ e^{\alpha_l^2/4x}$$

It is obvious that eq. 6.310 describes the shear wave that moves along the free surface with velocity c_l. Its wavefront is the plane defined approximately by the equation

$$k_l x+\sqrt{k_s^2-k_l^2}\ z=\text{const},$$

and the velocity of propagation in the direction perpendicular to the wavefront is c_s. Thus, we are dealing with the already familiar conical wave that is generated by a longitudinal wave in the same manner as in the case of the source situated at the free boundary. Thus, applying the stationary-phase and contour-integration methods, we have learned that along with the direct wave, the following waves arrive at the observation point:

1. The reflected longitudinal wave, which obeys Snell's law at the free boundary

$$\theta_l=\theta_i$$

2. The reflected shear wave, which appears at a different point of the boundary, and which has angles of incidence and reflection that satisfy Snell's law:

$$\frac{\sin\theta_l}{c_l}=\frac{\sin\theta_s}{c_s}$$

Note that the stationary-phase method allows us to obtain the leading terms of Debye's series, which describe wave amplitude in a zero approximation – that is, in this approximation, energy flux is preserved inside the ray tube.

3. The inhomogeneous longitudinal wave $\tilde{\varphi}_l$, which propagates along the x-axis with the velocity of shear wave c_s. It is generated by the shear wave, and the destructive interference of elementary longitudinal waves plays an important role.

4. The shear conical wave $\tilde{\psi}_l$, which appears as the result of constructive interference of elementary shear waves and moves along the x-axis with velocity c_l. The longitudinal wave is the moving source of this wave.

5. The shear wave, which may arise due to the inhomogeneous elementary waves radiated by the primary source and propagates with velocity c_s.

6. In addition, we observe the Rayleigh wave, which is described by integrals I_p and M_p around the pole $m = k_R$.

Direct SV **wave** Now we assume that the linear source of the SV wave is located at depth H below the free surface. As follows from eqs. 6.61, the complex amplitudes of scalar and vector potentials describing the secondary waves are

$$\tilde{\varphi} = -C_1 \int\limits_{-\infty}^{\infty} \frac{4im\,(2m^2 - k_s^2)\; e^{-m_l H}\; e^{-m_s H}}{(2m^2 - k_s^2)^2 - 4m^2 m_l\, m_s}\; e^{-m_l\, z}\; e^{i\, m\, x}\, dm \tag{6.311}$$

and
$$\tilde{\psi} = -C_1 \int\limits_{-\infty}^{\infty} \frac{(2m^2 - k_s^2)^2 + 4m^2 m_l\, m_s}{m_s[(2m^2 - k_s^2)^2 - 4m^2 m_l\, m_s]}\; e^{-2m_s H}\; e^{-m_s\, z}\; e^{i\, m\, x}\, dm$$

Here

$$C_1 = \frac{r_0^2\, \tau_{r\theta}^0}{4\,\mu}, \tag{6.312}$$

and $\tau_{r\theta}^0$ is the amplitude of shear stress at the surface of the linear source with a very small radius. At the same time, in accordance with eqs. 6.248 and 6.250, we have for the case of the direct P wave

$$\tilde{\varphi} = -C_0 \int\limits_{-\infty}^{\infty} \frac{(2m^2 - k_s^2)^2 + 4m^2 m_l\, m_s}{m_l[(2m^2 - k_s^2)^2 - 4m^2 m_l\, m_s]}\; e^{-2m_l H}\; e^{-m_l\, z}\; e^{i\, m\, x}\, dm \tag{6.313}$$

and $$\tilde{\psi} = C_0 \int\limits_{-\infty}^{\infty} \frac{4im\,(2m^2 - k_s^2)\; e^{-m_l H}\; e^{-m_s H}}{(2m^2 - k_s^2)^2 - 4m^2 m_l\; m_s}\; e^{-m_s z}\; e^{i\,m\,x} dm$$

The similarity between the two cases is obvious.

6.6 Linear source in the presence of the boundary: elastic medium and fluid

Suppose that the linear source of the P wave is situated in the fluid at distance H above the boundary, Fig. 6.7a. Since shear waves are absent in the fluid, the wavefields are described only by scalar potential φ_1. In contrast, in an elastic medium we need both scalar and vector potentials, φ_2 and ψ_2. At the boundary, the normal components of stress and displacement are continuous functions, whereas shear stress vanishes:

$$\tau_{zz}^{(1)} = \tau_{zz}^{(2)}, \qquad \tau_{xz}^{(2)} = 0 \qquad s_z^{(1)} = s_z^{(2)} \tag{6.314}$$

Taking into account results obtained in section 6.1, in place of eq. 6.314 we have

$$-\lambda_1 k_1^2 \tilde{\varphi}_1 = -\lambda_2\; k_l^2 \tilde{\varphi}_2 + 2\mu \left(\frac{\partial^2 \tilde{\varphi}_2}{\partial z^2} + \frac{\partial^2 \tilde{\psi}_2}{\partial x\, \partial z} \right),$$

$$2\,\frac{\partial^2 \tilde{\varphi}_2}{\partial x\, \partial z} + \frac{\partial^2 \tilde{\psi}_2}{\partial x^2} - \frac{\partial^2 \tilde{\psi}_2}{\partial z^2} = 0, \tag{6.315}$$

$$\frac{\partial \tilde{\varphi}_1}{\partial z} = \frac{\partial \tilde{\varphi}_2}{\partial z} + \frac{\partial \tilde{\psi}_2}{\partial x} \qquad \text{if} \qquad z = H$$

Here

$$\lambda_1 = c_1^2\; \rho_1, \qquad (\lambda_2 + 2\mu_2) = c_l^2\; \rho_2, \qquad \mu_2 = \rho_2 c_s^2 \tag{6.316}$$

As usual, the complex amplitudes of potentials can be written in the form

$$\tilde{\varphi}_1 = A_0 \left[\int\limits_{-\infty}^{\infty} \frac{e^{-m_1\,|z|}}{m_1} e^{i\,m\,x} dm + \int\limits_{-\infty}^{\infty} A_m\; e^{m_1 z}\; e^{i\,m\,x} dm \right], \tag{6.317}$$

$$\tilde{\varphi}_2 = A_0 \int\limits_{-\infty}^{\infty} B_m\; e^{-m_l\; z}\; e^{i\,m\,x} dm, \qquad \tilde{\psi}_2 = A_0 \int\limits_{-\infty}^{\infty} C_m\; e^{-m_s\; z}\; e^{i\,m\,x} dm,$$

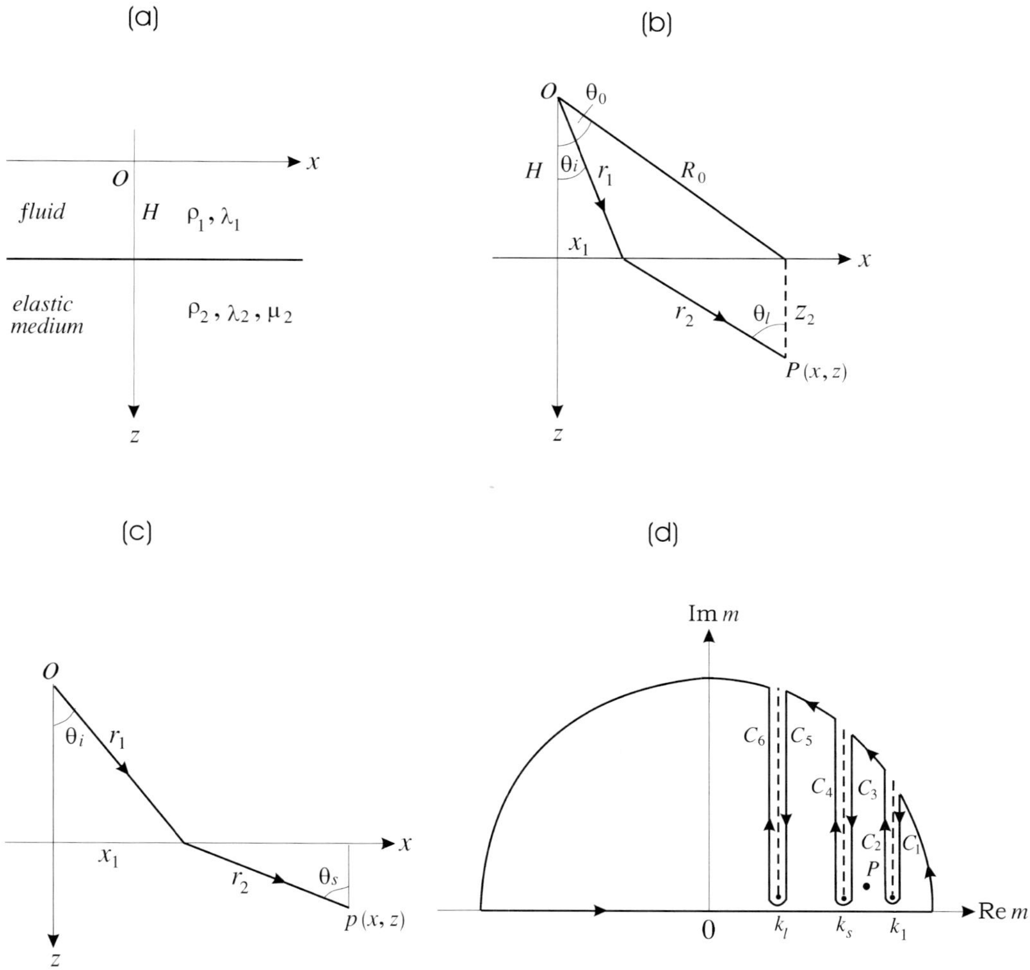

Figure 6.7: (a) Linear source in fluid above boundary with elastic medium (b) Illustration of wave φ_2 in far zone (c) Ray of shear transmitted wave $\psi_2(m_{01})$ (d) Integration along branch lines and around poles

where

$$m_1 = \sqrt{m^2 - k_1^2}, \qquad m_l = \sqrt{m^2 - k_l^2}, \qquad m_s = \sqrt{m^2 - k_s^2}$$

From eq. 6.9–6.15, it follows that

$$A_0 = \frac{\tau_{rr}^0}{2\rho_1 \ln r_0} \tag{6.318}$$

Substitution of eqs. 6.317 into eqs. 6.315 gives

$$-\lambda_1 k_1^2 \left(\frac{e^{-m_1 H}}{m_1} + A_m e^{m_1 H} \right) = -\lambda_2 \, k_l^2 B_m e^{-m_l H}$$

$$+ \, 2\mu_2 \left(m_l^2 B_m e^{-m_l H} - im \; m_s C_m e^{-m_s H} \right),$$

$$-2i \; m \; m_l \; e^{-m_l H} B_m - \; m^2 C_m \; e^{-m_s H} - m_s^2 e^{-m_s H} C_m = 0, \tag{6.319}$$

$$-e^{-m_1 H} + m_1 A_m \; e^{m_1 H} = -m_l B_m \; e^{-m_l H} + i \; m \; C_m \; e^{-m_s H}$$

The first two equations in set 6.319 can be slightly simplified. For the first we have

$$-\lambda_1 k_1^2 \left(\frac{e^{-m_1 H}}{m_1} - A_m \; e^{m_1 H} \right) = \mu_2 \left[\left(2m^2 - k_s^2\right) B_m \; e^{-m_l H} - 2i \; m \; m_s \; C_m \; e^{-m_s H} \right]$$

or

$$-n \; k_s^2 \left(\frac{e^{-m_1 H}}{m_1} + \; e^{m_1 H} A_m \right) = \left(2m^2 - k_s^2\right) B_m \; e^{-m_l H} - 2i \; m \; m_s \; C_m \; e^{-m_s H} \tag{6.320}$$

Also

$$2i \; m \; m_l \; e^{-m_l H} B_m + \left(2m^2 - k_s^2\right) C_m \; e^{-m_s H} = 0 \tag{6.321}$$

$$-e^{-m_1 H} + m_1 A_m \; e^{-m_1 H} = -m_l B_m \; e^{-m_l H} + im \; C_m \; e^{-m_s H} \tag{6.322}$$

Here

$$n = \frac{\rho_1}{\rho_2} \tag{6.323}$$

From eq. 6.321 we have

$$C_m = -\frac{2i\ m\ m_l}{(2m^2 - k_s^2)} e^{-m_l H} e^{m_s H}\ B_m \tag{6.324}$$

Then eq. 6.323 becomes

$$e^{-m_1 H} - m_1\ A_m\ e^{m_1 H} = \left(m_l - \frac{2m^2\ m_l}{2m^2 - k_s^2}\right)\ B_m\ e^{-m_l H}$$

or

$$e^{-m_1 H} - m_1 A_m\ e^{m_1 H} = -\frac{k_s^2\ m_l}{2m^2 - k_s^2}\ B_m\ e^{-m_l H} \tag{6.325}$$

Also, the right side of eq. 6.320 can be represented as

$$\left[(2m^2 - k_s^2) - \frac{4m^2 m_l\ m_s}{2m^2 - k_s^2}\right]\ e^{-m_l H} B_m = \frac{D}{2m^2 - k_s^2}\ e^{-m_l H} B_m,$$

where

$$D = \left(2m^2 - k_s^2\right)^2 - 4m^2 m_l\ m_s$$

Thus, we have a system for the determination of A_m and B_m:

$$-n\ k_s^2 \left(\frac{e^{-m_1 H}}{m_1} + e^{-m_1 H} A_m\right) = \frac{D}{2m^2 - k_s^2}\ e^{-m_l H}\ B_m \tag{6.326}$$

and $$e^{-m_1 H} - m_1 A_m\ e^{m_1 H} = -\frac{k_s^2\ m_l}{2m^2 - k_s^2}\ e^{-m_l H}\ B_m$$

Solution of the system gives

$$B_m = -\frac{2n\ k_s^2\ (2m^2 - k_s^2)\ e^{-m_1 H}\ e^{m_l H}}{D_1} \tag{6.327}$$

Here

$$D_1 = m_1 D + m_l\ n\ k_s^4 \tag{6.328}$$

Thus

$$C_m = \frac{4i\ m\ m_l\ n\ k_s^2\ e^{-m_1 H} e^{m_s H}}{D_1}, \tag{6.329}$$

and from eqs. 6.326 we have

$$A_m = \frac{1}{m_1} \frac{m_1 D - n\ k_s^4\ m_l}{D_1}\ e^{-2m_1 H} \tag{6.330}$$

In illustration, consider a case in which the second medium is also a fluid. Then $\mu_2 \to 0$, and therefore

$$c_s \to 0 \qquad \text{but} \qquad k_s \to \infty$$

Correspondingly, eqs. 6.327–6.330 give

$$B_m = \frac{2n\ e^{-m_1 H}\ e^{m_l H}}{m_1 + m_l\ n}, \qquad C_m = 0, \qquad A_m = \frac{m_1 - n\ m_l}{m_1 + n\ m_l}\ e^{-2m_1 H}$$

Now we will start to derive asymptotic formulas in the far zone, applying the stationary-phase method and then contour integration. To begin with, assume that distance x along the boundary is much greater than the wavelengths.

1. The stationary-phase method

First consider waves in a fluid. In accordance with eq. 6.317, the secondary potential is

$$\tilde{\varphi}_{1s} = A_0 \int\limits_{-\infty}^{\infty} \frac{1}{m_1} \frac{m_1 D - n\ k_s^4\ m_l}{m_1 D + n\ k_s^4\ m_l} e^{-m_1 z_1} e^{i\ m\ x} dm \tag{6.331}$$

Here

$$z_1 = 2H - z > 0$$

and

$$D = \left(2m^2 - k_s^2\right)^2 - 4m^2\ m_l\ m_s$$

We start from equality 6.243:

$$I(\alpha) = \int\limits_{-\infty}^{\infty} f(m)\ e^{i\ \alpha\ h\ (m)} dm = f(m_0)\ e^{i\ \alpha\ h\ (m_0)} \left(\frac{2\pi}{\alpha h''(m_0)}\right)^{1/2} e^{\pm\ i\pi/4}$$

Since the stationary point m_0 is absent when $m > k_1$, we have for the phase function

$$x\,h(m) = m\,x + \sqrt{k_1^2 - m^2}\,z_1 \tag{6.332}$$

and, correspondingly,

$$h(m) = m + \frac{z_1}{x}\sqrt{k_1^2 - m^2} \tag{6.333}$$

$$h'(m) = 1 - \frac{z_1}{x}\frac{m_0}{\sqrt{k_1^2 - m_0^2}} = 0$$

Thus

$$m_0 = \frac{k_1 x}{r_1} = k_1 \sin\theta_i \tag{6.334}$$

As before, m_0 depends on the wavenumber and the angle of incidence θ_i. Here

$$r_1 = \sqrt{x^2 + (2H - z)^2} \tag{6.335}$$

is the distance between an observation point and the mirror reflection of the origin, O, with respect to the interface. Since $m_0 < k_1$, the secondary wave is mainly defined by homogeneous elementary plane waves.

Taking derivatives of $h(m)$, we obtain

$$h''(m_0) = -\frac{z_1}{x\,k_1\cos^3\theta_i} = -\frac{1}{k_1\sin\theta_i\cos^2\theta_i} \tag{6.336}$$

For function $f(m_0)$, eq. 6.331, we have:

$$f(m_0) = \frac{A_0}{m_1}\frac{m_1 D - n\,k_s^4\,m_l}{m_1 D + n\,k_s^4\,m_l} = \frac{iA_0}{k_1\cos\theta_i}\frac{M_1}{N_1} \tag{6.337}$$

Here

$$M_1 = \cos\theta_i\left[\left(2\sin^2\theta_i - \frac{c_1^2}{c_s^2}\right)^2 - 4\sin^2\theta_i\left(\sin^2\theta_i - \frac{c_1^2}{c_l^2}\right)^{1/2}\left(\sin^2\theta_i - \frac{c_1^2}{c_s^2}\right)^{1/2}\right]$$

$$-\,n\,k_s^4\left(\frac{c_1^2}{c_l^2} - \sin^2\theta_i\right)^{1/2} \qquad \text{and} \tag{6.338}$$

$$N_1 = \cos\theta_i\left[\left(2\sin^2\theta_i - \frac{c_1^2}{c_s^2}\right)^2 - 4\sin^2\theta_i\left(\sin^2\theta_i - \frac{c_1^2}{c_l^2}\right)^{1/2}\left(\sin^2\theta_i - \frac{c_1^2}{c_s^2}\right)^{1/2}\right]$$

$$+ nk_s^4 \left(\frac{c_1^2}{c_l^2} - \sin^2 \theta_i \right)^{1/2}$$

Thus, the asymptotic expression for scalar potential in a fluid, $\tilde{\varphi}_{1s}$, is

$$\tilde{\varphi}_{1s} = A_0 \, \frac{M_1}{N_1} \sqrt{2\pi} \frac{1}{(k_1 r_1)^{1/2}} \, e^{i\pi/4} \, e^{\, ik_1 r_1} \tag{6.339}$$

This characterizes the reflected wave at the observation point, and the direction of its ray corresponds to Snell's law: $\theta_r = \theta_i$. As is seen from eqs. 6.338, if the angle of incidence does not exceed the critical angle

$$\sin \theta_i < \frac{c_1}{c_l}$$

the amplitude of the reflected wave at the interface is smaller than that of the direct wave. Beyond the critical angle for the shear wave, the amplitudes of the incident and reflected waves become equal at these points. However, there is a phase shift between them. This shows that total internal reflection has taken place.

Scalar potential, $\tilde{\varphi}_2$, in an elastic medium

From eqs. 6.317 and 6.327 we have

$$\tilde{\varphi}_2 = -2n \, k_s^2 \, A_0 \int\limits_{-\infty}^{\infty} \frac{(2m^2 - k_s^2) \; e^{-m_1 H} e^{-m_l \, z_2}}{m_1 D + m_l \, n \, k_s^4} \; e^{i \, m \, x} dm \tag{6.340}$$

Here $z_2 = z - H$. Suppose that

$$c_1 < c_s < c_l \qquad \text{or} \qquad k_1 > k_s > k_l \tag{6.341}$$

As will be shown, there are two stationary points. Consider first the contribution of point m_{01}, which is located within interval

$$0 \leq \; m_{01} < \; k_l \tag{6.342}$$

Then, the phase function is

$$x \; h(m) = x \; m + \sqrt{k_1^2 - m^2} H + \sqrt{k_l^2 - m^2} \; z_2$$

Correspondingly,

$$h(m) = m + \frac{H}{x}\sqrt{k_1^2 - m^2} + \frac{z_2}{x}\sqrt{k_l^2 - m^2} \tag{6.343}$$

and

$$h'(m) = 1 - \frac{H}{x}\frac{m}{\sqrt{k_1^2 - m^2}} - \frac{z_2}{x}\frac{m}{\sqrt{k_l^2 - m^2}} \tag{6.344}$$

Thus,

$$x = H\frac{m_{01}}{\sqrt{k_1^2 - m_{01}^2}} + z_2\frac{m_{01}}{\sqrt{k_l^2 - m_{01}^2}} \tag{6.345}$$

As is seen from Fig. 6.7b,

$$\frac{m_{01}}{\sqrt{k_1^2 - m_{01}^2}} = \tan\theta_i \qquad \text{and} \qquad \frac{m_{01}}{\sqrt{k_l^2 - m_{01}^2}} = \tan\theta_l,$$

whence

$$m_{01} = k_1 \sin\theta_i \qquad \text{and} \qquad m_{01} = k_l \sin\theta_l \tag{6.346}$$

Here θ_l is the angle of refraction of the longitudinal wave, and according to Snell's law,

$$\frac{\sin\theta_i}{c_1} = \frac{\sin\theta_l}{c_l}$$

Thus, we have proved the presence of a stationary point at the initial interval of integration, and, as follows from eqs. 6.342 and 6.346, the angle of incidence does not exceed the critical angle:

$$\sin\theta_i < \frac{k_l}{k_1} = \frac{c_1}{c_l}$$

Taking into account eqs. 6.343 and 6.346, the phase is equal to

$$x\, h\,(m_{01}) = k_1\, x_1 \sin\theta_i + k_1 H\cos\theta_i + k_l\,(x - x_1)\sin\theta_l + k_l\, z_2\cos\theta_l, \tag{6.347}$$

and x_1 is shown in Fig. 6.7b. Therefore,

$$e^{i\,x\,h\,(m_{01})} = e^{i\,k_1 r_1}\, e^{i\,k_l\,r_2}, \tag{6.348}$$

where

$$r_1 = \left(H^2 + x_1^2\right)^{1/2}, \qquad r_2 = \left[z_2^2 + (x - x_1)^2\right]^{1/2}$$

Eq. 6.348 shows that when the incident wave reaches the point of the boundary (x_1, H), it gives rise to the transmitted longitudinal wave. Correspondingly, the total change of phase is defined by eq. 6.347. Differentiation of eq. 6.344 yields

$$h''(m) = -\frac{k_1^2}{(k_1^2 - m^2)^{3/2}}\frac{H}{x} - \frac{k_l^2}{(k_l^2 - m^2)^{3/2}}\frac{z_2}{x} \tag{6.349}$$

and $$h''(m_{01}) = -\frac{1}{k_l}\left(\frac{H}{x}\frac{c_l}{c_1}\cos^{-3}\theta_i + \frac{z_2}{x}\cos^{-3}\theta_l\right)$$

Finally, for function $f(m_{01})$ we have

$$f(m_{01}) = \frac{2n\ i}{k_l}\frac{c_l^2}{c_s^2}A_1\frac{2\sin^2\theta_l - \dfrac{c_l^2}{c_s^2}}{N_2}, \tag{6.350}$$

where

$$N_2 = \left(\frac{c_l^2}{c_1^2} - \sin^2\theta_l\right)^{1/2}\left[\left(2\sin^2\theta_l - \frac{c_l^2}{c_s^2}\right)^2 + \sin^2\theta_l\cos\theta_l\left(\frac{c_l^2}{c_s^2} - \sin^2\theta_l\right)^{1/2}\right] \tag{6.351}$$

$$+\ n\frac{c_l^2}{c_s^2}\cos\theta$$

Thus, the asymptotic expression for a longitudinal wave in an elastic medium is

$$\tilde{\varphi}_2 = 2\sqrt{2\pi}\ n\ \frac{c_l^2}{c_s^2}A_0\frac{2\sin^2\theta_l - \dfrac{c_l^2}{c_s^2}}{N_2}e^{i\pi/4}\frac{1}{(k_l x)^{1/2}}e^{i(k_1 r_1 + k_l\ r_2)} \tag{6.352}$$

$$\times\frac{1}{\left(\dfrac{H}{x}\dfrac{c_l}{c_1}\cos^{-3}\theta_i + \dfrac{z_2}{x}\cos^{-3}\theta_l\right)^{1/2}},$$

and its amplitude represents a zero approximation of Debye's series. Eq. 6.352 is valid if

$$0 \le \theta_i \le \frac{c_1}{c_l}$$

In accordance with Snell's law, this means that the transmitted wave, $\tilde{\varphi}_2$, arrives at any point of the lower medium, Fig. 6.7b.

Next we will demonstrate the existence of the second stationary point, which is located within interval

$$k_l \le m_{02} \le k_1,$$

and evaluate its contribution. Since in this case the radical $(m^2 - k_l^2)^{1/2}$ is real, the phase function is

$$x\ h(m) = xm + \sqrt{k_1^2 - m^2}\ H$$

and

$$h(m) = m + \sqrt{k_1^2 - m^2}\ \frac{H}{x} \tag{6.353}$$

Hence

$$h'(m) = 1 - \frac{m}{\sqrt{k_1^2 - m^2}}\ \frac{H}{x}, \tag{6.354}$$

which gives

$$m_{02} = k_1 \sin \theta_0, \tag{6.355}$$

where θ_0 is shown on Fig. 6.7b. As follows from eq. 6.353, at the stationary point the phase of the integrand in eq. 6.340 is equal to

$$x\ h(m_{02}) = k_1 x \sin \theta_0 + k_1 H \cos \theta_0 = k_1 R_0 \tag{6.356}$$

Here R_0 is the distance between the primary source and the point of the boundary (x, H) located above the observation point (x, H). In accordance with eq. 6.340, we have

$$f(m_{02}) = -2n\ A_0\ \frac{c_1^2}{c_s^2}\ \frac{\left(2 \sin^2 \theta_0 - \frac{c_1^2}{c_s^2}\right)}{k_1 N_3}\ e^{-k_1 \sqrt{\sin^2 \theta_0 - c_1^2/c_l^2}\ z_2} \tag{6.357}$$

Here

$$N_3 = -i \cos \theta_0 \left[\left(2 \sin^2 \theta_0 - \frac{c_1^2}{c_s^2}\right)^2 - 4 \sin^2 \theta_0 \left(\sin^2 \theta_0 - \frac{c_1^2}{c_l^2}\right)^{1/2} \left(\sin^2 \theta_0 - \frac{c_1^2}{c_s^2}\right)^{1/2} \right]$$

$$+ n \left(\frac{c_1}{c_s}\right)^4 \left(\sin^2 \theta_0 - \frac{c_1^2}{c_l^2}\right)^{1/2} \tag{6.358}$$

Differentiation of eq. 6.354 gives

$$h''(m_{02}) = -\frac{H}{x\ k_1} \cos^{-3} \theta_0 \tag{6.359}$$

Thus, the complex amplitude of scalar potential associated with m_{02} is

$$\widetilde{\varphi}_2(m_{02}) = \frac{-2\sqrt{2\pi}\ n A_0 \left(\frac{c_1}{c_s}\right)^2 \left(2 \sin^2 \theta_0 - \frac{c_1^2}{c_s^2}\right)}{(k_1 x)^{1/2} \left(\frac{H}{x} \cos^{-3} \theta_0\right)^{1/2} N_3} \tag{6.360}$$

$$\times \exp\left[-k_1 z_2 \left(\sin^2\theta_0 - \frac{c_1^2}{c_l^2}\right)^{1/2} + i\,(k_1 R_0 - \pi/4)\right]$$

Since $m_{02} > k_l$, we have

$$\sin\theta_0 > \frac{c_1}{c_l}$$

In other words, the angle of incidence exceeds the critical angle, and function $\tilde{\varphi}_2\,(m_{02})$ describes the longitudinal wave, which exponentially decays with distance from the interface. As in the case of plane waves, we observe one of the features of total internal reflection, namely, the transmitted P wave becomes inhomogeneous. Its velocity of propagation along the boundary is now the function of x,

$$c = \frac{c_1}{\sin\theta_0}$$

and it varies within the range $c_1 \le c \le c_l$. From eq. 6.355 it follows that this wavefield appears at points of the boundary where

$$\frac{x}{(x^2 + H^2)^{1/2}} > \frac{c_1}{c_l}, \tag{6.361}$$

that is where the angle of incidence is greater the critical angle, $\theta_i > \theta_c$. Therefore,

$$\tilde{\varphi}_2 = \tilde{\varphi}_2\,(m_{01}) \tag{6.362}$$

if the x-coordinate of an observation point is such that

$$\frac{x}{\sqrt{x^2 + H^2}} < \frac{c_1}{c_l},$$

and

$$\tilde{\varphi}_2 = \tilde{\varphi}_2\,(m_{01}) + \tilde{\varphi}_2\,(m_{02}) \tag{6.363}$$

provided that x satisfies equality 6.361. In this last case, there is superposition of the transmitted wave, which obeys Snell's law, and the inhomogeneous wave. Both of them are caused by the incident wave, but they arise at different points of the boundary.

Vector potential, $\tilde{\psi}_2$

From eqs. 6.317 and 6.329, we have

$$\tilde{\psi}_2 = 4i\;n\;k_s^2\;A_0 \int\limits_{-\infty}^{\infty} \frac{m\;m_l\;e^{-m_1 H}e^{-m_s\;z_2}}{m_l D + m_l n\;k_s^4}\;e^{i\;m\;x}dm \tag{6.364}$$

As in the case of scalar potential $\widetilde{\varphi}_2$, there are two stationary points, m_{01} and m_{02}. Consider first the contribution of point m_{01}, which is located inside the interval

$$0 \le m_{01} \le k_s \tag{6.365}$$

Then the phase function is

$$x\ h(m) = m\ x + \sqrt{k_1^2 - m^2}\ H + \sqrt{k_s^2 - m^2}\ z_2 \tag{6.366}$$

Performing a differentiation, we obtain an equation for the first stationary point:

$$x = H\ \frac{m_{01}}{\sqrt{k_1^2 - m_{01}^2}} + z_2\ \frac{m_{01}}{\sqrt{k_s^2 - m_{01}^2}} \tag{6.367}$$

As is seen from Fig. 6.7c,

$$m_{01} = k_1 \sin\theta_i \qquad \text{or} \qquad m_{01} = k_s \sin\theta_s \tag{6.368}$$

Here θ_s is the angle of refraction. It is obvious that eq. 6.368 also describes Snell's law for the transmitted shear wave, and it allows us to represent the phase function as

$$x\ h(m_{01}) = k_1 x_1 \sin\theta_i + k_1 H \cos\theta_i + k_s\,(x - x_1) \sin\theta_s + k_s\ z_2 \cos\theta_s \tag{6.369}$$

Note that x_1 is the coordinate of the boundary point from which this wave arises and then arrives at the observation point. Of course, in eqs. 6.347 and 6.369, the values of x_1 differ from each other. From eq. 6.369 we have

$$e^{i\ x\ h\ (m_{01})} = e^{i\ k_1 r_1} e^{i\ k_s r_2} \tag{6.370}$$

Here r_1 and r_2 are shown in Fig. 6.7c. This means that function $\widetilde{\psi}_2(m_{01})$ describes the transmitted wave, which obeys Snell's law.

$$\frac{\sin\theta_i}{c_1} = \frac{\sin\theta_s}{c_s} = \frac{\sin\theta_l}{c_l} \tag{6.371}$$

By analogy with eq. 6.349, we have

$$h''\,(m_{01}) = -\frac{1}{k_s}\left(\frac{H}{x}\ \frac{c_1}{c_s}\cos^{-3}\theta_i + \frac{z_2}{x}\cos^{-3}\theta_s\right) \tag{6.372}$$

Function $f\,(m_{01})$ is equal to

$$f\,(m_{01}) = \frac{4n\ i\ A_0 \sin\theta_s \left(\sin^2\theta_s - \frac{c_s^2}{c_l^2}\right)^{1/2}}{k_s N_4} \tag{6.373}$$

where

$$N_4 = -i\left(\frac{c_s^2}{c_1^2} - \sin^2\theta_s\right)^{1/2}\left[\left(2\sin^2\theta_s - 1\right)^2 + 4i\sin^2\theta_s\cos\theta_s\left(\sin^2\theta_s - \frac{c_s^2}{c_l^2}\right)^{1/2}\right]$$

$$+\, n\left(\sin^2\theta_s - \frac{c_s^2}{c_l^2}\right)^{1/2} \tag{6.374}$$

Thus,

$$\tilde{\psi}_2\left(m_{01}\right) = \frac{4\,n\,i\,A_0\sqrt{2\pi}\,e^{-i\pi/4}\,e^{i\left(k_1 r_1 + k_s r_2\right)}}{\left(k_s\,x\right)^{1/2} N_4\left(\frac{H}{x}\frac{c_1}{c_s}\cos^{-3}\theta_i + \frac{z_2}{x}\cos^{-3}\theta_s\right)^{1/2}} \tag{6.375}$$

This wave arises at points of the boundary where

$$\sin\theta_i \leq \frac{c_1}{c_s},$$

and it exists everywhere in an elastic medium.

Now consider the contribution of the second stationary point m_{02}, which is situated within interval

$$k_s < m_{02} < k_1 \tag{6.376}$$

Then the function $h(m)$ is

$$h(m) = m + \frac{H}{x}\left(k_1^2 - m^2\right)^{1/2},$$

which coincides with eq. 6.353. Correspondingly,

$$m_{02} = k_1\sin\theta_0, \qquad h''\left(m_{02}\right) = -\frac{H}{x\,k_1}\cos^{-3}\theta_0 \tag{6.377}$$

and $\qquad xh\left(m_{02}\right) = k_1 R_0$

Also, we have

$$f(m_{02}) = 4n\,A_0\left(\frac{c_1}{c_s}\right)^2\frac{i}{k_1}\frac{\sin\theta_0\left(\sin^2\theta_0 - \frac{c_1^2}{c_l^2}\right)^{1/2}}{N_5}\,e^{-k_1\sqrt{\sin^2\theta_0 - c_1^2/c_s^2}\,z_2} \tag{6.378}$$

Here

$$N_5 = -i\cos\theta_0 \left[\left(2\sin^2\theta_0 - \frac{c_1^2}{c_s^2} \right)^2 - 4\sin^2\theta_0 \left(\sin^2\theta_0 - \frac{c_1^2}{c_l^2} \right)^{1/2} \left(\sin^2\theta_0 - \frac{c_1^2}{c_s^2} \right)^{1/2} \right]$$

$$+ n \left(\sin^2\theta_0 - \frac{c_1^2}{c_l^2} \right)^{1/2} \left(\frac{c_1}{c_s} \right)^4 \tag{6.379}$$

Thus, potential $\tilde{\psi}_2(m_{02})$ is

$$\tilde{\psi}_2(m_{02}) = \frac{4i\sqrt{2\pi} n A_0 c_1^2 \sin\theta_0 \left(\sin^2\theta_0 - \dfrac{c_1^2}{c_l^2} \right)^{1/2}}{(k_1 x)^{1/2}\ c_s^2\ N_5 \left(\dfrac{H}{x} \cos^{-3}\theta_0 \right)^{1/2}} \tag{6.380}$$

$$\times\ e^{-k_1 \sqrt{\sin^2\theta_0 - c_1^2/{c_s}^2}\ z_2 + i\,(k_1 R_0 - \pi/4)}$$

This potential may describe the evanescent shear wave that decays exponentially with distance from the boundary. As follows from eq. 6.377, velocity of propagation of this shear wave along the x-axis is

$$c = \frac{c_1}{\sin\theta_0}$$

and varies within the range $c_s < c < c_l$. In the same manner as in the case of the longitudinal wave, we have

$$\tilde{\psi}_2 = \tilde{\psi}_2(m_{01}) \qquad \text{if} \quad \frac{x_1}{r_1} < \sin\theta_c \tag{6.381}$$

$$\text{and} \qquad \tilde{\psi}_2 = \tilde{\psi}_2(m_{01}) + \tilde{\psi}_2(m_{02}) \qquad \text{if} \quad \frac{x_1}{r_1} \geq \sin\theta_c$$

Comparison of eqs. 6.360 and 6.380 shows that scalar and vector potentials $\tilde{\varphi}_2(m_{02})$ and $\tilde{\psi}_2(m_{02})$ characterize the wavefields, which produce both deformation and rotation of elementary volumes of a medium.

We see that the stationary-phase method allows us to describe reflection and transmission of waves obeying Snell's law as well as of inhomogeneous waves. In deriving asymptotic formulas for reflected and transmitted waves that obey Snell's law, we have paid attention to large distances along the boundary $(k\,x \gg 1)$. However, the same formulas are valid when either H or z, or both, exceed the wavelength, but offset x may be small. In essence, the magnitude of potentials describing these waves does not depend on x.

Next we will apply the second approximate method for seeing the presence of other waves.

2. Contour-integration method

In the same manner as earlier, we replace integration along the real axis of m with integration along branch lines and around poles, Fig. 6.7d. First consider wavefields associated with branch points in a fluid medium.

Scalar potential, $\widetilde{\varphi}_{1s}$

In accordance with eq. 6.331, we have

$$\widetilde{\varphi}_{1s} = A_1 \int_{-\infty}^{\infty} \frac{1}{m_1} \frac{m_1 D - n\, k_s^4\, m_l}{m_1 D + n\, k_s^4\, m_l}\, e^{-m_1 z_1}\, e^{imx} dm = A_1 \left(I_1 + I_s + I_l + I_p \right) \tag{6.382}$$

where I_i is an integral along a new path and

$$D = \left(2m^2 - k_s^2\right)^2 - 4m^2 m_l\, m_s, \qquad z_1 = 2H - z$$

Further we use several times the known integrals

$$\int_0^\infty x^{1/2}\, e^{-x\,t} dt = \frac{\sqrt{\pi}}{2x^{3/2}}, \qquad \int_0^\infty \frac{\cosh\left(\alpha\sqrt{t}\right)}{t^{1/2}}\, e^{-x\,t} dt = \sqrt{\frac{\pi}{x}} e^{\alpha^2/4x} \tag{6.383}$$

$$\int_0^\infty \sinh\left(\alpha\sqrt{t}\right) e^{-x\,t} dt = \frac{1}{2} \frac{\sqrt{\pi}}{x^{3/2}} \alpha\, e^{\alpha^2/4x},$$

$$\int_0^\infty t^{1/2} \cosh\left(\alpha\sqrt{t}\right)\, e^{-x\,t} dt = \frac{\sqrt{\pi}}{2x^{3/2}} \left(1 + \frac{\alpha^2}{2x}\right) e^{\alpha^2/4x}$$

Integral I_1 Since the radical m_1 changes sign around point k_1, integration along paths C_1 and C_2 gives

$$I_1 = \int_{k_1}^{k_1 + i\infty} \frac{1}{m_1} \left[\frac{m_1 D - n\, k_s^4\, m_l}{m_1 D + n\, k_s^4\, m_l}\, e^{-m_1 z_1} - \frac{m_1 D + n\, k_s^4\, m_l}{m_1 D - n\, k_s^4\, m_l}\, e^{m_1 z_1} \right] e^{i\,m\,x} dm$$

or

$$I_1 = -2 \int_{k_1}^{k_1+i\infty} \frac{A_1(m)}{m_1 B_1(m)} e^{i\, m\, x} dm \tag{6.384}$$

Here

$$A_1(m) = \left(m_1^2 D^2 + n^2\, k_s^8\, m_l^2\right) \sinh m_1 z_1 + 2n\, k_s^4\, m_1\, m_l\, D \cosh m_1 z_1 \tag{6.385}$$

and $\quad B_1(m) = m_1^2\, D^2 - n^2\, k_s^8\, m_l^2$

Introducing variable t: $m = k_1 + it$, we have $dm = i\, dt$. For small values of t,

$$m_1 = \sqrt{2i\, k_1}\, t^{1/2}, \qquad m_l = \sqrt{k_1^2 - k_l^2}, \qquad m_s = \sqrt{k_1^2 - k_s^2}$$

Then

$$D = k_1^4 \left[\left(2 - \frac{k_s^2}{k_1^2}\right)^2 - 4\left(1 - \frac{k_l^2}{k_1^2}\right)^{1/2}\left(1 - \frac{k_s^2}{k_1^2}\right)^{1/2}\right],$$

$$A_1(m) = n^2 k_s^8\, k_1^2 \left(1 - \frac{k_l^2}{k_1^2}\right) \sinh\left(\alpha\sqrt{t}\right) \quad \text{and} \quad B_1(m) = -n^2 k_s^8\, k_1^2 \left(1 - \frac{k_l^2}{k_1^2}\right) \tag{6.386}$$

Hence

$$I_1 = \frac{i\sqrt{2}\, e^{-i\pi/4}}{k_1^{1/2}}\, e^{i\, k_1 x} \int_0^\infty \frac{\sinh \alpha\sqrt{x}}{t^{1/2}}\, e^{-x\, t} dt$$

and

$$\alpha = \sqrt{2i\, k_1}\, z_1$$

The last integral is tabular and is equal to

$$\int_0^\infty \frac{\sinh \alpha\sqrt{t}}{t^{1/2}}\, e^{-x\, t} dt = 2\int_0^\infty \sinh(\alpha v)\, e^{-x\, v^2} dv = \sqrt{\frac{\pi}{x}}\, e^{\alpha^2/4x} \Phi\left(\frac{\alpha}{2\sqrt{x}}\right) \tag{6.387}$$

Here Φ is the probability integral, and for large values of k_1, i.e., small wavelengths, we have

$$\Phi\left(\frac{\alpha}{2\sqrt{x}}\right) = 1$$

Correspondingly,

$$I_1 = \frac{\sqrt{2\pi}}{(k_1 x)^{1/2}} \, e^{i\pi/4} \, e^{k_1 \left(x + z_1^2/2x\right)}$$

Thus the function

$$\widetilde{\varphi}_{1s}(k_1) = A_0 I_1 \tag{6.388}$$

describes the reflected wave, which was already studied by the stationary- phase method. Now let us evaluate integral I_s along the next two branch lines, C_3 and C_4, associated with the point k_s.

Integral I_s Taking into account that m_s changes sign around point k_s, we have

$$I_s = \int\limits_{k_s}^{k_s+i\infty} \frac{1}{m_1} \left[\frac{m_1 D\ (m_l, m_s) - n\ k_s^4\ m_l}{m_1 D\,(m_l, m_s) + n\ k_s^4\ m_l}\right.$$

$$\left. - \frac{m_1 D\ (m_l, -m_s) - n\ k_s^4\ m_l}{m_1 D\ (m_l, -m_s) + n\ k_s^4\ m_l}\right] e^{-m_1 z_1} e^{i\ m\ x} dm$$

or

$$I_s = -8n\ k_s^4 \int\limits_{k_s}^{k_s+i\infty} \frac{m^2 m_l^2\ m_s\ e^{-m_1 z_1}\ e^{i\ m\ x}\ dm}{\left[m_1 D\ (m_l, m_s) + n\ k_s^4\ m_l\right]\left[m_1 D\ (m_l, -m_s) + n\ k_s^4\ m_l\right]} \tag{6.389}$$

Using the variable t: $m = k_s + it$ and considering again only the vicinity of branch point k_s, we obtain

$$m^2 = k_s^2, \ m_l^2 = k_s^2\left(1 - \frac{k_l^2}{k_s^2}\right), \quad m_s = \sqrt{2ik_s}\ t^{1/2}, \ D = k_s^4, \ e^{-m_1 z_1} = e^{i\sqrt{k_1^2 - k_s^2}\ z_1}$$

Thus

$$I_s = -\frac{8\ i\ e^{i\pi/4}\ n\ \sqrt{2}\left(1 - \frac{c_s^2}{c_l^2}\right) e^{i\ k_s x}\ e^{i\sqrt{k_1^2 - k_s^2}\ z_1}}{k_s^{3/2}\left[n\left(1 - \frac{c_s^2}{c_l^2}\right)^{1/2} - i\left(\frac{c_s^2}{c_1^2} - 1\right)^{1/2}\right]^2} \int\limits_0^\infty t^{1/2}\ e^{-x\ t} dt$$

or

$$I_s = -\frac{4\,i\,e^{i\pi/4}\,n\,\sqrt{2\pi}\,\left(1-\dfrac{c_s^2}{c_l^2}\right)\,e^{i\left(k_s x+\sqrt{k_1^2-k_s^2}\,z_1\right)}}{(k_s x)^{3/2}\left[n\left(1-\dfrac{c_s^2}{c_l^2}\right)^{1/2}-i\left(\dfrac{c_s^2}{c_1^2}-1\right)^{1/2}\right]^2}, \tag{6.390}$$

and the potential related to point k_s is

$$\tilde{\varphi}_{1s} = A_0\, I_s \tag{6.391}$$

Eq. 6.391 characterizes a wave that is quite different from the reflected wave. First, the exponential function in eq. 6.390 depends on two wavenumbers, k_1 and k_s. This indicates that the wave path is located in a fluid and in an elastic medium, where the wave propagates with velocities c_1 and c_s, respectively. Also, as follows from eq. 6.390 the amplitude of the wave decreases as $1/\left(k_s\,x\right)^{3/2}$, that is, more rapidly than the amplitudes of the incident and reflected waves. Let us represent the exponent

$$k_s\,x + \sqrt{k_1^2-k_s^2}\,z_1$$

in the form

$$k_s\,x + \sqrt{k_1^2-k_s^2}\,z_1 = \omega\left(\frac{x}{c_s}+\frac{2H-z}{c_1}\sqrt{1-\frac{c_1^2}{c_s^2}}\right) = \omega\left(\frac{x}{c_s}+\frac{2H-z}{c_1}\cos\theta_c^s\right)$$

Here θ_c^s is the critical angle:

$$\sin\theta_c^s = \frac{c_1}{c_s}$$

Then we have

$$k_s\,x + \sqrt{k_1^2-k_s^2}\,z_1 = \omega\left[\frac{x-(2H-z)\tan\theta_c^s}{c_s}+\frac{2H-z}{c_1}\left(\cos\theta_c^s+\frac{c_1}{c_s}\tan\theta_c^s\right)\right]$$

$$= \omega\left[\frac{x-(2H-z)\tan\theta_c^s}{c_s}+\frac{2H-z}{c_1\cos\theta_c^s}\right] \tag{6.392}$$

This clearly shows that the wave propagates along path $OBCD$ (Fig. 6.8a). In fact, the length of the wavepath in the fluid is

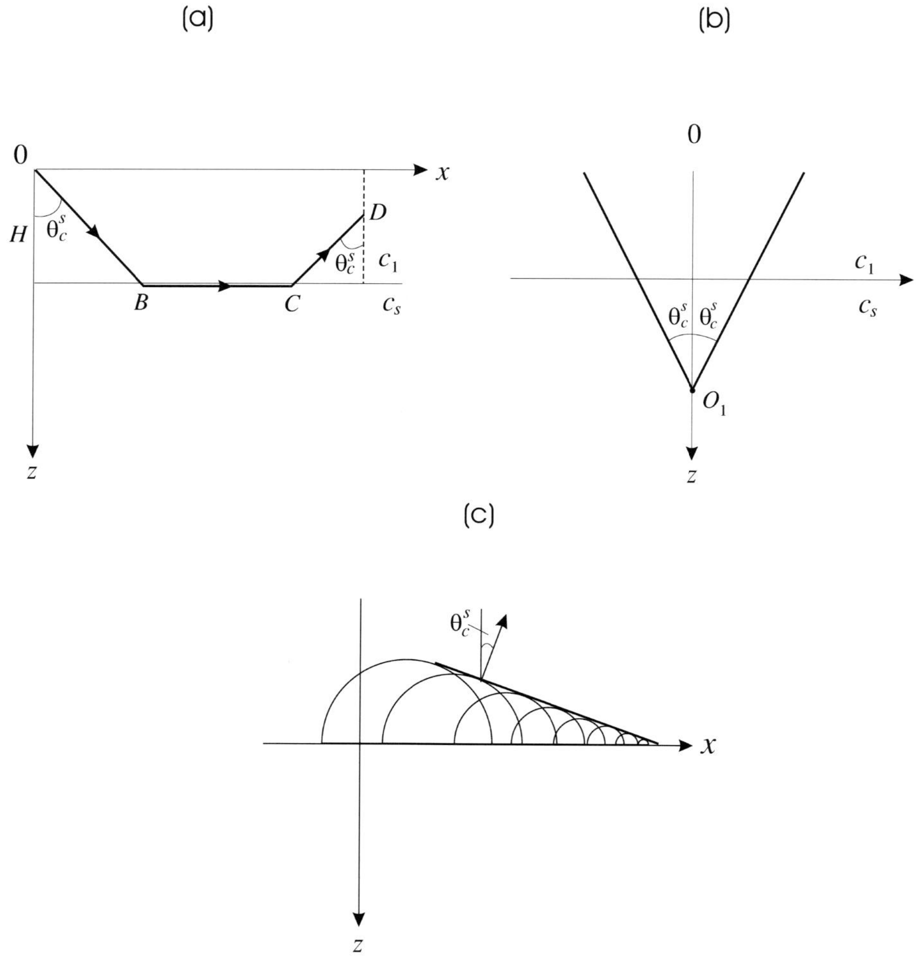

Figure 6.8: (a) Rays of PSP wave (b) Conical zone where PSP wave is absent (c) Huygen's principle and formation of PSP wave $(c_1 < c_s)$

$$OB + CD = \frac{H}{\cos\theta_c^s} + \frac{H-z}{\cos\theta_c^s} = \frac{2H-z}{\cos\theta_c^s},$$

which corresponds to the second term in the brackets of eq. 6.392. At the same time, the path length in the lower medium, BC, is equal to

$$BC = x - H\ \tan\theta_c^s - (H-z)\ \tan\theta_c^s = x - (2H-z)\ \tan\theta_c^s,$$

which coincides with the numerator of the first term.

Thus our assumption is correct, and we are dealing with a head wave (Part II) that is usually called PSP. This name can be easily explained. When the incident P wave reaches points of the interface where $\theta_i = \theta_c^s$, the shear wave S begins to move along this boundary, giving rise to the P wave in the upper medium. Thus, we twice observe a transition from one type of wave to another. In accordance with eq. 6.392, rays of the head waves are parallel to each other and they form with the z-axis the angle equal to θ_c^s. Correspondingly, the phase surfaces of this wave are planes. It is obvious that in the three-dimensional case these surfaces are conical and their apexes are located at the z-axis. This is why head waves are often called conical waves. Since scalar potential $\widetilde{\varphi}_{1s}$ depends on coordinate x, the wavefields vary on each phase surface and decrease with increased distance from the z-axis. This means that in general, motion of particles is characterized by elliptical polarization, which occurs in the case of reflected and transmitted waves. However, with an increase of distance x, this effect becomes weaker, and nearly linear polarization is observed.

Also, it is clear that the head wave is absent within the volume bounded by two planes, $\theta = \theta_c^s$, Fig. 6.8b. In order to visualize the appearance of this wave, it is useful to apply Huygen's principle (Part I). When the shear wave moves along the boundary, each of its points can be treated as the source of the secondary cylindrical P wave in a fluid. Since velocity c_s exceeds c_1, the elementary wavefronts overlap and form the envelope that represents the front of the head wave, Fig. 6.8c.

At the beginning, we assumed that $c_1 < c_s$. Now let us understand what happens in the opposite case $(c_1 > c_s)$. As follows from eq. 6.390, the exponential term can be written in the form

$$e^{i\,k_s x}\, e^{-\sqrt{k_s^2 - k_1^2}\, z_1}$$

This means that instead of a conical wave in fluid, we observe a evanescent wave that exponentially decays with distance from the boundary and propagates along the boundary with velocity c_s. In both cases $(c_1 < c_s$ or $c_1 > c_s)$, the shear wave in an elastic medium is the moving source of the wave, $\widetilde{\varphi}_{1s}(k_s)$, in fluid. However, in the latter case, due to destructive interference, an inhomogeneous wave is formed. In other words, the wavefronts of elementary waves do not have an envelope.

Next consider the contribution of integrals along branch lines C_5 and C_6, Fig. 6.7d.

Integral I_l In accordance with eq. 6.382, we have

$$I_l = \int_{k_l}^{k_l+i\infty} \frac{1}{m_1}\left[\frac{m_1 D(m_l,m_s) - nk_s^4\, m_l}{m_1 D(m_l,m_s) + nk_s^4\, m_l} - \frac{m_1 D(-m_l,m_s) + nk_s^4\, m_l}{m_1 D(-m_l,m_s) - nk_s^4\, m_l}\right] e^{-m_1 z_1 + imx} dm$$

or

$$I_l = -4n\, k_s^4 \int_{k_l}^{k_l+i\infty} \frac{m_l\left(2m^2 - k_s^2\right)^2 e^{-m_1 z_1}\, e^{i\,m\,x}\, dm}{\left[m_1 D(m_l,m_s) + nk_s^4\, m_l\right]\left[m_1 D(-m_l,m_s) - nk_s^4\, m_l\right]} \tag{6.393}$$

Applying the same procedure as before, we have

$$m_l = \sqrt{2i\, k_l}\; t^{1/2}, \qquad dm = idt, \qquad m_1 = -i\, k_l\left(\frac{k_s^2}{k_l^2} - 1\right)^{1/2},$$

$$\left(2m^2 - k_s^2\right)^2 = k_l^4\left(2 - \frac{k_s^2}{k_l^2}\right)^2, \quad D(m_l,m_s) = D(-m_l,m_s) = k_l^4\left(2 - \frac{k_s^2}{k_l^2}\right)^2$$

Thus

$$I_l = \frac{4n\sqrt{2}\; i\; e^{i\,\pi/4}\; e^{ik_l x}}{k_l^{3/2}\left(2 - \frac{k_s^2}{k_l^2}\right)^2\left(\frac{k_1^2}{k_l^2} - 1\right)}\left(\frac{k_s}{k_l}\right)^4 e^{i\sqrt{k_1^2 - k_l^2}\; z_1}\int_0^{\infty} t^{1/2}\, e^{-xt} dt$$

or

$$I_l = \frac{2\sqrt{2\pi}n\; i\; e^{i\,\pi/4}}{(k_l x)^{3/2}\left(2 - \frac{k_s^2}{k_l^2}\right)^2\left(\frac{k_1^2}{k_l^2} - 1\right)}\left(\frac{k_s}{k_l}\right)^4 e^{i\left(k_l x + \sqrt{k_1^2 - k_l^2}\; z_1\right)} \tag{6.394}$$

and the scalar potential is

$$\widetilde{\varphi}_{1s}(k_l) = A_1\, I_l \tag{6.395}$$

Comparison of eqs. 6.391 and 6.395 shows that the latter describes the PPP head wave. By analogy with the case of the PSP head wave, it is easy to show that the wave path consists of three elements. The first is the ray of the P direct wave with the angle of incidence equal to the critical angle

$$\theta_i = \theta_c^l = \sin^{-1}\frac{c_1}{c_l}$$

Second is the path in an elastic medium of the longitudinal wave which advances along the boundary with velocity c_l. Third is the ray of the head wave propagating from the boundary to an observation point; its reflection angle is also equal to the critical angle. As follows from eq. 6.394, the phase surface of this head wave is a plane, and its equation is

$$k_l\, x + \sqrt{k_1^2 - k_l^2}\; z_1 = \text{const} \tag{6.396}$$

The velocity of propagation along the x-axis is equal to c_l. This is because the head wave is caused by a longitudinal wave moving along the boundary. As follows from eq. 6.394, both head waves decrease with distance in the same manner. Since

$$\theta_c^l < \theta_c^s, \tag{6.397}$$

the PSP head wave appears at points of the boundary that are located at greater distances x than the PPP wave. Thus, the stationary-phase method and the contour-integration method allow us to describe the following secondary waves in fluid:

1. PP reflected wave.
2. PSP head wave, if $c_1 < c_s$.
3. PPP head wave.

Their paths are shown in Fig. 6.9a. Earlier we demonstrated that if $c_1 > c_s$, then instead of a PSP wave, we observe an evanescent wave. This group of waves does not include the boundary wave associated with the pole, which will be discussed later.

Next, we will derive asymptotic formulas for waves in an elastic medium.

Scalar potential $\tilde{\varphi}_{2s}$

As follows from eq. 6.340,

$$\tilde{\varphi}_2 = -2nk_s^2 A_1 \int\limits_{-\infty}^{\infty} \frac{(2m^2 - k_s^2)\; e^{-m_1 H} e^{-m_l\, z_2}}{m_1 D + m_l n\; k_s^4}\; e^{i\, m\, x} dm = A_1\,(L_1 + L_s + L_l + L_p) \tag{6.398}$$

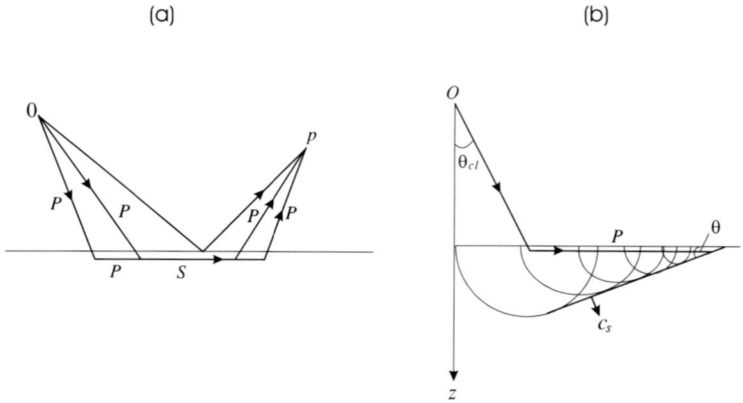

Figure 6.9: (a) Secondary waves in fluid (b) Conical shear wave in elastic medium

Here

$$z_2 = z - H > 0$$

and, as before, we assume that $k_1 > k_s > k_l$.

Integral L_1 Integration along paths C_1 and C_2 gives

$$L_1 = -2nk_s^2 \int_{k_1}^{k_1+i\infty} \left(2m^2 - k_s^2\right) e^{-m_l z_2} e^{imx} \left[\frac{e^{-m_1 H}}{m_1 D + n\, k_s^4\, m_l} + \frac{e^{m_1 H}}{m_1 D - n\, k_s^4\, m_l}\right] dm \tag{6.399}$$

Replacement of variable $m = k_1 + it$ gives

$$L_1 = \frac{4i\, k_1}{k_s^2} \frac{\left(2 - \frac{k_s^2}{k_1^2}\right)}{\left(1 - \frac{k_l^2}{k_1^2}\right)^{1/2}} e^{i\, k_1 x}\, e^{-\sqrt{k_1^2 - k_l^2}\, z_2} \int_0^\infty e^{-x\, t} \sinh \alpha_1 \sqrt{t} dt$$

where

$$\alpha_1 = \sqrt{2i\, k_1}\, H$$

Taking into account eq. 6.383, we have

$$L_1 = \frac{2\sqrt{2\pi}\, i\, e^{i\, \pi/4}}{(k_s\, x)^{3/2}} \left(\frac{k_1}{k_s}\right)^2 \frac{k_1 H \left(2 - \frac{k_s^2}{k_1^2}\right)}{\left(1 - \frac{k_l^2}{k_1^2}\right)^{1/2}} e^{i\, k_1 x}\, e^{i\frac{k_1 H^2}{2x}}\, e^{-\sqrt{k_1^2 - k_l^2}\, z_2} \tag{6.400}$$

and $\qquad \widetilde{\varphi}_2(k_1) = A_1 L_1$

It is clear that this function characterizes the wavefield that exponentially decays with distance from the boundary. Only in this sense does it resemble an inhomogeneous longitudinal wave. This field arises when the angle of incidence of the direct wave exceeds the critical angle θ_c^l.

Integral L_s Along paths C_3 and C_4 we have

$$L_s = -2\,n\,k_s^2 \int\limits_{k_s}^{k_s+i\infty} \left(2m^2 - k_s^2\right)\, e^{-m_l\, z_2} e^{-m_1 H} e^{i\,m\,x}$$

$$\times \left[\frac{1}{m_1 D\ (m_l, m_s) + n\,k_s^4\,m_l} - \frac{1}{m_1 D\ (m_l, -m_s) + n\,k_s^4\,m_l}\right] dm$$

or

$$L_s = -2nk_s \int\limits_{k_s}^{k_s+i\infty} \frac{m_1\left[D\ (m_l, -m_s) - D\ (m_l, m_s)\right]}{\left[m_1 D\,(m_l, m_s) + n\,k_s^4\,m_l\right]\left[m_1 D\ (m_l, -m_s) + n\,k_s^4\,m_l\right]}$$

$$\times \left(2m^2 - k_s^2\right)\, e^{-m_l\, z_2} e^{-m_1 H} e^{imx} dm$$

or

$$L_s = -16\,n\,k_s^2 \int\limits_{k_s}^{k_s+i\infty} \frac{m^2\,m_1\,m_l\,m_s\,\left(2m^2 - k_s^2\right)\, e^{-m_l\, z_2} e^{-m_1 H} e^{i\,m\,x} dm}{\left[m_1 D\ (m_l, m_s) + n\,k_s^4\,m_l\right]\left[m_1 D\ (m_l, -m_s) + n\,k_s^4\,m_l\right]} \qquad (6.401)$$

Applying the variable t: $m = k_s + it$, we obtain

$$L_s = -\frac{16\,n\,i\,e^{i\pi/4}\sqrt{2}\,e^{ik_s x}\,e^{-\sqrt{k_s^2 - k_1^2}\,H}\,e^{-\sqrt{k_s^2 - k_l^2}\,z_2}}{k_s^{3/2}\left[\left(1 - \dfrac{k_1^2}{k_s^2}\right)^{1/2} + n\left(1 - \dfrac{k_l^2}{k_s^2}\right)^{1/2}\right]^2} \int\limits_0^\infty t^{1/2}\, e^{-x\,t} dt$$

or

$$L_s = \frac{-8\,n\,i\sqrt{2\pi}\,e^{i\pi/4}\,e^{-\sqrt{k_s^2 - k_l^2}\,z_2}\,e^{i(k_s x + \sqrt{k_1^2 - k_s^2}\,H)}}{(k_s\,x)^{3/2}\left[\left(1 - \dfrac{k_1^2}{k_s^2}\right)^{1/2} + n\left(1 - \dfrac{k_l^2}{k_s^2}\right)^{1/2}\right]^2}$$

$$\text{and} \qquad \widetilde{\varphi}_2(k_s) = A_0 L_s \tag{6.402}$$

This expression describes an evanescent longitudinal wave that exponentially decays with distance from the boundary. For this reason it recalls the wave $\widetilde{\varphi}_2(k_1)$. However, there is a strong difference. First of all, in accordance with eq. 6.402, this wave moves along the boundary with the velocity of the shear wave, c_s. Also, let us introduce coordinate x_{cs} of the boundary points where the angle of incidence coincides with the critical angle, $\theta_i = \theta_c^s$, and represent the phase function in eq. 6.402 as

$$k_s x + \sqrt{k_1^2 - k_s^2}\ H = k_s\,(x - x_{cs}) + k_s x_{cs} + \sqrt{k_1^2 - k_s^2}\ H$$

Since

$$\frac{k_s}{k_1} = \sin\theta_c^s \qquad \text{and} \qquad r_c = \sqrt{x_{cs}^2 + H^2},$$

we have

$$k_s\ x + \sqrt{k_1^2 - k_s^2}\ H = k_s\,(x - x_{cs}) + k_1 r_c \tag{6.403}$$

Thus, appearance of wave $\widetilde{\varphi}_2(k_s)$ may be explained in the following way. When the direct P wave reaches the vicinity of points where $\theta_i = \theta_c^s$, its phase is equal to $k_1 r_c$, and it gives rise to the transmitted S wave. The latter moves along the boundary with velocity c_s and, correspondingly, a change of the phase at these points is equal to

$$k_s(x - x_{cs})$$

Propagation of the S wave causes vibration of the boundary and, as a result, elementary longitudinal waves arise in a fluid and in an elastic medium, where their velocities are c_1 or c_l, respectively. Because $c_1 < c_s$, constructive interference of elementary waves occurs in liquid, and we observe the PSP head wave. In contrast, in an elastic medium $c_l > c_s$, and superposition of elementary waves has a destructive character. Because of this, the evanescent longitudinal wave $\widetilde{\varphi}_2(k_s)$ appears.

Thus, the transmitted S wave propagating along the boundary is the moving source of wave described by eq. 6.402.

Integral L_l The integrals along paths C_5 and C_6 are

$$L_l = -2n\ k_s^2 \int\limits_{k_l}^{k_l + i\infty} (2m^2 - k_s^2)\ e^{-m_1 H} e^{i\ m\ x}$$

$$\times \left[\frac{e^{-m_l\, z_2}}{m_1 D\,(m_l, m_s) + n\; k_s^4\; m_l} - \frac{e^{m_l\, z_2}}{m_1 D\,(-m_l, m_s) - n\; k_s^4\; m_l} \right] dm$$

After changing the variable of integration and performing simplifications, we obtain:

$$L_l = -\frac{4n\; i\; k_s^2\; e^{ik_l x}\; e^{i\sqrt{k_1^2 - k_l^2}\; H}}{k_l^3 \left(1 - \dfrac{k_1^2}{k_l^2}\right)^{1/2} \left(2 - \dfrac{k_s^2}{k_l^2}\right)} \int\limits_0^\infty e^{-x\; t} \sinh \alpha_l \sqrt{t} dt \tag{6.404}$$

where

$$\alpha_l = \sqrt{2\; i\; k_l}\; z_2$$

Then the use of eqs. 6.383 gives

$$L_l = \frac{2\sqrt{2\pi}\; in\; e^{i\;\pi/4}}{(k_l\; x)^{1/2}} \left(\frac{k_s}{k_l}\right)^2 \frac{z_2}{x}\; e^{ik_l\; x}\; e^{i\sqrt{k_1^2 - k_l^2}\; H}\; e^{\dfrac{ik_l\; z^2}{2x}} \tag{6.405}$$

and $\quad \tilde{\varphi}_2\;(k_l) = A_0 L_l$

It is useful to represent the phase function in the following form

$$k_l\; x_{cl} + \sqrt{k_1^2 - k_l^2}\; H + k_l\,(x - x_{cl}) + \frac{k_l\; z_2^2}{2x}$$

Here x_{cl} is the coordinate of the boundary points where the angle of incidence of the direct wave is θ_c^l. Assuming that $x \gg x_{cl}$ and $z_2/x \ll 1$, the phase becomes equal to

$$k_1 r_c + \; k_l r_1$$

Here

$$r_c = \sqrt{x_{cl}^2 + H^2} \qquad \text{and} \qquad r_1 = \sqrt{(x - x_{cl})^2 + z_2^2}$$

Thus, we see that function $\tilde{\varphi}_2(k_l)$ describes the transmitted P wave near the boundary and that, therefore, its secondary sources are located in the vicinity of points where $\theta_i = \theta_c^l$. Again it is clear that the stationary-phase method allows us to obtain a better approximation of this wave.

Vector potential ψ_2

In accordance with eq. 6.364, we have

$$\tilde{\psi}_2 = 4ink_s^2 A_0 \int\limits_{-\infty}^{\infty} \frac{m\, m_l\, e^{-m_1 H} e^{-m_s\, z_2}}{m_1 D + n\, k_s^4\, m_l}\, e^{i\, m\, x} dm = A_0 \left(M_1 + M_s + M_l + M_p\right) \tag{6.406}$$

As before, we start our evaluation of integrals from branch point k_1.

Integral M_1 Integrals along paths C_1 and C_2 give

$$M_1 = 4\, i\, n\, k_s^2 \int\limits_{k_1}^{k_1 + i\infty} m\, m_l\, e^{-m_s z_2}\, e^{i\, m\, x} \left[\frac{e^{-m_1 H}}{m_1 D + m_l n\, k_s^4} + \frac{e^{m_1 H}}{m_1 D - m_l n\, k_s^4}\right] dm$$

or

$$M_1 = \frac{8\, k_1}{k_s^2} e^{i k_1 x}\, e^{-\sqrt{k_1^2 - k_s^2}\, z_2} \int\limits_0^{\infty} e^{-x\, t} \sinh \alpha_1 \sqrt{t} dt,$$

where

$$\alpha_1 = \sqrt{2\, i\, k_1}\, H$$

Thus

$$M_1 = \frac{4\sqrt{2\pi}\, e^{i\pi/4}}{(k_s\, x)^{3/2}} (k_1 H) \left(\frac{k_1}{k_s}\right)^{1/2} e^{-\sqrt{k_1^2 - k_s^2}\, z_2} e^{i\, k_1 \left(x + \frac{1}{2} \frac{H^2}{x}\right)} \tag{6.407}$$

$$\text{and} \qquad \tilde{\psi}_2(k_1) = A_0 M_1$$

It is obvious that this function characterizes a shear wavefield that exponentially decays with distance from the boundary and appears provided that $\theta_i > \theta_c^s$.

Integral M_s As follows from eq. 6.406,

$$M_s = 4ink_s^2 \int\limits_{k_s}^{k_s + i\infty} m\, m_l\, e^{-m_1 H} e^{i\, m\, x}$$

$$\times \left[\frac{e^{-m_s \, z_2}}{m_1 D \, (m_l, m_s) + n \, k_s^4 \, m_l} - \frac{e^{-m_s \, z_2}}{m_1 D \, (m_l, -m_s) + n \, k_s^4 \, m_l} \right] dm$$

or

$$M_s = \frac{8n \left(1 - \dfrac{k_l^2}{k_s^2}\right)^{1/2} e^{ik_s x} \, e^{i\sqrt{k_1^2 - k_s^2} \, H}}{k_s \left[\left(1 - \dfrac{k_1^2}{k_s^2}\right)^{1/2} + n \left(1 - \dfrac{k_l^2}{k_s^2}\right)^{1/2} \right]} \int\limits_0^\infty e^{-xt} \sinh \alpha_s \sqrt{t} \, dt,$$

where

$$\alpha_s = \sqrt{2 \, i \, k_s} H$$

Hence

$$M_s = \frac{4\sqrt{2\pi} e^{i\pi/4} \left(1 - \dfrac{k_l^2}{k_s^2}\right)^{1/2} e^{ik_s x} \, e^{i\sqrt{\, k_1^2 - k_s^2} \, H} \, e^{i\dfrac{k_s \, z_2^2}{2x}}}{(k_s \, x)^{1/2} \left[\left(1 - \dfrac{k_1^2}{k_s^2}\right)^{1/2} + n \left(1 - \dfrac{k_l^2}{k_s^2}\right)^{1/2} \right]} \, \frac{z_2}{x} \tag{6.408}$$

and $\quad \widetilde{\psi}_2(k_s) = A_0 \, M_s$

By analogy with function $\widetilde{\psi}_2(k_l)$, we conclude that eq. 6.408 describes the transmitted shear wave near the boundary, $z_2/x \ll 1$, and at sufficiently large distance from the z-axis.

Integral M_l The integral along paths C_5 and C_6 is

$$M_l = 4i \, n \, k_s^2 \int\limits_{k_l}^{k_1 + i\infty} m \, m_l \, e^{-m_1 H} e^{-m_s \, z_2} e^{imx}$$

$$\times \left[\frac{1}{m_1 D \, (m_l, m_s) + n \, k_s^4 \, m_l} + \frac{1}{m_1 D \, (-m_l, m_s) - n \, k_s^4 \, m_l} \right] dm$$

This gives

$$M_l = -\frac{4 \, n \, k_s^2 \sqrt{2i \, k_l} \, k_l}{(k_l^2 - k_1^2)^{1/2} \, (2k_l^2 - k_s^2)^2} \, e^{i\sqrt{k_1^2 - k_l^2} \, H} e^{i \, k_l x} \, e^{i\sqrt{k_s^2 - k_l^2} \, z_2} \int\limits_0^\infty t^{1/2} e^{-xt} dt$$

Therefore,

$$M_l = \frac{2n\sqrt{2\pi}e^{i\pi/4}e^{i\,k_l x}\,e^{i\sqrt{k_1^2-k_l^2}\,H}\,e^{i\sqrt{k_s^2-k_l^2}\,z_2}}{(k_l x)^{3/2}\left(1-\frac{k_1^2}{k_l^2}\right)^{1/2}\left(2-\frac{k_s^2}{k_l^2}\right)^2} \quad \text{and} \quad \overset{\sim}{\psi}_2(k_l) = A_1 M_l \tag{6.409}$$

To understand this wavefield, we represent its phase function in the form

$$k_l x_{cl} + \sqrt{k_1^2 - k_l^2}\,H + k_l\,(x - x_{cl}) + \sqrt{k_s^2 - k_l^2}\,z_2$$

As before, x_{cl} is the coordinate of the boundary points where $\theta_i > \theta_c^l$. Respectively, the sum of the two first terms is equal to $k_1 r_c$, and it defines the phase of the direct wave at points $(x_{cl}\,, H)$ and $r_c = \sqrt{x_{cl}^2 + H^2}$. The sum of the next terms remains constant at points of the plane:

$$k_l\,(x - x_{cl}) + \sqrt{k_s^2 - k_l^2}\,z_2 = \text{const} \tag{6.410}$$

Thus, function $\overset{\sim}{\psi}_2(k_l)$ describes a wave with plane phase surfaces that moves through an elastic medium with the velocity of a shear wave, c_s. At the same time, the velocity of propagation along the boundary is equal to c_l. As follows from the expression for the phase function, the wavefront and the boundary form angle θ, where

$$\sin\theta = \frac{c_s}{c_l}$$

The appearance of this wave was discussed earlier. When the longitudinal transmitted wave advances along the boundary, it also produces elementary shear waves. Since $c_s < c_l$, superposition of these waves has a constructive character. As before, applying Huygen's principle, it is easy to demonstrate that the phase surfaces of this wave are planes, Fig. 6.9b. We see that the transmitted P wave is the moving source of the conical wave, $\overset{\sim}{\psi}_2(k_l)$, and it is observed at points of an elastic medium where $x > x_{cl}$.

In summary, let us outline the main features of wavefields in the lower medium:

1. When the direct P wave reaches the boundary, we see the appearance at its points of the transmitted P and S waves, if $\theta_i < \theta_c^l$. We see only the S wave if $\theta_c^l \le \theta_i \le \theta_c^s$. Neither wave arises when $\theta_i > \theta_c^s$. These waves obey Snell's law and, correspondingly, they appear at any point of a medium. Their magnitude depends on frequency and distance as $1/\left(kr\right)^{1/2}$.

2. Within the next range of the angle of incidence

$$\theta_c^l \leq \theta_i \leq \theta_c^s,$$

besides the transmitted S wave, vibrations of the boundary cause a shear conical wave. It propagates through a medium with velocity c_s, and like the head wave, its magnitude depends on the wavenumber and distance as $1/(kr)^{3/2}$. Certainly, its magnitude is smaller than the magnitude of the transmitted waves. The phase surface of the conical wave is a plane that, with the boundary, forms angle θ, where

$$\sin\theta = \frac{c_s}{c_l}$$

Since the transmitted P wave generates a conical wave, its velocity of propagation along the boundary is equal to c_l. Besides, due to the primary wave, a longitudinal wavefield arises that exponentially decays the distance from the boundary. Use of the word "wave" in this case is hardly appropriate. In fact, the phase surface of this field is the vertical plane $x = \text{const}$, but its velocity continuously changes along the x-axis. This wavefield is usually called a "diffusive wave".

3. At points of the boundary where $\theta_i > \theta_c^s$, four wavefields appear, namely:

a. The shear conical wave.

b. Longitudinal and shear diffusive waves, which exponentially decay with the distance from the boundary.

c. The inhomogeneous longitudinal wave, which moves along the boundary with the velocity of the shear wave and also exponentially decreases with depth.

Our description of wavefields does not include the boundary Stoneley waves that arise at the interface between the fluid and elastic media. By definition, they are related to poles of integrands, which describe the scalar and vector potentials. In accordance with eq. 6.328, poles are roots of the equation

$$\sqrt{m^2-k_1^2}\left[\left(2m^2-k_s^2\right)^2-4m^2\sqrt{m^2-k_l^2}\,\sqrt{m^2-k_s^2}\right]+\sqrt{m^2-k_l^2}\;n\;k_s^4=0 \qquad (6.411)$$

Here $n = \rho_1/\rho_2$. Letting $\rho_1 = 0$, we arrive at the known Rayleigh equation

$$\left(2m^2-k_s^2\right)^2-4m^2\sqrt{m^2-k_l^2}\sqrt{m^2-k_s^2}=0$$

Applying the same approach as in this last case $(\rho_1 = 0)$, it is easy to show that for any parameters of a fluid and an elastic medium there is always one real root, m_p, of eq. 6.411, which slightly exceeds the maximal value of the wavenumbers:

$$m_p > k_1 \quad \text{if} \quad k_1 > k_s \qquad \text{and} \qquad m_p > k_s \quad \text{if} \quad k_s > k_1$$

Correspondingly, the surface wave related to pole m_p moves along the boundary a little more slowly than either the longitudinal wave in fluid or the S wave in the lower medium. Since radicals $\sqrt{m_p^2 - k_1^2}$, $\sqrt{m_p^2 - k_s^2}$, and $\sqrt{m_p^2 - k_l^2}$ are all real and positive, this wave exponentially decays with an increase of distance from the boundary. In a fluid, that wave is purely dilatational, whereas in an elastic medium the wave causes deformation and the rotation of elementary volumes. Of course, above and beneath the interface these waves advance with the same velocity

$$c_p = \frac{\omega}{m_p}$$

6.7 Point source of elastic waves in the presence of the free boundary

In the previous sections we assumed that the field is caused by a linear source oriented parallel to the boundary. Now we extend this study to the three-dimensional case and consider elastic waves generated by different types of point sources (Chapter 3).

1. Point source of the P wave Suppose that a very small spherical source is situated at distance d from the free boundary, Fig. 6.10a, and a change of its radius (pulsations) generates the P wave. As usual, in order to determine the wavefields we formulate the boundary value problem. Taking into account the axial symmetry of the direct wave and a medium, it is convenient to choose the cylindrical system of coordinates, r, φ, z, and place its origin at the boundary. By analogy with the two-dimensional case, we assume that the source generates a sinusoidal wave with frequency ω. Of course, the use of Fourier's integral permits us to obtain information about transient waves. The presence of the boundary creates both longitudinal and shear waves. Therefore it is natural to formulate the boundary value problem in terms of complex amplitudes of scalar and vector potentials $\tilde{\varphi}$ and $\tilde{\psi}$. Respectively, we have

$$\tilde{\varphi} = \tilde{\varphi}_i + \tilde{\varphi}_s \qquad \text{and} \qquad \tilde{\boldsymbol{\psi}} = \tilde{\boldsymbol{\psi}}_s \tag{6.412}$$

Here $\tilde{\varphi}_i$ and $\tilde{\varphi}_s$ are scalar potentials of the primary, $\tilde{\varphi}_i$, and secondary, $\tilde{\varphi}_s$, wavefields.

At regular points they obey the Helmholtz equations:

$$\nabla^2 \tilde{\varphi} + k_l^2\, \tilde{\varphi} = 0 \quad \text{and} \quad \nabla^2 \tilde{\boldsymbol{\psi}} + k_s^2\, \tilde{\boldsymbol{\psi}} = 0 \tag{6.413}$$

At the free surface the tangential, $\tilde{\tau}_{rz}$, and normal, $\tilde{\tau}_{zz}$, components of stress vanish,

$$\tilde{\tau}_{rz} = 0 \qquad \text{and} \qquad \tilde{\tau}_{zz} = 0, \qquad \text{if} \quad z = 0, \tag{6.414}$$

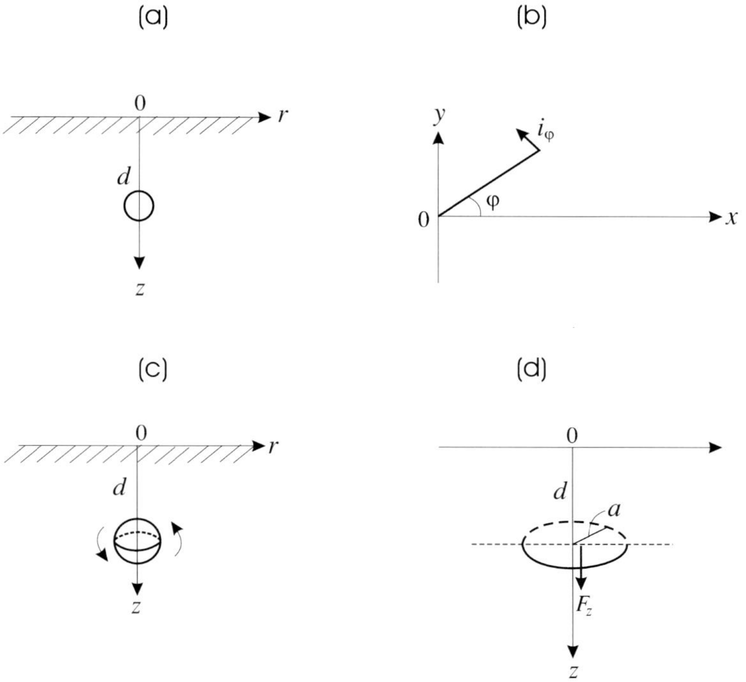

Figure 6.10: (a) Buried point explosive source of P wave (b) Illustration of eq. 6.427 (c) Buried shear wave point source (d) Vertical force

and potentials must also satisfy these two equations. Since

$$\tilde{\varphi}_i = C_l \, \frac{e^{i\, k_l\, R}}{R} \tag{6.415}$$

and potentials of the secondary fields have finite values everywhere, in approaching the source we have

$$\tilde{\varphi} \to \tilde{\varphi}_i \qquad \text{if} \qquad R \to 0 \tag{6.416}$$

Here

$$R = \sqrt{r^2 + (z-d)^2}$$

It is clear that the amplitude of the spherical wave caused by the source, as well as amplitudes of secondary waves, decrease with distance, and in the limit they obey the condition at infinity

$$\tilde{\varphi} \rightarrow 0 \qquad \text{and} \qquad \tilde{\boldsymbol{\psi}} \rightarrow 0 \qquad \text{if} \qquad R \rightarrow \infty \tag{6.417}$$

We have formulated the boundary value problem, and, as follows from the physical point of view and the theorem of uniqueness, only one field of displacement and stress satisfies these conditions. Since we have considered Dirichlet's boundary value problem (Part I), the potentials are also defined uniquely.

Solutions of the Helmholtz equation for scalar potential

First, we find a solution of the equation

$$\nabla^2 \tilde{\varphi} + k_l^2 \tilde{\varphi} = 0$$

Taking into account axial symmetry with respect to the z-axis, i.e., independence on the azimuthal coordinate, we have in the cylindrical system (Part I)

$$\frac{\partial^2 \tilde{\varphi}}{\partial r^2} + \frac{1}{r}\frac{\partial \tilde{\varphi}}{\partial r} + \frac{\partial^2 \tilde{\varphi}}{\partial z^2} + k_l^2 \, \tilde{\varphi} = 0 \tag{6.418}$$

Applying the method of separation of variables, potential $\tilde{\varphi}$ is written in the form

$$\tilde{\varphi}(r, z, \omega) = T(r) \; Z(z, \omega) \tag{6.419}$$

Substitution of eq. 6.419 into eq. 6.418 and division of both sides by the product TZ gives

$$\frac{1}{T}\frac{d^2T}{dr^2} + \frac{1}{rT}\frac{dT}{dr} + \frac{1}{Z}\frac{d^2Z}{dz^2} + k_l^2 = 0 \tag{6.420}$$

This equality indicates that the sum of the first two terms and the sum of the last two terms are constants that differ by sign only. Therefore, in place of eq. 6.420, we can write

$$\frac{1}{T}\frac{d^2T}{dr^2} + \frac{1}{r}\frac{1}{T}\frac{dT}{dr} = \pm m^2 \qquad \text{and} \qquad \frac{1}{Z}\frac{d^2Z}{dz^2} + k_l^2 = \mp m^2, \tag{6.421}$$

where m is an arbitrary number. Thus, instead of a partial differential equation, we arrive at two ordinary differential equations whose solutions are well known. Selecting the sign on the right side of eqs. 6.421, we have to take into account the fact that the wavefield has a finite amplitude everywhere except at the source location. Suppose that the sign "+" is chosen in the first equation of the set. It gives

$$\frac{d^2T}{dr^2} + \frac{1}{r}\frac{dT}{dr} - m^2 T = 0$$

As is well known (Part II), the solution of this equation is expressed in terms of modified Bessel functions of the zero order

$$I_0\,(mr) \qquad \text{and} \qquad K_0\,(mr)$$

Note that the same functions will be used to study wavefields inside the borehole (Chapter 7). Inasmuch as $I_0\,(mr)$ increases unlimitedly when $r \to \infty$, while $K_0\,(mr)$ becomes infinitely large at all points of the z-axis, neither of those functions can describe the wavefields. For this reason we choose "−" and "+" in the first and second equations of set 6.421, respectively. Therefore, this system becomes

$$\frac{d^2T}{dr^2} + \frac{1}{r}\frac{dT}{dr} + m^2T = 0 \qquad \text{and} \qquad \frac{d^2Z}{dz^2} - \left(m^2 - k_l^2\right) Z = 0 \tag{6.422}$$

The solutions of the first equation are Bessel functions (Part II),

$$J_0\,(mr) \quad \text{and} \quad Y_0\,(mr)\,,$$

or their combinations. Since function $Y_0\,(mr)$ is infinitely large at the z-axis, it cannot be used to describe the wavefield. The solutions of the second equation are

$$\exp\left[\pm\sqrt{m^2 - k_l^2}\; z\right]$$

Thus, in accordance with eq. 6.419, the partial solution is

$$\tilde{\varphi}_m = \left(A_m\, e^{-m_l\, z} + C_m\, e^{m_l\, z}\right)\; J_0\,(mr)\,, \tag{6.423}$$

where

$$m_l = \sqrt{m^2 - k_l^2}$$

Correspondingly, the general solution of the Helmholtz equation for scalar potential is

$$\tilde{\varphi}(r, z, \omega) = \int\limits_0^\infty \left[A_m\, e^{-m_l\, z} + C_m\, e^{m_l\, z}\right]\; J_0\,(mr)\;\, dm \tag{6.424}$$

Here A_m and C_m are unknown coefficients; they do not depend on coordinates r and z of the observation point.

Solution of Helmholtz equation for vector potential $\tilde{\boldsymbol{\psi}}$

Taking into account that displacement **s** has only two components, s_r and s_z, we assume that vector potential $\tilde{\boldsymbol{\psi}}$ can be described by the azimuthal component

$$\tilde{\boldsymbol{\psi}} = \tilde{\psi}\; \mathbf{i}_\varphi, \tag{6.425}$$

where $\mathbf{i}_\varphi$ is the unit vector along the φ-coordinate line, and its direction depends on point position. In order to obtain an equation for scalar component $\tilde{\boldsymbol{\psi}}$, we substitute eq. 6.425 into the second equation of set 6.413, which gives

$$\nabla^2\left(\tilde{\psi}\,\mathbf{i}_\varphi\right)+k_s^2\,\tilde{\psi}\,\mathbf{i}_\varphi=0 \qquad \text{or} \qquad \mathbf{i}_\varphi\nabla^2\,\tilde{\psi}+\tilde{\psi}\nabla^2\mathbf{i}_\varphi+k_s^2\,\tilde{\psi}\,\mathbf{i}_\varphi=0 \tag{6.426}$$

As is seen from Fig. 6.10b,

$$\mathbf{i}_\varphi=-\sin\varphi\,\mathbf{i}+\cos\varphi\,\mathbf{j} \tag{6.427}$$

Here $\mathbf{i}$ and $\mathbf{j}$ are unit vectors in the Cartesian system, and they are constant vectors. Since in the cylindrical system of coordinates

$$\nabla^2\mathbf{i}_\varphi=\frac{1}{r^2}\frac{\partial^2}{\partial\varphi^2}\left(-\sin\varphi\,\mathbf{i}+\cos\varphi\,\mathbf{j}\right),$$

we have

$$\nabla^2\mathbf{i}_\varphi=-\frac{\mathbf{i}_\varphi}{r^2} \tag{6.428}$$

Thus, eq. 6.426 becomes

$$\nabla^2\,\tilde{\psi}+\left(k_s^2-\frac{1}{r^2}\right)\tilde{\psi}=0 \tag{6.429}$$

Applying again the method of separation of variables

$$\tilde{\psi}=TZ,$$

we obtain as before the same equation for Z, but function T has to obey a different Bessel equation

$$\frac{d^2T}{dr^2}+\frac{1}{r}\frac{dT}{dr}+\left(m^2-\frac{1}{r^2}\right)T=0, \tag{6.430}$$

and its solutions are Bessel functions of the first order

$$J_1\,(mr) \qquad \text{and} \qquad Y_1\,(mr)$$

The latter has to be discarded, since it tends to infinity when $r\to 0$. Respectively, the general solution of eq. 6.429 has the form

$$\tilde{\psi}_s(r,z,\omega)=\int\limits_0^\infty\left(B_m\,e^{-m_s\,z}+D_m\,e^{m_s\,z}\right)\,J_1\,(mr)\;dm \tag{6.431}$$

Here

$$m_s = \sqrt{m^2 - k_s^2}$$

Note that functions $J_0(v)$ and $J_1(v)$ are related to each other:

$$J_1(x) = -J_0'(x) \tag{6.432}$$

It is clear that $\tilde{\varphi}$ and $\tilde{\psi}_s$ satisfy the corresponding Helmholtz equations, regardless of the values of the unknowns. In other words, as in the two-dimensional case, these equations have an infinite number of solutions. Our goal is to choose such values of $A_m, B_m, C_m,$ and D_m that potentials would also obey the other conditions of the boundary value problem. In order to accomplish this task, we represent the primary potential $\tilde{\varphi}_i$ in the same manner as potentials of secondary waves. This can be done with the help of the Sommerfeld integral,

$$C_l \frac{e^{i\,k_l\,R}}{R} = C_l \int\limits_0^\infty \frac{m}{m_l}\, e^{-m_l\,|z-d|}\; J_0(mr)\; dm, \tag{6.433}$$

where

$$R = \sqrt{r^2 + (z-d)^2}$$

Then, taking into account the condition at infinity, the potentials are written in the form

$$\tilde{\varphi} = C_l \int\limits_0^\infty \left[\frac{m}{m_l} e^{-m_l\,|z-d|} + A_m\, e^{-m_l\,z}\right] J_0(mr)\; dm \tag{6.434}$$

and

$$\tilde{\psi}_s = C_l \int\limits_0^\infty B_m\, e^{-m_s\,z} J_1(mr)\; dm \tag{6.435}$$

It is obvious that functions $\tilde{\varphi}$ and $\tilde{\psi}$ given by eqs. 6.434 and 6.435, obey the Helmholtz equations as well as conditions near the source and at infinity. Now, making use of eqs. 6.414, we will find unknown coefficients A_m and B_m.

Stress in the cylindrical system of coordinates and conditions at the free boundary

Due to axial symmetry, we can expect that the azimuthal component of displacement, s_φ, is equal to zero. Correspondingly, displacement vector $\mathbf{s}$ is

$$\mathbf{s} = s_r\ \mathbf{i}_r + s_z\ \mathbf{i}_z, \tag{6.436}$$

where $\mathbf{i}_r$ and $\mathbf{i}_z$ are unit vectors along coordinate lines. In this case the strains are

$$e_{rr} = \frac{\partial s_r}{\partial r},\quad e_{\varphi\varphi} = \frac{s_r}{r},\quad e_{zz} = \frac{\partial s_z}{\partial z},\quad e_{\varphi z} = 0,\ e_{rz} = \frac{\partial s_r}{\partial z} + \frac{\partial s_z}{\partial r},\quad e_{r\varphi} = 0 \tag{6.437}$$

Then, in accordance with Hooke's law, we have

$$\tau_{rr} = \lambda\ \text{div}\ \mathbf{s}{+}2\mu\ e_{rr},\quad \tau_{\varphi\varphi} = \lambda\ \text{div}\ \mathbf{s}{+}2\mu\ e_{\varphi\varphi},\qquad \tau_{zz} = \lambda\ \text{div}\ \mathbf{s}{+}2\mu\ e_{zz} \tag{6.438}$$

and

$$\tau_{rz} = \mu\ e_{rz}, \tag{6.439}$$

while

$$\tau_{\varphi z} = \tau_{r\varphi} = 0 \tag{6.440}$$

Thus, the boundary conditions, eqs. 6.414, can be written as

$$\lambda\ \text{div}\ \mathbf{s}{+}2\mu\ e_{zz} = 0 \qquad \text{and} \qquad e_{rz} = 0 \qquad \text{if} \qquad z = 0 \tag{6.441}$$

Since the boundary value problem is formulated with the help of potentials, it is necessary to express displacement in terms of functions $\widetilde{\varphi}$ and $\widetilde{\psi}$. By definition

$$\widetilde{\mathbf{s}} = \text{grad}\ \widetilde{\varphi} + \text{curl}\ \widetilde{\boldsymbol{\psi}}$$

Taking into account that

$$\text{grad}\ \widetilde{\varphi} = \frac{\partial\widetilde{\varphi}}{\partial r}\ \mathbf{i}_r + \frac{\partial\widetilde{\varphi}}{\partial z}\ \mathbf{i}_z \qquad \text{and} \qquad \text{curl}\ \widetilde{\boldsymbol{\psi}} = \frac{1}{r}\begin{vmatrix} \mathbf{i}_r & r\ \mathbf{i}_\varphi & \mathbf{i}_z \\ \dfrac{\partial}{\partial r} & \dfrac{\partial}{\partial \varphi} & \dfrac{\partial}{\partial z} \\ 0 & r\widetilde{\psi} & 0 \end{vmatrix},$$

we obtain

$$\widetilde{s}_r = \frac{\partial\widetilde{\varphi}}{\partial r} - \frac{\partial\widetilde{\psi}}{\partial z},\qquad\qquad \widetilde{s}_z = \frac{\partial\widetilde{\varphi}}{\partial z} + \frac{1}{r}\frac{\partial}{\partial r}\left(r\widetilde{\psi}\right) \tag{6.442}$$

Thus, eqs. 6.441 become

$$-\lambda\, k_l^2 \widetilde{\varphi} + 2\mu \left[\frac{\partial^2 \widetilde{\varphi}}{\partial z^2} + \frac{1}{r} \frac{\partial^2}{\partial r\, \partial z} \left(r \widetilde{\psi} \right) \right] = 0 \qquad (6.443)$$

and $$2\, \frac{\partial^2 \widetilde{\varphi}}{\partial r\, \partial z} - \frac{\partial^2 \widetilde{\psi}}{\partial z^2} + \frac{\partial}{\partial r} \left[\frac{1}{r} \frac{\partial}{\partial r} \left(r \widetilde{\psi} \right) \right] = 0, \qquad \text{if} \quad z = 0$$

since

$$\text{div grad } \widetilde{\varphi} = \nabla^2 \widetilde{\varphi} = -k_l^2 \widetilde{\varphi} \qquad \text{and} \qquad \text{div curl } \widetilde{\boldsymbol{\psi}} = 0$$

Further transformations are based on the following equalities:

$$\text{if} \quad \int\limits_0^\infty f(m)\, J_n(mr)\, dm = 0 \quad \text{then} \quad f(m) = 0 \qquad (6.444)$$

Here $J_n(mr)$ is the Bessel function of the n-order. This result is similar to that for Fourier's integral. Also

$$J_0'(v) = -J_1(v) \qquad \text{and} \qquad \frac{d}{dv}\left[v J_1(v) \right] = v J_0(v) \qquad (6.445)$$

Besides, letting $v = mr$, we have

$$\frac{\partial}{\partial r} \frac{1}{r} \left\{ \frac{\partial}{\partial r} \left[r\, J_1(mr) \right] \right\} = m^2 \frac{\partial}{\partial v} \frac{1}{v} \frac{\partial}{\partial v} \left[v J_1(v) \right] = -m^2 J_1(v)$$

Substitution of eqs. 6.434 and 6.435 into eq. 6.443 and using eqs. 6.444 and 6.445 gives

$$-\lambda\, k_l^2 \left(\frac{m}{m_l}\, e^{-m_l\, d} + A_m \right) + 2\mu \left(m\, m_l\, e^{-m_l\, d} + m_l^2\, A_m - m\, m_s B_m \right) = 0$$

and

$$2 \left(-m^2 e^{-m_l\, d} + m_l\, m\, A_m \right) - \left(2m^2 - k_s^2 \right)\, B_m = 0$$

or

$$-\left(2m^2 - k_s^2 \right)\, A_m + 2m\, m_s B_m = \frac{m}{m_l}\, e^{-m_l\, d}\, \left(2m^2 - k_s^2 \right) \qquad (6.446)$$

and $$2m\ m_l\ A_m - \left(2m^2 - k_s^2\right)\ B_m = 2m^2\ e^{-m_l\ d}$$

Solution of this system is

$$A_m = -\frac{m}{m_l}\,\frac{\left(2m^2 - k_s^2\right)^2 + 4m^2\ m_l\ m_s}{\left(2m^2 - k_s^2\right)^2 - 4m^2\ m_l\ m_s}\ e^{-m_l\ d} \tag{6.447}$$

and $$B_m = -\frac{4m^2\left(2m^2 - k_s^2\right)}{\left(2m^2 - k_s^2\right)^2 - 4m^2\ m_l\ m_s}\ e^{-m_l\ d}$$

Thus, we have found coefficients A_m and B_m, that stresses vanish at the free surface. Therefore, all conditions of the boundary value problem are satisfied. In particular, this means that the assumption about the azimuthal component of the vector potential was correct. From eqs. 6.434 and 6.435 we have

$$\tilde{\varphi}_s = -C_l \int_0^\infty \frac{m}{m_l}\,\frac{\left(2m^2 - k_s^2\right)^2 + 4m^2\ m_l\ m_s}{\left(2m^2 - k_s^2\right)^2 - 4m^2\ m_l\ m_s}\ e^{-m_l\ (z+d)}\ J_0(mr)\ dm \tag{6.448}$$

and $$\tilde{\psi}_s = -4C_l \int_0^\infty \frac{m^2\left(2m^2 - k_s^2\right)\ e^{-m_l\ d}e^{-m_s\ z}}{\left(2m^2 - k_s^2\right)^2 - 4m^2\ m_l\ m_s}\ J_1(mr)\ dm$$

The similarity with the two-dimensional case is obvious; for instance, the denominator of integrands in eqs. 6.448 is again described by the left side of the Rayleigh equation:

$$\left(2m^2 - k_s^2\right)^2 - 4m^2\ m_l\ m_s = 0 \tag{6.449}$$

2. Point source of S wave Suppose that a small spherical source rotates around the z-axis, Fig. 6.10c. As was shown in Chapter 3, the vector potential of a shear wave in a homogeneous medium is described by the z-component only. This potential can be represented in the form

$$\tilde{\psi}_i = C_s\frac{e^{i\ k_s R}}{R} = C_s \int_0^\infty \frac{m}{m_s}\ e^{-m_s\ |z-d|}\ J_0(mr)\ dm \tag{6.450}$$

Displacement carried out by the incident wave has the φ-component only, and the field possesses axial symmetry. It is natural to assume that even in the presence of the horizontal interface, secondary wave $\mathbf{s}$ has the same behavior, i.e.,

$$\mathbf{s} = s_\varphi \mathbf{i}_\varphi \qquad \text{and} \qquad \frac{\partial s_\varphi}{\partial \varphi} = 0$$

Correspondingly, the resultant potential can be written as

$$\tilde{\psi}_s = C_s \int_0^\infty \left[\frac{m}{m_s}\, e^{-m_s\,|z-d|} + A_m\, e^{-m_s\, z} \right] J_0(mr)\, dm \tag{6.451}$$

In this case the normal strains are absent:

$$\tilde{e}_{rr} = \tilde{e}_{\varphi\varphi} = \tilde{e}_{zz} = 0, \quad \text{but} \quad \tilde{e}_{\varphi z} = \frac{\partial \tilde{s}_\varphi}{\partial z}, \quad \tilde{e}_{rz} = 0, \quad \tilde{e}_{r\varphi} = \frac{\partial \tilde{s}_\varphi}{\partial r} - \frac{\tilde{s}_\varphi}{r} \tag{6.452}$$

Thus, at the free boundary we need to satisfy only one condition:

$$\frac{\partial \tilde{s}_\varphi}{\partial z} = 0 \qquad \text{if} \qquad z = 0 \tag{6.453}$$

By definition

$$\tilde{\mathbf{s}} = \text{curl}\, \tilde{\boldsymbol{\psi}} = \text{curl}\left(\tilde{\psi}\mathbf{i}_z\right),$$

and therefore

$$\tilde{s}_\varphi = -\frac{\partial \tilde{\psi}}{\partial r} \tag{6.454}$$

From eq. 6.454 we have

$$\tilde{s}_\varphi = C \int_0^\infty \left(\frac{m^2}{m_s}\, e^{-m_s\,|z-d|} - m\, A_m\, e^{-m_s\, z} \right) J_1(mr)\, dm, \tag{6.455}$$

and eq. 6.453 becomes

$$\int_0^\infty \left(m^2\, e^{-m_s d} - m\, m_s\, A_m \right) J_1(mr)\, dm = 0$$

This gives

$$A_m = \frac{m}{m_s}\, e^{-m_s d} \tag{6.456}$$

The vector potential of the secondary field is

$$\tilde{\psi}_s = C_s \int_0^\infty \frac{m}{m_s}\, e^{-m_s\,(z+d)}\, J_0(mr)\, dm$$

or

$$\tilde{\psi}_s = C_s\, \frac{e^{i\,k_s R_1}}{R_1} \tag{6.457}$$

where

$$R_1 = \sqrt{r^2 + (z+d)^2}$$

This shows that the reflected wave is also the shear wave, and its fictitious source is the mirror reflection of the real source with respect to the boundary. Thus, we have

$$\tilde{\psi} = C\left(\frac{e^{i\,k_s R}}{R} + \frac{e^{i\,k_s R_1}}{R_1}\right), \tag{6.458}$$

and in this case no additional waves are generated at the boundary.

3. Point vertical force Now we will consider a more complicated case, in which the vertical force per unit area

$$Z(t) = Z_0 e^{-i\,\omega\,t}$$

is applied in the vicinity of some point of the z-axis, Fig. 6.10d. It is natural to expect that wavefields generated by such a force produce deformation and rotation of elementary volumes of a medium. This means that the wavefield is described by both scalar and vector potentials. In order to find their expressions, suppose first that a medium is homogeneous and that vertical force F_z is constant within the disc of radius a situated at plane $z = 0$ (Fig. 6.10d):

$$F_z = F_z \quad \text{if} \quad z < a \qquad \text{and} \qquad F_z = 0 \quad \text{if} \quad z > a \tag{6.459}$$

Then, as follows from the theory of Bessel functions, $F_z(r, a)$ can be represented in the form

$$F_z = F_z\, a \int_0^\infty J_0(mr)\, J_1(ma)\, dm \tag{6.460}$$

In the limit, when the disc radius a tends to zero, we obtain

$$F_z = F_z \frac{a^2}{2} \int_0^\infty m \; J_0 (mr) \; dm, \tag{6.461}$$

since

$$J_1 (ma) \to \frac{ma}{2} \qquad \text{if} \quad ma \to 0$$

Thus

$$\frac{F_z}{\pi a^2} = \frac{F_z}{2\pi} \int_0^\infty m \; J_0 (mr) \; dm$$

or

$$Z_0 = \frac{F_z}{2\pi} \int_0^\infty m \; J_0 \; (mr) \; dm, \tag{6.462}$$

and it plays the same role as potentials of the incident wave for two other sources. By analogy with the case, when the source generates a P wave, we assume that the vector potential of the incident wave has only an azimuthal component. Then, due to axial symmetry, the potentials of this wave are

$$\tilde{\varphi}_i = \int_0^\infty C_m \, e^{-m_l \, z} \; J_0 (mr) \; dm, \quad \tilde{\psi}_i = \int_0^\infty D_m \, e^{-m_s \, z} \; J_1 (mr) \; dm, \quad \text{if} \quad z > 0 \tag{6.463}$$

and

$$\tilde{\varphi}_i = \int_0^\infty C'_m \, e^{m_l \, z} \; J_0 (mr) \; dm, \quad \tilde{\psi}_i = \int_0^\infty D'_m \, e^{m_s \, z} \; J_1 (mr) \, dm, \qquad \text{if} \;\; z < 0 \tag{6.464}$$

In order to determine the unknowns, we take into account that at plane $z = 0$, normal and shear stresses are discontinuous and continuous functions, respectively:

$$\tilde{\tau}^{+}_{zz} - \tilde{\tau}^{-}_{zz} = -Z_0^* \qquad \text{and} \qquad \tilde{\tau}^{+}_{rz} - \tilde{\tau}^{-}_{rz} = 0 \tag{6.465}$$

Here the "+" and "−" signs characterize stresses in the vicinity of plane $z = 0$ for positive and negative values of z. As follows from eqs. 6.438 and 6.439,

$$\tilde{\tau}_{zz} = -\lambda k_l^2 \tilde{\varphi} + 2\mu \left[\frac{\partial^2 \tilde{\varphi}}{\partial z^2} + \frac{1}{r} \frac{\partial^2}{\partial r \partial z} \left(r \tilde{\psi} \right) \right] \tag{6.466}$$

and $$\tilde{\tau}_{rz} = 2 \frac{\partial^2 \tilde{\varphi}}{\partial r \partial z} - \frac{\partial^2 \tilde{\psi}}{\partial z^2} + \frac{\partial}{\partial r} \left[\frac{1}{r} \frac{\partial}{\partial r} \left(r \tilde{\psi} \right) \right]$$

Substitution of eqs. 6.462-6.464, and 6.466 into eq. 6.465 gives,

$$\left(2m^2 - k_s^2\right) \ (C_m - C'_m) - 2m\, m_s \left(D_m + D'_m\right) = -\frac{F_z}{2\pi \mu}\, m \tag{6.467}$$

and $$2m\, m_l \left(C_m + C'_m\right) - \left(2m^2 - k_s^2\right) \ (D_m - D'_m) = 0$$

Besides, the radial and normal components of displacement are continuous functions everywhere in a medium, including at all points of plane $z = 0$. In accordance with eqs. 6.442, we have

$$\frac{\partial \tilde{\varphi}^+}{\partial r} - \frac{\partial \tilde{\psi}^+}{\partial z} = \frac{\partial \tilde{\varphi}^-}{\partial r} - \frac{\partial \tilde{\psi}^-}{\partial z} \tag{6.468}$$

and $$\frac{\partial \tilde{\varphi}^+}{\partial z} + \frac{1}{r} \frac{\partial}{\partial r} r \tilde{\psi}^+ = \frac{\partial \tilde{\varphi}^-}{\partial z} + \frac{1}{r} \frac{\partial}{\partial r} r \tilde{\psi}^-$$

or

$$-m \left(C_m - C'_m\right) + m_s \left(D_m + D'_m\right) = 0, \tag{6.469}$$

$$-m_l \left(C_m + C'_m\right) + m \left(D_m - D'_m\right) = 0$$

From both sets of equations, we have first

$$C'_m = -C_m, \qquad D'_m = D_m \tag{6.470}$$

and then

$$\left(2m^2 - k_s^2\right) C_m - 2m\, m_s D_m = -\frac{F_z m}{4\pi\mu}, \qquad - mC_m + m_s D_m = 0 \tag{6.471}$$

This gives

$$C_m = \frac{F_z\, m}{4\pi\mu\, k_s^2} \qquad \text{and} \qquad D_m = \frac{F_z\, m^2}{4\pi\mu\, m_s\, k_s^2} \tag{6.472}$$

Let us assume that force is applied at the point of plane $z = d$ and that the z-axis is still directed downward. Then it is obvious that potentials describing the direct wave are

$$\widetilde{\varphi}_i = -\frac{F_z}{4\pi\mu\, k_s^2}\int\limits_0^\infty m\, e^{-m_l\,|z-d|}\, J_0(mr)\, dm \qquad \text{if} \qquad z < d \tag{6.473}$$

and $$\widetilde{\varphi}_i = \frac{F_z}{4\pi\mu\, k_s^2}\int\limits_0^\infty m\, e^{-m_l\,|z-d|}\, J_0(mr)\, dm \qquad \text{if} \qquad z > d$$

Also

$$\widetilde{\psi}_i = \frac{F_z}{4\pi\mu\, k_s^2}\int\limits_0^\infty \frac{m^2}{m_s}\, e^{-m_s\,|z-d|}\, J_1(mr)\, dm \tag{6.474}$$

Using the Sommerfeld integral, we obtain

$$\widetilde{\varphi}_i = -\frac{F_z}{4\pi\mu\, k_s^2}\frac{\partial}{\partial z}\frac{e^{ik_lR}}{R} \qquad \text{and} \qquad \widetilde{\psi}_i = -\frac{F_z}{4\pi\mu\, k_s^2}\frac{\partial}{\partial r}\frac{e^{ik_sR}}{R} \tag{6.475}$$

Here

$$R = \left[r^2 + (z-d)^2\right]^{1/2}$$

Note that these expressions were derived in Chapter 3, where we considered the waves in a homogeneous medium.

Potentials of the secondary wave

Taking into account eqs. 6.474 and 6.475, expressions of potentials $\widetilde{\varphi}_s$ and $\widetilde{\psi}_s$ have the form

$$\widetilde{\varphi}_s = -\frac{F_z}{4\pi\mu\, k_s^2}\int\limits_0^\infty m\, A_m\, e^{-m_l\, z}\, J_0(mr)\, dm \tag{6.476}$$

and $$\tilde{\psi}_s = \frac{F_z}{4\pi\mu\, k_s^2} \int_0^\infty m B_m\, e^{-m_s\, z}\; J_1(mr)\; dm$$

Therefore

$$\tilde{\varphi} = -\frac{F_z}{4\pi\mu\, k_s^2} \int_0^\infty m \left[e^{-m_l\,|z-d|} + A_m\, e^{-m_l\, z} \right] \; J_0(mr)\; dm \tag{6.477}$$

and $$\tilde{\psi} = \frac{F_z}{4\pi\mu\, k_s^2} \int_0^\infty m \left[\frac{m}{m_s}\, e^{-m_s\,|z-d|} + B_m\, e^{-m_s\, z} \right] \; J_1(mr)\, dm$$

Then, making use of the boundary conditions at the free surface, $z = 0$, we obtain:

$$\left(2m^2 - k_s^2\right)\, A_m + 2m\, m_s B_m = -\left(2m^2 - k_s^2\right)\, e^{-m_l\, d} + 2m^2 e^{-m_s\, d} \tag{6.478}$$

and $$2m\, m_l\, A_m + \left(2m^2 - k_s^2\right) B_m = 2m\, m_l\, e^{-m_l\, d} - \frac{m}{m_s} \left(2m^2 - k_s^2\right)\, e^{-m_s\, d}$$

Solution of this system yields

$$A_m = -\frac{\left(2m^2 - k_s^2\right)^2 + 4m^2\, m_l\, m_s}{D} e^{-m_l\, d} + \frac{4m^2\left(2m^2 - k_s^2\right)}{D}\, e^{-m_s\, d} \tag{6.479}$$

and $$B_m = \frac{4m\, m_l\left(2m^2 - k_s^2\right)}{D}\, e^{-m_l\, d} - \frac{m}{m_s} \frac{\left(2m^2 - k_s^2\right)^2 + 4m^2 m_l m_s}{D}\, e^{-m_s\, d}$$

where D is the determinant of the system. It is clear that expressions of potentials are similar to those derived for the two-dimensional source.

Wave behavior

For illustration we derive formulas for displacement components in the far zone, $(k_l r \gg 1)$, when both the source of the P wave and an observation point are located in the vicinity of the free boundary. Then, letting $d = 0$ in eqs. 6.448 and using eqs. 6.442 we obtain

$$\tilde{s}_z = -2k_s^2\, C_l \int_0^\infty \frac{m\left(2m^2 - k_s^2\right)\; J_0(mr)}{\left(2m^2 - k_s^2\right)^2 - 4m^2 m_l\, m_s}\, dm \tag{6.480}$$

and $$\tilde{s}_r = 4k_s^2\, C_l \int\limits_0^\infty \frac{m^2 m_s\, J_1(mr)}{(2m^2 - k_s^2)^2 - 4m^2 m_l\, m_s}\, dm$$

From the physical point of view, it is clear that the reflected wave, which obeys Snell's law, is absent in this case. This is confirmed by the fact that integrands in eqs. 6.480 do not contain exponential terms and, therefore, the stationary-phase method cannot be used. Correspondingly, by analogy with the two-dimensional source, we apply contour integration to find approximate formulas for the wavefields. However, such an approach requires that the integrand in eqs. 6.480 vanish on the complex plane m, when the variable of integration tends to infinity (Jordan lemma). In order to meet this condition, we replace functions J_0 and J_1 by the Hankel functions (Part II). As follows from eqs. 6.480, displacements are expressed in terms of two types of integrals,

$$I = \int\limits_0^\infty F_1(m, m_l, m_s)\; J_0(mr)\; dm \tag{6.481}$$

and $$L = \int\limits_0^\infty F_2(m, m_l, m_s)\; J_1(mr)\; dm,$$

where F_1 and F_2 are analytical functions on the complex plane m except branch points and poles.

Integral I At the beginning consider integral I describing displacement s_z and use the equality

$$J_0(mr) = \frac{1}{2}\left[H_0^{(1)}(mr) + H_0^{(2)}(mr)\right] \tag{6.482}$$

Here $H_0^{(1)}(mr)$ and $H_0^{(2)}(mr)$ are Hankel's functions of the zero order and of the first and second kind. Note that the relationship eq. 6.482 follows from a definition of these functions of any order ν:

$$H_\nu^{(1)}(mr) = J_\nu(mr) + iY_\nu(mr) \quad \text{and} \quad H_\nu^{(2)}(mr) = J_\nu(mr) - iY_\nu(mr) \tag{6.483}$$

Substitution of eq. 6.482 into integral I gives

$$I = \frac{1}{2}\int\limits_0^\infty F_1(m)\; H_0^{(1)}(mr)\, dm + \frac{1}{2}\int\limits_0^\infty F_1(m)\; H_0^{(2)}(mr)\, dm \tag{6.484}$$

As was pointed out earlier, branch points and poles are located either in the first or in the third quadrants. Poles are

$$m_1 = k_s + i\,\varepsilon_s, \quad m_2 = -k_s - i\,\varepsilon_s, \quad m_3 = k_l + i\,\varepsilon_l, \quad m_4 = -k_l - i\,\varepsilon_l, \qquad (\,\varepsilon \ll 1)$$

The asymptotic expressions of functions $H_0^{(1)}(mr)$ and $H_0^{(2)}(mr)$ are

$$H_0^{(1)}(mr) \approx \sqrt{\frac{2}{\pi mr}} e^{i\,(m\,r\,-\,\pi/4)} \tag{6.485}$$

and $$H_0^{(2)}(mr) \approx \sqrt{\frac{2}{\pi mr}} e^{-i\,(m\,r\,-\,\pi/4)} \qquad \text{if } m\,r \gg 1$$

We see that these functions of the first and second kind decay exponentially with an increase of m, if $\operatorname{Im} m > 0$ and $\operatorname{Im} m < 0$, respectively. Applying the Cauchy theorem to the first integral in eq. 6.484, we have:

$$\oint_C F_1(m)\; H_0^{(1)}(mr)\,dm = \int_0^\infty F_1(m)\; H_0^{(1)}(mr)\,dm \tag{6.486}$$

$$+ \int_{C_1+C_2} F_1(m)\; H_0^{(1)}(mr)\,dm + \int_{C_3+C_4} F_1(m)\; H_0^{(1)}(mr)\,dm$$

$$+ \oint_{C_p} F_1(m)\; H_0^{(1)}(mr)\,dm + \int_{i\infty}^{0} F_1(m)\; H_0^{(1)}(mr)\,dm = 0$$

Here path C is situated in the first quadrant, Fig. 6.11, and singularities of the integrand are absent in the area surrounded by C. Also, due to the exponential decay of $H_0^{(1)}(mr)$ for large arguments, the integral along the portion of the path with an infinitely large radius, $\operatorname{Im} m \to \infty$, can be discarded (Jordan lemma). In accordance with eq. 6.486, integration along the real axis m is replaced by integration along the branch lines, around poles, and along the imaginary axis m. Since the last integral is unknown, we have to eliminate it. To do this consider the integral

$$\oint_{C_0} F_1(m)\; H_0^{(2)}(mr)\,dm,$$

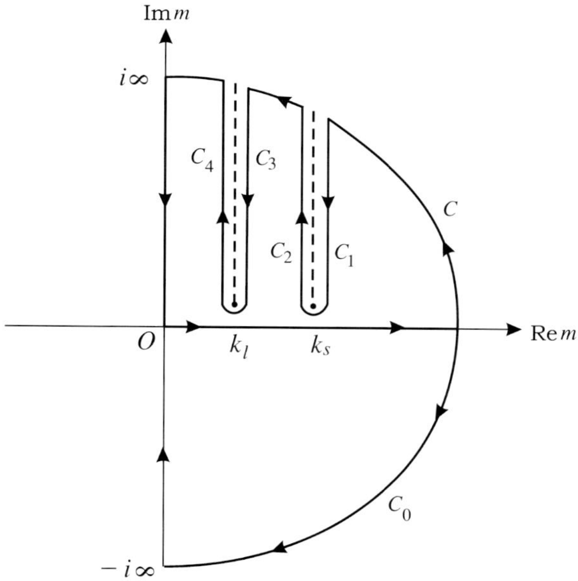

Figure 6.11: Contour of integration in eq. 6.486

where C_0 is located in the fourth quadrant, Fig. 6.11. Taking into account that $F_1(m)$ is the analytical function in all points of this quadrant, we have:

$$\oint_{C_0} F_1(m)\ H_0^{(2)}(mr)\,dm = \int_0^{\infty} F_1(m)\ H_0^{(2)}(mr)\,dm + \int_{-i\infty}^{0} F_1(m)\ H_0^{(2)}(mr)\,dm = 0$$

or

$$\int_0^{\infty} F_1(m)\ H_0^{(2)}(mr)\,dm = -\int_{-i\infty}^{0} F_1(m)\ H_0^{(2)}(mr)\,dm \tag{6.487}$$

The use of eqs. 6.484–6.487 gives for integral I

$$I = -\frac{1}{2}\left[\int_{C_1+C_2} F_1(m)\ H_0^{(1)}(mr)\,dm + \int_{C_3+C_4} F_1(m)\ H_0^{(1)}(mr)\,dm \right. \tag{6.488}$$

$$+\oint_{C_p} F_1(m)\; H_0^{(1)}(mr)\,dm + \int_{i\infty}^{0} F_1(m)\; H_0^{(1)}(mr)\,dm + \int_{-i\infty}^{0} F_1(m)\; H_0^{(2)}(mr)\,dm\Bigg]$$

Now we will demonstrate that the sum of integrals along the imaginary axis m vanishes. Introducing a new variable that differs from m only by sign, the last integral in eq. 6.488 becomes

$$-\int_{i\infty}^{0} F_1(-m)\; H_0^{(2)}(-mr)\,dm$$

Since

$$F_1(-m) = -F_1(m) \qquad \text{and} \qquad H_0^{(2)}(-mr) = -H_0^{(2)}(mr)$$

we have

$$-\int_{i\infty}^{0} F_1(-m)\; H_0^{(2)}(-mr)\,dm = -\int_{i\infty}^{0} F_1(m)\; H_0^{(1)}(mr)\,dm,$$

and eq. 6.488 yields

$$I = -\frac{1}{2}\Bigg[\int_{C_1+C_2} F_1(m)\; H_0^{(1)}(mr)\,dm \tag{6.489}$$

$$+\int_{C_3+C_4} F_1(m)\; H_0^{(1)}(mr)\,dm + \oint_{C_p} F_1(m)\; H_0^{(1)}(mr)\,dm\Bigg]$$

Thus, as in the case of the two-dimensional source, we have represented the vertical component of displacement as a sum of integrals along branch lines and around poles only.

Integral L In the same manner, consider the second integral in eq. 6.481, describing the radial component of displacement. As follows from eqs. 6.483,

$$J_1(mr) = \frac{1}{2}\left[H_1^{(1)}(mr) + H_1^{(2)}(mr)\right],$$

and therefore

$$L = \frac{1}{2}\left[\int_0^\infty F_2(m)\ H_1^{(1)}(mr)\,dm + \int_0^\infty F_2(m)\ H_1^{(2)}(mr)\,dm\right] \tag{6.490}$$

Taking into account the asymptotic behavior of Hankel's functions

$$H_1^{(1)}(mr) \approx \sqrt{\frac{2}{\pi m\, r}} e^{i\,(m\,r\, -\, 3\pi/4)} \quad \text{and} \quad H_1^{(2)}(mr) \approx \sqrt{\frac{2}{\pi m\, r}} e^{-i\,(m\,r\, -\, 3\pi/4)}, \tag{6.491}$$

we can again use closed paths C and C_0. By analogy with eq. 6.488, we get

$$L = -\frac{1}{2}\left[\int_{C_1+C_2} F_2(m)\,H_1^{(1)}(mr)\,dm + \int_{C_3+C_4} F_2(m)\,H_1^{(1)}(mr)\,dm\right.$$

$$\left. + \oint_{C_p} F_2(m)\,H_1^{(1)}(mr)\,dm + \int_{i\infty}^{0} F_2(m)\,H_1^{(1)}(mr)\,dm + \int_{-i\infty}^{0} F_2(m)\,H_1^{(2)}(mr)\,dm\right] \tag{6.492}$$

By definition, eq. 6.481, $F_2(m) = F_2(-m)$. Besides, as is well known, $H_1^{(2)}(-mr) = H_1^{(1)}(mr)$.

Replacement of the variable in the last integral yields

$$\int_{-i\infty}^{0} F_2(m)\ H_1^{(2)}(mr)\,dm = -\int_{i\infty}^{0} F_2(m)\ H_1^{(1)}(mr)\,dm,$$

and in place of eq. 6.492 we have

$$L = -\frac{1}{2}\left[\int_{C_1+C_2} F_2(m)\ H_1^{(1)}(mr)\,dm\right. \tag{6.493}$$

$$\left. + \int_{C_3+C_4} F_2(m)\ H_1^{(1)}(mr)\,dm + \oint_{C_p} F_2(m)\ H_1^{(1)}(mr)\,dm\right]$$

Now we are prepared to derive asymptotic formulas for displacement.

Vertical component s_z

As follows from eqs. 6.480 and 6.489,

$$\tilde{s}_z = -k_s^2\, C_l\,(I_s + I_l + I_p) \tag{6.494}$$

Integral I_s Integration along branch lines C_1 and C_2 yields

$$I_s = \int\limits_{k_s}^{k_s+i\infty} m\left(2m^2 - k_s^2\right)$$

$$\times \left[\frac{1}{\left(2m^2-k_s^2\right)^2 - 4m^2\, m_l\, m_s} - \frac{1}{\left(2m^2-k_s^2\right)^2 + 4m^2 m_l m_s}\right] H_0^{(1)}(mr)\ dm,$$

since the radical m_s has different signs on C_1 and C_2. Thus,

$$I_s = 8 \int\limits_{k_s}^{k_s+i\infty} \frac{m^3\, m_l\, m_s\,(2m^2 - k_s^2)}{(2m^2-k_s^2)^4 - 16m^4\, m_l^2\, m_s^2}\, H_0^{(1)}(mr)\ dm \tag{6.495}$$

Introducing a new variable, $t:\ m = k_s + it$, and assuming that the integral is defined by the initial part of an integration, we have

$$m = k_s, \qquad dm = i\ dt, \qquad m_s = \sqrt{2i\ k_s\ t} \tag{6.496}$$

Because $k_s r \gg 1$, we can use the asymptotic expression of Hankel's function, which gives

$$H_0^{(1)}(mr) = \sqrt{\frac{2}{\pi k_s r}}\, e^{i\,(k_s r - \pi/4)} e^{-r\ t} \tag{6.497}$$

Substituting eqs. 6.496 and 6.497 into eq. 6.495, we obtain

$$I_s = \frac{8\ i\left(1 - \dfrac{k_l^2}{k_s^2}\right)^{1/2}}{k_s^2\ r^2}\ e^{i\ k_s\ r}, \tag{6.498}$$

since

$$\int\limits_0^\infty t^{1/2}\ e^{-r\ t} dt = \frac{\sqrt{\pi}}{2r^{3/2}}$$

Correspondingly,

$$\tilde{s}_z(k_s) = -\frac{8C_l\, i\left(1-\dfrac{k_l^2}{k_s^2}\right)^{1/2} e^{ik_s r}}{r^2} \tag{6.499}$$

Eq. 6.499 describes the component of displacement, $\tilde{s}_z(k_s)$, related to two waves propagating along the boundary with velocity c_s. One is the shear wave moving through a medium; the other is the longitudinal wave propagating along the boundary, also with velocity c_s. The longitudinal wave arises in the same manner as in the two-dimensional case. When the shear wave moves through a medium, vibration of particles of the boundary gives rise to the evanescent longitudinal wave, which has the same velocity of propagation, c_s, along the boundary but exponentially decays with depth z.

Integral I_l Performing integration along paths C_3 and C_4, it is evident that the resultant integral I_l has the same integrand as that in eq. 6.495:

$$I_l = 8\int\limits_{k_l}^{k_l+i\infty} \frac{m^3 m_l\, m_s\,(2m^2-k_s^2)}{(2m^2-k_s^2)^4 - 16m^4 m_l^2\, m_s^2}\, H_0^{(1)}(mr)\; dm \tag{6.500}$$

Then, use of variable t, $m = k_l + it$, gives

$$I_l = \frac{8\left(1-\dfrac{k_l^2}{k_s^2}\right)^{1/2} e^{ik_l\, r}}{k_l^2\left(2-\dfrac{k_s^2}{k_l^2}\right)^3 r^2}, \tag{6.501}$$

and

$$\tilde{s}_z(k_l) = -8C_l\left(\frac{k_s}{k_l}\right)^2 \frac{\left(\dfrac{k_s^2}{k_l^2}-1\right)^{1/2}}{\left(2-\dfrac{k_s^2}{k_l^2}\right)^3}\, \frac{e^{i\,k_l\, r}}{r^2} \tag{6.502}$$

This portion of displacement, $\tilde{s}_z$, is also due to two waves. One is the P wave advancing through a medium with velocity c_l; it has a spherical wavefront. The second is the conical wave; its velocity of propagation is equal to that of the shear wave, c_s. Along the boundary, however, it moves with velocity c_l. The wavefront of this wave is the lateral surface of the cone; the apex is located on the z-axis. The appearance of this conical wave was described earlier when we considered wavefields caused by a linear source. Propagation of the longitudinal wave produces vibration of particles of the free boundary, which becomes the source of elementary shear waves. The conical wave is the result constructive interference by elementary shear waves. Thus, the sum

$$\widetilde{s}_z(k_s) + \widetilde{s}_z(k_l)$$

is associated with waves moving along the boundary with two different velocities, c_s and c_l, and each wave is a superposition of longitudinal and shear waves. Of course, this consideration is directly applied to function s_r.

Radial component s_r

In accordance with eqs. 6.480 and 6.493,

$$\widetilde{s}_r = -k_s^2 C_l \left(L_s + L_l + L_p\right) \tag{6.503}$$

Integral L_s Integration around branch point k_s gives

$$L_s = 2 \int\limits_{k_s}^{k_s+i\infty} m^2 m_s \left[\frac{1}{\left(2m^2 - k_s^2\right)^2 - 4m^2 m_l m_s} + \frac{1}{\left(2m^2 - k_s^2\right)^2 + 4m^2 m_l m_s} \right] H_1^{(1)}(mr)\, dm$$

or

$$L_s = 4 \int\limits_{k_s}^{k_s+i\infty} \frac{m^2 m_s \left(2m^2 - k_s^2\right)^2 H_1^{(1)}(mr)}{\left(2m^2 - k_s^2\right)^4 - 16m^4 m_l^2\, m_s^2}\, dm \tag{6.504}$$

Again performing a simple transformation, $m = k_s + it$, we obtain

$$L_s = \frac{4i\, k_s^2\, \sqrt{k_s}\sqrt{2i}\sqrt{2}\; e^{-i\,3\pi/4}\; e^{ik_s\, r}\sqrt{\pi}}{k_s^4\, \sqrt{\pi k_s\, r}\; r^{3/2}\; 2}$$

or

$$L_s = \frac{4\, e^{i\, k_s r}}{k_s^2\, r^2} \tag{6.505}$$

and

$$\widetilde{s}_r(k_s) = -\frac{4\, C_l\, e^{ik_s\, r}}{r^2} \tag{6.506}$$

Integral L_l In the same manner, for integrals along branch lines C_3 and C_4 we have

$$L_l = 2 \int_{k_l}^{k_l + i\infty} m^2 m_s \left[\frac{1}{(2m^2 - k_s^2)^2 - 4m^2 m_l\, m_s} - \frac{1}{(2m^2 - k_s^2)^2 + 4m^2 m_l m_s} \right] H_1^{(1)}(mr)\, dm$$

or

$$L_l = 16 \int_{k_l}^{k_l + i\infty} \frac{m^4 m_l\, m_s^2\, H_1^{(1)}(mr)\, dm}{(2m^2 - k_s^2)^4 - 16 m^4\, m_l^2\, m_s^2} \tag{6.507}$$

This gives

$$L_l = \frac{16i\, k_l^4 \sqrt{2i\, k_l}\, k_l^2 \left(1 - \dfrac{k_s^2}{k_l^2}\right) e^{ik_l\, r} \sqrt{2} \sqrt{\pi} e^{-i\, 3\pi/4}}{\sqrt{\pi k_l r}\, k_l^8 \left(2 - \dfrac{k_s^2}{k_l^2}\right)^4 2\, r^{3/2}}$$

and

$$L_l = - \frac{\left(\dfrac{k_s^2}{k_l^2} - 1\right) 16}{\left(2 - \dfrac{k_s^2}{k_l^2}\right)^4} \frac{e^{ik_l\, r}}{k_l^2\, r^2} \tag{6.508}$$

Thus

$$\tilde{s}_r(k_l) = \frac{16 C_l \left(\dfrac{k_s^2}{k_l^2} - 1\right)}{\left(2 - \dfrac{k_s^2}{k_l^2}\right)^4} \frac{e^{i\, k_l\, r}}{r^2} \tag{6.509}$$

Correspondingly, the radial component of displacement associated with branch points is

$$\tilde{s}_r(k_s) + \tilde{s}_r(k_l)$$

We have found asymptotic expressions of displacement at the free boundary caused by waves that propagate with either velocity c_s or velocity c_l. There is also a surface wave, and its velocity c_R $(c_R < c_s)$ is defined from the real root of the Rayleigh equation (eq. 6.449). In order to determine the magnitude of displacement caused by the Rayleigh wave, we have to calculate the residual at the pole, k_R, of the integrands in eq. 6.480. Since in the far zone $k_s r \gg 1$, asymptotic formulas for Hankel's functions can be used. This means that displacement components related to the surface wave decay as $1/\sqrt{r}$, – that is, much more slowly than the same components caused by the other waves.

Chapter 7

Propagation of elastic waves in borehole containing a fluid

In this chapter we will study wave propagation in a medium with a cylindrical interface and mostly at relatively large distances from the source. As in the horizontally layered medium, our attention is paid to reflected, transmitted, head, and surface waves, as well as to normal modes.

7.1 Solution of the boundary value problem

Suppose that a cylindrical borehole with radius a is filled by a fluid, and the elementary spherical source of the sinusoidal P wave is located on its axis, Fig. 7.1a. The surrounding medium is an elastic one, and the density and wave velocity inside and outside the borehole are ρ_1, c_1 and ρ_2, c_l, c_s, respectively. Taking into account the geometry of the model, we have chosen a cylindrical system of coordinates with the origin at point O, where the source is situated, and the z-axis coinciding with the axis of the cylinder, Fig. 7.1a. In order to describe the wavefields we again use scalar and vector potentials that, due to axial symmetry, depend on two coordinates, r and z, only. By definition, displacement $\mathbf{s}$ is related to the potentials as

$$\mathbf{s} = \text{grad}\ \varphi + \text{curl}\ \boldsymbol{\psi} \tag{7.1}$$

Correspondingly, vector $\mathbf{s}$ has two components, s_r and s_z, which are independent of the azimuthal coordinate. Since the borehole is filled by a fluid, the rotational waves are absent inside of it, and the wavefields are described by scalar potential only.

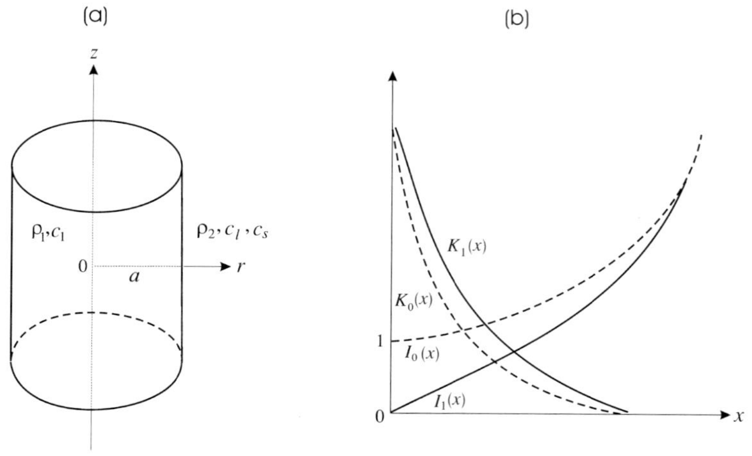

Figure 7.1: (a) Cylindrical borehole (b) Behavior of modified Bessel function

Next we will derive formulas for potentials inside the borehole and in an elastic medium, and with this purpose in mind we will formulate, as usual, the boundary value problem. First, we know that complex amplitudes of potentials obey the Helmholtz equations:

$$\nabla^2\tilde{\varphi}_1 + k_1^2\tilde{\varphi}_1 = 0 \qquad \text{if} \qquad r < a \tag{7.2}$$

and

$$\nabla^2\tilde{\varphi}_2 + k_l^2\tilde{\varphi}_2 = 0, \qquad \nabla^2\tilde{\boldsymbol{\psi}}_2 + k_s^2\tilde{\boldsymbol{\psi}}_2 = 0, \qquad \text{if} \qquad r > a \tag{7.3}$$

where

$$k_1 = \frac{\omega}{c_1}, \qquad k_l = \frac{\omega}{c_l}, \qquad k_s = \frac{\omega}{c_s} \tag{7.4}$$

Of course, eqs. 7.2 and 7.3 are invalid at the origin, where the source of the primary field is located, as well as at the interface $r = a$. At points of this boundary between a fluid and an elastic medium, the tangential component of displacement can be a discontinuous function. At the same time, the normal component of displacement and normal stress are continuous functions. Since shear stress is absent in the fluid, it is also equal to zero at points of the interface. Thus, the boundary conditions are

$$\tilde{s}_r^{(1)} = \tilde{s}_r^{(2)}, \qquad \tilde{\tau}_{rr}^{(1)} = \tilde{\tau}_{rr}^{(2)}, \qquad \tilde{\tau}_{zr}^{(2)} = 0 \qquad \text{if} \qquad r = a \tag{7.5}$$

It is convenient to represent the scalar potential inside the borehole as the sum

$$\tilde{\varphi}_1(r,z,\omega) = \tilde{\varphi}_0(r,z,\omega) + \tilde{\varphi}_s(r,z,\omega), \tag{7.6}$$

where $\tilde{\varphi}_0(r,z,\omega)$ is the scalar potential of a direct wave in a homogeneous medium with parameters ρ_1 and c_1. As we know,

$$\tilde{\varphi}_0 = C\,\frac{e^{ik_1 R}}{R} \tag{7.7}$$

Here C is the constant and R is the distance from the origin:

$$R = \left(r^2 + z^2\right)^{1/2}$$

Function $\tilde{\varphi}_s$ describes secondary waves caused by the presence of the boundary, and it is finite everywhere. It is clear that on approaching the source, the primary potential becomes dominant:

$$\tilde{\varphi}_1 \to \tilde{\varphi}_0 = C\,\frac{e^{ik_1 R}}{R} \qquad \text{if} \qquad R \to 0 \tag{7.8}$$

Finally, assuming the presence of attenuation, even it is very small, we can conclude that at very large distances from the source, wavefields vanish, and we have condition at infinity:

$$\tilde{\varphi}_1 \to 0, \qquad \tilde{\varphi}_2 \to 0, \qquad \text{and} \qquad \tilde{\psi}_2 \to 0, \qquad \text{if} \qquad R \to \infty \tag{7.9}$$

Thus, we have formulated the boundary value problem, and our goal is to find scalar and vector potentials that satisfy eqs. 7.2 and 7.3, as well as conditions 7.5, 7.8 and 7.9.

Solution of Helmholtz equation for scalar potential $\tilde{\varphi}$

As was shown in the previous chapter, in the cylindrical system of coordinates the Helmholtz equation is written in the form

$$\frac{\partial^2 \tilde{\varphi}}{\partial r^2} + \frac{1}{r}\frac{\partial \tilde{\varphi}}{\partial r} + \frac{\partial^2 \tilde{\varphi}}{\partial z^2} + k^2 \tilde{\varphi} = 0 \tag{7.10}$$

Here

$$\begin{aligned} k &= k_1 \qquad \text{if} \qquad r < a \\ k &= k_l \qquad \text{if} \qquad r > a \end{aligned}$$

Applying again the method of separation of variables, we have

$$\widetilde{\varphi}(r,z,\omega) = T\left(r,\omega\right)\ Z\left(z\right) \tag{7.11}$$

Substitution of eq. 7.11 into eq. 7.10 gives two ordinary differential equations of the second order,

$$\frac{1}{T}\frac{d^2T}{dr^2} + \frac{1}{rT}\frac{dT}{dr} + k^2 = \mp m^2 \qquad \text{and} \qquad \frac{1}{Z}\frac{d^2Z}{dz^2} = \pm m^2, \tag{7.12}$$

where m is an arbitrary variable of separation. In choosing a sign on the right side of eqs. 7.12, we assume, by analogy with function $\widetilde{\varphi}_0$, that the scalar potential in both media is an even function of coordinate z. Thus it is proper to choose minus sign on the right side of the second equation of set 7.12, and this gives

$$\frac{d^2Z}{dz^2} + m^2 Z = 0$$

Its partial solution is $\sin mz$ and $\cos mz$ with arbitrary constants, and the latter is an even function of Z. For this reason, $\cos mz$ is used to represent the complex amplitude of potential $\widetilde{\varphi}$. Correspondingly, the first equation of the set 7.12 becomes

$$\frac{d^2T}{dr^2} + \frac{1}{r}\frac{dT}{dr} - \left(m^2 - k^2\right)T = 0, \tag{7.13}$$

Its partial solutions are modified Bessel functions of the first and second kind,

$$I_0\left(\sqrt{m^2-k^2}\ r\right) \qquad \text{and} \qquad K_0\left(\sqrt{m^2-k^2}\ r\right),$$

but of the zero order. In particular, if the argument is real and positive, the behavior of these functions is very simple, Fig. 7.1b. For instance, their asymptotic expressions are

$$I_0\left(x\right) \to 1, \qquad K_0\left(x\right) \to -\ln x \qquad \text{if} \quad x \to 0 \tag{7.14}$$

$$\text{and} \qquad I_0\left(x\right) \to \frac{1}{\left(2\pi x\right)^{1/2}}\, e^x, \quad K_0\left(x\right) \to \left(\frac{\pi}{2x}\right)^{1/2} e^{-x} \qquad \text{if} \qquad x \to \infty$$

We will also use modified Bessel functions of the first order:

$$I_1\left(x\right) \qquad \text{and} \qquad K_1\left(x\right)$$

Their asymptotic formulas are

$$I_1(x) \to \frac{x}{2}, \qquad K_1(x) \to \frac{1}{x} \qquad \text{if} \quad x \to 0 \tag{7.15}$$

and $$I_1(x) \to \left(\frac{1}{2\pi x}\right)^{1/2} e^{x}, \qquad K_1(x) \to \left(\frac{\pi}{2x}\right)^{1/2} e^{-x} \qquad \text{if} \qquad x \to \infty$$

By definition, the sum

$$T_m(r,\omega) = A_m I_0(m_i r) + B_m K_0(m_i r) \tag{7.16}$$

satisfies eq. 7.13, and

$$m_i = m_1 = \sqrt{m^2 - k_1^2} \qquad \text{if} \qquad r < a$$
$$m_i = m_l = \sqrt{m^2 - k_l^2} \qquad \text{if} \qquad r > a$$

Taking into account the assumption about symmetry with respect to plane $z = 0$ and eq. 7.11, the partial solution of the Helmholtz equation is

$$\widetilde{\varphi}_m = [A_m I_0(m_i r) + B_m K_0(m_i r)] \cos mz \tag{7.17}$$

Therefore, for the general solution we obtain

$$\widetilde{\varphi}(r,z,\omega) = \int_0^\infty [A_m I_0(m_i r) + B_m K_0(m_i r)] \cos mz \; dm \tag{7.18}$$

As was pointed out earlier, secondary waves have finite value everywhere. For this reason, the secondary scalar potential $\widetilde{\varphi}_s$ is written as

$$\widetilde{\varphi}_s = \int_0^\infty A_m I_0(m_1 r) \cos mz \; dm, \qquad \text{if} \qquad r < a \tag{7.19}$$

since function $K_0(m_1 r)$ is infinitely large at the point of the borehole axis $(r = 0)$. In approaching the origin $(r \to 0, \; z \to 0)$, the secondary potential tends to a finite value. Correspondingly, the function $\widetilde{\varphi}_1$ satisfies the condition near the source. Taking into account that $I_0(m_l r)$ increases without limit with an increase of r, potential $\widetilde{\varphi}_2$ in an elastic medium is

$$\widetilde{\varphi}_2(r,z,\omega) = \int_0^\infty B_m K_0(m_l r) \cos mz \; dm \tag{7.20}$$

Thus functions $\tilde{\varphi}_1$ and $\tilde{\varphi}_2$ satisfy the Helmholtz equation as well as conditions near the source and at infinity when $r \to \infty$. Note that these functions also vanish with the increase of $|z|$. This happens because of the oscillating factor $\cos mz$ in integrals of eqs. 7.19 and 7.20.

Solution of Helmholtz equation for vector potential ψ

We will assume that the vector potential has only an azimuthal component. In this case the Helmholtz equation becomes

$$\nabla^2 \tilde{\psi} + \left(k_s^2 - \frac{1}{r^2} \right) \tilde{\psi} = 0$$

Then use of the method of separation of variables

$$\tilde{\psi} = T\ (r, \omega)\ Z\ (z)$$

gives the same equation for $Z(z)$ as before, and its solutions are the trigonometric functions

$$\sin mz \qquad \text{and} \qquad \cos mz$$

At the same time, function $T(r, \omega)$ has to obey a different Bessel equation

$$\frac{d^2T}{dr^2} + \frac{1}{r}\frac{dT}{dr} - \left(\frac{1}{r^2} + m_s^2 \right) T = 0, \tag{7.21}$$

where

$$m_s = \sqrt{m^2 - k_s^2}$$

Solutions of this equation are modified Bessel functions of the first order: $I_1(m_s r)$ and $K_1(m_s r)$. Since $I_1(m_s r)$ increases without limit with an increase of r, provided that m_s is real, this function cannot describe the potential $\tilde{\psi}$ outside the borehole. We will attempt to satisfy the boundary conditions assuming that the potential solution of the Helmholtz equation is the odd function of z, i.e.,

$$\tilde{\psi}_m(r, z, \omega, m) = C_m K_1(m_s r) \sin mz \tag{7.22}$$

Then the general solution for the vector potential may be written as

$$\tilde{\psi}(r, z, \omega,) = \int_0^\infty C_m K_1(m_s r) \sin mz\ dm \tag{7.23}$$

It is clear that this function also obeys the condition at infinity for any values of unknown C_m.

Boundary conditions and determination of unknowns

By analogy with a horizontally layered medium, we represent the scalar potential of the direct wave in terms of Bessel and trigonometric functions (Part II):

$$\frac{e^{i\,k_1 R}}{R} = \frac{2}{\pi}\int_0^\infty K_0\left(m_1 r\right)\cos mz\ dm \tag{7.24}$$

Correspondingly, the expressions of potentials satisfying the Helmholtz equations as well as conditions near the source and at infinity are

$$\tilde{\varphi}_1 = \frac{2}{\pi}C\int_0^\infty \left[K_0\left(m_1 r\right) + A_m I_0\left(m_1 r\right)\right]\cos mz\ dm \qquad \text{if} \qquad r < a \tag{7.25}$$

$$\text{and} \quad \tilde{\varphi}_2 = \frac{2}{\pi}C\int_0^\infty B_m K_0\left(m_l r\right)\cos mz\ dm,$$

$$\tilde{\psi}_2 = \frac{2}{\pi}C\int_0^\infty C_m K_1\left(m_s r\right)\sin mz\ dm \qquad \text{if} \qquad r > a$$

In order to determine unknown coefficients A_m, B_m, and C_m, we use the boundary conditions, eqs. 7.5. Since

$$\tilde{s}_r = \frac{\partial\tilde{\varphi}}{\partial r} - \frac{\partial\tilde{\psi}}{\partial z}, \qquad \tilde{s}_z = \frac{\partial\tilde{\varphi}}{\partial z} + \frac{1}{r}\frac{\partial}{\partial r}\left(r\tilde{\psi}\right), \tag{7.26}$$

$$\tilde{e}_{rr} = \frac{\partial\tilde{s}_r}{\partial r}, \qquad \tilde{e}_{rz} = \frac{\partial\tilde{s}_r}{\partial z} + \frac{\partial\tilde{s}_z}{\partial r},$$

$$\text{and} \qquad \tilde{\tau}_{rr} = \lambda\ \text{div}\ \tilde{\mathbf{s}} + 2\mu\frac{\partial\tilde{s}_r}{\partial r}, \qquad \tilde{\tau}_{zr} = \mu\ \tilde{e}_{zr},$$

eqs. 7.15 can be rewritten in terms of complex amplitudes of potentials:

$$\frac{\partial\tilde{\varphi}_1}{\partial r} = \frac{\partial\tilde{\varphi}_2}{\partial r} - \frac{\partial\tilde{\psi}_2}{\partial z}\ , \tag{7.27}$$

$$-\lambda_1 k_1^2 \widetilde{\varphi}_1 = -\lambda_2 k_l^2 \widetilde{\varphi}_2 + 2\mu_2 \left(\frac{\partial^2 \widetilde{\varphi}_2}{\partial r^2} - \frac{\partial^2 \widetilde{\psi}_2}{\partial r \partial z} \right)$$

and $$2 \frac{\partial^2 \widetilde{\varphi}_2}{\partial r \partial z} - \frac{\partial^2 \widetilde{\psi}_2}{\partial z^2} + \frac{\partial}{\partial r} \left(\frac{1}{r} \frac{\partial}{\partial r} \, r \widetilde{\psi}_2 \right) = 0 \qquad \text{if} \quad r = a$$

Here we use the known equalities

$$\text{div grad } \widetilde{\varphi} = \nabla^2 \widetilde{\varphi} = -k^2 \widetilde{\varphi} \quad \text{and} \quad \text{div curl } \widetilde{\boldsymbol{\psi}} \equiv 0$$

Let us recall that λ_1 is the bulk modulus of the fluid and

$$\lambda_1 = c_1^2 \rho_1, \qquad (\lambda_2 + 2\mu_2) = c_l^2 \rho_2, \qquad \mu_2 = c_s^2 \, \rho_2 \tag{7.28}$$

In solving system 7.27, we may apply the equalities

$$I_n' = I_{n-1} - \frac{n}{v} I_n, \qquad I_n' = I_{n+1} + \frac{n}{v} I_n, \qquad \frac{2n}{v} \, I_n = I_{n-1} - I_{n+1} \tag{7.29}$$

and $$-K_n' = K_{n-1} + \frac{n}{v} K_n, \qquad -K_n' = K_{n+1} - \frac{n}{v} K_n, \qquad -\frac{2n}{v} K_n = K_{n-1} - K_{n+1},$$

where v and n are the argument and order of the modified Bessel functions. Substitution of eqs. 7.25 into the first equation of set 7.27 yields:

$$-m_1 K_1 (m_1 a) + m_1 I_1 (m_1 a) A_m = -m_l \, K_1 (m_l a) B_m - m \, K_1 (m_s a) C_m$$

In the same manner, we have for the second equation

$$-\lambda_1 \, k_1^2 \left[K_0 (m_1 a) + A_m I_0 (m_1 a) \right] = -\lambda_2 \, k_l^2 \, B_m \, K_0 (m_l a)$$

$$-2\mu_2 \, m_l^2 B_m K_1' (m_l a) - 2\mu_2 m \, m_s C_m K_1' (m_s a)$$

Finally,

$$2m \, m_l K_1 (m_l a) B_m + m^2 C_m K_1 (m_s a) + m^2 C_m K_1 (m_s a) = 0,$$

since

$$\frac{\partial}{\partial r} r K_1 (m_s r) = \frac{\partial}{\partial v} v K_1 (v) = K_1 (v) + v K_1' (v) =$$

$$K_1(v) - v\left(K_0 + \frac{1}{v}K_1\right) = -vK_0(v)$$

and

$$\frac{\partial}{\partial r}\left[\frac{1}{r}\frac{\partial}{\partial r}rK_1(v)\right] = -\frac{\partial}{\partial r}m_sK_0(m_sa) = m_s^2K_1(m_sa)$$

Thus, the system of equations with respect to unknowns A_m, B_m, and C_m is

$$m_1I_1(m_1a)\ A_m + m_lK_1(m_la)\,B_m + mK_1(m_sa)\,C_m = m_1K_1(m_1a),$$

$$-\lambda_1k_1^2I_0(m_1a)\,A_m + \left[\lambda_2k_l^2K_0(m_la) + 2\mu_2\ m_l^2\ K_1'(m_la)\right]B_m \tag{7.30}$$

$$+2\mu_2m\ m_sK_1'(m_sa)\,C_m = \lambda_1k_1^2K_0(m_1a)$$

and $$2m\ m_lK_1(m_la)\,B_m + \left(2m^2 - k_s^2\right)K_1(m_sa)\,C_m = 0$$

Before we continue, it may be appropriate to make two comments:

1. The potentials $\widetilde{\varphi}$ and $\widetilde{\psi}$ are represented in the form of the Fourier cosine and sine transforms, respectively. This fact allows us to replace an equality of integrals with an equality of integrands.

2. At the beginning we assumed that expressions for potentials do not contain terms with either $\sin mz$ or $\cos mz$. However, the same result follows from eqs. 7.27 if we initially preserve these terms. In other words, our assumptions are justified.

From the last equation of set 7.30 we have

$$C_m = -\frac{2m\ m_lK_1(m_la)}{(2m^2 - k_s^2)\ K_1(m_sa)}\ B_m \tag{7.31}$$

Its substitution into the first equation of the set gives

$$m_1I_1(m_1a)\ A_m - \frac{k_s^2\ m_l\ K_1(m_la)}{(2m^2 - k_s^2)}\ B_m = m_1K_1(m_1a) \tag{7.32}$$

The second equation of the system becomes

$$-\lambda_1\ k_1^2\ I_0\ (m_1a)\ A_m + [\lambda_2\ k_l^2\ K_0(m_la) + 2\mu_2\ m_l^2K_1'(m_la) \tag{7.33}$$

$$-\frac{4\mu_2\ m^2m_l\ m_sK_1'(m_sa)\,K_1(m_la)}{(2m^2 - k_s^2)\,K_1(m_sa)}]\ B_m = \lambda_1k_1^2K_0(m_1a)$$

Consider the sum of the first two terms in the square brackets:

$$\lambda_2\, k_l^2 K_0\ (m_l a) + 2\mu_2\ m_l^2 K_1'\,(m_l a)$$

Since

$$K_1'(v) = -K_0\,(v) - \frac{1}{v} K_1\,(v)\,,$$

we have

$$\lambda_2\, k_l^2 K_0\,(m_l a) - 2\mu_2\ m_l^2 \left[K_0\,(m_l a) + \frac{1}{m_l a} K_1\,(m_l a)\right] = \left(\lambda_2\ k_l^2 - 2\mu_2\ m_l^2\right) K_0\,(m_l a)$$

$$-2\mu_2 \frac{m_l^2}{m_l\ a} K_1\,(m_l a) = -\mu_2 \left[\left(2m^2 - k_s^2\right) K_0\,(m_l a) + \frac{2m_l}{a} K_1\,(m_l a)\right]$$

Correspondingly, the expression in brackets becomes

$$-\mu_2 \left[\left(2m^2 - k_s^2\right) K_0\,(m_l a) + \frac{2m_l}{a} K_1\,(m_l a) + \frac{4m^2 m_l\ m_s K_1'\,(m_s a)\, K_1\,(m_l a)}{\left(2m^2 - k_s^2\right) K_1\,(m_s a)}\right] =$$

$$-\frac{\mu_2}{\left(2m^2 - k_s^2\right) K_1\,(m_s a)} \left\{\left(2m^2 - k_s^2\right)^2 K_0\,(m_l a)\, K_1\,(m_s a)\right. \tag{7.34}$$

$$+\frac{2m_l}{a}\left(2m^2 - k_s^2\right) K_1\,(m_l a)\, K_1\,(m_s a)$$

$$\left.-4m^2 m_l m_s \left[K_0\,(m_s a) + \frac{1}{m_s a} K_1\,(m_s a)\right] K_1\,(m_l a)\right\} = -\mu_2 P_m,$$

where

$$P_m = \frac{1}{\left(2m^2 - k_s^2\right) K_1\,(m_s a)} \left[\left(2m^2 - k_s^2\right)^2 K_0\,(m_l a)\, K_1\,(m_s a)\right. \tag{7.35}$$

$$\left. - \frac{2m_l\ k_s^2}{a} K_1\,(m_l a)\, K_1\,(m_s a) - 4m^2 m_l\ m_s K_0\,(m_s a)\, K_1\,(m_l a)\right]$$

It is interesting to note that function P_m contains terms like $(2m^2 - k_s^2)^2$ and $4m^2 m_l m_s$, which are also present in the Rayleigh equation (Chapter 6). Thus, in place of eq. 7.33, we have:

$$b\,k_s^2\,I_0\,(m_1 a)\;A_m + P_m B_m = -b\,k_s^2 K_0\,(m_1 a)\,, \tag{7.36}$$

because

$$\frac{\lambda_1 k_1^2}{\mu_2} = \frac{\rho_1 c_1^2\,k_1^2}{\rho_2\,c_s^2} = b\,k_s^2, \qquad b = \frac{\rho_1}{\rho_2}$$

Respectively, the system of equations with respect to A_m and B_m is

$$m_1 I_1\,(m_1 a)\,A_m - \frac{k_s^2 m_l K_1\,(m_l a)}{(2m^2 - k_s^2)} B_m = m_1 K_1\,(m_1 a) \tag{7.37}$$

and $\quad b\,k_s^2\,I_0\,(m_1 a)\,A_m + P_m B_m = -b\,k_s^2\,K_0\,(m_1 a)$

The solution of this set is

$$A_m = \frac{\begin{vmatrix} m_1 K_1\,(m_1 a) & \dfrac{-k_s^2\,m_l\,K_1\,(m_l\,a)}{2m^2 - k_s^2} \\ -k_s^2 K_0\,(m_1 a)\,b & P_m \end{vmatrix}}{D_m} \tag{7.38}$$

and

$$B_m = \frac{\begin{vmatrix} m_1 I_1\,(m_1 a) & m_1 K_1\,(m_1 a) \\ b\,k_s^2 I_0\,(m_1 a) & -b\,k_s^2 K_0\,(m_1 a) \end{vmatrix}}{D_m} = \tag{7.39}$$

$$-\,m\,b\,k_s^2 \frac{I_1\,(m_1 a)\,K_0\,(m_1 a) + I_0\,(m_1 a)\,K_1\,(m_1 a)}{D_m} = -\frac{b\,k_s^2}{a}\,\frac{1}{D_m},$$

since

$$I_1\,(v)\,K_0\,(v) + I_0\,(v)\,K_1\,(v) = \frac{1}{v} \tag{7.40}$$

As follows from eqs. 7.37, the determinant is equal to

$$D_m = m_1 I_1\,(m_1 a)\;P_m + \frac{b\,k_s^4\,m_l\,I_0\,(m_1 a)\;K_1\,(m_l a)}{(2m^2 - k_s^2)} \tag{7.41}$$

The coefficient C_m is defined from eq. 7.31.

We have derived formulas for scalar and vector potentials that can be used to perform numerical integration. However, they are hardly convenient for obtaining the asymptotic expressions for these functions. For this reason, we will begin by studying wave behavior in some special and much simpler cases.

7.2 Waves in an acoustic medium

We will start by considering a relatively simple case in which a medium surrounding a borehole, is also an acoustic medium $(\mu_2 = 0)$. Because shear waves are absent, wavefields caused by the point source, (Fig. 7.1a), are described by scalar potential only. Respectively, we have for the complex amplitudes of $\varphi(r, z, \omega)$ inside and outside the borehole (Part II)

$$\tilde{\varphi}_1(r, z, \omega) = C\frac{2}{\pi}\int_0^\infty [K_0(m_1 r) + A_m\, I_0(m_1 r)]\cos mz\, dm \qquad \text{if} \quad r < a \tag{7.42}$$

and $$\tilde{\varphi}_2(r, z, \omega) = C\frac{2}{\pi}\int_0^\infty B_m K_0(m_2 r)\cos mz\, dm \qquad \text{if} \quad r > a$$

where

$$m_1 = \sqrt{m^2 - k_1^2}, \qquad m_2 = \sqrt{m^2 - k_2^2}, \qquad \text{and} \qquad k_1 = \frac{\omega}{c_1}, \qquad k_2 = \frac{\omega}{c_2}$$

It is obvious that these functions satisfy the Helmholtz equations as well as conditions near the source and at infinity. Unlike in the general case, $(\mu_2 \neq 0)$, only two equations at the borehole surface describe continuity of normal stress and the normal component of displacement:

$$\tilde{\tau}_{rr}^{(1)} = \tilde{\tau}_{rr}^{(2)} \qquad \text{and} \qquad \tilde{s}_r^{(1)} = \tilde{s}_r^{(2)} \qquad \text{if} \quad r = a \tag{7.43}$$

or

$$\lambda_1\, k_1^2\tilde{\varphi}_1 = \lambda_2\, k_2^2\tilde{\varphi}_2, \qquad \frac{\partial\tilde{\varphi}_1}{\partial r} = \frac{\partial\tilde{\varphi}_2}{\partial r} \tag{7.44}$$

Taking into account that $\lambda = \rho c^2$, eq. 7.44 can be rewritten as

$$\rho_1\tilde{\varphi}_1 = \rho_2\tilde{\varphi}_2, \qquad \frac{\partial\tilde{\varphi}_1}{\partial r} = \frac{\partial\tilde{\varphi}_2}{\partial r} \qquad \text{if} \qquad r = a \tag{7.45}$$

Substitution of eqs. 7.42 into eqs. 7.45 gives

$$\rho_1[K_0(m_1 a) + A_m I_0(m_1 a)] = \rho_2 B_m K_0(m_2 a) \tag{7.46}$$

and $$m_1[-K_0(m_1 a) + A_m I_1(m_1 a)] = -m_2 B_m K_1(m_2 a)$$

Solution of this system gives (Part II)

$$A_m = \frac{m_1 K_0(m_2 a)\ K_1(m_1 a) - m_2\ b\ K_0(m_1 a)\ K_1(m_2 a)}{m_1 K_0(m_2 a)\ I_1(m_1 a) + m_2\ b\ I_0(m_1 a)\ K_1(m_2 a)} \tag{7.47}$$

and

$$B_m = \frac{b}{a[m_1 K_0(m_2 a)\ I_1(m_1 a) + m_2\ b\ I_0(m_1 a)\ K_1(m_2 a)]} \tag{7.48}$$

Here

$$b = \frac{\rho_1}{\rho_2}$$

In deriving the latter, eq. 7.48, we have used the equality

$$I_0\ (x)\ K_1(x) + I_1(x)\ K_0(x) = \frac{1}{x} \tag{7.49}$$

Also, it is useful to obtain eqs. 7.47 and 7.48 from formulas derived in the previous section. Assuming that $\mu_2 = 0$, we have $c_s \to 0$ and $k_s \to \infty$. Function P_m (eq. 7.35), then becomes

$$P_m \approx -k_s^2 K_0(m_l a)$$

Therefore, the determinant of the system D_m, eq. 7.39, is equal to

$$D_m \approx -k_s^2\left[m_1 I_1(m_1 a) K_0(m_l a) + b\ m_l I_0(m_1 a) K_1(m_l a)\right]$$

Thus, coefficient A_m, eq. 7.38, coincides with the coefficients for the acoustic medium, eq. 7.47, if $\mu_2 = 0$. In the same manner, we obtain coefficient B_m. It is clear that coefficient C_m, eq. 7.31, characterizing shear waves, vanishes if $\mu_2 = 0$.

Normal modes in the borehole

Next we will discuss wave behavior inside and outside of the borehole beginning with the normal modes. First, suppose that the primary source generates incident sinusoidal wave at sufficiently high frequencies that the wavelength is smaller than the borehole radius a $(\lambda_1 < a)$. This means that at each point of the boundary, reflection and transmission take place, as would also occur on the plane (Part II). At this point, the direct wave can be treated as a plane wave, and its incident angle increases with an increase of coordinate z. In order to understand the formation of normal modes we use axial symmetry and consider any plane that contains borehole axis z, Fig. 7.2a.

Let us assume that the incident wave reaches some point p of line 1 and forms angle θ_i with the normal to the boundary. The elementary reflected wave then appears and travels upward. It reaches line 2 and causes another reflected wave, which goes back to line 1. Reflections from both lines give rise to two families of waves, namely, downgoing and upgoing waves. From Fig. 7.2a we see that each wave of these sets makes the same angle θ_i with radius r. At a rather large distance from the source the resultant wavefield related to the normal modes can be approximately described by a system of plane waves propagating along the borehole and forming different angles with the boundary. This picture may suggest that each group of upgoing and downgoing waves undergoes multiple reflections at different points of the boundary and advances along the z-axis. The interference of these waves may be either destructive or constructive, and our goal is to examine the superposition of these waves when they interfere in a constructive way. Before we discuss this subject in some detail, it is appropriate to make two comments:

1. The incident wave simultaneously reaches all points of the boundary located at any plane $z = const$. The reflected wave also arises at these points at the same instant. Its rays are located at the lateral surface of the cone and at the wavefronts. For this reason, these reflected waves are called conical waves.

2. If the incident angle exceeds the critical angle, we observe total internal reflection, and the energy of these waves remains inside the borehole. However, when the incident angle is smaller than the critical angle a transmitted wave appears, and some energy penetrates into the surrounding medium. Thus, after each reflection, the waves inside the borehole become weaker. This shows that even when constructive interference takes place, rsulting mode usually rapidly decreases with distance z from the source. Often such a mode is called the leaking mode. Certainly, at large distances we observe the result of a constructive superposition of waves with the reflection angle exceeding the critical $\theta_i > \theta_c$.

We have qualitatively described an appearance of waves that move along the borehole. In other words, it is assumed that under certain conditions there is a constructive interference between reflected waves. Because of this a superposition of these waves – called normal modes – may exist in the borehole. From the mathematical point of view this means that each normal mode can itself satisfy, at some frequency, boundary conditions. It is also obvious that propagation of the normal mode inside the borehole causes vibration of the borehole surface $(r = a)$, and evanescent motion in the surrounding medium. Thus, every normal mode is accompanied by waves outside the borehole. As

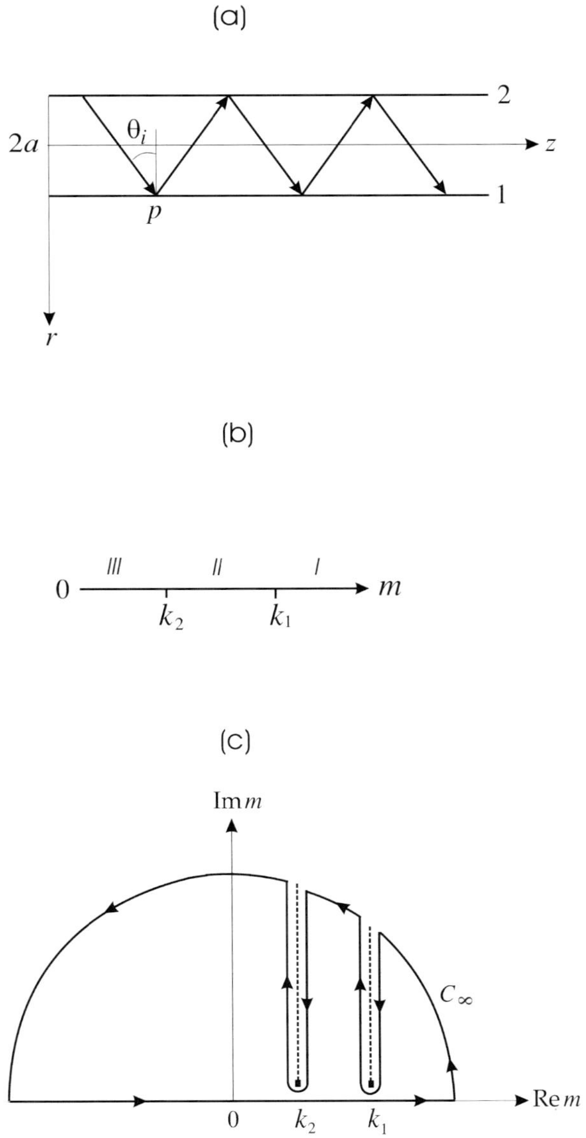

Figure 7.2: (a) Reflection of waves on borehole surface (b) Intervals of integration (c) Path of integration on complex plane

is well known from the theory of total internal reflection (Part II), the amount of energy moving into the surrounding medium during each period is equal to zero. This suggests that such waves are evanescent waves. The independence of each mode from the others allows us to find the conditions from which they arise, as well as their amplitudes and phase velocities.

Dispersion equation

In accordance with eqs. 7.42, scalar potentials of normal modes can be represented as

$$\tilde{\varphi}_1 = \int_0^\infty A_m I_0(m_1 r) \cos mz \, dm \qquad \text{if} \quad r < a \tag{7.50}$$

and

$$\tilde{\varphi}_2 = \int_0^\infty B_m K_0(m_2 r) \cos mz \, dm \qquad \text{if} \quad r > a$$

In eqs. 7.42 and in eqs. 7.50 we use the same notations for harmonic amplitudes, A_m and B_m. However, in the eqs. 7.50, they characterize the normal modes only. Substitution of eqs. 7.50 into set 7.44 gives a system of two homogeneous equations:

$$\rho_1 A_m I_0(m_1 a) - \rho_2 B_m K_0(m_2 a) = 0, \tag{7.51}$$

$$m_1 A_m I_1(m_1 a) + m_2 B_m K_1(m_2 a) = 0$$

Existence of the normal modes means that unknowns A_m and B_m differ from zero, which can happen if the determinant of this system is equal to zero:

$$D_m = \begin{vmatrix} \rho_1 I_0(m_1 a) & -\rho_2 K_0(m_2 a) \\ m_1 I_1(m_1 a) & m_2 K_1(m_2 a) \end{vmatrix} = 0$$

or

$$b\, m_2 I_0(m_1 a) K_1(m_2 a) + m_1 I_1(m_1 a) K_0(m_2 a) = 0 \tag{7.52}$$

Here

$$m_1 = \sqrt{m_n^2 - k_1^2} \qquad \text{and} \qquad m_2 = \sqrt{m_n^2 - k_2^2},$$

and m_n are roots of eq. 7.52. In order to demonstrate the presence of roots assume that $k_1 > k_2$. Correspondingly, three intervals of integration are shown in Fig. 7.2b. In the first one,

$$k_1 \leq m < \infty,$$

both radicals m_1 and m_2 are real and positive. Therefore values of the modified Bessel functions are also positive, Fig. 7.1b. This means that determinant D_m of system 7.46 does not have roots if $m > k_1$. In this light, it may be appropriate to notice that in this case, integrands in eqs. 7.42 describe elementary cylindrical waves that exponentially decay with an increase of distance from the boundary. In other words, only harmonics with $m > k_1$ may form surface waves. Since roots m_n are absent in the first interval, we conclude that this range of m does not produce normal modes. Within the second interval

$$k_2 \leq m \leq k_1$$

the radical m_2 still remains real and positive, while m_1 can be written as

$$m_1 = \left(m^2 - k_1^2\right)^{1/2} = -i\tilde{m}_1, \tag{7.53}$$

where $\tilde{m}_1$ is the positive number. Respectively, function D_m has the form

$$D_m = -i\tilde{m}_1\rho_2 I_1\left(-i\tilde{m}_1 a\right) K_0\left(m_2 a\right) + m_2\rho_1 I_0\left(-i\tilde{m}_1 a\right) K_1\left(m_2 a\right) \tag{7.54}$$

Taking into account that

$$I_0\left(-ix\right) = J_0\left(-ix\right), \quad I_1\left(-ix\right) = -iJ_1\left(x\right), \tag{7.55}$$

we have

$$D_m = -\tilde{m}_1\rho_2\, J_1\left(\tilde{m}_1 a\right) K_0\left(m_2 a\right) + m_2\rho_1 J_0\left(\tilde{m}_1 a\right) K_1\left(m_2 a\right) \tag{7.56}$$

Therefore the denominator becomes equal to zero at points m_n, where

$$\tilde{m}_1\rho_2\, J_1\left(\tilde{m}_1 a\right) K_0\left(m_2 a\right) = m_2\rho_1 J_0\left(\tilde{m}_1 a\right) K_1\left(m_2 a\right) \tag{7.57}$$

This is called a dispersion equation. Finally, within the last interval, $m < k_2$, we have:

$$D_m = -\tilde{m}_1\rho_2\, J_1\left(\tilde{m}_1 a\right) K_0\left(-i\tilde{m}_2 a\right) - i\tilde{m}_2\rho_1\, J_0\left(\tilde{m}_1 a\right) K_1\left(-i\tilde{m}_2 a\right) \tag{7.58}$$

Here

$$\tilde{m}_2 = \sqrt{k_2^2 - m^2}$$

Functions $K_0\left(-i\tilde{m}_2 a\right)$, and $K_1\left(-i\tilde{m}_2 a\right)$ have complex values, and so function D_m is also complex. At the same time, the presence of zeros requires that

$$\operatorname{Re} D_m = 0 \qquad \text{and} \qquad \operatorname{Im} D_m = 0$$

As calculations show, the real roots of the dispersion equation

$$D_m = 0$$

are absent if $m < k_2$. Let us assume that the real root m_n exists in this interval, i.e., m_2 is a purely imaginary number. Taking into account the asymptotic behavior of function $K_0(m_2 r)$, we have to conclude that in the surrounding medium there is a wave propagating away from the borehole. Correspondingly, the energy of the normal modes decreases, and they must vanish. Thus, we see again that roots of the determinant D_m are situated only within the second interval:

$$k_2 < m_n < k_1 \qquad \text{if} \qquad c_2 > c_1 \tag{7.59}$$

The term $\cos mz$ in the integrands of eqs. 7.50 characterizes propagation of the normal mode along the z-axis, and by definition the root m_n plays the role of the wavenumber. Correspondingly, the phase velocity, $c_{pn}(\omega)$, of the normal mode is

$$m_n = \frac{\omega}{c_{pn}(\omega)} \qquad \text{or} \qquad c_{pn}(\omega) = \frac{\omega}{m_n} \tag{7.60}$$

Thus, eq. 7.57 allows us to determine the wavenumber and phase velocity of the normal modes. In order to determine their amplitude, we apply the same approach as in the case of a medium with a plane interface.

Deformation of integration path

As usual, let us assume the presence of very small attenuation. Therefore, the singularities of the integrands in eqs. 7.42 are situated either slightly above the real axis of m in the first quadrant of the m-plane or a little beneath in the third quadrant. First, we represent the potential of the secondary waves in the borehole in a different way. Since

$$\cos mz = \frac{e^{i\,m\,z} + e^{-i\,m\,z}}{2},$$

we have

$$\tilde{\varphi}_s = \frac{C}{\pi}\int_0^\infty A_m I_0\left(m_1 r\right)\ e^{i\ m\ z}\ dm + \frac{C}{\pi}\int_0^\infty A_m I_0\left(m_1 r\right)\ e^{-i\ m\ z}\ dm$$

Since $A_m(m)$ and $I_0\left(m_1 r\right)$ are the even functions

$$A_m\left(-m\right) = A_m\left(m\right), \qquad I_0\left(m_1 r\right) = I_0\left(-m_1 r\right)$$

the last integral is written as

$$\int_0^\infty A_m\left(m\right) I_0\left(m_1 r\right)\ e^{-i\ m\ z}\ dm = -\int_0^{-\infty} A_m\left(-m\right) I_0\left(m_1 r\right)\ e^{i\ m\ z}\ dm$$

$$= \int_{-\infty}^0 A_m\left(m\right) I_0\left(m_1 r\right)\ e^{i\ m\ z}\ dm$$

Thus,

$$\tilde{\varphi}_s = \frac{C}{\pi}\int_{-\infty}^\infty A_m\left(m\right) I_0\left(m_1 r\right)\ e^{i\ m\ z}\ dm \tag{7.61}$$

Applying the Cauchy theorem to the closed path C_0 shown in Fig. 7.2c, we have

$$\oint_{C_0} \frac{N_m}{D_m} I_0\left(m_1 r\right)\ e^{imz}\ dm = \int_{-\infty}^\infty \frac{N_m}{D_m} I_0\left(m_1 r\right)\ e^{imz}\ dm + \tag{7.62}$$

$$\int_{C_\infty} \frac{N_m}{D_m} I_0\left(m_1 r\right)\ e^{imz}\ dm + M_p + M_b = 0$$

Here M_p and M_b are sums of integrals around poles and branch points, respectively, and

$$N_m = \left[m_1 \rho_2 K_0\left(m_2 a\right) K_1\left(m_1 a\right) - m_2 \rho_1 K_0\left(m_1 a\right) K_1\left(m_2 a\right)\right] \tag{7.63}$$

Since the integrand contains the exponential term $\exp\left(imz\right)$, we can use the Jordan lemma and discard the second integral in eq. 7.62. This gives

$$\int_{-\infty}^\infty A_m\left(m\right) I_0\left(m_1 r\right)\ e^{imz}\ dm = -M_p - M_b$$

and hence

$$\tilde{\varphi}_s = -\frac{C}{\pi}\,(M_p + M_b) \tag{7.64}$$

In a similar manner the potential $\tilde{\varphi}_2$ can be described in terms of integrals around poles and along branch lines. Since the poles m_n exceed the wavenumber k_2, the radical m_2 is positive, and, therefore the function $K_0\,(m_2 r)$ exponentially decays with distance r, if $k_2 r > 1$. As was already pointed out, this suggests that outside the borehole, the wavefields associated with normal modes behave like evanescent waves. They appear due to total internal reflection when destructive interference produces a rapidly weakening wave in the surrounding medium.

In accordance with eqs. 7.64 the wavefield in the borehole consists of three parts:

$$\tilde{\varphi}_1 = \tilde{\varphi}_0 + \tilde{\varphi}_p + \tilde{\varphi}_b \tag{7.65}$$

We focus now on the second term, $\tilde{\varphi}_p$, related to the poles. Because roots m_n of the dispersion equation $D_m = 0$ are poles of the integrands in eqs. 7.42, we can say that the normal modes are defined by behavior of the potential in the vicinity of the poles. This fundamental fact allows us to apply theory of the complex variables to determine the amplitude of the normal modes (Part II). At the same, the phase velocity $c_p\,(\omega)$ is calculated from the dispersion equation. Taking into account the residual theorem, we have

$$\tilde{\varphi}_p = 2Ci \sum_{n=1} \text{Res}\; F_n\; e^{i\;m_n\;z}, \tag{7.66}$$

where

$$F_n = \frac{N_m}{D_m}\; I_0\;(m_1 r)\,, \tag{7.67}$$

while D_m and N_m are given by eqs. 7.58 and 7.63, respectively. Taking into account that

$$\varphi_p\,(r,z,\omega) = \text{Re}\,\tilde{\varphi}_p e^{-i\;\omega\;t},$$

eq. 7.66 gives

$$\varphi_p\,(r,z,\omega) = \text{Re} \sum_n G_n\; e^{-i\,(\omega\;t\;-\;m_n\;z)} \tag{7.68}$$

Here

$$G_n = 2C\ i\ \mathrm{Res} F_n \tag{7.69}$$

The latter, eq. 7.69, clearly shows that each term of the sum characterizes a sinusoidal wave (normal mode) traveling along the borehole with phase velocity c_{pn}, eq. 7.60. At the same time, function G_n defines the complex amplitude of the normal modes, which depends on frequency, parameters of the medium, the distance r, and root m_n. From eq. 7.67 it follows that dependence of the normal mode amplitude of distance r is determined by the function

$$J_0\left(\sqrt{k_1^2 - m_n^2}\ r\right),$$

since

$$I_0(m_1 r) = J_0\left(\tilde{m}_1 r\right)$$

This means that each mode as a function of r represents a standing wave.

We will begin our study of normal modes with the simplest case, in which the borehole is surrounded by an ideally rigid medium.

Case 1 When the borehole is surrounded by an ideally rigid medium, function F_n is greatly simplified, and we have

$$F_n = -i\,\frac{K_1(m_1 a)}{J_1\left(\tilde{m}_1 a\right)}\,J_0\left(\tilde{m}_1 r\right) \tag{7.70}$$

Correspondingly, the dispersion equation is

$$J_1\left(\tilde{m}_1 a\right) = 0, \tag{7.71}$$

and its first zeros are given below:

n	0	1	2	3	4	5	6
$r_n = \tilde{m}_1 a$	0	3.83	7.02	10.17	13.32	16.47	19.64

Since with an increase of the argument the behavior of Bessel functions has a sinusoidal character, the difference between zeros, r_n, tends to π. Letting

$$\tilde{m}_1 a = r_n, \tag{7.72}$$

we obtain for each mode a relationship between wavenumber m_n and the frequency:

$$m_n = \frac{1}{a}\left(k_1^2 a^2 - r_n^2\right)^{1/2} \tag{7.73}$$

As follows from eq. 7.73, for all normal modes except $n = 0$, there is a nonzero frequency when

$$k_1 a = r_n \qquad \text{or} \qquad f_c = \frac{r_n\, c_1}{2\pi a} \tag{7.74}$$

This is called the cut-off frequency. In this case, wavenumber m_n is equal to zero and the phase velocity becomes infinitely large, eq. 7.60. With an increase of ω the wavenumber of the normal mode increases, eq. 7.73, and therefore phase velocity c_{pn},

$$c_{pn} = \frac{\omega\, a}{\left(k_1^2 a^2 - r_n^2\right)^{1/2}}, \tag{7.75}$$

gradually decreases. In the limit it tends to the wave velocity in the borehole fluid

$$c_{pn}(\omega) \to c_1 \qquad \text{if} \qquad \omega \to \infty \tag{7.76}$$

As was pointed out earlier, every normal mode is the result of the constructive interference of waves reflected from the cylindrical boundary. With an increase of frequency, the angle between the direction of propagation of these waves and the z-axis decreases, and it tends to zero when $\omega \to \infty$. Dispersion curves of phase velocity $c_{pn}(\omega)$ for the first several modes are shown in Fig. 7.3a. Unlike with the other modes, the phase velocity of the normal mode $n = 0$ is independent of frequency, and it is equal to c_1. In fact, since $r_0 = 0$, we have, eq. 7.75

$$m_0 = k_1 \qquad \text{and} \qquad c_{p0} = c_1 \tag{7.77}$$

Thus, regardless of frequency, this mode propagates along the borehole with constant velocity, and its cut-off frequency is equal to zero. From eq. 7.75 it follows that with an increase of the order of the normal mode, the cut-off frequency also increases. Correspondingly, there is a range of relatively low frequencies in which modes except $n = 0$ are absent. In the second range the first mode appears, so that there are two modes. In the next interval three modes exist, and so on. Note that the cut-off frequency of any mode is related to the pole $m_n = 0$. We see from eq. 7.74 that the normal mode arises when the smallest wavenumber is equal to

$$k_1 = \frac{r_n}{a} \qquad \text{or} \qquad \frac{a}{\lambda_1} = \frac{r_n}{2\pi} \tag{7.78}$$

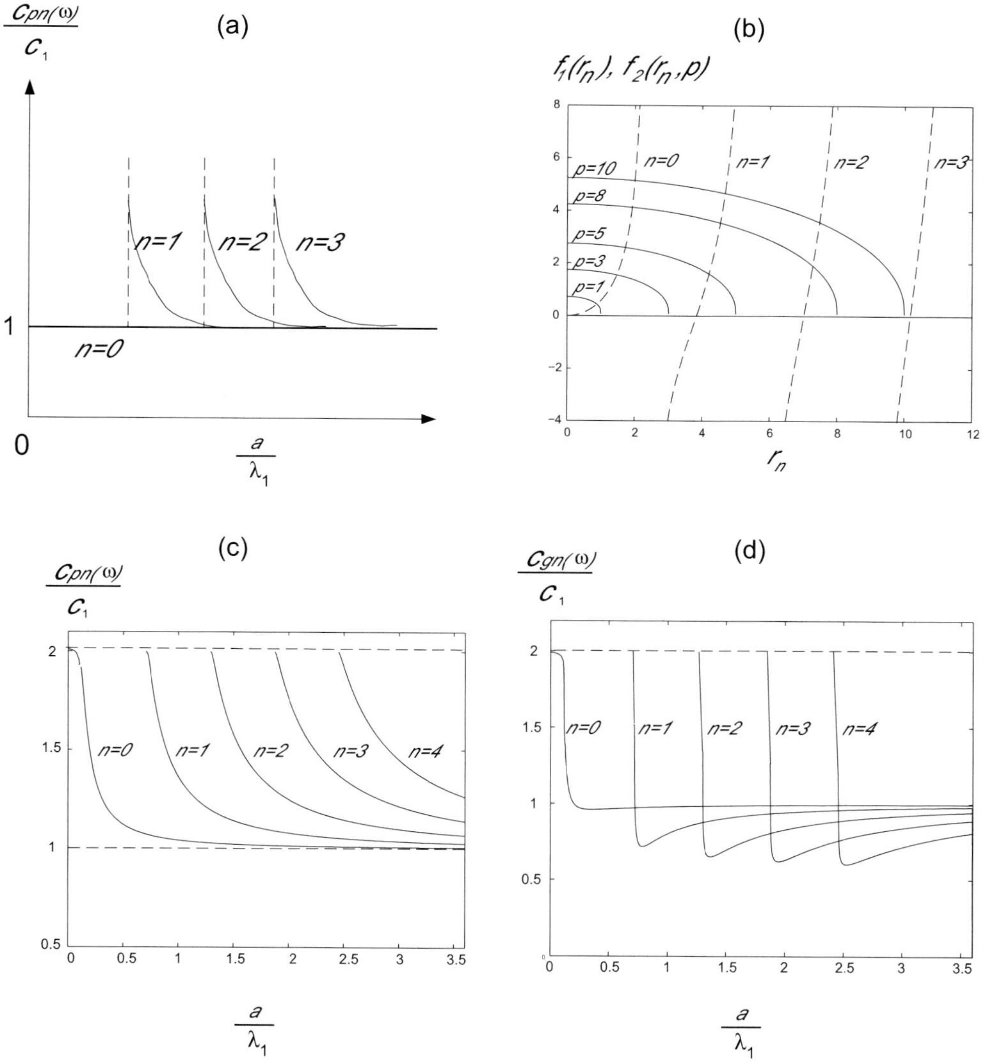

Figure 7.3: (a) Dispersion curves of phase velocity $(\rho_2 = \infty)$ (b) Functions $f_1(r_n)$ (solid lines), $f_2(r_n, p)$ (dashed lines) for different n and p (c) Dispersion curves of phase velocity (general case) (d) Dispersion curves of group velocity (general case)

For instance, if $n=1$, we have

$$\frac{a}{\lambda_1} \approx 0.6,$$

and the wavelength is slightly smaller than the borehole diameter. Of course, with an increase of the mode order, wavelength λ_c $(\lambda_c = c_1/f_c)$, decreases. (The normal mode of the zero order is an exception. The earliest "telephone" was based on air pipes, and communication took place due to propagation of normal mode $n=0$ with constant velocity c_1, regardless of frequency.)

In accordance with eq. 7.70, the residue of function F_n (Part II) is

$$\text{Res } F_n = i\frac{\tilde{m}_1 K_1(m_1 a)\ J_0\left(\tilde{m}_1 r\right)}{m_n a\ J_1'\left(\tilde{m}_1 a\right)}, \tag{7.79}$$

since

$$\frac{\partial}{\partial m}\ J_1'\left(\tilde{m}_1 a\right) = -\frac{m_n a}{\tilde{m}_1}\ J_1'\left(\tilde{m}_1 a\right)$$

Taking into account the equality

$$J_1'(x) = J_0(x) - \frac{J_1(x)}{x},$$

in place of eq. 7.79 we obtain

$$\text{Res } F_n = i\frac{\tilde{m}_1 K_1(m_1 a)\ J_0\left(\tilde{m}_1 r\right)}{m_n a\left[J_0\left(\tilde{m}_1 a\right) - \dfrac{J_1\left(\tilde{m}_1 a\right)}{\tilde{m}_1 a}\right]} \tag{7.80}$$

Now, making use of the asymptotic behavior of the Bessel functions

$$J_0(x) \to 1, \qquad J_1(x) \to \frac{x}{2}, \qquad K_1(x) \to \frac{1}{x}, \qquad \text{if} \quad x \to 0$$

we have for the normal mode of the zero order, when $m_0 = k_1$

$$\text{Res } F_n = -\frac{2}{k_1 a^2}$$

Therefore, function G_0, eq. 7.69, is

$$G_0 = -\frac{4iC}{k_1 a^2} \tag{7.81}$$

and

$$\widetilde{\varphi}_{0p} = -\frac{4iC}{k_1 a^2}\, e^{-i\,(\omega\, t - k_1 z)} \tag{7.82}$$

Proceeding from the equality

$$\widetilde{\tau}_{zz} = \widetilde{\tau}_{rr} = -\lambda_1\, k_1^2 \widetilde{\varphi}_{0p} = -\rho_1 \omega^2 \widetilde{\varphi}_{0p},$$

the expression for stress caused by the mode of the zero order is

$$\widetilde{\tau} = \frac{4\, i\, \rho_1 c_1 \omega}{a^2}\, C\, e^{-i\,(\omega t - k_1 z)} \qquad \text{if} \quad m_0 = k_1 \tag{7.83}$$

Thus, stress uniformly distributed over the borehole cross-section and, as we may expect, is inversely proportional to its area (πa^2). Since

$$\widetilde{s}_z = \frac{\partial \widetilde{\varphi}}{\partial z} \qquad \text{and} \qquad \widetilde{s}_r = \frac{\partial \widetilde{\varphi}}{\partial r},$$

we have

$$\widetilde{s}_r = \frac{4C}{a^2}\, e^{-i\,(\omega t - k_1 z)} \qquad \text{and} \qquad \widetilde{s}_r = 0 \qquad \text{if} \qquad m_0 = k_1 \tag{7.84}$$

That is, displacement has only the vertical component $\widetilde{s}_z$, which is independent of coordinate r.

Next let us discuss the general features of normal modes of the higher order. From eqs. 7.69 and 7.79, we have

$$G_n = -2C \frac{\widetilde{m}_1 K_1\,(m_1 a)\;\; J_0\left(\widetilde{m}_1 a\right)}{m_n a\;\; J_0\left(\widetilde{m}_1 a\right)}, \tag{7.85}$$

because

$$J_1\left(\widetilde{m}_1 a\right) = 0$$

Unlike in case $n = 0$, functions G_n depend on coordinate r. Since the complex amplitude of potential is

$$\widetilde{\varphi}_{np} = -2\, C\, \frac{\widetilde{m}_1 K_1\,(m_1 a)\;\; J_0\left(\widetilde{m}_1 r\right)}{m_n a\;\; J_0\left(\widetilde{m}_1 a\right)}\, e^{-i\,(\omega\, t - m_n\, z)}, \tag{7.86}$$

formulas for the stress and displacement components are

$$\tilde{\tau} = 2\, C \rho_1 \omega^2 \, \frac{\tilde{m}_1 K_1 (m_1 a) \; J_0 \left(\tilde{m}_1 r\right)}{m_n a \; J_0 \left(\tilde{m}_1 a\right)} \, e^{-i\,(\omega\, t - m_n\, z)},$$

$$\tilde{s}_z = -2\, C\, i \, \frac{\tilde{m}_1 K_1 (m_1 a) \; J_0 \left(\tilde{m}_1 r\right)}{a \; J_0 \left(\tilde{m}_1 a\right)} \, e^{-i\,(\omega\, t - m_n\, z)}, \tag{7.87}$$

$$\tilde{s}_r = 2\, C \, \frac{\tilde{m}_1^2 K_1 (m_1 a) \; J_1 \left(\tilde{m}_1 r\right)}{m_n a \; J_0 \left(\tilde{m}_1 a\right)} \, e^{-i\,(\omega\, t - m_n\, z)}$$

This clearly shows that with an increase of the order n, the wavefields vary more rapidly in the radial direction, forming standing waves and, correspondingly, a sequence of extensional and compressional zones. At each point of the borehole except its axis and the surface $r = a$, there are usually two components of displacement, and the phase shift between them is equal to $\pi/2$. For this reason the vector $\mathbf{s}$ is elliptically polarized. Note that the radial component s_r vanishes at the boundary since $J_1 \left(\tilde{m}_1 a\right) = 0$. This is obvious, because the surrounding medium is rigid.

Case 2 Consider the second limiting case, in which $\rho_2 = 0$, i.e., a cylinder containing a fluid is surrounded by free space. As follows from eq. 7.67,

$$F_n = -\frac{K_0 (m_1 a)}{J_0 \left(\tilde{m}_1 a\right)} \; J_0 \left(\tilde{m}_1 r\right) \tag{7.88}$$

Respectively, the dispersion equation is

$$J_0 (r_n) = 0, \tag{7.89}$$

where $r_n = \tilde{m}_1 a$. Values of r_n for the first six modes are given below:

n	1	2	3	4	5	6
r_n	2.40	5.52	8.65	11.79	14.93	18.07

As in the previous case $(\rho_2 \to \infty)$, with an increase of the order of a mode, the difference between neighboring r_n approaches π. The normal mode of the zero order is absent. Also, as before $(\rho_2 \to \infty)$, at the cut-off frequency phase velocity $c_{pn}(\omega)$ tends to an infinity, and with an increase of ω approaches c_1. In both limiting cases $(\rho_2 \to \infty$ and $\rho_2 \to 0)$, the value of r_n is independent of frequency. Since r_n can be represented as

$$r_n = \omega\sqrt{\frac{1}{c_1^2} - \frac{1}{c_{pn}^2}}, \tag{7.90}$$

we conclude that the radical in eq. 7.90 is inversely proportional to ω. In the same manner as in the first case, we find that the residue of function F_n is

$$\text{Res } F_n = -\frac{\widetilde{m}_1 K_0(m_1 a)}{m_n a \; J_1\left(\widetilde{m}_1 a\right)} J_0\left(\widetilde{m}_1 r\right) \tag{7.91}$$

Hence

$$G_n = -2\; C\; i\; \frac{\widetilde{m}_1 K_0(m_1 a)\;\; J_0\left(\widetilde{m}_1 r\right)}{m_n\; a\; J_1\left(\widetilde{m}_1 a\right)} \tag{7.92}$$

and

$$\widetilde{\varphi}_{np} = -2\; C\; i\; \frac{\widetilde{m}_1 K_0(m_1 a)\;\; J_0\left(\widetilde{m}_1 r\right)}{m_n a\; J_1\left(\widetilde{m}_1 a\right)}\; e^{-i\,(\omega\, t - m_n\, z)} \tag{7.93}$$

This demonstrates that boundary conditions are satisfied at the borehole surface, i.e., the strain is equal to zero at its points. Note that component s_z also vanishes if $r = a$. As a rule, inside the borehole the vector $\mathbf{s}$ is elliptically polarized, and we can also observe a regular change of the pressure sign in the radial direction (standing wave). As follows from eq. 7.93, the amplitude of the normal modes is independent of distance z. It also remains valid when the density of the surrounding medium has a finite nonzero value if $c_2 > c_1$.

General case When both media have nonzero and finite values of density, the wavenumbers of the normal modes m_n (poles of the integrands in eqs. 7.42) are defined from the dispersion equation 7.57

$$\widetilde{m}_1 J_1\left(\widetilde{m}_1 a\right)\;\; K_0(m_2 a) = m_2\; b\; J_0\left(\widetilde{m}_1 a\right)\;\; K_1(m_2 a), \tag{7.94}$$

where

$$b = \frac{\rho_1}{\rho_2}$$

Introducing notation

$$r_n = \widetilde{m}_1 a \qquad \text{or} \qquad r_n = \sqrt{k_1^2 - m_n^2}\; a, \tag{7.95}$$

we have for the wavenumber of the normal modes

$$m_n^2 = k_1^2 - \left(\frac{r_n}{a}\right)^2 \tag{7.96}$$

and, as was pointed out earlier:

$$r_n = \omega a \left[\frac{1}{c_1^2} - \frac{1}{c_{pn}^2(\omega)}\right]^{1/2}$$

Correspondingly,

$$m_2\, a = \sqrt{p^2 - r_n^2}, \tag{7.97}$$

where

$$p = \beta k_1 a \qquad \text{and} \qquad \beta = \left(1 - \frac{c_1^2}{c_2^2}\right)^{1/2} \tag{7.98}$$

Thus, eq. 7.94 becomes

$$\frac{r_n\, J_1(r_n)}{J_0(r_n)} = \frac{b\sqrt{p^2 - r_n^2}\; K_1\left(\sqrt{p^2 - r_n^2}\right)}{K_0\left(\sqrt{p^2 - r_n^2}\right)} \tag{7.99}$$

Unlike in the case of a medium, where $\rho_2 = \infty$ or $\rho_2 = 0$, the value of r_n depends on frequency. As in the presence of the plane interface, this value is defined numerically (Part II). The left side of eq. 7.99,

$$f_1(r_n) = \frac{r_n\, J_1(r_n)}{J_0(r_n)}, \tag{7.100}$$

is independent of parameter p; its behavior is shown in Fig. 7.3b. At small values of r_n, the function $f_1(r_n)$ is positive and decreases in proportion with r_n^2. At greater values of r, its behavior is dictated by the roots of equations

$$J_0(r_n) = 0 \qquad \text{and} \qquad J_1(r_n) = 0,$$

and in the limit we have the periodic function

$$f_1(r_n) \to r_n \tan\left(r_n - \frac{\pi}{4}\right) \to r_n \tan r_n \qquad \text{if} \quad r_n \to \infty \tag{7.101}$$

The right side of eq. 7.99

$$f_2(r_n, p, b) = b\sqrt{p^2 - r_n^2}\frac{K_1\left(\sqrt{p^2 - r_n^2}\right)}{K_0\left(\sqrt{p^2 - r_n^2}\right)} \tag{7.102}$$

for given b depends on two variables: p and r_n. Respectively, we can plot a system of curves describing this function, Fig. 7.3b. An intersection of graphs of $f_1(r_n)$ and $f_2(r)$ allows us to determine r_n and, correspondingly, the root m_n of the dispersion equation for each value of p and b. Inasmuch as functions K_0 and K_1 have complex values, if $r_n > p$, eq. 7.99 does not have real roots in this range. Let us assume that

$$r_n = p \tag{7.103}$$

Then the right side of eq. 7.99 becomes equal to zero, because

$$K_1\left(\sqrt{p^2 - r_n^2}\right) \to \frac{1}{\sqrt{p^2 - r_n^2}} \qquad \text{and} \qquad K_0\left(\sqrt{p^2 - r_n^2}\right) \to \infty$$

Therefore, eq. 7.99 takes place when

$$J_1(r_n) = 0 \tag{7.104}$$

The latter equation defines the roots, m_n, of the dispersion equation corresponding to the cut-off frequency. As was already pointed out, at smaller values of p_1 (frequencies), the real roots of eq. 7.99 are absent. Note that unlike the in limiting case when $\rho_2 \to \infty$, eq. 7.104 characterizes the roots for each mode only at the cut-off frequency. From eqs. 7.97 and 7.103, we obtain

$$m_n = k_2 \tag{7.105}$$

Thus, at the cut-off frequency the phase velocity coincides with that of the surrounding medium

$$c_{pn} = c_2 \qquad \text{if} \qquad f = f_c \tag{7.106}$$

Now suppose that parameter p, i.e., the frequency, tends to infinity and that $c_{pn}(\infty) \neq c_2$. By definition we have

$$r_n = \omega a\left(\frac{1}{c_1^2} - \frac{1}{c_{pn}^2}\right)^{1/2} \qquad \text{and} \qquad \left(p^2 - r_n^2\right)^{1/2} = \omega a\left(\frac{1}{c_{pn}^2} - \frac{1}{c_2^2}\right)$$

Since

$$K_0(x) \to K_1(x), \qquad \text{if} \quad x \to \infty$$

in place of eq. 7.99 we obtain

$$\frac{J_1(r_n)}{J_0(r_n)} = \frac{\left(\frac{1}{c_{pn}^2} - \frac{1}{c_2^2}\right)^{1/2}}{\left(\frac{1}{c_1^2} - \frac{1}{c_{pn}^2}\right)^{1/2}} \tag{7.107}$$

Here c_{pn} corresponds to the high-frequency limit. The solution of this equation gives some finite value of r_n, eq. 7.95, and, since ω tends to infinity, phase velocity c_{pn} should approach c_1:

$$c_{pn} \to c_1 \qquad \text{if} \quad \omega \to \infty \tag{7.108}$$

It is easy to show that eq. 7.99 does not have a solution, if $c_{pn}(\infty) = c_2$. Thus, the range of change of the phase velocity is

$$c_1 < c_{pn}(\omega) \leq c_2, \tag{7.109}$$

regardless of the order of the normal mode. Behavior of this function for several modes $(n = 0, 1, ..., 4)$ is shown in Fig. 7.3c. In this case, the fundamental mode exists for all frequencies.

From eqs. 7.98, it follows that parameter p can be written as

$$p = k_1 a \cos\theta_c, \tag{7.110}$$

where θ_c is the critical angle. As we know, if $\theta_i \geq \theta_c$ total internal reflection takes place and normal modes are formed that propagate without attenuation along the borehole.

Having solved the dispersion equation, we have found the wavenumbers of the normal modes and, therefore, their phase velocities $c_{pn}(\omega)$. Then, applying the residue theorem, we can calculate mode amplitudes at any point of the borehole cross-section. From the physical point of view, as well as eq. 7.67, it is obvious that amplitudes are independent of the z-coordinate. This is why at large distances from the sources, wavefields related to the normal modes play the dominant role. Also from eq. 7.67, we see that with an increase of the order of the normal modes, more modes of the standing wave are observed between the borehole axis and its surface. In the same manner we can evaluate the magnitude of evanescent motion in the surrounding medium that are associated with normal modes.

Transient modes

By analogy with a horizontally layered medium (Part II), study of the dispersion curves of $c_{pn}(\omega)$ allows us to discuss the main features of waves in the borehole. Suppose that a spectrum of the wave caused by the primary source is a continuous function of frequency. In illustration, let us consider just one normal mode with index n. Earlier it was shown that the primary wave with frequency f, which is smaller than the cut-off frequency $(n \neq 0)$

$$f < f_{cn},$$

does not cause the normal mode. The normal mode appears when $f = f_{cn}$. With an increase of frequency, the phase velocity c_{pn} decreases, varying between c_1 and c_2, eq. 7.109. Since each frequency component in this mode propagates at a different phase velocity, interference with this system of waves gives rise to different wave groups. Each group is characterized by a certain value of the dominant frequency and the group velocity c_{gn}. As is well known (Part II), phase and group velocities are related to each other in the following way:

$$\frac{1}{c_{gn}} = \frac{1}{c_{pn}} - \frac{\omega^2}{c_{pn}^2}\frac{dc_{pn}}{d\omega} \tag{7.111}$$

Because the function $c_{pn}(\omega)$ is known, the group velocity $c_{gn}(\omega)$ is easily calculated and its behavior is shown in Fig. 7.3d $(n = 0, 1, ..., 4)$. At the cut-off frequency, the group velocity coincides with the physical velocity c_2 of the surrounding medium. With an increase of frequency it becomes smaller, and at some frequency f_{nA} velocity c_{gn} reaches minimum value c_{gn}^A. With further increase of frequency, velocity c_{gn} begins to grow and asymptotically approaches c_1. Knowledge of the function $c_{gn}(\omega)$ permits us to describe qualitatively the behavior of the transient wave. Suppose that the source located at the origin is turned on at instant $t = 0$, and at some distance z the receiver measures the nth transient mode. During the time interval

$$0 < t < \frac{z}{c_2},$$

the wave is absent. At instant $t = z/c_2$ the wave group arrives, and its dominant frequency is equal to the cut-off frequencies, f_{cn}. With increased time, wave groups with higher frequencies begin to arrive. Within some frequency range

$$f_{cn1} < f_1 < f_n,$$

the dominant frequency of the wave group gradually increases with time. At an instant slightly greater than the ratio z/c_1, two wave groups arrive simultaneously at the observation point. Their group velocities are equal to each other and almost coincide with c_1. The dominant frequency of the first group equals f_1, while the second is characterized by much higher frequency. Such superposition of two wave groups is observed within the time interval

$$\frac{z}{c_1} < t < \frac{z}{c_{gn}^A},$$

and with increased time the dominant frequencies of both groups approach each other. Finally, starting from instant

$$t = \frac{z}{c_{gA}},$$

only one wave group exists, and its dominant frequency is equal to f_{nA}. It is customary to treat this last stage of the transient wave as the Airy phase. All other modes behave in a similar way. Note that dispersion causes change in the waveform with distance z. The waveform stretches in time and, in accordance with law of energy conservation, the amplitudes of the transient modes, unlike in the case of the stationary modes, decrease with z.

Normal modes and interference of conical waves

Let us represent the integrand in eq. 7.61, which describes the secondary wave inside the borehole, as

$$A_m \, J_0 \left(\tilde{m}_1 r\right) \, e^{i\, m\, z} \tag{7.112}$$

Here

$$\tilde{m}_1 = \sqrt{k_1^2 - m^2} > 0$$

We are interested in the high-frequency spectrum in which

$$\tilde{m}_1 r \gg 1$$

In this part of the spectrum, the Bessel function $J_0\left(\tilde{m}_1 r\right)$ is approximately equal to

$$J_0\left(\tilde{m}_1 r\right) \simeq \left(\frac{2}{\pi \tilde{m}_1 r}\right)^{1/2} \cos\left(\tilde{m}_1 r - \frac{\pi}{4}\right)$$

or

$$J_0\left(\tilde{m}_1 r\right) \simeq \left(\frac{1}{2\pi \tilde{m}_1 r}\right)^{1/2} \left(e^{-i\pi/4} e^{i\,\tilde{m}_1 r} + e^{i\pi/4} e^{-i\,\tilde{m}_1 r}\right) \quad \text{if} \quad \mathrm{r} \neq 0 \qquad (7.113)$$

This shows that the integrand, eq. 7.112, can be represented as a sum of two conical waves. One of them is proportional to

$$\frac{1}{\sqrt{r}} \exp\left(-i\tilde{m}_1 r + imz\right), \qquad (7.114)$$

and it moves toward the borehole axis. The other is

$$\frac{1}{\sqrt{r}} \exp\left(i\tilde{m}_1 r + imz\right), \qquad (7.115)$$

and it diverges from this axis. The wave front, i.e., the surface of the equal phase, coincides with the lateral surface of the cone. Its apex is located at the borehole axis, Fig. 7.4a, where

$$\tan\theta = \frac{m}{\tilde{m}_1} \qquad (7.116)$$

It is obvious that at each cross-section of the borehole, superposition of conical waves produces a standing wave.

In order to determine the relationship of borehole radius a, frequency ω, and angle θ, when the normal mode can be formed, consider the ray passing through point A of wave front $N'ON$, Fig. 7.4b. This ray intersects the borehole axis at point B and, after reflection at point C of the boundary, it moves toward the axis. It is essential that the ray remain normal to the phase surface of the convergent wave. At point D, it intersects the same wave front $N'ON$. The condition for the existence of a normal mode requires that the difference between the phase at point A and a change of phase along the ray must be equal $2\pi n$. In evaluating this difference, we assume that $\rho_2 = \infty$ and take into account three factors:

a. In accordance with eqs. 7.114 and 7.115, the conical waves propagate with the velocity of the borehole fluid, c_1.

b. The phase shift between the reflected and incident waves at points of the rigid boundary is equal to zero.

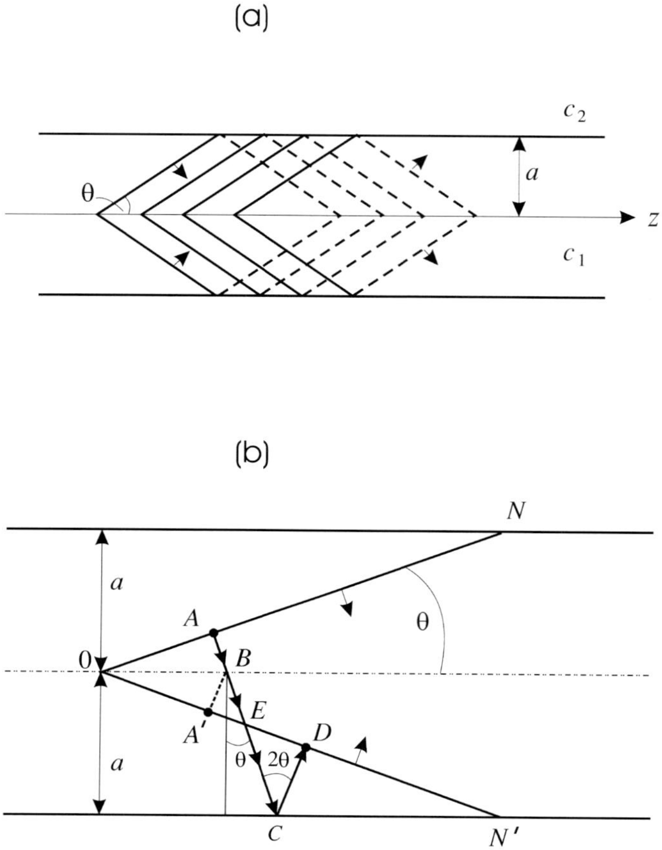

Figure 7.4: a) Conical wave fronts (b) Illustration of constructive interference, eq. 7.118

c. At each point of the borehole axis, the rays converge, and these points may be treated as foci. This means that the difference between phases of convergent and divergent waves in the vicinity of the axis is equal to $-\pi/2$.

Thus, the condition for constructive interference is

$$k_1 \left|ABECD\right| - \frac{\pi}{2} = 2\pi n$$

To find the ray length, we introduce point A', which is the mirror reflection of point A with respect to the borehole axis and which is located at the same wave front. It is clear that

$$AB = A'B$$

and hence

$$|ABECD| = |A'BECD|$$

or

$$|A'BECD| = A'B + BE + EC + CD,$$

where

$$A'B = BE\cos 2\theta, \qquad CD = EC\cos 2\theta, \qquad \text{and} \qquad BE + EC = \frac{a}{\cos\theta} \tag{7.117}$$

Note that points E and D are located at different wave fronts. In particular, point E belongs to the converging wave. Therefore

$$ABECD = a\,(1 + \cos 2\theta)\,\frac{1}{\cos\theta} \tag{7.118}$$

and, in place of eq. 7.117, we have

$$2k_1 a\cos\theta = \frac{\pi}{2} + 2\pi n$$

or

$$k_1 a\cos\theta = \frac{\pi}{2}\,(2n + 1) \tag{7.119}$$

Thus, permissible values of angle θ, when the normal mode is formed, are defined from eq. 7.119. The latter also shows that the minimal (cut-off) frequency of the mode occurs when $\theta = 0$, and it is equal to

$$f_{nc} = \frac{c_1\,(2n + 1/2)}{4a}$$

or

$$(k_1 a)_{\min} = \frac{\pi}{2}\,(2n + 1/2) \tag{7.120}$$

Regardless of the order of the normal mode $(n \neq 0)$, with an increase of frequency angle θ also increases; otherwise, constructive interference would not take place. Comparison *of* r_n (eq. 7.72) with values of $(k_1 a)_{\min}$ follows

n	1	2	3	4	5	6
r_n	3.83	7.02	10.17	13.32	16.47	19.64
$(k_1 a)_{\min}$	3.92	7.06	10.20	13.34	16.49	19.62

Next assume that the surrounding medium is free space $(\rho_2 = 0)$. Then the phase shift at the boundary between incident and reflected waves is equal to π, and in place of eq. 7.119 we obtain

$$k_1 a \cos\theta = \frac{\pi}{2}\left(2n - \frac{1}{2}\right) \tag{7.121}$$

Comparison of $(k_1 a)_{\min}$ with values of r_n (eq. 7.89) follows:

n	1	2	3	4	5	6
r_n	2.40	5.52	8.65	11.79	14.93	18.07
$(k_1 a)_{\min}$	2.35	2.49	8.63	11.80	14.91	18.05

The right sides in eqs. 7.121 and 7.122 define asymptotic values of roots of Bessel functions $J_1(r_n)$ and $J_0(r_n)$, respectively. As follows from eq. 7.116,

$$\sin\theta = \frac{m_n}{k_1} = \frac{c_1}{c_{pn}}, \tag{7.122}$$

i.e., the phase velocity of the normal mode is the apparent velocity of the conical wave along the borehole axis. Of course, the same result directly follows from Fig. 7.4b.

Now we will focus our attention an a different part of the wavefield, namely, head waves $(c_2 > c_1)$.

Head waves in the borehole $(c_2 > c_1)$

At the high-frequency spectrum, when the wavelength is smaller than the borehole radius, we can expect the appearance of the head wave, Fig. 7.5a. It arises at points of the borehole surface where the incident angle of the direct wave is close to the critical angle, θ_c. In order to find an asymptotic expression of the potential of the head wave, we evaluate the contribution of the integration along branch lines C_1 and C_2, Fig. 7.5b. In this light, it is convenient to represent the potential inside the borehole in the form (eq. 7.61)

$$\tilde{\varphi}_1 = C\,\frac{e^{ikR}}{R} + \frac{C}{\pi}\int_{-\infty}^{\infty} A_m\, I_0\,(m_1 r)\, e^{imz}\, dm \tag{7.123}$$

Here

$$R = \sqrt{r^2 + z^2}$$

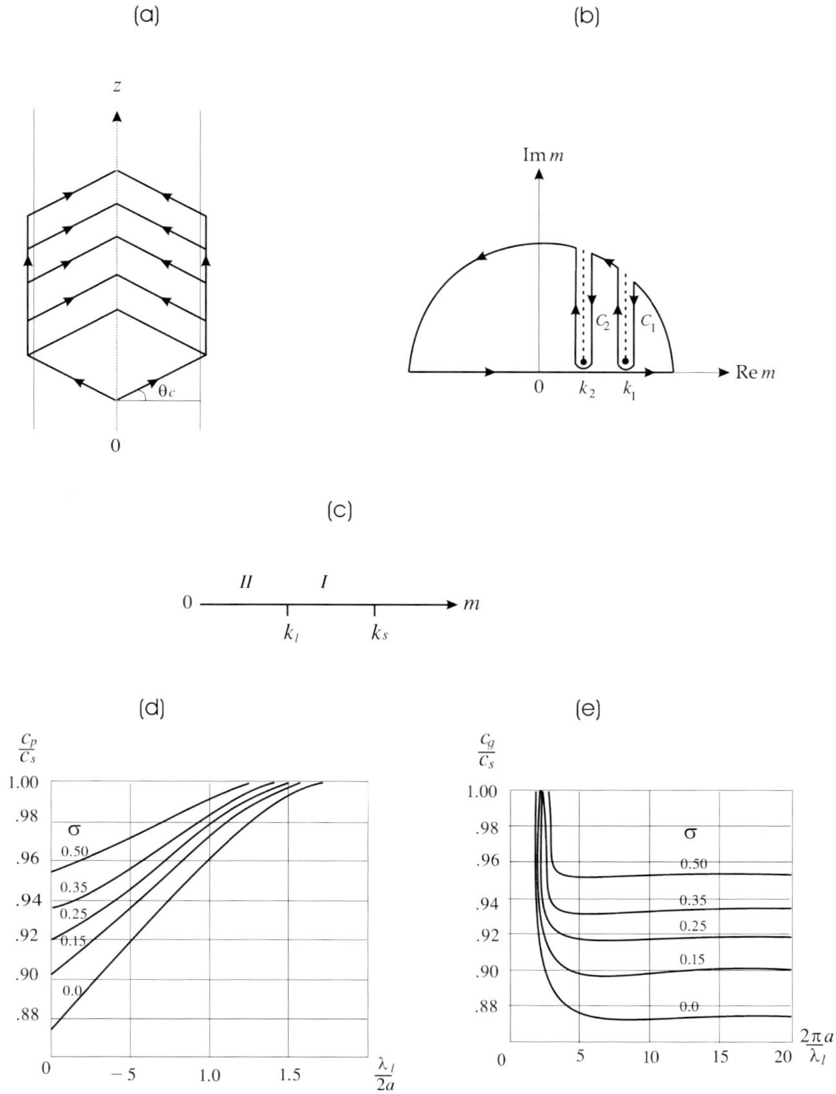

Figure 7.5: (a) Rays of head waves (b) Contour of integration along branch lines (c) Intervals of integration (d) Phase velocities of surface waves (e) Group velocities of surface waves. Numbers near curves are the values of Poisson's ratio σ. [After Biot, 1952]

and

$$A_m = \frac{m_1 K_0\ (m_2 a)\ K_1 (m_1 a) - m_2\ b\ K_0\ (m_1 a)\ K_1 (m_2 a)}{m_1 K_0\ (m_2 a)\ I_1 (m_1 a) + m_2\ b\ I_0\ (m_1 a)\ K_1 (m_2 a)} \tag{7.124}$$

To integrate along line C_1, we introduce a new variable,

$$m_1 = it, \tag{7.125}$$

where t alters from 0 to ∞ on one side of the branch line and from $-\infty$ to 0 on its other side, since after passing branch point k_1, radical m_1 changes sign. The variable of integration m along contour C_1 can be represented as

$$m = \left(m_1^2 + k_1^2\right)^{1/2} = \left(-t^2 + k_1^2\right)^{1/2} = i\left(t^2 - k_1^2\right)^{1/2}, \tag{7.126}$$

and, correspondingly,

$$dm = \frac{it\ dt}{\left(t^2 - k_1^2\right)^{1/2}} \qquad \text{and} \qquad m_2 = \left(-t^2 + k_1^2 - k_2^2\right)^{1/2} \tag{7.127}$$

Then the integral along branch line C_1 becomes

$$\int_0^\infty \left[\frac{i\ t\ K_0 (m_2 a)\ K_1\ (ita) - m_2\ b\ K_0\ (ita)\ K_1 (m_2 a)}{i\ t\ K_0 (m_2 a)\ I_1\ (ita) + m_2\ b\ I_0\ (ita)\ K_1 (m_2 a)} I_0 (itr)\right. \tag{7.128}$$

$$\left. - \frac{-i\ t\ K_0 (m_2 a) K_1 (-ita) - m_2\ b\ K_0 (-ita)\ K_1 (m_2 a)}{-i\ t\ K_0 (m_2 a) I_1 (-ita) + m_2\ b\ I_0 (-ita)\ K_1 (m_2 a)} I_0 (-itr)\right]$$

$$\times \frac{it}{\left(t^2 - k_1^2\right)^{1/2}}\ e^{-\sqrt{t^2 - k_1^2} z}\ dt$$

Making use of relations

$$I_0 (-ita) = I_0 (ita), \qquad I_1 (-ita) = -I_1 (ita) \tag{7.129}$$

and $\quad K_0 (-ita) = K_0 (ita) + i\pi I_0 (ita), \qquad K_1 (-ita) = -K_1 (ita) + i\pi I_1 (ita),$

we perform a transformation in eq. 7.128 and write the second fraction in parentheses in this equation as

$$\frac{-i\ t\ K_0 (m_2 a) \left[-K_1 (ita) + i\ \pi I_1 (ita)\right] - m_2 b K_1 (m_2 a) \left[K_0 (ita) + i\ \pi I_0 (ita)\right]}{i\ t K_0 (m_2 a)\ I_1 (ita) + m_2 b K_1 (ita)\ I_0 (ita)} \tag{7.130}$$

$$= \frac{i\ t\ K_0(m_2 a)\ K_1(ita) - m_2\ bK_0(ita)\ K_1(m_2 a)}{i\ tK_0(m_2 a)\ I_1(ita) + m_2\ bK_1(ita)\ I_0(ita)} - i\pi$$

The first term on the right side of eq. 7.130 coincides with the first term in parentheses in eq. 7.128. Therefore, the integral along branch line C_1 is equal to

$$-\pi \int_0^\infty \frac{t\ e^{-\sqrt{t^2 - k_1^2}\ z}}{\sqrt{t^2 - k_1^2}}\ I_0(itr)\ dt$$

or

$$-\pi \int_0^\infty \frac{t\ e^{-\sqrt{t^2 - k_1^2}\ z}}{\sqrt{t^2 - k_1^2}}\ J_0(tr)\ dt, \tag{7.131}$$

since

$$I_0(itr) = J_0(tr)$$

Taking into account that

$$\frac{e^{i\ k_1 R}}{R} = \int_0^\infty \frac{t}{\sqrt{t^2 - k_1^2}}\ e^{-\sqrt{t^2 - k_1^2}\ z}\ J_0(mr)\ dm, \tag{7.132}$$

we conclude that

$$\tilde{\varphi}_s(k_1) = -C\ \frac{e^{i\ k_1 R}}{R} \tag{7.133}$$

In other words, inside the borehole the sum of potentials due to the direct wave and to integration along branch line C_1 is equal to zero. Thus, the field in the borehole caused by the head wave is expressed only through the integral along branch line C_2

$$\tilde{\varphi}_{1b}(\omega, r, z) = \frac{C}{\pi} \int_{C_2} A_m\ I_0(m_1 r)\ e^{i\ m\ z}\ dt$$

or

$$\tilde{\varphi}_{1b}(\omega, r_1 z) = \frac{C}{\pi} \int_{k_2}^{i\infty} [A_m(m_1, m_2) - A_m(m_1, -m_2)]\ I_0(m_1 r)\ e^{i\ m\ z}\ dt, \tag{7.134}$$

and the integral can be written as

$$\int_{k_2}^{i\infty}\left[\frac{m_1 K_0(m_2 a) K_1(m_1 a) - m_2\ b\ K_0(m_1 a) K_1(m_2 a)}{m_1 K_0(m_2 a) I_1(m_1 a) + m_2\ b\ I_0(m_1 a) K_1(m_2 a)}\right. \tag{7.135}$$

$$\left. - \frac{m_1 K_0(-m_2 a) K_1(m_1 a) + m_2\ b\ K_0(m_1 a) K_1(m_2 a)}{m_1 K_0(-m_2 a)\ I_1(m_1 a) - m_2\ b\ I_0(m_1 a) K_1(-m_2 a)}\right] I_0(m_1 r)\ e^{i\ m\ z}\ dt$$

First, consider the numerator of the expression in parentheses, which is equal to

$$[m_1 K_0(m_2 a) K_1(m_1 a) - m_2 b K_0(m_1 a) K_1(m_2 a)]$$

$$\times [m_1 K_0(-m_2 a) I_1(m_1 a) - m_2 b I_0(m_1 a) K_1(-m_2 a)]$$

$$- [m_1 K_0(m_2 a) I_1(m_1 a) + m_2 b I_0(m_1 a) K_1(m_2 a)]$$

$$= [m_1 K_0(-m_2 a) K_1(m_1 a) + m_2\ b K_0(m_1 a) K_1(-m_2 a)]$$

The use of eqs. 7.129 greatly simplifies the numerator, which becomes

$$\frac{-b\ \pi i}{a^2} \tag{7.136}$$

Respectively, in place of eq. 7.134 we obtain

$$\tilde{\varphi}_{1b}(r, z, \omega) = -\frac{C\ i\ b}{a^2}\int_{k_2}^{i\infty} F(m_1, m_2)\ I_0(m_1 r)\ e^{i\ m\ z}\ dt \tag{7.137}$$

Here

$$F(m_1, m_2) = [m_1 K_0(m_2 a) I_1(m_1 a) + m_2 b I_0(m_1 a) K_1(m_2 a)]^{-1} \tag{7.138}$$

$$\times [m_1 K_0(-m_2 a) I_1(m_1 a) - m_2 b I_0(m_1 a) K_1(-m_2 a)]^{-1}$$

To evaluate this integral, it is useful to introduce variable q, whic is related to m in the following way:

$$m = k_2 + i\ \frac{q}{z} \tag{7.139}$$

It is clear that q varies within the range

$$0 \leq q < \infty \qquad \text{and} \qquad e^{i\,m\,z} = e^{i\,k_2\,z}\,e^{-q} \tag{7.140}$$

This means that the asymptotic value of the integral is determined by the interval of integration, where $q < 1$. For such values of q we have

$$m_1 \approx \left(k_2^2 - k_1^2 + 2\,i\,k_2\frac{q}{z}\right)^{1/2} \quad \text{and} \quad m_2 = \left(2ik_2\frac{q}{z}\right)^{1/2}, \qquad dm = \frac{i}{z}dq \tag{7.141}$$

Taking into account the behavior of functions K_0 and K_1 for small arguments:

$$K_0(x) \to \gamma_0 - \ln\frac{x}{2}, \qquad K_1(x) \to \frac{1}{x},$$

we have

$$F\left(m_1, m_2\right) \approx a^2\left[m_1 a\, I_1\left(m_1 a\right)\left(\gamma_0 + \ln\frac{m_2 a}{2}\right) + b\, I_0\left(m_1 a\right)\right]^{-1} \tag{7.142}$$

$$\times\left[m_1 a\, I_1\left(m_1 a\right)\left(\gamma + \ln\frac{-m_2 a}{2}\right) + b\, I_0\left(m_1 a\right)\right]^{-1},$$

where $\gamma_0 = 0.57722$ is Euler's constant. Because for small values of q

$$m_1 a \approx \sqrt{k_2^2 - k_1^2}\; a = \sqrt{k_1^2 - k_2^2}\; a\; e^{i\pi/2}, \qquad m_2 a \approx \sqrt{2k_2\frac{a^2}{z}}\; e^{i\pi/4},$$

$$\ln\frac{m_2 a}{2} \approx \frac{1}{2}\ln\frac{k_2 a^2 q}{z} - \frac{1}{2}\ln 2 + i\frac{\pi}{4}, \qquad \ln\frac{-m_2 a}{2} \approx \frac{1}{2}\ln\frac{k_2 a^2 q}{z} - \frac{1}{2}\ln 2 + i\frac{3\pi}{4},$$

and

$$I_0\left(i\sqrt{k_1^2 - k_2^2}\; a\right) = J_0\left(\sqrt{k_1^2 - k_2^2}\; a\right), \tag{7.143}$$

$$I_1\left(i\sqrt{k_1^2 - k_2^2}\; a\right) = i\; J_1\left(\sqrt{k_1^2 - k_2^2}\; a\right),$$

instead of eq. 7.137 we have

$$\tilde{\varphi}_{1b}\left(r, z, \omega\right) = \frac{C\, b}{z} e^{i\,k_2 z}\; J_0\left(\sqrt{k_1^2 - k_2^2}\; r\right) \tag{7.144}$$

$$\times \int_0^\infty \frac{e^{-q}\, dq}{\left(\alpha_1 + \beta_1 \ln \frac{k_2 q a^2}{z}\right)\left(\alpha_2 + \beta_2 \ln \frac{k_2 q a^2}{z}\right)}$$

where α_1, β_1 and α_2, β_2 are constants. At large values of z, when

$$\frac{k_2 a^2}{z} \ll 1,$$

the denominator rapidly varies with q. This fact does not allow us to take function F out of the integral. To overcome this difficulty we introduce a new variable, γ:

$$q = k_2\, z\, e^\gamma \tag{7.145}$$

whence

$$dq = k_2\, z\, e^\gamma\, d\gamma \qquad \text{and} \qquad \ln q = \ln k_2\, z + \gamma \tag{7.146}$$

Then the integral in eq. 7.144 becomes

$$k_2\, z \int_{-\infty}^{\infty} \frac{\exp\left(\gamma - k_2\, z\, e^\gamma\right)\, d\gamma}{(\alpha_1 + 2\beta_1 \ln k_2 a + \beta_1 \gamma)(\alpha_2 + 2\beta_2 \ln k_2 a + \beta_2 \gamma)} \tag{7.147}$$

It is clear that the integrand rapidly decreases when γ tends to either the upper or lower limit. In order to evaluate the integral, we use the stationary phase method. Taking the first derivative from the function

$$\gamma - k_2\, z\, e^\gamma,$$

we find that stationary point γ_s satisfies the equation

$$1 + k_2\, z\, e^{\gamma_s} = 0 \qquad \text{or} \qquad \gamma_s = -\ln k_2 z \tag{7.148}$$

The main contribution to the integral comes from the vicinity of stationary point γ_s. Since $\gamma_s < 0$ and $|\gamma_s| \gg 1$, the denominator of the integrand is a slowly varying function that can be taken out of the integral. This gives

$$\frac{k_2\, z}{(\alpha_1 + 2\beta_1 \ln k_2 a + \beta_1 \gamma_s)(\alpha_2 + 2\beta_2 \ln k_2 a + \beta_2 \gamma_s)} \int_{-\infty}^{\infty} \exp\left(\gamma - k_2\, z\, e^\gamma\right)\, d\gamma \tag{7.149}$$

Again introducing variable q, eq. 7.145, we find that the last integral is equal to $1/k_2 z$. Thus, eq. 7.144 becomes

$$\widetilde{\varphi}_{1b} = \frac{C\,b}{z}\, e^{ik_2 z}\, J_0\left(\sqrt{k_1^2 - k_2^2}\; r\right) \tag{7.150}$$

$$\times \frac{1}{(\alpha_1 + 2\beta_1 \ln k_2 a + \beta_1 \gamma_s)\,(\alpha_2 + 2\beta_2 \ln k_2 a + \beta_2 \gamma_s)}$$

The latter represents the leading term of the scalar potential of the head wave. As a function of z, $\widetilde{\varphi}_{1b}$, it is inversely proportional to $k_2 z \ln^2 k_2 z$ unless $J_1\left(\sqrt{k_1^2 - k_2^2}\; a\right) = 0$, and to $k_2\, z$ when $J_1\left(\sqrt{k_1^2 - k_2^2}\; a\right) = 0$. The dependence on r is given by Bessel function $J_0\left(\sqrt{k_1^2 - k_2^2}\; r\right)$. It is worthy to note that the product

$$e^{ik_2 z}\, J_0\left(\sqrt{k_1^2 - k_2^2}\; r\right)$$

at the right side of eq. 7.150 indicates that if $\sqrt{k_1^2 - k_2^2}\; r \gg 1$, the angle between the rays of the head wave and the borehole radius is equal to the critical angle, θ_c.

In our study of the wavefield inside the borehole at large distances from the source, we have found out that the wavefield consists of normal modes and head waves. The amplitude of normal modes is dominant, and the influence of the primary wave and of leaking modes is negligible. Besides, we have noted that normal modes behave as inhomogeneous waves in the surrounding medium. Also, outside the borehole, the transmitted wave propagates away from the boundary with velocity c_2. This wave arises at points of the interface $r = a$, where the incident angle is smaller than the critical angle. The interval of integration $0 \le m \le k_2$ defines the behavior of this wave.

7.3 Propagation of surface waves along a borehole

Next we will consider a more general model, when a medium surrounding a borehole is an elastic medium, ($\mu_2 \neq 0$). In this case wave behavior usually becomes more complicated. In particular, as in the case with the planar boundary, it is natural to expect the appearance of surface waves. By definition, surface waves have two main features, namely

a. Propagation along the boundary occurs without the support of other waves, and

b. The amplitude of potentials describing these waves exponentially decreases with increase of distance from the boundary.

Our first goal is to demonstrate the existence of such waves. To do this, we will use results obtained in section 7.1. In accordance with eqs. 7.25, scalar and vector potentials can be written in the form

$$\widetilde{\varphi}_1 = \int\limits_0^\infty A_m I_0\left(m_1 r\right) \cos mz \, dm \qquad \text{if} \quad r < a \tag{7.151}$$

and

$$\widetilde{\varphi}_2 = \int\limits_0^\infty B_m K_0\left(m_l r\right) \cos mz \, dm, \tag{7.152}$$

$$\widetilde{\psi}_2 = \int\limits_0^\infty C_m K_1\left(m_s r\right) \sin mz \, dm \qquad \text{if} \quad r > a$$

Here

$$m_1 = \sqrt{m^2 - k_1^2}, \qquad m_l = \sqrt{m^2 - k_l^2}, \qquad m_s = \sqrt{m^2 - k_s^2},$$

$$k_1 = \frac{\omega}{c_1}, \qquad k_l = \frac{\omega}{c_l}, \qquad k_s = \frac{\omega}{c_s},$$

and c_1, c_l, c_s are velocities in the borehole and in an elastic medium. As is well known, wavefields are such that at the boundary $(r = a)$, the normal components of displacement and stresses are continuous functions. Since shear stress in fluid vanishes, we have (eq. 7.5)

$$\widetilde{s}_r^{(1)} = \widetilde{s}_r^{(2)}, \qquad \widetilde{\tau}_{rr}^{(1)} = \widetilde{\tau}_{rr}^{(2)}, \qquad \widetilde{\tau}_{zr}^{(2)} = 0 \qquad \text{if} \qquad r = a \tag{7.153}$$

or, in terms of potentials (eq. 7.27),

$$\frac{\partial \widetilde{\varphi}_1}{\partial r} = \frac{\partial \widetilde{\varphi}_2}{\partial r} - \frac{\partial \widetilde{\psi}_2}{\partial z}, \tag{7.154}$$

$$-\lambda_1 k_1^2 \widetilde{\varphi}_1 = -\lambda_2 \, k_2^2 \widetilde{\varphi}_2 + 2\mu_2 \left(\frac{\partial^2 \widetilde{\varphi}_2}{\partial r^2} - \frac{\partial^2 \widetilde{\psi}_2}{\partial r \partial z} \right),$$

$$\text{and} \qquad 2\frac{\partial^2\widetilde{\varphi}_2}{\partial r\partial z} - \frac{\partial^2\widetilde{\psi}_2}{\partial z^2} + \frac{\partial}{\partial r}\left(\frac{1}{r}\frac{\partial}{\partial r}\, r\widetilde{\psi}_2\right) = 0$$

Substituting eqs. 7.151 and 7.152 into eqs. 7.154 and performing simple algebra with Bessel functions, we arrive at homogeneous system of equations with respect to A_m, B_m, and C_m:

$$m_1 I_1(m_1 a)\ A_m + m_l K_1(m_l a)\, B_m + m K_1(m_s a)\, C_m = 0,$$

$$-\lambda_1\, k_1^2\ I_0(m_1 a)\ A_m + \left[\lambda_2 k_l^2 K_0(m_l a) + 2\mu_2\ m_l^2\ K_1'(m_l a)\right] B_m \qquad (7.155)$$

$$+2\ \mu_2\ m\ m_s K_1'(m_s a)\, C_m = 0,$$

$$2m\ m_l K_1(m_l a)\, B_m + \left(2m^2 - k_s^2\right) K_1(m_s a)\, C_m = 0$$

Taking into account that the right side of this set is equal to zero, the solution of eqs. 7.155 exists only if the determinant of the system, D_m, obeys the equality

$$D_m = 0 \qquad (7.156)$$

However, the condition 7.156 is not sufficient for existence of surface wave. As was pointed out earlier, the potentials of these waves must decay exponentially with increased distance from the borehole surface. Taking into account the asymptotic behavior of the modified Bessel functions

$$I_0(x) \sim \frac{e^x}{\sqrt{x}} \qquad \text{and} \qquad K_0(x) \sim \frac{e^{-x}}{\sqrt{x}}, \qquad \text{if} \quad x \gg 1 \qquad (7.157)$$

we conclude that the solution of eq. 7.156 must obey the additional requirement

$$m > \max\ (k) \qquad (7.158)$$

It is necessary to distinguish two cases: $c_1 < c_s < c_l$ and $c_s < c_1 < c_l$. In place of eq. 7.158 we have, respectively,

$$m > k_1\ , \qquad m > k_s \qquad (7.159)$$

Our main attention will be paid to the existence of surface waves and their phase velocity. Wave amplitudes will be briefly discussed later.

The empty borehole

We will begin this study with the simplest case, in which the source of elastic waves has axial symmetry and is located in an elastic medium in the vicinity of the borehole. Along with the direct wave radiated by the source reflected P and S waves appear at points of the boundary when the corresponding incident angles do not exceed the critical angle. As these waves propagate along the interface, we may observe a conical shear wave caused by a longitudinal wave. The P evanescent wave appears at points of the boundary and is generated by the shear wave. By analogy with the plane interface, we observe a surface wave that causes deformation and rotation of elementary volumes of a medium. Its behavior is the subject of our study. Since waves are absent inside the borehole, i.e., the boundary is free, conditions 7.153 are simplified. They require only that both stresses vanish at points of the borehole surface. Correspondingly, letting $A_m = 0$, the last two equations of system 7.155 give

$$\left[\lambda_2 k_l^2 K_0(m_l a) + 2\,\mu_2\, m_l^2\, K_1'(m_l a)\right] B_m + 2\,\mu_2\, m\, m_s K_1'(m_s a)\, C_m = 0 \qquad (7.160)$$

and $$2m\, m_l K_1(m_l a)\, B_m + \left(2m^2 - k_s^2\right) K_1(m_s a)\, C_m = 0$$

Eliminating unknowns B_m and C_m and taking into account that the determinant of the system must be equal to zero, we obtain

$$-\left[\lambda_2\, k_l^2 K_0(m_l a) + 2\,\mu_2\, m_l^2 K_1'(m_l a)\right]\left(2m^2 - k_s^2\right)\frac{K_1(m_s a)}{2m\, m_l K_1(m_l a)} \qquad (7.161)$$

$$+\, 2\,\mu_2\, m\, m_s K_1'(m_s a) = 0$$

where

$$K_1'(x) = -\; K_0(x) - \frac{1}{x}\; K_1(x)$$

Thus, we have derived an equation with respect to the unknown wavenumber m. As numerical analysis shows, eq. 7.161 has the root $m_n(\omega)$, which obeys the condition

$$m_n(\omega) > k_s \qquad (7.162)$$

First of all, if the frequency is sufficiently high, we can expect the dispersion equation 7.161 to coincide with the Rayleigh equation. This happens because the wavelength becomes smaller than the borehole radius $(\lambda < a)$. To demonstrate this transition, we use eq. 7.157 and equality

$$K_1'(x) \sim - \quad K_0(x) \qquad \text{if} \quad x \gg 1$$

Replacing the modified Bessel functions by their asymptotic expressions, when

$$m_l a \gg 1 \qquad \text{and} \qquad m_s a \gg 1,$$

in place of eq. 7.161 we obtain the Rayleigh equation

$$\left(2m^2 - k_s^2\right)^2 - 4m^2 m_l m_s = 0$$

Therefore, at the high-frequency spectrum, the phase velocity of the surface wave approaches that of the Rayleigh wave, c_R, when the free boundary is planar.

Now we will discuss the behavior of phase velocity $c_p(\omega)$ at the low-frequency range. As is seen from Fig. 7.5c, in the first interval of integration, wavenumbers obey the condition

$$m \geq k_s = \frac{\omega}{c_s} \tag{7.163}$$

If the wavenumber is beyond this range $(m < k_s)$, the radical m_s becomes imaginary and, correspondingly, the integrand of the vector potential, $\tilde{\psi}_2$, does not demonstrate exponential decay. This suggests that there is a minimal (cut-off) frequency when the surface wave is still observed. Its phase velocity at this frequency coincides with that of the shear wave. In this case $(m_s \to 0)$, eq. 7.161 becomes a relationship between the cut-off frequency f_c or wavelength λ_c $(\lambda_c = c_l \,/\, f_c)$ and parameters of the medium. In fact, since

$$K_0(x) \to -\ln x, \qquad K_1(x) \to \frac{1}{x},$$

for the dispersion equation we have

$$\left[\lambda_2\, k_l^2 K_0(m_l a) + 2\mu_2\, m_l^2\, K_1'(m_l a)\right] \frac{1}{2m_l} + \frac{2\mu_2}{a} K_1(m_l a) = 0 \tag{7.164}$$

Here

$$k_l = \frac{\omega_c}{c_l}; \qquad m_l = \omega_c \left(\frac{1}{c_p^2} - \frac{1}{c_l^2}\right)^{1/2}$$

A graph of the ratio $c_p(\omega)/c_s$ as a function of $\lambda_l/2a$ is shown in Fig. 7.5d for different values of Poisson's ratio, σ. Here $\lambda_l = c_l/f$ is the wavelength of the longitudinal waves. We see that with an increase of the wavelength, phase velocity of the surface wave also gradually increases, and it tends to the same limit c_s regardless of the value of Poisson's ratio, σ. For example, when $\sigma = 1/4$, this asymptotic value is almost reached when the wavelength exceeds the borehole radius by about three. Note that unlike with Rayleigh waves, the phase velocity of the surface wave along the empty borehole depends on frequency, and its range of change is

$$c_R < c_p(\omega) < c_s \tag{7.165}$$

Finally, Fig. 7.5e illustrates the behavior of group velocity $c_g(\omega)/c_s$ as a function of parameter $2\pi a/\lambda_l$.

As follows from eqs. 7.152, with increased distance from the borehole, as well as increased frequency, both potentials display exponential decay. Also, because $c_p(\omega) < c_s$ and $c_l > c_s$, vector potential decreases more slowly than scalar potential. Since displacement $\mathbf{s}$ is related to the potentials as

$$\mathbf{s} = \text{grad}\,\varphi + \text{curl}\,\boldsymbol{\psi},$$

we again conclude that propagation of the surface wave is characterized by elliptical polarization of particle motion.

Next we will consider the behavior of surface waves in a more general case, when the borehole is filled by a fluid.

Stoneley waves

In Chapter 5, we discussed the behavior of surface waves propagating along the plane boundary between a fluid and an elastic medium. In fluid this wave is dilatational, whereas in an elastic medium it is accompanied by deformation and rotation of elementary volumes. In other words, this wave combines characteristics of dilatational and shear waves. It is essential that scalar and vector potentials describing the surface wave decrease with distance from the boundary. The velocity of propagation of this wave, which is usually called the Stoneley or Scholte wave, is independent of frequency. It is also natural to expect the appearance of a surface wave propagating along the cylindrical surface of the borehole. In order to prove the existence of this wave and to outline its main features, we proceed from the system of homogeneous equations 7.155. Our goal is to find such wavenumbers m that the determinant of set 7.155 becomes equal to zero. Besides, it is assumed that m obeys the inequality

$$m > \max(k_1, k_s)$$

Then it is clear that all three radicals – m_1, m_s, and m_l – are positive. Correspondingly, elementary cylindrical waves

$$A_m I_0(m_1 r)\cos mz, \qquad B_m K_0(m_l r)\cos mz\,, \qquad \text{and} \qquad C_m K_0(m_s r)\cos mz$$

become smaller with an increase of distance from the boundary $(r = a)$. To begin, we will consider two limiting cases, $\omega \to \infty$ and $\omega \to 0$.

The high-frequency spectrum

Suppose that frequencies are so high that the wavelength of longitudinal waves in an elastic medium is smaller than the borehole radius: $\lambda_l < a$. Correspondingly, this inequality takes place for the shear wave as well as well as in a fluid. Letting $c(\omega)$ be the velocity of propagation of the surface wave, we have

$$m_1 = \omega\left(\frac{1}{c^2} - \frac{1}{c_1^2}\right)^{1/2}, \qquad m_l = \omega\left(\frac{1}{c^2} - \frac{1}{c_l^2}\right)^{1/2}, \qquad m_s = \omega\left(\frac{1}{c^2} - \frac{1}{c_s^2}\right)^{1/2}$$

At the high-frequency spectrum, these radicals are large, and the modified Bessel functions are described by their asymptotic formulas:

$$I_0(x) = I_1(x) = \left(\frac{1}{2\pi x}\right)^{1/2} e^{x}, \qquad K_0(x) = K_1(x) = \left(\frac{\pi}{2x}\right)^{1/2} e^{-x}, \tag{7.166}$$

and $\qquad K_0'(x) \to -K_1(x) \qquad$ if $\quad x \gg 1$

Substitution of eqs. 7.166 into set 7.155 yields

$$\left(\frac{m_1}{\pi}\right)^{1/2} e^{m_1 a} A_m + (m_l \pi)^{1/2} e^{-m_l a} B_m + m\left(\frac{\pi}{m_s}\right)^{1/2} e^{-m_s a} C_m = 0,$$

$$\lambda_1 k_1^2 \left(\frac{1}{\pi m_1}\right)^{1/2} e^{m_1 a} A_m + \mu_2\left[\left(2m^2 - k_s^2\right) + \frac{2m_l}{a}\right]\left(\frac{\pi}{m_l}\right)^{1/2} e^{-m_l\, a} B_m \tag{7.167}$$

$$+\, 2\mu_2 m\, m_s \left(1 + \frac{1}{m_s a}\right)\left(\frac{\pi}{m_s}\right)^{1/2} e^{-m_s a} C_m = 0,$$

$$2m\, {m_l}^{1/2} e^{-m_l\, a} B_m + \frac{2m^2 - k_s^2}{{m_s}^{1/2}} e^{-m_s a} C_m = 0$$

The determinant of this system coincides with that for the case in which the boundary is a planar boundary. Thus, at sufficiently high frequencies $(\lambda_l \ll a)$, the surface wave arises and advances along the boundary of the borehole with a velocity close to that of the Stoneley wave at the plane interface. For instance, if

$$\frac{c_s}{c_1} = 1.5, \qquad \frac{\rho_2}{\rho_1} = 1, \qquad \sigma = \frac{1}{4},$$

we have

$$\frac{c\,(\infty)}{c_s} \approx 0.92,$$

i.e., this velocity is slightly smaller than the velocity of the shear wave. It is obvious that with an increase of the frequency, the wavefields, such as displacement or stress, are mainly concentrated in the vicinity of the borehole surface. At the same time, at sufficiently large distances r away from the borehole surface, wave propagation causes virtually no motion in the medium. Now consider the second limiting case, the low-frequency spectrum.

The low-frequency spectrum

Assuming that arguments of the Bessel functions are small, we can use the following approximations:

$$I_0\,(x) \to 1, \quad I_1\,(x) \to \frac{x}{2}, \qquad K_0\,(x) \to -\ln x, \quad K_1\,(x) \to \frac{1}{x},$$

$$\text{and} \qquad K_1' \to -\frac{1}{x^2} \qquad \text{if} \quad x \ll 1$$

After their substitution into set 7.155, we obtain

$$\frac{m_1^2\, a^2}{2} A_m + B_m + \frac{m}{m_s}\, C_m = 0,$$

$$\lambda_1 k_1^2\, a^2 A_m + 2\mu_2 B_m + 2\mu_2 \frac{m}{m_s}\, C_m = 0, \tag{7.168}$$

$$0 + 2m B_m + \frac{2m^2 - k_s^2}{m_s}\, C_m = 0 \qquad \text{if} \quad \omega \to 0$$

The dispersion equation in this case is

$$\begin{vmatrix} \dfrac{m_1^2 a^2}{2} & 1 & \dfrac{m}{m_s} \\ \lambda_1 k_1^2 a^2 & 2\mu_2 & \dfrac{2\mu_2 m}{m_s} \\ 0 & 2m & \dfrac{2m^2 - k_s^2}{m_s} \end{vmatrix} = 0 \qquad (7.169)$$

After doing simple algebra, we obtain

$$-m_1^2\,\mu_2 + k_1^2 \lambda_1 = 0$$

Therefore, wavenumbers m of the Stoneley wave at the low-frequency spectrum are defined as

$$m = \frac{\omega}{c\,(\omega)} = k_1 \left(\frac{\lambda_1 + \mu_2}{\mu_2} \right)^{1/2} \qquad (7.170)$$

Respectively, for phase velocity we have

$$c\,(\omega) \to \frac{c_1}{\left(1 + \dfrac{\lambda_1}{\mu_2}\right)^{1/2}} \qquad \text{if} \qquad \omega \to 0 \qquad (7.171)$$

Eq. 7.171 shows that the asymptotic value of $c\,(\omega)$ depends on the velocity in a fluid and on the parameter

$$\frac{\lambda_1}{\mu_2} = \frac{\rho_1\, c_1^2}{\mu_2}, \qquad (7.172)$$

which is the ratio of fluid compressibility λ_1 to shear modulus μ_2. Thus, measurements of the velocity of the Stoneley's wave inside the borehole, if $\omega \to 0$, allow us to evaluate the rigidity μ_2 of the elastic medium. Note that solution of system 7.155 shows that the Stoneley's wave does not have a cut-off frequency $(c_s > c_1)$. In other words, this wave can be observed at any frequency. As follows from eq. 7.151, the axial and radial components of displacement corresponding to elementary cylindrical harmonics are

$$\tilde{s}_z(m) = -m\; A_m I_0(m_1 r) \sin mz \qquad \text{and} \qquad \tilde{s}_r(m) = m_1 A_m I_1(m_1 r) \cos mz \qquad (7.173)$$

At the low-frequency range, when the wave is defined by small values of m, we have

$$\tilde{s}_z(m) = -m A_m \sin mz \qquad \text{and} \qquad \tilde{s}_r(m) = \frac{r}{2} m_1^2 A_m \cos mz, \qquad (7.174)$$

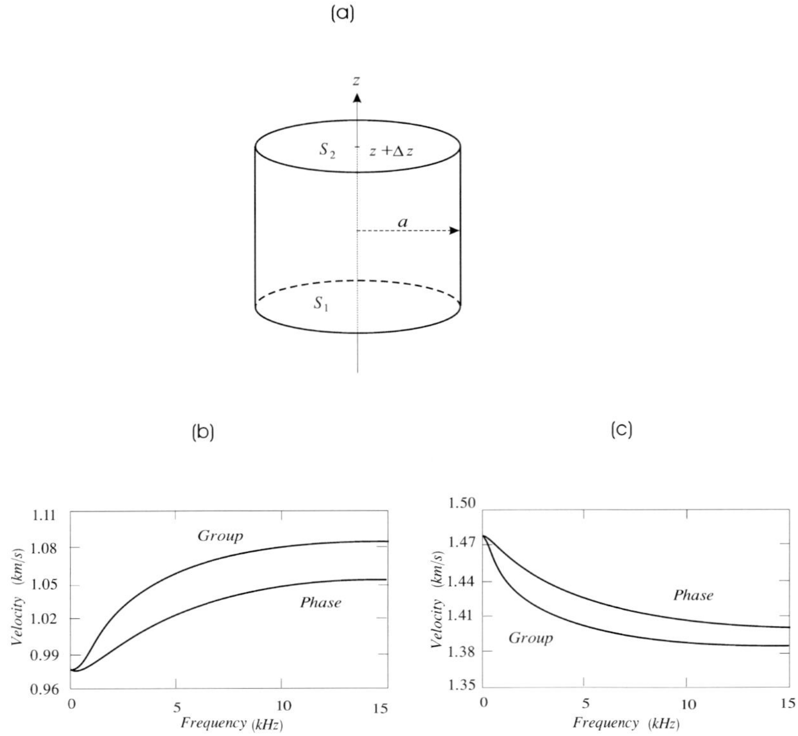

Figure 7.6: (a) Portion of borehole between two cross-sections S_1 and S_2 (b) Phase and group velocities when $c_s > c_1$ ($a = 0.1$ m, $c_l = 4.51$ km/s, $c_s = 2.41$ km/s, $\rho_1 = 1.2$ g/cm^3, $\rho_2 = 2.3$ g/cm^3) (c) Phase and group velocities when $c_s < c_1$ ($a = 0.1$ m, $c_l = 2.70$ km/s, $c_s = 1.20$ km/s, $\rho_1 = 1.2$ g/cm^3, $\rho_2 = 2.1$ g/cm^3). [After Paillet & Cheng, 1991]

i.e., the radial component of displacement linearly increases with r. At the same time, component $\tilde{s}_z$ remains constant at any cross-section of the borehole, although it varies with time and with coordinate z.

It is instructive to obtain eq. 7.171 in a completely different way. Let us consider a portion of the borehole bounded by its lateral surface, $r = a$, and by two cross-sections S_1 and S_2, located at distance Δz from each other, Fig. 7.6a. Note that

$$S = S_1 = S_2 = \pi a^2$$

The volume of this part

$$V_1 = \pi a^2 \, \Delta z$$

is not elementary in the axial direction, since it is characterized by a finite length a. Propagation of a surface wave along the borehole causes vibration of particles in a fluid and in the surrounding elastic medium. Because of symmetry, this motion is described by the axial, s_z, and radial, s_r, components of displacement. The azimuthal component, s_φ, is absent:

$$\mathbf{s} = s_r \mathbf{i}_r + s_z \mathbf{i}_z$$

Our main goal is to derive an equation of motion of volume V_1 and demonstrate that under certain conditions the motion equation coincides with the wave equation. Its solution describes a wave advancing along the borehole with the velocity given by eq. 7.171. In order to solve this task we make several assumptions that are obvious given that the fields slowly varying in the radial direction. For instance, in the frequency domain this means that the wavelength is much greater than the borehole radius. First, in agreement with eqs. 7.174, suppose that the axial component s_z and the additional pressure p are independent of coordinate r inside volume V_1, i.e.,

$$s_z = s_z(z, t) \qquad \text{and} \qquad p = p(z, t) \tag{7.175}$$

This volume moves along the borehole axis due to the external forces

$$F_z(z + \Delta z, t) \qquad \text{and} \qquad F_z(z, t),$$

which are caused by fluid located above and beneath volume V_1 and, correspondingly, have opposite directions. For instance, if force $F_z(z, t)$ produces compression and the additional pressure $p(z, t)$ becomes positive, we have for the resultant force

$$[-p(z + \Delta z, t) + p(z, t)] \; S$$

Therefore, in accordance with Newton's second law, the equation of motion of the volume is

$$\rho \, \frac{\partial^2 s_z}{\partial t^2} \, \pi a^2 \, \Delta z = -\frac{\partial p}{\partial z} \, \pi a^2$$

or

$$\rho \, \frac{\partial^2 s_z}{\partial t^2} = -\frac{\partial p}{\partial z}, \tag{7.176}$$

since it is assumed that the pressure varies linearly within Δz and that

$$-p\,(z+\Delta z,t)+p\,(z,t)=-\frac{\partial p\,(t)}{\partial z}\Delta z$$

whereas s_z is the axial component of displacement of the middle point of V_1. The equation of motion contains two unknowns, s_z and p, and we therefore must find another relation between them. Since a wave causes very small changes of any elementary volume V $(V<V_1)$, the principle of mass conservation gives (Part I)

$$\frac{\rho}{\rho_1}=-\frac{\Delta V}{V} \tag{7.177}$$

Here ρ and ΔV are variations of the original density ρ_1 and volume V. Inasmuch as deformation is also accompanied by small changes of pressure, we have

$$p=\alpha\,\rho, \tag{7.178}$$

and eq. 7.177 gives

$$p=-\lambda_1\frac{\Delta V}{V} \tag{7.179}$$

Here λ_1 is the bulk modulus of the fluid

$$\lambda_1=\alpha\,\rho_1=c_1^2\,\rho_1 \tag{7.180}$$

and pressure is constant inside volume V. By definition

$$\frac{\Delta V}{V}=\text{div }\mathbf{s},$$

and eq. 7.179, written as

$$p=-\lambda_1\text{div }\mathbf{s}, \tag{7.181}$$

describes the relationship between pressure and displacement in equilibrium. For instance, in the cylindrical system of coordinates we have

$$p=-\lambda_1\left(\frac{\partial s_z}{\partial z}+\frac{\partial s_r}{\partial r}+\frac{s_r}{r}\right), \tag{7.182}$$

since $s_\varphi=0$.

Next consider a change of volume V_1, which occurs for two reasons. One reason is axial motion, which is equal to

$$\pi a^2\,\frac{\partial s_z}{\partial z}\,\Delta z,$$

because $s_z(z)$ and $s_z(z+\Delta z)$ are displacements at the opposite faces of volume V_1. The other reason is radial expansion (compression), which is

$$2\pi a\ s_r\ \Delta z$$

Thus

$$\frac{\Delta V_1}{V_1} = \frac{\pi a^2\ \dfrac{\partial s_z}{\partial z}\ \Delta z + 2\pi a\ s_r(a)\ \Delta z}{\pi a^2\ \Delta z} = \frac{\partial s_z}{\partial z} + \frac{2}{a} s_r(a) \tag{7.183}$$

In accordance with the second equation of set 7.174, the radial component of displacement is the linear function of r, and it can be represented as

$$s_r(r) = \frac{s_r(a)}{a} r \tag{7.184}$$

Its substitution into eq. 7.181 shows again that

$$\frac{\Delta V_1}{V_1} = \text{div } \mathbf{s}$$

and, therefore,

$$p = -\lambda_1 \left[\frac{\partial s_z}{\partial z} + \frac{2}{a} s_r(a) \right], \tag{7.185}$$

where p is pressure at the middle point. Note that eq. 7.185 can be also derived from the Gauss formula:

$$\int\limits_{V_1} \text{div } \mathbf{s}\ dV_1 = \oint\limits_{S} \mathbf{s} \cdot d\mathbf{S}$$

In fact, making use of eq. 7.180 and assuming that $p =$ const, $\dfrac{\partial s_r(a)}{\partial z} = 0$, we obtain

$$-\lambda_1 p\ V = [s_z(z+\Delta z) - s_z(z)]\ \pi a^2 + s_r(a)\ 2\pi a\ \Delta z,$$

which coincides with eq. 7.185. Note that eqs. 7.176 and 7.185, derived for volumes that have different extensions Δz, contain three unknowns:

$$p, \quad s_z, \quad s_r$$

For this reason we need one more relationship, for instance, a relationship between p and s_r. Before solving this problem, let us consider a special case when s_r inside the borehole is equal to zero. In other words, fluid moves only along the z-axis. Then the system of eqs. 7.176 and 7.185 gives the wave equation

$$\frac{\partial^2 s_z}{\partial z^2} = \frac{1}{c_1^2} \frac{\partial^2 s_z}{\partial t^2},$$

which describes a wave moving with the velocity of the longitudinal wave, $c_1 = (\lambda_1/\rho_1)^{1/2}$. In reality, because the radial component s_r differs from zero, the elastic medium around the borehole influences wave velocity in the fluid. In establishing the linkage between s_r and p, we take into account the relatively slow change of these functions with time. This implies that both pressure and the radial component of displacement change almost synchronously at different points, even at large distances r from the borehole. With this limiting case in mind, we can expect the same relationship between p and s_r as we would expect in equilibrium. Let us consider now this subject of the static elasticity in detail.

Thick cylindrical shell

Suppose that a cylindrical shell of an arbitrary thickness is oriented along the z-axis. Its length is sufficiently large and we can ignore effects near the cylindrical ends. It is also assumed that the shell is under the action of fluid pressure p, which is uniformly distributed over the shell's internal surface. At the same time, at the external surface the pressure vanishes. Thus, we suppose that deformation is symmetrical about the z-axis and uniform along each generating line. Correspondingly, a cross-section remains plane after deformation, so that displacement along the z-axis is constant and can be zero. We can evaluate deformation inside the elastic cylinder as a two-dimensional boundary problem. It is related to the fact that pressure is distributed uniformly on the lateral surface of the shell. First of all, as was demonstrated in Chapter 1, the displacement $\mathbf{s}$ has to satisfy the equation

$$(\lambda_2 + \mu_2) \text{ grad div } \mathbf{s} + \mu_2 \nabla^2 \mathbf{s} = 0 \tag{7.186}$$

This is the condition of equilibrium when volume forces are ignored. Because the normal component of stress is a continuous function at the borehole surface between a fluid and an elastic medium, the first boundary condition is

$$\tau_{rr}(a) = -p, \tag{7.187}$$

where p is pressure caused by a fluid. Also, we have at the external surface of the cylinder

$$\tau_{rr}(b) = 0 \tag{7.188}$$

In solving eq. 7.186, it is natural to assume that displacement $\mathbf{s}$ has only the radial component

$$\mathbf{s} = s_r(r)\,\mathbf{i}_r, \tag{7.189}$$

which is a function of the single coordinate r. By definition,

$$\text{curl }\mathbf{s} = \frac{1}{r}\begin{vmatrix} \mathbf{i}_r & r\,\mathbf{i}_\varphi & \mathbf{i}_z \\ \dfrac{\partial}{\partial r} & \dfrac{\partial}{\partial \varphi} & \dfrac{\partial}{\partial z} \\ s_r & 0 & 0 \end{vmatrix},$$

i.e.,

$$\text{curl }\mathbf{s} = 0 \tag{7.190}$$

Inasmuch as

$$\text{curlcurl }\mathbf{s} = \text{graddiv }\mathbf{s} - \nabla^2\,\mathbf{s},$$

eq. 7.186 can be written as

$$(\lambda_2 + 2\mu_2)\,\nabla^2\mathbf{s} + \mu_2\text{ curlcurl }\mathbf{s} = 0$$

Making use of eq. 7.190, we obtain the Laplace equation for displacement $\mathbf{s}$:

$$\nabla^2\mathbf{s} = 0 \qquad \text{or} \qquad \mathbf{i}_r\nabla^2 s_r + s_r\nabla^2\,\mathbf{i}_r = 0 \tag{7.191}$$

The laplacian in the cylindrical system of coordinates is

$$\nabla^2 = \frac{1}{r}\frac{d}{dr}\left(r\frac{d}{dr}\right) + \frac{1}{r^2}\frac{d^2}{d\varphi^2}, \tag{7.192}$$

because the fields are independent of z. It is clear that

$$\mathbf{i}_r = \cos\varphi\;\mathbf{i} + \sin\varphi\;\mathbf{j}$$

Substituting the latter into eq. 7.192 and taking into account that $\mathbf{i}$ and $\mathbf{j}$ are constants, we obtain

$$\nabla^2\mathbf{i}_r = -\frac{1}{r^2}\,\mathbf{i}_r \tag{7.193}$$

Correspondingly, eq. 7.191 becomes

$$\frac{1}{r}\frac{d}{dr}\left(r\frac{ds_r}{dr}\right) - \frac{s_r}{r^2} = 0 \tag{7.194}$$

Thus, eq. 7.186 is greatly simplified. To determine displacement component s_r, we have to solve an ordinary differential equation. Suppose that its solution has the form

$$s_r(r) = Cr^m, \tag{7.195}$$

where C and m are unknowns. Substitution of eq. 7.195 into eq. 7.194 gives

$$m^2 - 1 = 0,$$

i.e., $m_1 = 1$ and $m_2 = -1$, and the general solution of eq. 7.194 is

$$s_r(r) = Ar + \frac{B}{r} \tag{7.196}$$

Here A and B are unknown constants that are determined from boundary conditions. As was shown earlier (Chapter 1), the diagonal elements of the strain tensor in the cylindrical system of coordinates are

$$e_{rr} = \frac{ds_r}{dr} \qquad \text{and} \qquad e_{\varphi\varphi} = \frac{s_r}{r}, \tag{7.197}$$

while

$$e_{zz} = 0,$$

since displacement s_z is independent of the z-coordinate. From eq. 7.196 we have

$$e_{rr} = A - \frac{B}{r^2} \qquad \text{and} \qquad e_{\varphi\varphi} = A + \frac{B}{r^2} \tag{7.198}$$

By definition, in the cylindrical system of coordinates

$$\text{div } \mathbf{s} = \frac{1}{r}\frac{d}{dr}(rs_r) = 2A, \tag{7.199}$$

which shows that at each point of any plane z=const, the relative change in elementary area is the same. This deformation causes the normal component of stress along the cylinder axis, and

$$\tau_{zz} = \text{const} \tag{7.200}$$

In accordance with Hooke's law (Chapter 1),

$$\tau_{rr}(r) = \lambda_2 \text{ div } \mathbf{s} + 2\mu_2 \, e_{rr}, \tag{7.201}$$

and using eq. 7.198 we have

$$\tau_{rr}(r) = 2\left[(\lambda_2 + \mu_2)\, A - \mu_2 \frac{B}{r^2}\right] \tag{7.202}$$

The boundary conditions give

$$-p = 2\left[(\lambda_2 + \mu_2)\, A - \mu_2 \frac{B}{a^2}\right], \qquad 0 = 2\left[(\lambda_2 + \mu_2)\, A - \mu_2 \frac{B}{b^2}\right] \tag{7.203}$$

Solution of this system is

$$A = \frac{a^2 p}{2\,(\lambda_2 + \mu_2)\,(b^2 - a^2)} \tag{7.204}$$

and

$$B = \frac{a^2 b^2 p}{2\mu_2\,(b^2 - a^2)} \tag{7.205}$$

Correspondingly,

$$\tau_{rr}(r) = \frac{p\, a^2\,(r^2 - b^2)}{r^2\,(b^2 - a^2)}, \tag{7.206}$$

which gradually changes with an increase of r. Now in eq. 7.196, letting $r = a$ we obtain

$$\frac{s_r(a)}{a} = \frac{a^2 p}{2\,(\lambda_2 + \mu_2)\,(b^2 - a^2)} + \frac{b^2\, p}{2\mu_2\,(b^2 - a^2)}$$

or

$$\frac{s_r(a)}{a} = \frac{p}{2M}, \tag{7.207}$$

where

$$M = \frac{\mu_2\,(\lambda_2 + \mu_2)\,(b^2 - a^2)}{\mu_2 a^2 + (\lambda_2 + \mu_2)\, b^2} \tag{7.208}$$

Thus we have solved our task and established the relationship between radial displacement of an elastic medium at the borehole surface and fluid pressure. Before we continue, let us note the following. The azimuthal component of stress is defined as

$$\tau_{\varphi\varphi}(r) = \lambda_2 \text{ div } \mathbf{s} + 2\,\mu_2\, e_{\varphi\varphi} \tag{7.209}$$

Taking into account eqs. 7.198 and 7.199, we have

$$\tau_{\varphi\varphi}(r) = 2\lambda_2\, A + 2\,\mu_2 \left(A + \frac{B}{r^2}\right)$$

or

$$\tau_{\varphi\varphi}(r) = \frac{a^2\,(r^2 + b^2)}{r^2\,(b^2 - a^2)}\; p \tag{7.210}$$

By definition (Chapter 1),

$$Ee_{zz} = \tau_{zz} - \sigma\,(\tau_{rr} + \tau_{\varphi\varphi})$$

Here E is the Young modulus and σ is Poisson' ratio. Inasmuch as strain is absent along the z-axis, we have

$$\tau_{zz} = \sigma\,(\tau_{rr} + \tau_{\varphi\varphi})\,, \tag{7.211}$$

and the use of eqs. 7.206–7.210 gives

$$\tau_{zz} = \frac{2\sigma\; a^2}{b^2 - a^2}\; p = \text{const} \tag{7.212}$$

Of course, this independence of coordinate r is expected. Also, it is interesting to mention that with an increase of the external radius r, the stress component τ_{zz} becomes smaller, and in the limit $(r \to \infty)$, it vanishes. Taking into account continuity of the normal component of displacement s_r and substituting eq. 7.207 into eq. 7.185, we obtain

$$p = -\lambda_1 \left(\frac{\partial s_z}{\partial z} + \frac{p}{M}\right)$$

or

$$p = -\frac{1}{\left(\frac{1}{\lambda_1} + \frac{1}{M}\right)}\,\frac{\partial s_z}{\partial z} \tag{7.213}$$

Then from eq. 7.213 and 7.176 we have

$$\rho\frac{\partial^2 s_z}{\partial t^2} = \frac{1}{\left(\frac{1}{\lambda_1} + \frac{1}{M}\right)}\,\frac{\partial^2 s_z}{\partial z^2} \tag{7.214}$$

This is the wave equation with respect to the vertical component of fluid displacement:

$$\frac{\partial^2 s_z}{\partial z^2} = \frac{1}{c^2}\frac{\partial^2 s_z}{\partial t^2}$$

Here

$$c = \frac{1}{\sqrt{\rho\left(\dfrac{1}{\lambda_1} + \dfrac{1}{M}\right)}} \tag{7.215}$$

or

$$c = c_1 \frac{1}{\left(1 + \dfrac{\lambda_1}{M}\right)^{1/2}} \tag{7.216}$$

In accordance with eq. 7.208, with an increase of external radius b, parameter M tends to the shear modulus

$$M \to \mu_2 \tag{7.217}$$

Respectively, eq. 7.216 coincides with eq. 7.171, which describes the velocity of Stoneley waves in the low-frequency spectrum.

It is also useful to consider the case in which radii a and b are close to each other, that is, the borehole is surrounded by a thin tube with thickness h:

$$b - a = h, \qquad b + a \approx 2a$$

Then, in place of eq. 7.208, we have

$$M \approx \frac{2\,\mu_2\,(\lambda_2 + \mu_2)\;h}{(\lambda_2 + 2\mu_2)\;a}$$

or

$$M \approx \frac{2\,\mu_2\,h}{\left(1 + \dfrac{\mu_2}{\lambda_2 + \mu_2}\right)\,a}, \tag{7.218}$$

and the velocity of propagation differs from that in the previous example $(M \sim \mu_2)$.

Phase and group velocities

We have studied phase velocity of the Stoneley wave in the low- and high-frequency spectrums. Next let us consider velocity dispersion, i.e., its dependence on frequency. Earlier it was demonstrated that the wavenumber of the Stoneley wave

$$m\,(\omega) = \frac{\omega}{c\,(\omega)}$$

is defined as the root of the characteristic equation 7.156 $D_m = 0$, provided that $m > \max(k_1, k_s)$.

As usual, it is convenient to consider the numerical solutions of this equation separately for two models of a medium, where either $c_s > c_1$ or $c_s < c_1$. In the first case, Fig. 7.6b, there is small dispersion of the phase and group velocities, and c_g is slightly higher than c_p. With an increase of frequency, values of both functions become larger and in the limit they approach the asymptote, which is a little smaller than wave velocity c_1 in the borehole fluid.

Before we discuss velocity dispersion in the second model $(c_s < c_1)$, let us use eq. 7.171,

$$c = \frac{c_1}{\left(1 + \dfrac{\lambda_1}{\mu_2}\right)^{1/2}}, \qquad \text{if} \quad \omega \to 0$$

which can also be represented in the form

$$\frac{c}{c_s} = \left(\frac{c_s^2}{c_1^2} + \frac{\rho_1}{\rho_2}\right)^{-1/2} \qquad \text{or} \qquad \frac{c_s^2}{c^2} = \frac{\rho_1}{\rho_2} + \frac{c_s^2}{c_1^2} \qquad \text{if} \quad \omega \to 0 \tag{7.219}$$

Inasmuch as existence of the Stoneley wave implies that $c < c_s$, it is natural to distinguish two cases in the second model $(c_s < c_1)$:

$$\frac{\rho_1}{\rho_2} + \frac{c_s^2}{c_1^2} > 1 \qquad \text{and} \qquad \frac{\rho_1}{\rho_2} + \frac{c_s^2}{c_1^2} < 1$$

An example of functions $c(\omega)$ and $c_g(\omega)$ in the first case is shown in Fig. 7.6c. Again there is small dispersion of phase and group velocities, and the latter is slightly smaller $(c_g < c)$. At the low-frequency limit we have, approximately,

$$\frac{c}{c_1} = \frac{c_g}{c_1} \simeq 0.72$$

With an increase of frequency phase and group velocities decrease a little and approach their corresponding asymptotic values. Note that the phase velocity of the Stoneley wave remains smaller than the shear velocity. In the second case, Fig. 7.7, the ratio c_s/c_1 is even smaller, which causes peculiar behavior of the Stoneley wave in the low-frequency spectrum. In accordance with eq. 7.219, we have

$$c_p(\omega) > c_s, \qquad \text{if} \qquad \omega \to 0$$

which means that a wave propagates in both the axial and radial directions. Because of this, its movement along the borehole axis is accompanied by attenuation. As is seen

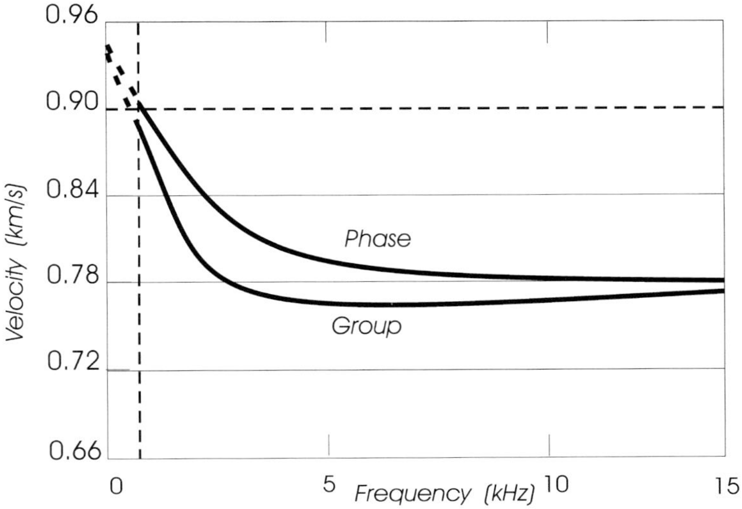

Figure 7.7: Dispersion curves of phase and group velocities ($a = 0.1$ m, $c_l = 2.40$ km/s, $c_s = 0.90$ km/s, $\rho_1 = 5.10$ g/cm^3, $\rho_2 = 2.1$ g/cm^3). [After Cheng & Toksöz, 1981]

from Fig. 7.7, when wavelength λ_1 is sufficiently large, the phase velocity of the surface wave exceeds the shear velocity. In this range the wave rapidly decays along the z-axis. However, with an increase of frequency, velocity $c_p(\omega)$ becomes smaller than c_s, and a Stoneley wave is observed. In other words, in the second case,

$$\frac{\rho_1}{\rho_2} + \frac{c_s^2}{c_1^2} > 1 \qquad ,$$

there is a cut-off frequency at which

$$c_p(f_c) = c_s$$

In addition, note the following:

1. The Stoneley wave propagates without attenuation along the borehole, whereas its wave amplitudes decay monotonically with increased distance r from the boundary, $r = a$.

2. At the beginning, this decrease is relatively slow. However, at large distances from the borehole surface, the amplitudes of displacement, strain, and stress decrease exponentially.

3. At the low-frequency spectrum, $a/\lambda_1 < 1$, the axial component of displacement s_z is almost constant inside the borehole, but the radial component s_r linearly increases as a function of r. At the borehole axis the latter is equal to zero, regardless of frequency.

4. As follows from boundary conditions, the component s_r is a continuous function at the borehole surface, but the axial component s_z has a discontinuity.

5. Propagation of the sinusoidal Stoneley wave is accompanied by elliptical partical motion in a fluid and in an elastic medium (elliptical polarization).

6. Wave amplitudes are defined by integration around the corresponding pole $m(\omega)$, where $m > \max(k_1, k_s)$.

7.4 Normal modes, head and transient waves

In the previous section, we studied the behavior of surface (Stoneley) waves propagating along the borehole. Certainly, this is a very important element of the wavefields that are caused by the source on the borehole axis.

As we know, scalar and vector potentials describing all of the waves inside and outside the borehole can be represented in the form:

$$\tilde{\varphi}_1 = C\,\frac{e^{i\,k_1 R}}{R} + \int\limits_0^\infty A_m I_0(m_1 r)\cos mz\; dm \qquad \text{if} \quad r < a \tag{7.220}$$

and $$\tilde{\varphi}_2 = \int\limits_0^\infty B_m K_0(m_l r)\cos mz\; dm,$$

$$\tilde{\psi}_2 = \int\limits_0^\infty C_m K_1(m_s r)\sin mz\; dm \qquad \text{if} \quad r > a$$

It is essential that all three functions, A_m, B_m, and C_m, have the same poles, which are roots of eq. 7.156:

$$D_m = 0 \tag{7.221}$$

In order to characterize the main features of waves, it is useful, as before, to separately consider two models of a medium, namely $c_1 < c_s < c_l$ (case 1) and $c_s < c_1 < c_l$ (case 2).

Case 1 Suppose that the ratio of the wavelength to the borehole radius is small ($\lambda_l \ll a$). Correspondingly, the wave behaves nearly in accordance with geometrical seismic postulates. In other words, we can use the concept of rays for P and S waves, as well as Snell's law. Earlier it was demonstrated that surface waves are absent when the borehole is surrounded by an acoustic medium, but the head wave and normal modes are present $(c_1 < c_2)$. It is natural to expect the appearance of these waves in the presence of an elastic medium. In describing wave behavior at the high-frequency approximation, we proceed from eqs. 7.220 and 7.221 and Fig. 7.8a,b. Energy of the direct wave caused by the primary source advances along the elementary ray tubes. In the tube oriented along the z-axis, scalar potential of the primary wave

$$\tilde{\varphi}_0 = C\,\frac{e^{i\,k_1 z}}{z}$$

decays inversely proportionally to distance z, and it might be observed alone at rather large distances from the origin.

Next let us consider the reflection and transmission of waves at the borehole surface using critical angles

$$\sin\theta_c^l = \frac{c_1}{c_l} \qquad \text{and} \qquad \sin\theta_c^s = \frac{c_1}{c_s} \tag{7.222}$$

Inasmuch as $c_l > c_s$, we have

$$\theta_c^l < \theta_c^s, \tag{7.223}$$

and the critical angle for longitudinal waves occurs at points of the borehole surface located relatively closer to the source. The coordinate z_l of these points is defined from the equality

$$\sin\theta_c^l = \frac{z_l}{\sqrt{a^2 + z_l^2}} = \frac{c_1}{c_l} \tag{7.224}$$

In the same manner, the place where we observe the critical angle for the shear wave is

$$\sin\theta_c^s = \frac{z_s}{\sqrt{z_s^2 + a^2}} = \frac{c_1}{c_s} \tag{7.225}$$

and

$$z_s > z_l \tag{7.226}$$

It is convenient to distinguish three intervals at the borehole surface:
they are $z < z_l$, $z_l < z < z_s$, $z > z_s$.
In the first range, $z < z_l$, the incident wave gives rise to the longitudinal reflected wave in the borehole fluid. The reflection angle obeys Snell's law and varies as

$$0 \le \theta_r < \theta_c^l$$

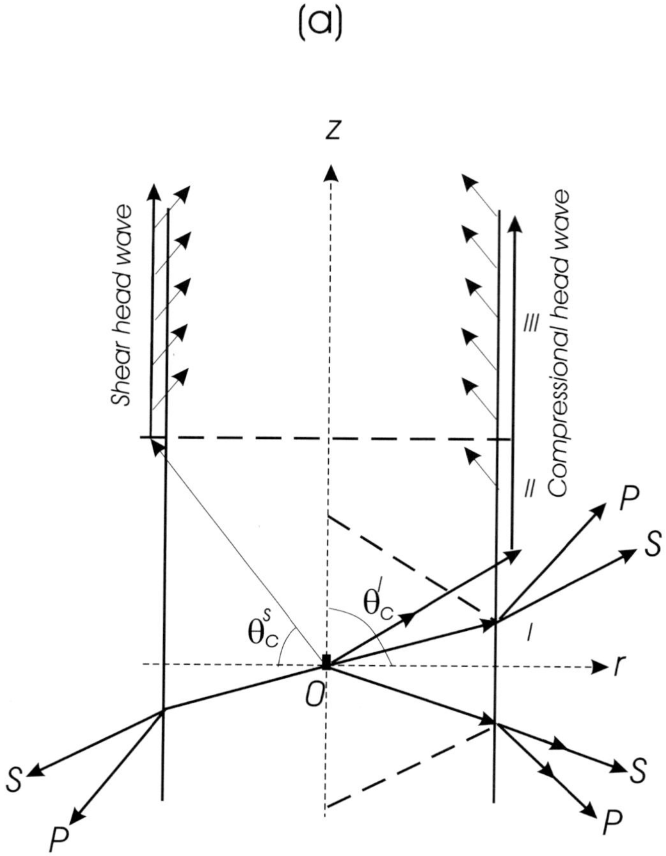

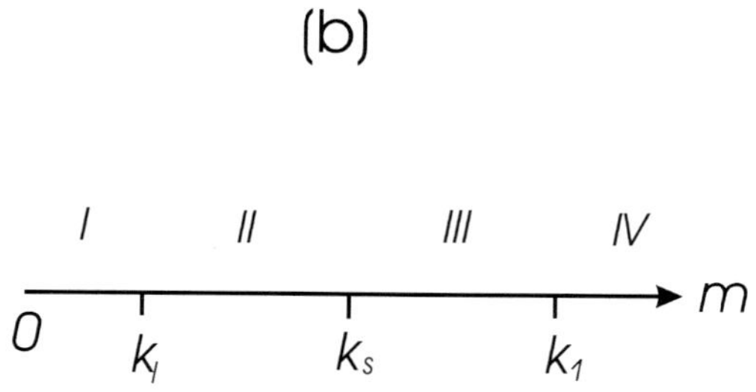

Figure 7.8: (a) Rays of reflected, transmitted, and head waves (b) Intervals of integration

At the same time, P and S transmitted waves appear in the elastic medium and move away from the boundary, $r = a$. As follows from Snell's law, transmission angles θ_2^l and θ_2^s obey the equality

$$\frac{\sin\theta_2^l}{c_l} = \frac{\sin\theta_2^s}{c_s}, \tag{7.227}$$

i.e.,

$$\theta_2^l > \theta_2^s \tag{7.228}$$

Certainly, the wavefields possess the same axial symmetry. In accordance with eq. 7.224,

$$\frac{z_l}{a} = \frac{c_1}{c_l}\left(1 - \frac{c_1^2}{c_l^2}\right)^{-1/2},$$

and z_l is usually smaller than the borehole radius. The longitudinal wave inside the borehole experiences reflections that may occur within the first interval as well as beyond it. In the case of constructive interference between reflections, modes are formed that propagate along the borehole. At any cross-section $z < z_l$, these modes are the standing waves. However, they rapidly attenuate with distance z, since after each reflection some part of the energy leaves the borehole and moves into elastic medium. This is why such waves are often called leaking modes, and with an increase of distance their role diminishes.

From eqs. 7.220, it follows that the initial interval of integration

$$m < k_l$$

entirely defines the P transmitted wave and makes a significant contribution to the S transmitted and reflected waves. In fact, the radicals

$$\sqrt{m^2 - k_1^2}, \qquad \sqrt{m^2 - k_s^2}, \qquad \text{and} \qquad \sqrt{m^2 - k_l^2}$$

are imaginary, and we are dealing with waves that propagate along rays (Snell's law). Note that in this interval the wavenumber of elementary cylindrical waves varies from zero to k_l. Therefore, phase velocity changes as $c_l < c < \infty$. When the angle of incidence θ_i of the direct wave approaches the critical angle, θ_c^l, the P transmitted wave starts to move in the vicinity of the borehole along its surface and generates a head wave in the fluid. This is a conical wave, and angle θ_c^l defines the orientation of the phase surface. At relatively large distances z from the source, the longest path of this wave is located outside the borehole, where it moves with velocity c_l. Correspondingly, the P head wave appears as the first arrival.

In the second range, $z_l < z < z_s$, we observe total internal reflection for the P wave. This means that the P evanescent wave appears in an elastic medium, and during each period energy flux of this wave in the radial direction is equal to zero. At the same time, the S transmitted wave propagates through the surrounding medium with velocity c_s, and it arises at each point of the second interval. Because of this, the amplitude of the reflected wave is still smaller than that of the incident wave, and it decreases after each reflection. Therefore, as with the first range, constructive interference of the reflected waves can produce only leaking modes. Of course, destructive interference is a second factor that also results in a decrease of wavefields.

Superposition of elementary cylindrical waves with wavenumbers $k_l < m < k_s$, i.e., the second interval of integration in eqs. 7.220, form leaking modes. At points of the borehole surface where $z_l < z < z_s$, the following waves appear:

a. The reflected wave inside the borehole,

b. The S transmitted wave in an elastic medium,

c. The P inhomogeneous wave, which rapidly decays with increased distance from the borehole, and

d. The P head wave, which originates near points of interference with coordinate z_l.

At the end of the second interval, $z \sim z_s$, the incident angle approaches θ_c^s, and the S transmitted wave begins to move along the boundary with the velocity of the shear wave, c_s. This creates the S head wave, which in the far zone comes after the P head wave, representing the second arrival of the transient wave. Note that the velocity of propagation of the evanescent wave associated with total internal reflection varies as

$$c_s \leq c \leq c_l$$

Since $m > k_l$ in the second interval of integration, Fig. 7.8b, corresponding harmonics (elementary cylindrical waves) make some contribution to the inhomogeneous wave. Finally, at points of the borehole surface where $z > z_s$, total internal reflection takes place for both P and S waves. This suggests that a reflected wave inside the borehole is accompanied by dilatational and rotational wavefields in an elastic medium. Besides, two head waves are propagating along the boundary with velocities c_l and c_s, respectively. Inasmuch as at these points of the boundary ($z > z_s$) a reflection does not cause leakage of energy into the surrounding medium, normal modes are formed, and they propagate along the borehole without attenuation. As we know, the same phenomenon takes place when the surrounding medium is acoustic and $c_2 > c_1$.

The harmonics of the third interval (eqs. 7.220),

$$k_s < m < k_1,$$

form normal modes and the corresponding evanescent wavefields. Fortunately, determination of the wavenumbers of these modes does not require integration of eqs. 7.220, because the former are roots of the characteristic equation 7.221. In this light, it is proper to note that wavenumbers of Stoneley's wave are also roots of eq. 7.221, but they correspond to the fourth interval of integration $(m > k_1)$, provided that $k_s < k_1$. This range of wavenumbers, primarily its initial part, contains information about surface waves. Thus, the roots of the characteristic equation describing the normal modes are within the third interval of integration. Correspondingly, their phase velocity changes as

$$c_1 < c < c_s$$

It may be useful to explain this fact in a different way. Consider the rays of the direct wave, which reach the borehole surface at points $z = z_s$. By definition, their incident angle coincides with the critical angle, θ_c^s:

$$\sin \theta_c^s = \frac{c_1}{c_s}$$

As follows from Snell's law, the reflected waves arising at these points are characterized by the same angle, θ_c^s, so that they form a conical wave. The latter causes the other reflected waves, which are still conical, and their reflection angle is still equal to θ_c^s. If frequency is such that their interference is constructive, they form a normal mode that propagates along the borehole axis. As in the case of the acoustic medium, the phase velocity of the normal mode is defined from elementary geometry and is equal to

$$c = \frac{c_1}{\sin \theta_c^s} = c_s$$

Next suppose that rays of the direct wave reach the interface at greater distances from the source, $z > z_s$. It is clear that their incident angle is also greater:

$$\theta_i(z) > \theta_c^s$$

As before, this wave gives rise to a new family of conical reflected waves with the same angle $\theta_i(z)$. Assuming that their superposition is constructive, we again observe wave propagation in the axial direction. Phase velocity becomes smaller, since $\theta_i(z) > \theta_c^s$, and

$$c = \frac{c_1}{\sin\theta(z)}$$

Thus, with an increase in the angle of incidence of the direct wave, phase velocity of the mode approaches that of the fluid

$$c \to c_1 \qquad \text{if} \qquad \theta_i \to \pi/2$$

Now let us discuss the relationship between wave velocities and frequency.

Dispersion curves of phase and group velocities

The dependencies of velocities $c_p(\omega)$ and $c_g(\omega)$ on frequency for the first two normal modes are shown in Fig. 7.9. Velocities are obtained by numerical solution of the characteristic equation 7.221. As we may expect, the behavior of these functions is similar to that in the case of an acoustic surrounding medium. First of all, the phase velocity for each normal mode has the same low-frequency asymptote, which is equal to the shear velocity, c_s. This occurs at the cut-off frequency that becomes higher with an increase of the mode order. Below these frequencies the normal mode cannot exist. As is seen from Fig. 7.9, with an increase of frequency phase velocity monotonically becomes smaller and asymptotically tends to the wave velocity in the fluid, regardless of the mode order. The behavior of group velocity is different, but its low-frequency asymptote is also equal to c_s. With an increase of frequency, it rapidly decreases and comes to the Airy phase, where $c_g(\omega) < c_1$. After passing the Airy phase group velocity approaches the high-frequency asymptote from below.

$$c_g(\omega) \to c_1 \qquad \text{if} \qquad \omega \to \infty$$

Thus, the normal modes are highly dispersive, which is a clear indication that they arise due to the constructive interference of reflected waves. For comparison, phase and group velocities of Stoneley waves are also shown in Fig. 7.9. These velocities are characterized by very minimal dispersion and do not have the cut-off frequency $(c_s > c_1)$. In this case, both velocities vary within the range

$$0.90 < \frac{c}{c_1} < 0.96$$

Thus, as in the case of the acoustic surrounding medium, there is an interval of relatively low frequencies in which normal modes are absent. Within this range only Stoneley and

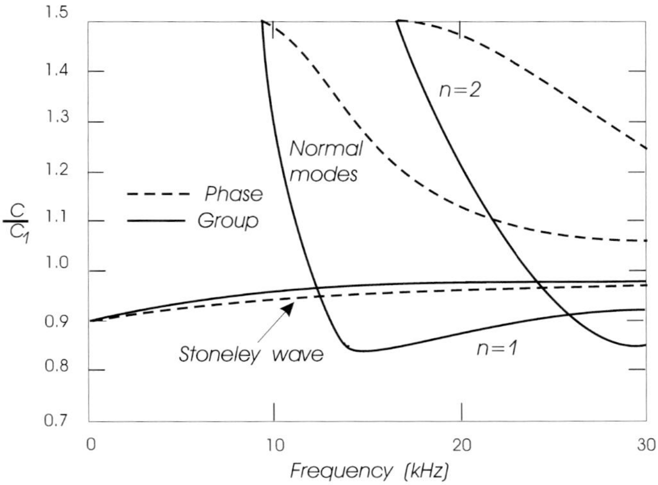

Figure 7.9: Phase and group velocities for the first two normal modes $(a = 0.1$ m, $c_l = 4.5$ km/s, $c_s = 2.2$ km/s, $\rho_1 = 1.2$ g/cm^3, $\rho_2 = 2.1$g/cm^3, $c_1 = 1.5$ km/s). [After Cheng & Toksöz, 1981]

head waves are present in the borehole $(z \gg a)$. With an increase of frequency, the first normal mode appears, then the second one, and so on. Thus, with an increase of frequency, the number of normal modes also increases. We can say that at each frequency, this part of the wavefield is the superposition of a finite number of normal modes moving with different phase velocities along the borehole. As is well known, each normal mode behaves like a standing wave at any cross-section of the borehole. With an increase of the mode order, the number of nodes, characterizing wave oscillations along the radius also increases. Applying the residual theorem, the amplitudes of Stoneley waves and normal modes can be found. One such evaluation is presented in Fig. 7.10. Near the cut-off frequency of the first normal mode there is a small range of frequencies, at which the amplitude of the Stoneley wave is greater. With a further increase of frequency, the field related to the normal modes prevails. Note that the normal modes are often called pseudo-Rayleigh waves as well as reflected conical waves.

Head waves

The procedure for deriving asymptotic formulas for P and S head waves is similar to that used in the case of the acoustic surrounding medium (section 7.2). The algebra related to this task is rather cumbersome, because function A_m in eqs. 7.220 is much more complicated. First of all, we represent the complex amplitude of scalar potential

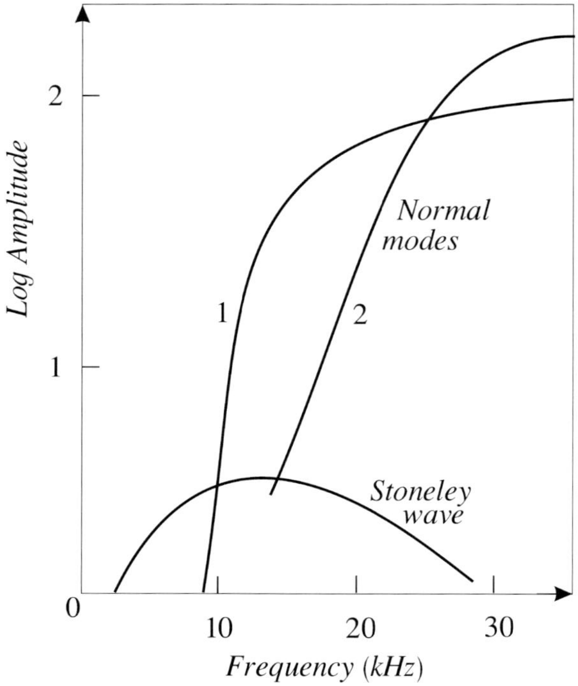

Figure 7.10: Amplitudes of normal modes and Stoneley wave

inside the borehole as

$$\tilde{\varphi}_s = \frac{1}{2} \int_{-\infty}^{\infty} A_m I_0(m_1 r)\, e^{i\, m\, z}\, dm$$

In the presence of an elastic medium around the borehole, there are three branch points of the integrand A_m,

$$k_1 + i\xi, \qquad k_s + i\xi, \qquad k_l + i\xi,$$

where ξ is a very small and positive number. Integration along branch cuts related to point $k_1 + i\xi$, as in the acoustic case, gives an expression of scalar potential that differs by sign only from that of the primary source in a uniform medium. This means that the influence of the primary wave is canceled, and we have to focus on the contribution of intervals around two other branch points. The asymptotic formulas for S and P head waves in the far zone are obtained by integration near branch points k_s and k_l,

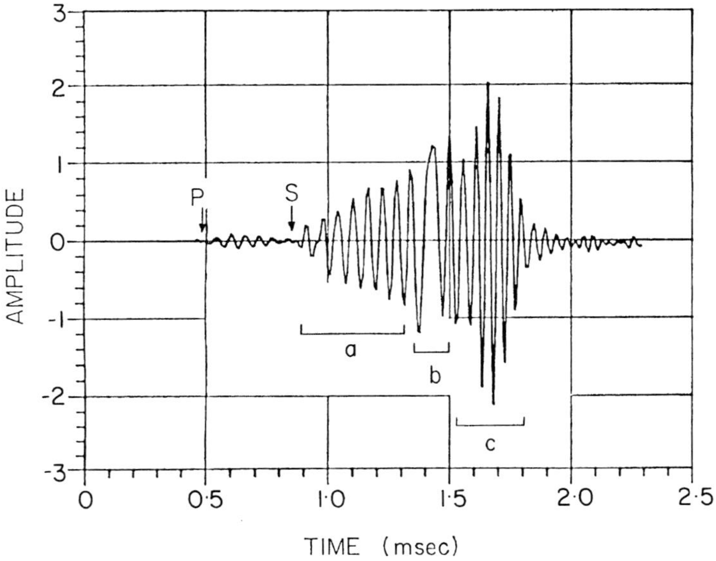

Figure 7.11: Transient wave at the borehole axis (a) Lowest-frequency part of normal mode (b) Stoneley wave (c) Airy phase. Here $L = 2.44$ m, $c_l = 5.94$ km/s, $\rho_1 = 1.2$ g/cm^3, $\rho_2 = 2.3$g/cm^3, $a = 10.2$ cm. [After Cheng & Toksöz, 1981]

respectively. In both cases, integration along opposite sides of the branch lines gives an integral with logarithmic singularity, and it has the known form

$$\int_{-\infty}^{\infty} \frac{e^{-mz}\, dm}{(a_1 + b_1 \ln m)\,(a_2 + b_2 \ln m)},$$

discussed in section 7.2. Correspondingly, both head waves usually decay either as $(k_s z) \ln k_s z$ or $(k_l z) \ln k_l z$. For this reason they are smaller in the far zone than the normal modes and Stoneley waves.

Transient waves

As an illustration of the transient wave at an observation point located at the borehole axis, consider the example shown in Fig. 7.11. Note that this theoretical response takes into account the first normal mode, while the influence of others is discarded. As we already know, the first arrival is due to the P head wave. The second arrival is caused by the S head wave. Inasmuch as both of these waves decrease with distance z, their magnitudes are sufficiently small. Then we observe portion "a" of the transient wave, which corresponds to the low-frequency range of the normal mode. As follows from Fig. 7.9 the group velocity in this range is higher than that of the Stoneley wave. Next is portion "b", which corresponds to Stoneley wave, and after that there is interval "c", which is formed by higher frequency normal modes and is called the Airy phase.

Chapter 8

Plane waves in a transversely isotropic medium

The conventional theory of elasticity assumes that a medium is continuous, that is an atomic structure of matter is not taken into account. This implies that any elementary volume contains practically unlimited number of atoms or molecules. From the macroscopic point of view this volume of a rock includes crystals, fluid, gas, as well as different amorphous solids. Crystals always demonstrate some kind of an anisotropy. In other words, their elastic parameters may vary with a direction. In general, an elementary volume may contain different kinds of crystals with more or less random orientation. If they are distributed evenly in a medium and their orientation is completely random, we may consider formation isotropic. In contrast, if there exists a preferable orientation along some metamorphic rocks, for instance, display a relatively significant anisotropy. Also, it can be caused by fracturing. There is another reason for such a behavior. Some sedimentary formations, and, first of all shales, have cleavage planes that are observed even inside small volumes, ($\sim 1\ \text{cm}^3$). At the same time this feature may characterize a layer of a great thickness and large horizontal extent. Correspondingly, the layer can be treated as a homogeneous but anisotropic solid.

Finally, the elastic properties of a system of horizontal layers often show symmetry with respect to a vertical axis, since all horizontal directions are equivalent. If the thicknesses of all layers are much smaller than the wavelength, then such a medium can be considered also as a homogeneous anisotropic one, called transversely isotropic. Its elastic properties are independent of direction in horizontal plane but differ from those in the vertical direction. Our goal is to investigate the behavior of plane waves in such a relatively simple medium. With this purpose in mind let us consider its elastic constants.

8.1 Stress-strain relations in a transversely isotropic medium

In general, Hooke's law relates the six independent stress components to six independent strain component and has form (Appendix E)

$$\begin{aligned}
\tau_{xx} &= c_{11}e_{xx} + c_{12}e_{yy} + c_{13}e_{zz} + c_{14}e_{yz} + c_{15}e_{xz} + c_{16}e_{xy} \\
\tau_{yy} &= c_{21}e_{xx} + c_{22}e_{yy} + c_{23}e_{zz} + c_{24}e_{yz} + c_{25}e_{xz} + c_{26}e_{xy} \\
\tau_{zz} &= c_{31}e_{xx} + c_{32}e_{yy} + c_{33}e_{zz} + c_{34}e_{yz} + c_{35}e_{xz} + c_{36}e_{xy} \\
\tau_{yz} &= c_{41}e_{xx} + c_{42}e_{yy} + c_{43}e_{zz} + c_{44}e_{yz} + c_{45}e_{xz} + c_{46}e_{xy}, \\
\tau_{xz} &= c_{51}e_{xx} + c_{52}e_{yy} + c_{53}e_{zz} + c_{54}e_{yz} + c_{55}e_{xz} + c_{56}e_{xy}, \\
\tau_{xy} &= c_{61}e_{xx} + c_{62}e_{yy} + c_{63}e_{zz} + c_{64}e_{yz} + c_{65}e_{xz} + c_{66}e_{xy},
\end{aligned} \tag{8.1}$$

where $c_{ij} = c_{ji}$.
As was demonstrated in the Appendix E only five coefficients c_{ij} differ from zero in the transversely isotropic medium and the matrix of elastic constants is

$$\begin{vmatrix}
c_{11} & c_{12} & c_{13} & 0 & 0 & 0 \\
c_{12} & c_{11} & c_{13} & 0 & 0 & 0 \\
c_{13} & c_{13} & c_{33} & 0 & 0 & 0 \\
0 & 0 & 0 & c_{44} & 0 & 0 \\
0 & 0 & 0 & 0 & c_{44} & 0 \\
0 & 0 & 0 & 0 & 0 & \dfrac{c_{11} - c_{12}}{2}
\end{vmatrix} \tag{8.2}$$

Thus, unlike isotropic media, which are defined by just two parameters:

$$c_{12} = c_{13} = \lambda \qquad \text{and} \qquad c_{44} = \mu,$$

with

$$c_{11} = c_{33} = \lambda + 2\mu,$$

for transversely isotropic media, the five independent coefficients:

$$c_{11}, \; c_{12}, \; c_{13}, \; c_{33}, \; c_{44}$$

describe the Hooke's law. Correspondingly, eqs. 8.1 reduce to

$$\tau_{xx} = c_{11}e_{xx} + c_{12}e_{yy} + c_{13}e_{zz}$$

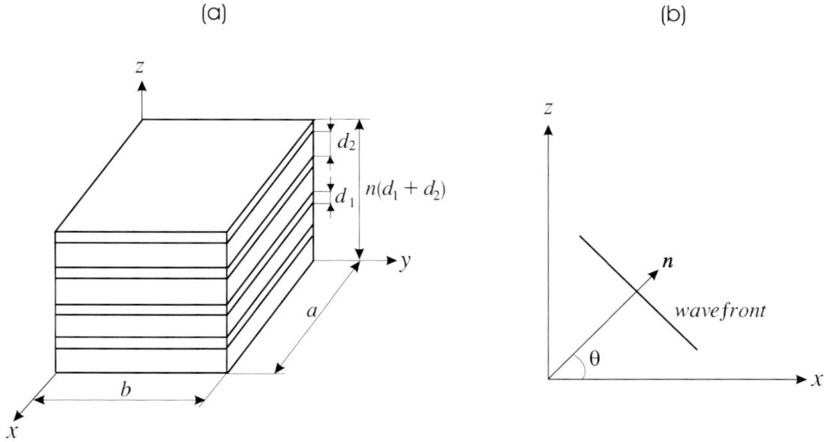

Figure 8.1: (a) Model of laminated medium (b) Orientation of of plane wavefront. [After Postma, 1952]

$$\tau_{yy} = c_{12}e_{xx} + c_{11}e_{yy} + c_{13}e_{zz}$$

$$\tau_{zz} = c_{13}e_{xx} + c_{13}e_{yy} + c_{33}e_{zz} \tag{8.3}$$

$$\tau_{yz} = c_{44}e_{yz}, \qquad \tau_{zx} = c_{44}e_{zx}, \qquad \tau_{xy} = \frac{c_{11} - c_{12}}{2}e_{xy}$$

Laminated solids

As an example of the transversely isotropic medium consider a periodic system of thin horizontal layers. Each of its element consists of two homogeneous isotropic layers with elastic parameters λ_1, μ_1 and λ_2, μ_2, respectively. Their thicknesses are d_1 and d_2. As is seen in Fig. 8.1a, the $x-$ and $y-$axes are parallel to the layer boundaries. Consider a parallelepiped with faces parallel to the coordinate planes and vertical and horizontal dimensions $n\,(d_1 + d_2)$, a, and b, where n is some large integer number. We assume that the wave length is much greater than the thickness of the elementary layers and by averaging replace this medium by a transversely isotropic one. First, suppose that the volume is subjected to normal stresses only. For instance, stress τ_{zz} acts on the faces perpendicular to the $z-$axis. At the same time, the normal stresses τ_{xx1} and τ_{xx2} are applied to the faces of elementary layers normal to the $x-$axis. By analogy, stresses τ_{yy1} and τ_{yy2} act on the faces perpendicular to the $y-$axis. Under the action of these stresses each elementary layer, with the thickness d_1 or d_1, experiences deformation.

Inasmuch as equilibrium is considered, normal stress τ_{zz} has the same value throughout the medium, that is,

$$\tau_{zz1} = \tau_{zz2} = \tau_{zz} \tag{8.4}$$

It is essential that a change of the length in any horizontal direction also has to be the same in both layers. Otherwise, we would observe a discontinuity of the corresponding components of the displacement. This means that strain elements e_{xx} and e_{yy} have to satisfy the condition

$$e_{xx1} = e_{xx2} = e_{xx}, \qquad e_{yy1} = e_{eyy2} = e_{yy} \tag{8.5}$$

Strains e_{zz1} and e_{zz2} however, may differ. Applying Hooke's law to the first isotropic layer we have:

$$\tau_{xx1} = (\lambda_1 + 2\mu_1)\, e_{xx} + \lambda_1 e_{yy} + \lambda_1 e_{zz1}$$

$$\tau_{yy1} = \lambda_1 e_{xx} + (\lambda_1 + 2\mu_1)\, e_{yy} + \lambda_1 e_{zz1} \tag{8.6}$$

$$\tau_{zz} = \lambda_1 e_{xx} + \lambda_1 e_{yy} + (\lambda_1 + 2\mu_1)\, e_{zz1}$$

In the same manner the use of the Hooke's law for the second layer gives

$$\tau_{xx2} = (\lambda_2 + 2\mu_2)\, e_{xx} + \lambda_2 e_{yy} + \lambda_2\, e_{zz2}$$

$$\tau_{yy2} = \lambda_2 e_{xx} + (\lambda_2 + 2\mu_2)\, e_{yy} + \lambda_2\, e_{zz2} \tag{8.7}$$

$$\tau_{zz} = \lambda_2 e_{xx} + \lambda_2 e_{yy} + (\lambda_2 + 2\mu_2)\; e_{zz2}$$

Now we perform an averaging of stresses weighted by the relative thickness of two types of layer. The mean stresses, acting on the faces perpendicular to the $x-$ and $y-$axes, are

$$\tau_{xx} = \frac{d_1\,\tau_{xx1} + d_2\,\tau_{xx2}}{d_1 + d_2} \quad \text{and} \quad \tau_{yy} = \frac{d_1\,\tau_{yy1} + d_2\,\tau_{yy2}}{d_1 + d_2} \tag{8.8}$$

In fact, the latter describe a replacement of two parallel forces by the resultant one. At the same time, as was pointed out earlier, the stress τ_{zz} has the same value at each layer and, therefore, its average value is equal to τ_{zz}. Making use of Hooke's law, eqs. 8.6 and 8.7, we find the relationship between the mean values of stresses and strains:

$$(d_1+d_2)\,\tau_{xx}=e_{xx}\left[d_1\left(\lambda_1+2\mu_1\right)+d_2\left(\lambda_2+2\mu_2\right)\right]$$

$$+e_{yy}\left(\lambda_1 d_1+\lambda_2 d_2\right)+e_{zz1}\lambda_1 d_1+e_{zz2}\,\lambda_2\,d_2,$$

$$(d_1+d_2)\,\tau_{yy}=e_{xx}\left(\lambda_1 d_1+\lambda_2\,d_2\right) \tag{8.9}$$

$$+e_{yy}[d_1\left(\lambda_1+2\mu_1\right)+d_2\left(\lambda_2+2\mu_2\right)]+e_{zz1}\lambda_1 d_1+e_{zz2}\,\lambda_2\,d_2,$$

$$(d_1+d_2)\,\tau_{zz}=e_{xx}\left(\lambda_1 d_1+\lambda_2\,d_2\right)+e_{yy}\left(\lambda_1 d_1+\lambda_2 d_2\right)$$

$$+e_{zz1}\,d_1\left(\lambda_1+2\mu_1\right)+e_{zz2}\,d_2\left(\lambda_2+2\mu_2\right)$$

In order to accomplish this process of averaging we introduce the mean strain e_{zz} as

$$(d_1+d_2)\;e_{zz}=d_1\,e_{zz1}+d_2\,e_{zz2} \tag{8.10}$$

Here e_{zz} is the average strain of the volume, which contains equal and large number of elementary layers with thicknesses d_1 and d_2. From the last equations of sets 8.6 and 8.7, we can express e_{zz1} and e_{zz2} in terms of the mean strains, giving

$$e_{zz1}=\frac{(d_1+d_2)\left(\lambda_2+2\mu_2\right)e_{zz}-\left(\lambda_1-\lambda_2\right)\left(e_{xx}+e_{yy}\right)d_2}{d_1\left(\lambda_2+2\mu_2\right)+d_2\left(\lambda_1+2\mu_1\right)} \tag{8.11}$$

and $$e_{zz2}=\frac{(d_1+d_2)\left(\lambda_1+2\mu_1\right)e_{zz}+\left(\lambda_1-\lambda_2\right)\left(e_{xx}+e_{yy}\right)d_1}{d_1\left(\lambda_2+2\mu_2\right)+d_2\left(\lambda_1+2\mu_1\right)}$$

Finally, substitution of eqs. 8.11 into the set 8.9 establishes the relationship between the mean normal stresses and strains:

$$\tau_{xx}=e_{xx}\frac{(d_1+d_2)^2\left(\lambda_1+2\mu_1\right)\left(\lambda_2+2\mu_2\right)+d_1 d_2\{[(\lambda_1+2\mu_1)-(\lambda_2+2\mu_2)]^2-\left(\lambda_1-\lambda_2\right)^2\}}{D}$$

$$+\,e_{yy}\frac{\lambda_1\lambda_2\left(d_1+d_2\right)^2+2\left(\lambda_1 d_1+\lambda_2 d_2\right)\left(\mu_2 d_1+\mu_1 d_2\right)}{D}$$

$$+e_{zz}\frac{(d_1+d_2)\left[\lambda_1 d_1(\lambda_2+2\mu_2)+\lambda_2 d_2(\lambda_1+2\mu_1)\right]}{D}$$

$$\tau_{yy}=e_{xx}\frac{\lambda_1\lambda_2(d_1+d_2)^2+2(\lambda_1 d_1+\lambda_2 d_2)(\mu_2 d_1+\mu_1 d_2)}{D} \tag{8.12}$$

$$+e_{yy}\frac{(d_1+d_2)^2(\lambda_1+2\mu_1)(\lambda_2+2\mu_2)+d_1 d_2\{[(\lambda_1+2\mu_1)-(\lambda_2+2\mu_2)]^2-(\lambda_1-\lambda_2)^2\}}{D}$$

$$+e_{zz}\frac{(d_1+d_2)^2\left[\lambda_1 d_1(\lambda_2+2\mu_2)+\lambda_2 d_2(\lambda_1+2\mu_1)\right]}{D}$$

$$\tau_{zz}=(e_{xx}+e_{yy})\,\frac{(d_1+d_2)\left[\lambda_1 d_1(\lambda_2+2\mu_2)+\lambda_2 d_2(\lambda_1+2\mu_1)\right]}{D}$$

$$+e_{zz}\frac{(d_1+d_2)^2(\lambda_1+2\mu_1)(\lambda_2+2\mu_2)}{D},$$

where

$$D=(d_1+d_2)\left[d_1(\lambda_2+2\mu_2)+d_2(\lambda_1+2\mu_1)\right] \tag{8.13}$$

Next assume that the volume is subjected to the action of shear stresses. First, consider the stress τ_{yz} acting on the faces perpendicular to the $z-$axis. Correspondingly, the deformation is characterized by the strains e_{yz1} and e_{yz2}, with their average value

$$(d_1+d_2)\,e_{yz}=d_1 e_{yz1}+d_2 e_{yz2} \tag{8.14}$$

Here

$$\mu_1 e_{yz1}=\tau_{yz}=\mu_2 e_{yz2} \tag{8.15}$$

The last two equations give

$$(d_1+d_2)\,e_{yz}=\left(\frac{d_1}{\mu_1}+\frac{d_2}{\mu_2}\right)\tau_{yz},\quad \text{or}\quad \tau_{yz}=\frac{(d_1+d_2)\,\mu_1\mu_2}{d_1\mu_2+d_2\mu_1}e_{yz} \tag{8.16}$$

In the same manner we find the relation between τ_{xz} and e_{xz}:

$$\tau_{xz}=\frac{(d_1+d_2)\,\mu_1\mu_2}{d_1\mu_2+d_2\mu_1}e_{xz} \tag{8.17}$$

Suppose that the shear forces $\tau_{x1}a\ d_1$ and $\tau_{x2}a\ d_1$ are applied to the faces normal to the $y-$axis of the layers with the thickness d_1 and d_2, respectively. Continuity of the displacement requires

$$e_{xy1} = e_{xy2} = e_{xy} \tag{8.18}$$

Therefore,

$$\tau_{xy1} = \mu_1 e_{xy} \qquad \text{and} \qquad \tau_{xy2} = \mu_2 e_{xy}, \tag{8.19}$$

and, for the average shear stress τ_{xy}, we have

$$\tau_{xy}\left(d_1 + d_2\right) = \tau_{xy1}d_1 + \tau_{xy2}d_2$$

or

$$\tau_{xy}\left(d_1 + d_2\right) = e_{xy}\left(\mu_1 d_1 + \mu_2 d_2\right) \tag{8.20}$$

Thus

$$\tau_{xy} = \frac{\mu_1 d_1 + \mu_2 d_2}{d_1 + d_2} e_{xy}, \tag{8.21}$$

We have thus established the relationships between the average values of the stresses and strains (Hooke's law).

Comparison of eqs. 8.12, 8.17, and 8.21 with eqs. 8.3 gives

$$c_{11} = \frac{1}{D}\{(d_1 + d_2)^2 (\lambda_1 + 2\mu_1)(\lambda_2 + 2\mu_2) + 4d_1 d_2 (\mu_1 - \mu_2)[(\lambda_1 + \mu_1) - (\lambda_2 + \mu_2)]\}$$

$$c_{12} = \frac{1}{D}\{(d_1 + d_2)^2 \lambda_1\lambda_2 + 2(\lambda_1 d_1 + \lambda_2 d_2)(\mu_2 d_1 + \mu_1 d_2)\}$$

$$c_{13} = \frac{1}{D}\{(d_1 + d_2)[\lambda_1 d_1 (\lambda_2 + 2\mu_2) + \lambda_2 d_2 (\lambda_1 + 2\mu_1)]\} \tag{8.22}$$

$$c_{33} = \frac{1}{D}(d_1 + d_2)^2 (\lambda_1 + 2\mu_1)(\lambda_2 + 2\mu_2)$$

$$c_{44} = \frac{(d_1 + d_2)\mu_1\mu_2}{d_1\mu_2 + d_2\mu_1} \qquad c_{66} = \frac{\mu_1 d_1 + d_2\mu_2}{d_1 + d_2}$$

Here

$$c_{66} = \frac{c_{11} - c_{12}}{2} \tag{8.23}$$

These formulas give the linkage between the five parameters of the transversely isotropic medium and the Lame constants of two elementary isotropic layers: λ_1, μ_1 and λ_2, μ_2, and their thicknesses d_1 and d_2. Applying algebra, we find that the elastic constants obey some inequalities. First, they are positive if $\lambda > 0$ and $\mu > 0$. Also,

$$c_{11} > c_{44}, \qquad c_{11} > c_{66}, \qquad c_{33} > c_{44} \tag{8.24}$$

In conclusion note that any laminated medium can be represented by a much more complicated periodic system of layers with different Lame constants and thicknesses.

Equations of a motion

As in an isotropic medium suppose that an elementary volume

$$dV = dx\ dy\ dz$$

is subjected to an action of the surface forces. Then, in accordance with the second Newton's law its motion is described by the following system (Chapter 1):

$$\begin{aligned}
&\frac{\partial \tau_{xx}}{\partial x} + \frac{\partial \tau_{xy}}{\partial y} + \frac{\partial \tau_{xz}}{\partial z} = \rho\, \frac{\partial^2 u}{\partial t^2}, \\
&\frac{\partial \tau_{yx}}{\partial x} + \frac{\partial \tau_{yy}}{\partial y} + \frac{\partial \tau_{yz}}{\partial z} = \rho\, \frac{\partial^2 v}{\partial t^2}, \\
&\frac{\partial \tau_{zx}}{\partial x} + \frac{\partial \tau_{zy}}{\partial y} + \frac{\partial \tau_{zz}}{\partial z} = \rho \frac{\partial^2 w}{\partial t^2}
\end{aligned} \tag{8.25}$$

Here

$$\mathbf{s} = u\mathbf{i} + v\mathbf{j} + w\mathbf{k} \tag{8.26}$$

is the displacement of the center of mass of an elementary volume, and, for the laminated medium, 8.1a,

$$\rho = \frac{\rho_1 d_1 + \rho_2 d_2}{d_1 + d_2} \tag{8.27}$$

Next, substitution of eqs. 8.3 into the set 8.25 and taking into account that

$$e_{xx} = \frac{\partial u}{\partial x}, \qquad e_{yy} = \frac{\partial v}{\partial y}, \qquad e_{zz} = \frac{\partial w}{\partial z},$$

$$e_{xy} = \frac{\partial u}{\partial y} + \frac{\partial v}{\partial x}, \qquad e_{xz} = \frac{\partial u}{\partial z} + \frac{\partial w}{\partial x}, \qquad e_{yz} = \frac{\partial v}{\partial z} + \frac{\partial w}{\partial y},$$

gives a system of equations with respect to the displacement components:

$$c_{11}\frac{\partial^2 u}{\partial x^2} + c_{66}\frac{\partial^2 u}{\partial y^2} + c_{44}\frac{\partial^2 u}{\partial z^2} + (c_{11} - c_{66})\frac{\partial^2 v}{\partial x \partial y} + (c_{13} + c_{44})\frac{\partial^2 w}{\partial x\, \partial z} = \rho\,\frac{\partial^2 u}{\partial t^2},$$

$$(c_{11} - c_{66})\frac{\partial^2 u}{\partial x\, \partial y} + c_{66}\frac{\partial^2 v}{\partial x^2} + c_{11}\frac{\partial^2 v}{\partial y^2} + c_{44}\frac{\partial^2 v}{\partial z^2} + (c_{13} + c_{44})\frac{\partial^2 w}{\partial y \partial z} = \rho\,\frac{\partial^2 v}{\partial t^2} \tag{8.28}$$

$$(c_{13} + c_{44})\frac{\partial^2 u}{\partial x\, \partial z} + (c_{13} + c_{44})\frac{\partial^2 v}{\partial y\, \partial z} + c_{44}\frac{\partial^2 w}{\partial x^2} + c_{44}\frac{\partial^2 w}{\partial y^2} + c_{33}\frac{\partial^2 w}{\partial z^2} = \rho\,\frac{\partial^2 w}{\partial t^2}$$

The counterpart for the equation in the isotropic medium is

$$\mu\,\nabla^2 \mathbf{s} + (\lambda + \mu)\ \text{grad div}\ \mathbf{s} = \rho\,\frac{\partial^2 \mathbf{s}}{\partial t^2}$$

8.2 Propagation of plane waves in a transversely isotropic medium

Now we demonstrate that the plane waves propagating through the transversely isotropic medium have velocities that relate in a certain manner to elastic constants and orientation of the phase surfaces. Let the axis of symmetry for the transversely isotropic medium be the z-axis. Suppose, with no loss of generality, that the Cartesian system of coordinates is such that the phase surfaces of these waves are parallel to the $y-$axis, that is the particle displacement is independent on the $y-$coordinate. Assuming that a plane wave exists, we can represent the components of the particle displacement as

$$u = u_0\ f\,(x\cos\theta + z\sin\theta - ct)$$

$$v = v_0\ f\,(x\cos\theta + z\sin\theta - ct) \tag{8.29}$$

$$w = w_0\ f\,(x\cos\theta + z\sin\theta - ct)\,,$$

with the angle θ is shown in Fig. 8.1b. Inasmuch as the derivatives with respect to y are equal to zero, the system of equations 8.28 simply to

$$c_{11}\frac{\partial^2 u}{\partial x^2}+c_{44}\frac{\partial^2 u}{\partial z^2}+(c_{13}+c_{44})\frac{\partial^2 w}{\partial x\partial z}=\rho\,\frac{\partial^2 u}{\partial t^2}$$

$$c_{66}\frac{\partial^2 v}{\partial x^2}+c_{44}\frac{\partial^2 v}{\partial z^2}=\rho\,\frac{\partial^2 v}{\partial t^2} \tag{8.30}$$

$$(c_{13}+c_{44})\,\frac{\partial^2 u}{\partial x\partial z}+c_{44}\frac{\partial^2 w}{\partial x^2}+c_{33}\frac{\partial^2 w}{\partial z^2}=\rho\,\frac{\partial^2 w}{\partial t^2}$$

Substitution of eq. 8.29 for the displacement into eqs. 8.30 gives:

$$\left(c_{11}\cos^2\theta+c_{44}\sin^2\theta\right)u_0+(c_{13}+c_{44})\sin\theta\cos\theta\;w_0=\rho\,c^2u_0$$

$$\left(c_{66}\cos^2\theta+c_{44}\sin^2\theta\right)v_0=\rho\,c^2v_0 \tag{8.31}$$

$$(c_{13}+c_{44})\sin\theta\cos\theta\;u_0+\left(c_{44}\cos^2\theta+c_{33}\sin^2\theta\right)\;w_0=\rho\,c^2w_0$$

Note that the system 8.31 does not contain the function

$$f\,(x\cos\theta+z\sin\theta-ct)\,,$$

so our results can be applied to any function of time; that is, independent of a frequency. Also, coefficients u_0 and w_0, characterizing motion in the plane $y-$const, are present in only the first and third equations of set 8.31. We thus have simple system of the homogeneous linear equations with respect to unknowns u_0, v_0, w_0, and c. Therefore, these parameters of the plane waves, $(u_0, v_0,$ and $w_0)$, cannot be uniquely determined. From the physical point of view this follows, because the primary source of the plane wave is not specified. However, existence of a plane wave in such a medium implies that these unknowns differ from zero. This is possible when the determinant of the system of homogeneous equations, 8.31, is equal to zero. This condition allows us, as with Rayleigh and Stoneley waves, to determine the velocity of propagation of the plane waves. This procedure is greatly simplified because the second equation of the set contains only the unknown v_0, and the first and third equations contain only u_0 and w_0. Thus, in place of this system we obtain two groups of equations, namely

$$\left(c_{66}\cos^2\theta+c_{44}\sin^2\theta-\rho\,c^2\right)\;v_0=0, \tag{8.32}$$

and

$$\left(c_{11}\cos^2\theta + c_{44}\sin^2\theta - \rho\, c^2\right)\; u_0 + \left(c_{13} + c_{44}\right)\sin\theta\cos\theta\; w_0 = 0, \qquad (8.33)$$

$$\left(c_{11} + c_{44}\right)\sin\theta\cos\theta\; u_0 + \left(c_{44}\cos^2\theta + c_{33}\sin^2\theta - \rho\, c^2\right)\; w_0 = 0$$

These sets describe different types of the plane waves and we begin their study from the simplest case.

1. Propagation along the z–axis (symmetry-axis direction)

Letting $\theta = \pi/2$, eqs. 8.32–8.33 become

$$\left(c_{44} - p\, c^2\right) v_0 = 0, \quad \left(c_{44} - p\, c^2\right)\; u_0 = 0, \quad \text{and} \quad \left(c_{33} - p\, c^2\right) w_0 = 0 \qquad (8.34)$$

Their solutions are the wave speeds

$$c_l = \sqrt{\frac{c_{33}}{\rho}} \qquad \text{and} \qquad c_s = \sqrt{\frac{c_{44}}{\rho}}, \qquad (8.35)$$

Therefore, in this direction we may observe the pure dilatational wave,

$$w\,(z,t) = w_0\; f\;\;(z - c_l t)\,, \qquad (8.36)$$

and pure shear waves,

$$u\,(z,t) = u_0\; f\;\;(z - c_s t) \qquad \text{and} \qquad v\,(z,t) = v_0\; f\;\;(z - c_s t) \qquad (8.37)$$

From the last inequality of set 8.24 it follows that

$$c_l > c_s \qquad (8.38)$$

Moreover, the particle displacement associated with each wave, is either normal or tangential to the phase surface which is defined as

$$z - c_l t = \text{const} \qquad \text{or} \qquad z - c_s t = \text{const} \qquad (8.39)$$

Inasmuch as

$$\frac{\partial w}{\partial z} = w_0\; f\,'\,(z - c_l t)\,, \qquad \frac{\partial w}{\partial x} = \frac{\partial w}{\partial y} = 0,$$

this wave causes compression (extension) only; for this reason it is called the pure dilatational wave. In contrast,

$$\frac{\partial v}{\partial y} = \frac{\partial v}{\partial x} = 0,$$

while

$$\frac{\partial v}{\partial z} = v_0 \, f\,'\,(z - c_s t)$$

Correspondingly, the wave $v\,(z,t)$ produces simple shear, which includes rotation, but no compression (extension). Thus,

$$v\,(z,t) = v_0 \, f\,'\,(z - c_s t)$$

is a pure shear wave, and likewise for $u(z,t)$. Because of the axial symmetry, both shear waves, $u\,(z,t)$ and $v\,(z,t)$ move with the same velocity.

2. Propagation along the x–axis

Suppose that plane waves may advance in the x–direction, orthogonal to the symmetry-axis direction. Then, taking into account that $\theta = 0$, eqs. 8.32 and 8.33 give

$$\left(c_{66} - \rho\, c^2\right) v_0 = 0, \qquad \left(c_{11} - \rho\, c^2\right) u_0 = 0 \tag{8.40}$$

and $\left(c_{44} - \rho\, c^2\right) w_0 = 0$

Three plane waves, with velocities

$$c_{SH} = \sqrt{\frac{c_{66}}{\rho}}, \qquad c_{SV} = \sqrt{\frac{c_{44}}{\rho}}, \qquad c_l = \sqrt{\frac{c_{11}}{\rho}}, \tag{8.41}$$

can propagate along the x–axis. They are

$$u\,(x,t) = u_0 \; f\,(x - c_l t)\,, \qquad v\,(x,t) = v_0 \; f\,(x - c_{SH} t)\,, \tag{8.42}$$

$$w\,(x,t) = w_0 \; f\,(x - c_{SV} t)$$

Certainly, the same waves can be observed in any plane containing the z–axis. System 8.42 includes one pure dilatational and two pure shear waves. Since

$$c_{11} > c_{66} \qquad \text{and} \qquad c_{11} > c_{44},$$

the velocity of the dilatational wave exceeds those of the shear ones:

$$c_l > c_{SH} \qquad \text{and} \qquad c_l > c_{SV} \tag{8.43}$$

Comparison of eqs. 8.35 and 8.41 shows that

$$c_s = c_{SV} \tag{8.44}$$

3. Propagation of the shear wave SH

Now we assume that θ is an arbitrary angle between the $x-$axis and the normal to the phase surface of the plane wave, Fig. 8.1b, and the wave has the single component v:

$$v\,(x,z,t)=v_0\ f\ \ (x\cos\theta+z\sin\theta-c_{SH}\ t) \tag{8.45}$$

In accordance with eq. 8.32, such a wave exists only if

$$c_{66}\cos^2\theta+c_{44}\sin^2\theta-\rho\ c_{SH}^2=0, \tag{8.46}$$

or which gives the expression for the velocity of propagation:

$$c_{SH}=\left(\frac{c_{66}\cos^2\theta+c_{44}\sin^2\theta}{\rho}\right)^{1/2} \tag{8.47}$$

Thus, the wave speed c_{SH} is a function of the angle θ, and varies within the range

$$c_{SH}\left(\frac{\pi}{2}\right)\leq c_{SH}(\theta)\leq c_{SH}(0),\qquad \text{for}\qquad c_{44}\leq c_{66} \tag{8.48}$$

As follows from eq. 8.45,

$$\frac{\partial v}{\partial y}=0,$$

but

$$\frac{\partial v}{\partial x}=v_0\ \cos\theta\ f\ '\,(x\cos\theta+z\sin\theta-c_{SH}\ t)$$

and $$\frac{\partial v}{\partial z}=v_0\ \sin\theta\ f\ '\,(x\cos\theta+z\sin\theta-c_{SH}\ t)$$

Therefore, for the displacement field

$$\mathbf{s}=v\mathbf{j},$$

we have

$$\begin{aligned}
\text{curl}_x\ \mathbf{s} &= -\frac{\partial v}{\partial z}=-v_0\ \sin\theta\ f\ '\,(x\cos\theta+z\sin\theta-c_{SH}\ t)\,,\\
\text{curl}_y\ \mathbf{s} &= 0,\\
\text{curl}_z\ \mathbf{s} &= \frac{\partial v}{\partial x}=v_0\ \cos\theta\ f\ '\,(x\cos\theta+z\sin\theta-c_{SH}\ t)
\end{aligned}$$

This means that the plane SH wave, regardless of the angle θ, is a pure shear wave, causing only the simple shear of an elementary volume around the $y-$axis.

4. Propagation of quasi-P and quasi-S waves

Next assume that the plane waves in general have particle displacement with two components:

$$\mathbf{s} = u\ \mathbf{i} + w\ \mathbf{k}, \tag{8.49}$$

while dispacements along the $y-$axis are absent. In accordance with system 8.33 such waves can exist only if the determinant of this system is equal to zero; that is,

$$\begin{vmatrix} c_{11}\cos^2\theta + c_{44}\sin^2\theta \ -\rho\, c^2 & (c_{13}+c_{44})\sin\theta\cos\theta \\ (c_{13}+c_{44})\sin\theta\cos\theta & c_{44}\cos^2\theta + c_{33}\sin^2\theta \ -\rho\, c^2 \end{vmatrix} = 0$$

Performing some algebra and introducing the notation

$$r = \rho c^2,$$

we have

$$r^2 - r\left(c_{44} + \frac{c_{11}+c_{33}}{2} + \frac{c_{11}-c_{33}}{2}\cos 2\theta\right) + \tag{8.50}$$

$$c_{44}\left(\frac{c_{11}+c_{33}}{2} + \frac{c_{11}-c_{33}}{2}\cos 2\theta\right) + \frac{1}{4}\left[(c_{11}-c_{44})(c_{33}-c_{44}) - (c_{13}+c_{44})^2\right]\sin^2\theta = 0$$

Making use of eq, 8.22 it is possible to show that

$$c_{44} < \frac{c_{11}+c_{33}}{2} + \frac{c_{11}-c_{33}}{2}\cos 2\theta,$$

the roots vary within the following limits:

$$c_{44} \le r_2 < r_1 \le \frac{c_{11}+c_{33}}{2} + \frac{c_{11}-c_{33}}{2}\cos 2\theta,$$

and they are distinct. Correspondingly, there can be two plane waves, propagating with the velocities c_1 and c_2, $(c_1 > c_2)$. We can write in the form,

$$u_1 = u_{01}\ f\left(x\cos\theta + z\sin\theta - c_1 t\right), \quad w_1 = w_{01}\ f\left(x\cos\theta + z\sin\theta - c_1 t\right), \tag{8.51}$$

$$u_2 = u_{02}\ f\left(x\cos\theta + z\sin\theta - c_2\ t\right), \quad w_2 = w_{02}\ f\left(x\cos\theta + z\sin\theta - c_2\ t\right) \tag{8.52}$$

For both waves, the two displacement components change synchronously and, therefore, vibrations of particles occur along a line (linear polarization). To illustrate the behavior of velocities c_1 and c_2 as functions of angle θ, consider an example of a laminated medium with parameters:

$$\begin{aligned} \rho_1 &= 2.7\cdot 10^3\ \mathrm{kg/m^3}, \quad \mu_1 = 2.5\cdot 10^{10}\ \mathrm{N/m^2}, \quad \lambda_1 = 3.0\cdot 10^{10}\ \mathrm{N/m^2}, \\ \rho_2 &= 2.3\cdot 10^3\ \mathrm{kg/m^3}, \quad \mu_2 = 0.6\cdot 10^{10}\ \mathrm{N/m^2}, \quad \lambda_2 = 0.8\cdot 10^{10}\ \mathrm{N/m^2} \end{aligned}$$

Correspondingly,

$$\begin{aligned} c_{l1} &= 5.40\ \mathrm{km/sec}, \quad c_{s1} = 3.04\ \mathrm{km/sec}, \\ c_{l2} &= 2.95\ \mathrm{km/sec}, \quad c_{s2} = 1.62\ \mathrm{km/sec} \end{aligned}$$

Also assume that

$$\frac{d_2}{d_1} = 3$$

Then the elastic constants, c_{ij}, in units of $\mathrm{N/m}^2$ are

$$\begin{aligned} c_{11} &= 3.36\cdot 10^{10}, \quad c_{33} = 2.46\cdot 10^{10}, \\ c_{12} &= 1.21\cdot 10^{10}, \quad c_{44} = 0.74\cdot 10^{10}, \\ c_{13} &= 0.97\cdot 10^{10}, \quad c_{66} = 1.08\cdot 10^{10}, \end{aligned}$$

and the average value of the density is $\rho = 2.4\cdot 10^3\ \mathrm{kg/m^3}$. Dependence of the wave speeds c_1, c_2, and c_{SH}, as functions of the angle θ is shown in Fig. 8.2. Note, that both plane waves, (u_1, w_1) and (u_2, w_2), become pure longitudinal and pure shear waves, respectively, when the angle θ is equal to $\pi/2$. Since $c_1(\pi/2) > c_2(\pi/2)$, these waves are often called the quasi-P and quasi-S waves. Because of axial symmetry ,

$$c_{SH}(\pi/2) = c_2(\pi/2)$$

As seen from Fig. 8.2, for $\theta = 0$ at the beginning with an increase of the angle θ $c_{SH} > c_2$. Then they become equal and, after that, $c_2 > c_{SH}$. Finally, for propagation along the z−axis these velocities coincide. The values of the longitudinal and shear velocities in each of the two elementary layers define a range of variation for c_1, c_2 and c_{SH}. We have:

$$c_{l2} < c_1(\theta) < c_{l1}, \qquad c_{s2} < c_2(\theta), \qquad c_{SH}(\theta) < c_{s1}, \qquad \text{for all values of } \theta.$$

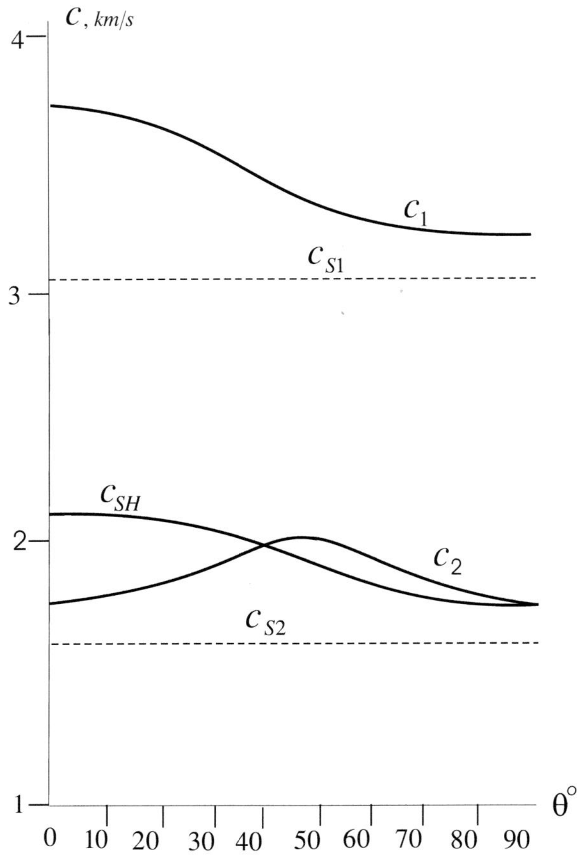

Figure 8.2: Velocities of propagation as functions of the angle θ

The variation of wave speed with angle θ for each wave type is an important feature of a wave propagation in an anisotropic medium.

Direction of particle motion

In general, the line of the particle vibrations for both waves, $(u_1, w_1$ and $u_2, w_2)$, do not coincide with the normal to the phase surface. To demonstrate this, we make use of eqs. 8.33, which give

$$\frac{w_0}{u_0} = \frac{\rho\, c_j - c_{11}\cos^2\theta - c_{44}\sin^2\theta}{(c_{13}+c_{44})\sin\theta\cos\theta} \tag{8.53}$$

or

$$\frac{w_0}{u_0} = \frac{(c_{13} + c_{44}) \sin\theta \cos\theta}{\rho\, c_2 - c_{44} \cos^2\theta - c_{33} \sin^2\theta} \tag{8.54}$$

Inasmuch as the determinant of this system is equal to zero, eqs. 8.53–8.54 give the same result. From eqs. 8.51–8.52 it follows that

$$\tan\alpha = \frac{w}{u} = \frac{1}{2} \frac{(c_{13} + c_{44}) \sin 2\theta}{\left(c_{44} \cos^2\theta + c_{33} \sin^2\theta - \rho\, c^2\right)}, \tag{8.55}$$

where α is the angle between the $z-$axis and the line of vibrations. It is different for the quasi-P and S waves and varies with the angle θ. A propagation of these waves is accompanied by a compression, (expansion) and a rotation of elementary volumes of a medium. In other words, the quasi-P or S waves are neither the dilatational or shear ones. In the case of quasi-P wave the vector of displacement is usually oriented close to the normal of the phase surface, whereas for the quasi-SV wave the vector of displacement is almost tangential to the phase surface. When quasi-SV wave propagates along the $z-$axis these waves become, respectively, pure dilatational and pure shear waves. To study this behavior in some detail let us compute the divergence and curl of the displacement. Since

$$\begin{aligned} u &= u_0\ f\,(x\cos\theta + z\sin\theta - ct)\,, \\ w &= w_0\ f\,(x\cos\theta + z\sin\theta - ct)\,, \end{aligned}$$

we have

$$\operatorname{div}\,\mathbf{s} = (u_0 \cos\theta + w_0 \sin\theta)\, f'\,(x\cos\theta + z\sin\theta - ct)$$

or

$$\operatorname{div}\,\mathbf{s} = s_n\ f\,'\,(x\cos\theta + z\sin\theta - ct) \tag{8.56}$$

Here

$$s_n = u_0 \cos\theta + w_0 \sin\theta \tag{8.57}$$

characterizes the displacement component, normal to the phase surface.

Again only the $y-$component of the $\operatorname{curl}\,\mathbf{s}$ differs from zero:

$$\operatorname{curl}_y\,\mathbf{s} = \frac{\partial u}{\partial z} - \frac{\partial w}{\partial x},$$

and rotation takes place about the $y-$axis. Therefore we have:

$$\text{curl}_y\ \mathbf{s} = s_t\ f\ '(x\cos\theta + z\sin\theta - ct)\,, \tag{8.58}$$

where

$$s_t = u_0\sin\theta - w_0\cos\theta \tag{8.59}$$

defines the displacement component, tangential to the phase surface of the plane wave. The ratio

$$\frac{\nabla \mathbf{s}}{|\nabla_y \times \mathbf{s}|} = \frac{\dfrac{u_0}{w_0} + \tan\theta}{\dfrac{u_0}{w_0}\tan\theta - 1}, \tag{8.60}$$

which show the relative roles of compression (extension) and rotation, depends on the angle θ, as well as on the type of the plane wave.

8.3 Rays and an energy flow

A ray may be considered to represent an elementary tube along which elastic energy flows. Correspondingly, in order to describe ray geometry, it is natural to proceed from the Poynting vector (Part I), the vector of flux density (Appendix E):

$$\mathbf{Y} = -\tau \cdot \dot{\mathbf{s}}, \tag{8.61}$$

tangential to the ray. Here τ is the stress tensor:

$$\tau = \begin{pmatrix} \tau_{xx} & \tau_{xy} & \tau_{xz} \\ \tau_{yx} & \tau_{yy} & \tau_{yz} \\ \tau_{zx} & \tau_{zy} & \tau_{zz} \end{pmatrix}$$

and

$$\dot{\mathbf{s}} = \dot{u}\mathbf{i} + \dot{v}\mathbf{j} + \dot{w}\mathbf{k}$$

is the particle velocity. In matrix notation, eq. 8.61 becomes

$$\mathbf{Y} = -\begin{pmatrix} \tau_{xx} & \tau_{xy} & \tau_{xz} \\ \tau_{yx} & \tau_{yy} & \tau_{yz} \\ \tau_{zx} & \tau_{zy} & \tau_{zz} \end{pmatrix} \begin{pmatrix} \dot{u} \\ \dot{v} \\ \dot{w} \end{pmatrix}$$

or

$$Y_x = -\tau_{xx}\dot{u} - \tau_{xy}\dot{v} - \tau_{xz}\dot{w}$$

$$Y_y = -\tau_{xy}\dot{u} - \tau_{yy}\dot{v} - \tau_{yz}\dot{w} \tag{8.62}$$

$$Y_z = -\tau_{xz}\dot{u} - \tau_{yz}\dot{v} - \tau_{zz}\dot{w},$$

since

$$\tau_{xy} = \tau_{yx}, \qquad \tau_{xz} = \tau_{zx}, \qquad \tau_{yz} = \tau_{zy}$$

Taking into account that any element of the ray,

$$d\,\mathbf{l} = dx\;\mathbf{i} + dy\;\mathbf{j} + dz\;\mathbf{k},$$

and the vector $\mathbf{Y}$ have the same directional cosines

$$\frac{dx}{dl} = \frac{Y_x}{Y}, \qquad \frac{dy}{dl} = \frac{Y_y}{Y}, \qquad \frac{dz}{dl} = \frac{Y_z}{Y},$$

the equation of a ray in the Cartesian system of coordinates is

$$\frac{dx}{Y_x} = \frac{dy}{Y_y} = \frac{dz}{Y_z} \tag{8.63}$$

Making use of the Hooke's law and eqs. 8.62 and 8.63, we consider several cases that illustrate ray behavior.

Case one. Plane SH wave in an isotropic medium Suppose that the plane wave moves away from the origin, with its phase surfaces are parallel to the $y-$axis. Then, the displacement

$$\mathbf{s} = v\;\mathbf{j}$$

is tangential to these surfaces, and for the single scalar component v we have

$$v = v_0\; f\;(x\cos\theta + z\sin\theta - c_s\;t) \tag{8.64}$$

Thus, the strains are

$$e_{xx} = e_{yy} = e_{zz} = 0, \tag{8.65}$$

and $$e_{xy} = \frac{\partial v}{\partial x} = v_0 \cos\theta \; f'\left(x\cos\theta + z\sin\theta - c_s t\right),$$

$$e_{yz} = \frac{\partial v}{\partial z} = v_0 \sin\theta \; f'\left(x\cos\theta + z\sin\theta - c_s t\right), \qquad e_{xz} = 0 \tag{8.66}$$

This means that the normal stresses:

$$\tau_{xx} = \lambda \text{div } \mathbf{s} + 2\mu \; e_{xx}, \quad \tau_{yy} = \lambda \text{div } \mathbf{s} + 2\mu \; e_{yy}, \quad \tau_{zz} = \lambda \text{div } \mathbf{s} + 2\mu \; e_{zz} \tag{8.67}$$

vanish, but the shear stresses are

$$\tau_{xy} = \mu \; e_{xy} = \mu \; v_0 \cos\theta \; f\,', \qquad \tau_{yz} = \mu \; e_{yz} = \mu \; v_0 \sin\theta \; f\,', \qquad \tau_{xz} = 0 \tag{8.68}$$

Before we continue, note that any phase surface of a wavefront of the plane wave is defined by the equation

$$x\cos\theta + z\sin\theta - c_s t = \text{const}; \tag{8.69}$$

i.e., the phase is the same at all points where the wave arrives simultaneously. In accordance with eqs. 8.62

$$Y_x = -\tau_{xy}\dot{v}, \qquad Y_y = 0, \qquad Y_z = -\tau_{yz}\dot{v} \tag{8.70}$$

or

$$Y_x = \mu \; v_0 c_s \cos\theta \; \left(f'\right)^2, \qquad Y_z = \mu \; v_0 \; c_s \sin\theta \; \left(f'\right)^2, \tag{8.71}$$

Substitution of eqs. 8.71 into eq. 8.63 gives:

$$\frac{dx}{\cos\theta} = \frac{dz}{\sin\theta} \qquad \text{or} \qquad \frac{dz}{dx} = \tan\theta \tag{8.72}$$

This demonstrates the known fact that rays are normal to the phase surface in an isotropic medium.

Case two. Plane P wave in an isotropic medium Next consider propagation of a longitudinal wave with displacement

$$\mathbf{s} = s_0 f\left(x\cos\theta + z\sin\theta - c_l \; t\right) \; \mathbf{n}$$

perpendicular to the phase surface. Correspondingly, components of the vector $\mathbf{s}$ along the coordinate axes are

$$u = s_0 \cos\theta \; f\left(x\cos\theta + z\sin\theta - c_l \; t\right), \tag{8.73}$$

$$w = s_0 \sin\theta \; f\left(x\cos\theta + z\sin\theta - c_l \; t\right)$$

Hence

$$e_{xx} = s_0 \cos^2\theta \; f', \qquad e_{yy} = 0, \quad e_{zz} = s_0 \sin^2\theta f', \tag{8.74}$$

$$e_{xy} = 0, \qquad e_{yz} = 0 \qquad \text{and} \qquad e_{xz} = \frac{\partial u}{\partial z} + \frac{\partial w}{\partial x} = 2s_0 \sin\theta\cos\theta \; f',$$

while

$$\text{div } \mathbf{s} = s_0 \; f'$$

The stresses are equal to

$$\tau_{xx} = \left(\lambda + 2\mu \; \cos^2\theta\right) \; s_0 \; f', \;\; \tau_{yy} = \lambda \; s_0 \; f', \;\; \tau_{zz} = \left(\lambda + 2\mu \; \sin^2\theta\right) \; s_0 \; f', \tag{8.75}$$

and

$$\tau_{xy} = 0, \qquad \tau_{yz} = 0, \qquad \tau_{xz} = \mu \; \sin^2\theta \; s_0 \; f' \tag{8.76}$$

From eq. 8.62 we obtain

$$Y_x = -\tau_{xx}\dot{u} - \; \tau_{xz}\dot{w}, \qquad Y_y = 0, \qquad Y_z = -\tau_{xz} \; \dot{u} - \tau_{zz}\dot{w} \tag{8.77}$$

Inasmuch as

$$\dot{u} = -c_l s_0 \cos\theta \; f', \qquad \dot{w} = -c_l s_0 \sin\theta f',$$

eqs. 8.77 give

$$Y_x = c_l \; s_0^2 \cos\theta \;\; (\lambda + 2\mu)\left(f'\right)^2, \quad Y_z = c_l \; s_0^2 \sin\theta \;\; (\lambda + 2\mu)\left(f'\right)^2 \tag{8.78}$$

Thus

$$\mathbf{Y} = c_l^3 \rho s_0^2 \left(f'\right)^2 \mathbf{n}, \tag{8.79}$$

because

$$\mathbf{n} = \cos\theta \; \mathbf{i} + \sin\theta \; \mathbf{k}, \qquad (\lambda + 2\mu) = \rho \; c_l^2,$$

and the ray is normal to the phase surface.

Case three. The plane SH wave in a transversely isotropic medium Again, as in the first case, the displacement has only the single component

$$v = v_0 \ f\left(x\cos\theta + z\sin\theta - c_{SH}\ t\right) \tag{8.80}$$

Therefore

$$e_{xx} = e_{yy} = e_{zz} = 0$$

and

$$e_{xy} = v_0 \cos\theta \ f'\left(x\cos\theta + z\sin\theta - c_{SH}\ t\right),$$

$$e_{yz} = v_0 \sin\theta \ f'\left(x\cos\theta + z\sin\theta - c_{SH}\ t\right), \tag{8.81}$$

$$e_{xz} = 0$$

Taking into account Hooke's law, eqs. 8.3:

$$\tau_{xx} = 0, \qquad \tau_{yy} = 0, \qquad \tau_{zz} = 0, \tag{8.82}$$

$$\tau_{yz} = c_{44}\ v_0 \sin\theta \ f', \qquad \tau_{xz} = 0, \qquad \tau_{xy} = \ c_{66}\ v_0 \cos\theta \ f',$$

and eqs. 8.62 become

$$Y_x = \ c_{66}\ v_0^2\ c_{SH} \cos\theta\,(xf')^2 \qquad\qquad Y_y = 0, \tag{8.83}$$

$$Y_z = \ c_{44}\ v_0^2\ c_{SH} \sin\theta\,(\ f')^2$$

Correspondingly, the ray equation is

$$\frac{dx}{c_{66}\cos\theta} = \frac{dz}{c_{44}\sin\theta}, \tag{8.84}$$

showing that the rays of the SH wave are still the straight lines in transversely isotropic media, but they are no longer normal to the phase surface, except when $\theta = 0$ and $\theta = \pi/2$. The angle formed by the ray and the $x-$axis is equal to

$$\tan\varphi = \frac{dz}{dx} = \frac{c_{44}}{c_{66}}\tan\theta$$

For instance, in the example considered above:

$$c_{44} = 0.74 \cdot 10^{10} \mathrm{N/m^2}, \qquad c_{66} = 1.08 \cdot 10^{10} \mathrm{N/m^2},$$

and we have

$$\tan\varphi = 0.69 \tan\theta,$$

that is the angle φ between the ray and the $x-$axis is smaller than θ.

Case four. Ray orientation for quasi-P and quasi-S waves These waves cause particle displacement that has two components:

$$u = u_0 \; f \;\; (x\cos\theta + z\sin\theta - c\; t)\,, \qquad w = w_0 \; f \;\; (x\cos\theta + z\sin\theta - c\; t) \tag{8.85}$$

Here c is the either c_1 or c_2. Correspondingly, the strains are

$$e_{xx} = u_0 \cos\theta f', \qquad e_{yy} = 0, \qquad e_{zz} = w_0 \sin\theta f', \qquad e_{xy} = 0, \qquad e_{yz} = 0,$$

$$e_{xz} = (u_0 \sin\theta \; + w_0 \cos\theta)\; f' \tag{8.86}$$

$$\text{and} \qquad \operatorname{div} \mathbf{s} = (u_0 \cos\theta \; + w_0 \sin\theta)\; f'$$

This gives for stresses

$$\tau_{xx} = (c_{11} \; u_0 \cos\theta \; + c_{13} \; w_0 \sin\theta)\; f',$$

$$\tau_{yy} = (c_{12} \; u_0 \cos\theta \; + c_{13} \; w_0 \sin\theta)\; f', \tag{8.87}$$

$$\tau_{zz} = (c_{13} \; u_0 \cos\theta \; + c_{33} \; w_0 \sin\theta)\; f' \;,$$

$$\text{and} \qquad \tau_{yz} = 0, \qquad \tau_{xy} = 0, \qquad \tau_{xz} = c_{44}\,(u_0 \cos\theta \; + w_0 \sin\theta) \;\; f'$$

Then, in accordance with eqs. 8.62, components of the Poynting vector are:

$$Y_x = \left[(c_{11} u_0 \cos\theta \; + c_{13} w_0 \sin\theta) \;\; u_0 + c_{44}\,(u_0 \sin\theta \; + w_0 \cos\theta)\, w_0\right] c\,(\; f')^2\,,$$

$$Y_y = 0 \;, \tag{8.88}$$

$$\text{and} \qquad Y_z = \left[c_{44}\,(u_0 \sin\theta \; + w_0 \cos\theta)\, u_0 + (c_{13} u_0 \cos\theta \; + c_{33} w_0 \sin\theta)\, w_0\right] c\,(\; f')^2$$

Clearly the angle between the normal $\mathbf{n}$ to the phase surface and the ray direction is nonzero. Since the ratio u_0/w_0 depends on the velocity of propagation, the rays of the quasi-P and quasi-S waves are oriented differently.

8.4 Phase and ray surfaces

By definition, the phase surface of the plane wave moves along its normal $\mathbf{n}$ with the velocity c_p, which differs for SH and quasi-P and S waves. At the same time an elastic energy propagates along rays. Because their orientation is characterized by the unit vector $\mathbf{r}_0$, which generally does not coincide with $\mathbf{n}$, it can be expected that the energy (ray) velocity c_r differs from c_p. In order to determine the former consider an elementary volume oriented along the ray, as shown in Fig. 8.3a. It has cross-section dS and length

$$dl = c_r dt \tag{8.89}$$

Here dt is a small time interval, and c_r is the energy velocity along the ray. Correspondingly, the amount of this energy inside the volume is

$$dW = e\ dS\ c_r dt, \tag{8.90}$$

where e is the density of the elastic energy. During the time interval dt all energy of the volume crosses dS, so it can be represented as

$$dW = Y\ dS\ dt \tag{8.91}$$

Here Y is the magnitude of the Poynting vector. Thus, we have:

$$e\ dS\ c_r dt = Y\ dS\ dt, \qquad \text{or} \qquad c_r = \frac{Y}{e}; \tag{8.92}$$

i.e., the ray velocity is equal to a relative change of the density per unit time. Consider quasi-P and S waves: As was shown earlier,

$$Y_x = -\tau_{xx}\dot{u} - \tau_{xz}\dot{w} \qquad \text{and} \qquad Y_z = -\tau_{xz}\dot{u} - \tau_{zz}\dot{w}$$

Also,

$$Y = \sqrt{Y_x^2 + Y_z^2} \tag{8.93}$$

Inasmuch as the densities of the potential and kinetic energies are equal to each other in the plane wave, and the latter is

$$\frac{1}{2}\rho\left(\dot{u}^2 + \dot{w}^2\right),$$

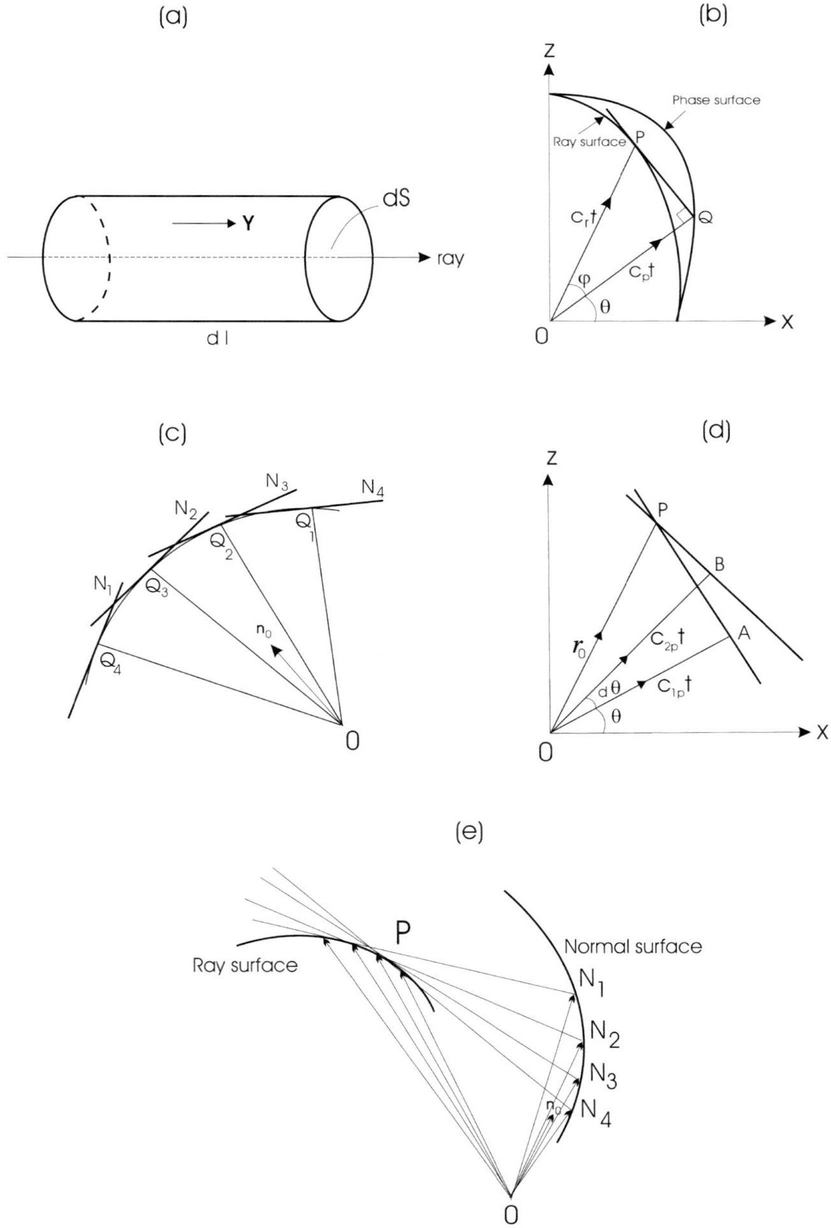

Figure 8.3: (a) Illustration of eq. 8.92 (b) Phase and ray surfaces (c) Interference of elementary waves in an isotropic medium (d) Interference of two plane waves in an anisotropic medium (e) Formation of ray surface by a system of plane waves in an anisotropic medium

the density of the elastic energy is equal to

$$e = \rho \left(\dot{u}^2 + \dot{w}^2 \right) \tag{8.94}$$

where

$$\dot{u} = \frac{\partial u}{\partial t} \qquad \text{and} \qquad \dot{w} = \frac{\partial w}{\partial t}$$

are scalar components of the particle velocity. Thus

$$c_r = \frac{\sqrt{Y_x^2 + Y_z^2}}{\rho \left(\dot{u}^2 + \dot{w}^2 \right)} \tag{8.95}$$

The unit vector $\mathbf{r}_0$, which defines a ray orientation, forms with the $z-$axis the angle φ

$$\tan \varphi = \frac{Y_x}{Y_z} = \frac{\tau_{xx} + \tau_{xz} \dfrac{\dot{w}}{\dot{u}}}{\tau_{xz} + \tau_{zz} \dfrac{\dot{u}}{\dot{w}}} \tag{8.96}$$

In transversely isotropic medium, as well as in more general cases, it is proper to distinguish the particle, phase, and ray velocities because they usually differ from each other by magnitude and direction.

In order to emphasize the difference between the phase and ray velocities we introduce, along with the phase (normal) surface, the concept of the wave or ray surface. Suppose that a source of an elastic wave is located at some point O of the transversely isotropic medium, and it starts to generate the wave at the instant $t = 0$. Let us plot along any straight line, drawn from O, a segment that is proportional to the phase velocity $c_p(\theta)$ in this direction:

$$l_p(\theta) = c_p(\theta)\, t,$$

where θ is the angle between the line and the $x-$axis. Connecting terminal points of the linear element, we obtain a position of the normal surface at the instant t. The shape of phase surfaces formed in this way is non-spherical and is independent of time; it is defined by parameters of the anisotropic medium. Similarly, we plot the segment along the same line that s proportional to the ray velocity

$$l_r(\theta) = c_r(\theta)\, t,$$

Its terminal points generate the wave (ray) surface, as seen in Fig. 8.3b. From the physical point of view, such a surface at the moment t is the boundary between the portion of a medium distorted by a wave and the portion that is still at rest. In other words, this surface is formed by points, where an energy arrives at the instant t, and, correspondingly, it plays the role of the wavefront. In an isotropic medium the ray and phase surfaces coincide and are spherical. There is a relationship between these surfaces. In order to describe this relationship, consider a wave that is a sinusoidal function of time. As is known, it is possible to represent a wave caused by a point source, as a superposition of an infinite number of plane harmonic waves, each of them with infinitely small amplitude. These elementary waves depend on time and distance from the source as

$$\sin\left[\omega t - \frac{\omega}{c_p}\left(x\cos\theta + z\sin\theta\right)\right] \qquad \text{or} \qquad \sin\left(\omega t - \frac{\omega}{c_p}\,\mathbf{r}\cdot\mathbf{n}\right), \tag{8.97}$$

where $\mathbf{r}$ is the radius-vector of any point, located on the plane phase surface:

$$\mathbf{r} = x\mathbf{i_1} + z\mathbf{i_2}, \qquad \text{while} \qquad \mathbf{n} = \cos\theta\ \mathbf{i_1} + \sin\theta\ \mathbf{i_2}$$

is the unit normal to this plane, such that $|\,\mathbf{n}| = 1$. In place of expression 8.97 we can write

$$\sin\left(\omega t - \mathbf{k}\cdot\mathbf{r}\right) \tag{8.98}$$

Here the wavenumber vector

$$\mathbf{k} = \frac{\omega}{c_p}\mathbf{n} \tag{8.99}$$

has a direction of the normal to the phase plane, with components

$$k_x = \frac{\omega}{c_p}\cos\theta, \qquad k_z = \frac{\omega}{c_p}\sin\theta \tag{8.100}$$

The vector

$$\mathbf{s} = \frac{\mathbf{n}}{c_p} \tag{8.101}$$

is called the slowness vector, and, correspondingly

$$\mathbf{k} = \omega\mathbf{s} \tag{8.102}$$

Now, consider a group of plane waves of the same frequency, with slightly different directions of propagation. One can imagine that unit normals $\mathbf{n}$ of these waves are inside a small cone with the apex O, characterized by a "mean wave vector $\mathbf{n}_0$", as depicted in Fig. 8.3c. Assume that constructive interference takes place between only those elementary waves whose normals are close to $\mathbf{n}_0$. By definition, at the instant $t = 0$ all these waves are in the phase in the vicinity of the source; i.e., the wave disturbance is maximal at the point O. Now we trace propagation of this maximum motion through the medium and focus on those wavefronts whose normals differs only slightly from $\mathbf{n}_0$. First, consider a simple case.

Isotropic medium

Since the velocity of propagation is independent of direction, at the instant t these plane waves advance at the same distance $c_p t$, as in Fig. 8.3c:

$$OQ_1 = OQ_2 = OQ_3 = OQ_4$$

The wavefronts of elementary plane waves are perpendicular to the corresponding directions of propagation. The phase of each wave at points of its the wavefront is the same as that at the initial instant $t = 0$ at the origin O. In other words, all wavefronts, N_1, N_2, N_3 and N_4, shown in Fig. 8.3c, have the same phase. A summation of wavefields at those points of the space where these wavefronts intersect each other thus has a constructive character. With a decrease of the angle between directions of propagation such points belong the envelope of the wavefronts, as illustrated in Fig. 8.3c. In isotropic media, for which $\partial c_p/\partial\theta = 0$, the latter is the arc of the circle, or a spherical segment in three dimensions. In this light let us make two comments:

1. In the presence of the dispersion, the radii of the wavefronts are functions of frequency, and are equal to $c_p(\omega)\, t$. The constructive interference between them leads to the formation of the wave groups, propagating with the group velocity c_g. The locus of points, where this interference occurs at time t is the arc of the circle, with radius $c_g t$.

2. The direction of propagation of each plane wave coincides with the direction of propagation of the energy of the wave group and is defined by the normal of the wavefront.

Anisotropic medium

First, consider only two plane waves with the same frequency, propagating in directions $\mathbf{n}_1$ and $\mathbf{n}_2$, as shown in Fig. 8.3d. Correspondingly, their phase velocities are equal to

$$c_p(\theta) \qquad \text{and} \qquad c_p(\theta + d\theta)$$

As before, at the initial moment $t = 0$, their phases are equal at the origin O and constructive interference is observed. During the time interval t, the wavefronts of the two waves advance in the directions of their normals at distances

$$l_{1p} = c_p(\theta)\ t \qquad \text{and} \qquad l_{2p} = c_p(\theta + d\theta)\ t,$$

respectively. The phases of these waves are the same at points A and B. Since these points are situated at different places, a superposition of waves with the same phase occurs at the point P, where wavefronts intersect, Fig. 8.3d. Thus, the interference maximum of the wave field, located at O when $t = 0$, moves to the point P during the time interval t. The distance OP is equal to $c_r t$, where c_r is the ray velocity along OP, and its direction is characterized by the unit vector $\mathbf{r}_0$. Along this ray energy propagates with its maximum at the point P, at time t. By analogy with the previous case consider a superposition of an infinite number of plane waves, moving in different directions. All of them have the same phase at the instant $t = 0$ at point O. Let us focus of those waves, whose normals are close to $\mathbf{n}_0$, as depicted in Fig. 8.3e. After the time interval t, the wavefront N, propagating in the direction $\mathbf{n}$, reaches a position N_1, so that the perpendicular, drawn from O to this plane wavefront is equal to $c_p t$. The amplitude of the group of these neighboring waves is largest provided that they reinforce each other (constructive interference). This happens where the wavefronts of the plane waves intersect; this defines a region in the vicinity of the envelope of these planes. Its position is characterized by the point P, where the energy arrives. The straight line between the points O and P is the ray along which energy travels with the velocity c_r. Different groups of elementary plane waves give rise to different points of the wave (ray) surface. A relationship exists between the point of the normal surface with the radius-vector $c_p\ t\ \mathbf{n}$ and the point of the wave surface characterized by the radius-vector $c_r\ t\ \mathbf{n}_0$. Because intersection of the phase planes, having almost the same orientation, defines the position of the point P of the wave surface where the constructive interference occurs, the coordinates of the point P can be derived from the condition that the first derivative of the phase of the plane waves with respect to the wavenumber $\mathbf{k}$ is zero. That is, any point of the wave surface is a stationary one. Taking into account that propagation of energy along the ray is accompanied by constructive interference of plane waves, the velocity c_r must coincide with the group velocity

$$c_g = \frac{\partial \omega}{\partial k} \tag{8.103}$$

This important result can be proved by different ways. For example, we can make use of eq. 8.103, which can be represented in the form

$$c_g = \frac{\partial}{\partial k}(kc_p) \tag{8.104}$$

Correspondingly, its components along the x and z coordinate axes are

$$c_{gx} = \frac{\partial}{\partial k_x}(kc_p), \qquad c_{gz} = \frac{\partial}{\partial k_z}(kc_p) \tag{8.105}$$

Since the phase velocity $c_p(\theta)$, as well as components of the wavenumber, are known, one can determine the group velocity and confirm that it coincides with c_r. By definition:

$$\mathbf{c}_g = c_{gx}\mathbf{i_1} + c_{gz}\mathbf{i_2}$$

or

$$\mathbf{c}_g = \text{grad}(kc_p), \tag{8.106}$$

where derivatives are taken with respect to k_x and k_z. The gradient is perpendicular to the level surface

$$kc_p = \text{const} \qquad \text{or} \qquad k = \frac{\text{const}}{c_p} \tag{8.107}$$

At the same time the plane slowness surface is defined from the condition

$$\frac{1}{c_p} = \text{const} \tag{8.108}$$

Thus, assuming that $c_p(\theta)$ is known, determination of values k, corresponding to the level surface, is equivalent to finding points of the slowness surface. This means that the vector of the group (ray) velocity is perpendicular to this surface. Taking into account that the wave surface is an envelope of the wavefronts of the plane waves, we can derive an equation of the ray surface. This task can be performed applying the conventional method of calculus (Part II). For instance, in the meridian section XOZ the wavefronts are straight lines and described by the equation

$$x\cos\theta + z\sin\theta = c_p(\theta)\,t \quad \text{for} \quad \text{constant } t \tag{8.109}$$

Here x and z are coordinates of any point of the line perpendicular to OQ in Fig. 8.3b, and θ is the angle between OQ and the $x-$axis. Taking the derivative with respect to θ we have

$$-x\sin\theta + z\cos\theta = \frac{\partial c_p(\theta)}{\partial\theta}t \tag{8.110}$$

Combining eqs. 8.109 and 8.110 to eliminate θ we obtain the equation of the envelope that describes the wave surface, which is tangential to PQ at the point P. Squaring and then adding eqs. 8.109 and 8.110 gives

$$l_r^2 = x^2 + z^2 = \left[c_p^2 + \left(\frac{\partial c_p}{\partial \theta}\right)^2\right] t^2 \tag{8.111}$$

Here l_r is the ray length, which is numerically equal to $l_r = c_r t$. In this light, note also that, from the triangle OPQ in Fig. 8.3b that

$$c_p(\theta) = c_r(x, z) \cos\varphi \tag{8.112}$$

This shows that the ray velocity along OP exceeds the phase velocity along the corresponding line OQ. Combining eqs. .8.111 and 8.112 shows that

$$\tan\varphi = \frac{1}{c_p}\frac{\partial c_p}{\partial \theta}, \tag{8.113}$$

where the φ characterizes the angle between the ray OP and the line OQ normal to the corresponding plane. Thus, knowledge of the angle φ and the distance l_r allows us to determine a position of a point P of the wave surface. Usually with an increase of the angle θ the ray OP approaches to the $z-$axis; i.e.,

$$\frac{d}{d\theta}(\theta + \varphi) > 0$$

However, it may happen that this derivative changes a sign, i.e. the angle $\theta + \varphi$ begins to decrease. In such a case the wave surface has a cusp. The condition for its appearance is

$$\frac{d}{d\theta}(\theta + \varphi) = 0,$$

or, taking into account eq. 8.113,

$$\frac{d}{d\theta}\left(\frac{1}{c_p}\frac{\partial c_p}{\partial \theta}\right) = -1 - \left(\frac{1}{c_p}\frac{\partial c_p}{\partial \theta}\right)^2 \tag{8.114}$$

The wave surface of the quasi-P wave has no such cusps.

Appendix A

Equations of motion of a rigid body

Resultant force and resultant torque

In order to find relationships between the motion of a rigid body and forces acting on it, we have to perform some transformations with forces. First, suppose that the external force $\mathbf{F}_e$ is applied to the body at some point p, Fig. A.1a. Then, imagine that two forces, $\mathbf{F}_e$ and $-\mathbf{F}_e$, also act at a different point q. It is obvious that they cancel each other and, correspondingly, the system of three forces, $\mathbf{F}_e(p)$, $\mathbf{F}_e(q)$ and $-\mathbf{F}_e(q)$, is equivalent to the single force $\mathbf{F}_e(p)$. As is well known, a combination of forces $\mathbf{F}_e(p)$ and $-\mathbf{F}_e(q)$ represents the couple with some torque $\boldsymbol{\tau}$

$$\boldsymbol{\tau} = [\mathbf{r}(p) - \mathbf{r}(q)] \times \mathbf{F}_e(p) \qquad \text{(A-1)}$$

Since the point of application of this vector is not important, assume that it acts at the point q. Let us also recall that the torque $\boldsymbol{\tau}$ coincides with the moment of the force $\mathbf{F}_e(p)$ with respect to the point q. Thus we described the rule, which allows us to replace the force $\mathbf{F}_e(p)$ by the same force $\mathbf{F}_e(q)$, but acting at different point, and the couple with the torque given by eq. A-1. If we have a distribution of external forces, applied at various points of the rigid body, the same procedure for each of them leads to two sums. The first one is the sum of external forces, the other is the sum of torques. It is essential that all these forces and torques act at the same point q, Fig. A.1b. Applying the known rule of a summation of vectors, we obtain the resultant force $\mathbf{F}_r(q)$ and the resultant torque $\boldsymbol{\tau}(q)$.

$$\mathbf{F}(q) = \sum \mathbf{F}_n(q), \qquad \boldsymbol{\tau} = \sum \boldsymbol{\tau}_n(q) \qquad \text{(A-2)}$$

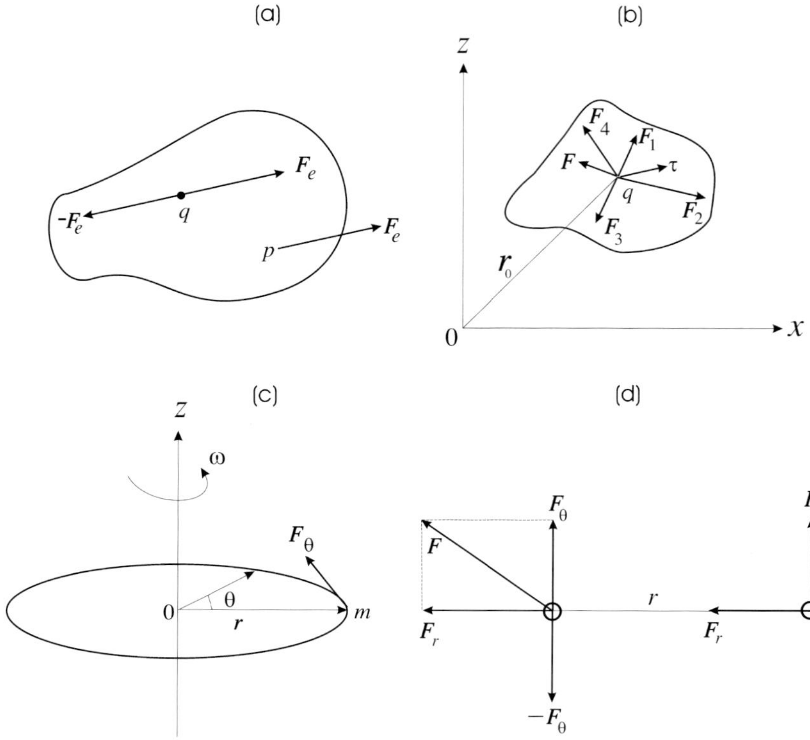

Figure A.1: (a) Replacement of the single force by the couple and the same force, applied at different point. (b) Superposition of forces and torques. (c) Rotation of an elementary mass (d) Torque of the force $\mathbf{F}_\theta$

Of course, in the case of a continuous distribution of forces, a summation has to be replaced by an integration. Note that a choice of the point q makes an influence on the resultant torque $\boldsymbol{\tau}$, but the resultant force $\mathbf{F}$ remains the same. Without any doubts, eqs. A-2 are of a great importance, because they suggest that any motion of a rigid body can be represented as the sum of two different types of motion. One of them is caused by the resultant force, while the other is due to the action of the couple of forces.

Further it is assumed that the point q coincided with the center of mass o, and, correspondingly, this motion is a superposition of a translation of the point o and a rotation around it. As is well known, (Part I), the motion of the center of mass with radius-vector $\mathbf{r}_0$ is described by the equation:

$$M\frac{d^2\,\mathbf{r}_0}{dt^2} = \mathbf{F} \tag{A-3}$$

It clearly demonstrates that the center of mass moves as if it where a particle with mass M, subjected to the resultant force, $\mathbf{F}$. Certainly, this is the remarkable feature of the center of mass. The last equation contains only one unknown, $\mathbf{r}_0(t)$, and its solution allows us to find a path of this point. Suppose that at the beginning of the motion the rigid body is at rest and then we apply the system of external forces, so that

$$\sum \mathbf{F}_e = \mathbf{F} = \mathbf{0}$$

In accordance with eq. A-3 the center of mass does not move, while the other parts of the body can be involved in a motion.

Next we obtain an equation of a rotation due to the resultant torque $\boldsymbol{\tau}$. This motion takes place about the axis of rotation and, by definition, all points do not move, including the center of mass. At the same time other particles move along circles with centers located on the axis of rotation. In general, the latter may change its orientation with time. As in the case of translation we begin from the simplest case of a rotation of an elementary mass about the fixed axis.

Example one Assume that the axis of rotation of mass m coincides with the z-axis, Fig. A.1c, and the origin of the cylindrical system is located at the point O. In accordance with eq. A.3, we have

$$mr\frac{d^2\theta}{dt^2} = F_\theta \tag{A-4}$$

For our purpose it is convenient to imagine that the mass and the axis are connected with a help of a massless rod. Bearing in mind that a rotation is caused by the torque, it is proper to transform eq. A-4 in such a way that instead of the force F_θ we would have its moment. After a multiplication of both sides of this equation by r we obtain

$$mr^2\frac{d^2\theta}{dt^2} = rF_\theta \tag{A-5}$$

Its right hand side is the magnitude of the moment of the force $\boldsymbol{\tau}$ with respect to the origin O, defined by

$$\boldsymbol{\tau} = \mathbf{r} \times \mathbf{F}$$

The force $\mathbf{F}$ has usually two components:

$$\mathbf{F} = F_r\ \mathbf{i}_r + F_\theta\ \mathbf{i}_\theta, \qquad \text{and} \qquad \mathbf{r} = r\ \mathbf{i}_r$$

By definition

$$|\mathbf{r} \times \mathbf{F}| = rF_\theta \tag{A-6}$$

It is obvious that $\boldsymbol{\tau}$ is directed along the axis of rotation, that is

$$\boldsymbol{\tau} = rF_\theta\ \mathbf{i}_z \tag{A-7}$$

The left hand side of eq. A-5 can be written in the form

$$mr^2\frac{d^2\theta}{dt^2} = \frac{d}{dt}(I\omega) \tag{A-8}$$

Here

$$I = mr^2 \tag{A-9}$$

is called the moment of inertia of mass with respect to the axis of rotation and

$$\omega = \frac{d\theta}{dt} \tag{A-10}$$

is the magnitude of the angular velocity, which characterizes a rate of a change of the angle θ, that is a turn of the mass m. The angular velocity, as the torque, is a vector, and it shows an orientation of the axis of rotation:

$$\boldsymbol{\omega} = \omega\mathbf{i}_z \tag{A-11}$$

Note that all particles of a rotating rigid body have the same angular velocity $\boldsymbol{\omega}$, that emphasizes an importance of this vector quantity. From Fig. A.1c we see that magnitudes of the linear and angular velocities are related as

$$v_\theta = r\omega$$

and in the vector form

$$\mathbf{v} = \boldsymbol{\omega} \times \mathbf{r} \tag{A-12}$$

This is the definition of the angular velocity. Also the product

$$\mathbf{L} = I\boldsymbol{\omega} \tag{A-13}$$

is called the angular momentum of a motion, and in this example vector $\mathbf{L}$ and $\boldsymbol{\omega}$ have the same direction.

Since each side of eq. A-5 represents the z-component of vectors $d\mathbf{L}/dt$ and $\boldsymbol{\tau}$, respectively, it can be rewritten as

$$\frac{d\mathbf{L}}{dt} = \boldsymbol{\tau} \tag{A-14}$$

Thus, the torque is equal to the rate of change with time of the angular momentum, $\mathbf{L}$. In this simplest case only its magnitude varies. In other words, the torque results in a change of the angular momentum. In particular, if $\boldsymbol{\tau} = 0$, the vector $\mathbf{L}$ remains constant. It is useful to note that eq. A-14 is also valid in the general case of a rigid body.

Let us compare eq. A-14 with Newton's second law:

$$m\,\mathbf{a} = \mathbf{F} \qquad \text{and} \qquad I\,\boldsymbol{\alpha} = \boldsymbol{\tau} \tag{A-15}$$

We took into account that the axis of rotation is fixed (the moment of inertia does not change with time), and

$$\frac{d\boldsymbol{\omega}}{dt} = \boldsymbol{\alpha} \tag{A-16}$$

is the angular acceleration. The analogy between these two equations of motion is obvious. The moment of inertia plays the same role as mass, while the linear and angular accelerations define a rate of a change of corresponding velocities. Both the force and torque cause a motion. Also there is some essential difference between m and I. The first one is independent of a position of the particle, while the moment of inertia rapidly varies with r. For instance, with an increase of the distance from the axis, it is more difficult to change the angular velocity ω.

Here it may be appropriate to make several comments:

1. In the Cartesian system of coordinates the moment of inertia, given by eq. A-9, has the form

$$I = m(x^2 + y^2) \tag{A-17}$$

2. Rotation of mass around point O, Fig. A.1c, is always accompanied by the presence of the radial component of force $\mathbf{F}_r$, and its physical meaning depends on the problem. For instance, it can be the gravitational, electrical or magnetic force. Also it may arise due to a deformation of the elastic rod, connecting the mass with the axis of rotation.

3. Since this centripetal force, $\mathbf{F}_r$, and the radius-vector $\mathbf{r}$ are opposite to each other, its torque is equal to zero:

$$\mathbf{r} \times \mathbf{F}_r = \mathbf{0}$$

4. As was mentioned above, we may treat the particle with mass m and massless rod, connecting it with the axis, as the rigid body. Then, forces $\mathbf{F}_r$ and $\mathbf{F}_\theta$, acting on the mass m, can be replaced by the resultant force, $\mathbf{F}$, applied at the point O, and the torque of the couple of forces, shown in Fig. A.1d. The action of the force $\mathbf{F}_r$ is compensated by the axis of rotation. Therefore, the motion occurs due to the couple of forces $\mathbf{F}_\theta$ and $-\mathbf{F}_\theta$ with the level r.

Example two Again consider a motion of an elementary mass in the plane around the z-axis, but, unlike the previous case, assume that the origin O is not located at the same plane, Fig. A.2a. This generalization is desirable, because it will help to take into account the fact that particles of the rotating rigid body move in different planes. We will proceed from Newton's second law

$$m\frac{d^2\mathbf{r}}{dt^2} = \mathbf{F}$$

Its vector multiplication by $\mathbf{r}$ gives

$$\mathbf{r}\times m\frac{d^2\mathbf{r}}{dt^2} = \mathbf{r} \times \mathbf{F} = \boldsymbol{\tau} \tag{A-18}$$

The right hand side is the torque about the point O, and it has both normal and tangential components with respect to the plane of motion. In approaching the point O to this plane the tangential component of $\boldsymbol{\tau}$ tends to zero, provided that $\mathbf{F}_z = 0$.

Now we demonstrate that eq. A-18 has the same form as A-14. With this purpose in mind consider the left hand side of eq. A-18, which can be written as:

$$\mathbf{r}\times m\frac{d^2\mathbf{r}}{dt^2} = \frac{d}{dt}(\mathbf{r}\times m\frac{d\mathbf{r}}{dt}) \tag{A-19}$$

In fact, performing a differentiation, we obtain

$$\frac{d}{dt}(\mathbf{r} \times m\frac{d\mathbf{r}}{dt}) = \frac{d\mathbf{r}}{dt} \times m\frac{d\mathbf{r}}{dt} + \mathbf{r}\times m\frac{d^2\mathbf{r}}{dt^2} = \mathbf{r}\times m\frac{d^2\mathbf{r}}{dt^2} \tag{A-20}$$

since vectors $d\mathbf{r}/dt$ and $m\,(d\mathbf{r}/dt)$ have the same direction. Thus, eq. A-18 becomes

$$\frac{d}{dt}(\mathbf{r} \times m\frac{d\mathbf{r}}{dt}) = \frac{d\mathbf{L}}{dt} = \boldsymbol{\tau} \tag{A-21}$$

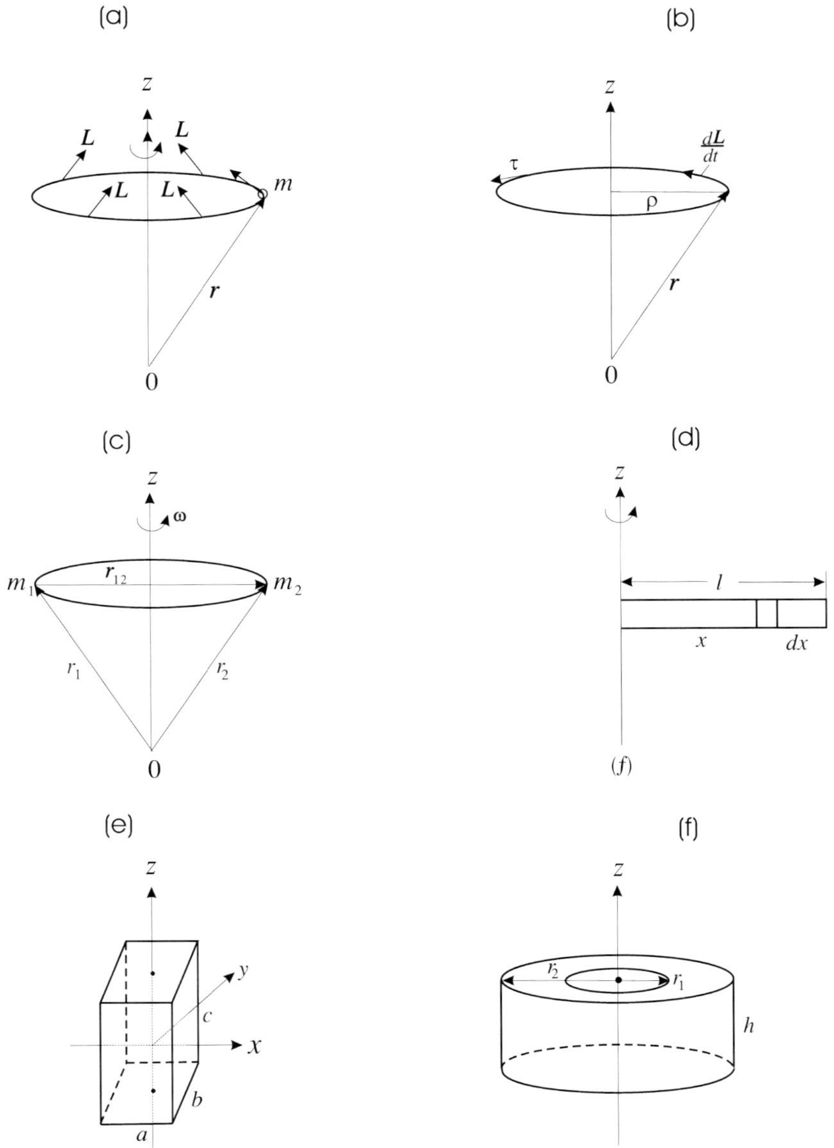

Figure A.2: (a) Illustration of eq. A-19. (b) Orientation of $\boldsymbol{\tau}$ and $d\mathbf{L}/dt$, when $F_\theta = 0$. (c) Illustration of eq. A-27 (d,e,f) Calculations of I_{zz} of the bar, rectangular, parallelepiped and the ring.

The expression in brackets is called the angular momentum $\mathbf{L}$, and it has several equivalent forms:

$$\mathbf{L} = \mathbf{r}\times\, m\frac{d\mathbf{r}}{dt} \tag{A-22}$$

or

$$\mathbf{L} = \mathbf{r}\times\, m\mathbf{v}, \tag{A-23}$$

where $\mathbf{v}$ is the particle velocity.

Taking into account eq. A-12, we also have

$$\mathbf{L} = \mathbf{r}\times m(\boldsymbol{\omega} \times \mathbf{r}) \tag{A-24}$$

It is useful to express the angular momentum in terms of the momentum $\mathbf{P} = m\mathbf{v}$, and it gives

$$\mathbf{L} = \mathbf{r} \times \mathbf{P} \tag{A-25}$$

Suppose that the origin O is situated at the plane of motion. Then both vectors $\mathbf{r}$ and $d\mathbf{r}/dt$ are located in this plane, and they are perpendicular to each other. As follows from eq. A-24 the magnitude of the angular momentum is equal to

$$mrv = mr^2\omega = I\omega,$$

and $\mathbf{L}$ is directed along the z-axis. In the same manner we see that the torque $\boldsymbol{\tau}$ has the component τ_z only and equals rF_θ . Correspondingly, eq. A-21 is greatly simplified, and it is transformed into eq. A-5. Returning to the general case we see from eq. A-22, that the angular momentum $\mathbf{L}$ is perpendicular to the plane, formed by vectors $\mathbf{r}$ and $\mathbf{v}$. Therefore, as the torque $\boldsymbol{\tau}$, the vector $\mathbf{L}$ has normal and tangential components with respect to the plane of motion. During a rotation of the mass this vector forms the conical surface. Let us first assume that the component F_θ is equal to zero and, correspondingly, the mass moves with the constant velocity. In this case the centripetal force gives rise to the torque:

$$\boldsymbol{\tau} = \mathbf{r} \times \mathbf{F}_r$$

which is located in the horizontal plane. At the same time the end of the vector $\mathbf{L}$ moves along the circle, located in the horizontal plane too, Fig. A.2b, since the magnitude of $\mathbf{L}$ remains the same. For this reason the rate of a change of the angular momentum,

$d\mathbf{L}/dt$, as well as the torque, does not have a normal component, and it is tangential to the circle, shown in Fig. A.2b. From the geometry it is a simple matter to see that

$$\frac{d\mathbf{L}}{dt} = \boldsymbol{\omega} \times \mathbf{L} \tag{A-26}$$

In approaching the point O to the plane of motion the torque $\boldsymbol{\tau}$ becomes smaller and in the limit it disappears. In the presence of the component $\mathbf{F}_\theta$ the normal component of $\boldsymbol{\tau}$ differs from zero, and the velocity of the particle varies. Because of this the magnitude of $\mathbf{L}$ also changes, and the end of vector moves along a more complicated path. It is not located in the horizontal plane, so that both $\mathbf{L}$ and $\boldsymbol{\tau}$ have the normal and tangential components. In particular, eq. A-21 can be separately written for each component.

Example three Suppose that two masses, m_1 and m_2, move around the z-axis, and the distance between them does not change, Fig. A.2c. Applying eq. A-21 for each mass we have

$$\frac{d}{dt}(\mathbf{r}_1 \times m\,\mathbf{r}_1) = \mathbf{r}_1 \times \mathbf{F}_1 \qquad \text{and} \qquad \frac{d}{dt}(\mathbf{r}_2 \times m\,\mathbf{r}_2) = \mathbf{r}_2 \times \mathbf{F}_2 \tag{A-27}$$

Forces $\mathbf{F}_1$ and $\mathbf{F}_2$, acting on the masses, in general, consist of external and internal forces:

$$\mathbf{F}_1 = \mathbf{F}_{1e} + \mathbf{F}_{12}, \qquad \mathbf{F}_2 = \mathbf{F}_{2e} + \mathbf{F}_{21} \tag{A-28}$$

Since internal forces, $\mathbf{F}_{12}$ and $\mathbf{F}_{21}$, are unknown, eqs. A-27 cannot be solved with respect to $\mathbf{r}_1(t)$ and $\mathbf{r}_2(t)$. By analogy with a translation (Part I), we make use of Newton's third law:

$$\mathbf{F}_{12} = -\mathbf{F}_{21} \tag{A-29}$$

In fact, $\mathbf{F}_{12}$ is the force, caused by mass m_2 and it acts on m_1, while $\mathbf{F}_{21}$ is applied to m_2, and it is generated by m_1. Now performing summation of eqs. A-27 we obtain

$$\frac{d}{dt}(\mathbf{L}_1+\mathbf{L}_2) = \sum_{n=1}^{2} \mathbf{r}_n \times \mathbf{F}_{ne} + (\mathbf{r}_1 \times \mathbf{F}_{12} - \mathbf{r}_2 \times \mathbf{F}_{12}) = \sum_{n=1}^{2} \boldsymbol{\tau}_{ne} + (\mathbf{r}_1 - \mathbf{r}_2) \times \mathbf{F}_{12} \tag{A-30}$$

As is seen from Fig. A.2c, vectors

$$\mathbf{r}_1 - \mathbf{r}_2 \qquad \text{and} \qquad \mathbf{F}_{12}$$

have the same direction and therefore

$$(\mathbf{r}_1 - \mathbf{r}_2) \times \mathbf{F}_{12} = \mathbf{0}.$$

Hence,

$$\frac{d\mathbf{L}}{dt} = \boldsymbol{\tau}_e, \tag{A-31}$$

where

$$\mathbf{L} = \mathbf{L}_1 + \mathbf{L}_2$$

is the sum of angular moments of this system of masses and

$$\boldsymbol{\tau}_e = \boldsymbol{\tau}_{1e} + \boldsymbol{\tau}_{2e} \tag{A-32}$$

is the sum of torques, caused by external forces only. An importance of eq. A-31 is obvious, since it does not contain the unknown internal forces. We are now ready to discuss a general case.

Equation of rotation of a rigid body

To slightly simplify a derivation, we choose the center of mass as the origin of coordinates and assume that the resultant external force equals zero. In other words, this point is at rest and, therefore, it belongs to the axis of rotation. Let us represent the rigid body as a system of N elementary masses, and for each of them we have

$$\frac{d\mathbf{L}_n}{dt} = \boldsymbol{\tau}_n = \mathbf{r}_n \times \mathbf{F}_{en} + \mathbf{r}_n \times \sum_{k=1}^{N} \mathbf{F}_{kn} \qquad k \neq n \tag{A-33}$$

Here $\mathbf{L}_n$ is the angular momentum of mass m_n, $\mathbf{r}_n$ is the radius-vector of this particle, $\mathbf{F}_{en}$ is the external force acting on m_n, and finally

$$\sum_{k=1} \mathbf{F}_{kn} \qquad k \neq n$$

is the total internal force at this point. Similarly, for an arbitrary mass m_m

$$\frac{d\mathbf{L}_m}{dt} = \mathbf{r}_m \times \mathbf{F}_{em} + \left(\mathbf{r}_m \times \sum_{k=1}^{N} \mathbf{F}_{km} \right) \qquad k \neq m \tag{A-34}$$

Sums in these equations contain terms

$$\mathbf{r}_n \times \mathbf{F}_{nm} \qquad \text{and} \qquad \mathbf{r}_m \times \mathbf{F}_{mn}$$

respectively.

By analogy with the previous example consider the sum of these terms. The use of the Newton's third law gives

$$\mathbf{r}_n \times \mathbf{F}_{nm} + \mathbf{r}_m \times \mathbf{F}_{mn} = (\mathbf{r}_n - \mathbf{r}_m) \times \mathbf{F}_{nm} = \mathbf{0},$$

since both vectors have the same direction. Performing a summation of eqs. A.60, written for all elementary masses, we eliminate an influence of internal forces on the torque, and this procedure gives again:

$$\frac{d\mathbf{L}}{dt} = \boldsymbol{\tau}_e, \tag{A-35}$$

where

$$\mathbf{L} = \sum_{k=1}^{N} \mathbf{r}_k \times m_k \mathbf{v}_k = \sum \mathbf{r}_k \times m_k (\boldsymbol{\omega} \times \mathbf{r}_k) \tag{A-36}$$

is the total angular momentum of the rigid body and

$$\boldsymbol{\tau}_e = \sum \boldsymbol{\tau}_{ek} \tag{A-37}$$

is the resultant torque due to the external forces only. Of course, in the limit, when elementary masses tend to zero, a summation is replaced by an integration and it yields

$$\mathbf{L} = \int_V \mathbf{r} \times \rho \, \mathbf{v} \, dV \tag{A-38}$$

Here ρ is the density of the rigid body, and, in general, it may vary.

Moment of inertia

Inasmuch as the angular momentum $\mathbf{L}$ is defined as the double-cross-product, eq. A-36, it is natural to represent this vector in terms of components in the Cartesian system of coordinates. First, by definition:

$$\mathbf{v} = \boldsymbol{\omega} \times \mathbf{r} = \begin{vmatrix} \mathbf{i} & \mathbf{j} & \mathbf{k} \\ \omega_x & \omega_y & \omega_z \\ x & y & z \end{vmatrix}$$

or

$$v_x = \omega_y z - \omega_z y, \qquad v_y = \omega_z x - \omega_x z, \qquad v_z = \omega_x y - \omega_y x$$

Correspondingly

$$\mathbf{r}_k \times m_k \mathbf{v}_k = m_k \begin{vmatrix} \mathbf{i} & \mathbf{j} & \mathbf{k} \\ x_k & y_k & z_k \\ v_{xk} & v_{yk} & v_{zk} \end{vmatrix}$$

whence

$$L_x = \sum_{k=1}^{N} m_k (y_k v_{zk} - z_k v_{yk}) = \sum_{k=1}^{N} m_k [y_k(\omega_x y_k - \omega_y x_k) - z_k(\omega_z x_k - \omega_x z_k)]$$

$$= \sum_{k=1}^{N} m_k (y_k^2 + z_k^2)\omega_x - \sum_{k=1}^{N} m_k x_k y_k \omega_y - \sum_{k=1}^{N} m_k x_k z_k \omega_z$$

or

$$L_x = I_{xx}\omega_x + I_{xy}\omega_y + I_{xz}\omega_z \tag{A-39}$$

In the same manner we obtain

$$L_y = I_{yx}\omega_x + I_{yy}\omega_y + I_{yz}\omega_z \tag{A-40}$$

$$L_z = I_{zx}\omega_x + I_{zy}\omega_y + I_{zz}\omega_z, \tag{A-41}$$

where

$$I_{xx} = \sum m_k (y_k^2 + z_k^2), \quad I_{xy} = -\sum m_k x_k y_k, \quad I_{xz} = -\sum m_k x_k z_k \tag{A-42}$$

and

$$I_{yx} = -\sum m_k y_k x_k, \quad I_{yy} = \sum m_k (x_k^2 + z_k^2), \quad I_{yz} = -\sum m_k y_k z_k,$$

$$I_{zx} = -\sum m_k z_k x_k, \quad I_{zy} = -\sum m_k z_k y_k, \quad I_{zz} = \sum m_k (x_k^2 + y_k^2)$$

The set of these nine quantities is called the moment of inertia and it represent the symmetrical tensor, (Appendix B):

$$I = \begin{pmatrix} I_{xx} & I_{xy} & I_{xz} \\ I_{yx} & I_{yy} & I_{yz} \\ I_{zx} & I_{zy} & I_{zz} \end{pmatrix}, \tag{A-43}$$

where

$$I_{\alpha\beta} = I_{\beta\alpha} \tag{A-44}$$

and α, β are any two indices x, y, z. Taking into account the rule of multiplication of a tensor by a vector (Appendix B), eqs. A-38-A-40 can be written as

$$L_\alpha = \sum_\beta I_{\alpha\beta}\omega_\beta, \tag{A-45}$$

where $\alpha, \beta = x, y, z$.

The right hand side is called a "tensor product" of the tensor I and the vector $\boldsymbol{\omega}$, and this operation gives the vector $\mathbf{L}$. The diagonal components, I_{xx}, I_{yy}, I_{zz} of the tensor are often called the "moments of inertia", while nondiagonal components: $I_{xy} = I_{yx}$, $I_{yz} = I_{zy}$, $I_{zx} = I_{xz}$ are "products of inertia". It is clear that I depends on the dimensions and shape of the rigid body, as well as its density. Moreover, I varies with a change of the axis of rotation. It is also useful to represent eq. A-45 in the form

$$\begin{pmatrix} L_x \\ L_y \\ L_z \end{pmatrix} = \begin{pmatrix} I_{xx} & I_{xy} & I_{xz} \\ I_{yx} & I_{yy} & I_{yz} \\ I_{zx} & I_{zy} & I_{zz} \end{pmatrix} \begin{pmatrix} \omega_x \\ \omega_y \\ \omega_z \end{pmatrix} \tag{A-46}$$

As an example suppose that the axis of rotation coincides with the z-axis, that is

$$\omega_x = \omega_y = 0 \qquad \text{and} \qquad \boldsymbol{\omega} = \omega_z \, \mathbf{k},$$

Then, in place of eqs. A-39–A-41, we obtain

$$L_x = I_{xz}\omega_z, \qquad L_y = I_{yz}\omega_z, \qquad L_z = I_{zz}\omega_z \tag{A-47}$$

and, as we already know, this indicates that the vector of the angular momentum is not usually directed along the axis of rotation.

In accordance with eq. A.42 the diagonal term of the tensor I_{zz} is positive

$$I_{zz} = \int_V \rho \, (x^2 + y^2) dx dy dz \tag{A-48}$$

Correspondingly, the vector $\mathbf{L}$ has the component L_z along the axis of rotation. However, nondiagonal terms, that is "products of inertia" may be equal to zero. For example, it happens in the case of a homogeneous rigid body, which is symmetrical with respect to the axis of rotation, ($\boldsymbol{\omega} = \omega_z \mathbf{k}$). In fact, the "products of inertia"

$$I_{xz} = -\rho \int_V xz \; dxdydz, \qquad I_{yz} = -\rho \int_V yz \; dxdydz$$

vanish, because there are always pairs of masses with the same coordinate z, while the other coordinate, x or y, differ by sign only. Then, instead of eq. A-47, we have

$$L_z = I_{zz}\omega_z \qquad \text{or} \qquad L = I\omega$$

For illustration let us derive an expression for I_{zz} in several simple cases. As follows from eq. A-48, we have to perform an integration of masses, which are multiplied by the square of their distance, $x^2 + y^2$, from the axis of rotation.

Example one Consider the rod with a very small cross-section, which rotates around the z-axis through one end, Fig. A.2d. Then we have

$$I_{zz} = \rho \int x^2 dxdydz = \rho \; dydz \int_0^l x^2 dx = \frac{\rho \; dydz \; l^3}{3}$$

Thus

$$I_{zz} = \frac{M \; l^2}{3} \tag{A-49}$$

The moment of inertia is directly proportional to the rod mass M, and the square of its length, l. If the axis of rotation passes through the center of mass, we obtain

$$I_{zz} = \rho \; dydz \int_{-l/2}^{l/2} x^2 dx = \frac{Ml^2}{12} \tag{A-50}$$

Correspondingly, the moment of inertia becomes four times smaller.

Example two Next, we take the rectangular parallelepiped with sides, a, b, c, which rotates about the z-axis, Fig. A.2e. From eq. A-48 we have

$$I_{zz} = \rho \int_{-c/2}^{c/2} dz \int_{-a/2}^{a/2} dx \int_{-b/2}^{b/2} (x^2 + y^2) dy \tag{A-51}$$

$$= \rho c \int_{-a/2}^{a/2} dx \left(x^2 b + \frac{b^3}{12} \right) = \rho \; cb\frac{a^3}{12} + \rho ca\frac{b^3}{12} = \frac{M}{12}(a^2 + b^2)$$

In particular, when $a = b$

$$I_{zz} = \frac{Ma^2}{6} \tag{A-52}$$

Example three Consider a ring with radii r_1 and r_2, which rotates around the z-axis, Fig. A.2f. Introducing the angle ϕ

$$x = r\cos\phi, \qquad y = r\sin\phi,$$

we have for the elementary volume in the cylindrical system of coordinates

$$dV = r dr dz d\phi$$

Therefore

$$I_{zz} = \rho \int_{-h/2}^{h/2} dz \int_0^{2\pi} d\phi \int_{r_1}^{r_2} r^3 dr = 2\pi\rho\, h\frac{r_2^4 - r_1^4}{4} = \frac{2\pi\rho h}{4}(r_2^2 - r_1^2)(r_2^2 + r_1^2)$$

Inasmuch as the volume of this body is equal to

$$V = h\int_0^{2\pi} d\phi \int_{r_1}^{r_2} r dr = 2\pi h\frac{r_2^2 - r_1^2}{2},$$

we obtain

$$I_{zz} = \frac{M(r_1^2 + r_2^2)}{2} \tag{A-53}$$

For instance if $r_2 \approx r_1$:

$$I_{zz} = Mr^2, \tag{A-54}$$

where

$$r = \frac{r_1 + r_2}{2}$$

Some of these expressions of the moment of inertia will be used in deriving the wave equation for several special cases.

Equations of motion of a rigid body

Now we return to eq. A-35, which can be written as

$$\frac{d}{dt}(I\boldsymbol{\omega})= \boldsymbol{\tau}_e \tag{A-55}$$

If the axis of rotation does not change its orientation, then the tensor of inertia I remains constant, and, therefore, eq. A-55 becomes

$$I\ \boldsymbol{\alpha} = \boldsymbol{\tau}_\mathbf{e} \tag{A-56}$$

As was pointed out earlier, there is a complete analogy with the equation, describing a translation of the rigid body:

$$M\frac{d^2\mathbf{r}_0}{dt^2} = \mathbf{F}_e \tag{A-57}$$

where $\mathbf{r}_0$ is the radius-vector of the center of mass with respect to the origin of the fixed system of coordinates. For instance, knowing the total mass and the resultant force, we can determine a position of the center mass as a function of time, provided that the initial location and velocity at this point are given. In the same manner we can solve eq. A-55, when the direction of the rotation axis is fixed. In such a case, the tensor of inertia is defined by integration, and it allows us to determine components of the angular acceleration, as well as other kinematic parameters of motion. On the other hand, if the axis of rotation changes its orientation with time, an analogy with eq. A-55 ceases, since the moment of inertia varies with time too. Then, eq. A-55 can be written as

$$\frac{d}{dt}(I\boldsymbol{\omega}) = \frac{dI}{dt}\boldsymbol{\omega}+I\ \boldsymbol{\alpha} = \boldsymbol{\tau}_\mathbf{e}\ , \tag{A-58}$$

where both coefficients I and dI/dt are unknowns. Thus, we have shown a motion of the rigid body consists of a translation of the center of mass and a rotation around it. Respectively, there are two equations

$$M\frac{d^2\mathbf{r}_0}{dt^2} = \mathbf{F}_e, \qquad \frac{d}{dt}(I\boldsymbol{\omega}) = \boldsymbol{\tau}_e\ , \tag{A-59}$$

or, in the Cartesian system of coordinates:

$$M\frac{d^2\mathbf{x}_0}{dt^2} = \mathbf{F}_{ex}, \qquad M\frac{d^2\mathbf{y}_0}{dt^2} = \mathbf{F}_{ey}, \qquad M\frac{d^2\mathbf{z}_0}{dt^2} = \mathbf{F}_{ez}$$

and

$$\frac{d}{dt}(I_{xx}\omega_x + I_{xy}\omega_y + I_{xz}\omega_z) = \tau_{ex}$$

$$\frac{d}{dt}(I_{yx}\omega_x + I_{yy}\omega_y + I_{yz}\omega_z) = \tau_{ey} \tag{A-60}$$

$$\frac{d}{dt}(I_{zx}\omega_x + I_{zy}\omega_y + I_{zz}\omega_z) = \tau_{ez}$$

As the special case, assume that both the resultant force and torque are equal to zero. Then we have

$$M\frac{d^2\mathbf{r}_0}{dt^2} = 0, \qquad I\boldsymbol{\omega} = const \tag{A-61}$$

They are conditions of equilibrium of the rigid body, when it is either at rest or moves with a constant velocity.

Appendix B

Matrix Algebra and Tensors

B.1 Matrix Algebra

Before we introduce the concept of matrices, it may be useful to discuss transformation of vectors in the simplest case when they are situated on a plane. Further, it is assumed that operations with determinants and methods of solving systems of linear equations are known.

Transformation of two-dimensional vectors

Suppose that $\boldsymbol{\alpha}$ is an operator which transforms a two-dimensional vector $\mathbf{u}$ into another vector $\mathbf{v}$. This transformation can be written as

$$\boldsymbol{\alpha}\,\mathbf{u} = \mathbf{v} \tag{B-1}$$

It is called unique if such operation produces only one vector $\mathbf{v}$. The operator $\boldsymbol{\alpha}$ is called regular when it transforms different vectors $\mathbf{u}$ into different $\mathbf{v}$, that is if

$$\mathbf{u}_1 \neq \mathbf{u}_2,$$

then

$$\boldsymbol{\alpha}\,\mathbf{u}_1 \neq \boldsymbol{\alpha}\,\mathbf{u}_2$$

If the equalities

$$\boldsymbol{\alpha}(c\,\mathbf{u}) = c\,(\boldsymbol{\alpha}\,\mathbf{u}) \quad \text{and} \quad \boldsymbol{\alpha}(\mathbf{u}+\mathbf{v}) = \boldsymbol{\alpha}\,\mathbf{u} + \boldsymbol{\alpha}\,\mathbf{v} \tag{B-2}$$

take place, the transformation $\boldsymbol{\alpha}$ is called linear. Here c is an arbitrary constant. We restrict ourselves to linear transformations only. In general, the operator $\boldsymbol{\alpha}$ acting on vector $\mathbf{u}$ produces a change of both its magnitude and direction. Consider one special case when the resulting vector $\mathbf{v}$ has the same direction as that of $\mathbf{u}$. In other words, the transformation $\boldsymbol{\alpha}$ changes only the magnitude of the vector $\mathbf{u}$. This property can be written in the form

$$\boldsymbol{\alpha}\,\mathbf{u} = \lambda\,\mathbf{u} \tag{B-3}$$

Here λ is a number which characterizes the change in the vector length. Usually, λ is called a eigenvalue of the transformation $\boldsymbol{\alpha}$.

Summation of two operators

Suppose that two operators $\boldsymbol{\alpha}$ and $\boldsymbol{\beta}$ are applied to the same vector $\mathbf{u}$. By definition, each of them produces new vector

$$\boldsymbol{\alpha}\,\mathbf{u} = \mathbf{v}_1 \quad \text{and} \quad \boldsymbol{\beta}\,\mathbf{u} = \mathbf{v}_2 \tag{B-4}$$

Then, the sum

$$\boldsymbol{\alpha}\,\mathbf{u} + \boldsymbol{\beta}\,\mathbf{u} = \mathbf{v}_1 + \mathbf{v}_2 = \mathbf{v}_3$$

is denoted as

$$(\boldsymbol{\alpha} + \boldsymbol{\beta})\,\mathbf{u} \tag{B-5}$$

The operator $\boldsymbol{\alpha} + \boldsymbol{\beta}$, that transforms the vector $\mathbf{u}$ into $\mathbf{v}_3$, is called the sum of operators $\boldsymbol{\alpha}$ and $\boldsymbol{\beta}$.

Product of two operators

Consider vector $\mathbf{u}$ and assume that it is subjected to two transformations, $\boldsymbol{\alpha}$ and $\boldsymbol{\beta}$, which follow one another. The one operator gives the vector $\mathbf{v}_1$:

$$\boldsymbol{\beta}\,\mathbf{u} = \mathbf{v}_1 \tag{B-6}$$

Then, the second transformation, applied to $\mathbf{v}_1$, produces the vector $\mathbf{v}_2$:

$$\boldsymbol{\alpha}\,\mathbf{v}_1 = \boldsymbol{\alpha}(\boldsymbol{\beta}\,\mathbf{u}) = \mathbf{v}_2 \tag{B-7}$$

If it is possible to obtain the vector $\mathbf{v}_2$ directly from $\mathbf{u}$ using a single operator $\boldsymbol{\gamma}$, the latter is called the product of operators $\boldsymbol{\alpha}$ and $\boldsymbol{\beta}$:

$$\boldsymbol{\gamma}\,\mathbf{u} = \boldsymbol{\alpha}(\boldsymbol{\beta}\,\mathbf{u}) \tag{B-8}$$

Respectively, we have

$$\boldsymbol{\gamma} = \boldsymbol{\alpha}\,\boldsymbol{\beta} \quad \text{and} \quad \boldsymbol{\gamma}\mathbf{u} = \mathbf{v}_2 \tag{B-9}$$

It is essential to notice that usually the operators $\boldsymbol{\alpha\beta}$ and $\boldsymbol{\beta\alpha}$ are different, because the order in which the transformations are performed is important. The difference

$$\boldsymbol{\alpha}\,\boldsymbol{\beta} - \boldsymbol{\beta}\,\boldsymbol{\alpha}$$

is often called the commutator of operators $\boldsymbol{\alpha}$ and $\boldsymbol{\beta}$. If the commutator of two operators is equal to zero, such transformations are commutative, that is,

$$\boldsymbol{\alpha}\,\boldsymbol{\beta} = \boldsymbol{\beta}\,\boldsymbol{\alpha} \tag{B-10}$$

Representation of transformations in terms of matrices

The operator $\boldsymbol{\alpha}$ in eq. B-1 plays a rather symbolic role because it does not show explicitly the operations with $\mathbf{u}$ that produce the vector $\mathbf{v}$. In order to overcome this problem, we introduce the rectangular coordinate system with the unit vectors $\mathbf{i}_1$, $\mathbf{i}_2$ and consider the components of both vectors $\mathbf{u}$ and $\mathbf{v}$:

$$\mathbf{u} = u_1\,\mathbf{i}_1 + u_2\,\mathbf{i}_2\,, \quad \mathbf{v} = v_1\,\mathbf{i}_1 + v_2\,\mathbf{i}_2 \tag{B-11}$$

Let us express the components of the vector $\mathbf{v}$ as a linear combinations of components of the vector $\mathbf{u}$. Then we arrive at two equalities

$$\begin{aligned} v_1 &= \alpha_{11}\,u_1 + \alpha_{12}\,u_2 \\ v_2 &= \alpha_{21}\,u_1 + \alpha_{22}\,u_2 \end{aligned} \tag{B-12}$$

Here α_{ik} are some numbers. Thus, the operator of transformation $\boldsymbol{\alpha}$, eq. B-1, is characterized by a table of coefficients

$$\boldsymbol{\alpha} = \begin{bmatrix} \alpha_{11} & \alpha_{12} \\ \alpha_{21} & \alpha_{22} \end{bmatrix}, \tag{B-13}$$

which is called the matrix of transformation $\boldsymbol{\alpha}$. It is clear that knowing the matrix and using eqs. B-12 allows us to determine the vector $\mathbf{v}$. For illustration, consider three matrices.

Case one Suppose that the matrix $\boldsymbol{\alpha}$ is

$$\boldsymbol{\alpha} = \begin{bmatrix} \alpha_{11} & 0 \\ 0 & \alpha_{22} \end{bmatrix}$$

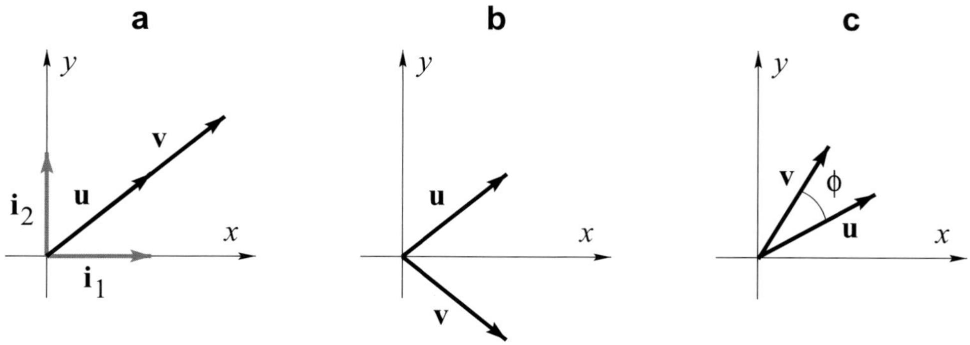

Figure B.1: Transformations of vectors.

Then, eqs. B-12 are greatly simplified and become

$$v_1 = \alpha_{11}\, u_1 \quad \text{and} \quad v_2 = \alpha_{22}\, u_2$$

In particular, when

$$\alpha_{11} = \alpha_{22} = a\,, \tag{B-14}$$

the matrix has the form

$$\boldsymbol{\alpha} = \begin{bmatrix} a & 0 \\ 0 & a \end{bmatrix},$$

and both vectors **u** and **v** have the same direction, Fig. B.1a.

Case two Consider the matrix

$$\boldsymbol{\alpha} = \begin{bmatrix} 1 & 0 \\ 0 & -1 \end{bmatrix}$$

As follows from eqs. B-12, the relationships between components of both vectors are

$$v_1 = u_1 \quad \text{and} \quad v_2 = -u_2 \tag{B-15}$$

Therefore, the matrix $\boldsymbol{\alpha}$ produces the vector **v** which is a mirror reflection of **u** with respect to the horizontal coordinate axis, Fig. B.1b.

Case three Now we assume that due to the transformation the unit vector **u** was rotated counter-clockwise by the angle ϕ, so the vector **v** is also the unit vector, Fig. B.1c. From the definition of the directional cosines we have

$$\mathbf{u} = \cos\alpha\, \mathbf{i}_1 + \cos\beta\, \mathbf{i}_2 \tag{B-16}$$

and

$$\mathbf{v} = \cos(\alpha + \phi)\,\mathbf{i}_1 + \cos(\beta + \phi)\,\mathbf{i}_2 \tag{B-17}$$

Therefore,

$$\begin{aligned} u_1 &= \cos\alpha\,, & u_2 &= \cos\beta\,, \\ v_1 &= \cos(\alpha + \phi)\,, & v_2 &= \cos(\beta + \phi), \end{aligned} \tag{B-18}$$

whence

$$v_1 = \cos\phi\,\cos\alpha - \sin\phi\,\sin\alpha$$

and

$$v_2 = \cos\phi\,\cos\beta - \sin\phi\,\sin\beta$$

Since

$$\beta = \frac{\pi}{2} - \alpha,$$

and taking into account eqs. B-18, we obtain

$$v_1 = \cos\phi\,u_1 - \sin\phi\,u_2$$

and

$$v_2 = \sin\phi\,u_1 + \cos\phi\,u_2$$

Comparison with eqs. B-12 shows that the transformation matrix is

$$\boldsymbol{\alpha} = \begin{bmatrix} \cos\phi & -\sin\phi \\ \sin\phi & \cos\phi \end{bmatrix}$$

Now we discuss some operations with matrices.

Summation of two matrices

Consider two transformations $\boldsymbol{\alpha}$ and $\boldsymbol{\beta}$, which are applied to the same vector $\mathbf{u}$. As a result, they give the vectors $\mathbf{v}$ and $\mathbf{w}$, respectively. The relationships between the vectors are

$$v_1 = \alpha_{11}\,u_1 + \alpha_{12}\,u_2\,, \tag{B-19}$$

$$v_2 = \alpha_{21}\, u_1 + \alpha_{22}\, u_2$$

and

$$w_1 = \beta_{11}\, u_1 + \beta_{12}\, u_2\,, \tag{B-20}$$

$$w_2 = \beta_{21}\, u_1 + \beta_{22}\, u_2$$

Summation of the corresponding components yields

$$v_1 + w_1 = (\alpha_{11} + \beta_{11})\, u_1 + (\alpha_{12} + \beta_{12})\, u_2\,,$$

(B-21)

$$v_2 + w_2 = (\alpha_{21} + \beta_{21})\, u_1 + (\alpha_{22} + \beta_{22})\, u_2$$

Therefore, in order to obtain the sum

$$\mathbf{v} + \mathbf{w}\,,$$

we can apply a single transformation, represented by the matrix $\boldsymbol{\gamma}$, with elements that are the sums of the corresponding elements in the matrices $\boldsymbol{\alpha}$ and $\boldsymbol{\beta}$.

The matrix $\boldsymbol{\gamma}$ is called the sum of matrices $\boldsymbol{\alpha}$ and $\boldsymbol{\beta}$:

$$\boldsymbol{\gamma} = \boldsymbol{\alpha} + \boldsymbol{\beta} \quad \text{and} \quad \mathbf{v} + \mathbf{w} = \boldsymbol{\gamma}\,\mathbf{u} \tag{B-22}$$

For example,

$$\begin{bmatrix} 1 & 2 \\ 4 & 5 \end{bmatrix} + \begin{bmatrix} 1 & -2 \\ 3 & -6 \end{bmatrix} = \begin{bmatrix} 2 & 0 \\ 7 & -1 \end{bmatrix}$$

Note that the rule of summation of matrices is the same as that for vectors, and this relation remains valid for several other operations.

Summation of matrices and multiplications by a number

Until now, we have considered a summation of two matrices. It is obvious that the same approach is applicable in the general case, when the number of matrices is arbitrary. Then, any jkth element of the matrix

$$\boldsymbol{\gamma} = \boldsymbol{\alpha}_1 + \boldsymbol{\alpha}_2 + \boldsymbol{\alpha}_3 + \ldots + \boldsymbol{\alpha}_n \tag{B-23}$$

is also a sum of the jkth elements of the matrices $\boldsymbol{\alpha}_i$. In particular, if all the matrices are equal, $\boldsymbol{\alpha}_i = \boldsymbol{\alpha}$, we have

$$\boldsymbol{\gamma} = n\,\boldsymbol{\alpha}\,, \tag{B-24}$$

For instance,

$$7\begin{bmatrix} 1 & -1 \\ 2 & 0 \end{bmatrix} = \begin{bmatrix} 7 & -7 \\ 14 & 0 \end{bmatrix}$$

Multiplication of two matrices

Suppose we perform two transformations characterized by the matrices

$$\boldsymbol{\alpha} = \begin{bmatrix} \alpha_{11} & \alpha_{12} \\ \alpha_{21} & \alpha_{22} \end{bmatrix} \qquad \text{and} \qquad \boldsymbol{\beta} = \begin{bmatrix} \beta_{11} & \beta_{12} \\ \beta_{21} & \beta_{22} \end{bmatrix}$$

Applying the transformation $\boldsymbol{\beta}$ to the vector $\mathbf{u}$, we obtain the vector $\mathbf{v}$:

$$\mathbf{v} = \boldsymbol{\beta}\,\mathbf{u} \tag{B-25}$$

Then, making use of the second transformation, we arrive at the vector $\mathbf{w}$:

$$\mathbf{w} = \boldsymbol{\alpha}\,\mathbf{v} = \boldsymbol{\alpha}\,(\boldsymbol{\beta}\,\mathbf{u}) \tag{B-26}$$

Now, let us determine the matrix $\boldsymbol{\gamma}$, which transforms the vector $\mathbf{u}$ into $\mathbf{w}$ directly.

By definition,

$$\begin{aligned} v_1 &= \beta_{11}\,u_1 + \beta_{12}\,u_2\,, \\ v_2 &= \beta_{21}\,u_1 + \beta_{22}\,u_2 \end{aligned} \tag{B-27}$$

and

$$\begin{aligned} w_1 &= \alpha_{11}\,v_1 + \alpha_{12}\,v_2\,, \\ w_2 &= \alpha_{21}\,v_1 + \alpha_{22}\,v_2 \end{aligned} \tag{B-28}$$

Substitution of eqs. B-27 into eqs. B-28 yields

$$\begin{aligned} w_1 &= (\alpha_{11}\,\beta_{11} + \alpha_{12}\,\beta_{21})\,u_1 + (\alpha_{11}\,\beta_{12} + \alpha_{12}\,\beta_{22})\,u_2\,, \\ w_2 &= (\alpha_{21}\,\beta_{11} + \alpha_{22}\,\beta_{21})\,u_1 + (\alpha_{21}\,\beta_{12} + \alpha_{22}\,\beta_{22})\,u_2 \end{aligned} \tag{B-29}$$

Thus, the matrix of the transformation

$$\mathbf{w} = \boldsymbol{\gamma}\,\mathbf{u} \tag{B-30}$$

is defined as

$$\begin{bmatrix} \gamma_{11} & \gamma_{12} \\ \gamma_{21} & \gamma_{22} \end{bmatrix} = \begin{bmatrix} \alpha_{11}\,\beta_{11} + \alpha_{12}\,\beta_{21} & \alpha_{11}\,\beta_{12} + \alpha_{12}\,\beta_{22} \\ \alpha_{21}\,\beta_{11} + \alpha_{22}\,\beta_{21} & \alpha_{21}\,\beta_{12} + \alpha_{22}\,\beta_{22} \end{bmatrix} \tag{B-31}$$

The matrix $\boldsymbol{\gamma}$ is called the product of two matrices $\boldsymbol{\alpha}$ and $\boldsymbol{\beta}$.

It is easy to recognize here the rule of formation of elements of the product of two determinants. To illustrate this operation, consider the following example:

$$\boldsymbol{\alpha} = \begin{bmatrix} 1 & 0 \\ 0 & -1 \end{bmatrix} \qquad \text{and} \qquad \boldsymbol{\beta} = \begin{bmatrix} 0 & 1 \\ -1 & 0 \end{bmatrix}$$

The product $\boldsymbol{\gamma} = \boldsymbol{\alpha}\,\boldsymbol{\beta}$ is

$$\boldsymbol{\gamma} = \begin{bmatrix} 1\cdot 0 + 0\cdot(-1) & 1\cdot 1 + 0\cdot 0 \\ 0\cdot 0 + (-1)\cdot(-1) & 0\cdot 1 + (-1)\cdot 0 \end{bmatrix} = \begin{bmatrix} 0 & 1 \\ 1 & 0 \end{bmatrix}$$

but

$$\boldsymbol{\beta}\,\boldsymbol{\alpha} = \begin{bmatrix} 0\cdot 1 + 1\cdot 0 & 0\cdot 0 + 1\cdot(-1) \\ (-1)\cdot 1 + 0\cdot 0 & (-1)\cdot 0 + 0\cdot(-1) \end{bmatrix} = \begin{bmatrix} 0 & -1 \\ -1 & 0 \end{bmatrix},$$

that is, the operators $\boldsymbol{\alpha}\,\boldsymbol{\beta}$ and $\boldsymbol{\beta}\,\boldsymbol{\alpha}$ are not equal to each other.

From eq. B-31 it is a simple matter to derive a rule for obtaining the product $\boldsymbol{\alpha}\,\boldsymbol{\beta}$. Every element of the row of the matrix $\boldsymbol{\alpha}$ is multiplied by the element of the column of the matrix $\boldsymbol{\beta}$, and those products are summed together. For instance, the element γ_{12} of the product is located on the intersection of the first row and the second column. It is the sum of products of the first element of the first row of the matrix $\boldsymbol{\alpha}$ times the first element of the second column of $\boldsymbol{\beta}$, and the second element of the same row of $\boldsymbol{\alpha}$ times the second element of the second column of $\boldsymbol{\beta}$, eq. B-31. We have considered the summation or subtraction and the multiplication of matrices, describing linear transformations of a plane.

Equality of two matrices

Two matrices $\boldsymbol{\alpha}$ and $\boldsymbol{\beta}$ are equal to each other if the transformation, defined by them and applied to an arbitrary vector $\mathbf{u}$, gives the same result. By definition, we have

$$\alpha_{11}\,u_1 + \alpha_{12}\,u_2 = \beta_{11}\,u_1 + \beta_{12}\,u_2$$

and

$$\alpha_{21}\,u_1 + \alpha_{22}\,u_2 = \beta_{21}\,u_1 + \beta_{22}\,u_2$$

Since the vector $\mathbf{u}$ is arbitrary, it is natural to conclude that two matrices coincide if their corresponding elements are equal.

Matrix representation for a vector

It is convenient to write the vector $\mathbf{u}$ in the form of matrix

$$\begin{bmatrix} u_1 \\ u_2 \end{bmatrix} \tag{B-32}$$

Certainly, the latter can be now treated as an operator. For instance, the summation of two vectors can be represented as

$$\begin{bmatrix} u_1 \\ u_2 \end{bmatrix} + \begin{bmatrix} v_1 \\ v_2 \end{bmatrix} = \begin{bmatrix} u_1 + v_1 \\ u_2 + v_2 \end{bmatrix}, \tag{B-33}$$

that is, the components of the total vector are sums of the corresponding components of $\mathbf{u}$ and $\mathbf{v}$.

Also a transformation of the vector $\mathbf{u}$ by the matrix $\boldsymbol{\alpha}$ can be written as the product of two matrices

$$\begin{bmatrix} \alpha_{11} & \alpha_{12} \\ \alpha_{21} & \alpha_{22} \end{bmatrix} \begin{bmatrix} u_1 \\ u_2 \end{bmatrix} = \begin{bmatrix} v_1 \\ v_2 \end{bmatrix} \tag{B-34}$$

Applying the rule of multiplication of matrices, we obtain again

$$v_1 = \alpha_{11}\,u_1 + \alpha_{12}\,u_2$$

$$v_2 = \alpha_{21}\,u_1 + \alpha_{22}\,u_2$$

Matrices in n-dimensional space

Our study of operations with matrices in the two-dimensional case is easily generalized to n-dimensional space. Suppose there is a vector $\mathbf{u}$ with components

$$u_1,\ u_2,\ \ldots,\ u_n$$

Then, a group of linear relationships

$$v_i = \sum_k \alpha_{ik}\, u_k \,, \tag{B-35}$$

where $i = 1, 2, \ldots, n$, describes the transformation from vector $\mathbf{u}$ to the new vector $\mathbf{v}$. The matrix of transformation has the form

$$\begin{bmatrix} \alpha_{11} & \cdots & \alpha_{1n} \\ \vdots & & \vdots \\ \alpha_{n1} & \cdots & \alpha_{nn} \end{bmatrix} \tag{B-36}$$

For instance, the vector $\mathbf{u}$ can be also written as the matrix which consists of one column

$$\begin{bmatrix} u_1 \\ u_2 \\ \vdots \\ u_{n-1} \\ u_n \end{bmatrix}$$

We denote the matrix with elements α_{ik} as

$$[\alpha_{ik}]$$

and also use a simplified notation $\boldsymbol{\alpha}$.

Equality of two matrices

As before, from the equality

$$v_i = w_i \quad \text{or} \quad \sum_k \alpha_{ik}\, u_k = \sum_k \beta_{ik}\, u_k \tag{B-37}$$

it follows that

$$\alpha_{ik} = \beta_{ik}$$

In other words, two matrices are equal,

$$[\alpha_{ik}] = [\beta_{ik}] \,, \tag{B-38}$$

if they define the transformations that produce the same vector.

Summation of two matrices

Now we assume that the n-component vector $\mathbf{u}$ is independently subjected to two transformations. Correspondingly, the components of new vectors $\mathbf{v}$ and $\mathbf{w}$ are

$$v_i = \sum_k \alpha_{ik}\, u_k \quad \text{or} \quad \mathbf{v} = \boldsymbol{\alpha}\, \mathbf{u} \tag{B-39}$$

and

$$w_i = \sum_k \beta_{ik}\, u_k \quad \text{or} \quad \mathbf{w} = \boldsymbol{\beta}\, \mathbf{u} \tag{B-40}$$

Their summation gives

$$v_i + w_i = \sum_k (\alpha_{ik} + \beta_{ik})\, u_k = \sum_k \gamma_{ik}\, u_k\,,$$

where

$$\gamma_{ik} = \alpha_{ik} + \beta_{ik} \tag{B-41}$$

or

$$\mathbf{v} + \mathbf{w} = \boldsymbol{\gamma}\, \mathbf{u} \tag{B-42}$$

Thus, the sum of vectors $\mathbf{v}$ and $\mathbf{w}$ can be obtained by one transformation, which is characterized by the matrix $[\gamma_{ik}]$, and its elements are the sums of proper elements of the matrices $\boldsymbol{\alpha}$ and $\boldsymbol{\beta}$, eq. B-41. This procedure is defined only for matrices that have the same number of rows and columns. For example,

$$\begin{bmatrix} 1 & 2 & 3 \\ 4 & 5 & 6 \end{bmatrix} + \begin{bmatrix} 1 & -2 & 0 \\ 3 & -6 & -10 \end{bmatrix} = \begin{bmatrix} 2 & 0 & 3 \\ 7 & -1 & -4 \end{bmatrix}$$

Summation of an arbitrary number of matrices is performed in the same manner. In particular, if there are ℓ equal matrices $\boldsymbol{\alpha}$, then, the result of their summation is the matrix $\boldsymbol{\gamma}$ such that

$$\gamma_{ik} = \ell\, \alpha_{ik} \quad \text{or} \quad \boldsymbol{\gamma} = \ell\, \boldsymbol{\alpha} \tag{B-43}$$

Multiplication of two matrices

Consider two sequential transformations of the vector $\mathbf{u}$. By definition, we have

$$v_l = \sum_k \beta_{lk} \, u_k \quad \text{and} \quad w_j = \sum_l \alpha_{jl} \, v_l \tag{B-44}$$

Now with the help of the transformation, defined by the matrix $\boldsymbol{\gamma}$, we obtain the vector $\mathbf{w}$ directly from $\mathbf{u}$.

Combining equations of the set B-44 yields

$$w_j = \sum_k \left(\sum_l \alpha_{jl} \, \beta_{lk} \right) u_k \tag{B-45}$$

Comparison with the relationship

$$w_j = \sum_k \gamma_{jk} \, u_k \tag{B-46}$$

allows us to establish the rule of multiplication of two matrices:

$$\gamma_{jk} = \sum_l \alpha_{jl} \, \beta_{lk} \tag{B-47}$$

or, in the shortened form,

$$\boldsymbol{\gamma} = \boldsymbol{\alpha} \, \boldsymbol{\beta} \tag{B-48}$$

Therefore, the elements of the matrix $\boldsymbol{\gamma}$ located at the intersection of the row j and the column k is obtained in the following way. We multiply the first term of the row j of the matrix $\boldsymbol{\alpha}$ by the first term of the column k of matrix $\boldsymbol{\beta}$, add the similar products of the second terms, then, the third ones and so on, as schematically shown in Fig. B.2. We see that the rule of multiplication of matrices describing linear transformations has a general character.

It may be proper to notice that applying the rule of multiplication of matrices, the transformation

$$\mathbf{v} = \boldsymbol{\alpha} \, \mathbf{u}$$

can be written as

$$\begin{bmatrix} v_1 \\ v_2 \\ \vdots \\ v_n \end{bmatrix} = \begin{bmatrix} \alpha_{11} & \alpha_{12} & \cdots & \alpha_{1n} \\ \vdots & & & \vdots \\ \vdots & & & \vdots \\ \alpha_{n1} & \alpha_{n2} & \cdots & \alpha_{nn} \end{bmatrix} \begin{bmatrix} u_1 \\ u_2 \\ \vdots \\ u_n \end{bmatrix} \tag{B-49}$$

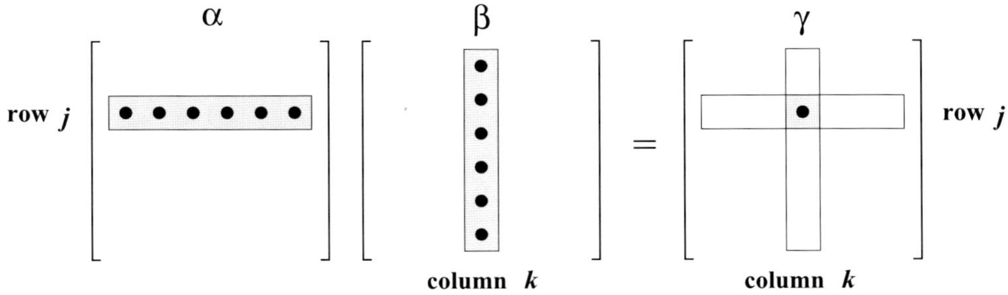

Figure B.2: Multiplication of matrices.

As follows from eq. B-48 the determinant of the matrix product is equal to the product of determinants, that is, if

$$\boldsymbol{\alpha}\,\boldsymbol{\beta} = \boldsymbol{\gamma}\,,$$

then,

$$|\boldsymbol{\alpha}|\,|\boldsymbol{\beta}| = |\boldsymbol{\gamma}| \tag{B-50}$$

Of course, the latter has meaning only in the case of square matrices.

It is obvious that one can form a product of two matrices if they are square or have just one row and one column. Also, this operation is possible for rectangular matrices, provided that the number of columns of the first matrix is equal to the number of rows of the second one.

Consider several types of matrices.

Symmetric and antisymmetric matrices

Suppose that the matrix $\boldsymbol{\alpha}$ is square:

$$\boldsymbol{\alpha} = \begin{bmatrix} \alpha_{11} & \alpha_{12} & \cdots & \alpha_{1k} & \cdots & \alpha_{1n} \\ \vdots & \vdots & & \vdots & & \vdots \\ \alpha_{i1} & \alpha_{i2} & \cdots & \alpha_{ik} & \cdots & \alpha_{in} \\ \vdots & \vdots & & \vdots & & \vdots \\ \alpha_{n1} & \alpha_{n2} & \cdots & \alpha_{nk} & \cdots & \alpha_{nn} \end{bmatrix}$$

Its element α_{ik} is situated on the intersection of the row i and the column k. The elements $\alpha_{11}, \alpha_{22}, \ldots, \alpha_{nn}$, located along the downgoing diagonal from the left to right, form the main diagonal of the matrix $\boldsymbol{\alpha}$. If the elements, situated symmetrically with respect to the main diagonal, are equal to each other,

$$\alpha_{ik} = \alpha_{ki}\,, \tag{B-51}$$

the matrix is called symmetric. When such elements have the same magnitudes but the opposite signs,

$$\alpha_{ik} = -\alpha_{ki}\,, \tag{B-52}$$

the matrix is antisymmetric. All its diagonal elements then are equal to zero.

Diagonal matrices

A matrix with all off-diagonal elements equal to zero

$$\alpha_{ik} = 0 \quad \text{if} \quad i \neq k \tag{B-53}$$

is called diagonal. It is easy to see that the product of two diagonal matrices is also a diagonal matrix.

Identity and zero matrices

The identity matrix $[\mathbf{1}]$ is a special case of diagonal matrices whose diagonal elements are all equal to one. It plays the same role in matrix algebra as the number 1 in algebra of numbers. The identity matrix transforms any vector into itself

$$[\mathbf{1}]\,\mathbf{u} = \mathbf{u} \tag{B-54}$$

The zero matrix $[\mathbf{0}]$ is such that all its elements are zeros. It is equivalent to the number 0 in algebra.

Order and rank of matrix

Suppose a square matrix has n rows and columns. Then, n is called the order of this matrix. A minor is a determinant obtained by removing from the matrix the same number of rows and columns. Consider the case, when all minors of the order higher than r, which can be formed from the matrix, are equal to zero. At the same time, there is at least one minor of the order r which differs from zero. Then the number r is called the rank of this matrix. In other words, the rank of the matrix is equal to the highest order of its minors (determinants) which are not equal to zero.

Let us notice that the matrix is called singular when its determinant is zero. For instance, the matrix

$$\begin{bmatrix} 3 & 2 & 1 & -1 \\ 9 & 6 & 3 & -3 \\ -6 & -4 & -2 & 2 \\ 0 & 0 & 0 & 0 \end{bmatrix}$$

is singular. Its order is 4 but the rank is equal to 1.

Transposed matrix

Let us replace the rows of the matrix $\boldsymbol{\alpha}$ with its columns. Then, the new matrix $\tilde{\boldsymbol{\alpha}}$, where

$$\tilde{\alpha}_{ik} = \alpha_{ki} \tag{B-55}$$

is called the transposed of $\boldsymbol{\alpha}$. For example, the matrix

$$\tilde{\boldsymbol{\alpha}} = \begin{bmatrix} 2 & 3 & 7 \\ 4 & 2 & 4 \\ 5 & 8 & 1 \end{bmatrix}$$

is the transpos of the matrix

$$\boldsymbol{\alpha} = \begin{bmatrix} 2 & 4 & 5 \\ 3 & 2 & 8 \\ 7 & 4 & 1 \end{bmatrix}$$

It is easy to see that the matrix, transposed to the product $\boldsymbol{\alpha}\,\boldsymbol{\beta}$, is equal to the product of the transposed matrices $\tilde{\boldsymbol{\alpha}}$ and $\tilde{\boldsymbol{\beta}}$ taken in the opposite order

$$(\widetilde{\boldsymbol{\alpha}\,\boldsymbol{\beta}}) = \tilde{\boldsymbol{\beta}}\,\tilde{\boldsymbol{\alpha}} \tag{B-56}$$

In fact, the matrix $(\widetilde{\boldsymbol{\alpha}\,\boldsymbol{\beta}})$ is obtained first by multiplication of row elements of $\boldsymbol{\alpha}$ by column elements of $\boldsymbol{\beta}$ and then by replacement of its rows by columns. The same result follows if we multiply the column elements of $\boldsymbol{\beta}$, that is, the rows of $\tilde{\boldsymbol{\beta}}$ by the row elements of $\boldsymbol{\alpha}$, that is, the columns of $\tilde{\boldsymbol{\alpha}}$. The same rule is also applied to a product of any number of matrices

$$(\widetilde{\boldsymbol{\alpha}\boldsymbol{\beta}\ldots\boldsymbol{\omega}}) = \tilde{\boldsymbol{\omega}}\ \ldots\ \tilde{\boldsymbol{\beta}}\,\tilde{\boldsymbol{\alpha}} \tag{B-57}$$

Dot product of two vectors

Consider two vectors

$$\mathbf{u} = u_1\,\mathbf{i}_1 + u_2\,\mathbf{i}_2 + \ldots + u_n\,\mathbf{i}_n$$

and

$$\mathbf{v} = v_1\,\mathbf{i}_1 + v_2\,\mathbf{i}_2 + \ldots + v_n\,\mathbf{i}_n$$

Then, by definition of the dot product, we have

$$\mathbf{u}\cdot\mathbf{v} = u_1\,v_1 + u_2\,v_2 + \ldots + u_n\,v_n \qquad \text{(B-58)}$$

If these vectors are given in the form of matrices

$$\begin{bmatrix} u_1 \\ u_2 \\ \vdots \\ u_n \end{bmatrix} \quad \text{and} \quad \begin{bmatrix} v_1 \\ v_2 \\ \vdots \\ v_n \end{bmatrix},$$

the dot product can be written as

$$\mathbf{u}\cdot\mathbf{v} = [u_1\ u_2\ \ldots\ u_n] \begin{bmatrix} v_1 \\ v_2 \\ \vdots \\ v_n \end{bmatrix}$$

or

$$\mathbf{u}\cdot\mathbf{v} = \tilde{\mathbf{u}}\,\mathbf{v} \qquad \text{(B-59)}$$

In particular, the square of the magnitude of the vector $\mathbf{u}$ is equal to

$$u_1^2 + u_2^2 + \ldots + u_n^2 = \tilde{\mathbf{u}}\,\mathbf{u} \qquad \text{(B-60)}$$

Inverse matrix

Suppose that the matrix $\boldsymbol{\alpha}$ transforms the vector $\mathbf{u}$ into vector $\mathbf{v}$

$$\mathbf{v} = \boldsymbol{\alpha}\,\mathbf{u}$$

This means that there are n linear relationships

$$v_i = \sum_k \alpha_{ik} u_k \tag{B-61}$$

Here $i = 1, 2, \ldots, n$.

Let us find the matrix $\boldsymbol{\beta}$ that determines the inverse transformation from the vector $\mathbf{v}$ to the original one

$$\mathbf{u} = \boldsymbol{\beta}\,\mathbf{v} \tag{B-62}$$

In other words, we want to find the table of coefficients β_{jl} for the linear relationships

$$u_j = \sum_k \beta_{jl} v_l \tag{B-63}$$

and $j = 1, 2, \ldots, n$. The matrix $\boldsymbol{\beta}$ is called the inverse of $\boldsymbol{\alpha}$ and is denoted as $\boldsymbol{\alpha}^{-1}$. The coefficients β_{jl} can be obtained from the system B-61, solving it with respect to u_1, u_2, $\ldots$, u_n. For instance, they can be found using the formula

$$\beta_{jl} = \frac{A_{lj}}{\Delta} \tag{B-64}$$

known as the Cramer rule. Here Δ is the determinant of the matrix $\boldsymbol{\alpha}$ and A_{lj} is the algebraic addition of the element α_{lj}. In other words, A_{lj} is the determinant obtained by removing the row l and the column j from the matrix $\boldsymbol{\alpha}$. Besides, this determinant is multiplied by the term $(-1)^{l+j}$. Calculation of the matrix $\boldsymbol{\alpha}^{-1}$ usually may consist of the following steps.

1. Matrix $\tilde{\boldsymbol{\alpha}}$, transposed to $\boldsymbol{\alpha}$, is written down.

2. Every element of $\tilde{\boldsymbol{\alpha}}$ is replaced by the determinant, which is obtained by removing the row and column where the given element is situated.

3. The sign of this determinant is changed to the opposite if the sum of indexes $j + l$ is odd.

4. The last matrix is divided by Δ.

As an example, consider the matrix

$$\boldsymbol{\alpha} = \begin{bmatrix} 1 & 2 & 3 \\ 4 & 5 & 6 \\ 2 & 8 & 9 \end{bmatrix}$$

The transposed matrix is

$$\tilde{\boldsymbol{\alpha}} = \begin{bmatrix} 1 & 4 & 2 \\ 2 & 5 & 8 \\ 3 & 6 & 9 \end{bmatrix}$$

Next we replace each element with the determinant obtained by removing the corresponding row and column. This gives

$$\begin{bmatrix} -3 & -6 & -3 \\ 24 & 3 & -6 \\ 22 & 4 & -3 \end{bmatrix}$$

Then we change the signs of the elements with the odd sums of indexes and obtain

$$\begin{bmatrix} -3 & 6 & -3 \\ -24 & 3 & 6 \\ 22 & -4 & -3 \end{bmatrix}$$

Finally, the division of the last matrix by $\Delta = 15$ produces the inverse matrix

$$\tilde{\boldsymbol{\alpha}}^{-1} = \begin{bmatrix} -1/5 & 2/5 & -1/5 \\ -8/5 & 1/5 & 2/5 \\ 22/15 & -4/15 & -1/5 \end{bmatrix}$$

Calculating the inverse matrix, we assume the original matrix is not singular and, correspondingly, the determinant Δ differs from zero. In the opposite case the inverse matrix does not exist. For instance, this happens if the matrix $\tilde{\boldsymbol{\alpha}}$ is not square. In fact, such a matrix can be made square by adding a certain number of zeros but then its determinant also becomes zero.

As follows from the Cramer rule, the inverse of a diagonal matrix is also diagonal with the elements which are reciprocal to the elements of the given matrix. For instance, the inverse of

$$\boldsymbol{\alpha} = \begin{bmatrix} a & 0 & 0 & 0 \\ 0 & b & 0 & 0 \\ 0 & 0 & c & 0 \\ 0 & 0 & 0 & d \end{bmatrix}$$

is

$$\boldsymbol{\alpha}^{-1} = \begin{bmatrix} 1/a & 0 & 0 & 0 \\ 0 & 1/b & 0 & 0 \\ 0 & 0 & 1/c & 0 \\ 0 & 0 & 0 & 1/d \end{bmatrix}$$

By definition, we have

$$\mathbf{v} = \boldsymbol{\alpha}\,\mathbf{u} \quad \text{and} \quad \mathbf{u} = \boldsymbol{\alpha}^{-1}\,\mathbf{v},$$

whence

$$\mathbf{v} = \left(\boldsymbol{\alpha}\,\boldsymbol{\alpha}^{-1}\right)\,\mathbf{v}\,, \tag{B-65}$$

that is, the product of a matrix and its inverse is always equal to the identity matrix

$$\boldsymbol{\alpha}\,\boldsymbol{\alpha}^{-1} = [\mathbf{1}] \tag{B-66}$$

Comparison of operations with matrices and determinants

As was shown earlier, the rule of multiplication of matrices is the same as that for the determinants. However, this is the only case when the rules coincide. In particular, the rules of multiplication by a number and summation of matrices and determinants are different. In fact,

$$\ell \begin{bmatrix} a_1 & b_1 & c_1 \\ a_2 & b_2 & c_2 \\ a_3 & b_3 & c_3 \end{bmatrix} = \begin{bmatrix} \ell\, a_1 & \ell\, b_1 & \ell\, c_1 \\ \ell\, a_2 & \ell\, b_2 & \ell\, c_2 \\ \ell\, a_3 & \ell\, b_3 & \ell\, c_3 \end{bmatrix}$$

whereas

$$\ell \begin{vmatrix} a_1 & b_1 & c_1 \\ a_2 & b_2 & c_2 \\ a_3 & b_3 & c_3 \end{vmatrix} = \begin{vmatrix} \ell\, a_1 & b_1 & c_1 \\ \ell\, a_2 & b_2 & c_2 \\ \ell\, a_3 & b_3 & c_3 \end{vmatrix}$$

Also

$$\begin{bmatrix} a_1 & b_1 & c_1 \\ a_2 & b_2 & c_2 \\ a_3 & b_3 & c_3 \end{bmatrix} + \begin{bmatrix} a_1' & b_1 & c_1 \\ a_2' & b_2 & c_2 \\ a_3' & b_3 & c_3 \end{bmatrix} = \begin{bmatrix} a_1 + a_1' & 2\,b_1 & 2\,c_1 \\ a_2 + a_2' & 2\,b_2 & 2\,c_2 \\ a_3 + a_3' & 2\,b_3 & 2\,c_3 \end{bmatrix}$$

but

$$\begin{vmatrix} a_1 & b_1 & c_1 \\ a_2 & b_2 & c_2 \\ a_3 & b_3 & c_3 \end{vmatrix} + \begin{vmatrix} a_1' & b_1 & c_1 \\ a_2' & b_2 & c_2 \\ a_3' & b_3 & c_3 \end{vmatrix} = \begin{vmatrix} a_1 + a_1' & b_1 & c_1 \\ a_2 + a_2' & b_2 & c_2 \\ a_3 + a_3' & b_3 & c_3 \end{vmatrix}$$

Application of matrices for solving the systems of linear equations

We mentioned earlier that systems of linear equations can be solved by applying the Cramer rule. It turns out that the matrix notation allows us to write the system in compact form and, correspondingly, it becomes more convenient to deal with the set of equations. For instance, this approach facilitates determination of some group of unknowns, while calculation of others can be avoided. Consider a system of equations

$$\begin{aligned} y_1 &= \alpha_{11}\, x_1 + \ldots + \alpha_{1n}\, x_n \\ &\;\;\vdots \\ y_n &= \alpha_{n1}\, x_1 + \ldots + \alpha_{nn}\, x_n \end{aligned} \tag{B-67}$$

Introducing the notation

$$\mathbf{y} = \begin{bmatrix} y_1 \\ \vdots \\ y_n \end{bmatrix} \quad \text{and} \quad \mathbf{x} = \begin{bmatrix} x_1 \\ \vdots \\ x_n \end{bmatrix},$$

in place of the set B-67 we have

$$\mathbf{y} = \boldsymbol{\alpha}\, \mathbf{x}, \tag{B-68}$$

where $\boldsymbol{\alpha}$ is the matrix of coefficients. Suppose that we want to solve this system only with respect to the first k unknowns $x_1, x_2, \ldots, x_k$. The matrix $\boldsymbol{\alpha}$ can be written as

$$\boldsymbol{\alpha} = \left[\begin{array}{ccc|ccc} \alpha_{11} & \cdots & \alpha_{1k} & \alpha_{1,k+1} & \cdots & \alpha_{1n} \\ \vdots & & \vdots & \vdots & & \vdots \\ \alpha_{k1} & \cdots & \alpha_{kk} & \alpha_{k,k+1} & \cdots & \alpha_{kn} \\ \hline \alpha_{k+1,1} & \cdots & \alpha_{k+1,k} & \alpha_{k+1,k+1} & \cdots & \alpha_{k+1,n} \\ \vdots & & \vdots & \vdots & & \vdots \\ \alpha_{n1} & \cdots & \alpha_{nk} & \alpha_{n,k+1} & \cdots & \alpha_{kn} \end{array}\right] = \left[\begin{array}{c|c} A_1 & A_2 \\ \hline A_3 & A_4 \end{array}\right] \tag{B-69}$$

The latter represents the combination of four block matrices A_1, A_2, A_3 and A_4. Similar approach is applied to matrices $\mathbf{y}$ and $\mathbf{x}$:

$$\mathbf{y} = \begin{bmatrix} y_1 \\ y_2 \\ \vdots \\ y_k \\ \hline y_{k+1} \\ y_{k+2} \\ \vdots \\ y_n \end{bmatrix} = \begin{bmatrix} Y_1 \\ \hline Y_2 \end{bmatrix} \quad \text{and} \quad \mathbf{x} = \begin{bmatrix} x_1 \\ x_2 \\ \vdots \\ x_k \\ \hline x_{k+1} \\ x_{k+2} \\ \vdots \\ x_n \end{bmatrix} = \begin{bmatrix} X_1 \\ \hline X_2 \end{bmatrix}$$

Therefore, the system B-67 is written in the form

$$\begin{bmatrix} Y_1 \\ Y_2 \end{bmatrix} = \begin{bmatrix} A_1 & A_2 \\ A_3 & A_4 \end{bmatrix} \begin{bmatrix} X_1 \\ X_2 \end{bmatrix} \tag{B-70}$$

Formulae of matrix multiplication show that the matrices A_1, A_2, A_3, A_4, X_1, X_2, Y_1 and Y_2 can be considered as the elements of matrices in eq. B-70 and, therefore, the last system consists of two equations

$$\begin{aligned} Y_1 &= A_1\,X_1 + A_2\,X_2 \\ Y_2 &= A_3\,X_1 + A_4\,X_2 \end{aligned} \tag{B-71}$$

Let us eliminate one group of unknowns, X_2, from this system. The second equation gives

$$A_4\,X_2 = Y_2 - A_3\,X_1\,,$$

whence

$$X_2 = A_4^{-1}\,(Y_2 - A_3\,X_1) \tag{B-72}$$

Substitution of the latter into the first equation of the set B-71 yields

$$Y_1 = A_1\,X_1 + A_2\,A_4^{-1}\,(Y_2 - A_3\,X_1)$$

Finally,

$$Y_1 - A_2\,A_4^{-1}\,Y_2 = \left(A_1 - A_2\,A_4^{-1}\,A_3\right)X_1 \tag{B-73}$$

This is a group of linear equations

$$\mathbf{Y} = \boldsymbol{\alpha}\,\mathbf{X}$$

which does not contain $x_{k+1}, x_{k+21}, \ldots, x_n$.

Also it may be proper to notice that formally the system B-67 can be solved very quickly. In matrix notation, $\mathbf{Y} = \boldsymbol{\alpha}\,\mathbf{X}$, therefore,

$$\mathbf{X} = \boldsymbol{\alpha}^{-1}\,\mathbf{Y}\,,$$

and the calculations are reduced to the determination of the inverse matrix $\boldsymbol{\alpha}^{-1}$, which can be found using Cramer rule.

Until now we have studied matrices with real elements. Next generalization is related to the case when the matrix elements are complex numbers. In other words, the vectors are considered in the n-dimensional complex space, where their components are complex. This space essentially differs from the complex plane, used in the theory of complex numbers, where the real numbers, representing a complex number were plotted along the x- and y-axes.

It is useful to define two new types of matrices.

Hermitian matrices

The matrix $\boldsymbol{\alpha}$ is called Hermitian if its elements, located symmetrically with respect to the main diagonal are complex conjugate numbers:

$$\alpha_{kj} = \alpha^*_{jk} \tag{B-74}$$

For instance, the matrix

$$\begin{bmatrix} 2 & 2+3\,i & i \\ 2-3\,i & 4 & 3 \\ -i & 3 & 1 \end{bmatrix}$$

is the Hermitian. The elements on the main diagonal of such a matrix are always real. In particular, a real Hermitian matrix is symmetric.

Hermitian conjugate matrices

This new matrix is obtained from the Hermitian one in two steps. First, the transposed matrix $\tilde{\boldsymbol{\alpha}}$ is constructed. Then, its elements are replaced with their complex conjugate. This matrix is denoted as $\boldsymbol{\alpha}^+$, and by definition we have

$$\alpha^+_{ik} = \alpha^*_{jk} \tag{B-75}$$

Eigenvalues, eigenvectors and characteristic equation

Suppose that the matrix $\boldsymbol{\alpha} = [\alpha_{ik}]$ and the nonzero vector $\mathbf{u}$ are given. The latter can be also described by the matrix $[u_i]$. If the vector $\mathbf{u}$ is such that the transformation produced by $\boldsymbol{\alpha}$ causes only a change of its length, $|\mathbf{u}|$, it is called an eigenvector. The coefficient λ, characterizing the change of its length, is named the eigenvalue. Considering transformations in two dimensions, we already mentioned those concepts. By definition, we have

$$\boldsymbol{\alpha}\,\mathbf{u} = \lambda\,\mathbf{u}$$

or

$$\sum_i \alpha_{ki}\,u_i = \lambda\,u_k \tag{B-76}$$

Here $k = 1, 2, \ldots, n$. The set of equations B-76 can be also written as

$$\begin{aligned}
(\alpha_{11} - \lambda)\,u_1 + \alpha_{12}\,u_2 + \ldots + \alpha_{1n}\,u_n &= 0\\
\alpha_{21}\,u_1 + (\alpha_{22} - \lambda)\,u_2 + \ldots + \alpha_{2n}\,u_n &= 0\\
&\vdots\\
\alpha_{n1}\,u_1 + \alpha_{n2}\,u_2 + \ldots + (\alpha_{nn} - \lambda)\,u_n &= 0
\end{aligned} \tag{B-77}$$

Inasmuch as at least one component of the vector $\mathbf{u}$ differs from zero, the determinant of this system of linear equations is equal to zero:

$$\Delta(\lambda) = \begin{vmatrix} \alpha_{11} - \lambda & \alpha_{12} & \ldots & \alpha_{1n} \\ \vdots & & & \vdots \\ \alpha_{n1} & \alpha_{n2} & \ldots & (\alpha_{nn} - \lambda) \end{vmatrix} = 0 \tag{B-78}$$

This gives the so-called characteristic equation of the matrix $\boldsymbol{\alpha}$ for calculating the eigenvalues λ. The root λ, substituted into the system of equations B-76, allows us to determine the direction of eigenvector, which corresponds to this particular root.

Example Consider the matrix

$$\begin{bmatrix} 11 & -6 & 2 \\ -6 & 10 & -4 \\ 2 & -4 & 6 \end{bmatrix}$$

Its characteristic equation is

$$\begin{vmatrix} 11-\lambda & -6 & 2 \\ -6 & 10-\lambda & -4 \\ 2 & -4 & 6-\lambda \end{vmatrix} = -\lambda^3 + 27\,\lambda^2 - 180\,\lambda + 324 = 0$$

The solution of this equation gives three roots

$$\lambda_1 = 18\,, \quad \lambda_2 = 6\,, \quad \lambda_3 = 3$$

First, we determine the eigenvector corresponding to λ_3. Substitution of λ_3 into the system B-76 gives

$$\begin{aligned} 8\,u_1 - 6\,u_2 + 2\,u_3 &= 0\,, \\ -6\,u_1 + 7\,u_2 - 4\,u_3 &= 0\,, \\ 2\,u_1 - 4\,u_2 + 3\,u_3 &= 0 \end{aligned}$$

The determinant of this system is zero. This means that the equations are linearly dependent. For this reason, we can, for example, discard the last equation and solve the first two equations with respect to two unknowns. Letting $u_1 = c_3$, we have

$$\begin{cases} -6\,u_2 + 2\,u_3 = -8\,c_3 \\ 7\,u_2 - 4\,u_3 = 6\,c_3 \end{cases}$$

The latter gives the vector

$$\mathbf{u}^{(3)} = c_3 \begin{bmatrix} 1 \\ 2 \\ 2 \end{bmatrix} \qquad \text{or} \qquad \mathbf{u}^{(3)} = c_3\,\mathbf{i}_1 + 2\,c_3\,\mathbf{i}_2 + 2\,c_3\,\mathbf{i}_3\,,$$

which defines the direction of the eigenvector. In the same manner we find the eigenvectors for $\lambda_2 = 6$ and $\lambda_1 = 18$:

$$\mathbf{u}^{(2)} = c_2 \begin{bmatrix} 2 \\ 1 \\ -2 \end{bmatrix} \qquad \text{and} \qquad \mathbf{u}^{(1)} = c_1 \begin{bmatrix} 2 \\ -2 \\ 1 \end{bmatrix}$$

B.2 Tensors

At the beginning we introduced the matrix $\boldsymbol{\alpha}$ as the operator, which transforms one vector into another, and this transformation is linear one. For instance, if the matrix α is applied to the vector $\mathbf{u}$ with components u_1, u_2, u_3, then we arrive at the new vector $\mathbf{v}$ with components v_1, v_2 and v_3 in the same system of coordinates. It is essential that each component of $\mathbf{v}$ is a linear function of $\mathbf{u}$. Now consider a behavior of the matrix α, when the system of coordinates is changed. First, suppose that there are two vectors: $\mathbf{u}$ and $\mathbf{v}$ and the relationship between them is

$$\alpha\mathbf{u} = \mathbf{v}, \tag{B-79}$$

where components of vectors are given in the Cartesian system of coordinates x, y, z and α is the matrix with elements:

$$\alpha = \begin{pmatrix} \alpha_{11} & \alpha_{12} & \alpha_{13} \\ \alpha_{21} & \alpha_{22} & \alpha_{23} \\ \alpha_{31} & \alpha_{32} & \alpha_{33} \end{pmatrix} \tag{B-80}$$

Applying the rule of multiplication of the matrix by the vector, in place of eq. B-79 we can write

$$\begin{aligned} \alpha_{11}u_1 + \alpha_{12}u_2 + \alpha_{13}u_3 &= v_1 \\ \alpha_{21}u_1 + \alpha_{22}u_2 + \alpha_{23}u_3 &= v_2 \\ \alpha_{31}u_1 + \alpha_{32}u_2 + \alpha_{33}u_3 &= v_3 \end{aligned} \tag{B-81}$$

This clearly shows that α performs the linear transformation of $\mathbf{u}$ into $\mathbf{v}$.

Next assume that there is another system of coordinates, x', y', z' with the same origin. Certainly, in this system the magnitude and direction of vectors $\mathbf{u}$ and $\mathbf{v}$ remain the same, but their components vary. By definition, if in the new system of coordinates, obtained by a rotation of the old one, we arrive at the system, like eq. B-81, then the matrix α is called the tensor. In other words, in such the case the linearity of transformation is preserved. Let us notice that often the physical considerations allow us to conclude that the matrix represents the tensor.

Since components of vectors $\mathbf{u}$ and $\mathbf{v}$ are different in the new system we can expect that elements of the tensor α also change. Our goal is to find the relationship between its elements in both system, and preliminary it is useful to consider two topics.

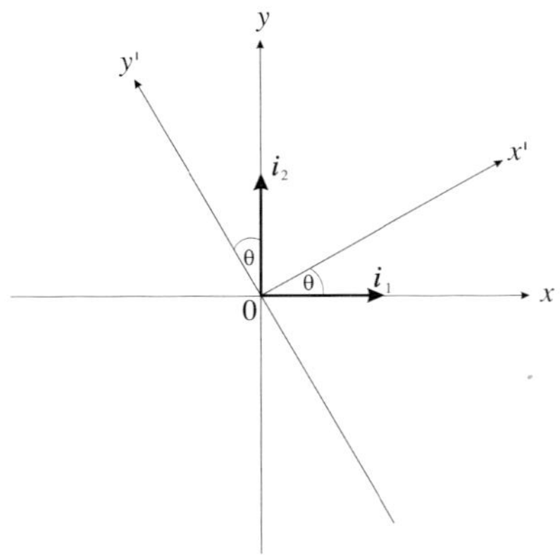

Figure B.3: Transformation of coordinates

Transformation of scalars and vectors

As is well known, there are two groups of scalars. One of them consists of scalars, which change with a rotation of the Cartesian system of coordinates, and they are called the variant scalars. Components of vector and coordinates of the point are examples of such scalars. At the same time scalars, like temperature, pressure and density of sources are independent of the orientation of coordinate axis, and they represent invariant scalars. Naturally, we are interested in variant quantities and, first, study the simplest case, when scalars are coordinates of a point. Consider the Cartesian system of coordinates, X with unit vectors $\mathbf{i}_1$, $\mathbf{i}_2$ and $\mathbf{i}_3$, and the origin at the point O, Fig. B.3. The radius-vector $\mathbf{r}$, characterizing a position of some point p is

$$\mathbf{r} = x_1 \, \mathbf{i}_1 + x_2 \, \mathbf{i}_2 + x_3 \, \mathbf{i}_3 \tag{B-82}$$

Here x_1, x_2 and x_3 are coordinates of the point p. It is convenient to introduce the notation

$$\delta_{jk} = 1 \qquad \text{if} \qquad j = k \tag{B-83}$$

$$\delta_{jk} = 0 \qquad \text{if} \qquad j \neq k$$

Since

$$\mathbf{i}_j \cdot \mathbf{i}_k = \delta_{jk}, \tag{B-84}$$

the coordinate of the point in the system x is

$$x_k = \mathbf{r} \cdot \mathbf{i}_k \tag{B-85}$$

Next assume that we also have the new Cartesian system X', which is obtained from the old one by a rotation of its coordinate axis about the origin, and $\mathbf{i}'_1$, $\mathbf{i}'_2$, $\mathbf{i}'_3$ are unit vectors of the system X'. It is clear that the vector $\mathbf{r}$ remains the same in both systems, that is

$$x_1\ \mathbf{i}_1 + x_2\ \mathbf{i}_2 + x_3\ \mathbf{i}_3 = x'_1\ \mathbf{i}'_1 + x'_2\ \mathbf{i}'_2 + x'_3\ \mathbf{i}'_3 \tag{B-86}$$

The latter allows us to find a relationship between coordinates in system X and X'. Multiplying both sides of eq. B-86 by unit vector $\mathbf{i}'_j$ and taking into account eq. B-85 we obtain

$$x'_j = \mathbf{r} \cdot \mathbf{i}'_j = x_1\ \mathbf{i}_1 \cdot \mathbf{i}'_j + x_2\mathbf{i}_2 \cdot \mathbf{i}'_j + x_3\mathbf{i}_3 \cdot \mathbf{i}'_j \tag{B-87}$$

Here $j = 1, 2, 3$. Thus, every coordinate of the system X' is the linear function of coordinates of the system X. Coefficients

$$\gamma_{jk} = \mathbf{i}'_j \cdot \mathbf{i}_k \tag{B-88}$$

are directional cosines of the angles, formed by the axis $\mathbf{i}'_j$ and $\mathbf{i}_k$. For example, γ_{12} is the directional cosine of the angle between the coordinate axis x' and y.

We see that a rotation of the Cartesian system leads to a change of coordinates of the point, and it is described by the linear transformation

$$x'_j = \sum_{k=1}^{3} \gamma_{jk} x_k, \tag{B-89}$$

or

$$\begin{aligned} x'_1 &= \gamma_{11}x_1 + \gamma_{12}x_2 + \gamma_{13}x_3 \\ x'_2 &= \gamma_{21}x_1 + \gamma_{22}x_2 + \gamma_{23}x_3 \\ x'_3 &= \gamma_{31}x_1 + \gamma_{32}x_2 + \gamma_{33}x_3 \end{aligned} \tag{B-90}$$

The matrix

$$\gamma = \begin{pmatrix} \gamma_{11} & \gamma_{12} & \gamma_{13} \\ \gamma_{22} & \gamma_{23} & \gamma_{23} \\ \gamma_{31} & \gamma_{32} & \gamma_{33} \end{pmatrix} \tag{B-91}$$

consists of nine elements and each of them is the directional cosine of the angle, formed by axis of systems X and X'. It is a simple matter to see that these elements are related to each other. In other words, they obey the certain condition, which follows from the fact that the distance from the origin O to the point p, that is $|\mathbf{r}|$, is independent on the direction of coordinate axis:

$$\sum_{j=1}^{3}(x_j)^2 = \sum_{j=1}^{3}(x'_j)^2 \tag{B-92}$$

Before we use this equality, let us determine the relationship between γ_{ik}.

First, suppose that the vector $\mathbf{r}$ coincides with the unit vector $\mathbf{i}_1$. Then its components in the old system are

$$x_1 = 1, \qquad x_2 = 0, \qquad x_3 = 0,$$

where

$$x_1 = x, \qquad x_2 = y, \qquad x_3 = z$$

In accordance with eqs. B-90 components of the unit vector $\mathbf{i}$ in the new system of coordinates are

$$x'_1 = \gamma_{11}, \qquad x'_2 = \gamma_{21}, \qquad x'_3 = \gamma_{31} \tag{B-93}$$

Therefore, in this system the vector $\mathbf{i}_1$ can be expressed as

$$\mathbf{i}_1 = \gamma_{11}\,\mathbf{i}'_1 + \gamma_{21}\,\mathbf{i}'_2 + \gamma_{31}\,\mathbf{i}'_3 \tag{B-94}$$

Now suppose that the vector $\mathbf{r}$ coincides with the second unit vector $\mathbf{i}_2$ and after it with $\mathbf{i}_3$. By analogy we have

$$\mathbf{i}_2 = \gamma_{12}\,\mathbf{i}'_1 + \gamma_{22}\,\mathbf{i}'_2 + \gamma_{32}\,\mathbf{i}'_3, \qquad \mathbf{i}_3 = \gamma_{13}\,\mathbf{i}'_1 + \gamma_{23}\,\mathbf{i}'_2 + \gamma_{33}\,\mathbf{i}'_3 \tag{B-95}$$

From eqs. B-94–B-95 we see that columns of the matrix γ is composed of components of the unit vectors $\mathbf{i}_1$, $\mathbf{i}_2$ and $\mathbf{i}_3$ in the new system of coordinates. Inasmuch as these vectors, as well as $\mathbf{i}'_1$, $\mathbf{i}'_2$, $\mathbf{i}'_3$, are orthogonal to each other, the matrix γ has several important features. First, forming the dot product of eqs. B.94–95 we find that

$$\gamma_{11}\gamma_{12} + \gamma_{21}\gamma_{22} + \gamma_{31}\gamma_{32} = 0$$

$$\gamma_{12}\gamma_{13} + \gamma_{22}\gamma_{23} + \gamma_{32}\gamma_{33} = 0 \qquad \text{(B-96)}$$

$$\gamma_{13}\gamma_{11} + \gamma_{23}\gamma_{21} + \gamma_{33}\gamma_{31} = 0$$

This indicates that columns of γ are orthogonal to each other. Next, forming the dot product of $\mathbf{i}_1 \cdot \mathbf{i}_1$, $\mathbf{i}_2 \cdot \mathbf{i}_2$ and $\mathbf{i}_3 \cdot \mathbf{i}_3$ we obtain

$$\gamma_{11}^2 + \gamma_{21}^2 + \gamma_{31}^2 = 1$$

$$\gamma_{12}^2 + \gamma_{22}^2 + \gamma_{32}^2 = 1 \qquad \text{(B-97)}$$

$$\gamma_{13}^2 + \gamma_{23}^2 + \gamma_{33}^2 = 1$$

It turns out that rows of the matrix are also orthogonal and their magnitude is equal to unit. Such matrixes are called orthogonal ones. To illustrate last properties consider the two dimensional case. As it seen from Fig. B.3.

$$\gamma_{11} = \cos\theta \qquad \gamma_{21} = -\sin\theta \qquad \gamma_{12} = \sin\theta \qquad \gamma_{22} = \cos\theta$$

Correspondingly, the matrix $\boldsymbol{\gamma}$ has the form

$$\gamma = \begin{pmatrix} \cos\theta & \sin\theta \\ -\sin\theta & \cos\theta \end{pmatrix}, \qquad \text{(B-98)}$$

and the orthogonality of its columns and rows is obvious. Of course, the determinant is equal to unity. Also this figure shows that rows of the matrix are formed by components of unit vectors $\mathbf{i}'_1$ and $\mathbf{i}'_2$ in the old system of coordinates:

$$\begin{aligned} \mathbf{i}'_1 &= \cos\theta\, \mathbf{i}_1 + \sin\theta\, \mathbf{i}_2 \quad \text{and} \\ \mathbf{i}'_2 &= -\sin\theta\, \mathbf{i}_1 + \cos\theta\, \mathbf{i}_2 \end{aligned}$$

The latter clearly demonstrates that rows of the matrix are also orthogonal. It is useful to replace six equations, given by eqs. B-96–B-97, by one. As an example, consider the equality

$$\gamma_{11}\gamma_{12} + \gamma_{21}\gamma_{22} + \gamma_{31}\gamma_{32} = 0$$

which can be represented as

$$\sum_{j=1}^{3} \gamma_{ji}\gamma_{jk} = 0 \qquad \text{if} \qquad i \neq k \tag{B-99}$$

Here i and k are either 1 or 2, or 3. It is obvious that eq. B-99 describes two other equations of the set B-96. Moreover, in place of eq. B-97 we have

$$\sum_{j=1}^{3} \gamma_{ji}\gamma_{jk} = 1 \qquad \text{if} \qquad i = k \tag{B-100}$$

Thus, combining the last two equations and making use of notations of eq. B-83, we obtain

$$\sum_{j=1}^{3} \gamma_{ji}\gamma_{jk} = \delta_{ik} = \begin{cases} 1 & , \quad i = k \\ 0 & \quad\;\; i \neq k \end{cases} \tag{B-101}$$

which expresses the condition of orthogonality in the compact form. The same result follows from eq. B-92. In fact, we have

$$\sum_{j=1}^{3}(x_j')^2 = \sum_{j=1}^{3}\left(\sum_{i=1}^{3}\gamma_{ji}x_i\right)\left(\sum_{k=1}^{3}\gamma_{jk}x_k\right) = \sum_{i=1}^{3}\sum_{k=1}^{3} x_i x_k \sum_{j=1}^{3}\gamma_{ji}\gamma_{jk}$$

Since the latter is equal to $\mathbf{r}^2$ we again arrive at eq. B-101.

Bearing in mind that the determinant of the matrix $\boldsymbol{\alpha}$ is not equal to zero, the coordinates x_k can be expressed in term of x_k', and, by analogy with eq. B-89, we have

$$x_k = \sum_{j=1}^{3} \gamma_{jk} x_j' \tag{B-102}$$

Next suppose that $\mathbf{M}$ is an arbitrary vector

$$\mathbf{M} = \sum_{k=1}^{3} M_k \; \mathbf{i}_k = \sum_{k=1}^{3} M_k' \; \mathbf{i}_k' \tag{B-103}$$

Any component $\mathbf{M}_j'$ of this vector in the system X' is given by the equation

$$M_j' = \mathbf{M} \cdot \mathbf{i}_j' = \sum_{k=1}^{3} M_k \; \mathbf{i}_k \cdot \mathbf{i}_j' = \sum_{k=1}^{3} \gamma_{jk} M_k \tag{B-104}$$

Thus, components of the vector are transformed as coordinates of the point, when the Cartesian system of coordinates is rotated. As we know, every vector is characterized by three scalar components. However, it does not mean that any three scalars, M_1, M_2 and M_3 can be treated as the vector components. It happens, if they are transformed as coordinates of the point.

Finally, let us represent eqs. B-81 in the compact form:

$$v_j = \sum_{k=1}^{3} \alpha_{jk} u_k \tag{B-105}$$

Assuming that the transformation is invariant with respect to the rotation, i.e. α is the tensor, we have to obtain:

$$v_i' = \sum_{l=1}^{3} \alpha_{il}' \, u_l' \tag{B-106}$$

Here $i = 1, 2, 3$. Multiplication of eq. B-105 by γ_{ij} and a summation by the index j gives

$$\sum_{l=1}^{3} \gamma_{ij} \, v_j = \sum_{j=1}^{3} \sum_{k=1}^{3} \gamma_{ij} \, \alpha_{jk} u_k \tag{B-107}$$

On the other hand

$$v_i' = \sum_{j=1}^{3} \gamma_{ij}' \, v_j, \qquad u_k = \sum_{l=1}^{3} \gamma_{lk}' \, u_l' \tag{B-108}$$

Therefore, eq. B-106 is written as

$$v_i' = \sum_{l=1}^{3} \left(\sum_{j=1}^{3} \sum_{k=1}^{3} \gamma_{ij} \, \gamma_{lk} \, \alpha_{jk} \right) u_l' = \sum_{l=1}^{3} \alpha_{il}' \, u_l', \tag{B-109}$$

where

$$\alpha_{il}' = \sum_{j=1}^{3} \sum_{k=1}^{3} \gamma_{ij} \, \gamma_{lk} \, \alpha_{il}, \tag{B-110}$$

and we found the relationship between tensor elements in both system of coordinates.

Appendix C

Stress tensor

Volume and surface forces in an ideal fluid

In studying propagation of acoustic waves in an ideal fluid, it was shown that there are two types of forces: volume forces and surface forces. The classical example of the former is the gravitational force. For instance, in the case of an elementary volume ΔV this force is equal to

$$\mathbf{F}(p) = \mathbf{f}(p)\Delta V = \rho(p)\mathbf{g}(p)\Delta V, \tag{C-1}$$

where $\mathbf{f}(p)$ is the vector, characterizing the density of volume forces,

$$\mathbf{f}(p) = \rho(p)\ \mathbf{g}(p) \tag{C-2}$$

Here $\rho(p)$ is the mass density and $\mathbf{g}(p)$ is the gravitational field, caused by all masses, except $\Delta m(p)$:

$$\Delta m(p) = \rho(p)\Delta V \tag{C-3}$$

Of course, masses of the Earth are main sources of the gravitational field.

The surface forces may arise differently. Consider an elementary plane surface, $da(p)$, inside an ideal fluid, Fig. C.1a. A medium, situated at the right side of $da(p)$ and at its vicinity, acts on the medium, located at the left side of this elements, with the force

$$\mathbf{F}(p) = \mathbf{t}(p)da \tag{C-4}$$

The vector $\mathbf{t}$ is the density of surface forces, and in an ideal fluid it is normal to the surface da:

$$\mathbf{t} = P\mathbf{n} \tag{C-5}$$

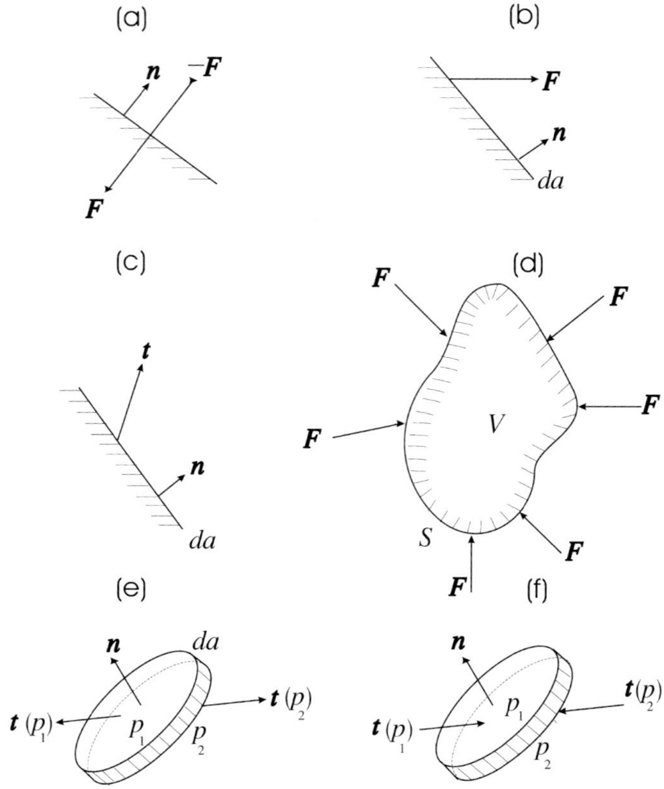

Figure C.1: (a) Orientation of surface forces inside an ideal fluid (b) Orientation of surface forces inside an elastic medium (c) Traction vector **t** (d) Arbitrary volume of elastic medium (e) Tractions at opposite sides of a disk

where $\mathbf{n}$ is the unit vector, normal to the surface, and P is the pressure. The first remarkable feature of surface forces is the fact that they act only in the direction perpendicular to the surface. In other words, the tangential components of these forces are absent. This means that an action (pull or push) in the direction, tangential to the element da, does not have any influence on the ideal fluid, located on the other side of this surface. In accordance with Newton's third law a medium, situated at the left side of da, also exerts a force across this element, and it is equal to

$$\mathbf{F} = -\mathbf{t}\,(p)\,da \tag{C-6}$$

Thus, in the vicinity of any point p of the surface element there are two surface forces with equal magnitude and opposite directions. They are applied at two points, located at different sides of the surface and infinitely close to each other. The second outstanding feature of these forces is independence of their magnitude on an orientation of the surface da. A change of the direction of the unit vector $\mathbf{n}$ does not make an influence on $|\mathbf{F}|$. As is well known, this allows us to characterize a force distribution in the ideal fluid by the single scalar function only, which is called the pressure.

Surface forces inside an elastic medium

Completely different behavior of surface forces is observed in an elastic medium. First of all, both the normal and tangential components of the force are transmitted through the surface. This means that, in general, the force $\mathbf{F}$, acting on some element da, can be arbitrary oriented with respect to the normal $\mathbf{n}$, Fig. C.1b. As in the case of an ideal fluid, media located at both sides of the surface act on each other with forces $\mathbf{F}(p)$ and $-\mathbf{F}(p)$, respectively. The second feature of these forces in an elastic medium is the fact that a change of an orientation of the element da results in a change of the force, exerted across it. We can imagine infinite number of orientations of the element da and, correspondingly, an unlimited number of different forces, acting at the same point of an elastic medium. Because of this the following question arises. How can we characterize such a distribution of forces? We attempt to find one quantity, which will allow us to determine the force density, $\mathbf{t}$, acting on the element da, regardless of its orientation. We have already performed a similar task then we studied studying scalar fields, (Part I). In principle, at each point there is always an infinite number of the directional derivatives of such field, and, in order to calculate them, the gradient of the scalar field was introduced. It turns out that a behavior of forces $\mathbf{t}$ as a function of an orientation of the elementary surface at the same point is also described by single quantity, which is called the stress tensor. Before we demonstrate this fact, let us make some comments about volume and surface forces.

a. In the absence of external forces a body is not deformed and its atoms are in a stable equilibrium. Correspondingly, forces of interaction are equal to zero. Because of a deformation a relative position of atoms changes and the internal forces arise. They try to return atoms to their original position.

b. The volume force, acting on elementary mass, $\Delta m(p)$, is caused by masses inside and outside an elastic body. Also these forces may have electric or magnetic origin.

The traction vector $\mathbf{t}$

Let us take a small element of the surface da inside an elastic body. We consider the force, transmitted through the element da and caused by a medium, which is situated at a certain side of da. In order to specify this portion of the body, we draw the normal $\mathbf{n}$ toward it, $(\mathbf{da} = da\,\mathbf{n})$. In other words, a direction of $\mathbf{n}$ defines a medium, which produces the surface force. As in the case of an ideal fluid, the density of the surface force is defined by eq. C-4

$$\mathbf{t} = \frac{\mathbf{F}}{da},$$

and it implies that the force $\mathbf{F}$ is uniformly distributed over the element da. The vector $\mathbf{t}$ is called the traction across this surface at the point p, Fig. C.1c. The dimension of $\mathbf{t}$ is

$$[\mathbf{t}] = \text{kg m}^{-1}\ \text{sec}^{-2},$$

and, by definition, $\mathbf{t}$ components along the Cartesian coordinate axes are

$$t_x = \mathbf{t} \cdot \mathbf{i} = t\cos(\mathbf{t}, \mathbf{i}), \qquad t_y = \mathbf{t} \cdot \mathbf{j} = t\cos(\mathbf{t}, \mathbf{j}), \qquad t_z = \mathbf{t} \cdot \mathbf{k} = t\cos(\mathbf{t}, \mathbf{k}) \tag{C-7}$$

Here $\cos(\mathbf{t}, \mathbf{i})$, $\cos(\mathbf{t}, \mathbf{j})$, $\cos(\mathbf{t}, \mathbf{k})$ are directional cosines of the vector $\mathbf{t}$.

It is a simple matter to find the normal and tangential components of the traction with respect to the plane element da. For instance, the scalar component along the normal $\mathbf{n}$ is

$$t_n(p) = \mathbf{t}(p) \cdot \mathbf{n} = t(p)\cos(\mathbf{t}, \mathbf{n}), \tag{C-8}$$

where t is the traction magnitude. If $t_n(p)$ is negative, it is called the pressure. In the opposite case, $t_n > 0$, this component is called the tension. For instance, when the fluid is at rest, directions of the vector $\mathbf{t}$ and the normal $\mathbf{n}$ are exactly opposite to each other. In an elastic medium the traction can be at any angle to the normal $\mathbf{n}$, Fig. C.1c.

Equations of equilibrium in integral form

In order to understand a distribution of internal forces it is very useful to consider the case when an elastic body is in a state of static equilibrium. The latter is provided by a system of external forces. This means that all particles of the body are at rest and, in particular, wave propagation is absent. Consider an arbitrary volume V of the elastic medium, surrounded by the surface S, Fig. C.1d. Since the body is in equilibrium the resultant external force, $\mathbf{F}$, and the resultant torque, $\mathbf{M}$, have to be equal to zero, (Appendix A):

$$\mathbf{F} = 0 \qquad \text{and} \qquad \mathbf{M} = 0 \tag{C-9}$$

Earlier we pointed out that the force $\mathbf{F}$ consists of the external surface and volume forces. For instance, the former is caused by elements of the medium, located at the external side of the surface S. They act on the neighboring elements near the internal side of S. As in the case of the ideal fluid we will use the concept of the density of volume forces, $\mathbf{f}$, and the traction, $\mathbf{t}$. Therefore, the elementary volume, dV, and the elementary surface, da, are subjected to the action of forces:

$$\mathbf{dF} = \mathbf{f}dV \qquad \text{and} \qquad \mathbf{dF} = \mathbf{t}da \tag{C-10}$$

As we already know, such presentation means that the volume and surface forces are uniformly distributed over dV and da, respectively. Now we are prepared to write down conditions of an equilibrium when both translation and rotation are absent. Making use of eqs. C-9 and the principle of superposition we obtain

$$\int\limits_V \mathbf{f}dV + \oint\limits_S \mathbf{t}\, da = 0 \tag{C-11}$$

and

$$\int\limits_V (\mathbf{r} \times \mathbf{f})\, dV + \oint\limits_S (\mathbf{r} \times \mathbf{t})\, da = 0 \tag{C-12}$$

Here $\mathbf{r}$ is the radius–vector drawn from an arbitrary chosen origin to any element of the volume V or the surface S (Appendix A). The first equality shows that the volume V does not experience translation, while the second one guarantees that this body is not involved in rotation. In both cases it is assumed that at the initial instant the body was at rest. The two equations represent conditions of equilibrium in integral form, since the volume V may have arbitrary dimensions.

Because our purpose is to find out relationships between surface forces on the vicinity of any point p inside an elastic body, we replace eqs. C-11–C-12 by their differential form. This task can be solved at least by two ways, related to each other. The first approach is based on an assumption that the volume V is very small. Correspondingly, points of the surface S are close to the point p, located at the middle of the volume V. This allows us to expand components of the traction $\mathbf{t}$ at points of the surface in the Taylor series around the point p. Also we assume that these components linearly change within the volume V. For this reason, terms of the series, which contain the second and higher order derivatives, are discarded. Then it turns out that after an integration over S, eq. C-11, it becomes possible to express the first condition of an equilibrium in terms of the traction $\mathbf{t}$ and the density of volume forces $\mathbf{f}$, at the point p.

The same equation of an equilibrium with respect to a translation can be obtained slightly differently, and the second approach follows from the Gauss divergence theorem, (Part I):

$$\int\limits_V \operatorname{div} \mathbf{M}\, dV = \oint\limits_S \mathbf{M} \cdot \mathbf{da}$$

where $\mathbf{da} = da\, \mathbf{n}$, and $\mathbf{n}$ is the unit vector, directed outward the volume V. This orientation is in agreement with a direction of the traction $\mathbf{t}$. It emphasizes the fact that a medium, surrounding the volume, generates a force, acting on V. In other words, these forces are external. In the Cartesian system of coordinates we have

$$\mathbf{n} = n_x \mathbf{i} + n_y \mathbf{j} + n_z \mathbf{k} \tag{C-13}$$

and n_x, n_y and n_z are directional cosines of the normal $\mathbf{n}$ with respect to coordinate axes.

Vectors X, Y, Z

In order to obtain the differential form of eq. C-11 it is very useful to introduce three vectors: $\mathbf{X}$, $\mathbf{Y}$ and $\mathbf{Z}$. By definition

$$\mathbf{X} = X_x \mathbf{i} + X_y \mathbf{j} + X_z \mathbf{k}, \qquad \mathbf{Y} = Y_x \mathbf{i} + Y_y \mathbf{j} + Y_z \mathbf{k}, \qquad \mathbf{Z} = Z_x \mathbf{i} + Z_y \mathbf{j} + Z_z \mathbf{k} \tag{C-14}$$

There vectors obey the following rule. The dot product of each vector and the normal $\mathbf{n}$ of the elementary surface da gives the corresponding component of the traction $\mathbf{t}$ on the coordinate axes

$$t_x = \mathbf{X} \cdot \mathbf{n}, \qquad t_y = \mathbf{Y} \cdot \mathbf{n}, \qquad t_z = \mathbf{Z} \cdot \mathbf{n} \tag{C-15}$$

For instance, the dot product $\mathbf{X} \cdot \mathbf{n}$ defines the projection of the vector $\mathbf{X}$ on the normal $\mathbf{n}$, and it is equal to the x–component of the traction. As follows from eq. C-15

$$\mathbf{t} = (\mathbf{X} \cdot \mathbf{n})\, \mathbf{i} + (\mathbf{Y} \cdot \mathbf{n})\, \mathbf{j} + (\mathbf{Z} \cdot \mathbf{n})\, \mathbf{k}$$

or

$$\mathbf{t} = X_n\mathbf{i} + Y_n\mathbf{j} + Z_n\mathbf{k}, \tag{C-16}$$

where X_n, Y_n and Z_n are projections of vectors **X**, **Y** and **Z** on the normal **n**. Respectively, the normal component of the traction can be written as

$$t_n = \mathbf{t} \cdot \mathbf{n}$$

or

$$t_n = X_n n_x + Y_n n_y + Y_z n_z \tag{C-17}$$

Let us rewrite eqs. C-15 in the form

$$t_x = X_x n_x + X_y n_y + X_z n_z, \quad t_y = Y_x n_x + Y_y n_y + Y_z n_z, \quad t_z = Z_x n_x + Z_y n_y + Z_z n_z \tag{C-18}$$

or in the compact form

$$\mathbf{t} = \begin{pmatrix} X_x & X_y & X_z \\ Y_x & Y_y & Y_z \\ Z_x & Z_y & Z_z \end{pmatrix} \mathbf{n} \tag{C-19}$$

Here t_x, t_y and t_z are the Cartesian components of the traction, acting on the plane element with the normal **n**.

Equations C-18 or C-19 can be treated as a transformation of the normal **n** into the vector **t**. However, they have much more important meaning and in order to understand it we consider three special orientations of the elementary surface, da, at the point p. First, suppose that this element is perpendicular to the x–axis, $(\mathbf{n} = \mathbf{i})$, that is

$$n_x = 1, \qquad n_y = n_z = 0$$

Then, as follows from eq. C-18

$$t_x(p) = X_x(p), \qquad t_y(p) = Y_x(p), \qquad t_z(p) = Z_x(p) \tag{C-20}$$

Comparison with eq. C-19 shows that the first column of the matrix characterizes the traction $\mathbf{t}(p)$, when the element da is normal to the x–axis. At the same time, X_x, Y_x, Z_x are components of the vector $\mathbf{t}$:

$$\mathbf{t} = X_x\mathbf{i} + Y_x\mathbf{j} + Z_x\mathbf{k}$$

In the second case the element $da(p)$ is normal to the y–axis, and correspondingly

$$n_x = 0, \qquad n_y = 1, \qquad n_z = 0$$

Then we have

$$t_x = X_y, \qquad t_y = Y_y, \qquad t_z = Z_y$$

and

$$\mathbf{t}(p) = X_y\mathbf{i} + Y_y\mathbf{j} + Z_y\mathbf{k}$$

We see that the second column of the matrix defines the traction $\mathbf{t}$ at the same point, when the element da is perpendicular to the y–axis. In a similar manner we find that the last column represents the vector $\mathbf{t}$, if the element $da(p)$ is normal to the z–axis and

$$\mathbf{t}(p) = X_z\mathbf{i} + Y_z\mathbf{j} + Z_z\mathbf{k}$$

Cauchy formulas

Thus, the matrix, (eq. C-19), contains information about the traction $\mathbf{t}$ for three mutually perpendicular positions of the element, $da(p)$. It is essential, that each time the normal $\mathbf{n}$ and one of the unit vectors of the Cartesian system coincide. Assume that components of vectors $\mathbf{X}$, $\mathbf{Y}$ and $\mathbf{Z}$ are given. In other words, we know the vector $\mathbf{t}$ for three orientations of the element da, corresponding to the coordinate planes, (Part I). Then, an importance of eqs. C-15 or C-18 becomes clear. In fact, they allow us to calculate the traction $\mathbf{t}$ at the same point for any orientation of the element $da(p)$ and these relationships are called Cauchy formulas. One can say that we have solved our main task and found out that the matrix, given by eq. C-19, is the desired quantity, which completely describes the traction $\mathbf{t}$ for an arbitrary orientation of the surface element da.

The first condition of an equilibrium in the differential form

To understand better some properties of this matrix we should return to conditions of an equilibrium and obtain their differential form. It is natural to start from eq. C-11. First, consider this equation for the x–component of vectors $\mathbf{f}$ and $\mathbf{t}$. It is clear that

$$\int_V f_x dV + \oint_S t_x da = 0 \qquad \text{(C-21)}$$

Substitution of the first equality of the set. C-15 into eq. C-21 yields

$$\int_V f_x dV + \oint_S \mathbf{X} \cdot \mathbf{n}\, da = 0 \qquad \text{(C-22)}$$

The integrand of the surface integral is represented as the flux of the vector $\mathbf{X}$ through the element da, and, therefore, we can make use of the Gauss theorem. As was already mentioned, this was one of the reasons for introduction of vectors $\mathbf{X}$, $\mathbf{Y}$, $\mathbf{Z}$. Correspondingly, in place of eq. C-22 we obtain

$$\int_V f_x dV + \int_V \mathrm{div}\mathbf{X}\, dV = 0$$

or

$$\int_V (f_x + \mathrm{div}\mathbf{X})\, dV = 0$$

Since this equality takes place regardless of dimensions and shape of the volume V, we conclude that the integrand is also equal to zero

$$f_x + \mathrm{div}\mathbf{X} = 0 \qquad \text{(C-23)}$$

By analogy, applying the same approach to components f_y, t_y and f_z, t_z, we have

$$f_y + \mathrm{div}\mathbf{Y} = 0, \qquad f_z + \mathrm{div}\mathbf{Z} = 0 \qquad \text{(C-24)}$$

Thus, eqs. C-23–C-24 represent the differential form of eq. C-11, and they show that an elementary volume around some point p does not experience a translation. It is obvious that the left hand side of these equations describe the resultant force, acting on the unit volume. Also it may be proper to notice the following. By definition, the divergence is a sum of the first derivatives, for instance

$$\mathrm{div}\ \mathbf{X} = \frac{\partial X_x}{\partial x} + \frac{\partial X_y}{\partial y} + \frac{\partial X_z}{\partial z}$$

or

$$\mathrm{div}\ \mathbf{X} = \lim \frac{\oint \mathbf{X} \cdot \mathbf{n}\, da}{\Delta V} \qquad \Delta V \to 0$$

Respectively, a calculation of divergence implies that the elementary volume ΔV has to be so small that functions $\mathbf{X}$, $\mathbf{Y}$ and $\mathbf{Z}$ change almost linearly inside of it. At the same time the density of volume forces remains the same. Of course, with a decrease of the volume a variation of each component of these vectors also tends to zero. As we know (Part I), an equilibrium does not take place instantly and it is always preceded by the dynamic stage. Suppose that at some instant the constant external forces are applied to the surface S, surrounding an elastic body. At the same moment a wave begins to propagate through the volume and ultimately it provides an equilibrium of each its portion. For illustration consider two examples of an elementary volume inside a body.

Example one Suppose that a volume is a very thin disk with elementary surfaces $da(p_1)$ and $da(p_2)$,

$$da(p_1) = da(p_2) = da$$

Its lateral surface is so small that one can neglect the forces acting on it. At the same time, we assume that forces, exerted on surfaces $da(p_1)$ and $da(p_2)$ are distributed uniformly over them. Therefore, a distribution of these forces is characterized by the tractions $\mathbf{t}(p_1)$ and $\mathbf{t}(p_2)$. Suppose that the wave approaches to the face $da(p_1)$ of the disk and produces its expansion. Then, the traction $\mathbf{t}(p_1)$ is directed towards the surrounding medium, as well as the normal $\mathbf{n}(p_1)$. In accordance with Newton's third law the traction $\mathbf{t}(p_2)$ has the opposite direction on this volume, Fig. C.1e. This means that vector components of $\mathbf{t}(p_1)$ and $\mathbf{t}(p_2)$ in the direction, which is either normal or tangential to the disk, are also opposite to each other.

In particular, in the state of an equilibrium

$$\mathbf{t}(p_1) = -\ \mathbf{t}(p_2),$$

provided that we can neglect the volume forces. If the wave produces a compression of an elementary disk, a direction of tractions is given in Fig. C.1f. It is essential that such an orientation of the traction at opposite faces of an elementary volume is always observed.

Example two Now consider an elementary parallelepiped, shown in Fig. C.2a. The sides of this volume are equal to Δx, Δy and Δz, and the middle point p has coordinates x, y, z. As in the first example, because of the wave, the volume, ΔV, is subjected to an action of forces, caused by a deformation of the surrounding medium. These surface forces are uniformly distributed over each face of ΔV, but they may have different magnitudes and directions at different faces. First assume that the wave moves along the x-axis and produces a compression. Therefore, the vector component of the traction

$$\mathbf{t}_x(x - \frac{\Delta x}{2},\ y, z)$$

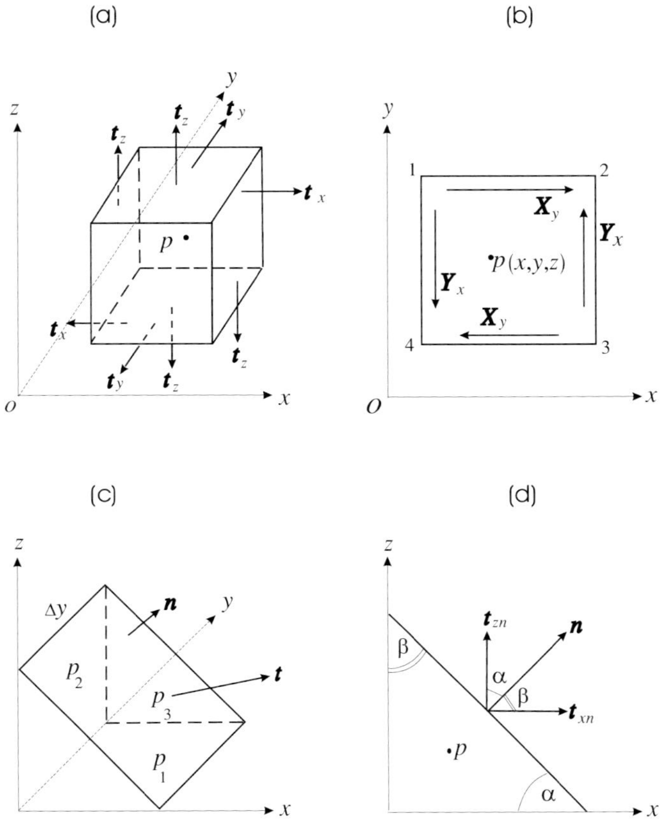

Figure C.2: (a,b) Tractions on faces of an elementary parallelepiped (c) Derivation of Cauchy formulas

is directed along the x–axis. When this wave reaches the opposite face, the force acts on a medium, which is in front of the volume, ΔV. As follows from Newton's third law, the traction, caused by this medium,

$$\mathbf{t}_x(x+\frac{\Delta x}{2},\ y,z)$$

has an opposite direction. If the wave is accompanied by tangential components of the traction, $\mathbf{t}_y$ and $\mathbf{t}_z$, then, applying the same law, we find that

$$\mathbf{t}_y(x-\frac{\Delta x}{2},\ y,z), \qquad\qquad \mathbf{t}_y(x+\frac{\Delta x}{2},\ y,z)$$

and

$$\mathbf{t}_z(x - \frac{\Delta x}{2},\ y, z) \qquad \text{and} \qquad \mathbf{t}_z(x + \frac{\Delta x}{2},\ y, z)$$

also have opposite directions. The same behavior of vector components of the traction is observed at other faces of the volume V, Fig. C.2a. At the same time each scalar component of vectors $\mathbf{X}$, $\mathbf{Y}$ and $\mathbf{Z}$ has the same sign at the opposite faces. In fact, by definition, we have

$$t_x(x - \frac{\Delta x}{2},\ y, z) = \mathbf{X} \cdot \mathbf{n} = -\mathbf{i} \cdot \mathbf{X}(x - \frac{\Delta x}{2},\ y, z) = -X_x(x - \frac{\Delta x}{2},\ y, z) \qquad \text{(C-25)}$$

For instance, in the case of the compressional wave, the scalar component $t_x(x - \frac{\Delta x}{2},\ y,\ z)$ is positive. Therefore, we conclude that

$$X_x(x - \frac{\Delta x}{2},\ y, z) < 0$$

In the opposite face we have

$$t_x(x + \frac{\Delta x}{2},\ y, z) = \mathbf{i} \cdot \mathbf{X}(x + \frac{\Delta x}{2},\ y, z) = X_x(x + \frac{\Delta x}{2},\ y, z) \qquad \text{(C-26)}$$

In accordance with Newton's third law, the component t_x is negative. Correspondingly, as on the back face:

$$X_x(x + \frac{\Delta x}{2},\ y, z) < 0$$

It is a simple matter to demonstrate that all other scalar components of vectors $\mathbf{X}$, $\mathbf{Y}$ and $\mathbf{Z}$ do not change sign at opposite faces of the elementary volume.

Flux of the vector X

Next we derive again eq. C-23 in more explicit way. With this purpose let us calculate the flux of the vector $\mathbf{X}$ through the closed surface, surrounding the volume ΔV, Fig. C.2a. Our goal is to simplify eq. C-22, when this volume is very small. It is clear that the flux through both faces, perpendicular to the x–axis, is

$$\left[X_x(x + \frac{\Delta x}{2},\ y, z) - X_x(x - \frac{\Delta x}{2},\ y, z)\right] \Delta y \Delta z$$

or

$$\frac{\partial X_x(x, y, z)}{\partial x}\ \Delta x\ \Delta y\ \Delta z = \frac{\partial X_x}{\partial x} \Delta V$$

The pair of faces, normal to the y–axis, gives

$$\left[X_y(x, y + \frac{\Delta y}{2},\ z) - X_y(x, y - \frac{\Delta y}{2}, z)\right]\ \Delta x\ \Delta z$$

or

$$\frac{\partial X_y}{\partial y}\Delta x \Delta y \Delta z = \frac{\partial X_y}{\partial y}\Delta V$$

Finally the flux through opposite faces, normal to the z–axis, is equal to

$$X_z[X_x(x, y, z + \frac{\Delta z}{2}) - X_z(x, y, z - \frac{\Delta z}{2})]\Delta x \Delta y$$

or

$$\frac{\partial X_z}{\partial z}\ \Delta V$$

Thus, the total flux is

$$\oint \mathbf{X} \cdot \mathbf{n}\ da = \left(\frac{\partial X_x}{\partial x} + \frac{\partial X_y}{\partial y} + \frac{\partial X_z}{\partial z}\right)\ \Delta V$$

As before, assuming that the density of volume forces is constant inside ΔV, we again obtain eq. C-23. The same approach gives the flux of vectors $\mathbf{Y}$ and $\mathbf{Z}$:

$$\oint \mathbf{Y} \cdot \mathbf{n}\ da = \left(\frac{\partial Y_x}{\partial x} + \frac{\partial Y_y}{\partial y} + \frac{\partial Y_z}{\partial z}\right)\ \Delta V$$

and

$$\oint \mathbf{Z} \cdot \mathbf{n}\ da = \left(\frac{\partial Z_x}{\partial x} + \frac{\partial Z_y}{\partial y} + \frac{\partial Z_z}{\partial z}\right)\ \Delta V,$$

and, correspondingly, eqs. C-24. Now it is appropriate to make several comments.

1. The last three equations allow us to express the flux through a surface, surrounding an elementary volume, in terms of the first derivatives of scalar components of $\mathbf{X}$, $\mathbf{Y}$ and $\mathbf{Z}$ at the middle point p. In other words, eqs. C-23–C-24 establish relationships between these components and the density of the volume force around the same point p.

2. We use values of functions at all faces of an elementary volume, but in the limit obtain formulas, which characterize a behavior of vectors $\mathbf{X}$, $\mathbf{Y}$, $\mathbf{Z}$ and $\mathbf{f}$ at one point, p.

3. As was mentioned earlier we assume that each component of these vectors linearly changes between opposite faces. This implies that a difference between values of any component at the middle point and at a face is directly proportional to the distance, $(\Delta x/2, \Delta y/2$ or $\Delta z/2)$, i.e., values of each scalar component, for instance, X_y, differ only slightly at the opposite faces. However, corresponding vector components have opposite directions.

The second condition of an equilibrium

We continue a study of vectors **X**, **Y** and **Z**, and with this purpose in mind consider eq. C-12. Its left hand side describes the resultant moment, and in order to provide an equilibrium, it has to be equal to zero. Respectively, each its component also vanishes. For instance, in the case of the x–component we have

$$\int_V (yf_z - zf_y)dV + \oint_S (yt_z - zt_y)da = 0 \tag{C-27}$$

To obtain its differential form we make use of eqs. C-15, and it gives

$$\int_V (yf_z - zf_y)dV + \oint_S (y\,\mathbf{Z} - z\mathbf{Y}) \cdot \mathbf{n}\; da = 0 \tag{C-28}$$

Applying again the Gauss's theorem we replace the surface integral by a volume integral, and eq. C-28 becomes

$$\int_V [yf_z - zf_y + \mathrm{div}(y\,\mathbf{Z} - z\,\mathbf{Y})]\; dV = 0 \tag{C-29}$$

By analogy with the first condition of an equilibrium, we take into account that eq. C-29 is valid for an arbitrary volume. This means that integrand is equal to zero, too:

$$yf_z - zf_y + \mathrm{div}(y\,\mathbf{Z} - z\,\mathbf{Y}) = 0 \tag{C-30}$$

This is the differential form of eq. C-28, and it shows that the x–component of the torque is zero. It is a relationship between components f_y, f_z and vectors **Y** and **Z** in the vicinity of any point. This equality contains extremely important information about scalar components of vectors **Y** and **Z**. To describe these new features we perform some simplifications in eq. C-30 As is well known from vector analysis,

$$\mathrm{div}\;(y\,\mathbf{Z}) = \mathbf{Z}\;\mathrm{grad}\;y + y\;\mathrm{div}\;\mathbf{Z} \qquad \text{and} \qquad \mathrm{div}\;(z\mathbf{Y}) = \mathbf{Y}\;\mathrm{grad}\;z + z\;\mathrm{div}\;\mathbf{Y} \tag{C-31}$$

Since $\mathrm{grad}\;y = \mathbf{j}$ and $\mathrm{grad}\;z = \mathbf{k}$, instead of eq. C-30 we have

$$yf_z - zf_y + \mathbf{Z} \cdot \mathbf{j} + y\;\mathrm{div}\;\mathbf{Z} - \mathbf{Y}\;\mathbf{k} - z\;\mathrm{div}\;\mathbf{Y} = 0$$

or

$$yf_z - zf_y + Z_y - Y_z + y\;\mathrm{div}\;\mathbf{Z} - z\;\mathrm{div}\;\mathbf{Y} = 0$$

Now, making use of the first condition of an equilibrium, eqs. C-23–C-24, we discover that

$$Z_y = Y_z \tag{C-32}$$

In the same manner, considering the y and z components of the resultant torque:

$$\int_V (z\, f_x - x\, f_z)dV + \oint_S (z\, t_x - x\, t_z)da = 0$$

and

$$\int_V (xf_y - yf_x)dV + \oint_S (x\, t_y - y\, t_x)da = 0,$$

we see that

$$Z_x = X_z \qquad \text{and} \qquad Y_x = X_y \tag{C-33}$$

In essence eqs. C-32–C-33 represent the second condition of an equilibrium of an elementary volume when its dimensions tend to zero. Thus, from both conditions of an equilibrium we found out that some elements of the matrix, eq. C-19, are equal to each other:

$$X_y(p) = Y_x(p), \qquad X_z(p) = Z_x(p), \qquad Y_z(p) = Z_y(p) \tag{C-34}$$

Taking into account an importance of these equalities, let us discuss them in some details. With this purpose consider an elementary cube, $(\Delta x = \Delta y = \Delta z)$ and its cross-section in the plane XOY, Fig. C.2b. First, we pay attention to tangential components of vectors $\mathbf{X}$ and $\mathbf{Y}$, which act on faces 1–2 and 2–3. Applying again the Taylor series we have

$$X_y(x, y + \frac{\Delta y}{2}, z) = X_y(p) + \frac{\partial X_y(p)}{\partial y}\frac{\Delta y}{2} - \ldots \tag{C-35}$$

and $$Y_x(x + \frac{\Delta x}{2}, y, z) = Y_x(p) + \frac{\partial Y_x(p)}{\partial x}\frac{\Delta x}{2} - \ldots$$

The traction components t_x and t_y, associated with X_y and Y_x, try to rotate the cube in opposite directions. As follows from eqs. C-35, in the limit, when the volume becomes infinitely small, their resultant torque vanishes, if $X_y(p) = Y_x(p)$. However, at faces 1–2 and 2–3 of the elementary volume, we may have:

$$X_y(x, y + \frac{\Delta y}{2}, z) \neq Y_x(x + \frac{\Delta x}{2}, y, z)$$

The same components at the opposite faces of the cube, 1–4 and 4–3, also form torques. As before, they have opposite directions and in the limit, when $\Delta x \to 0$ we again obtain that $X_y(p)$ is equal to $Y_x(p)$. Similarly, studying all tangential components of vectors **X**, **Y** and **Z**, we again arrive at eqs. C-34. This consideration also shows that in a state of an equilibrium these components are not usually equal at opposite faces. For example,

$$X_z(x, y, z + \frac{\Delta z}{2}) \neq X_z(x, y, z - \frac{\Delta z}{2})$$

Now it is proper to make several comments.

1. The set of equalities C-34 describes relationships between tangential components of a traction at point p. They act on elements of coordinate planes, which are equal to

$$da_x = dydz, \qquad da_y = dxdz, \qquad da_z = dxdy$$

2. As was demonstrated, eqs. C-34 remain valid, regardless of the volume force density, $\mathbf{f}$.

3. If we assume at the beginning that eqs. C-34 take place, then the second condition of an equilibrium is not independent and it follows from the first one. This approach is very useful, and it will be used later in deriving equations of motion.

4. In general, an equilibrium of an elementary volume depends on both the volume and surface forces. For instance, if the former can be neglected, the first condition is greatly simplified and we obtain

$$\text{div}\mathbf{X} = 0, \quad \text{div}\mathbf{Y} = 0, \quad \text{div}\mathbf{Z} = 0 \tag{C-36}$$

Stress tensor

As was already pointed out the matrix

$$\begin{pmatrix} X_x & X_y & X_z \\ Y_x & Y_y & Y_z \\ Z_x & Z_y & Z_z \end{pmatrix} \tag{C-37}$$

transforms the unit vector $\mathbf{n}$ of an arbitrary surface element into the traction, $\mathbf{t}(p)$, eq. C-19. This matrix is the tensor, and the relationship between $\mathbf{t}$ and $\mathbf{n}$ remains linear in new system of Cartesian coordinates, obtained by a rotation from the old one. Correspondingly, elements of the tensor in the new system can be calculated, applying formulas, derived in the previous Appendix B. By definition, nine scalar elements of the tensor, eq. C-37, are called stresses, and they allow us to find forces, acting on any element da. Its diagonal elements

$$X_x, \qquad Y_y \text{ and } \qquad Z_z$$

are called the normal stresses, since they characterize forces, which are perpendicular to corresponding coordinate planes. The other elements are shear stresses, and it is understandable, because they define tangential components of forces, exerted on the same coordinate planes.

In accordance with eqs. C-34 the stress tensor is symmetrical, and, therefore, it is defined by six elements only. There are different notations for tensor elements and one of them is given above, eq. C-37. It clearly shows the meaning of each element. For instance, X_y describes the force at the point p, directed along the x–axis and applied to the surface element $da(p)$, which is perpendicular to the y–axis.

The second notation uses one letter only for all elements, and it has a form

$$\begin{array}{lll} X_x = T_{11} & X_y = T_{12} & X_z = T_{13} \\ Y_x = T_{21} & Y_y = T_{22} & Y_z = T_{23} \\ Z_x = T_{31} & Z_y = T_{32} & Z_z = T_{33} \end{array} \tag{C-38}$$

and in place of eq. C-37 we have

$$\begin{pmatrix} T_{11} & T_{12} & T_{13} \\ T_{21} & T_{22} & T_{23} \\ T_{31} & T_{32} & T_{33} \end{pmatrix} \tag{C-39}$$

Respectively, the first index defines the component of the force, while the second characterizes a direction of the normal to the surface element. For instance, T_{32} describes the z–component of the traction, which acts on the surface element, perpendicular to the y–axis. It is obvious that

$$T_{21} = T_{12}, \qquad T_{13} = T_{31} \qquad \text{and} \qquad T_{32} = T_{23}$$

Also the stress tensor is sometimes written as

$$\begin{pmatrix} p_{xx} & p_{xy} & p_{xz} \\ p_{yx} & p_{yy} & p_{yz} \\ p_{zx} & p_{zy} & p_{zz} \end{pmatrix} \tag{C-40}$$

Comparison with eq. C-37 easily defines the meaning of each element. Finally, in order to emphasize a difference between the normal and shear stresses the following notations are used, too:

$$\begin{aligned} p_{xx} &= \sigma_x, \qquad p_{yy} = \sigma_y, \qquad p_{zz} = \sigma_z \\ \text{and} \qquad p_{xy} &= \tau_{xy}, \qquad p_{xz} = \tau_{xz}, \qquad p_{yz} = \tau_{yz} \end{aligned}$$

Therefore, the stress tensor is

$$\begin{pmatrix} \sigma_x & \tau_{xy} & \tau_{xz} \\ \tau_{yx} & \sigma_y & \tau_{yz} \\ \tau_{zx} & \tau_{zy} & \sigma_z \end{pmatrix} \tag{C-41}$$

In general, the latter is a function of a point and represents the example of a tensor field.

Cauchy formulas and an equilibrium

Earlier we obtained Cauchy formulas by simply introducing vectors $\mathbf{X}$, $\mathbf{Y}$ and $\mathbf{Z}$, eqs. C-15. It is also very fruitful to derive them, eqs. C-18, proceeding from the integral form of the first condition of an equilibrium. In other words, we again demonstrate, that the stress tensor, given at some point p, allows us to determine the traction, $\mathbf{t}(p)$, acting on the elementary surface, $da(p)$, arbitrary oriented with respect to coordinate planes. Solving this task it is convenient to deal with two different elementary volumes.

Case one: two-dimensional model Consider an elementary volume inside an elastic medium, which has a shape of the wedge, Fig. C.2c. Before we use the first condition of an equilibrium it is proper to notice the following. With a decrease of the wedge volume the surface forces decrease proportional to the area of its faces, that is as a square of linear dimensions. At the same time the volume force, for instance, the gravitational one, decays more rapidly; as a cube, since it is directly proportional to mass. For this reason we can neglect this force, that is

$$|\mathbf{f}|dV \ll |\mathbf{t}|da \tag{C-42}$$

Also we assume that the volume width, Δy, is very small, and forces, acting on two faces, perpendicular to the y–axis, cancel each other. This simplifies the first condition of an equilibrium, too, eq. C-11, and it can be written as

$$\mathbf{t}(p_3)\Delta y\Delta l + \mathbf{t}(p_2)\Delta y\Delta z + \mathbf{t}(p_1)\Delta x\Delta y = 0, \tag{C-43}$$

where

$$\Delta l = \sqrt{(\Delta x)^2 + (\Delta z)^2}$$

Respectively, for the x and z–components we have

$$t_{xn}(p_3)\Delta l + t_x(p_2)\Delta z + t_x(p_1)\Delta x = 0 \tag{C-44}$$

$$\text{and} \qquad t_{zn}(p_3)\Delta l + t_z(p_2)\Delta z + t_z(p_1)\Delta x = 0$$

By definition,

$$t_x(p_1) = -\mathbf{X}(p_1)\cdot\mathbf{k} = -X_z(p_1)$$

$$\text{and} \qquad t_x(p_2) = -\mathbf{X}(p_2)\cdot\mathbf{i} = -X_x(p_2),$$

since at both faces the normal has a direction, opposite to the corresponding unit vector. Then, the first equation of the set C-44 becomes

$$t_{xn}(p_3) = X_x(p_2)\frac{\Delta z}{\Delta l} + X_z(p_2)\frac{\Delta x}{\Delta l}$$

As is seen from Fig. C.2d

$$\frac{\Delta z}{\Delta l} = \cos\beta = n_x, \qquad \frac{\Delta x}{\Delta l} = \cos\alpha = n_z$$

are directional cosines of the normal $\mathbf{n}$. Thus, we have

$$t_{xn}(p_3) = X_x(p_2)n_x + X_z(p_1)n_z \tag{C-45}$$

It is clear that

$$t_z(p_1) = -\mathbf{Z}(p_1)\cdot\mathbf{k} = -Z_z(p_1)$$

$$\text{and} \qquad t_z(p_2) = -\mathbf{Z}(p_2)\cdot\mathbf{i} = -Z_z(p_2)$$

Therefore, the second equation of the set C-44 gives

$$t_{zn}(p_3) = Z_x(p_2)n_x + Z_z(p_1)n_z \tag{C-46}$$

Here $t_{xn}(p_3)$ and $t_{zn}(p_3)$ are the $z-$ and the x–components of the traction $\mathbf{t}(p_3)$ at the elementary plane with the normal $\mathbf{n}$.

As follows from eqs. C-45–C-46, they describe relationships between stresses at different points p_1, p_2 and p_3. However, with a decrease of the wedge volume, all faces approach to the same point p. In the limit these stresses characterize forces, exerted on three elementary surfaces, which have a common point p. Comparison with Cauchy formulas, eqs. C-18, shows that eqs. C-45–C-46 represent their special case, when an influence of forces, acting on faces, perpendicular to the y–axis, can be neglected.

Until now we found the Cartesian components, t_{xn} and t_{zn}, in terms of the stress tensor:

$$\begin{pmatrix} X_x & X_z \\ Z_x & Z_z \end{pmatrix}$$

It is also a simple matter to determine the normal and shear components of the traction $\mathbf{t}$ at the same point p_3. As is seen from Fig. C.2d

$$t_{nn}(p_3) = t_{xn}\cos\beta + t_{zn}\cos\alpha, \qquad t_{sn}(p_3) = -t_{xn}\cos\alpha + t_{zn}\cos\beta$$

or

$$t_{nn}(p_3) = t_{xn}n_x + t_{zn}n_z, \qquad t_{sn}(p_3) = -t_{xn}n_z + t_{zn}n_x \tag{C-47}$$

Here t_{nn} and t_{sn} are the normal and shear components of the traction at the point p_3. Substitution of eqs. C-45–C-46 into the set C-47 yields

$$t_{nn}(p_3) = n_x(X_x n_x + X_z n_z) + n_z(Z_x n_x + Z_z n_z)$$

or

$$t_{nn}(p_3) = n_x^2 X_x + n_x n_z X_z + n_x n_z Z_x + n_z^2 Z_z \tag{C-48}$$

Similarly, for the shear component we have

$$t_{sn}(p_3) = n_x(Z_x n_x + Z_z n_z) - n_z(X_x n_x + X_z n_z)$$

or

$$t_{sn}(p_3) = n_x^2 Z_x + n_x n_z Z_z - n_x n_z X_x - n_z^2 X_z \tag{C-49}$$

As we know, from the second condition of an equilibrium it follows that $Z_x = X_z$. Assume that the system of coordinates, x and z, is rotated about the y–axis, and the new axis x' is directed along the normal $\mathbf{n}$, while the z' axis is tangential to the elementary surface around p_3. Then, $t_{nn}(p_3)$ and $t_{sn}(p_3)$ represent the stress elements, $X_{x'}(p_3)$ and $Z_{x'}(p_3)$ in the new system of coordinates. Therefore, eqs. C-48–C-49 perform a transformation of two elements of the stress tensor, caused by a rotation of the coordinate system. Considering the surface element, normal to the old one, $(\Delta y, \Delta l)$, we can determine the stress $Z_{z'}$. As concerns $X_{z'}$, it is equal to $Z_{x'}$. It is proper to notice that the same result, eqs. C-48–C-49, follows from expressions derived in the Appendix B.

Case two: three-dimensional model Next we study an equilibrium of a tetrahedron, shown in Fig. C.3a. Three of its faces coincide with corresponding elements of the coordinate planes. The areas of all plane faces of this body are related to each other in the following way:

$$da_{xy} = da \cos(\mathbf{k}, \mathbf{n}), \qquad da_{xz} = da \cos(\mathbf{j}\ \mathbf{n})$$

and

$$da_{yz} = da \cos(\mathbf{i}, \mathbf{n})$$

Here da is the area of the face with the normal $\mathbf{n}$.

Our goal is to determine Cartesian components of the traction $\mathbf{t}$ at this oblique element of the closed surface. Applying again the first condition of an equilibrium in the integral form, we see that

$$\mathbf{t}\ da + \mathbf{t}(p_1) da_{yz} + \mathbf{t}(p_2) da_{xz} + \mathbf{t}(p_3) da_{xy} = 0 \tag{C-50}$$

Considering the Cartesian components of this equality and making use of the set C-15, we again arrive at the Cauchy formulas, eqs. C-18. At the same time, eq. C-50 does not relate to each other tensor elements, but it establishes a relationship between the traction $\mathbf{t}$ at an oblique surface element and the stress tensor. In addition note the following:

a. Earlier we derived Cauchy formulas without the use of the condition of an equilibrium, and, therefore, the volume forces.

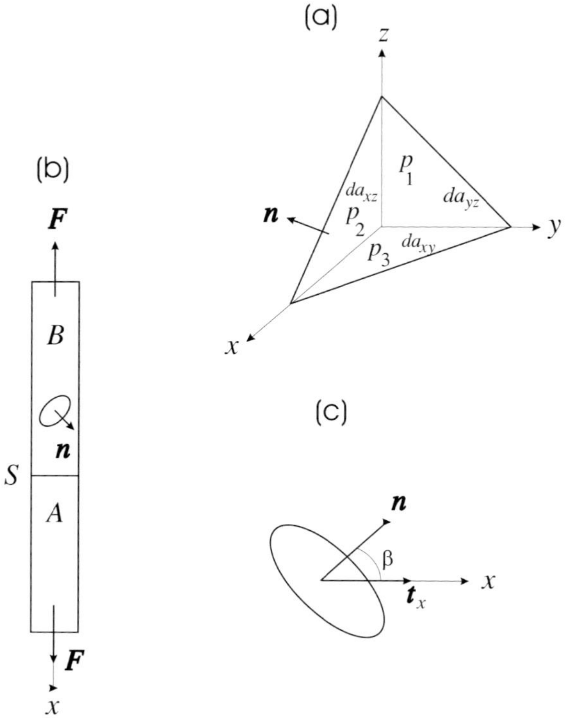

Figure C.3: (a) Cauchy formula for three dimensional cases (b) Equilibrium of bar (c) Illustration of eq. C-54

b. As was demonstrated in the first case, Cauchy formulas permit us to find elements of the stress tensor in the new Cartesian system of coordinates. Certainly, the same is correct in the three-dimensional case, (Appendix B)

Stress behavior and an equilibrium

Next we proceed from Cauchy formulas and illustrate a behavior of the traction components in an elastic medium. With this purpose in mind consider two examples, assuming that a body is in an equilibrium.

Example one Suppose that the stress tensor at some point of a medium is

$$X_x = Y_y = Z_z = -P \qquad \text{and} \qquad Y_z = Z_x = X_y = 0 \tag{C-51}$$

By definition, for an arbitrary oriented element da with the normal $\mathbf{n}$ we have

$$t_x = \mathbf{X} \cdot \mathbf{n} = -P\cos(\mathbf{i}, \mathbf{n}) = -P\ n_x$$

$$t_y = \mathbf{Y} \cdot \mathbf{n} = -P\cos(\mathbf{j}, \mathbf{n}) = -P\ n_y$$

$$t_z = \mathbf{Z} \cdot \mathbf{n} = -P\cos(\mathbf{k}, \mathbf{n}) = -P\ n_z$$

Respectively, the traction $\mathbf{t}$, acting on this surface element, is

$$\mathbf{t} = -P\ \mathbf{n}, \tag{C-52}$$

that is the vector $\mathbf{t}$ has a direction, which is opposite to the normal $\mathbf{n}$. Its magnitude is equal to the pressure P. As we know, such a behavior is observed in the ideal fluid when an equilibrium takes place.

Example two Consider an elastic bar, oriented along the x–axis, and assume that two forces, $\mathbf{F}$ and $-\mathbf{F}$, applied at bar ends, provide an equilibrium, Fig. C.3b. Because of these forces an extension occurs and internal forces arise. In order to find their distribution we mentally draw a cross- section S in any place of the bar. Its portions, A and B, are located at both sides of this surface. Inasmuch as the bar is an equilibrium, parts A and B are at rest, too. Therefore, the internal force, acting on S and caused by the portion A, is equal to $\mathbf{F}$. In other words, the resultant force, exerted on B, is equal to zero. Otherwise, it would be in a state of motion. Changing a position of the cross-section S and bearing in mind that the force is distributed uniformly over it, we conclude that $\mathbf{F}$ is the same at all points of the bar. Besides, this force is perpendicular to the section S, that is the traction has only the normal component, equal to

$$t_x = X_x = \frac{F}{S} \tag{C-53}$$

As was shown, this stress element provides an equilibrium, while the others are equal to zero:

$$T = \begin{pmatrix} \sigma_x & 0 & 0 \\ 0 & 0 & 0 \\ 0 & 0 & 0 \end{pmatrix}$$

Here it is proper to note that an influence of volume forces is ignored, and only surface forces are able to sustain this state of the bar. In the same manner we can consider the internal force, acting on the boundary S of the portion A. It is clear that this force differs from $\mathbf{F}$ by a sign only (Newton's third law).

Now we take an arbitrary oriented surface element inside the bar, Fig. C.3b. In accordance with Cauchy formulas, eq. C-18, the traction has only the component along the x–axis, and it is equal to

$$t_x = X_x \cos(\mathbf{n}, \mathbf{i}), \tag{C-54}$$

while $t_y = t_z = 0$ at all points of the bar. In particular, if the normal $\mathbf{n}$ is perpendicular to the x–axis, the traction is equal to zero. In general, there are both the normal and tangential components of the traction. Indeed, as is seen from Fig. C.3c

$$t_{nn} = t_x \cos\beta = X_x \cos^2\beta = X_x n_x^2 \tag{C-55}$$

and $$t_{sn} = -t_x \sin\beta = -X_x \sin\beta \cos\beta$$

Thus, the component t_{nn} gradually decreases with an increase of the angle β, while the shear component has a maximum, when $\beta = \pi/4$, and it is equal to

$$t_{sn} = \frac{X_x}{2} \tag{C-56}$$

Stress equations of motion

As is well known, a motion of an elastic body can be represented as a superposition of a translation and a rotation around its center of mass (Appendix A). Of course, there is also a deformation, and this phenomenon will be studied later. Equation describing translation is

$$M\mathbf{a}_0 = \mathbf{F}, \tag{C-57}$$

Here M is the total mass of the body, $\mathbf{a}_0$ is an acceleration of the center of mass, and $\mathbf{F}$ is the resultant of external forces. By definition, we have

$$M\mathbf{a}_0 = \int\limits_V \rho\, \mathbf{a}\, dV \qquad \text{and} \qquad \mathbf{F} = \oint\limits_V \mathbf{f}\, dV + \int\limits_S \mathbf{t}\, dS$$

Correspondingly, the first equation of motion in the integral form is

$$\int\limits_V \rho\, \mathbf{a}\, dV = \int\limits_V \mathbf{f}\, dV + \int\limits_S \mathbf{t}\, dS, \tag{C-58}$$

where $\mathbf{a}$ is the acceleration of an elementary mass, $\rho\, dV$. To obtain the differential form of this equation, consider, as in the case of an equilibrium, any component of this equality. For instance, the x-component is equal to:

$$\int_V \rho\, a_x dV = \int_V f_x\, dV + \oint_S \mathbf{X} \cdot \mathbf{n}\, dS \qquad \text{(C-59)}$$

Replacing the last integral by the volume one, we have

$$\int_V (\rho\, a_x - f_x - \operatorname{div} \mathbf{X}) dV = 0 \qquad \text{(C-60)}$$

Inasmuch as eq. C-60 is valid for any arbitrary volume, we conclude that the integrand is equal to zero, too:

$$\rho\, a_x = f_x + \operatorname{div} \mathbf{X} \qquad \text{(C-61)}$$

By analogy we obtain

$$\rho\, a_y = f_y + \operatorname{div} \mathbf{Y} \qquad \text{(C-62)}$$

and

$$\rho\, a_z = f_z + \operatorname{div} \mathbf{Z} \qquad \text{(C-63)}$$

The last three equations are the differential form of eq. C-58. For example, eq. C-62 shows that an acceleration of an elementary volume, ΔV, along the y-axis is defined by the volume force $f_y\, dV$ and the resultant of surface forces:

$$\operatorname{div}\mathbf{Y}\, dV$$

In fact, by a definition of the divergence, this product can be replaced as

$$\oint_S \mathbf{Y} \cdot \mathbf{dS},$$

and the integral describes the total surface force caused by a surrounding medium.

In most cases, discussed in this monograph, an influence of volume forces can be neglected. Therefore, in place of eq. C-58 we have

$$\int_V \rho\, \mathbf{a} dV = \oint_S \mathbf{t} dS \qquad \text{(C-64)}$$

As follows from eqs. C-61–C-63 in the Cartesian system of coordinates

$$\rho\, a_x = f_x + \frac{\partial X_x}{\partial x} + \frac{\partial X_y}{\partial y} + \frac{\partial X_z}{\partial z}$$

$$\rho\, a_y = f_y + \frac{\partial Y_x}{\partial x} + \frac{\partial Y_y}{\partial y} + \frac{\partial Y_z}{\partial z} \tag{C-65}$$

$$\rho\, a_z = f_z + \frac{\partial Z_x}{\partial x} + \frac{\partial Z_y}{\partial y} + \frac{\partial Z_z}{\partial z}$$

Here

$$a_x = \frac{\partial^2 s_x}{\partial t^2}, \qquad a_y = \frac{\partial^2 s_y}{\partial t^2}, \qquad a_z = \frac{\partial^2 s_z}{\partial t^2}$$

and

$$\mathbf{s} = s_x\mathbf{i} + s_y\mathbf{j} + s_z\mathbf{k}$$

is the displacement of the center of mass which in general takes place due to a translation of the volume, as a rigid body, and its deformation. Usually the set C-65 is called the stress equations. It may be proper to notice that they contain too many unknowns and this shortcoming will be removed later. Now we consider several examples, illustrating a stress behavior of a moving body.

Example one Suppose that at some instant $t = 0$ the constant force is applied to one end of the bar and it is directed along the x-axis, Fig. C.4a. As is well known, (Part I), at the beginning we observe waves, propagating between bar ends, and its different elements move with different velocities. It is essential that within this time interval a deformation changes. Then, after some time an influence of waves becomes negligible, and it happens due to an attenuation. Correspondingly, each elementary mass starts to move with the same acceleration, and we can apply the second Newton's law to any part of the bar. It is obvious that with a decrease of its length the first time interval becomes smaller and in the limit it tends to zero. Let us mentally draw the cross-section S of the bar and consider a portion A, Fig. C.4a. Since it moves with the same acceleration as the whole bar, the internal force, F_i, acting at points of S is defined from the equality

$$\frac{F}{\rho\, l\, S} = \frac{F_i}{\rho\, l_A S} \qquad \text{or} \qquad F_i = \frac{l_A}{l} F \tag{C-66}$$

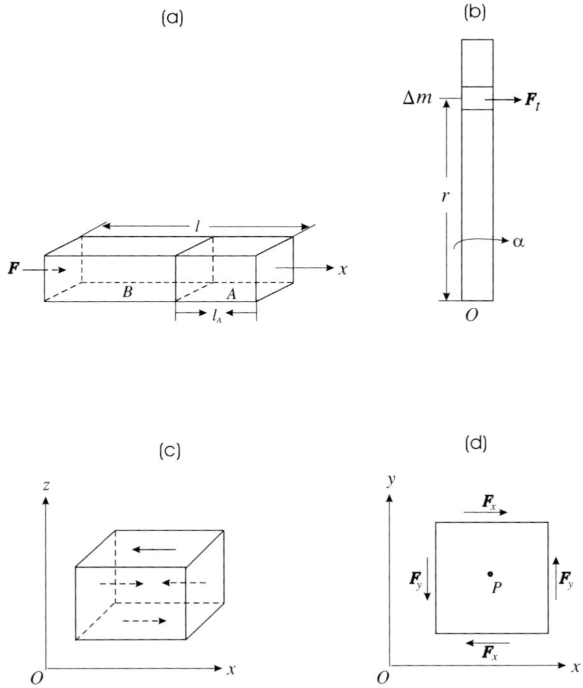

Figure C.4: (a) Distribution of internal forces in the bar when $\mathbf{F}$ =const (b) Rotation of the bar with the constant acceleration (c) Translation of elementary volume (d) Rotation of elementary volume

Thus, the internal force linearly decreases towards the front end of bar. Applying the principle of superposition we can determine a distribution of the internal forces, when both ends of the bar are under action of forces.

Example two Consider a bar rotating around its end with the constant acceleration α, Fig. C.4b. As follows from the second equation of motion the force, normal to the bar, is defined as

$$\mathbf{r} \times \mathbf{F}_t = I \ \boldsymbol{\alpha}, \tag{C-67}$$

Here I is the moment of inertia and r is the distance from an elementary mass, Δm, to the axis of rotation. In this case, eq. C-67 is simplified and we obtain

$$F_t \, r = \Delta m \, r^2 \alpha \qquad \text{or} \qquad F_t = \Delta m \, r\alpha \tag{C-68}$$

Thus, the shearing force linearly decreases with the distance r. Of course, there is also the centripetal force, directed along the bar.

Example three. Translation of an elementary volume Consider an elementary volume of an elastic medium and assume that the wave propagates along the x-axis, Fig. C.4c. In addition, we suppose that the force, associated with wave, has only the component $\mathbf{F}_x$. As soon as the wave reached the face with the coordinate: $x - \Delta x/2$, the left portion of the volume starts to move, while the other part remains at rest. Of course, during this motion we also observe a deformation. When the wave approaches the front face, $x+\Delta x/2$, the force $\mathbf{F}_x$ begins to act on a medium in front of the volume. In accordance with the third Newton's law, the force, caused by the surrounding medium and acting on this face:

$$\mathbf{F}_x(x + \frac{\Delta x}{2}, y, z, t)$$

has the same magnitude but opposite direction.

Thus, the total force, exerted on the opposite faces of the volume, is

$$\mathbf{F}_x(x - \frac{\Delta x}{2}, y, z, t) + \mathbf{F}_x(x + \frac{\Delta x}{2}, y, z, t) \tag{C-69}$$

Inasmuch as the x-component of the traction is related to the vector $\mathbf{X}$ as

$$t_x = \mathbf{X} \cdot \mathbf{n},$$

the sum in C-69 can be represented in the form

$$\left[X_x(x + \frac{\Delta x}{2}, y, z, t) - X_x(x - \frac{\Delta x}{2}, y, z, t)\right] \Delta y \Delta z \ \mathbf{i} \tag{C-70}$$

Here X_x is the normal stress. Now it is appropriate to point out that at the opposite faces the forces have opposite directions but stress values differ only slightly from each other. Because of this we assume that the function X_x changes linearly between these faces, the difference in C-70 is written as

$$\frac{\partial X_x(x, y, z, t)}{\partial x} \Delta x \Delta y \Delta z \ \mathbf{i}, \tag{C-71}$$

where X_x is the stress value at the middle point of the volume. The total force also includes the volume one

$$\frac{\partial X_x}{\partial x} + f_x = \rho \frac{\partial^2 s_x}{\partial t^2}$$

Geometry of the wave, propagating along the x-axis, allows us to assume that vectors **Y** and **Z** are absent. At the same time stresses X_y and X_z at opposite faces, perpendicular to the Y and Z-axes, are the same

$$\frac{\partial X_y}{\partial y} = \frac{\partial X_z}{\partial z} = 0$$

In such a case the set of equations of motion, C-65, is reduced to the single equation. If at each instant of time

$$X_x(x - \frac{\Delta x}{2}, y, z, t) = X_x(x + \frac{\Delta x}{2}, y, z, t) \qquad \text{and} \qquad f_x = 0,$$

then a deformation of the volume does not change, and it experiences a translation with the constant velocity. In a more general case, when these stresses are different, the volume is involved in a more complicated motion, including a vibration around its center of mass.

Example four. Rotation of elementary volume Now we investigate rotation of an elementary volume and, for simplicity, assume that it has a shape of the cube, Fig. C.4d. Its sides are equal to h. Since we are interested in rotation, an influence of normal stresses are not considered. As before, assume that the wave propagates along the x-axis, but unlike the first example, it produces the force, perpendicular to the x-axis. At the beginning, consider an action of its vector component $\mathbf{F}_y$. At points of the back face of the cube we have

$$\mathbf{F}_y(x - \frac{h}{2}, y, z, t) = t_y h^2 \mathbf{j} = (\mathbf{Y} \cdot \mathbf{n}) h^2 \mathbf{j}$$

or

$$\mathbf{F}_y(x - \frac{h}{2}, y, z, t) = -Y_x(x - \frac{h}{2}, y, z, t) h^2 \mathbf{j} \tag{C-72}$$

As soon as the wave front passes this face, the left portion of the volume begins to move along the y-axis, but the other portion is at rest. Such motion causes a deformation of the volume. Besides the normal and tangential components of the force appear at faces, perpendicular to the y-axis. When the wave approaches the face, $x + h/2$, the medium in front of the volume acts with the force

$$\mathbf{F}_y = t_y(x + \frac{h}{2}, y, z, t) h^2 \mathbf{j}$$

or

$$\mathbf{F}_y = Y_x(x + \frac{h}{2}, y, z, t) h^2 \mathbf{j} \tag{C-73}$$

It is essential that forces, given by eqs. C-72–C-73, have opposite directions, but their magnitudes differ only slightly from each other.

Forces, acting on faces, normal to the y-axis, display the same behavior, and we have:

$$\mathbf{F}_x(x, y + \frac{h}{2}, z, t) = X_y(x, y + \frac{h}{2}, z, t)h^2\ \mathbf{i} \tag{C-74}$$

and $$\mathbf{F}_x(x, y - \frac{h}{2}, z, t) = -X_y(x, y - \frac{h}{2}, z, t)h^2\ \mathbf{i}$$

It is clear that all four forces may cause a rotation of the volume about the z-axis, and it is described by the equation

$$M_z = I\alpha_z \tag{C-75}$$

Here M_z is the z-component of the torque, I is moment of inertia and α_z is the component of the angular acceleration.

As is seen from Fig. C.4d, the forces, acting at opposite faces of the volume, produce the torques in the same direction. For this reason, evaluating the total torque, we have to add their magnitudes. It is important to emphasize that torques due to forces, oriented along the x and y axes, have opposite directions. Otherwise we would not be able to observe an equilibrium. As follows from eqs. C-72–C-73 the magnitude of the first pair of torques is equal to

$$M_1 = \left[Y_x(x + \frac{h}{2}, y, z, t) + Y_x(x - \frac{h}{2}, y, z, t)\right]\frac{h^3}{2}, \tag{C-76}$$

since the level arm is $h/2$.

In the same manner the torque magnitude of the second pair is

$$M_2 = \left[X_y(x + \frac{h}{2}, z, t) + X_y(x, y - \frac{h}{2}, z, t)\right]\frac{h^3}{2} \tag{C-77}$$

Next we expand stresses at each face of the cube in the Taylor series around the middle point of the volume, $p(x, y, z)$. Discarding terms of the order h^3 and higher, we obtain

$$M_1 = \left[Y_x(p, t) + \frac{\partial^2 Y_x(p, t)}{\partial x^2}\frac{h^2}{8}\right]h^3 \tag{C-78}$$

and $$M_2 = \left[X_y(p, t) + \frac{\partial^2 X_y(p, t)}{\partial y^2}\frac{h^2}{8}\right]h^3 \qquad \text{if } h \to 0$$

It is proper to notice that the Taylor series of the magnitude of each torque contains the term with the first derivatives. However, fortunately they are absent in expressions for M_1 and M_2. As follows from eqs. C-78 the magnitude of the resultant torque around the z-axis is

$$M = M_1 - M_2 = \left[(Y_x - X_y) + \left(\frac{\partial^2 Y_x}{\partial x^2} - \frac{\partial^2 X_y}{\partial y^2} \right) \frac{h^2}{8} \right] h^3 \tag{C-79}$$

In accordance with eq. A-51 the moment of inertia of the cube is equal to

$$I = \frac{Mh^2}{6} = \frac{\rho h^5}{6} \tag{C-80}$$

that is it has the same order with respect to h, as the second term in eq. C-79. Substitution of eqs. C-77–C-80 into eq. C-75 yields

$$Y_x(p,t) - X_y(p,t) + \left(\frac{\partial^2 Y_x}{\partial x^2} - \frac{\partial^2 X_y}{\partial y^2} \right) \frac{h^2}{8} = \frac{\rho h^2}{6} \alpha_z \tag{C-81}$$

Here Y_x and X_y and their derivatives are taken at the point p. Since the acceleration can not be infinitely large, we conclude that

$$X_y(p,t) = Y_x(p,t) \tag{C-82}$$

By analogy, considering a rotation about the x and y axes it follows that

$$X_z = Z_x \qquad \text{and} \qquad Y_z = Z_y \tag{C-83}$$

The set of equations C-82–C-83 is very important result, because it shows that, as in the case of an equilibrium, the stress tensor is also symmetrical in the dynamic stage.

Besides from eq. C-81 it follows that the acceleration α_z is defined by the difference of second derivatives of stress components and naturally, it is independent on h. In fact, we have

$$\alpha_z = \frac{3}{4\rho} \left(\frac{\partial^2 Y_x}{\partial x^2} - \frac{\partial^2 X_y}{\partial y^2} \right) \tag{C-84}$$

In the same manner, considering a rotation of the volume around two other axes, we can obtain expressions for α_x and α_y.

Let us make some comments.

a. During a propagation of the wave through a given elementary volume, eq. C-75 is applied to its portion, located behind the wave front

b. If the shearing forces are such that

$$\frac{\partial^2 Y_x}{\partial x^2} = \frac{\partial^2 X_y}{\partial y^2},$$

the elementary volume rotates around the z-axis with the constant velocity ω, $(h \to 0)$.

Relationship between the second and first equations of motion

Next we pay attention to the second equation of motion, which describes rotation. As is well known, (Appendix A), this equation has the form

$$\frac{\partial \mathbf{L}}{\partial t} = \mathbf{M} \tag{C-85}$$

Here $\mathbf{M}$ is the resultant torque, caused by the external forces and, by definition, it is equal to

$$\mathbf{M} = \int_V (\mathbf{r} \times \mathbf{f}) dV + \oint \mathbf{r} \times \mathbf{t} \, dS \tag{C-86}$$

and $\mathbf{r}$ is the radius-vector, characterizing a position of the point of a volume with respect to the center of mass. The left hand side of eq. C-85 describes the rate of a change of the total angular momentum $\mathbf{L}$,

$$\mathbf{L} = \int_V \mathbf{r} \times \rho \mathbf{v} \, dV$$

Respectively

$$\frac{\partial \mathbf{L}}{\partial t} = \int_V \mathbf{r} \times \rho \mathbf{a} \, dV, \tag{C-87}$$

since

$$\frac{\partial}{\partial t}(\mathbf{r} \times \rho \mathbf{v}) = \mathbf{v} \times \rho \mathbf{v} + \mathbf{r} \times \rho \mathbf{a} = \mathbf{r} \times \rho \mathbf{a}$$

Thus, the second equation of motion in the integral form is

$$\int_V \rho \, \mathbf{r} \times \mathbf{a} \, dV = \int_V \mathbf{r} \times \mathbf{f} \, dV + \oint_S \mathbf{r} \times \mathbf{t} \, dS \tag{C-88}$$

It turns out that this equation is not independent, but it follows from the first equation of a motion. To demonstrate this very important fact, we consider at the beginning the x-component of eq. C-85, which is equal to

$$\int_V \rho(y \, a_z - z \, a_y) dV = \int_V (y f_z - z f_y) dV + \oint_S (y \, t_z - z \, t_y) dS \tag{C-89}$$

It is proper to notice that the same approach was used in the case of an equilibrium. Taking into account eqs. C-61–C-63 in place of eq. C-89 we have

$$\int\limits_V [y(f_z + \text{div } \mathbf{Z}) - z(f_y + \text{div } \mathbf{Y})]dV = \int\limits_V (yf_z - zf_y)dV + \oint\limits_S y(\mathbf{Z}\cdot\mathbf{n})dS - \oint\limits_S z(\mathbf{Y}\cdot\mathbf{n})dS$$

or

$$\int\limits_V (y \text{ div } \mathbf{Z} - z \text{ div } \mathbf{Y})dV = \int\limits_S \text{div}(y\mathbf{Z})dV - \int\limits_S \text{div}(z\mathbf{Y})dV \tag{C-90}$$

As was shown earlier

$$\text{div } y\mathbf{Z} = y \text{ div } \mathbf{Z} + Z_y \qquad \text{and} \qquad \text{div } z\mathbf{Y} = z \text{ div } \mathbf{Y} + Y_z$$

Therefore, eq. C-89 is greatly simplified, and we obtain

$$\int\limits_V (Z_y - Y_z)dV = 0 \tag{C-91}$$

Considering two other components of eq. C-85, we have

$$\int_V (X_z - Z_x)dV = 0 \qquad \text{and} \qquad \int\limits_V (X_y - Y_x)dV = 0 \tag{C-92}$$

These equalities can be interpreted in two different ways. First of all, we earlier proved that the stress tensor is symmetrical and, therefore, the left hand side of eqs. C-91–C-92 is equal to zero. This means that the second equation of motion follows from the first one. In other words, all information about both types of motion contains in eqs. C-61–C63. Certainly, it is important result, which greatly simplifies a study of wave fields. At the same time in some cases, when a translation is absent, the use of eq. C-85 can be more preferable. Also, eqs. C-91–C-92 may serve as another proof of the stress tensor symmetry. This follows from the fact that these equalities are valid for any volume, and, correspondingly, integrands are also equal to zero, that is

$$Z_y = Y_z, \qquad X_z = Z_x, \qquad X_y = Y_x$$

Appendix D

Deformation and strain tensor

The surface and volume forces applied to an ideally rigid body cause a motion, which can be in general described as a superposition of a translation, when all particles of a body have the same displacement, and a rotation about the center of mass. In both cases a distance between any two points of the body does not change. Since in reality every medium is an elastic one, we also observe a change of a relative position of different portions of the body. As a result its shape changes, as well as distance between points, and this phenomenon is called a deformation, Fig. D.1a. For instance, a straight line (1) of an arbitrary length becomes after a deformation a rather complicated line (2), Fig. D.1b. This suggests that under an action of different forces the same straight line can be transformed in unlimited number of lines of different shapes. Certainly, such a behavior makes a study of a deformation a very difficult task. For this reason the conventional approach considers a displacement, vector of an elementary segment of a curve, which can be treated as a straight line. In other words, we are going to study a change of a relative position of terminal points of this line, located close to each other. Thus, as was first established by Helmholtz, a motion of an elementary volume, in particular, the linear segment, can be represented as a sum of three components, namely
a. translation, b. rotation, c. deformation.
Our goal is to describe main features of a displacement, caused by a deformation.

Displacement and relative displacement

Consider some point p of the body. Its position with respect to the origin O is characterized by the radius-vector $\mathbf{r}$. Under an action of forces the medium around this point is moved, and after a deformation it is located around the point p' with the

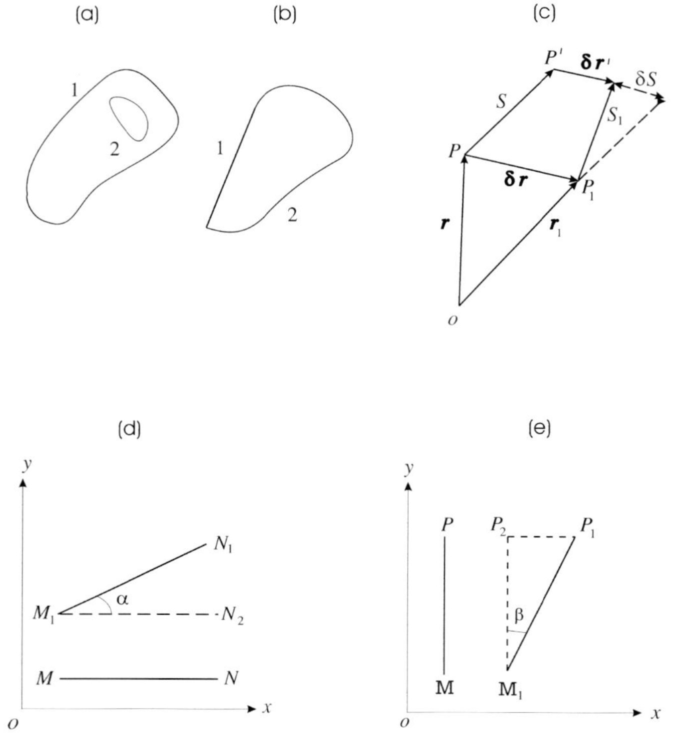

Figure D.1: (a) Deformation of an elementary volume (b) Change of a line due to deformation (c) Illustration of eqs. D-1–D-6 (d) Deformation of segment MN, parallel to the $x-$axis (e) Deformation of segment MP, parallel to the $y-$axis

radius-vector $\mathbf{r}'$, Fig. D.1c. Write

$$\mathbf{r}' = \mathbf{r} + \mathbf{s} \tag{D-1}$$

where $\mathbf{s}$ is the displacement of the point p, and, in general, it depends on coordinates of this point. Further we imply that the function $\mathbf{s}(\mathbf{r})$ is a continuous. Next, we take a neighboring point p_1 with the radius-vector $\mathbf{r}_1$. A deformation causes also a displacement of a medium around this point to the point p_1', and, as is seen from Fig. D.1c:

$$\mathbf{r}_1' = \mathbf{r}_1 + \mathbf{s}_1 \tag{D-2}$$

It is usually $\mathbf{s}_1 \neq \mathbf{s}$. From eqs. D-1 and D-2 it follows that

$$\delta\mathbf{r}' = \delta\mathbf{r} + \delta\mathbf{s} \tag{D-3}$$

where

$$\delta\mathbf{r} = \mathbf{r}_1 - \mathbf{r} \qquad \text{and} \qquad \delta\mathbf{r}' = \mathbf{r}_1' - \mathbf{r}' \tag{D-4}$$

characterize a relative position of two points (p and p_1) before and after a deformation. At the same time

$$\delta\mathbf{s} = \mathbf{s}(\mathbf{r}_1) - \mathbf{s}(\mathbf{r})$$

or

$$\delta\mathbf{s} = \mathbf{s}(\mathbf{r} + \delta\mathbf{r}) - \mathbf{s}(\mathbf{r}) \tag{D-5}$$

By definition, eq. D-3,

$$\delta\mathbf{s} = \delta\mathbf{r}' - \delta\mathbf{r}, \tag{D-6}$$

and it describes a change of a relative position of points p and p_1, Fig. D.1c. In other words, the vector $\delta\mathbf{s}$ is a measure of a deformation around some point p. As is seen from Fig. D.1c, vectors $\delta\mathbf{r}'$ and $\delta\mathbf{r}$ usually differ from each other by a magnitude and a direction. In particular, it may happen that the length of both vectors, $\delta\mathbf{r}'$ and $\delta\mathbf{r}$, remains the same, that is in such a case a displacement is not accompanied by a deformation. This indicates that the vector $\delta\mathbf{s}$ does not necessarily describe only a deformation, and this question will be studied in detail. Bearing in mind that $\delta\mathbf{s}$ represents a difference of the vector field $\mathbf{s}(\mathbf{r})$ at two neighboring points, it is natural to express this vector, $\delta\mathbf{s}$, in terms of the partial derivatives of the field $\mathbf{s}(\mathbf{r})$. Introducing the curvilinear orthogonal system of coordinates, x_1, x_2 and x_3, we have

$$\mathbf{s}(\mathbf{r}) = u(\mathbf{r})\, \mathbf{i}_1 + v(\mathbf{r})\, \mathbf{i}_2 + w(\mathbf{r})\, \mathbf{i}_3 \tag{D-7}$$

Here $\mathbf{i}_1$, $\mathbf{i}_2$ and $\mathbf{i}_3$ are unit vectors along coordinate lines, and $u(\mathbf{r})$, $v(\mathbf{r})$ and $w(\mathbf{r})$ are scalar components of the vector $\mathbf{s}$. Therefore, components of the vector $\delta\mathbf{s}$ are

$$\delta u = u(\mathbf{r}+\delta\mathbf{r}) - u(\mathbf{r})$$

$$\delta v = v(\mathbf{r}+\delta\mathbf{r}) - v(\mathbf{r}) \tag{D-8}$$

$$\delta w = w(\mathbf{r}+\delta\mathbf{r}) - w(\mathbf{r}),$$

First, consider the component $u(\mathbf{r})$ of the displacement $\mathbf{s}(\mathbf{r})$, which can be treated as a scalar field. Then its directional derivative at the point p is written as

$$\frac{\partial u}{\partial(\delta r)} = \mathbf{i}\cdot \text{grad } u \tag{D-9}$$

where $\mathbf{i}$ is the unit vector directed along $\delta\mathbf{r}$. Respectively, a change of the function u between points p and p_1 is equal to

$$\delta u = \frac{\partial u}{\partial(\delta r)}\delta\mathbf{r} = \text{grad } u \cdot \delta\mathbf{r} \tag{D-10}$$

By analogy, a change of two other components of the vector $\delta\mathbf{s}$ is

$$\delta v = \frac{\partial v}{\partial(\delta r)}\delta r = \text{grad } v \cdot \delta\mathbf{r} \quad \text{and} \quad \delta w = \frac{\partial w}{\partial(\delta r)}\delta r = \text{grad } w \cdot \delta\mathbf{r} \tag{D-11}$$

As was demonstrated in Part I the operator of gradient has a form

$$\text{grad} = \frac{\partial}{h_1\partial x_1}\mathbf{i}_1 + \frac{\partial}{h_2\partial x_2}\mathbf{i}_2 + \frac{\partial}{h_3\partial x_3}\mathbf{i}_3 \tag{D-12}$$

Here h_1, h_2 and h_3 are metric coefficients. Taking into account that

$$\delta\mathbf{r} = dx_1\,\mathbf{i}_1 + dx_2\,\mathbf{i}_2 + dx_3\,\mathbf{i}_3 \tag{D-13}$$

a change of scalar components of the field $\mathbf{s}$, eqs. D-10–D-11, can be represented in the form:

$$\begin{aligned}
\delta u &= \frac{1}{h_1}\frac{\partial u}{\partial x_1}dx_1 + \frac{1}{h_2}\frac{\partial u}{\partial x_2}dx_2 + \frac{1}{h_3}\frac{\partial u}{\partial x_3}dx_3 \\
\delta v &= \frac{1}{h_1}\frac{\partial v}{\partial x_1}dx_1 + \frac{1}{h_2}\frac{\partial v}{\partial x_2}dx_2 + \frac{1}{h_3}\frac{\partial v}{\partial x_3}dx_3 \\
\delta w &= \frac{1}{h_1}\frac{\partial w}{\partial x_1}dx_1 + \frac{1}{h_2}\frac{\partial w}{\partial x_2}dx_2 + \frac{1}{h_3}\frac{\partial w}{\partial x_3}dx_3
\end{aligned} \tag{D-14}$$

Before we study this set of equations let us perform some operations. Multiplication of eqs. D-10–D-11 by corresponding unit vectors and then a summation yields

$$\delta\mathbf{s} = (\mathbf{i}_1 \text{ grad } u + \mathbf{i}_2 \text{ grad } v + \mathbf{i}_3 \text{ grad } w)\cdot\delta\mathbf{r}$$

or

$$\delta \mathbf{s} = (\delta \mathbf{r} \cdot \text{grad})\ \mathbf{s} \tag{D-15}$$

For instance, the latter gives

$$\delta u = (\delta \mathbf{r} \cdot \text{grad}) u \tag{D-16}$$

or

$$\delta u = \frac{1}{h_1}\frac{\partial u}{\partial x_1}dx_1 + \frac{1}{h_2}\frac{\partial u}{\partial x_2}dx_2 + \frac{1}{h_3}\frac{\partial u}{\partial x_3}dx_3,$$

that coincides with the first equation of the set D-14

Tensor of deformation

It is convenient to represent eq. D-14 as

$$\delta \mathbf{s} = S\delta \mathbf{r} \tag{D-17}$$

Here

$$S = \begin{pmatrix} \frac{1}{h_1}\frac{\partial u}{\partial x_1} & \frac{1}{h_2}\frac{\partial u}{\partial x_2} & \frac{1}{h_3}\frac{\partial u}{\partial x_3} \\ \frac{1}{h_1}\frac{\partial v}{\partial x_1} & \frac{1}{h_2}\frac{\partial v}{\partial x_2} & \frac{1}{h_3}\frac{\partial v}{\partial x_3} \\ \frac{1}{h_1}\frac{\partial w}{\partial x_1} & \frac{1}{h_2}\frac{\partial w}{\partial x_2} & \frac{1}{h_3}\frac{\partial w}{\partial x_3} \end{pmatrix} \tag{D-18}$$

is the matrix which transforms the vector $\boldsymbol{\delta}\mathbf{r}$ into the vector $\boldsymbol{\delta}\mathbf{s}$. As was shown in the Appendix B this means that S is the tensor. Its nine elements are derivatives of scalar components of the field s with respect to displacements along the coordinate lines. In accordance with eq. D-17, if the tensor S is given, then we can determine a change of the relative position of two neighboring points, $\delta\mathbf{s}$, (eq. D-6), regardless of an orientation of $\delta\mathbf{r}$. It may be appropriate to notice that the tensor S plays the similar role, as the stress tensor, which allows us to find the traction $\mathbf{t}$, acting on arbitrary surface element. Let us make several comments:

a. The matrix S is an example of the tensor field, since its elements usually depend on coordinates of a point.

b. By definition of S, derivatives of the displacement s allow us to study a deformation, which can be caused by a motion of particles of a medium.

c. In principle, performing an integration of the tensor elements it is possible to determine the vector field $\mathbf{s}$ itself. The system of eqs. D-14 implies that we deal with very small displacements; for instance, the scalar component u at the point p_1, Fig. D.1c, is

$$u(\mathbf{r}+d\mathbf{r}) \quad \text{or} \quad u(x+dx,\, y+dy,\, z+dz)$$

Expanding this function in the Taylor's series around the point p, we have

$$u(\mathbf{r}+\delta\mathbf{r}) = u(\mathbf{r}) + \frac{\partial u}{\partial x}dx + \frac{\partial u}{\partial y}dy + \frac{\partial u}{\partial z}dz + \frac{1}{2}\frac{\partial^2 u}{\partial x^2}(dx)^2 + ... \tag{D-19}$$

Comparison with the first equation of the set D-8 clearly shows that eqs. D-14 are based on an assumption that terms of the series of the second and higher orders are neglected. In other words, the field $\mathbf{s}$ changes linearly within an elementary volume, where both vectors, $\delta\mathbf{r}$ and $\delta\mathbf{r}'$, Fig. D.1c, are situated. Because of this a study of a deformation is greatly simplified.

Homogeneous deformation

Earlier we introduced dx_1, dx_2 and dx_3 as relative coordinates of one point with respect to the other, while δu, δv and δw characterize its relative displacement. To emphasize this fact let us choose the Cartesian system of coordinates and change notations in the set D-14 in the following way

$$dx_1 \to x, \qquad dx_2 \to y, \qquad dx_3 \to z$$

and

$$\delta u \to u, \qquad \delta v \to v, \qquad \delta w \to w$$

Then we have

$$u = a_{11}x + a_{12}y + a_{13}z$$

$$v = a_{21}x + a_{22}y + a_{23}z \tag{D-20}$$

$$w = a_{31}x + a_{32}y + a_{33}z$$

Here a_{jk} are elements of the tensor

$$S=\begin{pmatrix} \dfrac{\partial u}{\partial x} & \dfrac{\partial u}{\partial y} & \dfrac{\partial u}{\partial z} \\ \dfrac{\partial v}{\partial x} & \dfrac{\partial v}{\partial y} & \dfrac{\partial v}{\partial y} \\ \dfrac{\partial w}{\partial x} & \dfrac{\partial w}{\partial y} & \dfrac{\partial w}{\partial z} \end{pmatrix}, \tag{D-21}$$

since in the Cartesian system of coordinates

$$h_1 = h_2 = h_3 = 1 \tag{D-22}$$

Thus, in our approximation the relative displacements, $(u,\ v,\ w)$, are linear functions of coordinates. Such a deformation is called the homogeneous one, and it may be observed in a volume of an arbitrary size. At the same time, a deformation of an elementary volume is always homogeneous because its dimensions are small. This follows from the fact that terms of Taylor's series, proportional to $(dx)^k$, $(dy)^k$ and $(dz)^k$, are negligible, if $k > 1$.

It is convenient to assume that the point p, Fig. D.1c, coincides with the origin of coordinates and its displacement is equal to zero. Correspondingly, u, v and w are scalar components of the displacement of the point p_1 with coordinates: x, y, z. Comparison with eq. D-6 shows that in this case

$$\delta \mathbf{r} = x\ \mathbf{i} + y\ \mathbf{j} + z\ \mathbf{k} \qquad \text{and} \qquad \delta \mathbf{s} = \mathbf{s}(p_1) = u\ \mathbf{i} + v\ \mathbf{j} + w\ \mathbf{k} \tag{D-23}$$

and both vectors, $\delta\mathbf{r}$ and $\delta\mathbf{s}$, have the common point p.

As was mentioned earlier the homogeneous deformation may take place also in a volume of finite dimensions, when components of the displacement, $\mathbf{s}$, are linear functions of a point inside a volume. In other words, derivatives of these components of the second and higher order vanish, and elements of the tensor S are independent of coordinates of a point. Considering an elementary volume we demonstrated that a small deformation transforms the straight segment, $\delta\mathbf{r}$, also into the straight element, $\delta\mathbf{s}$. It turns out that it is one of properties of the linear transformation, given by the set D-20. After these comments we describe the main features of the homogeneous deformation.

Transformation of a plane

Let us imagine a plane inside a volume. As is well known, its equation can be written in the form:

$$A\ x + B\ y + C\ z + D = 0 \tag{D-24}$$

Here x, y, z are coordinates of any point, situated on the plane, while A, B, C are scalar components of the vector $\mathbf{N}$, perpendicular to the plane:

$$\mathbf{N} = A\ \mathbf{i} + B\ \mathbf{j} + C\ \mathbf{k} \tag{D-25}$$

and, finally, D is constant.

Under an action of external forces the volume becomes deformed. Correspondingly, each point of the plane experiences a displacement, and as a result, new surface is formed. In order to obtain its equation, we have to perform in eq. D-24 the following replacements

$$x \to x + u, \qquad y \to y + v, \qquad z \to z + w$$

Making use of the set D-20, we obtain

$$A\ (x + a_{11}x + a_{12}y + a_{13}z) + B\ (y + a_{21}x + a_{22}y + a_{23}z)$$

$$+C\ (z + a_{31}x + a_{32}y + a_{33}z) + D = 0$$

or

$$(A + a_{11}A + Ba_{21} + Ca_{31})x + \tag{D-26}$$

$$(B + a_{22}B + a_{12}A + a_{32}C)y + (C + a_{33}C + a_{13}A + a_{23}B)z + D = 0$$

This is the equation of a plane, and, therefore, the homogeneous deformation transforms the plane into a plane. This is the first important feature of such a deformation.

Next, consider two parallel planes. Their equations are

$$A_1x + B_1y + C_1z + D_1 = 0 \qquad \text{and} \qquad A_2x + B_2y + C_2z + D_2 = 0 \tag{D-27}$$

Inasmuch as vectors $\mathbf{N}_1$ and $\mathbf{N}_2$, eq. D-25, are parallel to each other too we have

$$\frac{A_2}{A_1} = \frac{B_2}{B_1} = \frac{C_2}{C_1} = k, \tag{D-28}$$

where k is a constant. Substitution of eq. D-28 into the second equation of the set D-27 yields

$$k(A_1x + B_1y + C_1z) + D_2 = 0$$

or

$$A_1x + B_1y + C_1z + D_3 = 0 \tag{D-29}$$

Comparison with the first equation of the set D-27 shows that both planes are characterized by the same vector $\mathbf{N}_1$, which is perpendicular to both planes. Thus, two parallel planes remain parallel to each other and this is the second feature of the homogeneous deformation. As illustration consider a rectangular parallelepiped. It is clear that after a small deformation the opposite faces are still parallel, and a new volume has in general a shape of an oblique parallelepiped.

Transformation of a straight line

It is convenient to treat the straight line as an intersection of two planes. As was shown, the latter remain plane after a deformation. Correspondingly, the line of their intersection is the straight one. Thus, the set D-20 transforms the straight line into the straight one too, and this is another feature of the homogeneous deformation. Applying the same approach as in the case of planes, one can show that if two straight lines are parallel to each other, then they remain parallel in the strained volume. For instance, after a small deformation opposite sides of a parallelogram are still parallel.

Transformation of a sphere

Consider the spherical surface:

$$x^2 + y^2 + z^2 = r^2$$

Because of the deformation its points are shifted and a new surface is described by the equation

$$(x + u)^2 + (y + v)^2 + (z + w)^2 = r^2$$

After a substitution of displacements, eqs. D-20, we arrive at the equation

$$b_{11}\, x^2 + b_{22}\, y^2 + b_{33}\, z^2 + 2b_{12}\, x\, y + 2b_{13}\, x\, z + 2b_{23}\, y\, z - b = 0, \tag{D-30}$$

which describes the surface of the second order. In general, it is ellipsoid, and this is also an important feature of such a deformation. Let us notice that there are cases when a sphere is transformed into a hyperboloid.

Components of strain tensor

Next we study a geometrical meaning of the tensor components, eq. D-21, provided that the rate of a change of the displacement, **s**, is very small. As was pointed out we deal with the small deformation only. With this purpose in mind it is convenient to rewrite eqs. D-14 in the Cartesian system of coordinates. This gives

$$u_1 = u + \frac{\partial u}{\partial x}\,dx + \frac{\partial u}{\partial y}\,dy + \frac{\partial u}{\partial z}\,dz$$

$$v_1 = v + \frac{\partial v}{\partial x}\,dx + \frac{\partial v}{\partial y}\,dy + \frac{\partial v}{\partial z}\,dz \qquad \text{(D-31)}$$

$$w_1 = w + \frac{\partial w}{\partial x}\,dx + \frac{\partial w}{\partial y}\,dy + \frac{\partial w}{\partial z}\,dz$$

where

$$u = u(x, y, z), \qquad v = v(x, y, z), \qquad w = w(x, y, z)$$

and

$$u_1 = u(x + dx,\ y + dy,\ z + dz),$$

$$v_1 = v(x + dx,\ y + dy,\ z + dz),$$

$$w_1 = w(x + dx,\ y + dy,\ z + dz)$$

Diagonal elements: $\dfrac{\partial u}{\partial x}, \dfrac{\partial v}{\partial y}, \dfrac{\partial w}{\partial z}$

First, suppose that the linear segment MN is initially parallel to the x-axis. Due to the small deformation a medium around points M and N is moved, and we obtain a new straight line, M_1N_1. In general it differs from MN by a length and an orientation, Fig. D.1d. As follows from eqs. D-13, we have

$$MN = dx \qquad \text{and} \qquad dy = dz = 0$$

or

$$\mathbf{MN} = dx\ \mathbf{i} \qquad \text{(D-32)}$$

and

$$\delta u = \frac{\partial u}{\partial x}\, dx, \quad \delta v = \frac{\partial v}{\partial x}\, dx, \quad \delta w = \frac{\partial w}{\partial x}\, dx$$

or

$$\delta \mathbf{s} = \frac{\partial u}{\partial x} dx\ \mathbf{i} + \frac{\partial v}{\partial x} dx\ \mathbf{j} + \frac{\partial w}{\partial x} dx\ \mathbf{k} \tag{D-33}$$

By definition, Fig. D.1c,

$$\mathbf{M}_1\mathbf{N}_1 = \mathbf{MN} + \boldsymbol{\delta}\mathbf{s}$$

or

$$\mathbf{M}_1\mathbf{N}_1 = \left(1 + \frac{\partial u}{\partial x}\right)\ dx\ \mathbf{i} + \frac{\partial v}{\partial x}\ dx\ \mathbf{j} + \frac{\partial w}{\partial x}\ dx\ \mathbf{k} \tag{D-34}$$

Unlike $\mathbf{MN}$, the vector $\mathbf{M}_1\mathbf{N}_1$ may have all three components, and its length is defined from the equality:

$$M_1N_1 = MN\left[\left(1 + \frac{\partial u}{\partial x}\right)^2 + \left(\frac{\partial v}{\partial x}\right)^2 + \left(\frac{\partial w}{\partial x}\right)^2\right]^{1/2} \tag{D-35}$$

Inasmuch as we are interested by small deformation only, terms: $(\partial v/\partial x)^2$, $(\partial w/\partial x)^2$, as well as $(\partial u/\partial x)^2$, can be neglected. In particular, it implies that the length of M_1N_1 does not change when the point N_1 slightly moves in parallel to either the y or z–axes, Fig. D.1d. Thus, eq. D-35 gives

$$M_1N_1 = \left(1 + \frac{\partial u}{\partial x}\right) MN \tag{D-36}$$

or

$$e_{xx} = \frac{\partial u}{\partial x} = \frac{M_1N_1 - MN}{MN}, \tag{D-37}$$

and it describes the relative change of the length of MN. By analogy, two other diagonal elements:

$$e_{yy} = \frac{\partial v}{\partial y} \qquad \text{and} \qquad e_{zz} = \frac{\partial w}{\partial z}$$

characterize a relative change of the length of the straight segments, which are parallel to either the y or z–axes. It is essential that the diagonal elements of the tensor T differ from zero only in those cases when there is a change of the distance between points of a medium. In other words, they describe so called the pure deformation.

Nondiagonal elements

Suppose that at the beginning the linear segment MN is oriented along the x–axis, but after a deformation it is transformed into the linear segment M_1N_1, Fig. D.1d. They form the angle α, which in general differ from zero and characterizes a turn of MN during a deformation:

$$\sin\alpha = \frac{N_1N_2}{M_1N_1} \tag{D-38}$$

By definition, N_1N_2 is the difference of displacements at points M and N along the y–axis. As follows from the second equation of the set D-31 we have

$$N_1N_2 = v_1 - v = \frac{\partial v}{\partial x}dx \tag{D-39}$$

Substitution of eqs. D-36 and D-39 into eq. D-38 gives

$$\sin\alpha = \frac{\dfrac{\partial v}{\partial x}}{1 + \dfrac{\partial u}{\partial x}}$$

Since α and $\partial u/\partial x$ are very small, the latter is greatly simplified, and we obtain

$$\alpha = \frac{\partial v}{\partial x} \tag{D-40}$$

Thus, the first element in the second row of the tensor T defines an orientation change of the linear segment, MN, which was initially parallel to the x–axis.

It is interesting to derive the same result differently. Making use of the dot product of vectors $\mathbf{MN}$ and $\mathbf{M_1N_1}$ the same angle between them is defined as

$$\cos\alpha = \frac{\mathbf{MN}\cdot\mathbf{M_1N_1}}{MN\ M_1N_1} \tag{D-41}$$

From eqs. D-32 and D-36 it follows that

$$\cos\alpha = \frac{1 + \dfrac{\partial u}{\partial x}}{\left[(1 + \dfrac{\partial u}{\partial x})^2 + (\dfrac{\partial v}{\partial x})^2\right]^{\frac{1}{2}}}, \tag{D-42}$$

because $w = 0$.

In spite of the fact, that the term $(\partial v/\partial x)^2$ is very small, it has to be presented. Expanding the left and right hand side of eq. D-42 in a series and discarding all terms, except the first two, we have

$$1 - \frac{\alpha^2}{2} = 1 - \frac{1}{2}\frac{\left(\frac{\partial v}{\partial x}\right)^2}{\left(1 + \frac{\partial u}{\partial x}\right)^2} \tag{D-43}$$

or

$$\alpha = \frac{\partial v}{\partial x},$$

that coincides with eq. D-40. Certainly the geometrical approach in deriving this angle is simple.

Next assume that the linear segment MP is initially parallel to the y–axis, and due to a deformation it becomes M_1N_1. The distortion angle β is defined from the triangle $M_1P_2P_1$, Fig. D.1e

$$\sin\beta = \frac{P_1P_2}{M_1P_1} \tag{D-44}$$

The numerator is a difference of the displacement of points M and N along the x–axis:

$$P_1P_2 = u_1 - u = \frac{\partial u}{\partial y}dy$$

Whence

$$\beta = \frac{\partial u}{\partial y}, \tag{D-45}$$

since $\sin\beta \sim \beta$.

Therefore, the second element of the first row of the tensor T characterizes a change of the direction of the vector $\mathbf{MN}$, oriented at the y–axis. In the same manner the tensor elements

$$\frac{\partial u}{\partial z} \quad \text{and} \quad \frac{\partial w}{\partial x}, \quad \text{as well as} \quad \frac{\partial w}{\partial y} \quad \text{and} \quad \frac{\partial v}{\partial z}$$

describe the distortion angles in planes XOZ and YOZ, respectively. It is obvious that each element of the tensor of a deformation, T, can be either positive or negative or equal to zero. If we introduce a notation

$$\mathbf{s} = \sum s_j\mathbf{i}_j,$$

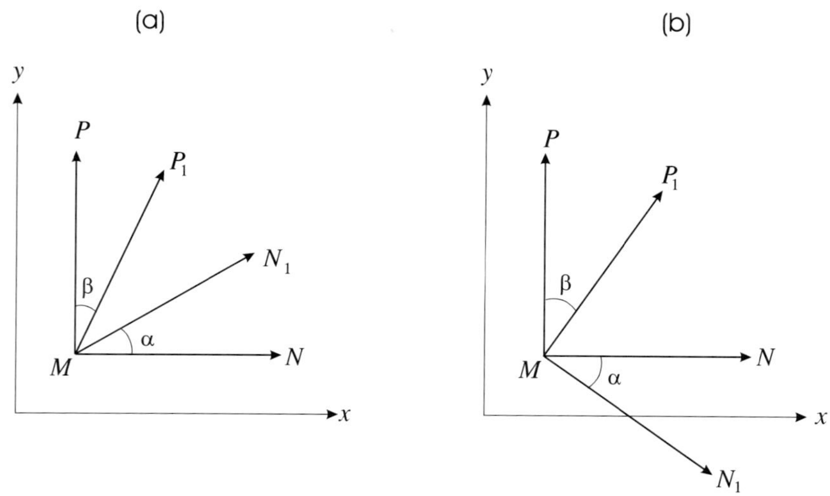

Figure D.2: Relationship between the distortion angles and nondiagonal elements of the strain tensor

then the nondiagonal element $\dfrac{\partial s_j}{\partial x_k}$ is positive when the component s_j increases with an increase of x_k. Thus, all nondiagonal elements of T characterize a change of a direction of the straight segments, oriented initially along the coordinate axes.

In general, the distortion angle may arise for two reasons, namely, a pure deformation and a rotation. In order to study the first factor consider the right angle between the linear segments MN and MP, having the common point M, Fig. D.2a. Before a deformation they are parallel to the $x-$ and y–axes, respectively. As follows from eqs. D-40 and D-45, after a deformation this angle decreases by a sum of angles

$$\alpha+\beta=\frac{\partial v}{\partial x}+\frac{\partial u}{\partial y}$$

This quantity is called the shearing strain, and it is denoted as

$$e_{xy}=\frac{\partial v}{\partial x}+\frac{\partial u}{\partial y}$$

In the same manner we can observe a decrease of the right angle in the XOZ plane:

$$e_{xz}=\frac{\partial w}{\partial x}+\frac{\partial u}{\partial z},$$

as well as in the YOZ plane

$$e_{yz}=\frac{\partial w}{\partial y}+\frac{\partial v}{\partial z}$$

Thus, we obtain six quantities, which are formed by elements of the tensor T, and they are

1.Three unit elongations:

$$e_{xx} = \frac{\partial u}{\partial x}, \qquad e_{yy} = \frac{\partial v}{\partial y}, \qquad e_{zz} = \frac{\partial w}{\partial z} \tag{D-46}$$

They describe a relative change of the length of the linear segments, which were initially parallel to the coordinate axes.

2.Three shearing strains

$$e_{xy} = \frac{\partial v}{\partial x} + \frac{\partial u}{\partial y}, \qquad e_{xz} = \frac{\partial w}{\partial x} + \frac{\partial u}{\partial z}, \qquad e_{yz} = \frac{\partial w}{\partial y} + \frac{\partial v}{\partial z}, \tag{D-47}$$

characterizing a decrease of the right angle between the linear segments, oriented along coordinate axes before a deformation. These quantities, eqs. D.46–D.47, are called components of a small deformation, and as will be shown later, they describe only the pure deformation.

Influence of translation and rotation on tensor T

At the beginning we pointed out that in general the small displacement, $\mathbf{s}(x, y, z)$, can be represented as a sum of three displacements, caused by a translation, a rotation and the pure deformation. In the last case the distance between points usually varies. Of course, it is very useful to know how elements of the tensor T (eq. D-21) are sensitive to every type of motion. It is natural to start from the simplest case, that is a translation, when all particles have the same displacement

$$u(x, y, z) = C_1, \qquad v(x, y, z) = C_2, \qquad w(x, y, z) = C_3,$$

where C_1, C_2 and C_3 are constants. Since derivatives of these functions with respect to coordinates are equal to zero, we make an obvious conclusion: translation does not make any influence on the tensor T.

Next, suppose that an elementary volume of a medium is involved in a rotation only, and $M(\mathbf{r})$ and $N(\mathbf{r}_1)$ are two its arbitrary points. As before, $\mathbf{r}$ and $\mathbf{r}_1$ are radius-vectors, characterizing a position of these points with respect to the origin O, located inside the volume. After a rotation the linear segment $\mathbf{MN} = \delta\mathbf{r}$ is transformed into the element $\mathbf{M}_1\mathbf{N}_1 = \delta\mathbf{r}_1$, and in accordance with eq. D-3

$$\mathbf{M}_1\mathbf{N}_1 = \mathbf{MN} + \delta\mathbf{s} \qquad \text{or} \qquad \delta\mathbf{r}' = \delta\mathbf{r} + \delta\mathbf{s} \tag{D-48}$$

Here

$$\delta\mathbf{s} = \mathbf{s}(\mathbf{r}_1) - \mathbf{s}(\mathbf{r})$$

First of all, it is useful to show that a small rotation is not accompanied by a change of the length of MN. To demonstrate this important fact, we recall that at each instant of time a rotation is a motion about some axis. In other words, the displacement $\mathbf{s}$ is located in the plane, perpendicular to this axis. Moreover, the magnitude of the vector $\mathbf{s}$ is proportional to the distance from the axis of rotation, (Appendix A). Because of this it is proper to represent the vector $\mathbf{s}$ as

$$\mathbf{s}(\mathbf{r}_1) = \mathbf{b} \times \mathbf{r}_1 \qquad \text{and} \qquad \mathbf{s}(\mathbf{r}) = \mathbf{b} \times \mathbf{r}, \tag{D-49}$$

where $\mathbf{b}$ is the vector, directed along the axis of rotation. Therefore, for the vector $\delta\mathbf{s}$ we have

$$\delta\mathbf{s} = \mathbf{b} \times \mathbf{r}_1 - \mathbf{b} \times \mathbf{r} = \mathbf{b} \times \delta\mathbf{r} \tag{D-50}$$

Substituting into eq. D-48 and taking the square from both its sides we have

$$(\delta\mathbf{r}')^2 = (\delta\mathbf{r} + \delta\mathbf{s})^2 = (\delta\mathbf{r} + \mathbf{b} \times \boldsymbol{\delta}\mathbf{r})^2 = (\delta\mathbf{r})^2 + 2(\mathbf{b} \times \delta\mathbf{r}) \cdot \delta\mathbf{r}, \tag{D-51}$$

since the term $(\delta\mathbf{s})^2$ is very small and it can be neglected. Because

$$(\mathbf{b} \times \delta\mathbf{r}) \cdot \delta\mathbf{r} = \mathbf{b} \cdot (\delta\mathbf{r} \times \delta\mathbf{r}) = 0,$$

eq. D-51 gives

$$|\delta\mathbf{r}'| = |\delta\mathbf{r}| \tag{D-52}$$

Thus, a small rotation does not change a length of the linear segment. In particular, it can be initially oriented along coordinate axes, Fig. D.1d,e, that is

$$M_1 N_1 = MN \qquad \text{and} \qquad MP_1 = MP \tag{D-53}$$

This means that if only a rotation takes place, the diagonal elements of the deformation tensor are equal to zero:

$$\frac{\partial u}{\partial x} = \frac{\partial v}{\partial y} = \frac{\partial w}{\partial z} = 0 \tag{D-54}$$

Suppose again that there is only a rotation and determine a relationship between the vector $\mathbf{b}$ and nondiagonal elements of the tensor T. First, we rewrite eq. D-50 in the form

$$\delta u\,\mathbf{i}+\delta v\,\mathbf{j}+\delta w\,\mathbf{k}=\begin{vmatrix}\mathbf{i} & \mathbf{j} & \mathbf{k}\\ b_x & b_y & b_z\\ dx & dy & dz\end{vmatrix} \tag{D-55}$$

Making use of eqs. D-31 and eq. D-55 we have

$$\begin{aligned}&\frac{\partial u}{\partial y}dy+\frac{\partial u}{\partial z}dz=b_y dz-b_z dy\\[2ex]&\frac{\partial v}{\partial x}dx+\frac{\partial v}{\partial z}dz=b_z dx-b_x dz\\[2ex]&\frac{\partial w}{\partial x}dx+\frac{\partial w}{\partial y}dy=b_x dy-b_y dx\end{aligned} \tag{D-56}$$

Inasmuch as dx, dy and dz are independent and arbitrary quantities, coefficients in front of each of them should be equal, that is

$$b_y=\frac{\partial u}{\partial z}\quad b_z=-\frac{\partial u}{\partial y},\qquad b_z=\frac{\partial v}{\partial x}\quad b_x=-\frac{\partial v}{\partial z},\qquad b_x=\frac{\partial w}{\partial y}\quad b_y=-\frac{\partial w}{\partial x} \tag{D-57}$$

or

$$b_x=-\frac{\partial v}{\partial z}=\frac{\partial w}{\partial y} \tag{D-58}$$

$$b_y=\frac{\partial u}{\partial z}=-\frac{\partial w}{\partial x} \tag{D-59}$$

$$b_z=-\frac{\partial u}{\partial y}=\frac{\partial v}{\partial x} \tag{D-60}$$

These equalities have very simple geometrical meaning. In fact, each derivative shows a change of an orientation of the linear segment, which was initially parallel to one of coordinate axes. For illustration, consider a rotation about the z–axis, Fig. D.2b. As we already know the distortion angle of the segment MP is equal to $\partial u/\partial y$, while the same angle of the segment MN is defined as $-\partial v/\partial x$. The minus is related to the fact that the scalar component of the displacement, v, is negative. As is seen from

Fig. D.2b in the case of a rotation the distortion angles are equal to each other and, correspondingly, derivatives $\partial u/\partial y$ and $\partial v/\partial x$ have the same magnitude but differ by a sign, eq. D-60. In other words, the right angle, formed by linear segments MP and MN, which are parallel to the coordinate axes, remains unchanged. Certainly, this result is easily expected. In the similar manner, eqs. D-58–D-59 describe a rotation about the $x-$ and y–axes.

Bearing in mind that during a rotation the right angle between the linear segments is preserved, we conclude that in this case the shearing strains are equal to zero. Also it follows from eqs. D-58–D-60, which give

$$e_{yz} = \frac{\partial v}{\partial z} + \frac{\partial w}{\partial y} = 0, \qquad e_{xz} = \frac{\partial u}{\partial z} + \frac{\partial w}{\partial x} = 0, \qquad e_{yx} = \frac{\partial v}{\partial x} + \frac{\partial u}{\partial y} = 0$$

Thus, we found out that neither the unit elongations, nor shearing strains are sensitive to a small rotation.

Now we present the vector of rotation, $\mathbf{b}$, in the form, which allows us to treat the tensor of deformation as a sum of two tensors, describing either the pure deformation or a rotation. In order to solve this task, let us first make use of the set D-57. Performing a summation of equalities for the same component of the vector $\mathbf{b}$, we obtain

$$b_x = \frac{1}{2}\left(\frac{\partial w}{\partial y} - \frac{\partial v}{\partial z}\right), \quad b_y = \frac{1}{2}\left(\frac{\partial u}{\partial z} - \frac{\partial w}{\partial x}\right), \quad b_z = \frac{1}{2}\left(\frac{\partial v}{\partial x} - \frac{\partial u}{\partial y}\right) \tag{D-61}$$

or

$$\mathbf{b} = \frac{1}{2}\,\text{curl}\;\mathbf{s} \tag{D-62}$$

Correspondingly, any change of the displacement, caused by a small rotation, eq. D-50, can be written in the form

$$\delta\mathbf{s} = \frac{1}{2}\,(\text{curl}\;\mathbf{s}\times\delta\mathbf{r}) \tag{D-63}$$

The latter can be treated as a transformation of the vector $\delta\mathbf{r}$. By definition, it can be also done with a help of some tensor B:

$$\delta\mathbf{s} = B\;\delta\mathbf{r}, \tag{D-64}$$

where

$$B = \begin{pmatrix} b_{11} & b_{12} & b_{13} \\ b_{21} & b_{22} & b_{23} \\ b_{31} & b_{32} & b_{33} \end{pmatrix} \tag{D-65}$$

Since

$$\mathbf{b}\times\delta\mathbf{r} = B\ \delta\mathbf{r},$$

we have

$$b_y dz - b_z dy = b_{11}\ dx + b_{12}\ dy + b_{13}\ dz$$

$$b_z dx - b_x dz = b_{21}\ dx + b_{22}\ dy + b_{23}\ dz$$

$$b_x dy - b_y dx = b_{31}\ dx + b_{32}\ dy + b_{33}\ dz$$

Its solution is

$$b_{11} = 0 \qquad b_{12} = -b_z \qquad b_{13} = b_y$$

$$b_{21} = b_z \qquad b_{22} = 0 \qquad b_{23} = -b_x \tag{D-66}$$

$$b_{31} = -b_y \qquad b_{32} = -b_x \qquad b_{33} = 0$$

Therefore, we conclude that the small rotation can be described by the antisymmetric tensor:

$$B = \begin{pmatrix} 0 & b_{12} & b_{13} \\ -b_{12} & 0 & b_{23} \\ -b_{13} & -b_{23} & 0 \end{pmatrix} \tag{D-67}$$

In accordance with eqs. D-61–D-66

$$b_{12} = \frac{1}{2}\left(\frac{\partial u}{\partial y} - \frac{\partial v}{\partial x}\right), \qquad b_{13} = \frac{1}{2}\left(\frac{\partial u}{\partial z} - \frac{\partial w}{\partial x}\right), \qquad b_{23} = \frac{1}{2}\left(\frac{\partial v}{\partial z} - \frac{\partial w}{\partial y}\right) \tag{D-68}$$

Certainly, the vector $\mathbf{b}$ and tensor B both allow us equally to describe a small rotation of an elementary volume.

Tensor of pure deformation, E

By analogy with the tensor of rotation, B, we introduce the tensor E, which characterizes the pure deformation only:

$$E = \begin{pmatrix} \varepsilon_{11} & \varepsilon_{12} & \varepsilon_{13} \\ \varepsilon_{21} & \varepsilon_{22} & \varepsilon_{23} \\ \varepsilon_{31} & \varepsilon_{32} & \varepsilon_{33} \end{pmatrix}, \tag{D-69}$$

where its diagonal elements

$$\varepsilon_{11} = \frac{\partial u}{\partial x}, \qquad \varepsilon_{22} = \frac{\partial v}{\partial y}, \qquad \varepsilon_{33} = \frac{\partial w}{\partial z} \tag{D-70}$$

are unit elongations, while the nondiagonal elements are two times smaller than the corresponding shearing strains:

$$\varepsilon_{12} = \varepsilon_{21} = \frac{1}{2}\left(\frac{\partial u}{\partial y} + \frac{\partial v}{\partial x}\right),$$

$$\varepsilon_{13} = \varepsilon_{31} = \frac{1}{2}\left(\frac{\partial u}{\partial z} + \frac{\partial w}{\partial x}\right), \tag{D-71}$$

$$\varepsilon_{23} = \varepsilon_{32} = \frac{1}{2}\left(\frac{\partial v}{\partial z} + \frac{\partial w}{\partial y}\right)$$

As follows from eqs. D-58–D-60, this tensor is not subjected to an influence of a rotation, as well as a translation. Comparison of tensors, T, E and B, clearly shows that T can be represented as a sum of tensors:

$$T = E + B \tag{D-72}$$

As was pointed out earlier, one of them, E, characterizes only the pure deformation, while the other, B, gives information about the angle of turn of an elementary volume, as a rigid body, with respect to coordinate axes

Superposition of small deformations

By definition, a homogeneous deformation is described by a linear transformation of the straight segment with terminal points, located close to each other. Because of this linearity we can expect that a sum of small deformations is also the homogeneous one. In particular, it may suggest that an arbitrary homogeneous deformation can be represented as a superposition of simpler deformations. Certainly, this fact may greatly simplify a study of small deformations. As we know, a deformation is defined by a relative change of a position of one point, N, with respect to the other point M. It is essential that coordinates of the latter do not participate in equations, describing a deformation. This allows us to place the origin of coordinates at the point M, which moves during a deformation. In other words, coordinates of the point M remain equal to zero. It is convenient to introduce the following notations:

a. x, y, z and x_1, y_1 z_1 are coordinates of the point N before and after a deformation, respectively.

b. u, v, and w are scalar components of the displacement $\mathbf{s}$ of the point N, that is

$$x_1 = x + u, \qquad y_1 = y + v, \qquad z_1 = z + w$$

Therefore, eqs. D-31 can be written in the form:

$$u = x_1 - x = \frac{\partial u}{\partial x}x + \frac{\partial u}{\partial y}y + \frac{\partial u}{\partial z}z$$

$$v = y_1 - y = \frac{\partial v}{\partial x}x + \frac{\partial v}{\partial y}y + \frac{\partial v}{\partial z}z \tag{D-73}$$

$$w = z_1 - z = \frac{\partial w}{\partial x}x + \frac{\partial w}{\partial y}y + \frac{\partial w}{\partial z}z$$

Now we are ready to consider a superposition of two small deformations, following one after another. Due to the first deformation a medium around the point N is moved, and it is located in the vicinity of point N_1. Its coordinates are defined by eqs. D-73

$$x_1 = \left(1 + \frac{\partial u}{\partial x}\right)x + \frac{\partial u}{\partial y}y + \frac{\partial u}{\partial z}z$$

$$y_1 = \frac{\partial v}{\partial x}x + \left(1 + \frac{\partial v}{\partial y}\right)y + \frac{\partial v}{\partial z}z \tag{D-74}$$

$$z_1 = \frac{\partial w}{\partial x}x + \frac{\partial w}{\partial y}y + \left(1 + \frac{\partial w}{\partial z}\right)z$$

After the second deformation a particle in the vicinity of the point N_1 is situated around some point N_2 with coordinates x_2, y_2 and z_2:

$$x_2 = \left(1 + \frac{\partial u_1}{\partial x_1}\right) x_1 + \frac{\partial u_1}{\partial y_1} y_1 + \frac{\partial u_1}{\partial z_1} z_1$$

$$y_2 = \frac{\partial v_1}{\partial x_1} x_1 + \left(1 + \frac{\partial v_1}{\partial y_1}\right) y_1 + \frac{\partial v_1}{\partial z_1} z_1 \qquad \text{(D-75)}$$

$$z_2 = \frac{\partial w_1}{\partial x_1} x_1 + \frac{\partial w_1}{\partial y_1} y_1 + \left(1 + \frac{\partial w_1}{\partial z_1}\right) z_1$$

Substitution of x_1, y_1, z_1 from eqs. D-74 into eqs. D-75 gives us relationships between coordinates of points N_2 and N. First equation of this set yields

$$x_2 = \left(1 + \frac{\partial u_1}{\partial x_1}\right)\left(x + \frac{\partial u}{\partial x} x + \frac{\partial u}{\partial y} y + \frac{\partial u}{\partial z} z\right)$$

$$+ \frac{\partial u_1}{\partial y_1}\left(\frac{\partial v}{\partial x} x + y + \frac{\partial v}{\partial y} y + \frac{\partial v}{\partial z} z\right) + \frac{\partial u_1}{\partial z_1}\left(\frac{\partial w}{\partial x} x + \frac{\partial w}{\partial y} y + \frac{\partial w}{\partial z} z + z\right)$$

Performing a multiplication and discarding terms with a product of derivatives we obtain

$$x_2 = \left(1 + \frac{\partial u}{\partial x} + \frac{\partial u_1}{\partial x_1}\right) x + \left(\frac{\partial u}{\partial y} + \frac{\partial u_1}{\partial y_1}\right) y + \left(\frac{\partial u}{\partial z} + \frac{\partial u_1}{\partial z_1}\right) z \qquad \text{(D-76)}$$

In the same manner we have

$$y_2 = \left(\frac{\partial v}{\partial x} + \frac{\partial v_1}{\partial x_1}\right) x + \left(1 + \frac{\partial v}{\partial y} + \frac{\partial v_1}{\partial y_1}\right) y + \left(\frac{\partial v}{\partial z} + \frac{\partial v_1}{\partial z_1}\right) z \qquad \text{(D-77)}$$

$$z_2 = \left(\frac{\partial w}{\partial x} + \frac{\partial w_1}{\partial x_1}\right) x + \left(\frac{\partial w}{\partial y} + \frac{\partial w_1}{\partial y_1}\right) y + \left(1 + \frac{\partial w}{\partial z} + \frac{\partial w_1}{\partial z_1}\right) z$$

Thus, a superposition of small deformations is a result of a summation of unit elongations as well as that of the distortion angles, corresponding to each small deformation. In other words, small homogeneous strains obey the principle of superposition. As was mentioned earlier this means that a complicated but a small deformation can be treated as a combination of rather simple strains, and their summation can be carried out in any order. We illustrated the principle of superposition in the case of two subsequent deformations. Of course, it is valid for any number of strains.

Appendix E

Relationship between stress and strain

Hooke's law

As was demonstrated in two previous appendices a distribution of surface forces at each point of an elastic medium is characterized by six elements of the symmetrical tensor:

$$T = \begin{pmatrix} \tau_{11} & \tau_{12} & \tau_{13} \\ \tau_{21} & \tau_{22} & \tau_{23} \\ \tau_{31} & \tau_{32} & \tau_{33} \end{pmatrix} \tag{E-1}$$

Also they can be written in the form:

$$X_x, \quad Y_y, \quad Z_z, \quad Y_z, \quad Z_x, \quad X_y \tag{E-2}$$

where

$$X_x = \tau_{xx}, \quad Y_y = \tau_{yy}, \quad Z_z = \tau_{zz}, \quad X_y = \tau_{xy}, \quad X_z = \tau_{xz}, \quad Y_z = \tau_{yz} \tag{E-3}$$

At the same time, a deformation is also defined by six elements of the symmetrical tensor

$$E = \begin{pmatrix} \varepsilon_{11} & \varepsilon_{12} & \varepsilon_{13} \\ \varepsilon_{21} & \varepsilon_{22} & \varepsilon_{23} \\ \varepsilon_{31} & \varepsilon_{32} & \varepsilon_{33} \end{pmatrix} \tag{E-4}$$

Often different notations are used:

$$\varepsilon_{xx} = \varepsilon_{11}, \quad \varepsilon_{yy} = \varepsilon_{22}, \quad \varepsilon_{zz} = \varepsilon_{33}, \quad \varepsilon_{yz} = \varepsilon_{23}, \quad \varepsilon_{xz} = \varepsilon_{13}, \quad \varepsilon_{xy} = \varepsilon_{12} \tag{E-5}$$

Here

$$\varepsilon_{xx} = \frac{\partial s_x}{\partial x}, \quad \varepsilon_{yy} = \frac{\partial s_y}{\partial y}, \quad \varepsilon_{zz} = \frac{\partial s_z}{\partial z} \tag{E-6}$$

and

$$\varepsilon_{yz} = \frac{1}{2}(\frac{\partial s_y}{\partial z} + \frac{\partial s_z}{\partial y}), \quad \varepsilon_{xz} = \frac{1}{2}(\frac{\partial s_x}{\partial z} + \frac{\partial s_z}{\partial x}), \quad \varepsilon_{xy} = \frac{1}{2}(\frac{\partial s_x}{\partial y} + \frac{\partial s_y}{\partial x}) \tag{E-7}$$

and the particle displacement $\mathbf{s}$ is

$$\mathbf{s} = s_x\mathbf{i}+s_y\mathbf{j}+s_z\mathbf{k} \tag{E-8}$$

Since due to a deformation the internal surface forces arise, it is natural to assume that there is a relationship between stress and strains. This dependence can be written as

$$
\begin{aligned}
X_x &= f_1(\varepsilon_{xx},\ \varepsilon_{yy},\ \varepsilon_{zz},\ \varepsilon_{zy},\ \varepsilon_{zx},\ \varepsilon_{xy}) \\
Y_y &= f_2(\varepsilon_{xx},\ \varepsilon_{yy},\ \varepsilon_{zz},\ \varepsilon_{zy},\ \varepsilon_{zx},\ \varepsilon_{xy}) \\
Z_z &= f_3(\varepsilon_{xx},\ \varepsilon_{yy},\ \varepsilon_{zz},\ \varepsilon_{zy},\ \varepsilon_{zx},\ \varepsilon_{xy}) \\
Y_z &= f_4(\varepsilon_{xx},\ \varepsilon_{yy},\ \varepsilon_{zz},\ \varepsilon_{zy},\ \varepsilon_{zx},\ \varepsilon_{xy}) \\
Z_x &= f_5(\varepsilon_{xx},\ \varepsilon_{yy},\ \varepsilon_{zz},\ \varepsilon_{zy},\ \varepsilon_{zx},\ \varepsilon_{xy}) \\
X_y &= f_6(\varepsilon_{xx},\ \varepsilon_{yy},\ \varepsilon_{zz},\ \varepsilon_{zy},\ \varepsilon_{zx},\ \varepsilon_{xy})
\end{aligned} \tag{E-9}
$$

Suppose that if a medium is not deformed, stresses are absent, that is

$$f_i(0,0,0,0,0,0) = 0 \qquad (i = 1,2,3...6) \tag{E-10}$$

We also imply that these functions are continuous and have first derivatives. Then expanding f_i in Maclaurin series and discarding all terms except the first one we obtain

$$X_x = c_{11}\varepsilon_{xx} + c_{12}\varepsilon_{yy} + c_{13}\varepsilon_{zz} + c_{14}\varepsilon_{zy} + c_{15}\varepsilon_{zx} + c_{16}\varepsilon_{xy}$$

$$Y_y = c_{21}\varepsilon_{xx} + c_{22}\varepsilon_{yy} + c_{23}\varepsilon_{zz} + c_{24}\varepsilon_{zy} + c_{25}\varepsilon_{zx} + c_{26}\varepsilon_{xy}$$

$$Z_z = c_{31}\varepsilon_{xx} + c_{32}\varepsilon_{yy} + c_{33}\varepsilon_{zz} + c_{34}\varepsilon_{zy} + c_{35}\varepsilon_{zx} + c_{36}\varepsilon_{xy}$$

$$Y_z = c_{41}\varepsilon_{xx} + c_{42}\varepsilon_{yy} + c_{43}\varepsilon_{zz} + c_{44}\varepsilon_{zy} + c_{45}\varepsilon_{zx} + c_{46}\varepsilon_{xy} \tag{E-11}$$

$$Z_x = c_{51}\varepsilon_{xx} + c_{52}\varepsilon_{yy} + c_{53}\varepsilon_{zz} + c_{54}\varepsilon_{zy} + c_{55}\varepsilon_{zx} + c_{56}\varepsilon_{xy}$$

$$X_y = c_{61}\varepsilon_{xx} + c_{62}\varepsilon_{yy} + c_{63}\varepsilon_{zz} + c_{64}\varepsilon_{zy} + c_{65}\varepsilon_{zx} + c_{66}\varepsilon_{xy}$$

These linear functions with respect to strains are called Hooke's law and they describe a relationship between stresses and strains. By definition, each coefficient c_{ij} defines the first derivative of the stress tensor element with respect to a corresponding strain. Speaking strictly, the derivative is calculated when this element of the strain tensor is equal to zero. For instance

$$c_{34} = \frac{\partial Z_z}{\partial \varepsilon_{zy}} \qquad \text{if } \varepsilon_{zy} \to 0$$

In accordance with eq. E-11 the Hooke's law contains 36 coefficients. However, we will demonstrate that some of them are equal to each other, and in general, this law is described by 21 independent parameters. Propagation of elastic waves is usually accompanied by a deformation with extremely small strains, which has order 10^{-6} and much smaller. This fact allows us to neglect terms of the second and higher orders in the Maclaurin series, eq. E-11. At the same time elastic constants, c_{ij} could be very large, and their dimension is the same as stresses, since ε_{ij} are dimensionless. Let us also notice that the linear theory of elasticity is based on Hooke's law, while eqs. E-9 are a foundation of the nonlinear theory. In order to study elastic constants in different types of a medium it is useful to derive expressions of the work, performed by stresses, as well as the potential energy of a deformed medium and the elastic potential. In particular, this approach allows us to show that the number of coefficients c_{ij} does not exceed 21.

The work of forces and potential energy of a deformed body

Consider some volume V of an elastic body, subjected to an action of the surface and volume forces. They cause a change in a relative position of particles of a body and a deformation takes place. Let us pay attention to a very small time interval during which these forces remain constant. Variation of the work, performed by them, is equal to

$$\delta A = \int\limits_V \mathbf{f} \cdot \boldsymbol{\delta}\mathbf{s}\, dV + \oint\limits_S \mathbf{t} \cdot \boldsymbol{\delta}\mathbf{s}\, dS \tag{E-12}$$

Here $\mathbf{f}$ is the density of volume forces, $\mathbf{t}$ is a traction, and $\boldsymbol{\delta}\mathbf{s}$ is a change of the displacement of particles, S is the surface, surrounding the volume V. Now we carry out some transformatious, which permit us to express this work δA in terms of stresses and strains at points inside the volume only. With this purpose in mind consider the surface integral in eq. E-12. As was shown in Appendix C the traction $\mathbf{t}$ can be represented in the form

$$\mathbf{t} = (\mathbf{X}\cdot\mathbf{n})\,\mathbf{i} + (\mathbf{Y}\cdot\mathbf{n})\,\mathbf{j} + (\mathbf{Z}\cdot\mathbf{n})\,\mathbf{k}, \tag{E-13}$$

where $\mathbf{n}$ is the unit vector normal to the surface S and directed outward. Therefore

$$\mathbf{t}\cdot\boldsymbol{\delta}\mathbf{s} = \delta s_x\,\mathbf{X}\cdot\mathbf{n} + \delta s_y\,\mathbf{Y}\cdot\mathbf{n} + \delta s_z\,\mathbf{Z}\cdot\mathbf{n} \tag{E-14}$$

Applying Gauss theorem

$$\oint_S \mathbf{M}\cdot\mathbf{dS} = \int_V \text{div}\,\mathbf{M}\,dV$$

we obtain

$$\oint_S \mathbf{t}\cdot\boldsymbol{\delta}\mathbf{s}\,dS = \oint_S [\delta s_x\,\mathbf{X}\cdot\mathbf{dS} + \delta s_y\,\mathbf{Y}\cdot\mathbf{dS} + \delta s_z\,\mathbf{Z}\cdot\mathbf{dS}] \tag{E-15}$$

$$= \int_V \text{div}\,(\mathbf{X}\,\delta s_x + \mathbf{Y}\,\delta s_y + \mathbf{Z}\,\delta s_z)dV$$

Correspondingly, eq. E-12 becomes

$$\delta A = \int_V [f_x\delta s_x + f_y\delta s_y + f_z\delta s_z + \text{div}\,(\mathbf{X}\,\delta s_x + \mathbf{Y}\,\delta s_y + \mathbf{Z}\,\delta s_z)]dV \tag{E-16}$$

The integrand can be greatly simplified by making use of the identities

$$\text{div}\,(\mathbf{X}\delta s_x) = \delta s_x\,\text{div}\,\mathbf{X} + \mathbf{X}\cdot\text{grad}\,\delta s_x$$

$$\text{div}\,(\mathbf{Y}\delta s_y) = \delta s_y\,\text{div}\,\mathbf{Y} + \mathbf{Y}\cdot\text{grad}\,\delta s_y \tag{E-17}$$

$$\text{div}\,(\mathbf{Z}\delta s_z) = \delta s_z\,\text{div}\,\mathbf{Z} + \mathbf{Z}\cdot\text{grad}\,\delta s_z$$

Note that this type of equalities was already used to obtain equations of a motion and an equilibrium, (Appendix C). In deriving an expression of the work we distinguish two cases.

Case one First, assume that a displacement of particles takes place very slowly, that is their velocity is negligible, and it is possible to neglect a change of kinetic energy. This allows us to apply a relation between the surface and volume forces for an equilibrium. As was demonstrated in Appendix C in this case

$$\text{div } \mathbf{X} + f_x = 0, \qquad \text{div } \mathbf{Y} + f_y = 0, \qquad \text{div } \mathbf{Z} + f_z = 0 \tag{E-18}$$

Substitution of eqs. E-17–E-18 into eq. E-16 yields

$$\delta A = \int_V (\mathbf{X} \cdot \text{grad } \delta s_x + \mathbf{Y} \cdot \text{grad } \delta s_y + \mathbf{Z} \cdot \text{grad } \delta s_z) dV \tag{E-19}$$

The integrand is a sum of three terms, and it can be written as

$$\begin{aligned} & X_x \frac{\partial}{\partial x}\delta s_x + X_y \frac{\partial}{\partial y}\delta s_x + X_z \frac{\partial}{\partial z}\delta s_x \\ & +Y_x \frac{\partial}{\partial x}\delta s_y + Y_y \frac{\partial}{\partial y}\delta s_y + Y_z \frac{\partial}{\partial z}\delta s_y \\ & +Z_x \frac{\partial}{\partial x}\delta s_z + Z_y \frac{\partial}{\partial y}\delta s_z + Z_z \frac{\partial}{\partial z}\delta s_z \end{aligned} \tag{E-20}$$

Taking into account that $X_y = Y_x$, $X_z = Z_x$, $Y_z = Z_y$ and the derivative of a difference is equal to a difference of derivatives, the last sum is equal to

$$X_x\, \delta e_{xx} + Y_y\, \delta e_{yy} + Z_z\, \delta e_{zz} + Y_z\, \delta e_{yz} + Z_x\, \delta e_{xz} + X_y\, \delta e_{xy} \tag{E-21}$$

Here

$$e_{xx} = \varepsilon_{xx} = \frac{\partial s_x}{\partial x}, \qquad e_{yy} = \varepsilon_{yy} = \frac{\partial s_y}{\partial y}, \qquad e_{zz} = \varepsilon_{zz} = \frac{\partial s_z}{\partial z} \tag{E-22}$$

and

$$e_{yz} = 2\varepsilon_{yz} = \frac{\partial s_y}{\partial z} + \frac{\partial s_z}{\partial y}, \qquad e_{zx} = 2\varepsilon_{zx} = \frac{\partial s_x}{\partial z} + \frac{\partial s_z}{\partial x}, \qquad e_{xy} = 2\varepsilon_{xy} = \frac{\partial s_x}{\partial y} + \frac{\partial s_y}{\partial z}$$

Thus, in place of eq. E-19 we have

$$\delta A = \int_V (X_x\, \delta e_{xx} + Y_y\, \delta e_{yy} + Z_z\, \delta e_{zz} + Y_z\, \delta e_{yz} + Z_x\, \delta e_{xz} + X_y\, \delta e_{xy})dV, \qquad \text{(E-23)}$$

and the work, performed by external forces during a short interval of time, is expressed in terms of stresses inside a volume, as well as a change of strains. In deriving eq. E-23 we neglected variations of the kinetic energy. Moreover, let us assume that the heat remains the same, that is no heat is gained or lost by any element of the body. Such adiabatic change can be expected, since particles of a medium are usually involved in rapid and small vibrations. Under these conditions the work δA results in only an increase of the potential (intrinsic) energy, U. In a fact, the external forces produce a work and a elastic body becomes deformed. If the body is allowed to return to its unstrained state, it gives back all the work, performed by external forces. Correspondingly, this work can be treated as an energy, stored in a body and is called the strain (potential) energy. Then, eq. E-23 can be rewritten as

$$\delta A = \delta U = \int \delta u_0\, dV, \qquad \text{(E-24)}$$

where δu_0 is a change of the density of the potential energy, while δU is a change of this energy of the deformed body. From a comparison of eqs. E-23 and E-24 we conclude that

$$\delta u_0 = X_x\, \delta e_{xx} + Y_y\, \delta e_{yy} + Z_z\, \delta e_{zz} + Y_z\, \delta e_{yz} + Z_x\, \delta e_{xz} + X_y\, \delta e_{xy} \qquad \text{(E-25)}$$

This clearly shows that during a very small time interval, δt, a change of the density of the strain energy is defined by a product of stresses and a variation of the corresponding strains.

Case two Next we demonstrate that eq. E-25 is still valid even when there are variations of the kinetic energy. With this purpose in mind eq. E-25 will be derived in a slightly different way. In accordance with the principle of conservation of energy, the work, performed by external forces during a unit time, results in a change of the kinetic and potential energy, as well as heat, δQ. This can be written as

$$\delta A = \delta K + \delta U + \delta Q$$

or

$$\frac{\partial A}{\partial t} = \frac{\partial K}{\partial t} + \frac{\partial U}{\partial t} + \frac{\partial Q}{\partial t}, \qquad \text{(E-26)}$$

that is the total mechanical energy

$$\delta U + \delta K$$

is not in general equal to the work, δA, done by external forces.

As was shown in Appendix A, the kinetic energy of an elementary volume is defined as

$$\frac{1}{2}\rho\ (v_x^2 + v_y^2 + v_z^2), \tag{E-27}$$

provided that displacements are small. Here v_x, v_y, and v_z are components of the particle velocity along the coordinate axes. Respectively, the kinetic energy of a deformed body is

$$K = \int\limits_V \frac{\rho}{2}(v_x^2 + v_y^2 + v_z^2)dV, \tag{E-28}$$

and its rate of a change with time is

$$\frac{\partial K}{\partial t} = \int\limits_V \rho(v_x a_x + v_y a_y + v_z a_z)dV \tag{E-29}$$

where $a_x = \partial v_x/\partial t$, $a_y = \partial v_y/\partial t$, $a_z = \partial v_z/\partial t$ are components of an acceleration. As is seen from eq. E-12 the rate, at which the work is done, is equal to

$$\frac{\partial A}{\partial t} = \int\limits_V \mathbf{f}\cdot\mathbf{v}\ dV + \oint\limits_S \mathbf{t}\cdot\mathbf{v}\ dS$$

or

$$\frac{\partial A}{\partial t} = \int\limits_V (f_x v_x + f_y v_y + f_z v_z)dV + \oint\limits_S (t_x v_x + t_y v_y + t_z v_z)dS$$

Making use of E-15, we obtain

$$\frac{\partial A}{\partial t} = \int\limits_V [f_x v_x + f_y v_y + f_z v_z + \text{div}\ (\mathbf{X}\ v_x + \mathbf{Y}\ v_y + \mathbf{Z}\ v_z)]dV \tag{E-30}$$

It almost coincides with eq. E-16 and can be represented as

$$\frac{\partial A}{\partial t} = \int\limits_V [(f_x + \text{div}\ \mathbf{X})v_x + (f_y + \text{div}\ \mathbf{Y})v_y + (f_z + \text{div}\ \mathbf{Z})v_z \tag{E-31}$$

$$+\mathbf{X}\cdot \text{grad } v_x + \mathbf{Y}\cdot \text{grad } v_y + \mathbf{Z}\cdot \text{grad } v_z]dV$$

In Appendix C we demonstrated that

$$f_x + \text{div } \mathbf{X} = \rho\ a_x, \qquad f_y + \text{div } \mathbf{Y} = \rho\ a_y, \qquad f_z + \text{div } \mathbf{Z} = \rho\ a_z \tag{E-32}$$

Substitution of eqs. E-32 into eq. E-31 gives

$$\frac{\partial A}{\partial t} = \int\limits_V \rho(a_x v_x + a_y v_y + a_z v_z)dV \tag{E-33}$$

$$+\frac{\partial}{\partial t}\int\limits_V (\mathbf{X}\cdot \text{grad } s_x + \mathbf{Y}\cdot \text{grad } s_y + \mathbf{Z}\cdot \text{grad } s_z)]dV$$

Taking into account eqs. E-29 and E-33, in place of eq. E-26 we have

$$\int\limits_V [\rho(a_x v_x + a_y v_y + a_z v_z) + \frac{\partial}{\partial t}\int\limits_V (\mathbf{X}\cdot \text{grad } s_x + \mathbf{Y}\cdot \text{grad } s_y + \mathbf{Z}\cdot \text{grad } s_z)]dV$$

$$= \int\limits_V \rho(v_x\ a_x + v_y\ a_y + v_z\ a_z)dV + \frac{\partial U}{\partial t} + \frac{\partial Q}{\partial t}$$

or

$$\frac{\partial}{\partial t}\int\limits_V (\mathbf{X}\cdot \text{grad } s_x + \mathbf{Y}\cdot \text{grad } s_y + \mathbf{Z}\cdot \text{grad } s_z)]dV = \frac{\partial U}{\partial t} + \frac{\partial Q}{\partial t}, \tag{E-34}$$

or

$$\int\limits_V (\mathbf{X}\cdot \text{grad } \delta s_x + \mathbf{Y}\cdot \text{grad } \delta s_y + \mathbf{Z}\cdot \text{grad } \delta s_z)]dV = \delta U + \delta Q \tag{E-35}$$

Assuming that a change is adiabatic, $\delta Q = 0$, we again arrive at eq. E-25. In this light it may be appropriate to notice that an adiabatic compression of a gas increases its temperature. Also if a metal is adiabatically compressed, there is an increase of a temperature too, but it is quite small. In principle it is possible to remove a portion of heat and restore an original temperature. Such procedure slightly changes a strain, that is a difference between the adiabatic and isothermal elastic parameters is very small and it is usually much less than one percentage. As follows from eqs. E-24–E-25, the work, causing an elementary change of strains in the unit volume, is equal to

$$dA_s = X_x de_{xx} + Y_y de_{yy} + Z_z de_{zz} + Y_z de_{yz} + Z_x de_{xz} + X_y de_{xy} \tag{E-36}$$

Since the work is transformed into the internal energy small variations of strains are replaced by the full differentials. Let us note that considering the work we did not make any assumptions about a relationship between the stress and strain.

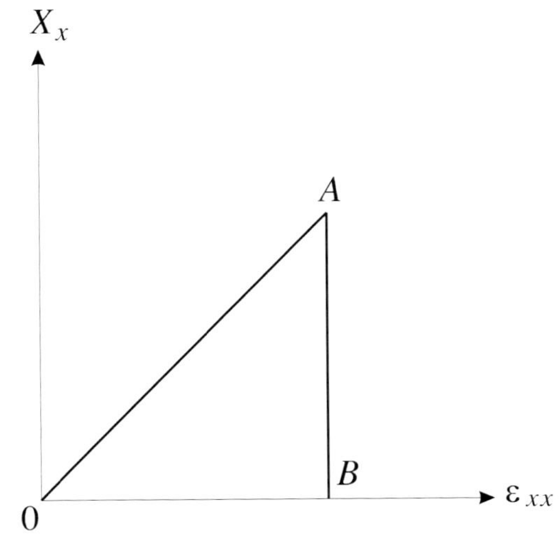

Figure E.1: Illustration of Eq. E-38

The work of stresses and Hooke's law

Now suppose that Hooke's law is valid and find an expression for the work A_s. By definition, we have

$$A_s = \int (X_x de_{xx} + Y_y de_{yy} + Z_z de_{zz} + Y_z de_{yz} + Z_x de_{xz} + X_y de_{xy}) \tag{E-37}$$

At the beginning consider the first integral

$$\int_0^{e_{xx}} X_x \, de_{xx}$$

Its evaluation is illustrated in Fig. E.1. Since the stress X_x linearly depends on the strain, e_{xx} (Hooke's law), this relation is described by the straight line OA. Correspondingly, the area of the triangle OAB defines the integral, that is

$$\int_0^{e_{xx}} X_x(e_{xx}) de_{xx} = \frac{1}{2} X_x e_{xx}, \tag{E-38}$$

where X_x is the function of the final value of the strain.

Applying the same approach to other integrals in eq. E-37 we arrive at the expression of the work

$$A_s = \frac{1}{2}(X_x e_{xx} + Y_y e_{yy} + Z_z e_{zz} + Y_z e_{yz} + Z_x e_{xz} + X_y e_{xy}) \tag{E-39}$$

For illustration consider an elementary volume of an elastic medium.

Example one Assume that each element of the stress tensor is the same at the opposite faces of the volume. In other words, the derivatives of stresses with respect to coordinates are equal to zero. We first calculate the work, performed by the force, related to the stress X_x. Introduce the following notation: $ds_x(x, y, z)$ is the displacement of the middle point of the volume along the x-axis. Then $ds_x(x-dx/2, y, z)$ and $ds_x(x+dx/2, y, z)$ are displacements of the back and front faces of the volume, respectively.

As was demonstrated in Appendix C, the x-components of forces, acting on these faces, are equal to

$$F_x(x - \frac{dx}{2}, y, z) = t_x dydz = (\mathbf{X} \cdot \mathbf{n}_1)dydz = -X_x dydz$$

and

$$F_x(x + \frac{dx}{2}, y, z) = t_x dydz = (\mathbf{X} \cdot \mathbf{n}_2)dydz = X_x dydz,$$

since X_x =const and $\mathbf{n}_2 = -\mathbf{n}_1 = \mathbf{i}$. Therefore, the work of these forces, is

$$X_x d\left[s_x(x + \frac{dx}{2}, y, z) - s_x(x - \frac{dx}{2}, y, z)\right] dydz \tag{E-40}$$

Expanding s_x in the Taylor's series and discarding all terms, except the first and second ones, eq. E-40 becomes

$$X_x d\left(\frac{\partial s_x}{\partial x}\right) dV = X_x de_{xx} dV$$

Its integration within the interval: $0 - e_{xx}$ gives the first term of eq. E-39, if $dV = 1$.

In the same manner, we derive expressions of the work, which is done by two other normal stresses and they are $\frac{1}{2}\, Y_y\, e_{yy}\, dV$ and $\frac{1}{2}\, Z_z\, e_{zz}\, dV$. Next, consider a contribution of tangential components of forces, directed along the x-axis and acting on faces of the volume, perpendicular to the z-axis. As a result, these faces experience displacements: $ds_x(x, y, z + dz/2)$ and $ds_x(x, y, z - dz/2)$. Since

$$F_x(x, y, z + \frac{dz}{2}) = (\mathbf{X} \cdot \mathbf{n}_2)dxdy = X_z(x, y, z)dxdy$$

and

$$F_x(x, y, z - \frac{dz}{2}) = (\mathbf{X} \cdot \mathbf{n}_1)dxdy = -X_z(x, y, z)dxdy$$

and

$$\mathbf{n}_2 = -\mathbf{n}_1 = \mathbf{k},$$

the work, performed by these forces, is

$$X_z d\left[s_x(x, y, z + \frac{dz}{2}) - s_x(x, y, z - \frac{dz}{2})\right] dxdy = X_z d\frac{ds_x}{dz} dV \tag{E-41}$$

By analogy, the work of the forces, directed along the z–axis and acting on faces, normal to the x–axis, is equal to

$$Z_x d\left[s_z(x + \frac{dx}{2}, y, z) - s_z(x - \frac{dx}{2}, y, z)\right] dydz$$

or

$$Z_x d\left(\frac{\partial s_z}{\partial x}\right) dV \tag{E-42}$$

Hence $X_z = Z_x$ a sum of works, given by eqs. E-41–E-42, is

$$Z_x d\left(\frac{\partial s_x}{\partial x} + \frac{\partial s_z}{\partial x}\right) dV = Z_x de_{xz} dV \tag{E-43}$$

Here de_{xz} is an elementary change of the strain e_{xz}.

After an integration of eq. E-43 from 0 to e_{xz} we obtain the corresponding term of eq. E-39. Similarly, considering the work of other shearing forces, all terms of the sum in eq. E-39 can be found.

Example two Next consider a more complicated case, when the volume forces are present and stresses vary linearly inside the volume dV. Making use of results, derived in the first example, it is clear that the elementary work of the force F_x, acting on faces, normal to the x–axis, is equal to

$$\left[X_x(x + \frac{dx}{2}, y, z) ds_x(x + \frac{dx}{2}, y, z) - X_x(x - \frac{dx}{2}, y, z) ds_x(x - \frac{dx}{2}, y, z)\right] dydz$$

or

$$\frac{\partial}{\partial x}(X_x ds_x) dV \tag{E-44}$$

In the same manner the elementary work, associated with forces F_x, acting on faces, perpendicular to the y– and z–axes, is

$$\frac{\partial}{\partial y}(X_y ds_x) dV \quad \text{and} \quad \frac{\partial}{\partial z}(X_z ds_x) dV \tag{E-45}$$

Thus, the total elementary work, caused by the surface forces F_x becomes

$$\left[\frac{\partial}{\partial x}(X_x ds_x) + \frac{\partial}{\partial y}(X_y ds_x) + \frac{\partial}{\partial x}(X_z ds_x)\right] dV$$

or

$$\left[X_x d\frac{\partial s_x}{\partial x} + X_y d\frac{\partial s_x}{\partial y} + X_z d\frac{\partial s_x}{\partial z} + ds_x(\frac{\partial X_x}{\partial x} + \frac{\partial X_y}{\partial y} + \frac{\partial X_z}{\partial z})\right] dV$$

The latter can be also written as

$$\left[X_x d\frac{\partial s_x}{\partial x} + X_y d\frac{\partial s_x}{\partial y} + X_z d\frac{\partial s_x}{\partial z} + ds_x \text{ div } \mathbf{X}\right] dV \tag{E-46}$$

The elementary work due to the x–component of the volume force f_x is

$$f_x ds_x dV \tag{E-47}$$

Adding eqs. E-45–E-46 we obtain

$$\left[X_x d\frac{\partial s_x}{\partial x} + X_y d\frac{\partial s_x}{\partial y} + X_z d\frac{\partial s_x}{\partial z} + (\text{div } \mathbf{X} + f_x) ds_x\right] dV$$

Taking into account the first equation of the set E-18, the elementary work, related to the x–component of the surface and volume forces, is defined, as in the previous example, by the sum

$$\left(X_x d\frac{\partial s_x}{\partial x} + X_y d\frac{\partial s_x}{\partial y} + X_z d\frac{\partial s_x}{\partial z}\right) dV$$

By analogy, the y– and z–component of the forces produce the work:

$$\left(Y_x d\frac{\partial s_y}{\partial x} + Y_y d\frac{\partial s_y}{\partial y} + Y_z d\frac{\partial s_y}{\partial z}\right) dV$$

and

$$(Z_x d\frac{\partial s_z}{\partial x} e_{xz} + Z_y d\frac{\partial s_z}{\partial y} e_{yz} + Z_z d\frac{\partial s_z}{\partial z}) dV$$

After a summation of all these elementary work and an integration, we again obtain eq. E-39.

Strain potential and Hooke's law

In accordance with eq. E-25 a change of the density of the potential energy is

$$du_0 = X_x de_{xx} + X_y de_{yy} + Z_z de_{zz} + Y_z de_{yz} + Z_x de_{xz} + X_y de_{xy}$$

Correspondingly, the density u is defined as

$$u_0 = \int (X_x de_{xx} + X_y de_{yy} + Z_z de_{zz} + Y_z de_{yz} + Z_x de_{xz} + X_y de_{xy}), \qquad \text{(E-48)}$$

and it represents the potential energy, stored in the unit volume, due to a deformation. The function u is called the elastic (strain) potential or strain energy function. From the physical point of view it is obvious that the density u_0 is a function of strains, that is

$$u_0 = u_0 \left(e_{xx}, \quad e_{yy}, \quad e_{zz}, \quad e_{yz}, \quad e_{xz}, \quad e_{xy}\right) \qquad \text{(E-49)}$$

For this reason, its small variation, du, can be represented in terms of small changes of strains. With this purpose in mind we expand the function u_0 in the power series and preserve only terms with the first derivatives. Since

$$u_0(0,0,0,0,0,0) = 0,$$

we have

$$du_0 = \frac{\partial u_0}{\partial e_{xx}} de_{xx} + \frac{\partial u_0}{\partial e_{yy}} de_{yy} + \frac{\partial u_0}{\partial e_{zz}} de_{zz} + \frac{\partial u_0}{\partial e_{yz}} de_{yz} + \frac{\partial u_0}{\partial e_{xz}} de_{xz} + \frac{\partial u_0}{\partial e_{xy}} de_{xy} \qquad \text{(E-50)}$$

Thus, we obtain two expressions of the same function, du_0, which characterize a change of the elastic potential, eqs. E-25 and E-50. Their comparison gives Green's formulas

$$X_x = \frac{\partial u_0}{\partial e_{xx}}, \qquad Y_y = \frac{\partial u_0}{\partial e_{yy}}, \qquad Z_z = \frac{\partial u_0}{\partial e_{zz}}, \qquad \text{(E-51)}$$

and $$Y_z = \frac{\partial u_0}{\partial e_{yz}}, \qquad Z_x = \frac{\partial u_0}{\partial e_{xz}}, \qquad X_y = \frac{\partial u_0}{\partial e_{xy}}$$

It is a very important result, because it allows us to express components of the stress tensor as derivatives of the elastic potential with respect to strains. Thus, the function u contains an information about surface forces and strains. It is essential that with a help of the potential u_0 we can demonstrate that some elastic constants in the Hooke's law are equal to each other. Later, considering some special types of a medium, it will be shown that certain elastic constants vanish. At the beginning we take the first two equations of the set E-11. Making use of eqs. E-51 and relationships between ε and e, these equations can be written as

$$X_x = \frac{\partial u_0}{\partial e_{xx}} = c_{11}e_{xx} + c_{12}e_{yy} + c_{13}e_{zz} + \frac{c_{14}}{2}e_{zy} + \frac{c_{15}}{2}e_{zx} + \frac{c_{16}}{2}e_{xy} \tag{E-52}$$

and $$Y_y = \frac{\partial u_0}{\partial e_{yy}} = c_{21}e_{xx} + c_{22}e_{yy} + c_{23}e_{zz} + \frac{c_{24}}{2}e_{zy} + \frac{c_{25}}{2}e_{zx} + \frac{c_{26}}{2}e_{xy}$$

Since strains are independent arguments of stresses X_x, we have

$$\frac{\partial X_x}{\partial e_{yy}} = \frac{\partial^2 u_0}{\partial e_{yy}\partial e_{xx}} = c_{12} \qquad \text{and} \qquad \frac{\partial Y_y}{\partial e_{xx}} = \frac{\partial^2 u_0}{\partial e_{xx}\partial e_{yy}} = c_{21} \tag{E-53}$$

The principle of conservation of energy requires that the work cannot depend on an order in which forces are applied, but only on their final magnitudes. Otherwise, it is possible to gain energy when a deformation has a complete cycle and a body returns to the original state. Correspondingly, we obtain

$$\frac{\partial^2 u_0}{\partial e_{xx}\partial e_{yy}} = \frac{\partial^2 u_0}{\partial e_{yy}\partial e_{xx}}, \tag{E-54}$$

that is $c_{21} = c_{12}$. Considering all pairs of stresses we find

$$c_{ij} = c_{ji} \tag{E-55}$$

where $i, j = 1, 2, 3...6$. Thus, in general, the Hooke's law is defined by 21 elastic constants, shown below:

$$\begin{matrix} c_{11} & c_{12} & c_{13} & c_{14} & c_{15} & c_{16} \\ & c_{22} & c_{23} & c_{24} & c_{25} & c_{26} \\ & & c_{33} & c_{34} & c_{35} & c_{36} \\ & & & c_{44} & c_{45} & c_{46} \\ & & & & c_{55} & c_{56} \\ & & & & & c_{66} \end{matrix} \tag{E-56}$$

Elastic potential in terms of strains and elastic constants

Earlier we represented u_0, as the integral, eq. E-48. In order to perform an integration we make use of the Hooke's law and eq. E-55. Substitution of eqs. E-11 into eq. E-48 and on integration gives

$$u_0 = \frac{1}{2}c_{11}e_{xx}^2 + c_{12}e_{xx}e_{yy} + c_{13}e_{xx}e_{zz} + c_{14}e_{xx}e_{yz} + c_{15}e_{xx}e_{xz} + c_{16}e_{xx}e_{xy} + c_{13}e_{xx}e_{zz}$$

$$+\frac{1}{2}c_{22}e_{yy}^2 + c_{23}e_{yy}e_{zz} + c_{24}e_{yy}e_{yz} + c_{25}e_{yy}e_{zx} + c_{26}e_{yy}e_{xy}$$

$$+\frac{1}{2}c_{33}e_{zz}^2 + c_{34}e_{zz}e_{zy} + c_{35}e_{zz}e_{zx} + c_{36}e_{zz}e_{xy} \tag{E-57}$$

$$+\frac{1}{2}c_{44}e_{zy}^2 + c_{45}e_{zy}e_{zx} + c_{46}e_{zy}e_{xy}$$

$$+\frac{1}{2}c_{55}e_{zx}^2 + c_{56}e_{zx}e_{xy} + \frac{1}{2}c_{66}e_{xy}^2$$

This shows that the elastic potential, u_0, is a function of the second order with respect to strains. As was pointed out the potential u is very useful to study elastic constants in the Hooke's law for the isotropic and anisotropic media. With this purpose in mind consider a change of strains with a rotation of coordinate axes.

Transformation of strain elements

Earlier in the Appendix B we briefly discussed a transformation of the tensor elements when a direction of coordinate axes of the Cartesian system changes. Now we consider this question in some detail. Suppose that six components of the strain tensor are known at the original system x, y, z. Then our goal is to find these elements in the new system x', y', z'. The position of this system with respect to the original one is defined by nine cosines, given in the table

	x	y	z
x'	l_1	m_1	n_1
y'	l_2	m_2	n_2
z'	l_3	m_3	n_3

For instance $l_1 = \cos(\mathbf{i}_1, \mathbf{i})$ is the cosine of the angle, formed by the x and x' axes, but $m_3 = \cos(\mathbf{k}_1, \mathbf{j})$ and so on. Now we use the following notations for the displacement components in both systems

$$\begin{matrix} u & v & w \\ u' & v' & w' \end{matrix}$$

By definition, components of the strain in the new system of coordinates are

$$e_{x'x'} = \frac{\partial u'}{\partial x'} \qquad e_{y'z'} = \frac{\partial w'}{\partial y'} + \frac{\partial v'}{\partial z'}$$

$$e_{y'y'} = \frac{\partial v'}{\partial y'} \qquad e_{z'x'} = \frac{\partial u'}{\partial z'} + \frac{\partial w'}{\partial x'} \tag{E-58}$$

$$e_{z'z'} = \frac{\partial w'}{\partial z'} \qquad e_{x'y'} = \frac{\partial v'}{\partial x'} + \frac{\partial u'}{\partial y'}$$

In order to determine relationships between strains in the new and old systems we, first, make use of simple formulas, which relate components of the displacement. Since the new and old systems have the same origin we can write for the displacement vector

$$u\,\mathbf{i} + v\,\mathbf{j} + w\,\mathbf{k} = u'\mathbf{i}_1 + v'\mathbf{j}_1 + w'\mathbf{k}_1$$

Its multiplication by unit vectors of the new system gives three important equations

$$u' = u\,l_1 + v\,m_1 + w\,n_1, \quad v' = u\,l_2 + v\,m_2 + w\,n_2, \quad w' = u\,l_3 + v\,m_3 + w\,n_3 \tag{E-59}$$

The next useful relation was derived in Part I and it can be written as

$$\frac{\partial \varphi}{\partial s} = \text{grad}\ \varphi\ \mathbf{i}_s \tag{E-60}$$

Here φ is an arbitrary function and $\partial\varphi/\partial s$ is the directional derivative along the line which unit vector is $\mathbf{i}_s$. By definition:

$$\mathbf{i}_s = \cos(\mathbf{i}_s, \mathbf{i})\ \mathbf{i} + \cos(\mathbf{i}_s, \mathbf{j})\ \mathbf{j} + \cos(\mathbf{i}_s, \mathbf{k})\ \mathbf{k} \tag{E-61}$$

In particular, the vector $\mathbf{i}_s$ may coincide with unit vectors of the new system of coordinates. Bearing in mind that

$$\text{grad}\varphi = \frac{\partial \varphi}{\partial x}\ \mathbf{i} + \frac{\partial \varphi}{\partial y}\ \mathbf{j} + \frac{\partial \varphi}{\partial z}\ \mathbf{k},$$

it is convenient to rewrite eq. E-60 in the form

$$\frac{\partial}{\partial s}(\,) = \frac{\partial}{\partial x}(\,)\cos(s,x) + \frac{\partial}{\partial y}(\,)\cos(s,y) + \frac{\partial}{\partial z}(\,)\cos(s,z) \tag{E-62}$$

Here $(\,)$ means any function φ.

First, suppose that $s = x'$ and $\varphi = u' = l_1 u + m_1 v + n_1 w$. Then eq. E-62 becomes

$$e_{x'x'} = \frac{\partial u'}{\partial x'} = (l_1 \frac{\partial}{\partial x} + m_1 \frac{\partial}{\partial y} + n_1 \frac{\partial}{\partial z})(l_1 u + m_1 v + n_1 w)$$

$$= l_1^2 \frac{\partial u}{\partial x} + m_1^2 \frac{\partial v}{\partial y} + n_1^2 \frac{\partial w}{\partial z} + (\frac{\partial w}{\partial y} + \frac{\partial v}{\partial z}) m_1 n_1 + (\frac{\partial u}{\partial z} + \frac{\partial w}{\partial x}) n_1 l_1 + +(\frac{\partial v}{\partial x} + \frac{\partial u}{\partial y}) l_1 m_1$$

whence

$$e_{x'x'} = l_1^2 e_{xx} + m_1^2 e_{yy} + n_1^2 e_{zz} + m_1 n_1 e_{yz} + n_1 l_1 e_{zx} + l_1 m_1 e_{xy} \tag{E-63}$$

Now assuming that $s = y'$ and $\varphi = v'$ or $s = z'$ and $\varphi = w'$ we arrive at the following equations:

$$e_{y'y'} = l_2^2\, e_{xx} + m_2^2\, e_{yy} + n_2^2\, e_{zz} + m_2 n_2 e_{yz} + n_2 l_2 e_{zx} + l_2 m_2\, e_{xy} \tag{E-64}$$

and $$e_{z'z'} = l_3^2\, e_{xx} + m_3^2\, e_{yy} + n_3^2\, e_{zz} + m_3 n_3 e_{yz} + n_3 l_3 e_{zx} + l_3 m_3\, e_{xy}$$

The last three equations describe a transformation of the diagonal elements of the strain tensor. Applying the same approach we have

$$e_{y'z'} = \frac{\partial w'}{\partial y'} + \frac{\partial v'}{\partial z'} = (l_2 \frac{\partial}{\partial x} + m_2 \frac{\partial}{\partial y} + n_2 \frac{\partial}{\partial z})(l_3 u + m_3 v + n_3 w)$$

$$+(l_3 \frac{\partial}{\partial x} + m_3 \frac{\partial}{\partial y} + n_3 \frac{\partial}{\partial z})(l_2 u + m_2 v + n_2 w) = 2 l_2 l_3 \frac{\partial u}{\partial x} + 2 m_2 m_3 \frac{\partial v}{\partial y} + 2 n_2 n_3 \frac{\partial w}{\partial z}$$

$$+(\frac{\partial w}{\partial y} + \frac{\partial v}{\partial z})(m_2 n_3 + m_3 n_2) + (\frac{\partial u}{\partial z} + \frac{\partial w}{\partial x})(n_2 l_3 + n_3 l_2) + (\frac{\partial v}{\partial x} + \frac{\partial u}{\partial y})(l_2 m_3 + l_3 m_2)$$

or

$$e_{y'z'} = 2(l_2 l_3\, e_{xx} + m_2 m_3\, e_{yy} + n_2 n_3\, e_{zz}) + e_{yz}(m_2 n_3 + m_3 n_2) \tag{E-65}$$

$$+ e_{zx}(n_2 l_3 + n_3 l_2) + e_{xy}(l_2 m_3 + l_3 m_2)$$

By analogy we have

$$e_{z'x'} = 2(l_1 l_3\, e_{xx} + m_1 m_3\, e_{yy} + n_1 n_3\, e_{zz}) + e_{yz}(m_3 n_1 + m_1 n_3)$$

$$+e_{zx}(n_3 l_1 + n_1 l_3) + e_{xy}(l_3 m_1 + m_3 l_1) \qquad \text{and} \qquad \text{(E-66)}$$

$$e_{x'y'} = 2(l_1 l_2\, e_{xx} + m_1 m_2\, e_{yy} + n_1 n_2 e_{zz}) + e_{yz}(m_1 n_2 + m_2 n_1)$$

$$+e_{zx}(n_1 l_2 + n_2 l_1) + e_{xy}(l_1 m_2 + l_2 m_1)$$

Thus, our task is solved and we can determine strain elements in the new system of coordinates, as soon as directional cosines are known. Formulas of strain transformation have different applications. Consider one of them, which allows us to find relationships between strains and stresses in an isotropic medium. Adding eqs. E-63–E-64 we obtain

$$e_{x'x'} + e_{y'y'} + e_{z'z'} = e_{xx}(l_1^2 + l_2^2 + l_3^2) + e_{yy}(m_1^2 + m_2^2 + m_3^2) + e_{zz}(n_1^2 + n_2^2 + n_3^2)$$

$$+e_{yz}(m_1 n_1 + m_2 n_2 + m_3 n_3) + e_{zx}(n_1 l_1 + n_2 l_2 + n_3 l_3) + e_{xy}(l_1 m_1 + l_2 m_2 + l_3 m_3)$$

Taking into account the known relations of directional cosines:

$$l_1^2 + l_2^2 + l_3^2 = 1 \qquad m_1 n_1 + m_2 n_2 + m_3 n_3 = 0$$

$$m_1^2 + m_2^2 + m_3^2 = 1 \qquad n_1 l_1 + n_2 l_2 + n_3 l_3 = 0$$

$$n_1^2 + n_2^2 + n_3^2 = 1 \qquad m_1 l_1 + m_2 l_2 + m_3 l_3 = 0,$$

we have

$$e_{x'x'} + e_{y'y'} + e_{z'z'} = e_{xx} + e_{yy} + e_{zz} \qquad \text{(E-67)}$$

The latter is the first invariant of this transformation. This result is obvious, since the relative volume extension, (dilatation)

$$\text{div}\,\mathbf{s} = \Theta = e_{xx} + e_{yy} + e_{zz}$$

is independent of the system of coordinates. Simple but cumbersome algebra shows that there are two more invariants, which are

$$e_{yy}e_{zz} + e_{zz}e_{xx} + e_{xx}e_{yy} - \frac{1}{4}(e_{yz}^2 + e_{zx}^2 + e_{xy}^2) \qquad \text{(E-68)}$$

$$\text{and} \qquad e_{xx}e_{yy}e_{zz} + \frac{1}{4}(e_{yz}e_{zx}e_{xy} - e_{xx}e_{yz}^2 - e_{yy}e_{zx}^2 - e_{zz}e_{xy}^2)$$

Now we are ready to find expressions for the Hooke's law at different media and start from the simplest but very important case.

Isotopic medium

In such a medium all physical parameters do not depend on a direction along which they are considered. It is natural that it is also applied to the strain tensor. This means that its elements remain the same, regardless of an orientation of coordinate axes. Inasmuch as the elastic potential, u, is a function of strains, we conclude that it is an invariant with respect to a rotation of coordinate axes too. In order to observe such behavior the right hand side of eq. E-57 should be represented as a combination of the strain invariants, eqs. E-67–E-68. Bearing in mind that the strain potential is a homogeneous function of the second order but the last invariant is of the third order, we make use of the first two invariants. This allows us to write u_0 as

$$u_0 = \frac{1}{2}[a(e_{xx} + e_{yy} + e_{zz})^2 + b(e_{yz}^2 + e_{zx}^2 + e_{xy}^2 - 4e_{yy}e_{zz} - 4e_{zz}e_{xx} - 4e_{xx}e_{yy})], \quad \text{(E-69)}$$

where a and b are elastic parameters of a medium.

In other words, elastic constants c_{ij} are related to each other so that the potential u_0 is described by only two parameters. It is conventional in place of a and b to use Lamé constants, λ and μ, which are introduced in the following way

$$a = \lambda + 2\mu \qquad \text{and} \qquad b = \mu \quad \text{(E-70)}$$

Respectively, we have

$$u_0 = \frac{\lambda + 2\mu}{2}(e_{xx} + e_{yy} + e_{zz})^2 + \frac{\mu}{2}(e_{yz}^2 + e_{xz}^2 + e_{xy}^2 - 4e_{yy}e_{zz} - 4e_{zz}e_{xx} - 4e_{xx}e_{yy}) \quad \text{(E-71)}$$

Now, making use of Green's formulas, eqs. E-51, it is easy to determine components of the stress tensor. Performing a differentiation of the potential u with respect to strains we obtain

$$\begin{aligned} X_x &= \lambda\Theta + 2\mu e_{xx}, & Y_z &= \mu\, e_{yz} \\ Y_y &= \lambda\Theta + 2\mu e_{yy}, & X_z &= \mu\, e_{xz} \\ Z_z &= \lambda\Theta + 2\mu e_{zz}, & X_y &= \mu\, e_{xy}, \end{aligned} \quad \text{(E-72)}$$

where $\Theta = e_{xx} + e_{yy} + e_{zz}$ is the dilatation. These equations establish a relationship between stresses and strains. It is interesting to notice that the normal stresses are related to the diagonal elements of the strain tensor only and depend on two elastic parameters λ and μ. At the same time, the shearing stresses are functions of the corresponding strain and the single parameter μ, which is usually called rigidity. Formulas, given by the set E-72, represent the Hooke's law when a medium is isotropic, and they were derived by Cauchy.

Relationships between elastic moduli of an isotropic medium

By definition elastic parameters, λ and μ, are independent of a distribution of stresses. For this reason, it is very useful to consider the simplest case when

$$X_x \neq 0, \qquad Y_y = Z_z = Y_z = X_z = X_y = 0 \tag{E-73}$$

For instance, if a thin bar experiences an extension we have

$$X_x = E\, e_{xx} \tag{E-74}$$

Here E is the Young modulus, and it always has a very large value. Taking into account the condition E-73 the set E-72 becomes

$$\lambda\Theta + 2\mu\, e_{xx} = X_x, \qquad \lambda\Theta + 2\mu\, e_{yy} = 0, \qquad \lambda\Theta + 2\mu\, e_{zz} = 0 \tag{E-75}$$

Adding eqs. F-75 we obtain

$$(3\lambda + 2\mu)\Theta = X_x \tag{E-76}$$

The latter allows us to express the relationship between X_x and e_{xx} in terms of λ and μ. In fact, we have

$$\Theta = \frac{X_x}{3\lambda + 2\mu} \tag{E-77}$$

and its substitution into the first equation of the set E-75 yields

$$\frac{\lambda}{3\lambda + 2\mu} X_x + 2\mu\, e_{xx} = X_x$$

or

$$X_x = \frac{\mu(3\lambda + 2\mu)}{\lambda + \mu} e_{xx} \tag{E-78}$$

Comparison with eq. E-74 gives for Young modulus

$$E = \frac{\mu(3\lambda + 2\mu)}{\lambda + \mu}, \tag{E-79}$$

and it depends on both Lamé parameters. In order to find the relation between the Poisson ratio, σ, and Lamé parameters consider the last two equations of the set E-75, which give

$$e_{yy} = e_{zz} = -\frac{\lambda\,\Theta}{2\mu}$$

Taking into account eq. E-76, we obtain

$$e_{yy} = e_{zz} = -\frac{\lambda}{2\mu}\frac{X_x}{(3\lambda + 2\mu)}$$

or, making use of E-78, the latter gives

$$e_{yy} = e_{zz} = -\frac{\lambda}{2(\lambda + \mu)} e_{xx} \tag{E-80}$$

By definition $e_{yy} = e_{zz} = -\sigma e_{xx}$. Whence the Poisson ratio can be represented as

$$\sigma = \frac{\lambda}{2(\lambda + \mu)} \tag{E-81}$$

Thus, pair of elastic parameters, E and σ, are expressed in terms of Lamé constants. Now, let us resolve eqs. E-79 and E-81 with respect to λ and μ. The first equation can be written in the form:

$$E = \frac{\mu(2\lambda + 2\mu)}{\lambda + \mu} + \frac{\mu\lambda}{\lambda + \mu} = 2\mu + \mu\, 2\sigma$$

Therefore

$$\mu = \frac{E}{2(1 + \sigma)} \tag{E-82}$$

Its substitution into eq. E-81 yields

$$\lambda = \frac{E\,\sigma}{(1 + \sigma)(1 - 2\sigma)} \tag{E-83}$$

It is instructive to consider the case when the normal stresses are equal to each other, but the shearing stresses are absent:

$$X_x = Y_y = Z_z = -P \qquad\qquad X_y = X_z = Y_z = 0 \tag{E-84}$$

Respectively, the set E-72 becomes

$$\lambda\Theta + 2\mu\, e_{xx} = -P, \qquad \lambda\Theta + 2\mu\, e_{yy} = -P, \qquad \lambda\Theta + 2\mu\, e_{zz} = -P \tag{E-85}$$

and

$$e_{yz} = e_{xz} = e_{xy} = 0$$

Summation of eqs. E-85 gives

$$(3\lambda + 2\mu)\Theta = -3P$$

Therefore, the relationship between the pressure, P, and dilatation is

$$P = -(\lambda + \frac{2}{3}\mu)\ \Theta \tag{E-86}$$

As was shown in Part I

$$P = -M\Theta, \tag{E-87}$$

where M is the bulk modulus, which characterizes a compression or an expansion of an elementary volume. Comparison of eqs. E-86 and E-87 gives

$$M = \lambda + \frac{2}{3}\ \mu \tag{E-88}$$

Replacing λ and μ by E and σ we obtain

$$M = \frac{E\sigma}{(1+\sigma)(1-2\sigma)} + \frac{E}{3(1+\sigma)}$$

or

$$M = \frac{E}{3(1-2\sigma)} \tag{E-89}$$

Anisotropic medium

As was pointed out earlier the strain-energy-function, eq. E-57, and, therefore, the Hooke's law is in general defined by 21 elastic constants. Now we consider several models of an anisotropic medium, where the number of these constants is greatly reduced.

Case one Suppose that a medium is such that at each point there is a plane of symmetry with respect to elastic properties. This means that two forces with equal magnitudes but opposite direction, normal to the plane, cause the same strain. Let us assume that the plane of symmetry coincides with the coordinate plane XOY. Then the elastic potential has to remain the same, when a direction of the z–axis, perpendicular to the plane of symmetry, changes. Correspondingly, a sign of z and w_z changes, too, as well as that of strains

$$e_{zx} = \frac{\partial w}{\partial x} + \frac{\partial u}{\partial z}, \qquad e_{yz} = \frac{\partial w}{\partial y} + \frac{\partial v}{\partial z}$$

Because of symmetry, terms containing e_{zx} and e_{zy} in the first power have to vanish. At the same time, terms with the product $e_{zx}e_{yz}$ remain. This gives

$$c_{14} = c_{15} = c_{24} = c_{25} = c_{34} = c_{35} = c_{46} = c_{56} = 0, \tag{E-90}$$

and the number of different elastic constants becomes 13.

Case two Assume that at each point of an elastic medium there are three mutually perpendicular planes of symmetry. It is convenient to treat them as coordinate planes. Therefore, an expression of an elastic potential, eq. E-57, does not change, if an orientation of coordinate axes becomes opposite. Since this transformation changes signs of e_{xy}, e_{xz}, e_{yz}, terms of the sum in eq. E-57, which contain e_{xy}, e_{xz}, e_{yz} in the first degree, should be equal to zero. It happens, if along with the condition E-90, we also have

$$c_{16} = c_{26} = c_{36} = c_{45} = 0 \tag{E-91}$$

The number of elastic constants is reduced to 9, and they usually characterize a deformation of an elementary volume, which is crystallized as the rectangular parallelepiped.

Case three In this medium, as before, at each point there are three mutually perpendicular planes of symmetry and, moreover, elastic properties are the same with respect to each of them. This means that the potential u does not change, if the x–axis is replaced by either the y or z–one. Therefore, the expression of the strain potential remains unchanged, if mutual replacement of the following quantities takes place:

$$e_{xx}, e_{yy}, e_{zz} \qquad \text{or} \qquad e_{xy}, e_{xz}, e_{yz}$$

It happens if

$$c_{11} = c_{22} = c_{33}, \qquad c_{44} = c_{55} = c_{66}, \qquad c_{23} = c_{12} = c_{13}, \tag{E-92}$$

provided that conditions E-90–E-91 are met. Then, the potential u_0 can be written as

$$u_0 = \frac{c_{11}}{2}(e_{xx}^2 + e_{yy}^2 + e_{zz}^2) + c_{12}(e_{yy}e_{zz} + e_{zz}e_{xx} + e_{xx}e_{yy}) + \frac{c_{44}}{2}(e_{yz}^2 + e_{zx}^2 + e_{xy}^2), \tag{E-93}$$

and it is characterized by three elastic constants. Such a deformation usually occurs in a medium with the cubical crystals.

Case four Now we assume that a medium is isotropic and derive again an expression for the strain potential. It is clear that this function has to be the same for any orientation of coordinate axes. For simplicity let us turn the original system, (x, y, z) by a small angle φ around the z–axis. This gives a new system, (x', y', z'), and, in accordance with Table 1, their directional cosines, are

$$\begin{array}{lll} l_1 = 1 & m_1 = \varphi & n_1 = 0 \\ l_2 = -\varphi & m_2 = 1 & n_2 = 0 \\ l_3 = 0 & m_3 = 0 & n_3 = 1 \end{array} \qquad \text{(E-94)}$$

Here we used the fact, that φ is very small and

$$\sin\varphi = \varphi, \qquad \cos\varphi = 1 - \frac{\varphi^2}{2} \approx 1$$

Applying formulas of the strain transformation, eqs. E-63–E-66 and discarding small terms with φ^2, we obtain components of the strain in new system of coordinates:

$$e_{x'x'} = e_{xx} + \varphi\, e_{xy}, \qquad e_{y'y'} = e_{yy} - \varphi\, e_{xy}, \qquad e_{z'z'} = e_{zz}, \qquad \text{(E-95)}$$

and $$e_{y'z'} = e_{yz} - \varphi\, e_{zx}, \qquad e_{z'x'} = e_{zx} + \varphi\, e_{zy}, \qquad e_{x'y'} = e_{xy} + 2\varphi(e_{yy} - e_{xx})$$

It is clear that any plane in an isotropic medium is a plane of symmetry and elastic properties are independent on a direction. For this reason, we can use eq. E-93, which can be written in new system of coordinates:

$$u_1 = \frac{c_{11}}{2}(e^2_{x'x'} + e^2_{y'y'} + e^2_{z'z'}) + c_{12}(e_{y'y'}e_{z'z'} + e_{z'z'}e_{x'x'} + e_{x'x'}e_{y'y'}) \qquad \text{(E-96)}$$

$$+\frac{c_{44}}{2}(e^2_{y'z'} + e^2_{z'x'} + e^2_{x'y'})$$

Now we represent u_1 as a sum of u_0 and terms, depending on φ in the first power. Substituting eqs. E-95 into eq. E-96 and preserving terms with φ of the first power we obtain:

$$u_1 = u + \varphi(e_{yy} - e_{xx})(2c_{44} + c_{12} - c_{11})\, e_{xy} \qquad \text{(E-97)}$$

Since $u_1 = u_0$ we conclude that elastic constants are related to each other and it gives

$$2c_{44} + c_{12} - c_{11} = 0 \qquad \text{or} \qquad c_{11} = 2c_{44} + c_{12} \tag{E-98}$$

Correspondingly eq. E-93 becomes:

$$2u = c_{12}(e_{xx} + e_{yy} + e_{zz})^2 + 2c_{44}(e_{xx}^2 + e_{yy}^2 + e_{zz}^2) + c_{44}(e_{yz}^2 + e_{zx}^2 + e_{xy}^2),$$

that coincides with eq. E-71, if we let $c_{44} = \mu$ and $c_{12} = \lambda$

Case five Next consider a transversely isotropic medium, which is of a great importance in a seismology. Suppose that a distribution of elastic parameters possesses axial symmetry around the z–axis. Correspondingly, any plane with this line is the plane of symmetry. In order to find the elastic potential in such a medium we undertake several steps. First, let us consider the plane XOZ. Since a change of the direction of the x–axis does not make an influence on the potential, eq. E-57, we obtain, as in the first example:

$$c_{15} = c_{16} = c_{25} = c_{26} = c_{35} = c_{36} = c_{45} = c_{46} = 0$$

Now we choose the second plane of symmetry, YOZ. For the same reason its symmetry gives:

$$c_{14} = c_{16} = c_{24} = c_{26} = c_{34} = c_{36} = c_{45} = c_{56} = 0$$

Because of two last conditions, eq. E-57 becomes

$$u_0 = \frac{1}{2}c_{11}e_{xx}^2 + c_{12}e_{xx}e_{yy} + c_{13}e_{xx}e_{zz} + \frac{1}{2}c_{22}e_{yy}^2 + c_{23}e_{yy}e_{zz} \tag{E-99}$$

$$+\frac{1}{2}c_{33}e_{zz}^2 + \frac{1}{2}c_{44}e_{zy}^2 + \frac{1}{2}c_{55}e_{zx}^2 + \frac{1}{2}c_{66}e_{xy}^2$$

Thus, a number of elastic constants is already reduced from 21 to 9. Next simplification is related to the axial symmetry of a medium. For this reason we can perform a replacement of the following strains: e_{xx}, e_{yy} and e_{xz}, e_{yz}. This means that $c_{11} = c_{22}$, $c_{13} = c_{23}$, $c_{44} = c_{55}$, and eq. E-99 can be written as

$$u_0 = \frac{1}{2}c_{11}(e_{xx}^2 + e_{yy}^2) + c_{12}e_{xx}e_{yy} + c_{13}(e_{xx} + e_{yy})e_{zz} + \frac{1}{2}c_{33}e_{zz}^2 \tag{E-100}$$

$$+\frac{1}{2}c_{44}(e_{zy}^2 + e_{zx}^2) + \frac{1}{2}c_{66}e_{xy}^2$$

At this stage we expressed the strain potential with help of six elastic constants. It turns out that only five of them are independent. To demonstrate this fact suppose, as in the case of an isotropic medium, that a system of coordinates, x, y, z, is turned by a small angle φ about the z–axis. Then, in the new system, x', y', z' eq. E-100 becomes

$$u_0 = \frac{1}{2}c_{11}(e^2_{x'x'} + e^2_{y'y'}) + c_{12}e_{x'x'}e_{y'y'} + c_{13}(e_{x'x'} + e_{y'y'})e_{z'z'} \tag{E-101}$$

$$+\frac{1}{2}c_{33}e^2_{z'z'} + \frac{1}{2}c_{44}(e^2_{z'y'} + e^2_{z'x'}) + \frac{1}{2}c_{66}\ e^2_{x'y'}$$

In the same manner as in the previous example, we substitute eqs. E-95 into eq. E-101 and discard terms with φ in the power exceeding one. This gives

$$u_1 = u_0 + \varphi[c_{11}(e_{xx} - e_{yy})e_{xy} + c_{12}(e_{yy} - e_{xx})e_{xy} + 2c_{66}(e_{yy} - e_{xx})e_{xy}]$$

or

$$u_1 = u_0 + \varphi(-c_{11} + c_{12} + 2c_{66})(e_{yy} - e_{xx})e_{xy} \tag{E-102}$$

Since $u_1 = u_0$, we arrive at one more condition for elastic constants:

$$c_{11} = c_{12} + 2c_{66}$$

Thus, in place of eq. E-100 we have

$$2u_0 = (c_{12} + 2c_{66})(e^2_{xx} + e^2_{yy}) + 2c_{12}\ e_{xx}e_{yy} + 2c_{13}(e_{xx} + e_{yy})e_{zz} + c_{33}\ e^2_{zz} \tag{E-103}$$

$$+c_{44}(e^2_{zy} + e^2_{zx}) + c_{66}\ e^2_{xy} = c_{12}(e_{xx} + e_{yy})^2 + 2c_{13}(e_{xx} + e_{yy})\ e_{zz} + c_{33}\ e^2_{zz}$$

$$+c_{44}(e^2_{zy} + e^2_{zx}) + c_{66}(e^2_{xy} + 2e^2_{xx} + 2e^2_{yy}),$$

and the potential u_0 is defined by five parameters.

References

K. Aki and P. G. Richards. Quantitative seismology: theory and methods, vol. I. W. Freeman and Co., 1980.

A. Angot. Compliments de Mathematiques. Paris, 1957.

M. A. Biot. Propagation of elastic waves in a cylindrical borehole containing a fluid. J. Appl. Phys. 23, pp. 997-1109, 1952.

M. Born and E. Wolf. Principles of optics: electromagnetic theory of propagation, interference and diffraction of light. Permagon Press, 1993.

L. M. Brekhovskikh and V. Goncharov. Mechanics of continua and wave dynamics. Springer-Verlang, 1985.

C. H. Cheng and M. N. Toksöz. Elastic wave propagation in fluid-filled borehole and synthetic acoustic logs. Geophysics, 46, pp.1042–1053, 1981.

C. H. Dix. Seismic prospecting for oil. Boston: International Human Resources Development Corp., 1981.

M. Ewing, W. Jardetzky, and F. Press. Elastic waves in layered media. McGraw-Hill Book Co., 1957.

F. L. Fedorov. Theory of elastic waves in crystals. Plenum Press, 1968.

G. A. Gamburtsev. On waves caused by a moving source in an elastic medium. Izvestia AN SSSR, ser. geogr. & geofiz., 10, n. 1, pp. 31–44, (in Russian), 1946.

F. S. Grant and G. F. West. Interpretation theory in applied geophysics. McGraw-Hill Book Co., 1965.

A. A. Kaufman and A. L. Levshin. Acoustic and elastic wave fields in geophysics, I. Elsevier, 2000.

A. A. Kaufman, A. L. Levshin, and K. Larner. Acoustic and elastic wave fields in geophysics, II. Elsevier, 2002.

H. Kolsky. Stress waves in solids. Dover Publications, 1963.

H. Lamb. On the propagation of tremors over the surface of an elastic solid. Phil. Trans. Roy. Soc., A, vol. 203, pp. 1-42, 1904.

H. Lamb. Statics. Cambridge University Press., 1960.

E. R. Lapwood. The disturbance due to a line source in a semiinfinite elastic medium. Phil. Trans. Roy. Soc. A 841, vol. 142, pp. 63–100, 1949.

L. S. Leibenson. Course of the theory of elasticity. OGIZ, Moscow (in Russian), 1947.

A. E. H. Love. The propagation of wave motion in an isotropic elastic solid medium. Proc. Math. Soc., 1, 1904.

A. E. H. Love. A treatise on the mathematical theory of elasticity. Dover Publications, 1944.

F. L. Paillet and C.H. Cheng. Acoustic waves in boreholes. CRC Press, 1991.

G. W. Postma. Wave propagation in a stratified medium. Geophysics, 20, n. 4, pp.780–806, 1955.

Lord Rayleigh. On waves propagated along the plane surface of an elastic solid. Proc. Math. Soc., 17, 4, 1885.

G. G. Stokes. Dynamical theory of diffraction. Trans. Phil. Soc., 9, 1849.

R. Stoneley. Elastic waves at the surface of separation of two solids. Proc. Roy. Soc., A., vol. 106, pp. 416–428, 1924.

S. P. Timoshenko and J. N. Goodler. Theory of elasticity. McGraw-Hill Book Co., 1951.

L. Tsang. Time harmonic solution of the elastic head wave problem incorporating the influence of Rayleigh poles. J. Acoust. Soc. Am., 63, n. 5, 1978.

I. Tsvankin. Seismic signatures and analysis of reflection data in anisotropic media. Pergamon Press, 2001.

J. E. White. Underground sound. Application of seismic waves. Elsevier, 1983.

Index